Control of Semiconductor
Surfaces and Interfaces

MATERIALS RESEARCH SOCIETY
SYMPOSIUM PROCEEDINGS VOLUME 448

Control of Semiconductor Surfaces and Interfaces

Symposium held December 2–5, 1996, Boston, Massachusetts, U.S.A.

EDITORS:

S.M. Prokes
Naval Research Laboratory
Washington, D.C., U.S.A.

O.J. Glembocki
Naval Research Laboratory
Washington, D.C., U.S.A.

S.K. Brierley
Raytheon Company
Andover, Massachusetts, U.S.A.

J.M. Gibson
University of Illinois, Urbana
Urbana, Illinois, U.S.A.

J.M. Woodall
Purdue University
West Lafayette, Indiana, U.S.A.

MATERIALS
RESEARCH
SOCIETY

PITTSBURGH, PENNSYLVANIA

This work was supported in part by the Office of Naval Research under Grant Number ONR: N00014-97-1-0051. The United States Government has a royalty-free license throughout the world in all copyrightable material contained herein.

Single article reprints from this publication are available through University Microfilms Inc., 300 North Zeeb Road, Ann Arbor, Michigan 48106

CODEN: MRSPDH

Published by:

Materials Research Society
9800 McKnight Road
Pittsburgh, Pennsylvania 15237
Telephone (412) 367-3003
Fax (412) 367-4373
Website: http://www.mrs.org/

Library of Congress Cataloging in Publication Data

Control of semiconductor surfaces and interfaces : symposium held December
    2–5, 1996, Boston, Massachusetts, U.S.A. / editors, S.M. Prokes,
    O.J. Glembocki, S.K. Brierley, J.M. Gibson, J.M. Woodall
        p. cm—(Materials Research Society symposium proceedings ; v. 448)
        Includes bibliographical references and index.
        ISBN 1-55899-352-5
        1. Semiconductors—Surfaces—Congresses. 2. Semiconductors—
    Junctions—Congresses. 3. Crystals—Growth—Congresses. 4. Surface
    chemistry—Congresses. I. Prokes, S.M. II. Glembocki, O.J. III. Brierley, S.K.
    IV. Gibson, J.M. V. Woodall, J.M. VI. Series: Materials Research Society
    symposium proceedings ; v. 448.
QC611.6.S9C68 1997   97-13302
621.384'134—dc21   CIP

Manufactured in the United States of America

CONTENTS

PART I: <u>CHEMICAL MODIFICATION OF SURFACES</u>

*Invited Paper

*Invited Paper

PART III: <u>CONTROL OF GROWTH II: NANOSTRUCTURE FORMATIONS/SELF-ASSEMBLY</u>

PART IV: <u>CONTROL OF GROWTH III: SELECTED AREA EPITAXY</u>

PART V: <u>DIELECTRIC AND SEMICONDUCTOR INTERFACES</u>

*Invited Paper

PART VI: <u>METAL/SEMICONDUCTOR INTERFACES:
STRUCTURAL AND ELECTRICAL PROPERTIES</u>

*Invited Paper

PREFACE

These proceedings consist of refereed papers presented at the symposium on "Control of Semiconductor Surfaces and Interfaces," held as a part of the 1996 MRS Fall Meeting in Boston, MA. Semiconductor surfaces and interfaces play a vital role in modern-day electronic devices. This is especially true as device dimensions shrink. The properties of clean surfaces and chemically processed surfaces can also have a significant impact on the properties of subsequently grown layers. These surfaces and interfaces may exhibit modified structural, electronic and optical properties, so it is important to understand their effects on subsequent growth, processing and device fabrication.

In this symposium topics include the structure of surfaces, control of surface defects and properties through chemical etching and passivation, modification of surfaces for growth and processing, nucleation on semiconductor surfaces and self-assembly, the effects of surfaces and interfaces on subsequent growth, and the properties of semiconductor/ dielectric and semiconductor/metal interfaces. In addition, *in situ* and *ex situ* monitoring of these properties, using various electrical and optical techniques, was also presented.

We would like to thank all of the contributors and participants for enhancing our understanding of this important area, the invited speakers for providing current and interesting reviews, and all of the reviewers for their timely and conscientious reviews. We would also like to thank the Office of Naval Research for support of this symposium.

S.M. Prokes
O.J. Glembocki
S.K. Brierley
J.M. Gibson
J.M. Woodall

May 1997

MATERIALS RESEARCH SOCIETY SYMPOSIUM PROCEEDINGS

MATERIALS RESEARCH SOCIETY SYMPOSIUM PROCEEDINGS

Volume 444— Materials for Mechanical and Optical Microsystems, M.L. Reed, M. Elwenspoek, S. Johansson, E. Obermeier, H. Fujita, Y. Uenishi, 1997, ISBN: 1-55899-348-7

Volume 445— Electronic Packaging Materials Science IX, P.S. Ho, S.K. Groothuis, K. Ishida, T. Wu, 1997, ISBN: 1-55899-349-5

Volume 446— Amorphous and Crystalline Insulating Thin Films—1996, W.L. Warren, J. Kanicki, R.A.B. Devine, M. Matsumura, S. Cristoloveanu, Y. Homma, 1997, ISBN: 1-55899-350-9

Volume 447— Environmental, Safety, and Health Issues in IC Production, R. Reif, A. Bowling, A. Tonti, M. Heyns, 1997, ISBN: 1-55899-351-7

Volume 448— Control of Semiconductor Surfaces and Interfaces, S.M. Prokes, O.J. Glembocki, S.K. Brierley, J.M. Woodall, J.M. Gibson, 1997, ISBN: 1-55899-352-5

Volume 449— III–V Nitrides, F.A. Ponce, T.D. Moustakas, I. Akasaki, B.A. Monemar, 1997, ISBN: 1-55899-353-3

Volume 450— Infrared Applications of Semiconductors—Materials, Processing and Devices, M.O. Manasreh, T.H. Myers, F.H. Julien, 1997, ISBN: 1-55899-354-1

Volume 451— Electrochemical Synthesis and Modification of Materials, S.G. Corcoran, P.C. Searson, T.P. Moffat, P.C. Andricacos, J.L. Deplancke, 1997, ISBN: 1-55899-355-X

Volume 452— Advances in Microcrystalline and Nanocrystalline Semiconductors—1996, R.W. Collins, P.M. Fauchet, I. Shimizu, J–C. Vial, T. Shimada, A.P. Alivisatos, 1997, ISBN: 1-55899-356-8

Volume 453— Solid–State Chemistry of Inorganic Materials, A. Jacobson, P. Davies, T. Vanderah, C. Torardi, 1997, ISBN: 1-55899-357-6

Volume 454— Advanced Catalytic Materials—1996, M.J. Ledoux, P.W. Lednor, D.A. Nagaki, L.T. Thompson, 1997, ISBN: 1-55899-358-4

Volume 455— Structure and Dynamics of Glasses and Glass Formers, C.A. Angell, T. Egami, J. Kieffer, U. Nienhaus, K.L. Ngai, 1997, ISBN: 1-55899-359-2

Volume 456— Recent Advances in Biomaterials and Biologically–Inspired Materials: Surfaces, Thin Films and Bulk, D.F. Williams, M. Spector, A. Bellare, 1997, ISBN: 1-55899-360-6

Volume 457— Nanophase and Nanocomposite Materials II, S. Komarneni, J.C. Parker, H.J. Wollenberger, 1997, ISBN: 1-55899-361-4

Volume 458— Interfacial Engineering for Optimized Properties, C.L. Briant, C.B. Carter, E.L. Hall, 1997, ISBN: 1-55899-362-2

Volume 459— Materials for Smart Systems II, E.P. George, R. Gotthardt, K. Otsuka, S. Trolier-McKinstry, M. Wun-Fogle, 1997, ISBN: 1-55899-363-0

Volume 460— High–Temperature Ordered Intermetallic Alloys VII, C.C. Koch, N.S. Stoloff, C.T. Liu, A. Wanner, 1997, ISBN: 1-55899-364-9

Volume 461— Morphological Control in Multiphase Polymer Mixtures, R.M. Briber, D.G. Peiffer, C.C. Han, 1997, ISBN: 1-55899-365-7

Volume 462— Materials Issues in Art and Archaeology V, P.B. Vandiver, J.R. Druzik, J. Merkel, J. Stewart, 1997, ISBN: 1-55899-366-5

Volume 463— Statistical Mechanics in Physics and Biology, D. Wirtz, T.C. Halsey, J. van Zanten, 1997, ISBN: 1-55899-367-3

Volume 464— Dynamics in Small Confining Systems III, J.M. Drake, J. Klafter, R. Kopelman, 1997, ISBN: 1-55899-368-1

Volume 465— Scientific Basis for Nuclear Waste Management XX, W.J. Gray, I.R. Triay, 1997, ISBN: 1-55899-369-X

Volume 466— Atomic Resolution Microscopy of Surfaces and Interfaces, D.J. Smith, R.J. Hamers, 1997, ISBN: 1-55899-370-3

Prior Materials Research Society Symposium Proceedings available by contacting Materials Research Society

Part I

Chemical Modification of Surfaces

Chemically Stable Semiconductor Surface Layers Using Low-Temperature Grown GaAs

D. B. JANES *, S. HONG **, V. R. KOLAGUNTA *, D. McINTURFF *, T.-B. NG *, R. REIFENBERGER **, S. D. WEST *, J. M. WOODALL *

* NSF MRSEC for Technology Enabling Heterostructure Materials and School of Electrical and Computer Engineering Purdue University West Lafayette, IN 47907, janes@ecn.purdue.edu

** Department of Physics, Purdue University, West Lafayette, IN 47907

ABSTRACT

The chemical stability of a GaAs layer structure consisting of a thin (10 nm) layer of low-temperature-grown GaAs (LTG:GaAs) on a heavily n-doped GaAs layer, both grown by molecular beam epitaxy, is described. Scanning tunneling spectroscopy and X-ray photoelectron spectroscopy performed after atmospheric exposure indicate that the LTG:GaAs surface layer oxidizes much less rapidly than comparable layers of stoichiometric GaAs. There is also evidence that the terminal oxide thickness is smaller than that of stoichiometric GaAs. The spectroscopy results are used to confirm a model for conduction in low resistance, nonalloyed contacts employing comparable layer structures. The inhibited surface oxidation rate is attributed to the bulk Fermi level pinning and the low minority carrier lifetime in unannealed LTG:GaAs. Device applications including low-resistance cap layers for field-effect transistors are described.

INTRODUCTION

In stoichiometric n-type GaAs, a significant layer of oxide (approximately 25 Å in thickness) forms rapidly upon exposure to atmosphere. In order to avoid this rapid oxidation, a number of passivation procedures have been employed, including As or S cap layers [1, 2]. Further evidence for the instability of most GaAs surfaces is given by the special preparation procedures employed in previous scanning tunneling microscope (STM) spectroscopy studies to avoid surface oxidation and the associated loss of STM resolution. Cleaved (110) surfaces have been prepared by either *in-situ* cleaving in an ultra high vacuum (UHV) STM system or *ex-situ* cleaving followed by sulfide passivation [2-5]. GaAs (001) surfaces passivated with As cap layers have been studied in UHV STM experiments following removal of the As layer by heating in the STM vacuum system [1]. STM spectroscopy has been performed on an unannealed layer of LTG:GaAs (225 °C) capped with a layer of GaAs grown at 350 °C. Characterization of a (110) surface of the LTG:GaAs layer exposed by cleaving in UHV identified a band of midgap states associated with the excess arsenic [3-5]. For heavily doped n-type (n+) layers, this band of states was located above the valence band edge of the material. The rapid surface oxidation in stoichiometric GaAs plays a major role in *ex-situ* ohmic contact and Schottky barrier characteristics.

This paper describes the surface oxidation characteristics and related device applications of low-temperature grown GaAs (LTG:GaAs), i.e. layers of GaAs grown by molecular beam epitaxy (MBE) at substrate temperatures of 250-300 °C. LTG:GaAs materials exhibit a number of interesting electronic properties associated with the excess arsenic incorporated during growth [3,6]. The studies reported in this work involve as-grown LTG:GaAs material, in which the excess arsenic results in a large concentration (approximately $1 \times 10^{20} cm^{-3}$) of point defects, primarily as arsenic antisites. The large concentration of point defects results in very short minority carrier lifetimes (< 1 ps) and pinning of the bulk Fermi level near midgap.

Recently, *ex-situ* low resistance, nonalloyed contacts to n and p-type GaAs have been demonstrated using a structure consisting of a thin layer of LTG:GaAs (2-5 nm) on a highly doped layer of normal growth temperature GaAs, both grown by MBE [7]. Specific contact resistances as low as $2 \times 10^{-7} \Omega cm^2$ have been reported on n-type GaAs layers. The conduction model for the contact structure consisted of defect assisted tunneling through the LTG:GaAs layer and tunneling through the space charge region in the heavily doped layer. Since

3

LTG:GaAs layers as thin as 2 nm were effective in providing low-resistance ohmic contacts following air exposure and an oxide etch, it can be inferred that the oxidation rate and oxide thickness in the LTG:GaAs layer are lower than those typically observed in stoichiometric GaAs following air exposure. This paper describes several recent experiments which provide direct evidence that as-grown LTG:GaAs surface layers are more stable during air exposure than stoichiometric GaAs. The inhibited oxidation, along with the effective surface potential control, indicate that thin layers (2-5 nm) of this material can provide effective surface passivation. Device applications of the stable LTG:GaAs layers will also be briefly described.

EXPERIMENT

Two spectroscopic studies, namely scanning tunneling spectroscopy and X-ray photoelectron spectroscopy (XPS), have been performed on the layer structure shown in Fig. 1. While the spectroscopic studies were performed in ultra high vacuum (UHV), the samples were exposed to atmosphere during transfer between the growth chamber and the UHV characterization systems. The semiconductor layer structures used in this study were grown in a Varian Gen II MBE on epi-ready n+ GaAs (100) substrates, with a growth temperature of 580°C for all layers except the not intentionally doped (nid) LTG:GaAs top layer, which was grown at 250°C. The growth rate for all layers was 1μm/hour. A silicon filament was used for the n-type doping, allowing doping concentrations of at least an order of magnitude higher than possible with conventional effusion cells. It has been shown that the surface Fermi level is pinned near midgap during MBE growth of GaAs [8]. As a consequence, Si dopant atoms located within the surface carrier depletion region are incorporated primarily at donor sites, even for Si doping concentrations approaching the solid solubility limit [7]. In stoichiometric GaAs, this n++ layer cannot be exploited in device applications involving *ex-situ* processing, since this layer rapidly oxidizes upon exposure to air and is subsequently removed during oxide etching steps. The LTG:GaAs cap layer provides a controlled surface potential and, as will be shown in the following sections, prevents the surface from oxidizing significantly. The presence of the LTG:GaAs layer therefore maintains the high space charge density in a thin region at the top of the heavily doped layer and therefore maintains the high concentration of activated donors within the region. The LTG:GaAs layer therefore preserves the n++ doped layer even following air exposure and enables device applications such as the non-alloyed contact structure.

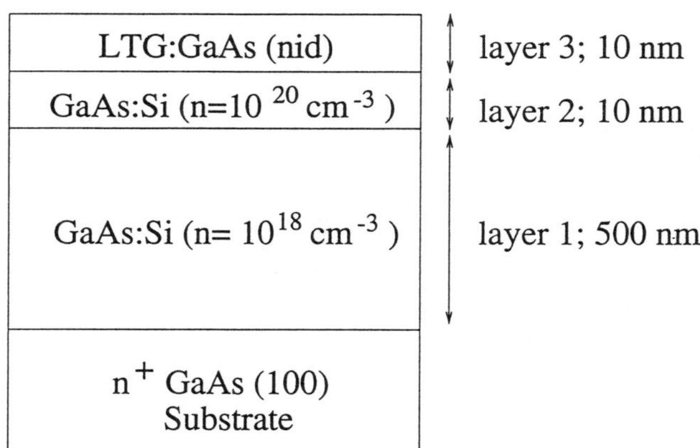

Fig. 1.   A diagram of the GaAs structure investigated in STM and XPS studies.

Scanning Tunneling Spectroscopy

In order to characterize the stability of LTG:GaAs after air exposure, spectroscopic STM studies were performed on the layer structure shown in Fig. 1. The tunneling current versus tip to substrate voltage was measured at various spatial locations on the surface. While the experiment actually probes the electronic characteristics of the surface layer, it also provides an indication of the amound of surface oxidation, since the observation of features such as a midgap band of states and significant tunnel current near zero bias requires a surface with very little oxide. The base pressure of the stainless steel vacuum chamber is below 4×10^{-10} torr. The tips are etched Pt/Ir wires cleaned in the STM chamber by field emission prior to use. Details of the measurement system have been reported elsewhere [9,10]. While the STM measurements were performed under UHV, the sample was exposed to the atmosphere for a period of approximately 20 minutes during transfer from the MBE system to the STM chamber. After initial measurements were made, the sample was stored in a nitrogen filled desiccator for ~ 25 hours to further study the effect of ambient on the electronic properties of the LTG:GaAs.

In STM spectroscopy, the density of states (DOS) is calculated from the measured current (I) versus voltage (V) relationship by [11]

$$\text{DOS} \propto \frac{dI}{dV} \times \frac{V}{I}$$

Dividing the differential conductivity dI/dV by I/V serves to remove the slowly varying transmission function inherent in all I(V) data, resulting in a quantity that more closely mirrors the desired DOS. As suggested by Martensson and Feenstra, smoothing the conductance (I/V) to reduce DOS features provides a better approximation to the tunneling transmission function [11]. Satisfactory results were obtained by smoothing the conductance with a one pole, low-pass Fourier filter algorithm with a pole frequency specified by $1/\delta = 1/1.6\ V^{-1}$. By choosing δ larger than E_g, the semiconductor gap (1.43 eV for GaAs), the low-pass Fourier filter suppresses DOS features within the band gap. As a result, the transmission function (which should not depend on the gap structure) can be recovered to a better approximation.

Fig. 2 shows the normalized conductance obtained after 20 minutes exposure to ambient air. The dashed line is a representative scan at a specific spatial location, while the solid line is the average of the scans at 100 spatial locations in 50x70 nm area. The effective conduction and valence band edges (marked by E_c and E_v), along with a band of gap states near the valence band edge are observed. In order to have a well-defined criterion for locating the band edges, the inflection point determined from the second derivative of I(V) was used to define an effective band edge. Although this procedure probably overestimates the size of the band-gap, it does provide a reliable way to compare band-gaps from I(V) data obtained at different locations. Following this procedure, the measured effective gap was found to be 1.58eV in Fig. 2, a value slightly larger than the band gap of bulk GaAs (1.43eV). From Fig. 2, the gap states are centered at a sample bias voltage near -0.64V. Using the valence band edge as a reference, this translates into a state located 0.54 eV above the effective valence band edge. The location of this feature is similar to the one observed in UHV-cleaved, n-doped LTG:GaAs [4, 5]. The observation of a band gap, along with a band of midgap states above the valence band edge, indicates that the LTG:GaAs surface layer does not significantly oxidize during atmospheric exposure and confirms the defect-assisted tunneling model for the contact structure. From the array of I(V) data taken at 100 uniformly spaced points in a 50 nm x 70 nm area, it is also possible to assess the integrity of the LTG:GaAs layer. In 88 of the 100 scans taken at different spatial locations, evidence for a clear gap state peak is found. In the other 12 scans, noisy data resembling the GaAs band gap was observed. The spatial distribution of midgap state density is consistent with the previous reports of defect densities in LTG:GaAs [3-5].

During the course of the measurements described above, the sample was stored in the UHV chamber for a period of ~ 3 weeks. During this time, no significant degradation of the mid-gap states was detected. In order to further assess the stability of the LTG:GaAs layer, the sample was removed from the UHV chamber and stored in a nitrogen filled desiccator for 25 hours. The GaAs band edges as well as the gap states can be readily resolved without dramatic change from the initial measurements. This data shows that the electronic properties of the structure are stable and supports the claim that LTG:GaAs does not rapidly oxidize upon exposure to air.

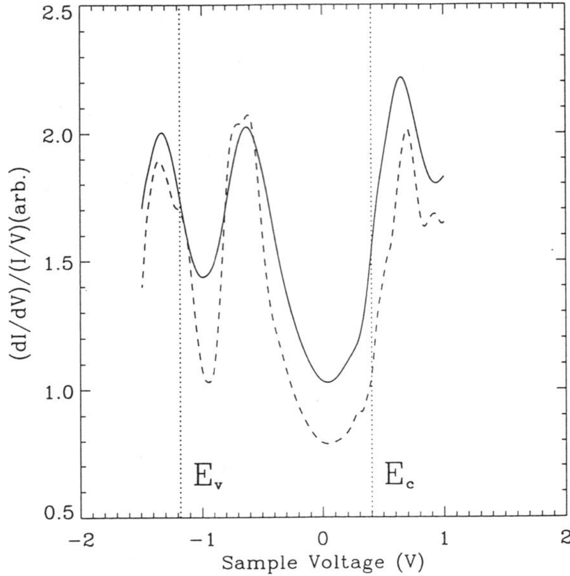

Fig. 2.   Normalized conductance of LTG:GaAs as a function of sample voltage after 20 minutes exposure to ambient. The dashed line is a typical set of data obtained at an arbitrary location of the Pt/Ir tip. The solid line is the average of data obtained from 100 different spatial locations in 50nmx70nm area.

X-Ray Photoelectron Spectroscopy

In order to provide a more detailed study of the oxidation characteristics of LTG:GaAs after air exposure, X-ray photoelectron spectroscopy (XPS) studies were performed on the layer structure shown in Fig. 1. A control sample identical to the first sample, but without the LTG:GaAs surface layer, was prepared using the same growth conditions and was exposed to atmosphere simultaneously with the LTG:GaAs sample. XPS measurements were performed on the samples following exposure to atmosphere for i) a period of approximately 60 minutes during transfer from the MBE system to the XPS chamber and ii) an additional 31 hours under illumination of fluorescent light in a clean air laminar flow chemical hood. Details of the UHV XPS system, the measurement conditions and the interpretation of specific peaks have been provided elsewhere [12].

The XPS measurement detected peaks corresponding to the oxygen 1s level and the $2p_{3/2}$ and 3d levels for both Ga and As. The horizontal axis of the spectra represents binding energy. Higher binding energy corresponds to lower kinetic energy for the emitted electrons, which translates to shorter escape depth from the material and hence better surface sensitivity. Based on published photoelectron escape depths for Ga and As levels [13, 14] and the 45° take-off angle used in this study, the depth sensitivity of the XPS measurement is estimated to be two to three monolayers for the $2p_{3/2}$ levels and six monolayers for the 3d levels

Fig. 3 presents the XPS spectra for the LTG:GaAs and the control samples after approximately 60 minutes of exposure to atmosphere. For regions where multiple peaks overlap,

Gaussian peaks used to fit the data are shown along with the measured curves. Fig. 3a shows the oxygen 1s peak at 531 eV. The peak area for the LTG:GaAs is only about 20% of that for the control sample, indicating a significantly reduced level of surface oxidation in the LTG:GaAs sample. The effect is also evident in Figs. 3b and 3c, which show the As and Ga $2p_{3/2}$ level peaks. The As $2p_{3/2}$ peak of the control sample shows a well resolved Ga-bonded As peak at 1322.6eV and an oxide peak at 3.1 eV towards higher binding energy, similar to cases reported for GaAs substrate surface after exposure to air [15, 16] and also after chemical preparations for MBE growth [17]. This has been identified as As bonded to oxygen in a chemisorbed phase which relates to surface Fermi level pinning [14]. In contrast, the LTG:GaAs has a sharp Ga-bonded As peak and only a small shoulder at about 2.5 eV towards the higher energy side. The Ga $2p_{3/2}$ peak for LTG:GaAs can be curve fitted by a Gaussian peak at 1116.9 eV plus another small Gaussian peak at 1.3 eV higher binding energy which consists of only about 8% of the total area under the curves. Using the same peak positions, curve fitting of the Ga $2p_{3/2}$ peak for the control sample shows the second peak to consist of about 40% of the total area.

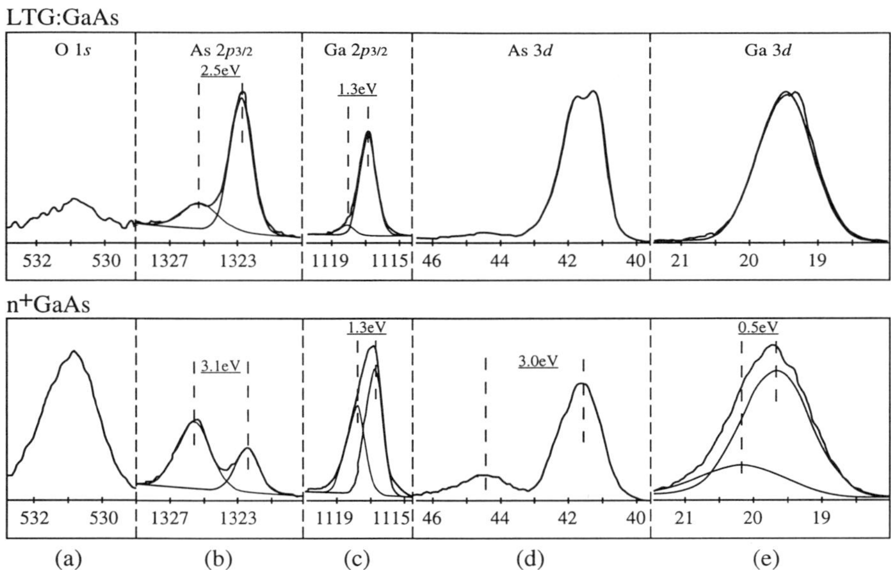

Fig. 3.   X-ray photoelectron spectra for LTG:GaAs sample and control sample, following atmospheric exposure of approximately 60 minutes.
a) O 1s peak
b) As $2p_{3/2}$ peak
c) Ga $2p_{3/2}$ peak
d) As 2d peak
e) Ga 2d peak

Figs. 3d and 3e show the more bulk sensitive As and Ga 3d level peaks. The As peak on the control sample again shows the characteristic oxide peak shifted by about 3.0 eV from the Ga-bonded As at 41.5 eV binding energy [13, 17, 18]. This feature is hardly discernible on the LTG:GaAs. In Fig. 3e, the Ga 3d peak for LTG:GaAs can be reasonably well approximated by a single Gaussian peak at 19.5 eV, indicating relative absence of components other than the

As-bonded Ga. In contrast, the corresponding peak for the control sample is wider and can be shown to consist of two Gaussian peaks 0.5 eV apart.

Fig. 4 illustrates the XPS spectra taken under the same conditions and parameters following the 31 hour air exposure. The oxygen 1s peaks for the LTG:GaAs and the control sample, shown in Fig. 4a, have become comparable in intensity and FWHM (1.8 eV), and can both be very well approximated by single Gaussian peaks. The observation strongly suggests that equal amounts of oxygen have been incorporated into the surface of both samples. This is also supported by the changes in the Ga $2p_{3/2}$, Ga 3d and As 3d peaks. There is also evidence that the As has evolved to a higher oxidation state, e.g. As_2O_5 [12]. From this behavior, it is apparent that the LTG:GaAs layer does oxidize to some extent when exposed to atmosphere under illumination. The XPS is essentially a surface probe, so the thickness of the oxidation layer cannot be directly determined. Other experiments, including the nonalloyed contact study, indicate that this layer is less than 2 nm thick even following prolonged air exposure.

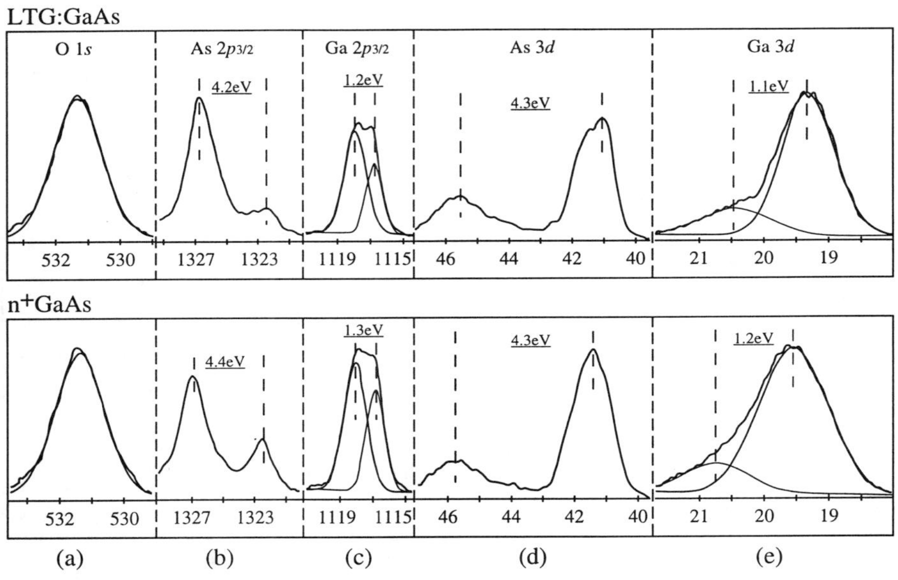

Fig. 4. X-ray photoelectron spectra for LTG:GaAs sample and control sample, following atmospheric exposure of approximately 31 hours.
a) O 1s peak
b) As $2p_{3/2}$ peak
c) Ga $2p_{3/2}$ peak
d) As 2d peak
e) Ga 2d peak

RESULTS

Oxidation Mechanism

A qualitative picture of the oxidation mechanisms can explain the difference in oxidation characteristics between the LTG:GaAs sample and the control sample. The oxidation is primarily via photo-oxidation, in which 6 holes are required to oxidize a single GaAs pair [19].

The rate of surface oxidation should therefore increase with increasing steady-state density of excess holes generated within a minority carrier (hole) diffusion length of the surface. In n-type stoichiometric GaAs, the minority carrier lifetime and diffusion length are relatively large, so the surface rapidly oxidizes during exposure to air under illumination. This process continues until the surface oxide thickness and midgap surface Fermi level pinning are sufficient to stabilize the surface during air exposure [19]. Kink sites, which are present on this surface, may also play a role in the decomposition of O_2 and H_2O at the surface [20] In unannealed LTG:GaAs, the bulk Fermi level is pinned approximately 0.4 eV below the conduction band edge [21], so the layer is lightly n-type. However, the short minority carrier lifetime (100 fs) and corresponding short minority carrier diffusion length in the LTG:GaAs material result in a lower minority carrier concentration near the surface than would be found in stoichiometric material for the same illumination intensity, and therefore a lower photo-oxidation rate. While midgap surface Fermi level pinning may occur, the associated band bending will be small in both energy and spatial extent, due to the bulk Fermi level pinning and the high space charge density associated with the antisite defects. There is also evidence that the (100) surface of LTG:GaAs does not reconstruct and therefore that kink sites do not form on this surface [22]. Since the XPS technique is only sensitive to atoms located within a few monolayers of the surface, it is not possible to determine from this measurement whether the thickness of the oxide layer formed on the LTG:GaAs sample after long air exposure is different from that of stoichiometric GaAs. However, scanning ellipsometry studies on comparable samples following air exposure indicate that the total oxide thickness on the LTG:GaAs surface is significantly less than that of an n-type surface [23].

While this study involved a 10 nm thick layer of LTG:GaAs on a heavily n-doped GaAs layer, the inhibited oxidation is associated with a bulk effect in the LTG:GaAs layer and therefore should occur in most semiconductor structures which are capped by a nonalloyed LTG:GaAs layer of sufficient thickness to provide bulk behavior, i.e., a region in the LTG:GaAs in which flat-band conditions exist. Based on the fact that layers of LTG:GaAs as thin as 2 nm are effective in providing low-resistance ohmic contact layers to n-type GaAs [7], the minimum thickness of LTG:GaAs required to achieve this bulk behavior is believed to be less than 2 nm.

Device Applications

In addition to the nonalloyed contact structure discussed in the Introduction, preliminary studies of other device applications have been performed. The first application involves patterning the LTG:GaAs cap layer using chemical etching techniques. As noted previously, thin surface layers of LTG:GaAs can provide effective passivation for semiconductor surfaces, i.e. they remain unoxidized during exposure to air and provide a controlled effective surface potential. It is therefore possible to fabricate thin n+ regions capped using the LTG:GaAs and make ohmic contact to the structure. Controlled stripping of the 2-5 nm thick cap layer will eliminate this surface passivation effect and will result in a surface with a surface depletion layer comparable to that found in stoichiometric GaAs samples. The removal of the LTG:GaAs cap will also lead to the oxidation of the n++ layer immediately below the cap, making that layer ineffective in device applications.

Patterning of the LTG:GaAs passivation layer has been demonstrated by shallow etching using chemical oxidation/etching techniques. The layer structure used for this etching experiment consists of a 10 nm thick heavily doped ($\sim 10^{20} cm^{-3}$) layer grown at normal growth temperature, capped by a 3.5 nm thick LTG:GaAs layer grown at 250 °C, all grown on a semi-insulating GaAs substrate. Transmission line method (TLM) patterns, consisting of 20-60μm x 120μm Ti/Au pads separated by 3-90μm, were defined on ths structure. The measured sheet resistance of the as-grown layer structure was R_{sh}= 633 Ω/□ with a contact resistance ρ_c= $7 \times 10^{-7} \Omega cm^2$. The variation across the sample was less than 15%. The TLM sample was then exposed to a chemical oxidation/ oxide etch sequence, in which each iteration consisted of an oxidization in H_2O_2 for 1 minute, followed by chemical stripping of the oxide in 1:8::HCl:DI for 30 seconds and a DI water rinse. Each such iteration allows controlled stripping of 3-4 nm of GaAs material. Although not shown in the figure, the resistance between the pads was measured after each iteration. Fig. 5 illustrates the measured resistance versus pad separation for various etch iterations, including the "as-grown" film (no etch iterations). The resistance between the pads exceeded few tens of KΩ after three such iterations. The drastic changes in the sheet resistance between successive etch iterations beyond the first iteration are associated with the re-pinning of the surface at midgap for the underlying n+ layer and associated depletion of the doped layer. It is also possible that the etch rate of the stoichiometric GaAs is greater than that of the protective LTG:GaAs cap, so that later etch steps remove more material per iteration. Preliminary experiments indicate that it is also possible to pattern the structure using photo-oxidation. The oxidation of LTG:GaAs layers requires higher illumination levels and/or longer times than would be required to oxidize comparable thicknesses of stoichiometric GaAs. This observation is consistent with the model of inhibited oxidation discussed earlier. The ability to effectively pattern semiconductor active areas by etching less than 5 nm of material could provide a useful capability for the realization of nanometer scale electronic devices. The shallow etching techniques eliminate the need to employ highly energetic beams and result in nearly planar surfaces.

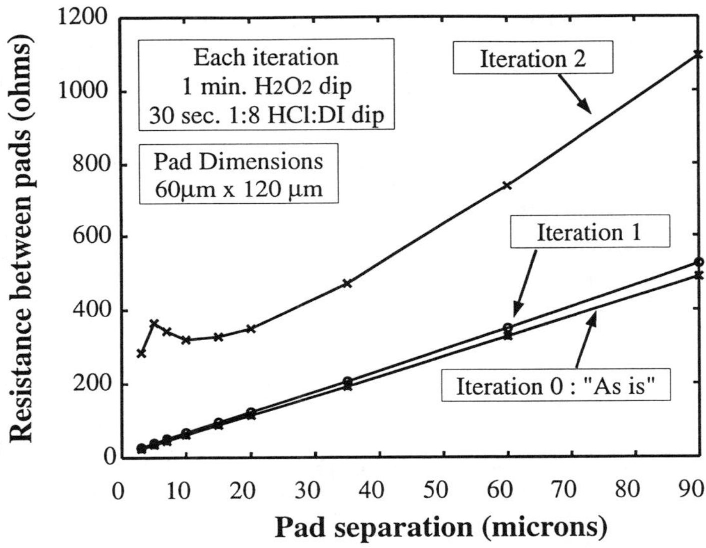

Fig. 5   Measured resistance versus TLM pad separation after various chemical oxidation/oxide strip iterations

The second application is for cap layers for field-effect transistors (FETs) such as GaAs metal-semiconductor FETs (MESFETs) or high-electron mobility transistors (HEMTs). By employing a LTG:GaAs top layer and an underlying n++ layer, a single structure can provide both low-resistance, nonalloyed ohmic contacts to the channel and a cap layer to provide high sheet conductivity and controlled surface potential in the region between the source/drain electrodes and the gate recess. Comparable cap layers employing thicker structures are commonly employed to reduce parasitic source and drain resistances in FETs.

MESFETs have been fabricated using a structure grown on a semi-insulating GaAs substrate and comprised of (in order of growth) i) a 150 nm thick n-doped channel layer ($\sim 2 \times 10^{17} cm^{-3}$) ii) a 5.0 nm n-doped ($\sim 1 \times 10^{16} cm^{-3}$) spacer layer iii) a cap layer consisting of 10 nm of n++ GaAs ($\sim 1 \times 10^{20} cm^{-3}$) and a 3.5 nm thick LTG:GaAs surface layer. Following definition of source and drain metallizations using liftoff, the devices were isolated using a $H_2SO_4:H_2O_2:DI$ etch solution. TLM patterns were simultaneously fabricated on the same sample to characterize the source/drain contacts and the sheet resistance of the channel and cap layer structure. Curve A of Fig. 6 shows the measured source to drain resistance, normalized to device width, versus source to drain spacing. The sheet resistance and specific contact resistance obtained from this data were $R_{sh} \sim 237$ $\Omega/\square$ and $\rho_c \sim 7 \times 10^{-7} \Omega cm^2$, respectively. A single photoresist step was used to pattern the gate recess etch and gate metallization. Before metallization, gate recesses were etched by exposing the samples to 3-5 chemical oxidation and etch iterations as described in the earlier section. The oxidation and etch procedure causes the stripping of controlled thicknesses of the LTG:GaAs layer and the underlying n+ GaAs layer thus permitting the formation of a good Schottky gate. The gate length in most of the MESFET devices fabricated were about 2μm with drain-source lengths of 4-8μm. A Ti/Au metallization (300-1000Å Ti/2000-3500Å Au), defined using liftoff, was used for both the source/drain contacts and the gate contacts. The only difference between the formation of the two different types of contacts is the pretreatment of the LTG:GaAs surface before the metallization.

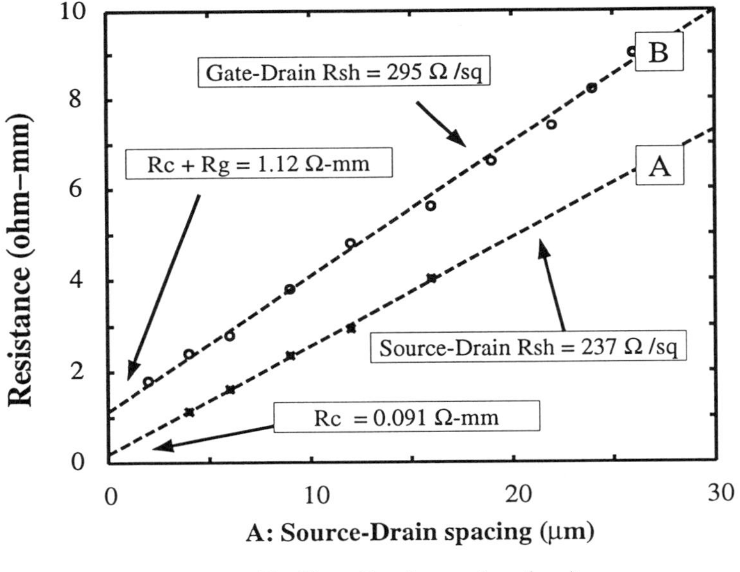

A: Source-Drain spacing (μm)

B: Gate-Drainspacing (μm)

Fig. 6   Source to drain resistance versus drain to source spacing for ungated TLM pattern (curve A) and gate to drain resistance versus gate to drain spacing for MESFET in floating gate configuration (curve B). In each case, the resistance is normalized to device width.

It is believed that the use of this process would allow one to fabricate MESFETs with a self-aligned sub-micron gate and with low drain/source access resistance due to the surface pinning and n++ layer activation associated with the LTG:GaAs in the regions external to the gate. This is evidenced by the floating gate measurement of gate to drain resistance for MESFETS, i.e. the ratio between the gate to drain voltage and the source to drain current with the gate floating. In the floating gate case, the gate to drain resistance is the sum of a sheet resistance term and a gate to channel resistance, presumed to be constant for a constant gate dimension. Curve B in Fig. 6 presents the measured gate to drain resistance versus gate to drain spacing. The slope of this curve can be used to determine the sheet resistance of the region between the gate and the drain metallization in the MESFET configuration. The measured sheet resistance of the capped region external to the gate region is $R_{sh} \sim 295\ \Omega/\square$, indicating that the cap layer is intact. Devices in which the gate recess etch consisted of 5 oxidation/etch iterations (~ 15 nm etch depth) show low gate leakage currents and well behaved common source characteristics (drain current versus drain-source voltage). For a device with a gate length of $\sim 2\mu m$ and a spacing between the drain source metallizations of $\sim 6\mu m$, the measured I_{DSS} of the device is ~ 232 mA/mm with peak transconductance of $g_m \sim 105$ mS/mm. While the MESFET performance has not been optimized, these results indicate that the LTG:GaAs cap layers can provide low resistance access layers while maintaining low gate leakage.

CONCLUSIONS

Measurements including STM spectroscopy and XPS have shown that LTG:GaAs surface layers oxidize less rapidly than layers of stoichiometric GaAs during air exposure. In addition, the midgap band of defect states responsible for midgap Fermi level pinning in LTG:GaAs are observable using STM spectroscopy following air exposure. Layers of LTG:GaAs with thicknesses of 2-5 nm therefore represent chemically stable surfaces and provide effective surface passivation due to the inhibited oxidation and bulk Fermi level pinning in the material. A qualitative model for the oxidation mechanism in LTG:GaAs has also been presented. Finally, device applications including low-resistance non-alloyed ohmic contacts, patterning of active areas using shallow etching and FET cap layers have been described.

ACKNOWLEDGMENTS

This work was partially supported by the NSF MRSEC program under Grant 9400415-DMR, the Army Research Office URI program under Contract DAAL03-G-0144 and by DARPA/ARO under grant DAAH04-96-1-0437. We would like to thank T. P. Chin, Prof. R. L. Gunshor and Prof. M. R. Melloch for technical assistance and helpful discussions.

REFERENCES

1. M. D. Pashley, K. W. Haberern, W. Friday, J. M. Woodall and P. D. Kirchner, Phys. Rev. Lett. **60,** 2176 (1988)

2. S. Gwo, K. -J. Chao, A. R. Smith, C. K. Shih, K. Sadra and B. G. Streetman, J. Vac. Sci. Technol. **B11,** 1509 (1993).

3. M. R. Melloch, J. M. Woodall, E. S. Harmon, N. Otsuka, F. H. Pollak, D. D. Nolte, R. M. Feenstra and M. A. Lutz, Annu. Rev. Mater. Sci **25,** 547 (1995).

4. R. M. Feenstra, J. M. Woodall and G. D. Pettit, Phys. Rev. Lett. **71,** 1176 (1993).

5. R. M. Feenstra, A. Vaterlaus, J. M. Woodall and G. D. Pettit, Appl. Phys. Lett. **63,** 2528 (1993).

6. M. R. Melloch, D. D. Nolte, J. M. Woodall, J. C. P. Chang, D. B. Janes and E. S. Harmon, CRC Critical Rev. in Solid State and Mat. Sci. **21,** 189 (1996).

7. M. P. Patkar, T. P. Chin, J. M. Woodall, M. S. Lundstrom and M. R. Melloch, Appl. Phys. Lett., **66,** 1412 (1995).

8. A. D. Katnani, P. Chiaradia, H. W. Sang, Jr. and R. S. Bauer, J. Vac. Sci. Technol. B **2,** 471

(1984).

9. S. Hong, D. B. Janes, D. McInturff, R. Reifenberger and J. M. Woodall, Appl. Phys. Lett. **68,** 2258 (1996).

10. M. Dorogi, J. Gomez, R. Osifchin, R. P. Andres and R. Reifenberger, Phys. Rev. B **52,** 9071 (1995).

11. P. Martensson and R. M. Feenstra, Phys. Rev. B **39,** 7744 (1988).

12. T.-B. Ng, D. B. Janes, D. McInturff and J. M. Woodall, to appear in Appl. Phys. Lett.

13. E. Huber and H. L. Hartnagel, Solid State Electron. **27,** 589 (1984).

14. W. E. Spicer, I. Lindau, P. Pianetta, P. W. Chye and C. M. Garner, Thin Solid Films **56,** 1 (1979).

15. T. Ishikawa and H. Ikoma, Jpn. J. Appl. Phys. **31,** 3981 (1992).

16. H. Ohno, M. Motomatsu, W. Mizutani and H. Tokumoto, Jpn. J. Appl. Phys. **34,** 1381 (1995).

17. J. Massies and J. P. Contour, J. Appl. Phys. **58,** 806 (1985); J. P. Contour, J. Massies and A. Salettes, Jpn. J. Appl. Phys. **24,** L563 (1985).

18. W. Storm, D. Wolany, F. Schroder, G. Becker, B. Burkhardt, L. Wiedmann and A. Benninghoven, J. Vac. Sci. Technol. **B12,** 147 (1994).

19. H. Gerischer, J. Vac. Sci. Technol. **15,** 1422 (1978).

20. W. A. Goddard III, J. J. Barton, A. Redondo and T. C. McGill, J. Vac. Sci. Technol. **15,** 1274 (1978).

21. A. C. Warren, J. M. Woodall, P. D. Kirchner, X. Yin, X. Guo, F. H. Pollak and M. R. Melloch, J. Vac. Sci. Technol. **B,** 10, 1904 (1992); H. Shen, F. C. Rong, R. Lux, J. Pamulapati, M. Taysing-Lara, M. Dutta, E. H. Poindexter, L. Calderon and Y. Lu, Appl. Phys. Lett. **61,** 1585 (1992).

22. M. R. Melloch, D. C. Miller and B. Das, Appl. Phys. Lett. **54,** 943 (1989).

23. F. Pollak, private communication.

SURFACE RECONSTRUCTION AND MORPHOLOGY OF HYDROGEN SULFIDE TREATED GaAs (001) SUBSTRATE

Jun SUDA*, Yoichi KAWAKAMI, Shizuo FUJITA, Shigeo FUJITA
Department of Electronic Science and Engineering, Kyoto University, Kyoto 606-01, Japan
*suda@kuee.kyoto-u.ac.jp

ABSTRACT

We report several new results in hydrogen sulfide (H_2S) treatment of a GaAs (001) substrate. Surface reconstruction and morphology were investigated by *in situ* reflection high energy electron diffraction (RHEED) and *ex situ* atomic force microscopy (AFM) in terms of the annealing temperature and the H_2S irradiation sequence. A (4×3) GaAs surface was obtained by annealing the substrate under H_2S irradiation (4×10^{-7} Torr). The surface was atomically flat, i.e., large terraces with monolayer steps were clearly observed. A (2×6) S-terminated GaAs surface was obtained by irradiation H_2S at 300°C on a Ga-terminated surface, which was formed by annealing at 580°C in high vacuum. The molecular beam epitaxy (MBE) growth of ZnSSe-based semiconductors on the (4×3) surface results in high quality structures such as a novel ZnSSe/ZnMgSSe tensile-strained quantum well (QW).

INTRODUCTION

Sulfur treatment of GaAs is very useful not only for device application but also for substrate modification for subsequent crystal growth. For example, ZnSe based II-VI compound semiconductors, which are promising materials for short wavelength light emitting devices, are generally grown on GaAs substrates, since high quality ZnSe substrates are not provided commonly. To obtain a high quality ZnSe layer on GaAs, it is very important to control the surface condition of the GaAs substrates.

As one of the solutions, a two chamber molecular beam epitaxy (MBE) system is used to grow a GaAs epitaxial buffer layer and to control the surface stoichiometry. Recently, it was reported that (2×4) As-stabilized GaAs surface followed by pre-irradiation of Zn brought layer-by-layer growth of ZnSe and very high quality ZnSe epilayers[1].

As other approaches, surface treatment techniques such as ammonium-sulfide treatment[2] and hydrogen plasma cleaning[3] have been proposed. These techniques, if they contribute to high quality ZnSe, are more handy and convenient, and will be applicable to regrowth of II-VI on patterned III-V layers for future novel opto-electronic integrated devices.

Recently, we proposed *in situ* surface treatment of GaAs substrates by using hydrogen sulfide (H_2S) for the MBE growth of ZnSe[4]. In order for the development of this new technique, it is very important to investigate relation between the surface structure (reconstruction, morphology, etc.) and the treatment conditions. In this paper, surface reconstruction and morphology of H_2S treated GaAs substrates are investigated by *in situ* reflection high energy electron diffraction (RHEED) and *ex situ* atomic force microscopy (AFM) in terms of annealing temperature and H_2S irradiation sequence. In addition,

Mat. Res. Soc. Symp. Proc. Vol. 448 © 1997 Materials Research Society

successful application of this technique to MBE growth of ZnSe-based semiconductors, including a new tensile-strained quantum well structure, is also mentioned.

EXPERIMENT

Substrates used in the experiments were Zn-doped p^+-GaAs (001) just oriented wafers. The substrates were prepared by the standard cleaning and etching procedures. Then, they were mounted onto a molybdenum holder by indium welding, and loaded into the exchanging chamber immediately.

The main chamber is pumped by a diffusion pump with a Vacuum Generators CCT-150 liquid nitrogen trap. It is equipped with gas-cells and RHEED system. The background pressure is lower than 1×10^{-9} Torr after chamber baking.

H_2S gas (99.99% purity) was introduced through the gas cell whose temperature was kept at 100°C to avoid condensation of the gas. At this temperature, H_2S does not readily thermally decompose. Therefore the species impinging onto the sample surface are uncracked H_2S molecules. The flow rate of H_2S was controlled by a mass flow controller. Beam pressure was measured by a movable ion gauge at the substrate position.

The substrate surface was characterized by *in situ* RHEED (10 keV) and *ex situ* atomic force microscope (AFM) observations. In order for the AFM observation, the substrate heater and H_2S irradiation were switched off when a desirable surface was obtained, then after cooling down to 300°C the sample was loaded to the exchanging chamber and transferred to the AFM system in the air as soon as possible. It is thought that the surface morphology is basically unchanged during the cooling down and in the air.

RESULTS

Annealing of GaAs substrates under H_2S irradiation

Afer loading into the main chamber, a GaAs substrate was heated up to 300°C. Then H_2S irradiation was started and the substrate temperature was increased at a rate of 10°C/min. *In situ* RHEED patterns and *ex situ* AFM images of a GaAs surface with increase of substrate temperature under H_2S irradiation are summarized in Fig. 1. In this experiment, the H_2S beam pressure was set at 4×10^{-7} Torr, which is much smaller than that employed in previous studies of H_2S treatment[5, 6].

Below 500°C, a halo pattern originated from the surface oxide layer was observed. Then, faint streaks began to appear with increasing the substrate temperature. At about 590°C, the intensity of RHEED greatly increased and an arrow head-like spotty pattern appeared as shown in Fig. 1(a). Desorption of the surface oxide seems to begin at this temperature. Fig. 1(a) shows an *ex situ* AFM image correspond to this surface. A wavy surface, which may be due to partial oxide desorption, was observed.

After 10–20 seconds, the RHEED pattern changed into sharp streaks with (4×3) reconstruction as shown in Fig. 1(b). Laue zone reflections were clearly observed for $[1\bar{1}0]$ direction. Fig. 1(b) shows an AFM image for this surface. The surface was atomically flat with monolayer height (~ 3 Å) steps. Obtaining such a flat surface only by annealing the GaAs substrate under H_2S irradiation is the new finding and will be useful not only for a preparation for subsequent growth but also for other applications. A (4×3) reconstruction

was known for a GaAs surface thermally annealed with selenium irradiation[7]. Although
an atomic arrangement of the (4×3) structure in ref. [7] is not clearly understood, it
was attributed to partial Se-adsorption on a GaAs surface. The (4×3) reconstruction
observed in our experiment will be the same kind of structure except for replacing Se with
S. In the view point of the growth of ZnSe on GaAs, such a flat surface is attractive since
inhomogeneous nucleation will be prevented. Also a *partially* S-adsorbed GaAs surface
will be better for beginning the ZnSe growth compared with a *completely* S-covered GaAs
surface, because mixing of Se-Ga and Zn-As bondings is required for charge valance at a
ZnSe/GaAs heterovalent interface[8].

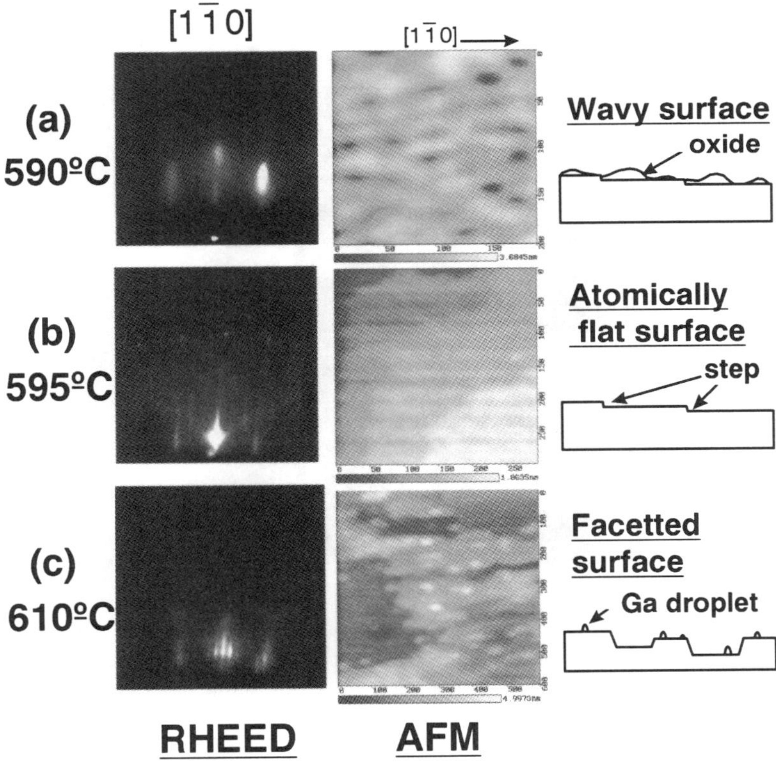

Fig. 1   *In situ* RHEED patterns and *ex situ* AFM images of a GaAs surface with
increase of substrate temperature under H_2S irradiation. (a) 590°C, (b)
595°C and (c) 610°C.

If the substrate temperature is further increased up to 610°C, the RHEED pattern
changed into arrow head-like streaks as shown in Fig. 1(c). Although surface reconstruction seems to depend on H_2S pressure, (4×6)-like reconstruction, which may originate

from a Ga-terminated GaAs surface[9], was generally observed. If the substrate was cooled down to 300°C under H_2S irradiation, the reconstruction changed to (2×6). This change is due to S-adsorption as mentioned below. An AFM image for this surface is shown in Fig. 1(c). The surface was faceting, i.e., composed of small atomically flat terraces developed along $[1\bar{1}0]$ direction. It is thought that these terraces were formed by etching effect of H_2S. Many protrusions (50Å in height) were also observed.

H_2S irradiation on Ga-terminated GaAs surface

If a GaAs substrate is annealed carefully in high vacuum without As pressure, a Ga-terminated surface is obtained after surface oxide removal. We also investigate the effect of H_2S irradiation on this Ga-terminated surface.

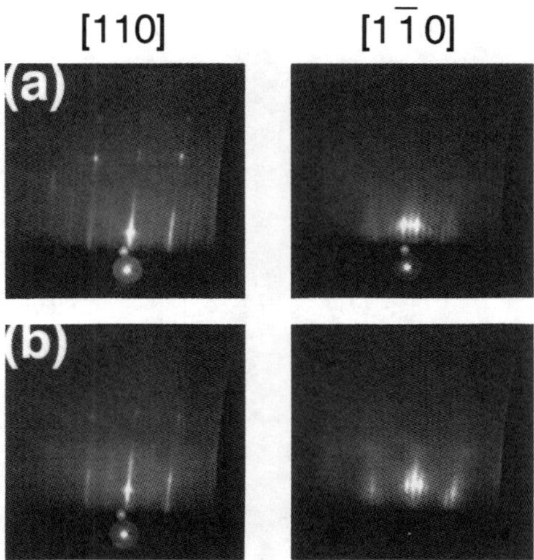

Fig. 2　RHEED patterns for GaAs substrates (a) before and (b) after 10-min irradiation of H_2S at 300°C. (a) A mixture of (3×6) and (4×6) reconstructions which are attributed to Ga-terminated surfaces. (b) A (2×6) reconstruction which is attributed to the S-terminated surface.

At first, a GaAs substrate was thermally annealed in high vacuum ($< 1 \times 10^{-9}$ Torr) to remove surface oxide. To avoid contamination, the main chamber was baked out before the experiment. After the annealing, a surface oxide layer was removed and mixture of (3×6) and (4×6) Ga-terminated surface reconstructions [9] appeared as shown in Fig. 2(a). After cooling down to 300°C, H_2S irradiation was started. As shown in Fig. 2(b), the surface reconstruction changed to (2×6) within a few minutes after the start of the irradiation. Auger electron spectroscopy (AES) shows that the surface is covered by S. Therefore, this (2×6) reconstruction is thought to be the same as the S-terminated (2×6) GaAs surface reported by Tsukamoto and Koguchi[10]. This (2×6) S-terminated surface is proposed to be energetically more stable than conventionally reported (2×1)

one. Very stable surface is attractive to being applied to surface passivation. The (2×6) GaAs surface has generally achieved by exposing a Ga-stabilized GaAs surface to S vapor. Obtaining the (2×6) surface by using H_2S is the new finding and is the more convenient technique.

<u>Application of H_2S treatment to the MBE growth of ZnSe-based semiconductors</u>

Recently, we proposed ZnSSe/ZnMgSSe tensile-strained quantum well structures for the study of light-hole excitons in the ZnSe-based system. It is known that a tensile-strained layer is more easily relaxed by introducing misfit dislocations compared to compressive one[11]. Thus, it is very important to prepare very smooth and high quality underlying layers for the growth of tensile strained systems. We applied the hydrogen sulfide treatment of GaAs, which results in atomically flat (4×3) surface, prior to the growth of tensile-strained ZnSSe/ZnMgSSe multiple quantum well (MQW) structure. As a result, TM-mode lasing due to light-hole exciton was realized at 24 K as shown in Fig. 3. This result indicates successful growth of high quality tensile-strained ZnSSe/ZnMgSSe MQW structure. Although detailed studies such as structural characterization by transmission electron microscopy (TEM) are under investigation, the present results encourage the usefulness of this H_2S treatment technique for the growth of ZnSe systems.

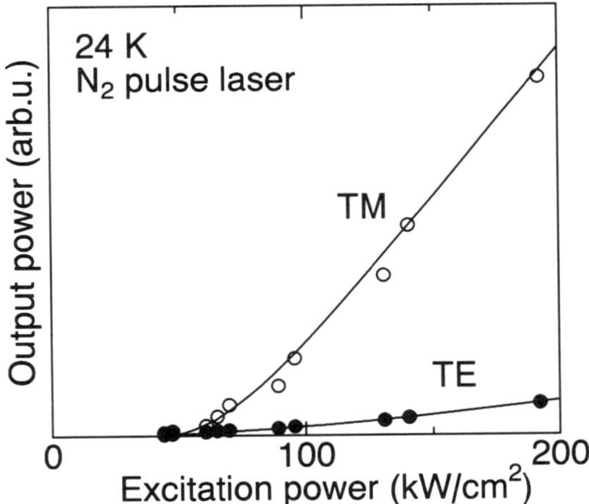

Fig. 3 Output power from cleaved edge vs. excitation power for ZnSSe/ZnMgSSe MQW structure. TM-mode lasing due to light-hole exciton occures at 428 nm. Threshould power is about 80 kW/cm^2.

CONCLUSIONS

An atomically flat GaAs surface with (4×3) reconstruction was obtained by annealing a GaAs substrate under H_2S irradiation. Successful growth of a tensile-strained ZnSSe/ZnMgSSe MQW structure was achieved on this (4×3) GaAs surface. The (2×6)

S-terminated GaAs surface was obtained by H_2S irradiation on Ga-terminated GaAs surface. The energetically stable (2×6) surface is promising for reduction of surface states.

ACKNOWLEDGEMENTS

The authors acknowledged Dr. Shiro Tsukamoto at National Research Institute for Metals for valuable discussion. We also wish to thank Masahiro Ogawa and Keiichiro Sakurai for optical characterization of ZnSSe/ZnMgSSe MQWs. One of the authors (J.S.) is supported by Research Fellowships of the Japan Society for the Promotion of Science for Young Scientists. This work was partly supported by a Grant in Aid for Scientific Research from the Ministry of Education, Science, Sports and Culture and by the Hōsō Bunka Foundation.

REFERENCES

1. L. H. Kuo, L. Salamanca-Riba, B. J. Wu, G. Hofler, J. M. DePuydt and H. Cheng, Appl. Phys. Lett. **67**, 3298 (1995).

2. Y. H. Wu, Y. Kawakami, Sz. Fujita and Sg. Fujita, Jpn. J. Appl. Phys. **29**, L1062 (1990).

3. C. M. Rouleau and R. M. Park, J. Vac. Sci. Technol. A **11**, 1792 (1993).

4. J. Suda, R. Tokutome, Y. Kawakami, Sz. Fujita and Sg. Fujita, J. Cryst. Growth to be published.

5. J. Massies, F. Dezaly and N. T. Linh, J. Vac. Sci. Technol. **17**,1134 (1980).

6. H. Kawanishi, Y. Sugimoto and K. Akita, J. Vac. Sci. Technol. B **9**, 1535 (1991).

7. S. Takatani, T. Kikawa and M. Nakazawa, Phys. Rev. B **45**, 8494 (1992).

8. H. H. Farrell, M. C. Tamargo and J. L. de Miguel, Appl. Phys. Lett. **58**, 355 (1991).

9. L. Daweritz and R. Hey, Surf. Sci. **236**, 15 (1988).

10. S. Tsukamoto and N. Koguchi, Appl. Phys. Lett. **65**, 2199 (1994).

11. K. Ozasa, M. Yuri, S. Tanaka and H. Matsunami, J. Appl. Phys. **68**, 107 (1990).

IMPROVEMENT OF InGaP/GaAs HETEROINTERFACE QUALITY BY CONTROLLING AsH_3 FLOW CONDITIONS

Yoshino K. Fukai, Fumiaki Hyuga, Takumi Nittono, Kazuo Watanabe and Hirohiko Sugahara,
NTT System Electronics Laboratories, 3-1 Morinosato Wakamiya Atsugi-shi, Kanagawa, 243-01 JAPAN, fukai@aecl.ntt.co.jp

ABSTRACT

This paper examines the conduction band offset, ΔEc, and interface charge density, σ, of disordered InGaP and GaAs heterointerfaces by controlling the AsH_3 cover time and flow rate at the growth interval from GaAs to InGaP. Short AsH_3 cover time (0.05 min) creates high ΔEc of 0.2 eV and low σ of $6.3 \times 10^{10} cm^{-2}$. Extending the AsH_3 cover time by 50 times cuts ΔEc to almost 0 eV and increases σ by one order. Interface morphology for long-AsH_3-cover-time samples observed by atomic force microscopy shows a terrace structure on GaAs surface, which means the surface is As rich. These results suggest that an As-poor GaAs surface is essential to achieving high-quality InGaP/GaAs heterointerfaces.

INTRODUCTION

InGaP has a low recombination velocity [1] and a low concentration of deep levels [2]. The InGaP/GaAs heterostructure, thus represents an alternative to the AlGaAs/GaAs heterostructure and is now being applied for electrical and optical devices [3], [4]. However, various values of energy band offset, ΔEc, and a large concentration of the interface charge, σ, have been reported for this structure, and the reason for this ΔEc variation and the origin of the interface charges have not been elucidated.

In this study, we fabricated the InGaP/GaAs heterostructure by controlling the gas flow sequence at the growth interval from GaAs to InGaP. We characterized ΔEc and σ by capacitance-voltage (C-V) measurements and observed interfacial GaAs morphology by atomic force microscopy (AFM). As a result, we clarified that ΔEc and σ change depending on the growth conditions and that reducing the AsH_3 cover time is the most effective way to create a high-quality interface with a large band offset and low interface charge density.

TARGET DEVICE STRUCTURE

A schematic cross section of the GaAs/InGaP/(In)GaAs heterostructure MESFET we developed for analog-microwave-ICs is shown in Figure 1 [5]. The (In)GaAs channel is heavily Si-doped ($\sim 5 \times 10^{18} cm^{-3}$) and the remaining layers are undoped. The source and drain regions were formed by Si ion implantation. The GaAs cap layer prevents InGaP from reacting with the refractory WSiN gate during the activation anneal process for implanted Si-ions. The n^+-(In)GaAs channel surface is passivated by InGaP which we provides a higher

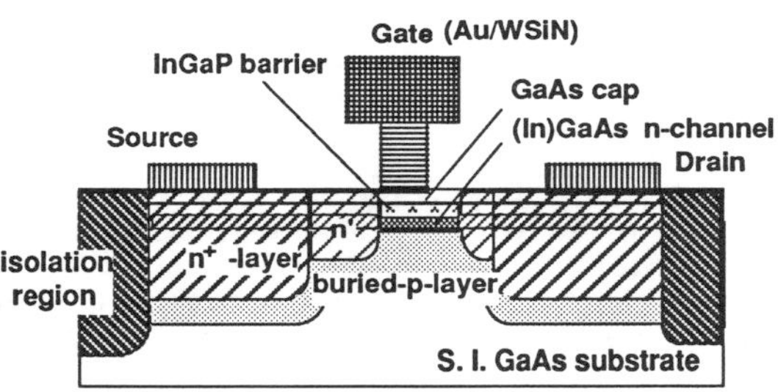

Fig. 1 Cross-section of the target device: self-alighned gate GaAs/InGaP/(In)GaAs heterostructure MESFETs.

Mat. Res. Soc. Symp. Proc. Vol. 448 © 1997 Materials Research Society

Schottky barrier. Thus, the InGaP/(In)GaAs interface quality is quite important to get high device performance.

EXPERIMENTAL

GaAs and InGaP layers were grown on 3-inch-diameter (100) GaAs substrate by low pressure (60 Torr) metal organic chemical vapor deposition (MOCVD) at a growth temperature of 500 °C. Both layers were 3000-A thick and doped to 1 to 2 x 10^{17} cm $^{-3}$. InGaP layers were grown in the disordered state to obtain high ΔEc [6] and their energy gap at room temperature was 1.9 eV, which was determined by photoluminescence (PL) measurement. InGaP layers were lattice matched to GaAs layers and the In-composition determined from X-ray diffraction measurements was 0.48.

The gas flow sequence at the growth interval was as follows, (1) AsH_3 was continuously supplied to cover GaAs after the Ga source was switched off. (2) After the AsH_3 flow was stopped, PH_3 purge gas was switched on to remove AsH_3 from around the sample. (3) After that, InGaP source gases were supplied. The timing of these gas changed at the growth interval is shown in Fig. 2. In this study, the AsH_3 cover time and flow rate were changed to investigate the influence of As on the interface characteristics. The PH_3 purge time was fixed at 2.5 min and the PH_3 flow late was fixed at 300 cc/min.

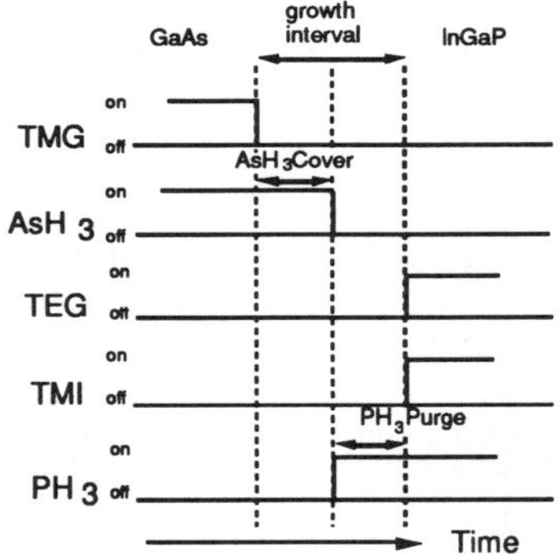

Fig. 2 Gas flow sequence for growth of the InGaP/GaAs structure. GaAs surface was covered with AsH_3 and subsequently purged by PH_3 at the GaAs-to-InGaP growth interval. PH_3 flow conditions were kept constant (300 cc/min and 2.5 min).

Schottky diodes with a 280 μm x 280 μm Ti/Au Schottky electrode encircled by a AuGe/Ni ohmic electrode were fabricated to measure the electrical characteristics. We measured the carrier concentration profile around the interface using C-V measurement. To avoid break down at the deep bias condition, we also fabricated diodes after etching off about half the thickness of the InGaP layer.

Since the interface morphology can not be observed in situ at the time of MOCVD, we observed the GaAs surface by AFM after removing the InGaP layers selectively by the HCl. The observation was performed in air using the tapping mode of Digital Instrument's Nanoscope III.

RESULTS
AsH_3 cover time dependence

AsH_3 cover time dependence on the heterointerface parameters was compared between 2 samples whose AsH_3 cover times were 0.05 (#2174) and 2.5 min (#2172). The AsH_3 flow rate was fixed at 100 cc/min. The carrier concentrations around the InGaP/GaAs interface for these samples are respectively shown in Figs. 3(a) and (b), by solid lines. These apparent carrier concentration profiles are quite different and the difference is related to the AsH_3 cover time. The electron depletion at the interface region in the InGaP layer is annihilated in the sample with the 2.5 min cover time.

We determined the conduction band offset, ΔEc, and interface charge density, σ, at the interfaces from these carrier concentration profiles using Kroemer's method [7]. Although the interface position is important in this method, it has some ambiguity because of the growth rate fluctuation. We defined the distance of the interface from the InGaP surface, x_i,

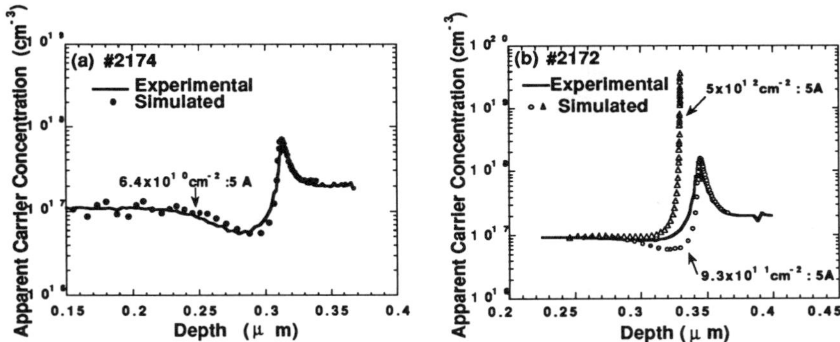

Fig. 3 Apparent carrier concentration obtained by capacitance-voltage measurement with (solid line) and by two-dimensional simulation. Sample #2174, (a), was grown with AsH$_3$ flow of 100 cc/min for 0.05 min and #2172, (b), was grown with AsH$_3$ flow of 100 cc/min for 2.5 min.

at the maximum carrier concentration position. The reason for doing so is that electrons accumulate at the interface due to the band bending and interface charges. The x_i dependence of ΔEc and σ for #2174 are shown in Fig. 4. The interface position is 0.314 μm and the obtained ΔEc and σ are 0.20 eV and 6.4×10^{10} cm^{-2}, respectively. The same process was performed for #2172 and obtained values were 0.05 eV and 9.3×10^{11} cm^{-2}. When the AsH$_3$ cover time was increased by 50 times, the conduction band offset decreased to 1/4 and the interface charge density increased by 10 times. The positive sign of the interface charge means that the charges are likely to be donors.

The apparent carrier concentration profile was reconstructed by two-dimensional simulation using the above ΔEc and σ values for sample #2174. In the simulation it was assumed that the interface charges are shallow donors and they exist at the interface in the InGaP layer to a thickness of 5A. The experimental curve and simulation, shown in Fig. 3(a) by solid circles, are in good agreement.

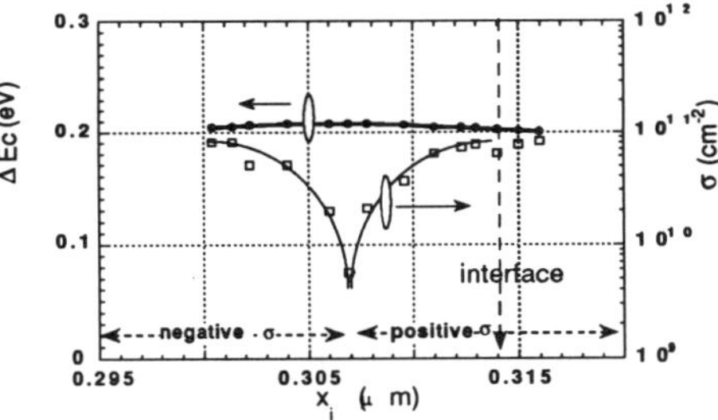

Fig. 4 Conduction band offset ΔEc and interface charge density σ of #2174 calculated by Kroemer's method using the carrier profile in Fig. 3(a).

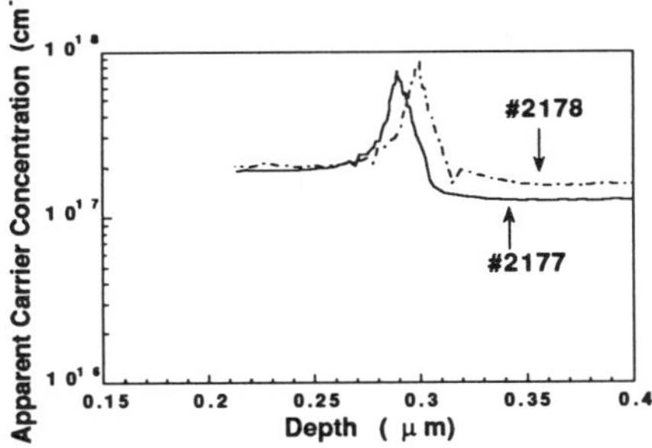

Fig. 5 Apparent carrier concentration obtained by capacitance-voltage measurement. Sample #2178 was grown with AsH$_3$ flow of 10 cc/min for 2.5 min and #2177 was grown with AsH$_3$ flow of 200 cc/min for 2.5 min.

The small ΔEc (about 0 eV) for #2172 is confirmed by 2 dimensional simulation. If the excess charge at the interface shaded the carrier depletion at the InGaP, modulation of the interface charge model would fit the reconstructed profile to experimental results with $\Delta Ec = 0.2$ eV. The interface conditions in the simulation were interface charge densities of 9.3×10^{11} and 5×10^{12} cm^{-2} for the interface charge layer thickness of 5A. The simulated carrier concentration profile is shown in Fig. 3(b). The assumption of a large interface charge density did not yield the correct profile. This means that it is not the interface charge, but the change of the band line up that is resposible for the annihilation on the carrier depletion.

AsH₃ flow rate dependence

AsH$_3$ flow rate dependence was examined with the AsH$_3$ cover time fixed at 2.5 min. The flow rates were 10 (#2178) and 200 cc/min (#2177). The results are shown in Fig.5. ΔEc and σ were determined in the manner mentioned above. The obtained values are -0.04 eV and 8.5×10^{11} cm^{-2} for #2178 and -0.01 eV and 7.3×10^{11} cm^{-2} for #2177.

The obtained parameters for the four samples are listed in Table 1. The ΔEc values for all flow rates with the long cover time are close to 0 eV and the σ values are on the order of 10^{11} cm^{-2}. This result shows that the flow rate change does not influence the interface parameters. The dominant parameter for controlling interface ΔEc and σ is AsH$_3$ cover time.

DISCUSSIONS

Samples whose InGaP layer was replaced by GaAs to form GaAs/GaAs homointerface were grown to compare it with the InGaP interface. Two samples, #2386 and #2387, were grown under the same AsH$_3$ cover condition as #2174 and #2172 at the growth interval without PH$_3$ purge. A third sample, #2407, was treated with a 2.5-min AsH$_3$ cover and 2.5-min PH$_3$ purge. The apparent carrier concentration profiles of these three samples are shown in Fig. 6. The profiles are flat over the total epitaxial growth layer. If the gas flow treatment at the growth interval had affected the GaAs layer, carrier accumulation at the interface would be expected. The flat profiles mean neither excess As nor P affect GaAs. No interface charges were created in the GaAs layer. Therefore, at the InGaP/GaAs interface, As

AsH$_3$ cover time (min)		
	0.05	2.5
10		#2178 -0.04 8.5E11
100	#2174 0.20 6.4E10	#2172 0.05 9.3E11
200		#2177 -0.01 7.3E11

(Row labels in the left column read: AsH$_3$ flow rate (cc/min))

Table. 1 The AsH$_3$ flow condition dependence of conduction band offset ΔEc (upper) and interface change density σ (lower).

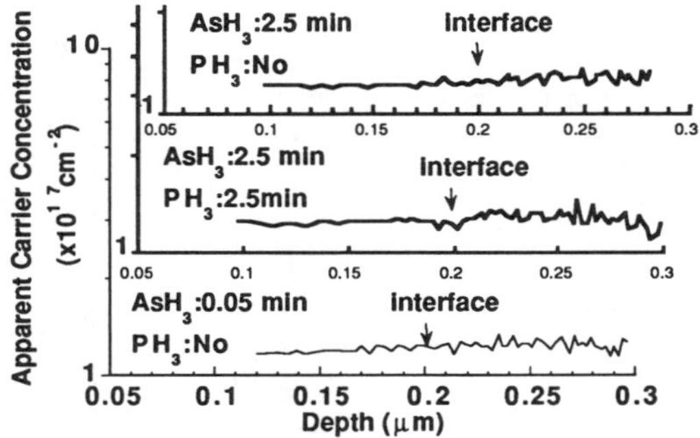

Fig. 6 Apparent carrier concentration of the GaAs/GaAs structures obtained by capacitance-voltage measurement. The interface lies 0.2 μm from the surface (depth = 0 μm). Total epitaxial-growth layer thickness is 0.4 μm.

24

is taken into the InGaP layer at the beginning of InGaP growth.

There is a possibility that a new thin interfacial layer made of In, Ga, As, and P, i.e. the quarternary $In_xGa_{(1-x)}As_yP_{(1-y)}$, is created when As affects the first stage of InGaP growth. If this is true it means the interface charge has some other origin. The photoluminescence spectrum measured at 4.2 K in Fig. 7 has a bulk InGaP line and bulk GaAs lines related to carbon and silicon impurities. The $In_xGa_{(1-x)}As_yP_{(1-y)}$ signal was not detected, although there were many possibility of energy gaps because of the unfixed atomic composition rate of x and y. Furthermore, the thin layer would show the quantum confinement effect. The InGaAsP related emission, that is frequently observed at the GaAs/InGaP interface at around 880 nm [8] was not detected. This indicates

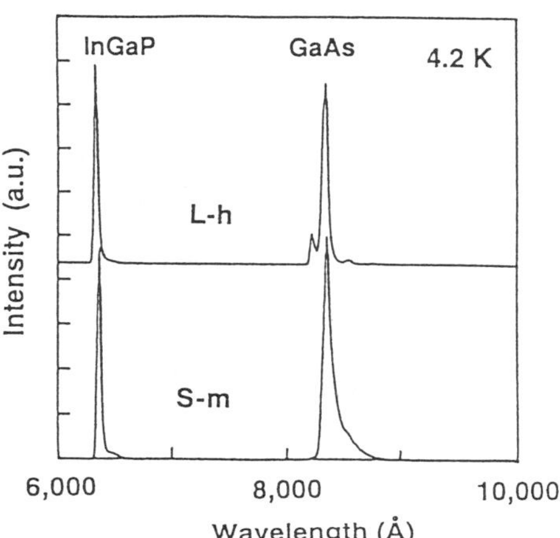

Fig. 7. Photoluminescence spectrum of #2174 (AsH₃ flow: 100 cc/min; 0.05 min) and #2177 (AsH₃ flow: 200 cc/min; 2.5 min) measured at 4.2 K.

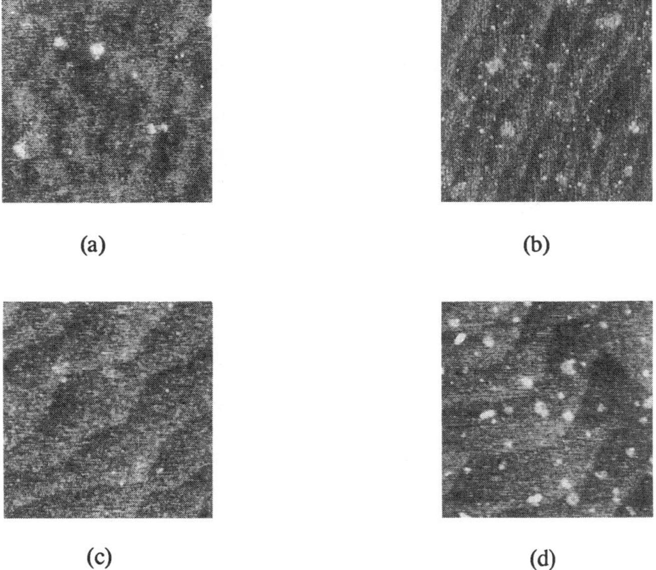

Fig. 8. GaAs surface morphology of a 1 μm x 1 μm segment of #2174 (a), #2172 (b), #2178 (c) and #2177 (d) observed by AFM. The long-AsH₃ cover-time samples (#2172, #2178, #2177) have a terrace structure on the GaAs surface.

that As/P intermixing did not occur at the InGaP/GaAs interfaces in spite of the long PH_3 purge time of 2.5 min.

Arsenic could be the origin of the interface charge, but its function is not clear. It can be donar like defects such as As+ ions or the origin of the lattice strain. The observed GaAs surface morphology is shown in Fig. 8. A terrace structure was observed for the long-AsH_3-cover-time samples and this may be related to the small ΔEc. The terraces may not directly affect the band line up of the heterostructure; however, they might trigger variations in the interface characteristics, e.g. they might induce ordering at the InGaP interface. The ΔEc reduction related to the long AsH_3 cover time might be explained if the ordering of InGaP [9], [10] at the interface due to the GaAs terrace or the interface strain [11] could be confirmed.

SUMMARY

We studied the dependence of the AsH_3 gas flow condition at the growth interval on the change in the band offset and the interface charge density of the InGaP/GaAs heterointerface. The ΔEc is reduced from 0.2 to about 0 eV when AsH_3 cover time is increased from 0.05 to 2.5 min, while σ increased from 6×10^{10} to about $8\times10^{11}cm^{-2}$. A two-dimensional simulation revealed the interface charge was found to be shallow donor type impurity. A terrace structure was observed for long-AsH_3-cover-time samples by AFM.

The variation of interface morphology of GaAs arises from excess As created by the long AsH_3 cover time at growth interval. The excess As and the morphology of GaAs probably induced the variation of ΔEc and σ. Although the function of As in InGaP is not clear at this stage, it might be As^+ ions or the origin of the lattice strain which makes donar like defects. The reason for the ΔEc reduction might be the ordering of InGaP at the interface caused by the GaAs terraces or by the interface strain.

In conclusion, reducing the AsH_3 cover time rather than controlling the AsH_3 flow rate was found to be the dominant factor in controlling ΔEc and σ and obtaining a high quality interface.

ACKNOWLEDGMENTS

The authors would like to thank T. Enoki for the two-dimensional simulation and for useful discussions and G. Araki and H. Tomaru for epitaxial film growth. We are indebted to K. Yamasaki for his encouragement throughout this work.

REFERENCES

1. J. Chen, J. R. Sites, I. L. Spaun, M. J. Hafich and G. Y. Robinson, Appl. Phys. Lett. **58**, 744 (1991).
2. K. Kitahara, M. Hoshino and M. Ozeki, Jpn. J. Appl. Phys. **27**, L110 (1988).
3. N. Nakasha, T. Miyata, Y. Watanabe, H. Ochimizu, S. Kuroda and M. Takikawa, IEEE 1994 CICC Dig. p 391 (1994).
4. T. Takahashi, S. Sasa, A. Kawano, T. Iwai and T. Fujii, IEEE IEDM94 Dig. p 191 (1994).
5. S. Sugitani, Y. Yamane, T. Nittono, H. Yamazaki and K. Yamasaki, 1994 GaAs IC Symp. Dig., pp. 123-126.
6. R. P. Schneider, Jr., E. D. Jones and D. M. Follstaedt, Appl. Phys. Lett. **65**, 587 (1994).
7. H. Kroemer et al., Appl. Phys. Lett. **36**, 295(1980).
8. F. E. G. Guimarass et al., J. Cryst. Growth **124** , 199 (1992).
9. P. Ernst, C. Geng, F.Scholz, H. Schweizer, Y. Zhang and A. Mascarenhas, Appl. Phys. Lett. **67**, 2347 (1995).
10. S. Froyen, A. Zunger and A. Mascarrenhas, Appl. Phys. Lett. **68**, 2852 (1996).
11. K. Ozasa, M. Yuri, S. Tanaka and H. Matsunami, J. Appl. Phys. **68**, 107 (1990).

PROCESSING OF InP AND GaAs SURFACES BY HYDROGEN AND OXYGEN PLASMAS: IN SITU REAL TIME ELLIPSOMETRIC MONITORING

M. LOSURDO, P. CAPEZZUTO and G. BRUNO
Centro di Studio per la Chimica dei Plasmi - CNR - Dipartimento di Chimica - Università di Bari - via Orabona, 4 - 70126 Bari, Italy

ABSTRACT

Remote radiofrequency H_2 and O_2 plasma processing of InP and GaAs surfaces was investigated by *in situ* real time spectroscopic ellipsometry. Hydrogen plasmas were used for the native oxide removal and the defect passivation of III-V surfaces. The effect of hydrogen exposure time and of crystallographic orientation (GaAs (100), (110), (111)) on the chemistry and kinetics of oxygen removal and of phosphorus/arsenic depletion was investigated. Oxygen plasma anodization was used to grow oxide films on GaAs (100), (110) and (111) substrates. The effect of bias voltage and UV-light irradiation on the chemistry and kinetics of oxidation process and on the oxide properties was studied. The composition and morphology of the InP and GaAs surfaces resulting from these plasma treatments was described.

INTRODUCTION

Recently, remote plasma treatment of III-V semiconductor surfaces has drawn the attention of many researchers due to the facilities of low processing temperatures and low material damage. Among others, hydrogen plasmas have been demonstrated to be effective in the *in situ* native oxide removal [1] and in the defect passivation [2,3] of III-V materials, and oxygen plasmas have been reported capable of producing GaAs oxidation; the resulting GaAs oxides have shown properties comparable to those of thermally grown oxides. In both cases, however, the knowledge of the involved chemistries and kinetics is still far from being complete. A common approach in the investigation of III-V plasma processing consists in exposing the surface to hydrogen and/or oxygen plasmas for an arbitrary time, and in the aposteriori evaluation of the material quality and morphology. This "trial and error" approach has produced contradictory results on the applicability of plasma processing to InP and GaAs materials. As an example, many authors report that hydrogen plasmas are not effective in the treatment of InP and GaAs materials, as the exposed surfaces loose their stoichiometry because of the surface depletion of phosphorus [4] and arsenic [5], respectively. Even more lively is the debate on the GaAs plasma oxidation process, although widely investigated in these last years. Specifically, while the oxidation kinetics has been well described and different models have been proposed [6,7], few attempts have been reported to correlate the oxidation kinetics with the resulting oxide chemistry and properties.

In this paper, we report on the use of *in situ* real time spectroscopic ellipsometry to monitor and control the dry cleaning and oxidation plasma processes of InP and GaAs substrates, and several questions concerning their chemistry and kinetics are addressed.

EXPERIMENT

The remote-plasma metalorganic chemical vapor deposition (RP-MOCVD) apparatus, used for the growth and treatment of III-V materials and described in detail previously [8], consists of a stainless steel reactor, equipped with a r.f. (13.56 MHz) plasma source and an in situ phase modulated spectroscopic ellipsometer (UVISEL-ISA Jobin Yvon). Samples used in the present

Mat. Res. Soc. Symp. Proc. Vol. 448 © 1997 Materials Research Society

study were commercially available SI (100) InP wafers and SI GaAs wafers with the three different orientations (100), (110) and (111)A. The samples were introduced, through a load lock chamber, in the MOCVD reactor without any ex situ pre-treatment, and were positioned on a molybdenum susceptor whose temperature is monitored by a K-thermocouple. All the hydrogen treatments were performed at a r.f. power of 60 watt, a pressure of 1 torr, a H_2 flow rate of 800 sccm, and the substrate temperature was changed in the range 25-350°C. For r.f. plasma oxidation, the plasma was fed by O_2:Ar=10:2 sccm mixture, the r.f. power was 100 watt, the pressure was varied in the range 7-70 mtorr, the surface temperature was 130°C, and a dc bias between 0 and +70 V was applied to the substrates. A Hg lamp, with maximum emission at 250 nm, was used to irradiate the substrate surface in order to investigate the effect of UV light on the oxidation kinetics and oxide chemistry. Ellipsometric spectra were recorded in the range 1.5 - 5.5 eV at room temperature before and after plasma treatments and were modelled by using the Bruggemann Effective Medium Approximation (BEMA) [9]. As for the process kinetics, the previous plasma treatments were monitored in real time by single wavelenght ellipsometry (SWE). The wavelength corresponding to the E_2 interband critical point (see fig. 1) was used to investigate the cleaning process of InP and GaAs, due to the high sensitivity to the surface state at E_2. To study the GaAs oxidation kinetics, the wavelenght of 359 nm, where the ellipsometric measurement is insensitive to temperature variation, was used.

A Perkin-Elmer 5300 spectrometer was used for the ex situ XPS analysis by using a standard Mg source (1253.6 eV).

RESULTS AND DISCUSSION

<u>Hydrogen plasma treatment for native oxide removal</u>

Surface temperature and exposure time are crucial parameters in hydrogen plasma treatment of III-V surfaces. The effect of surface temperature on the cleaning kinetics, reported by us in a previous paper [10], has evidenced that a substrate temperature of 230°C is the optimum to selectively remove oxygen from InP surfaces. In addition, Gottscho reported [2] that H-atom cleaning at room temperature is able to reduce As-oxide but not Ga-oxide, whereas the complete GaAs native oxide reduction can be achieved at temperature of about 200°C.

Figure 1 shows SE spectra for (100) InP and GaAs before and after hydrogen cleaning performed at T=230°C. To evaluate the quality of the surface resulting from the present hydrogen treatment, the reference SE spectra for c-InP and c-GaAs provided by Aspnes are also shown. The recorded spectra of the pseudodielectric function, $\langle \varepsilon \rangle = \langle \varepsilon_r \rangle + \langle \varepsilon_i \rangle$, have been modeled using BEMA theory and, as demonstrated previously [10], the best-fit optical model for the hydrogen plasma cleaned III-V surfaces is the one-layer model also shown in fig. 1. From this model, it comes out that the hydrogen cleaning of III-V surfaces results in a less dense top layer, including voids, due to the selective oxygen removal from GaAs and InP lattices.

Figure 2 shows the time evolution of the $\langle \varepsilon_i \rangle$ value at the E_2 point, $\langle \varepsilon_i(E_2) \rangle$, during the hydrogen exposure at T=230°C of (100) InP and of GaAs surfaces with (100), (110) and (111) crystallographic orientation. All these substrates have a 25Å-thick native oxide layer. At the plasma onset, the $\langle \varepsilon_i \rangle$ increases, so indicating that oxygen is removed from the surface. By prolonged exposure of the InP surface to hydrogen atoms, the trend reverses and the $\langle \varepsilon_i(E_2) \rangle$ value starts to decrease. This is indicative that some damage is occurring on the InP surface. In fact, the ellipsometric spectra and AFM and XPS measurements [10] have shown that, when oxygen has been completely removed, H atoms react selectively with phosphorus via PH_3 formation

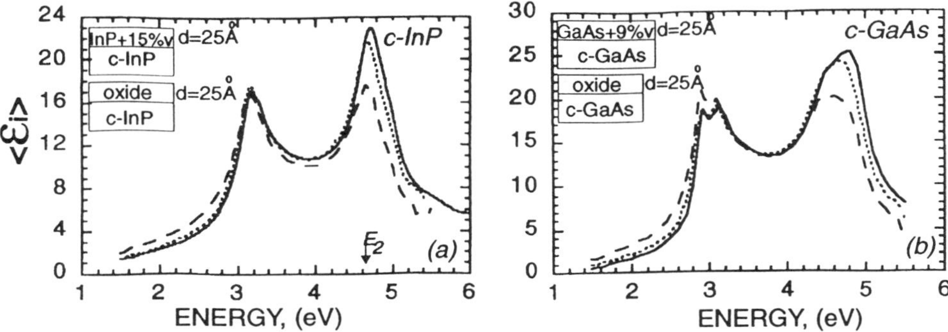

FIGURE 1. Imaginary part of the pseudodielectric function, $\langle \varepsilon_i \rangle$, of (a) InP and (b) GaAs substrates. Spectra before (dashed line) and after (dotted line) hydrogen treatment are shown. The c-InP and c-GaAs database from Aspnes are also reported. The inset shows the best-fit BEMA models.

which desorbes leaving an indium-rich surface. When the cleaning is performed at temperature as high as 350°C, the $\langle \varepsilon_i(E_2) \rangle$ value increases at first but soon after plummets because of a very high phosphorus depletion rate.

The situation for GaAs is rather different with respect to InP. For (100) GaAs surface, in fact, it is observed (see fig. 2b) that after a fast $\langle \varepsilon_i \rangle$ increase, an almost constant value is reached even at prolonged H-atom exposure. However, much longer hydrogen exposure leads to a further slight increase of $\langle \varepsilon_i \rangle$ value for (110) and (111) GaAs surfaces. This crystallographic orientation dependence of the $\langle \varepsilon_i \rangle$ kinetic profile can be explained on the basis of an etching process of GaAs surface, occurring after the oxygen removal, which slowly decreases the thickness of the outmost porous layer (see model in fig.1).

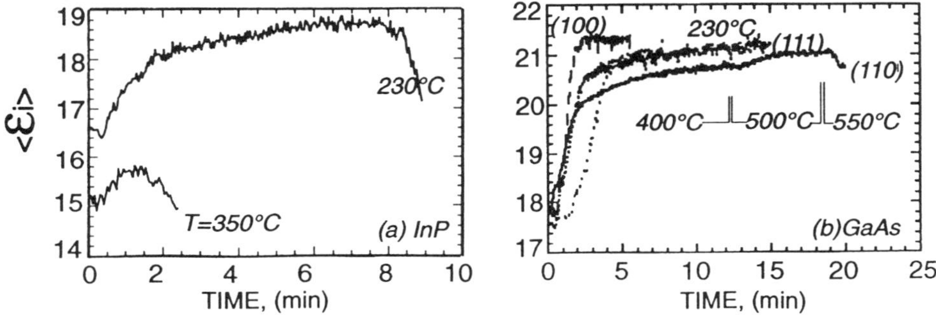

FIGURE 2. $\langle \varepsilon_i \rangle$ profiles at the E_2 interband critical point of (a) InP and (b) GaAs samples with different crystallographic orientation during the H-atom exposure at different temperatures; temperature range 400 - 550°C refers to (110) GaAs.

In fact, it has been reported [11] that GaAs etching by H atoms occurs in the temperature range 200-400°C without any evident Ga-enrichment, and that the etching rate is different for (111) and (100) crystallographic planes also depending on substrate temperature. The constancy of $\langle \varepsilon_i \rangle$ value for prolonged H-atom treatment of (100) GaAs at the substrate temperature of 230°C indicates that no increase/decrease of roughness is observed for (100) samples. At high substrate temperature (>400°C), the etching rate increases as evidenced by the increased slope of the $\langle \varepsilon_i \rangle$ profile in fig. 2b. On the contrary, data obtained during the H-atom treatment of a clean GaAs surface at room temperature (not reported in the figure), shows a decrease of the $\langle \varepsilon_i \rangle$ so indicating that As atoms are selectively depleted from GaAs surface and that the desorption rates of Ga-etch products GaH_3/Ga_2H_6 is lower than that of AsH_3.

From these results it comes out that the interaction of H-atoms with InP and GaAs surfaces, at T=230°C, is a two-step process: during the first step, hydrogen atoms selectively react with oxygen; in the second step, once the oxygen has been completely removed, H atoms react preferentially with phosphorus in InP material giving degradation, while the etching process occurs in GaAs so reducing the surface roughness resulting from the oxide removal. Thus, the statement "the exposure to H atoms leads to depletion of As on an oxide-free GaAs surface", as reported by many authors, is not generally true because it depends on the surface temperature. In fact, succesful III-V surface cleaning can be achieved by operating a remote hydrogen plasma treatment at T=230°C.

<u>Oxygen plasma anodization for GaAs oxide growth</u>

The plasma oxidation kinetics, for oxide thickness larger than 300Å, has been described by many authors, but the kinetics and chemistry of the growth of the first oxide layers, the so called "initial oxidation regime", are still not well known. Also, some other important effects such as crystallographic orientation and light irradiation of the surface during plasma oxidation remain to be examined. To investigate all these aspects, the plasma oxidation process of GaAs surfaces has been peformed at different bias voltages, with and without UV light irradiation, under the same plasma conditions. The oxide growth has been followed by SWE and the recorded ellipsometric trajectories have been transformed in the corresponding time dependence of the oxide thickness by using, as BEMA model, a GaAs-oxide single layer on GaAs substrate.

Figure 3 shows the oxidation kinetics for two (100) GaAs samples oxidized with (sample A) and without (sample B) UV light irradiation. The UV irradiation of the GaAs surface induces both an increase of the oxidation rate and a modification of the oxidation kinetic law. In fact, a linear oxidation kinetics applies when no UV light irradiates the surface, according to what expected on the basis of the linear-parabolic model already reported in literature. On the contrary, a non-linear time dependence of the oxide thickness is observed for UV light assisted plasma oxidation. In this case, as the oxide thickness increases, the photo-enhancement effect decreases, i.e. the oxidation rate decreases and approaches the value measured without UV light irradiation (sample B). This result suggests that the UV-photoeffect on the oxidation rate depends on the thickness of the oxide layer itself. In fact, the UV light is increasingly absorbed by the oxide layer and, according to the Lambert-Beer law, it is found that the oxide thickness has a logarithmic time dependence. Thus, the UV light activation occurs at the GaAs/oxide interface rather than at the bulk of the oxide or at the oxide surface.

At first sight, in the thickness profile, and particularly in that of sample B, different oxidation regimes can be distinguished: an early oxidation regime, (EO-regime) which correspond to the growth of the first 30 Å of oxide, a transition regime (TO-regime), where the oxide thickness increases with the square root of time and a late regime (LO-regime) where the oxide thickness depends linearly on time. In the EO-regime, the oxidation appears to be an easy process as the oxidation of the first

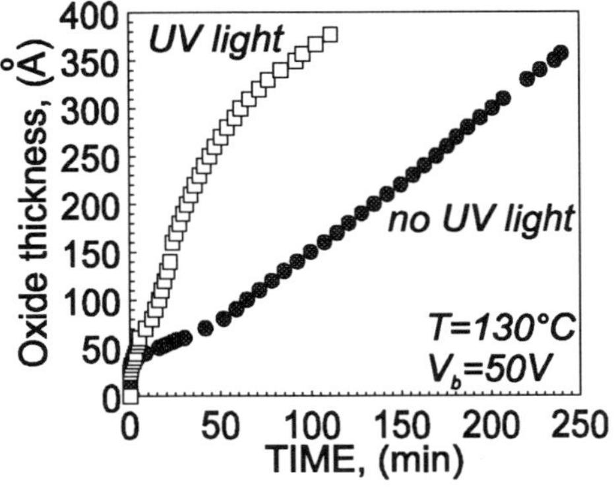

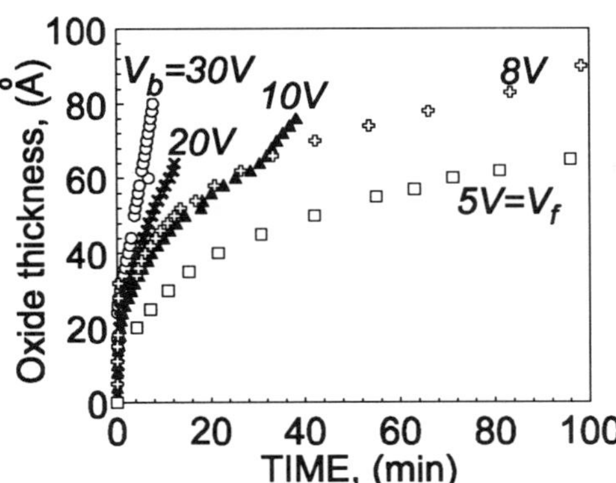

FIGURE 3. Oxide layer thickness *vs* oxidation time for GaAs samples with (□) and without (●) UV-light irradiation during plasma oxidation.

FIGURE 4. Thickness of the grown oxide layer *vs* oxidation time at different bias voltages under UV-light irradiation of GaAs surfaces.

$30\mathring{A}$ occurs in a relatively short time (<10 sec). The existence of this EO regime is related to the presence of the outmost porous layer ($\sim30\mathring{A}$) induced by hydrogen plasma pre-cleaning. Thus, the oxidation takes place throughout its entire volume due to its open porosity and the easy penetration of oxygen atoms into the pores. The key feature of the EO-regime is that it seems to be not affected by the bias voltage, while the GaAs morphology and chemical nature determine the initial oxidation rate. In the TO-regime, it is assumed that the electric field starts to built up in the thin oxide layer and the further oxidation is due to both in-diffusion of oxygen species and out-migration of ionic gallium and arsenic. Thus, in the TO-regime, the oxidation is not a surface process but takes place over a reaction volume zone. The time duration of this regime depends on bias, i.e. the higher the bias, the shorter the TO-regime (see fig. 4). In the LO-regime, the electric field in the oxide layer is well established and the oxide growth is controlled by the drift of Ga and As ions. Here, an increase of the bias voltage results in an increase of the oxidation rate. Figure 4 shows the oxidation kinetics for different bias voltages, Vb, applied to GaAs substrates during the UV photoassisted oxidation. It is evident that the overall oxidation rate strongly increases with increasing bias voltages.

The chemical composition of the grown oxide layers has been analyzed by XPS. Figure 5 shows the Ga3d and As3d photoelectron peaks of the two samples of fig. 3. For the As3d peak, two fitting components assigned to As_2O_5 and As_2O_3 oxides have been considered. In particular, the UV-oxides evidence that most of the oxidized As is present as As_2O_5, whereas As_2O_3 is mainly observed in the oxide layer grown without UV-light irradiation. By comparing the Ga3d it results that some elemental gallium is present in UV-grown oxides. However, we have experimented that the relative ratio As_2O_5/As_2O_3 and the amount of elemental gallium are strictly related to the oxidation kinetics under UV light irradiation, i.e. the lower the oxidation rate, the higher the As_2O_5/As_2O_3 ratio and the lower the gallium amount. The ratio of the areas of Ga3d versus As3d is 1.2 for the sample (A) and 1.1 for sample (B), respectively, indicating that GaAs stoichiometry is almost preserved in the oxide layer.

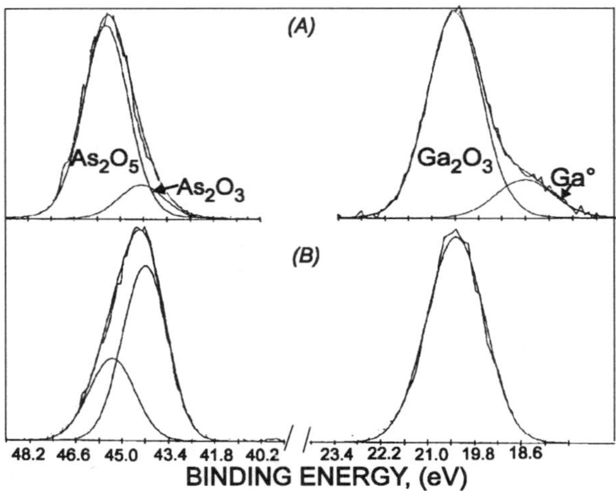

FIGURE 5. XPS spectra of Ga3d and As3d photoelectron peaks for GaAs oxide layers grown with (sample A) and without (sample (B) UV-light irradiation. Other experimental conditions: P = 70 mtorr, r.f. power = 100 watt, Vb = 50 V, O_2 flux = 10 sccm.

CONCLUSIONS

An in situ real time investigation of InP and GaAs surfaces during treatments in H_2 and O_2 remote plasmas has been presented. Hydrogen plasmas are active for native oxide removal (the cleaning process) from both InP and GaAs surfaces and, respectively, for P-depletion and GaAs etching. All these processes are found to strongly depend on surface temperature. Typically, at the optimal temperature of 230°C, a complete removal of oxide layer without surface damage is achieved. Oxygen plasmas are capable of oxidizing GaAs surface to grow GaAs oxide layers. The oxidation rate strongly increases with the bias voltage applied to the GaAs substrate and is strongly enhanced by the UV-light irradiation of the GaAs surface. The UV-light irradiation modifies the kinetics and chemistry at the GaAs/oxide interface and, hence, a stoichiometric mixture of Ga_2O_3 and As_2O_5 mainly is obtained.

REFERENCES

1. M. Losurdo, G. Bruno, P. Capezzuto, J. Vac. Sci. Technol. B, **14**, p. 691 (1996).
2. R.A. Gottscho, B.L. Preppenau, S.J. Pearton, A.B. Emerson, J. Appl. Phys.**68**(2) p.440 (1990)
3. E.M. Omeljanovsky, A.V. Pakhomov, A.Y. Polyakov, Semicond. Sci. Technol. **4**, p.947 (1989)
4. C.W. Tu, R.P.H. Chang, A.R. Schlier, Appl. Phys. Lett. **41**, p. 80 (1982).
5. M.C. Chuang, J.W. Colburn, J. Appl. Phys. **67**, p. 4372 (1990).
6. N. Cabrera, N.F. Mott, Rep. Prog. Phys. **12**, p. 163 (1948).
7. Y. Wang, Y.Z. Hu, E.A. Irene, J. Vac. Sci. Technol. B, **14**(3) p. 1687 (1996).
8. G. Bruno, M. Losurdo, P. Capezzuto, J. Vac. Sci. Technol. A, **13**, p. 349 (1995).
9. D.A.G. Bruggemann, Ann. Phys. (Liepzig) **24**, p. 636 (1935).
10. G. Bruno, P. Capezzuto, M. Losurdo, Phys. Rew. B, **54** (23) p. 1 (1996).
11. L.M. Weegels, T. Saitou, H. Kanbe, Appl. Phys. Lett. **66**(21) p. 2870 (1995).

CHARACTERIZATION OF GaAs SURFACES SUBJECTED TO A Cl_2/Ar HIGH DENSITY PLASMA ETCHING PROCESS

C.R. EDDY, JR.[*][+], O.J. GLEMBOCKI[*], V.A. SHAMAMIAN[*], D. LEONHARDT[*][‡], R.T. HOLM[*], J.E. BUTLER[*], B.D. THOMS[**], S.W. PANG[***], K.K. KO[***], E.W. BERG[***], AND C.E. STUTZ[%]

[*]U.S. Naval Research Laboratory, 4555 Overlook Ave., SW, Washington, DC 20375-5345
[‡] NRC Postdoctoral Fellow
[**]Dept. of Physics and Astronomy, Georgia State University, Atlanta, GA 30303
[***]University of Michigan, EECS Bldg., Ann Arbor, MI 48109-2122
[%]Wright Patterson Laboratories, Dayton, OH 45433
[+]e-mail address: eddy@ccfsun.nrl.navy.mil

ABSTRACT

High density plasma etching of III-V compound semiconductors is critically important to the development of advanced optoelectronic and high frequency devices. Unfortunately, the surface chemistry of these processes is not well understood. In an effort to monitor surface processes and their dependence on process conditions in a realistic etching environment, we have applied mass spectroscopic techniques for the study of GaAs etching in Cl_2/Ar chemistry. Etch product chlorides were monitored, together with optical measurement of the surface temperature by diffuse reflectance spectroscopy, as pressure (neutral flux), microwave power (ion flux) and rf bias of the substrate (ion energy) were varied. Observations from the spectroscopic techniques were correlated with ex situ surface damage assessments of unpassivated surfaces by photoreflectance spectroscopy. As a result, insights are made into regions of process conditions that are well suited to anisotropic, low damage etching.

INTRODUCTION

As critical dimensions of device structures approach the nanometer-scale regime, the need for fine control of semiconductor surface properties increases. This is especially true for plasma processing of semiconductor surfaces, either deposition of insulating/passivating films or device feature delineation by plasma etching. In this work we attempt to examine the surface processes involved in the high density plasma etching of GaAs and to correlate our observations with ex situ measurement of surface damage.   The ultimate goal is to gain insight into regions of parameter space best suited to low damage, anisotropic pattern transfer.

EXPERIMENT

Gallium arsenide samples used to examine surface chemistry in this study were as-received, bulk-grown GaAs wafers. Samples used for damage assessment were grown by MBE and consisted of a thick (~ 1 μm), heavily doped ($\sim 10^{18}$ cm^{-3}) p-type (UP) or n-type (UN) film with a thin (~ 1500Å) undoped layer grown on top. These structures were needed to evaluate surface damage by the photoreflectance spectroscopic technique discussed below.

All plasma processing experiments were carried out in an electron cyclotron resonance microwave plasma source configured for reactive ion etching and described in detail elsewhere[1]. Standard etching conditions were: 25% Cl_2 in Ar (by flow), 10 sccm total flow, 1.0 mTorr total pressure, 300 W coupled microwave power, -100 V rf-induced dc bias on the substrate, a substrate/ECR condition separation of 48-52 cm and substrate temperatures of $\sim 20°$C. For surface chemistry experiments, the traditional etching stage was replaced with an Extranuclear ELQ400 quadrupole mass spectrometer with a custom designed front cap/sampling aperture as shown in

Mat. Res. Soc. Symp. Proc. Vol. 448 © 1997 Materials Research Society

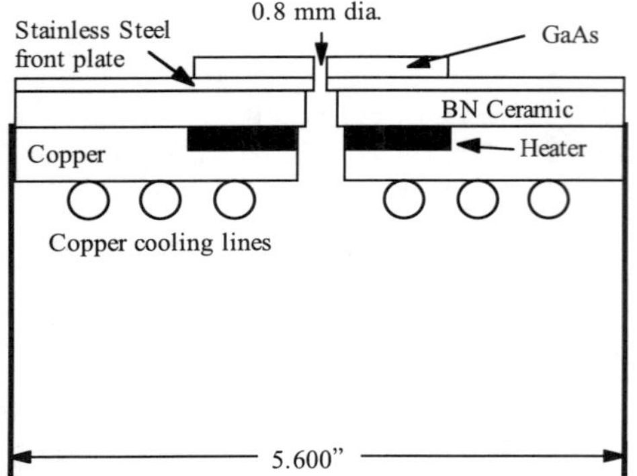

Figure 1. Schematic diagram of mass spectrometer front cap/sampling aperture.

Figure 1. The front cap is designed to emulate a traditional etching stage with full temperature control (both heating and cooling capabilities). When examining the surface chemistry, process conditions were varied as: total pressure = 0.5-5 mTorr, coupled microwave power = 200-500 W, and rf-induced dc bias = 0-400 V.

The mass spectrometer is calibrated using both perfluorotributlyamine (which contains major peaks at 69, 131 and 219) and etch product samples of $AsCl_3$ and $GaCl_3$. These efforts resulted in excellent calibration over the range of etch product masses (100-190 amu).

A critical process parameter in investigating surface chemistry issues in a high density plasma etching process is the substrate temperature. Accurate measurement of this parameter is important to assigning thermal and ion-assisted chemical etch processes occurring at the surface. As anisotropic pattern transfer is desired, ion-assisted chemistry is preferred with minimal variation in thermal chemistry. In this work, diffuse reflectance spectroscopy [2] is used to determine the temperature of the substrate itself. In this technique, the shift in the position of the absorption edge of the semiconductor is measured. This shift corresponds to the change in the band gap which is a well-defined function of temperature[3]. Application of this technique to the highly emissive environment of a plasma processing reactor requires substantial efforts to achieve good signal-to-noise ratios. A schematic of our experimental configuration is shown in Figure 2. In this application, both lock-in amplification and post sample wavelength selection are employed to improve signal to noise. The technique gives 6Å/°C resolution and signal-to-noise ratios of 10:1.

Surface damage is evaluated by measuring (with photoreflectance) the position of the surface Fermi level before and after the sample is subjected to plasma etching. The photoreflectance technique is described in full detail elsewhere [4]. In short, the built-in electric field ($\underline{E}_{bi}$) at the surface is measured by the change in reflectance (at wavelengths just above the band gap) of the semiconductor surface as hv>Eg radiation is applied to that surface. The built-in electric field is defined as:

$$\underline{E}_{bi} = \frac{|E_F - E_{c/v}|}{d} \qquad (1)$$

where E_F is the Fermi level at the surface, $E_{c/v}$ is the conduction band or valence band energy levels for UN or UP samples, respectively, and d is the thickness of the undoped layer. The etch process can effect both d and E_F so it is necessary to measure d

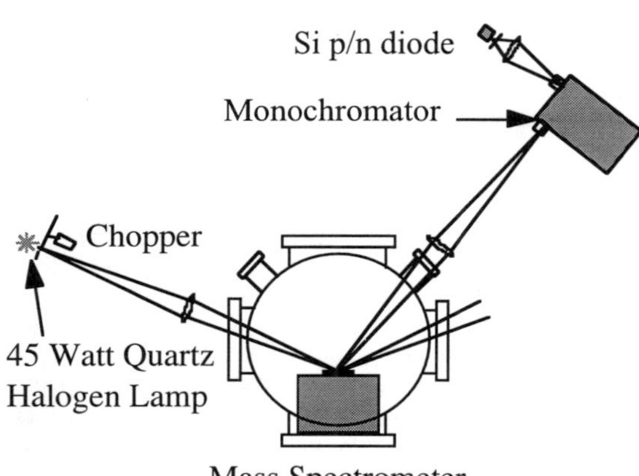

Figure 2. Schematic diagram of diffuse reflectance spectroscopy apparatus used to monitor substrate temperature

using independent step height measurements (performed here using a Tencor Alphastep 250).

RESULTS

The typical mass spectrum for the GaAs:Cl_2/Ar etch system consists of primarily AsCl, $AsCl_2$, $AsCl_3$ and to a much lesser extent $GaCl_2$ and GaCl. Gallium and arsenic monomers and dimers were not detectable as the peaks were either below our detection sensitivity or in the shoulders of molecular chlorine peaks. Of the peaks in the spectra, the $AsCl_2$ and $GaCl_2$ are the dominant etch product peaks for each element in the semiconductor. In the discussions that follow below the AsCl, $AsCl_2$, $AsCl_3$ and $GaCl_2$ peaks are monitored as the plasma etching environment is altered. Experiments were conducted in an effort to determine the origin of the product peaks detected in the mass spectrometer since it is entirely likely that plasma chemistry can generate the species as well as surface chemistry. These experiments involved taking mass spectra with the samples located near the sampling aperture and with them located only on the periphery of the front cap (~7 cm away) and the spectra were compared. When the sample was near the aperture, there was clear evidence for the etch product peaks mentioned above. However, when the samples were on the periphery, etch product peaks were at least four orders of magnitude smaller. Thus, the etch product peaks observed in the mass spectrometer are mapped to surface chemistry products through convolution with a degree of plasma chemistry.

Ion flux to the substrate is largely controlled by the input microwave power as plasma density is shown to increase with microwave power[5]. Thus, by varying the coupled microwave power from 200 to 500W the ion flux is monotonically increased. The etch product peak signals in the mass spectrum shows a commensurate increase from 200-300 W, but then saturate for coupled microwave powers greater than 300 W. This implies that the necessary ion and atomic neutral fluxes required at the substrate to maximize etching, namely Cl and Cl^+, Ar^+, are attained at relatively low microwave powers. Greater microwave powers do not result in additional etch product formation.

Neutral flux to the substrate is largely controlled by the total pressure in the process chamber during etching. As the total pressure is varied from 0.5 to 5 mTorr, the product mass peaks first increase, maximize at 1.0 mTorr and then gradually decrease with increasing pressure. The lower etch rate at lower pressures can be attributed to an insufficient neutral chlorine flux, as mass spectral characterizations [6] of the flux in absence of the substrate reveal Cl^+ and Ar^+ to dominate the spectrum. These same characterizations show that as the pressure is increased recombination of both ions and neutrals become dominant and the flux is largely made up of neutral Ar and Cl_2. Thus, a sufficient flux of neutral chlorine atoms and ions are necessary to maximize the etch rate and are available at a pressure of 1.0 mTorr. Lower total pressures are consistent with more anisotropic etching due to reduced ion collisions in the sheath and pre-sheath regions, therefore, 1.0 mTorr or lower pressures are desirable.

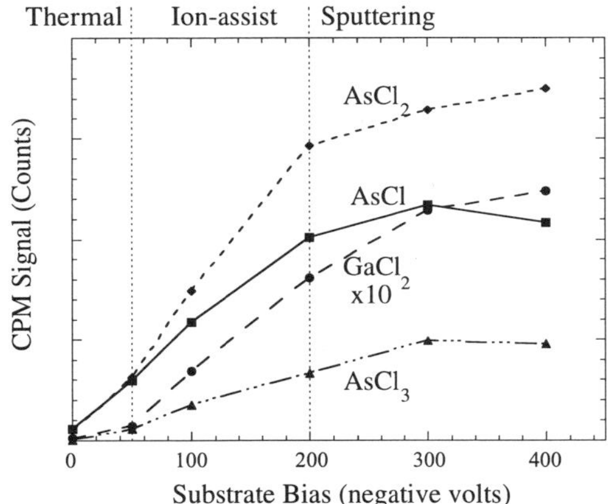

Figure 3. Etch product mass peak variation with substrate bias level.

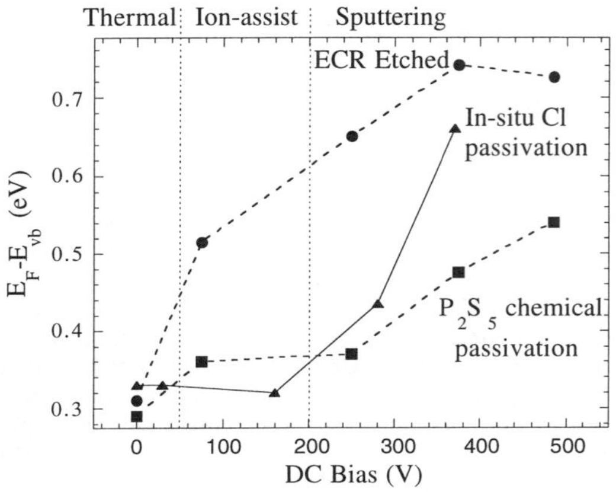

Figure 4. Surface Fermi level variations with ECR etching and subsequent passivation as a function of ion energy.

The most interesting parametric effect on the formation of etch products is the ion energy. Figure 3 shows the behavior of the various etch product desorption intensities with increasing incident ion energy. We postulate that there are three different regions of etching in these curves. The first, for ion energies < 50 eV, is attributed to largely chemical etching with chlorine atoms with a small amount of ion assist. In the range from 50-200 eV there is an abrupt change in the slope of the curves indicative of a substantial increase in the ion-assist to the chemical etching mechanism. For ion energies >200 eV the product peaks saturate. Measured etch rates of GaAs under the same conditions show the etch rate monotonically increasing over this ion energy range. Thus, the saturation above 200 eV can be explained as a change from ion-assisted chemical etching to more direct sputtering since product formation does not increase. Unfortunately, elemental Ga and As monomer and dimer peaks are hard to extract from the data due to larger reactant and product peaks nearby.

When the mass spectral characterizations of ion energy effects are compared to ex situ damage measurements, there is an excellent correlation. The UN sample is largely unaffected by the etch process and the Fermi level remains trapped at midgap throughout. The UP samples, however, do show a marked change in the position of the surface Fermi level with increasing ion energy (see Figure 4). In fact, ion energies greater than 100 eV result in considerable damage to the surface. As is shown in Figure 4, when the surface is in situ passivated with atomic chlorine from a chlorine ECR plasma, the Fermi level returns to the control sample level for ion energies <200 eV. Further, ex situ chemical passivation with P_2S_5 recovers the control sample surface Fermi level for ion energies <250 eV. These behaviors result from the mechanisms by which surface oxides form on the sample after removal from the etching reactor and are explained in greater detail elsewhere[7]. The end result is that for ion energies <200 eV, the damage is manifested largely as surface stoichiometry variations. However, for energies >200 eV, the inability to recover the surface Fermi level implies sub-surface damage, consistent with sputtering as proposed from the mass spectral characterizations.

CONCLUSIONS

The effect of Cl_2/Ar high density plasma etching of the surfaces of gallium arsenide has been investigated from the surface chemistry viewpoint by mass spectrometry and the surface damage viewpoint by photoreflectance spectroscopy. The combined results from these techniques infer that low microwave power (≤ 300 W), low pressure (≤ 1.0 mTorr) discharges with ion energies between 75 and 200 eV are best suited to anisotropic, low damage pattern transfer.

ACKNOWLEDGMENTS

This work is supported by the Office of Naval Research. D.L. gratefully acknowledges the support of National Research Council.

REFERENCES

1. C.R. Eddy, Jr., E.A. Dobisz, J.R. Meyer and C.A. Hoffman, J. Vac. Sci. Technol. **A11**, 1763 (1993).

2. T.P. Pearsall, S.R. Saban, J. Booth, B.T. Beard, Jr. and S.R. Johnson, Rev. Sci. Instr. **66**, 4977 (1995).

3. D.E. Aspenes in <u>Properties of Gallium Arsenide (2nd edition)</u>, EMIS Datareviews Series No. 2, (INSPEC, The Institute of Electrical Engineers, London, England, 1990), p. 153.

4. O.J. Glembocki, J. A. Tuchman, K.K. Ko, S.W. Pang, A. Giordana, R. Kaplan and C.E. Stutz, Appl. Phys. Lett. **66**, 3054 (1995) and references therein.

5. S.M. Gorbatkin, L.A. Berry, and J.B. Roberto, J.Vac. Sci. Technol. **A8**, 2893 (1990).

6. C.R. Eddy, Jr. and S.R. Douglass in <u>Diagnostic Techniques for Semiconductor Materials Processing II,</u> S.W. Pang, O.J. Glembocki, F.H. Pollak, F. Celii and C.M.S. Torres eds. (Mater. Res. Soc. Proc. 406, Pittsburgh, PA, 1996), p. 45-50.

7. O.J. Glembocki, K.K. Ko, E.W. Berg, S.W. Pang and C.E. Stutz, submitted to Appl. Phys. Lett.

Cl_2 PLASMA ETCHING OF Si(100): SURFACE CHEMISTRY AND DAMAGE

N. LAYADI, V. M. DONNELLY, and J. T. C. LEE
Bell Laboratories, Lucent Technologies, 700 Mountain Avenue, Murray Hill, NJ 07974, USA.

ABSTRACT

The interaction of a Cl_2 plasma with a Si(100) surface has been investigated by angle resolved x-ray photoelectron spectroscopy (XPS) and spectroscopic ellipsometry. From XPS, it was found that the amount of chlorine incorporated at the Si surface increases with ion energy. Chlorine is present as $SiCl_x$ (x = 1-3) with average relative coverages (integrated over depth) of $[SiCl]:[SiCl_2]:[SiCl_3] \cong 1:0.33:0.1$. These relative coverages don't depend strongly on ion energy between 40 and 280 eV. Real-time spectroscopic ellipsometry measurements showed that the layer present during etching is stable when the plasma is extinguished and the gas pumped away. In addition, the equivalent thickness of damaged silicon and silicon-chloride within the surface layer increases with ion energy.

INTRODUCTION

Low pressure, high density Cl_2 plasmas are widely used to etch Si and other materials for Si integrated circuits [1]. The overall mechanism for Si etching in a Cl_2 plasma is well established: Cl_2 and Cl react with the Si surface to form a chlorinated layer that is removed by ion bombardment. The nature of that chlorinated surface layer (thickness, stoichiometry, and structure) are, however, largely unknown. Both the chemical nature (chlorine content, surface stoichiometry) and physical nature (depth and degree of damage) of the chlorinated silicon surface layer must be clarified to better understand the chemical and physical etching mechanisms in high density plasmas and suggest new directions for etching processes and plasma reactor designs. In this paper, angle-resolved XPS and spectroscopic ellipsometry are used to study respectively the surface chemistry and damage of the Si surface during and after Cl_2 plasma etching over a range of ion energies and densities. More detailed accounts of these studies are published elsewhere [2,3].

EXPERIMENTAL PROCEDURE

The chemical nature of the Si surface layer etched with Cl_2 plasma was investigated in a helical resonator reactor coupled to an x-ray photoelectron spectrometer with a vacuum transfer chamber [2,4]. The physical nature was studied *in situ* by spectroscopic ellipsometry mounted on a helicon reactor [3]. Monocrystalline Si (100) wafers (p-type, 15-25 Ω-cm) were stripped of their native oxide before etching. The reactors were evacuated to a pressure of 10^{-6} Torr before the plasma was ignited. Two etching modes were evaluated: 1) reactive ion etching (RIE) with radiofrequency (rf) power applied only

Mat. Res. Soc. Symp. Proc. Vol. 448 © 1997 Materials Research Society

to the wafer stage and 2) a high density (HD) plasma with and without rf power on the stage.

RESULTS AND DISCUSSION

After etching under steady state conditions, low resolution (1.3 eV) XPS spectra were recorded as a function of take-off angle θ (with respect to the wafer plane), and the integrated intensities of the core-levels peaks were used to determine the total chlorine coverage (Fig. 1). The highest ion energy (RIE, -240 V DC bias) leads to the highest chlorine coverage (37% at $\theta = 20°$). On the other hand, low ion energy (HD, 0 V DC bias) produces the lowest chlorine coverage (28% at $\theta = 20°$). RIE and HD at the same ion energy (-75 V DC bias) produce surface layers with nearly identical Cl contents, indicating that the degree of dissociation and ionization of Cl_2 (much higher for the HD etching) is relatively unimportant.

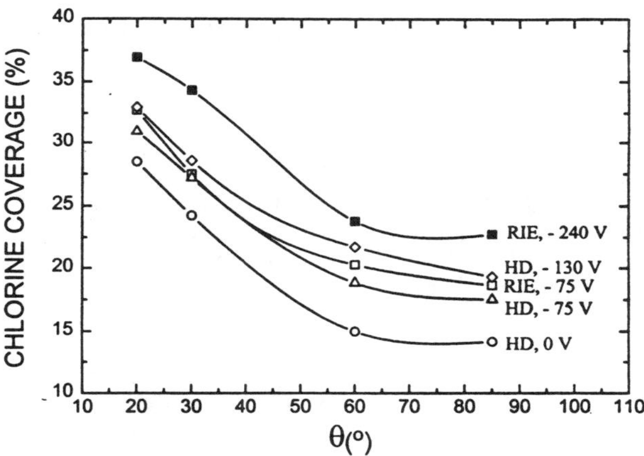

Figure 1. Chlorine coverage versus take off angle θ in the near-surface region of Si samples after etching for 30 s in Cl_2 plasma under different conditions [3].

Higher resolution (0.6 eV) spectra of the Si(2p) region were also recorded to determine the stoichiometry of the $SiCl_x$ layer. Fig. 2(a) presents a spectrum recorded after etching (RIE, -240 V DC bias), as well as spectrum for air-exposed, HF-cleaned, unetched Si. The latter spectrum consists of an asymmetric peak at a binding energy of 99.4 eV due to the partially resolved $2p_{3/2}$ and $2p_{1/2}$ components of bulk Si. After etching in a Cl_2 plasma under RIE conditions (-240 V DC bias), the integrated intensity of the Si peak at 99.4 eV decreases due to inelastic scattering of electrons by chlorine on the surface. In addition, a broad tail appears, extending from 100 to 104 eV, and is attributed to SiCl, $SiCl_2$, and $SiCl_3$. Fig. 2(b) is an expanded, processed version of the Si(2p) region of this latter spectrum. In addition to Si(bulk) at 99.4 eV, three features ascribed to SiCl, $SiCl_2$, and

SiCl$_3$ were identified at 100.2, 101.2, and 102.3 eV, respectively. An additional, weak peak appearing as a shoulder on the low binding energy side of the Si peak was attributed to Si·. Its binding energy was found to be the same (98.8 eV) as that measured for Si· after Ar$^+$ sputtering, providing strong evidence for the assignment. The integrated intensities of deconvoluted SiCl, SiCl$_2$, SiCl$_3$, and Si· peaks (normalized to the Si(bulk) integrated peak intensity) were recorded as a function of θ, for a range of processing conditions. SiCl, SiCl$_2$, and SiCl$_3$ decrease with θ and so with depth, for all conditions in agreement with the decrease of the chlorine content. SiCl$_2$ and SiCl$_3$ fall off more rapidly with θ than does SiCl, indicating that they are more confined to the surface, and that the Cl penetrating more deeply into Si is present almost entirely as SiCl. To gain more insight into the SiCl$_x$ layer thickness and composition, we computed angle resolved XPS intensities from assumed composition versus depth profiles. This procedure, described more thoroughly elsewhere, allows us to derive a fairly detailed picture of the Si surface layer schematically shown in Fig.3. The top monolayer of the near-surface region is predominantly SiCl$_3$ and SiCl$_2$, while below the surface, mainly disordered Si and SiCl are distributed over a depth of about 25 Å for ion energies of about 280 eV, and not much thicker than 13 Å at the lowest ion energies (40 eV). The layer thickness and total Cl content are in reasonable agreement with the molecular dynamic simulations by Barone and Graves [5], and especially with the recent extensions of these simulations by Barone [6].

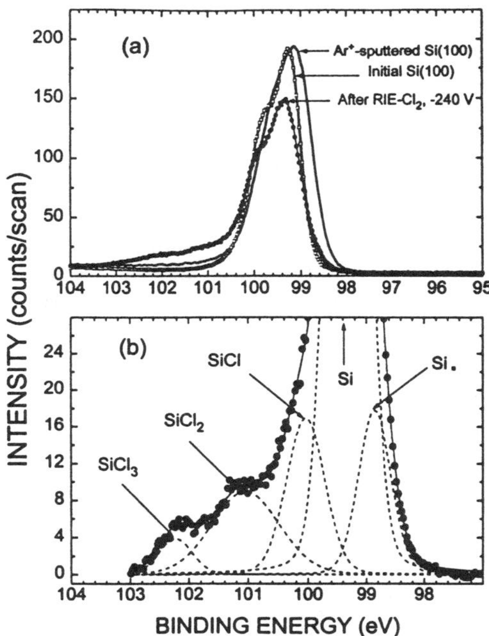

Figure 2. (a) High resolution Si(2p) spectra of air exposed (unetched) Si and plasma etched (RIE, -240 V DC bias) Si at θ = 30°. (b) High resolution Si(2p) spectrum of plasma etched (RIE, -240 V DC bias) Si at θ = 30° with the background substracted off and the 2p$_{1/2}$ components removed. The fit (solid lines) includes five peaks (dashed lines) with 85% gaussian, 15% lorentzian line shapes. The relative areas SiCl:SiCl$_2$:SiCl$_3$ are 1:0.67:0.33. These values (θ = 30°, more surface sensitive) are, as expected, higher than those integrated over the whole layer (1:0.33:0.10).

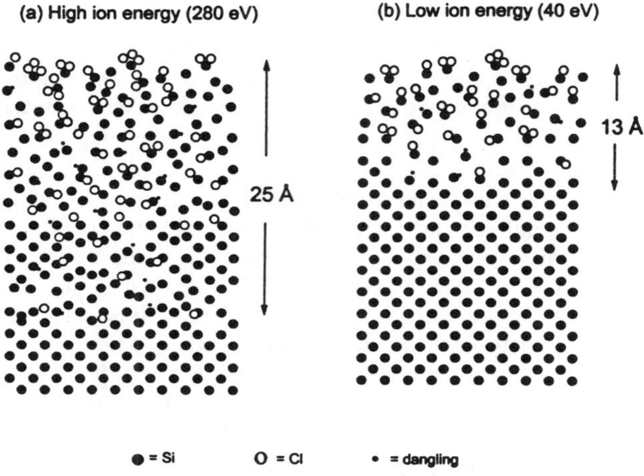

Figure 3. Schematic depiction of Si(100) surface after etching in a Cl_2 plasma, based on the results reported in [3]. (a): HD, 0 V DC bias (40 eV peak ion energy) and (b): RIE, -240 V DC bias (280 eV peak ion energy).

Spectroscopic ellipsometry was used to study the interaction of a Cl_2 plasma with Si(100) during and after etching in the helicon plasma under HD conditions, as well as under RIE conditions [3]. Fig. 4 displays the real-time traces of the ellipsometric angle Δ at 4.5 eV during Cl_2 etching of Si(100) at different DC-biases (i.e, ion energies). This photon energy corresponds to a more surface-sensitive region of the spectrum where light penetrates into c-Si no more than 100 Å. The initial value of Δ before the plasma is ignited is identical for all three samples indicating that the native oxide removal step is reproducible. As the plasma is switched on (at t = 10 s), Δ decreases and reaches a stable value within the first 4 s. This constant value becomes more negative with increasingly negative DC bias. Simulations at different photon energies [3] indicate that this decrease in Δ is the result of the creation of a damaged surface layer with a thickness that increases with negative DC-bias. Δ remains unchanged when the plasma is switched off (t = 130 s) and also when the gas flow is stopped (t = 145 s). Ψ (not shown) also remains constant after plasma extinction. The surface stability after plasma extinction (shown here for only one photon energy) was also confirmed at energies between 2 and 4.5 eV. From this, we conclude that the layer that forms quickly upon initiation of the plasma is stable when the plasma is extinguished and the Cl_2 is pumped away, under vacuum conditions.

To interpret the ellipsometric data, insights derived from the XPS study of the chemical nature of the surface were considered. To account for the graded composition of the surface layer found in that study, the damaged surface layer was assumed to be composed of two layers: i) an interface layer composed of undamaged crystalline Si (c-Si), damaged (i.e. amorphous) silicon (a-Si) and a transparent silicon-chloride material ($SiCl_x$) and ii) a top surface layer containing a-Si and $SiCl_x$. The optical properties of $SiCl_x$ were obtained using the Cauchy formula applied to the experiment carried out under conditions that maximized the $SiCl_x$ content in the layer (RIE, -300V), and that matches with the XPS study which provided the average thicknesses and fraction of $SiCl_x$ in the two assumed layers. The refractive index of $SiCl_x$ (~1.66) was found to be close to that of SiO_2 (~1.45) [3]. The thickness and composition of the damaged surface layer was computed as a function of exposure time and processing conditions through the use of the Bruggeman effective medium approximation (BEMA) and linear regression analysis [3]. The results in terms of the surface layer thickness are plotted in Fig. 5.

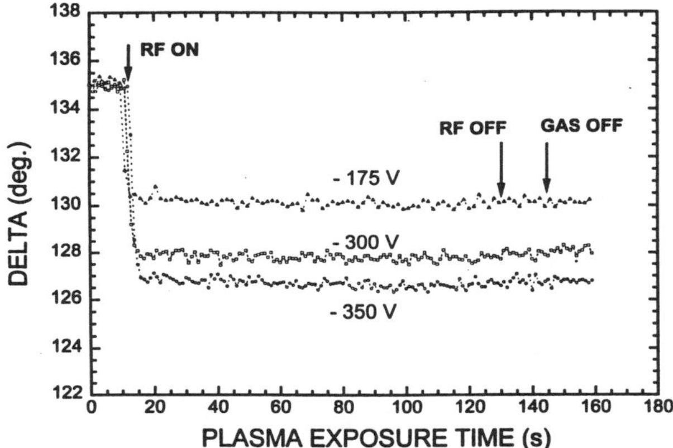

Figure 4. Real-time traces of the ellipsometric angle Δ with time, at 4.5 eV, during etching of Si(100) under RIE conditions (- 175, - 300, and - 350 V).

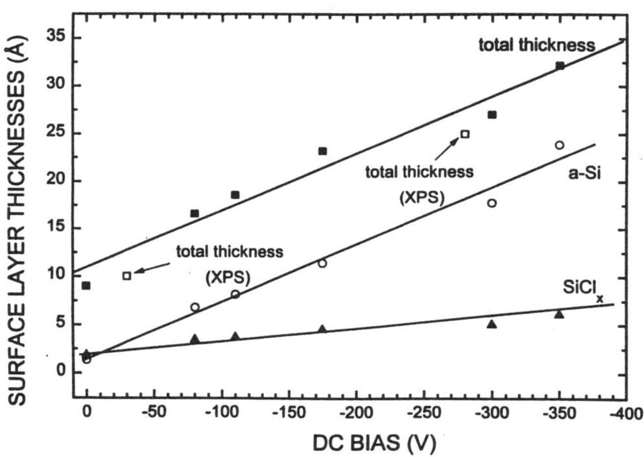

Figure 5. Surface layer thicknesses (deduced from ellipsometry) versus DC-bias. Two thicknesses derived from XPS study are also plotted for comparison.

As expected, energetic ion bombardment increases the depth of damage in the surface layer. When the DC-bias increases from 0 V to - 350 V, the thickness of the surface layer increases from 9 to 32 Å. Ion energy also increases the equivalent thicknesses of a-Si and $SiCl_x$ within the surface layer.

CONCLUSION

Angle-resolved x-ray photoelectron spectroscopy and spectroscopic ellipsometry were used to study the Si surface etched in a Cl_2 plasma. The steady-state chlorine coverage increases with the energy of Cl_2^+ and Cl^+ ions bombarding the Si surface during etching. The surface layer contains SiCl, $SiCl_2$ and $SiCl_3$. For high ion energies (280 eV), the near surface region contains 3.0×10^{15} Cl/cm^2, with $SiCl_2$ and $SiCl_3$ enriched at the surface, and SiCl and disordered Si distributed over a depth of ~ 25 Å. For low ion energies (40 eV), the layer contains 1.7×10^{15} Cl/cm^2, with relatives coverages of $SiCl_x$ smaller than those formed at higher energy and distributed in a much thinner layer (~ 13 Å). From full ellipsometry spectra recorded after etching, we found that a damaged silicon surface layer is created quickly after Cl_2 plasma exposure. The thickness of that layer increases linearly with DC-bias from 9 Å at 0 V (~ 20 eV peak ion energy) to 32 Å at -350 V (~ 370 eV peak ion energy). Ion energy also increases the equivalent thicknesses of both $SiCl_x$ and a-Si within the surface layer. Moreover, real-time spectroscopic ellipsometry measurements showed that the layer present during etching is stable when the plasma is extinguished and the gas pumped away, indicating that post-plasma surface diagnostic measurements such as x-ray photoelectron spectroscopy (XPS) reflect the surface that is present during etching.

REFERENCES

1. D. L. Flamm, V. M. Donnelly, and D. E. Ibbotson, in VLSI Electronics Microstructure Science, edited by N. G. Einspruck and D. M. Brown (Academic Press, NY, 1984), pp. 189-251.

2. N. Layadi, V. M. Donnelly, and J. T. C. Lee, J. Appl. Phys., (1997), to be published.

3. N. Layadi, V. M. Donnelly, J. T. C. Lee, and F. P. Klemens, J. Vac. Sci. Technol. A, (1997), in press.

4. C. C. Cheng, K. V. Guinn, V. M. Donnelly, and I. P. Herman, J. Vac. Sci. Technol. A **12**, 2630 (1994).

5. M. E. Barone and D. B. Graves, J. Appl. Phys. **78**, 6604 (1995).

6. M. E. Barone, 43rd AVS Symp., Oct. 14-18, 1996, Philadelphia, PA, paper PS-ThM4.

Theory of Reactive Adsorption on Si(100)

D.J. DOREN*, A. ROBINSON BROWN, R. KONECNY
Department of Chemistry and Biochemistry, University of Delaware, Newark, DE 19716
*doren@udel.edu

ABSTRACT

Density functional calculations on cluster models of Si(100)-2x1 have been used to predict reaction mechanisms and energetics for several reactive adsorption processes. Dissociative adsorption of H_2, H_2O, BH_3, and SiH_4 will be described and compared to available experimental data. Based on these examples, a qualitative theory of mechanisms for dissociative adsorption of hydrides on silicon surfaces will be proposed. These reactions can largely be understood in terms of the electron density distributions in the molecule and surface dimer. On a buckled dimer, there are both electron-rich and electron-deficient sites, which have different chemical interactions with adsorbates. The role of this difference is illustrated in a novel surface Diels-Alder reaction, for which symmetric addition to an unbuckled surface dimer is allowed by orbital symmetry. This reaction creates new reactive surface sites that may be useful for subsequent chemical surface modification.

INTRODUCTION

Many important processes in film growth on Si surfaces begin with a dissociative adsorption reaction. These include dissociation of silane and disilane, or boron and phosphorous hydrides used as sources of dopants. Dissociation of water is the initial step in oxide growth. Some of these adsorption reactions have significant activation barriers, while others are unactivated. This paper describes first-principles theoretical studies of several reactions on the Si(100) surface. These examples provide a general qualitative understanding of the differences between activated and unactivated reactions, and a detailed picture of the mechanisms of these reactions at an atomistic scale.

The specific reactions discussed here are the dissociative adsorption of H_2, H_2O, SiH_4 and BH_3. A reaction between cyclohexadiene (C_6H_8) and Si(100) is described as well. The latter reaction belongs to a novel class of surface reactions that may permit new techniques for controlling surface chemistry. It is also of interest as a possible example of a reaction mechanism that does not occur in the other examples studied.

THEORY

The results described here are obtained from first-principles calculations, based on density functional theory (DFT) using cluster models of the surface. Unless otherwise noted, the energies reported here are calculated with the nonlocal exchange functional developed by Becke[1] and the nonlocal correlation functional of Lee, Yang and Parr.[2] Basis sets are atom-centered triple-zeta gaussian bases, or double-zeta numerical bases, with polarization functions on all atoms. The surface is modeled with a Si_9H_{12} cluster (Figure 1). This model includes one surface dimer (two dangling bonds). The H atoms terminate subsurface bonds that would be Si-Si bonds in the solid. All coordinates in this model are optimized without constraints. Second derivatives are calculated at all critical points to verify that minima have no negative force constants and that transition states have exactly one negative force constant.

The small cluster model used here is computationally efficient and reproduces the main features of the surface structure and transition states. While it neglects interactions with the electronic states of neighboring sites and the forces from neighboring surface atoms, the errors appear to be on the order of 0.1-0.2 eV, as determined by comparisons to calculations with larger clusters or with periodic boundary conditions.[3,4] Other sources of error of similar

Mat. Res. Soc. Symp. Proc. Vol. 448 © 1997 Materials Research Society

magnitude come from basis set effects and the DFT method itself. The conclusions of this work are not affected by errors of this order.

RESULTS

Bare Surface Structure

The Si(100) surface displays a 2×1 reconstruction with surface atoms paired up into rows of "dimers." It is now generally agreed that the dimers have an asymmetric structure, being tilted with one atom of the dimer higher than the other. This structure is illustrated in the cluster model of Figure 1. This structure is slightly more stable than a symmetric structure. The stabilization energy is predicted to be no more than 0.2 eV, though the precise value depends on details of the theory (a brief review of the literature on this subject is included in Ref. 4). Using larger cluster models or periodic boundary conditions generally results in even greater asymmetry than shown in Figure 1. Many transition states display a similar asymmetry.

Tilting the surface dimers causes an asymmetry in the surface charge distribution, so that the "up" atom has a higher electron density than the "down" atom. Mulliken charges for the two atoms illustrating this effect are shown in Figure 1a. An important consequence of this charge asymmetry is that the two atoms in a dimer are chemically inequivalent. The electron-deficient "down" atom can be expected to react as an electrophile, preferring to interact with the electron-rich regions of incident molecules. On the other hand, the electron-rich "up" atom of the dimer is expected to react as a nucleophile, preferring to interact with electron-poor regions of incident molecules. The asymmetry appears in the highest occupied molecular orbital (HOMO) of the bare surface (Figure 1b), where the two dangling bond orbitals have different spatial extents. The asymmetry increases with the tilting angle. Thus, tilting the surface dimers establishes two chemically inequivalent sites, even on well-ordered surfaces of this elemental semiconductor. The consequences of this effect are illustrated in the following examples.

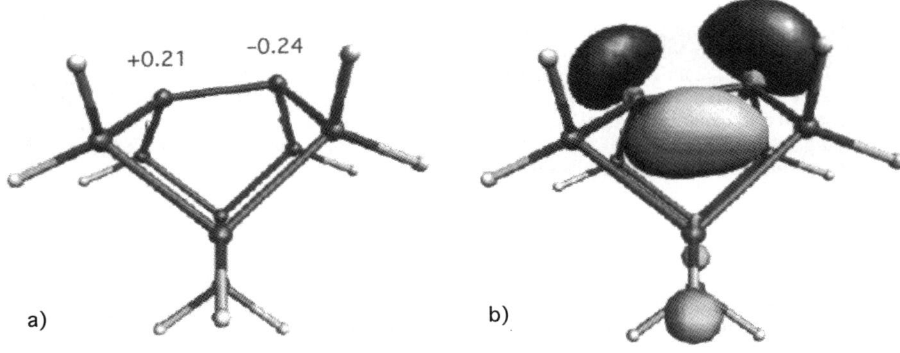

Figure 1. a) The Si_9H_{12} cluster model of the bare surface b) The HOMO of the bare surface.

Hydrogen (H_2)

Hydrogen dissociates with a very low probability (10^{-7}) to form Si–H bonds. The reaction mechanism for adsorption and desorption of H_2 remains controversial. Two reviews of the issues involved have recently been published.[5,6] Some workers have suggested that reaction is much more likely at defect sites. For present purposes, it is valuable examine the reaction path that involves the majority sites since it establishes a comparison with the mechanisms for reactions of other adsorbates at these sites. The transition state structure calculated with the cluster model is shown in Figure 2a. The energy of this configuration is 1.0 eV above the energy of the isolated H_2 and bare surface, consistent with a small sticking probability. A more detailed discussion and comparisons to other calculations (which find similar transition state structures) are given in Ref. 3.

At the transition state, the H_2 molecule is located with its bond directly over the electrophilic (buckled-down) surface dimer atom. The two H atoms are nearly equidistant from this surface atom. The H–H bond is stretched to about 1.0 Å, though the electron density between the H atoms indicates the molecular bond is not yet broken. The transition state geometry suggests that there is little interaction of the incident molecule with the nucleophilic (buckled-up) side of the dimer. This interpretation is consistent with the structure of the transition state HOMO, which is essentially an unperturbed dangling bond on the nucleophilic atom.

This transition state can be given a simple interpretation. The H–H bond region is the most polarizable region of the molecular electron density, and it is located to interact with the electrophilic surface site. Plots of the electron density show that electrons in the H–H bond are polarized toward the surface site, though there is no indication of bonding to the surface. Because the electron density in H_2 is low, the interaction with the electron-poor surface atom is not strongly favorable. Indeed, the surface is buckled even more strongly at the transition state than on the clean surface, making the surface atom more electrophilic to enhance the interaction with the molecule. The net effect is that the transition state can only be reached at the cost of some energy, so that adsorption is activated.

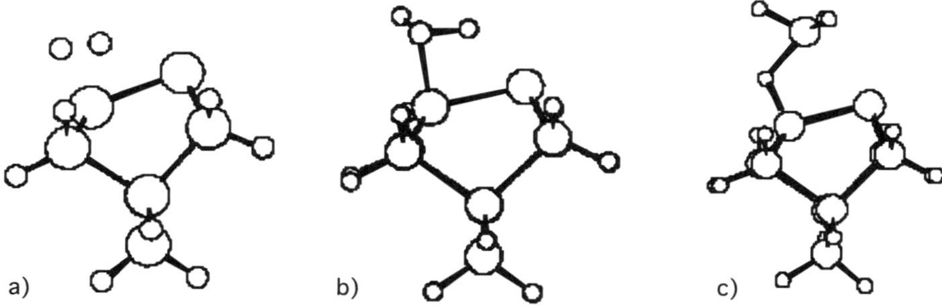

Figure 2. Transition state structures for adsorption of a) H_2 b) H_2O c) SiH_4.

Water (H_2O)

Water dissociates to leave hydroxyl (OH) and H fragments bound to the surface. In contrast to hydrogen molecules, water has a sticking probability close to unity, indicating that there is no barrier to adsorption. The experimental work on water adsorption on Si(100), as well as the details of our own work, are reviewed in Ref. 4. The geometry of the transition state to adsorption of water is shown in Figure 2b. The energy of the transition state is below that of the isolated molecule and clean surface: the structure shown corresponds to a saddle point between a weakly bound molecular precursor and the dissociatively chemisorbed state. The energy of this configuration is only slightly (< 0.2 eV) above the energy of the precursor state, a minor bump in a reaction path that is downhill nearly all the way from the reactant state to the chemisorbed state.

As with H_2, the water molecule interacts primarily with the electrophilic side of the dimer, though the interaction with the electron-rich oxygen atom is much more favorable than that with the bond in H_2. At the transition state, the O–H bond directed toward the nucleophilic side of the dimer is stretched to 1.2 Å (compared to 1.0 Å in an isolated water molecule). The electron density indicates that the O–H bond is not broken, but there is a Si–O bond forming that stabilizes the transition state. While the nucleophilic side of the dimer does not have strong interactions with the incident molecule, the HOMO shows some overlap between the nucleophilic dangling bond and the dissociating H atom, further stabilizing the transition state.

Thus, while H_2 and H_2O both adsorb on the electrophilic side of the dimer, several factors account for the large difference in activation energies. Water contains a polarizable, electron-rich atom that can react with the electrophilic surface site before the O–H bond is broken and the geometry of the transition state allows a stabilizing interaction with the nucleophilic side of the dimer. Finally, because oxygen can be polarized so easily, additional surface distortion is not required to enhance the interaction of the electrophilic Si with the electron-rich O. In fact, at the transition state for water adsorption, the surface is actually buckled less than the bare surface, which enhances the overlap of the Si orbitals with O.

Silane (SiH$_4$)

Silane dissociates to leave silyl (SiH$_3$) and H fragments bound to the surface.[7] The sticking probability of silane is low (10^{-5}),[7] though not as small as that of H_2. Measurements of film growth rates as a function of surface temperature have found an apparent activation energy for silane sticking of less than 0.2 eV on a clean surface, though the absolute sticking probability remains low at all temperatures studied.[8] On the other hand, measurements of sticking as a function of incident silane translational energy[9] show that sticking is increased by two orders of magnitude over that of a thermal gas if the molecules are given an initial translational energy on the order of 1 eV. This suggests that part of the barrier to adsorption is associated with molecular degrees of freedom which cannot be activated by heating the surface alone.

The geometry of the calculated transition state to adsorption is shown in Figure 2c. The energy of this state is 0.6-0.8 eV (depending on the basis set used) above the energy of the isolated silane molecule and clean surface. Again the primary interaction is with the electrophilic side of the dimer, though it corresponds to transfer of a hydrogen from silane to the surface. In fact, H is more electronegative than Si, so that an H atom will be more strongly attracted than Si to the electron-deficient surface atom. At the transition state, the dissociating H atom is closer to the surface Si (1.62 Å) than to the silane Si (1.79 Å). The electron density indicates that the silane Si–H bond is broken and a surface Si–H bond is being formed at the transition state. This is consistent with the picture that a large part of the activation energy is associated with molecular degrees of freedom, in this case a Si–H stretch. The nucleophilic side of the dimer shows little interaction with the incident silane, with the HOMO being essentially an unperturbed dangling bond. However, the silyl fragment is oriented to facilitate formation of a bond between silyl and the nucleophilic side of the dimer.

Silane adsorption has several features in common with hydrogen adsorption. Like H_2, silane has no easily polarizable, electron-rich atoms. Electrophilic attack can occur at an H atom, but it is not nearly as favorable as attack on an atom like oxygen, so adsorption of silane is activated. The barrier to adsorption is lower than for hydrogen, largely because the Si–H bond in silane is weaker than the H–H bond. There may be also be a weak stabilizing interaction between the silyl fragment and the nucleophilic side of the dimer, analogous to that seen with water. As with H_2, the surface is distorted at the silane transition state, making the "down" side of the dimer more electrophilic. This surface distortion corresponds to an increase of 0.2 eV in the surface energy. This corresponds closely to the observed effective activation barrier corresponding to the surface coordinates. However, the calculations show that both surface and molecular coordinates must be excited to cross the saddle point.

<u>Nucleophilic Attack Mechanisms: Silane and Borane (BH$_3$)</u>

All of the reactions described to this point are initiated by an interaction at the electrophilic side of the dimer. There are good reasons to expect that silane might adsorb by another mechanism in which the nucleophilic side of the dimer forms a bond with the Si in silane, followed by formation of a surface Si–H bond on the opposite side of the dimer. An earlier theoretical study predicted that silyl radical could attack silane in an analogous gas phase reaction to form a Si–Si bond and eliminate a hydrogen radical.[10] On the surface, this path would presumably be stabilized by formation of a Si–H bond instead of formation of a gas phase radical. Moreover, in this transition state the silane Si would be five-fold coordinated, a configuration that occurs in stable compounds of Si. Despite persistent efforts to search for such a saddle point, none has been found. The difference between the surface and gas phase reactions may reflect the difference between Si–H bonds in silyl radical and Si–Si bonds on the surface, with the latter being more effective in stabilizing the dangling bond, thus making the surface relatively less reactive as a nucleophile than silyl.

Adsorption of borane is a clear example of a reaction that can begin at the nucleophilic dimer atom. BH$_3$ is a strong electron acceptor, and adsorption occurs by forming a Si–B bond on the nucleophilic side of the dimer, leaving BH$_2$ and H fragments. This work is still in progress, but it appears that there is little or no activation barrier to borane adsorption.

<u>Symmetric Addition: Cyclohexadiene (C$_6$H$_8$)</u>

The dissociative adsorption reactions described here all proceed through asymmetric transition states. In some cases, it is easy to argue using qualitative molecular orbital theory that this must be the case. For example, Figure 3a indicates schematically the orbitals that would be involved in adsorption of H$_2$ along a symmetric path. To break the H–H bond, the anti-bonding lowest unoccupied orbital (LUMO) of H$_2$ must be populated by overlap with a filled surface state. To form Si–H bonds, the H$_2$ orbitals must overlap with the surface dangling bond states (the surface HOMO) with a proper phase relation for bonding. But the H$_2$ LUMO and surface HOMO have opposite parity, so that it is not possible to have positive overlap between the surface dangling bonds and the molecular antibonding orbital on a symmetric path.

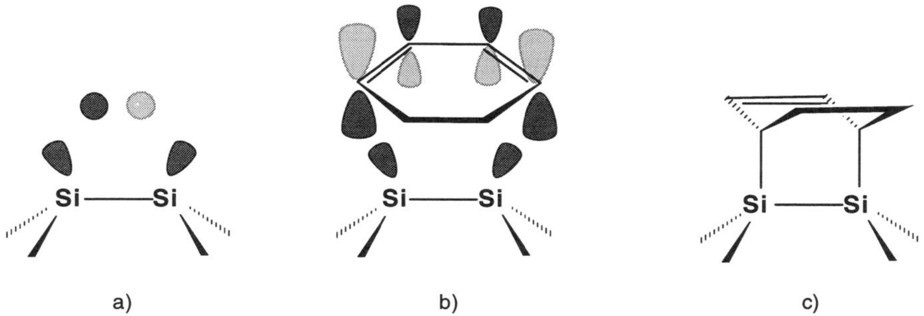

Figure 3. a) Orbitals involved in symmetric adsorption path for H$_2$ b) Orbitals involved in the Diels-Alder reaction of cyclohexadiene on Si(100) c) Product of the Diels-Alder reaction.

This kind of argument leads to a different conclusion for molecules such as cyclohexadiene. The relevant orbitals are shown in Figure 3b. It is clear that the surface

HOMO and molecular LUMO may have positive overlap on a symmetric path. The result of populating the molecular LUMO will be to form the product shown in Figure 3c, with two C–C π-bonds broken, and a new one formed, in addition to the Si-C bonds. The reaction is a surface analog of a common reaction in organic synthesis, the Diels-Alder reaction. While a symmetric reaction path is not prohibited by this simple argument, this need not be the lowest energy path. Preliminary calculations indicate that there is a lower energy path that is somewhat asymmetric, with the C–Si bonds formed sequentially rather than simultaneously. The distinction is nearly meaningless, however, as the reaction has little or no activation barrier on either path. The product is stable by 1.6 eV with respect to the reactants.

This reaction is a prototype of a large class of reactions, analogs of which have been explored thoroughly in the organic chemistry literature. The most direct relatives are other dienes, possibly including substituents. In such cases, the product contains a C–C double bond which offers the possibility of doing further controlled chemical reactions to modify the adsorbate layer.

CONCLUSIONS

Several examples of reactive adsorption have been described which illustrate the chemical features of the Si(100) surface. In all the reactions described here, both dangling bonds of the surface dimer play a role in the reaction. Dissociation of hydrides may begin at either the electrophilic or nucleophilic side, depending on the adsorbate. Indeed, the presence of both electron-rich and electron-poor sites in close proximity is an advantage in dissociative adsorption since both molecular fragments will be able to bind to a favorable site: the more electron-donating fragment binds to the electrophilic surface site, and the fragment that is the better electron acceptor binds to the nucleophilic site. In the examples shown, hydrogen can act as either type of fragment, binding to the electrophilic or nucleophilic surface site. However, reactions that proceed by electrophilic attack on H are activated, because the electron density around H is so low. On the other hand, electrophilic attack on an electron-rich atom (such as O) can proceed with no activation barrier, as can nucleophilic attack on an electron-deficient compound (such as BH_3) or addition of a diene.

ACKNOWLEDGMENTS

This material is based on work supported by the National Science Foundation under Grant No. CHE-9401312. We wish to thank Dave Allara for suggesting the surface Diels-Alder reaction.

REFERENCES

1. A.D. Becke, Phys. Rev. A **38**, 3098 (1988).
2. C. Lee, W. Yang and R.G. Parr, Phys. Rev. B **37**, 785 (1988).
3. S. Pai and D.J. Doren, J. Chem. Phys. **103**, 1232 (1995).
4. R. Konecny and D.J. Doren, J. Chem. Phys. (in press).
5. K. Kolasinski, Int. J. Mod. Phys. B **9**, 2753 (1995).
6. D.J. Doren, Adv. Chem. Phys. **95**, 1 (1996).
7. S.M. Gates, C.M. Greenlief, D.B. Beach and P.A. Holbert, J. Chem. Phys. **92**, 3144 (1990).
8. R.J. Buss, P. Ho, W.G. Breiland and M.E. Coltrin, J. Appl. Phys. **63**, 2808 (1988).
9. M.E. Jones, L.-Q. Xia, N. Maity and J.R. Engstrom, Chem. Phys. Lett. **229**, 401 (1994).
10. K.D. Dobbs and D.J. Doren, J. Am. Chem. Soc. **115**, 3731 (1993).

Nonradiative recombination on Si surfaces during anodic oxidation in fluoride solution

J. Rappich*, V. Yu. Timoshenko** and Th. Dittrich***

*   Hahn-Meitner-Institut, Abt. AP, Rudower Chaussee 5, D-12489 Berlin, Germany
**   Moscow State University, Physics Department, 119899 Moscow, Russia
*** Technische Universität München, Physik-Department E16, D-85747 Garching, Germany

Abstract

The anodic oxidation of p-Si(100) in aqueous NH_4F solution is investigated in-situ by photoluminescence (PL) with short N_2-laser pulses as a function of pH at constant potential. The PL intensity depends sensitively on the modification of the Si surface and on the change of the oxidation rate during current oscillations. The interruption of the anodic oxidation at the maximum of a current oscillation peak leads to a current transient which shows typical features of oxide removal and hydrogenation for two different oxide thicknesses. This makes it possible to realize an intermediate state of the Si surface with a coexisting hydrophilic (oxide) and hydrophobic (hydrogenated) part.

Introduction

The silicon/fluoride electrolyte system is of great interest due to the possibility of the formation of structured or smooth surfaces as used for micromachining and electronic devices. There exist some surprising effects like a current transient at the end of the dissolution of an oxide covered silicon electrode in fluoride solutions [1-4]. The hydrogenation of the Si surface is obtained when the current transient tends to decay and is completed when the current levels out [4,5].

We use in-situ photoluminescence (PL) measurements to investigate the creation and passivation of nonradiative recombination centers at the silicon / electrolyte interface during anodic oxidation reactions at p-type Si(100) in the oscillating regime. The results are discussed from the point of view of non-radiative defect generation and passivation, which is kinetically controlled by the charge transfer and the oxidation and etching rate, respectively. The stroboscopic probing of the PL was chosen to suppress light induced electrochemical reactions.

Experimental

P-type Si(100) samples with a specific resistivity of about 1 Ωcm are used in the experiments. The Si sample is placed at the center of a quartz tube and serves as working electrode. Pt wire and 1 M KCl/AgCl/Ag are used as counter and reference

Mat. Res. Soc. Symp. Proc. Vol. 448 © 1997 Materials Research Society

electrodes, respectively. The electrolyte is continuously pumped through the quartz tube during the experiments. The electrode current or potential is controlled by a galvanostat/potentiostat (Jaissle IMP 88 PC). The electrolytes are made from p.a. NH_4F, triply distilled water and H_2SO_4 for pH adjustment. The PL is excited by a N_2-laser (LTB-MSG200, wavelength 337 nm, pulse width 0.5 ns). The light intensity is about 0.5 mJ/cm^2. The PL transients are detected at a wavelength of 1.05 µm using a prism monochromator, a Si-photodiode with a high impedance preamplifier (EMM) and a digital oscilloscope (HP 54510 A). The experimental setup is described elsewhere [6].

Results

Fig.1a shows a typical current voltage curve of p-type Si(100) in the acidic aqueous fluoride solution starting from the hydrogenated surface at -0.5 V. With increasing potential the current increases leading to current peak at which the formation of oxide sets on [7].

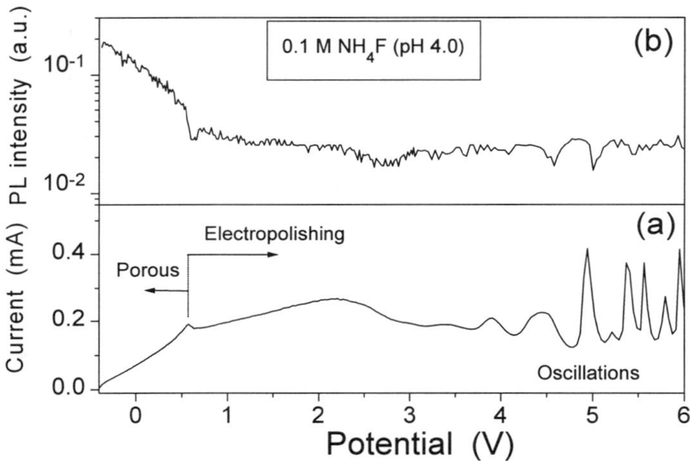

Fig.1 Current-potential dependence of p-Si(100) in 0.1 M NH4F (pH 4.0); scanrate 20 mV/s.

The formed oxide is simultaneously etched-back by the acidic fluoride solution resulting in the well-known electropolishing behavior. Current oscillations occur in the potential region above about +3.5 V in our case. Simultaneously, the PL intensity (fig.1b) decreases with increasing potential due to the formation of porous silicon in this potential regime [9]. At the current maximum, a sudden drop in the PL intensity occurs indicating to a change in the surface modification, the hydrogenated surface becomes hydroxylated and further oxidized [7]. A second, less pronounced drop can be seen after the second broad current maximum. Up to now, it is not known what kind of changes occur at the interface or in the oxide itself at this potential condition. At high anodic potentials the PL intensity oscillates anticorrelated to the oscillating current [8] which is suppressed when the current shows chaotic behavior above +5 V in our case.

Fig.2 shows the current-time (thin solid line) and PL-time (thick solid line) curves during electropolishing at a fixed potential of +8 V at different pH values (2.5, 3.5 and 4.0; fig. 2a,b and c, respectively). The mean current decreases while the mean PL intensity is nearly constant with increasing pH. The anticorrelation of the PL intensity to the oscillating current is seen at higher pH values due to a lower etchrate of the oxide by the electrolyte which increases the oscillation period and the oxide thickness [6]. The sharper the current peak of an oscillation the stronger the decrease of the PL intensity, up to one order in magnitude, which indicates to a sudden formation of a large density of nonradiative recombination centers if the change in the oxidation rate is fast. The damping of the current oscillations is also reflected by a damping of the PL intensity (fig.2b).

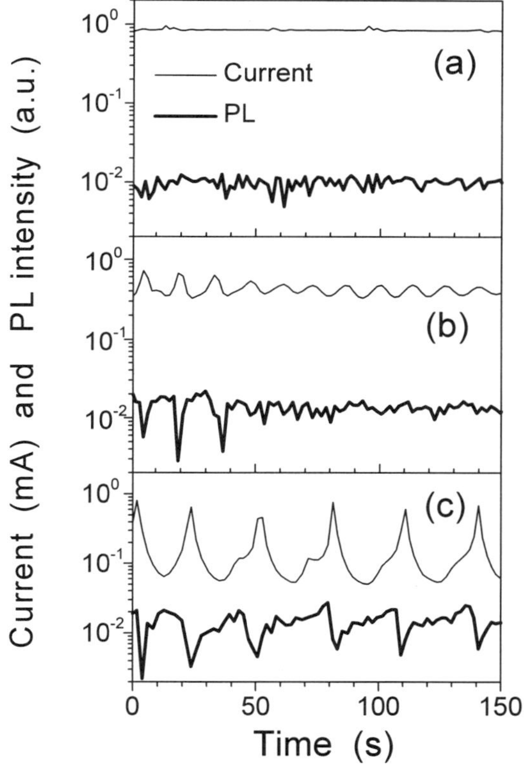

Fig:2 Current-time (thin solid line) and PL-time (thick solid line) curves for electropolishing at +8 V and hydrogenation at -0.4 V in 0.1 M NH_4F (a: pH 2.5, b: pH 3.5 and c: pH 4.0)

The PL intensity at +8 V (pH 4) and the deviation of the current from fig.2c are plotted in figure 3. There is a pronounced correlation of the PL intensity with the deviation of the current. The PL intensity increases slightly during the increase of the current, but when the current suddenly drops down (negative dI/dt) the PL intensity decreases heavily and cures with time until the next oscillation peak occurs.

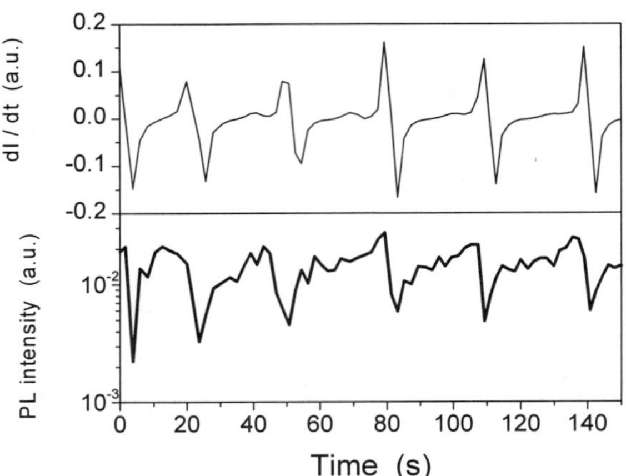

Fig.3 PL intensity and deviation of the current (dI/dt) from fig.2 (0.1M NH$_4$F, pH 4.0) as a function of time.

Fig.4 shows the influence of the etch-back of an anodic oxide by the acidic fluoride solution on the current transient by switching the potential from +8 V to -0.4 V either at

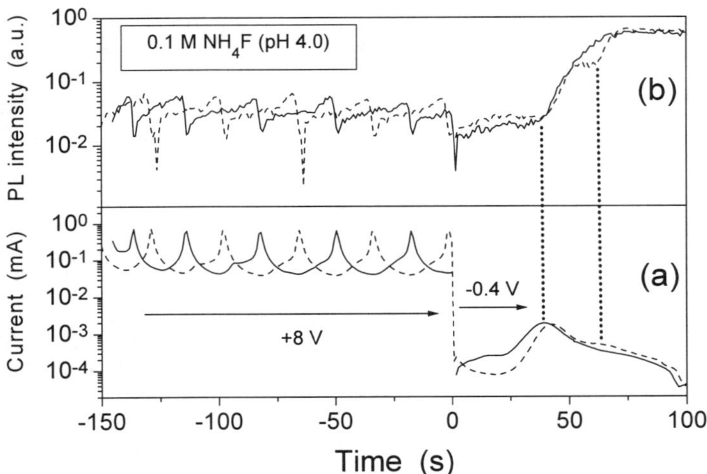

Fig.4 Current (a) and PL intensity (b) during electropolishing at +8 V and switching the potential to -0.4 V at the maximum (dashed line) or the minimum (solid line) of an oscillation.

the minimum (solid line) or at the maximum (dashed line) of a current peak (the timebase of the current and PL intensity is scaled to zero at the interruption of the current). There is no difference of the current amplitude and oscillation period at +8 V but the current transient at -0.4 V differs (fig.4a). The maximum of the current transient after interruption at the current oscillation maximum occurs somewhat later in time and shows a second small current peak with increasing time. The anticorrelation of the PL intensity to the current oscillation and the curing of the damaged interface by a slow increase of the PL intensity during the current minimum are clearly seen at +8 V in fig.4b. Switching the potential to -0.4 V leads to an increase of the PL intensity during the hydrogenation of the Si surface which begins at the current transient maximum and is completed when the current levels out. Interrupting the oxidation process at the maximum of the current oscillation leads not only to a second small peak at the current transient but also to a relative plateau at the PL intensity. When the second current peak begins to decay, the PL intensity increases to the value after interruption at the minimum of the current oscillation.

Discussion

The measured PL intensity of the crystalline silicon is controlled by the nonradiative recombination in the bulk and at the Si surface. The PL quenching in our experiments is related to the concentration of nonradiative defects at the Si surface, while the bulk recombination is unchanged during the electrochemical treatments at room temperature. The defect concentration depends on the morphology of the interface region Si/anodic oxide, which is controlled by the oxidation rate. As seen from fig.1, the in-situ PL measurement can be used as a probe for changes of the interface modification (hydrated, hydroxylated and oxidized respectively). Our experimental findings show that the nonradiative recombination, i.e. the defect concentration at the Si surface, is about 10 to 100 times higher during anodic oxidation than at the hydrogenated Si surface which correlates well with the finding from ex-situ surface photovoltage measurements on n-Si(111) [10]. Although the current maximum is less during anodic oxidation at higher pH´s (fig.2b and 2c, 0.5 mA) than at pH 2.5 (fig.2a, 1 mA) the mean PL intensity remains nearly constant. When the change of the oxidation rate (dI/dt) is to fast as observed during oscillations (fig.3) the oxidation process is obviously kinetically blocked by a limitation of the charge transfer into the electrolyte or by the transport of O^{2-} ions through the quickly formed passivating oxide layer. As a result of the higher etchrate of the oxide at higher pH the oxidation process leads to a thinner oxide layer where no limitation of the charge transfer into the electrolyte could be observed. It is known that dangling bonds at Si atoms with three Si backbonds, that means the dangling bonds closest to the bulk silicon, are very efficient nonradiative recombination centers [11]. The PL intensity during oxidation is the highest just before the current increases for the oscillation which points to a curing of the defective oxidized Si species with time (passivation of dangling bonds, increase of the oxidation state). The curing of the PL and the higher PL intensity during the minimum of a current oscillation points to a better passivation of the interface during this oxidation process and may be a hind why the oscillation treatment leads to smooth and electric well passivated surfaces [9,12]. Furthermore, a recent published analyses of the current transients indicates to the existence of mainly two different oxide thicknesses if the current transient shows a double-peak-like structure [8]. This behavior is confirmed by our in-situ PL measurement. The occurrence of a plateau and a further increase of the PL intensity during the second current peak indicates to a partially hydrogenated Si surface at the first plateau where the remaining oxide is etched back during the second current peak. The PL intensity of the hydrogenated Si surface is nearly identical for both interruption processes.

Conclusions

In situ PL measurements are used for the characterization of the p-type Si(100) surface during anodic oxidation. The fast change in the oxidation rate leads to a blocking of the passivation of Si dangling bonds at the interface. The thickness of the thin anodic oxide is most inhomogeneous at the maximum of the current oscillation where the PL intensity reaches the minimum. The etch-back of this inhomogeneous oxide layer leads to a Si surface state where hydrophobic (oxide) and hydrophilic (hydrogenated) part coexist.

Acknowledgment

J.R. is grateful to the support of this work by the Bundesministerium fuer Bildung und Forschung BMBF for financial support.

References

1.	M. Matsumura and S.R. Morrison; J. Electrochem. Soc. **147**, 157 (1983)

2.	H. Gerischer and M. Lübke; Ber. Bunsenges. Phys. Chem., **92**, 573 (1988)

3.	F. Ozanam, J.-N. Chazaviel; A. Radi and M. Etman; Ber. Bunsenges. Phys. Chem. **95**, 98 (1991)

4.	T. Bitzer, M. Gruyters, H.J. Lewerenz and K. Jacobi; Appl. Phys. Lett. **63**, 397 (1993)

5.	J. Rappich and H. J. Lewerenz; J. Electrochem. Soc.**142**, 1234 (1995)

6.	J. Rappich, Th. Dittrich and V. Yu. Timoshenko; J. Electrochem. Soc. in press

7.	F. Ozanam and J.-N. Chazalviel; J. Electron Spectrosc. Relat. Phenom. **64/65**, 395 (1993)

8.	J. Rappich, Th. Dittrich and V. Yu. Timoshenko; Ber. Bunsenges. Phys. Chem., in press

9.	J. Rappich, H. Jungblut, M. Aggour and H.J. Lewerenz; J. Electrochem. Soc. **141**, L99 (1994)

10.	Th. Dittrich, S. Rauscher, T. Bitzer, M. Aggour, H. Flietner and H.J. Lewerenz; J. Electrochem. Soc. **142**, 2411 (1995)

11.	H. Flietner, Proc. 7th Conf. Insulating Films on Semiconductors (INFOS) ed. by W. Eccleston, M. Uren (Hilger, Bristol 1991) pp. 151-154 and references therein.

12.	S. Rauscher, Th. Dittrich, M. Aggour, J. Rappich, H. Flietner and H.J. Lewerenz; Appl. Phys. Lett. **66**, 3018 (1995)

ADVANCED LITHOGRAPHY FOR NANOFABRICATION

Frank Y.C. Hui and Gyula Eres
Oak Ridge National Laboratory, P.O. Box 2008
Solid State Division, Bldg. 3150, MS 6056, Oak Ridge, TN 37831-6056

ABSTRACT

A novel method for generating lateral features by patterning the naturally forming surface hydride layer on Si is described. Because of the relatively strong chemical bonding between silicon and hydrogen, the hydride layer acts as a robust passivation layer with essentially zero surface mobility at ordinary temperatures. A focused electron beam from a scanning electron microscope was used for patterning. Upon losing the hydrogen passivation the silicon surface sites become highly reactive. Ideally, the lifetime of such a pattern in a clean environment should be infinite. Deliberate exposure of the entire wafer to a suitable gas phase precursor results in selective area film growth on the depassivated pattern. Linewidths and feature sizes of silicon dioxide on silicon below 100 nm were achieved upon exposure to air. The silicon dioxide is robust and allows effective pattern transfer by anisotropic wet-chemical etching. In this paper, the mechanism of hydrogen desorption and subsequent pattern formation, and the factors that govern the ultimate pattern resolution will be discussed.

INTRODUCTION

Electron beam lithography (EBL), X-ray, ion beam, and focused ion-beam (FIB) lithography are some of the candidates for the next generation microfabrication processes (sub-100 nm feature size) beyond optical lithography. However, none of the above processes satisfy the requirements for both high yield and high resolution. EBL and FIB have high ultimate resolution but suffer from slow speed, while the other methods support high yield processes but provide limited resolution. In terms of resolution, STM lithography can produce the finest structures. Linewidths down to 1 nm have been demonstrated on hydrogen passivated silicon surfaces [1]. However, because of the extremely slow writing speed and the narrow field of view, it is unlikely that STM patterning will become useful for high volume device fabrication. In addition, the STM fabricated ultrahigh resolution patterns were found to be too frail to survive the subsequent pattern transfer steps [2]. EBL is attractive for novel patterning methods because other than STM, it provides the highest resolution among all patterning tools and has a well established track record in device fabrication for research purposes [3].

In this paper, we report results concerning a novel resistless e-beam lithography process that involves electron beam induced patterning without the use of organic polymer resist such as PMMA. The patterning medium is the surface hydride layer on silicon. The function of the electron beam is to alter the chemical reactivity of the passivated surface. The exposed patterns are transferred by subsequent processing. In this work, we employed anisotropic wet chemical etching of the substrate. The advantages of resistless EBL are total elimination of all the toxic chemicals associated with the resist processes and a higher ultimate resolution resulting from the absence of line broadening by electron scattering processes in the resist. In this report we describe the linewidth dependence on the electron beam energy, the electron exposure dose, and the substrate thickness.

Mat. Res. Soc. Symp. Proc. Vol. 448 © **1997 Materials Research Society**

EXPERIMENTAL

The scanning electron beam lithography (SEBL) system consists of a commercial SEM, Amray 1400, configured for external scanning. The external scanning is driven by a 16 bit D/A converter controlled by a 486-50MHz personal computer. With a writing field of 100 μm, the D/A converter has a pixel resolution of 1.5 nm which is much less than both the beam diameter (~5-10 nm with a LaB$_6$ emitter) and the electrical noise (S/N ratio of ~120 dB) from the scanning coils. Larger writing fields with lower pixel resolution are also possible. Typical beam currents used were on the order of 5 to 100 pA. The electron dose for our application was 1-4 μC/cm. For comparison, the typical PMMA EBL dose is only a few nC/cm. The pressure in the sample chamber is in the low 10^{-6} to high 10^{-7} Torr range when both the turbo pump and ion pump are used. RGA monitoring of the sample chamber did not show detectable contamination. The base line performance capabilities of the EBL were determined using conventional resist based EBL. Minimum linewidths of 50 nm were routinely achieved using PMMA and gold liftoff. Note that we used relatively thick PMMA (~250 nm) and substrate (~300 μm) for the gold liftoff. Higher resolution is possible with better noise isolation and both thinner resist and substrate.

Following standard solution cleaning using trichloroethylene, acetone, methanol, and DI water rinse, the Si substrates were oxidized in a UV photoreactor to remove the remaining hydrocarbons and to oxidize any metallic impurities. The silicon hydride layers were prepared by a dilute (~10%) HF dip of the UV oxidized samples at room temperature. After the HF dip the surface is passivated by a uniform silicon hydride layer. Hydride passivation of the surface was confirmed by scanning Auger electron microprobe analysis. No surface contaminants such as carbon or oxygen were observed.

Upon electron beam irradiation, the hydrogen on the Si surface is desorbed and a highly reactive Si surface is produced that is the basis for pattern formation. The exposed area can either serve as a positive or a negative patterning mask depending on subsequent processes. One can oxidize the exposed area followed by CVD on the unexposed area. On the other hand, one can inject a source gas into the sample chamber concurrently with electron irradiation. In this work, we used wet chemical etching to transfer the oxide pattern to the Si substrate. A 10% tetramethyl-ammonium-hydroxide (TMAH) at 70°C is used as a Si etch. It is a highly anisotropic etch that etches Si(100) and Si(110) three times as fast as Si(111); for Si(110) the etch rate is 6.8 nm/s while for Si(111) it is 2.2 nm/s. In addition, the etch rate for SiO$_2$ is at least an order of magnitude slower than for Si. To elucidate the mechanism of pattern formation, the effects of electron energy, electron dose and the substrate thickness on the resulting linewidth were investigated.

RESULTS

Figure 1 shows an AFM image of a pattern on Si(100) transferred by TMAH etching following a 30 kV electron beam exposure at a dose of 1.14 μC/cm. The line spacings from top to bottom (same set of lines from left to right) are 0.25, 0.50, 1.00, 2.00, and 4.0 μm. Since the pressure in the sample chamber of the SEM was only in the low 10^{-6} Torr range, oxidation occurred in-situ during actual exposure. The oxide thickness was estimated to be around 1.5 nm. Attempts were made at imaging immediately following electron beam exposure, but due to the poor contrast between Si and SiO$_2$ we were unable to observe the oxide pattern in the SEM. Figure 2 shows a close-up view of the sample shown in Fig. 1 with line spacing of 0.5 μm. The lines are about 90 nm tall, and the width of the lines at the top is 25 nm.

Figure 3 shows a plot of the linewidth as a function of the electron dose at an electron energy of 30 kV. The plot shows a threshold dose of 0.75 μC/cm for electron beam induced oxidation. No exposure was found with a dose less than 0.75 μC/cm at 30 kV. The linewidth rises sharply until the dose reaches 1.9 μC/cm and saturates at a linewidth of 0.2 μm. The inset in the plot shows optical images of the corresponding test patterns. Note that the dose for the first pattern (which was not observable under the microscope) was plotted as corresponding to zero linewidth. This dose dependence is in contrast to the one reported by Kramer *et al.* [4] which showed no dose dependence in electron beam induced oxidation in UHV. However, it is possible that the lowest dose in their work has already exceeded the saturation dose.

In order to understand the mechanism of electron beam induced hydrogen desorption, we used different electron energies to test the role of back scattered (BSE) and secondary electrons (SE). Figure 4 depicts two patterns on the same substrate exposed at 10 kV and 30 kV using the same dose. At 10 kV the lines were much wider with less defined edges. This result strongly suggests that either the BSE and/or the SE are contributing to hydrogen desorption. The R_{max} given in Figure 4 is the Bethe range [5] for the given electron energies in Si. Figure 5 shows the measured linewidth as a function of electron energy at 5, 10, 30, and 40 kV. The substrate is Si(100) and the patterns were transferred using a 10 s TMAH etch. The linewidth decreases sharply from 5 kV and saturates at 30 kV. A minimum linewidth of ≤ 0.1 μm was obtained at around 30 kV.

To differentiate the role of BSE and SE, exposures on a substrate with a tapered edge were performed. The upper left picture in Figure 6 is a backscattered electron image showing a 140 μm hole in Si(100). The thickness is gradually increasing away from the edge as shown by the BSE image where bright areas indicate high intensities of BSE (thick substrate) and dark areas indicate low BSE (thin substrate). An optical image of two sets of line exposures (at 30 kV) with the lines written from the thin part of the substrate toward the thick part are depicted in the lower left of Figure 6. Close up images taken at (a) and (b) are shown at top right (thin substrate) and

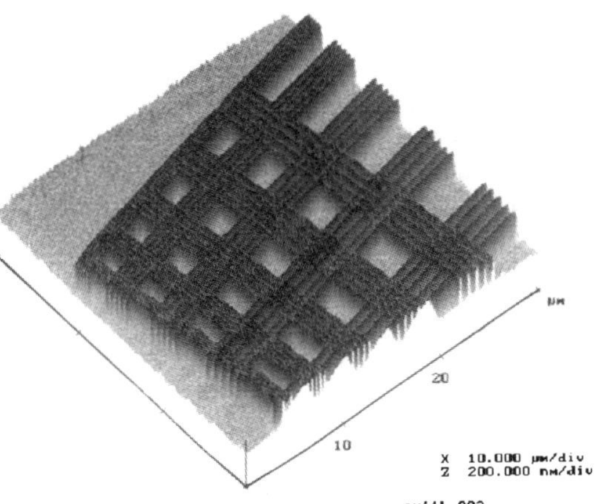

Figure 1

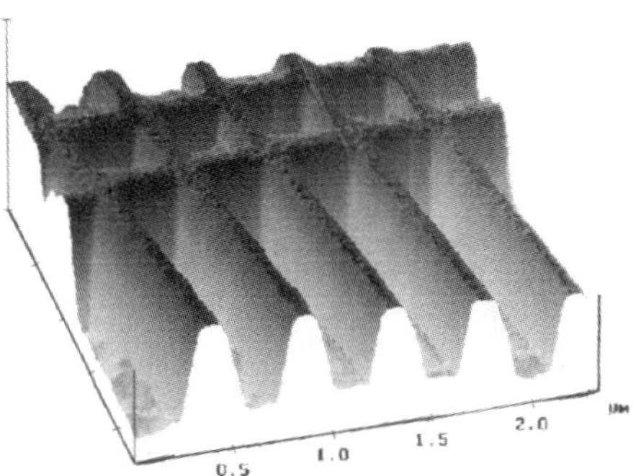

Figure 2

bottom right (thick substrate). The SEM images on the right hand side of Figure 6 show that the lines on the thick part of the substrate (b) are only slightly wider than the lines on the thin part. No significant change in the linewidth was found as the substrate thickness changed from about 1 μm to 5 μm. At 30 kV, the Bethe electron range is 8.5 μm. A significant reduction in BSE that would be expected for a substrate thickness that is less than half the Bethe range has not been observed. This finding suggests that BSE do not play a major role in determining the linewidth. It further implies that BSE exposure is not the mechanism that governs hydrogen desorption.

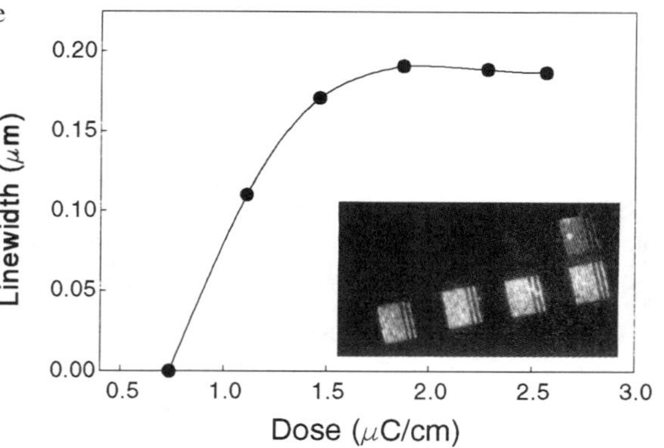

Figure 3

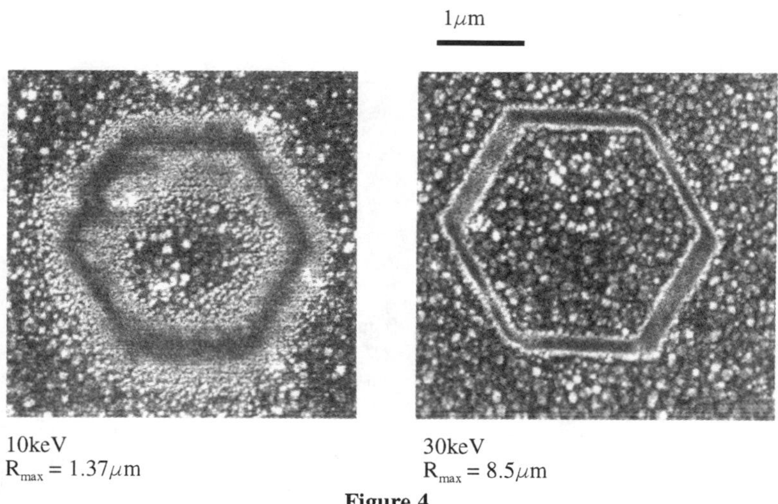

10keV
$R_{max} = 1.37 \mu$m

30keV
$R_{max} = 8.5 \mu$m

Figure 4

SUMMARY

In conclusion, the passivating hydride layer on silicon was used as a prototype for exploring the feasibility of electron beam patterning of surface adsorption layers for nanofabrication applications. The patterns obtained by this resistless hydride lithography were continuous with a minimum linewidth of 0.1 μm or better. At present, no special effort was made to obtain the ultimate resolution of this patterning method. Systematic studies were performed to explore the correlation between the exposure parameters such as the electron energy, dose, substrate thickness and the resulting linewidth. Further comprehensive investigations are needed to fully understand the interplay between the linewidth and the mechanism of hydrogen desorption.

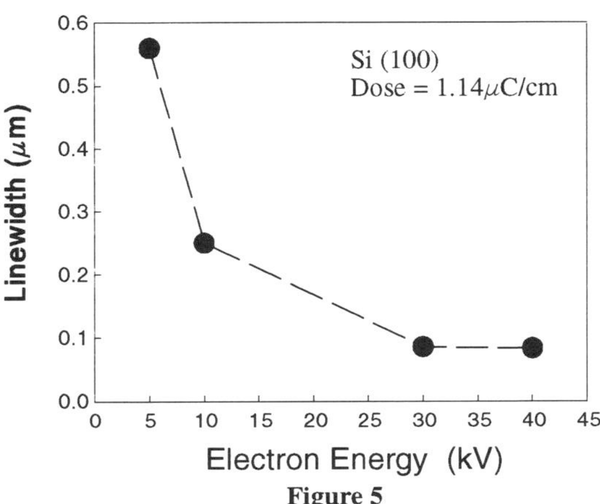

Figure 5

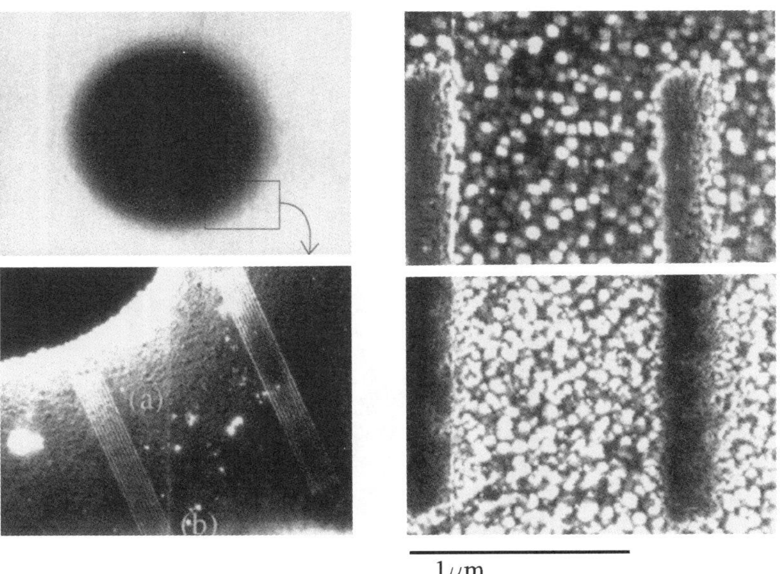

1μm

Figure 6

ACKNOWLEDGMENT

We acknowledge Professor David Joy for providing in depth understanding of electron scattering in solids and for obtaining the BSE images.   AFM images were taken with the help of Dr. Thomas Thundat of the Health Sciences Research Division of ORNL.   AES microprobe measurements were carried out by Dr. Jae-Won Park of the Materials Science and Engineering Department of the University of Tennessee.

REFERENCES

1.	J.W. Lyding, G.C. Abeln, T.C. Shen, C. Wang, and J.R. Tucker, J. Vac. Sci. Technol. B **12**, 3735 (1994).
2.	P. Fay, R.T. Brockenbrough, G. Abeln, P. Scott, S. Agarwala, I. Adesida, and J.W. Lyding, J. Appl. Phys. **75**, 7545 (1994).
3.	See IBM Journal of Research and Development, **32**, (1988).
4.	N. Kramer, J. Jorritsma, H. Birk, and C. Schönenberger, J. Vac. Sci. Technol. B **13**, 805 (1995).
5.	J.I. Goldstein, D.E. Newbury, P. Echlin, D.C. Joy, A.D. Romig, Jr., C.E. Lyman, C. Fiori, and E. Lifshin, *Scanning Electron Microscopy and X-ray Microanalysis: A Text for Biologists, Materials Scientists, and Geologists,* 2 nd ed. (Plenum, New York, 1992) p. 88.

IN SITU INFRARED OBSERVATION OF HYDROGENATION, OXIDATION, AND ADSORPTION ON SILICON SURFACES IN SOLUTIONS

YOSHIHIRO SUGITA and SATORU WATANABE
Fujitsu Laboratories ltd.,
10-1 Morinosato-Wakamiya Atsugi, 243-01 Japan, sugita@flab.fujitsu.co.jp

ABSTRACT

Fourier transform infrared attenuated total reflection method was employed to observe Si surfaces during wet chemical treatments. We observed the time evolution of the surface chemical structure during the oxidizing of hydrogenated Si surfaces in such oxidants as ozonized water and hydrogen peroxide using (100) and (111) surfaces. We also examined the adsorption of surfactants which were introduced in an HF solution. The interaction of adsorbates at the interface and with molecules in a liquid phase was discussed based on our *in situ* observations.

INTRODUCTION

In Si device fabrication, HF acid is widely used to produce oxide-free hydrogenated Si surfaces [1]. However, it is well known that the surface may become oxidized during the subsequent rinse in water containing dissolved oxygen [2]. The existence of any oxide on the Si surface degrades the electrical properties, when low resistance metal contacts and high dielectric capacitors formed on the surface. Recently, it was reported that thin gate oxides formed out of hydrogenated surfaces were more reliable than that out of chemical oxides formed by conventional cleaning procedures [3]. Precise understanding of the initial oxidation of hydrogenated surfaces is important. It is widely accepted, based on experimental evidence obtained by *ex situ* techniques, that the oxidation of a hydrogenated Si surface begins with oxygen attaching to the back bond of surface Si atoms when oxidation occurs in pure water containing dissolved oxygen and in a dry oxygen atmosphere [4-8].

In this paper, we investigated surface oxidation in an H_2O_2 solution and in ozonized water to examine the affects of the redox potential of each. Both solutions are nearly neutral but are different in redox potential. In addition, oxidants such as H_2O_2 and O_3 are intentionally added to solutions used for surface cleaning processes to dissolve some metallic contaminants and to passivate the surface with chemical oxide. Little of the oxidation chemistry is known regarding solutions with a large redox-potential.

The progress of oxidation and the consumption of the Si-hydrides structure were monitored using *in situ* infrared spectroscopy. We used Si (100) and (111), whose major surface structures are di-hydride and mono-hydride, respectively, and examined the relationship between the oxidation reaction and the surface structure. The adsorption of surfactants in an HF solution is also described in this paper.

EXPERIMENT

Non-doped, FZ-Si prism-shaped wafers which were polished on both sides were used for Fourier transform infrared attenuated total reflection (FT-IR ATR) analysis. We used both (100) and (111). The size of the ATR prism was 53 x 53 mm, and the thicknesses were 1.0 mm for (100) and 0.5 mm for (111). IR analysis was performed using a Nicolet 740 FT-IR spectrometer

Mat. Res. Soc. Symp. Proc. Vol. 448 © 1997 Materials Research Society

and an HgCdTe detector cooled with liquid N_2. The measurement resolution was 4 cm-1. Polarization experiments were performed using wire grid polarizer. We developed a specially designed chemical cell to examine the Si surface in solution [9].

After mounting a conventionally pre-cleaned sample in the chemical cell with ATR geometry, we introduced a 0.5% HF solution into the cell. The surface oxide was then removed and the surface Si bonds were terminated with hydrogen atoms. An oxidizing solution was then introduced for a specified time. To allow sufficient time for spectrum accumulation (typically 1000 times for 10 min), we exchanged the oxidizing solution with water to pose the oxidation reaction. This procedure with the oxidizing solution and water was repeated. We used a 3% H_2O_2 solution (pH 5.5, NHE. +0.5~0.7 V) and a 2 ppm ozonized water (pH 6-7, NHE. +1.0~1.2 V) as the oxidizing solutions.

An electron device industry grade surfactant added HF solution (Morita Chemical Industries Company Ltd. ; SHF) was used to examine affects of the surfactant on surface chemical state [10].

RESULTS

Figure 1 shows IR-ATR spectra of water near the Si surface. The spectra were recorded with a chemically oxidized Si (100) prism. The reference was air, that is, there was no water in the cell. Strong absorption originating from O-H stretching at about 3400 cm-1 and H-O-H bending at about 1650 cm-1 were observed. A weak absorption seen at 2210 cm-1 is attributed to an water related absorption. The region below 1450 cm-1 were undetectable because of the strong absorption of the Si substrate itself.

1. Oxidation of H/Si in solution

The IR-ATR technique is sensitive enough to reveal the structural changes of hydrogenated Si surfaces. About 1% structural modification of the monolayer is detectable with our equipment. And the chemical and physical adsorption from the atmosphere is negligible. Due to potential instability caused by *ex-situ* observation, it was quite difficult to evaluate the early oxidation stage based on the detection of $O_xSi_y\text{-}H_z$ (back bond oxidized Si-hydrides, x = 1 - 3, x + y + z = 5) [11]. In order to reveal the nature of the oxidation process of hydrogenated Si in solution, *in situ* IR measurements were required.

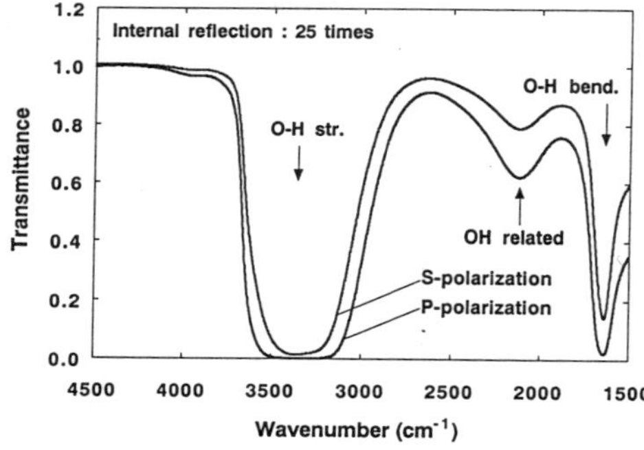

Fig. 1. ATR spectra of water near Si surface. Spectra were recorded with a native-oxide formed Si (100) prism. Vertical axis shows transmission when the cell was filled with pure water. The reference spectra were recorded without water in the cell.

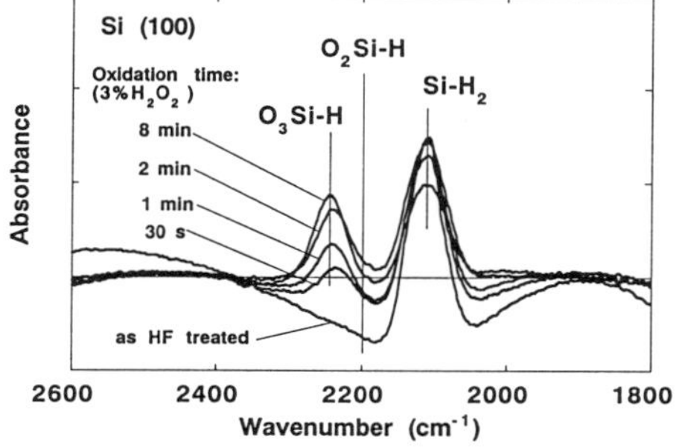

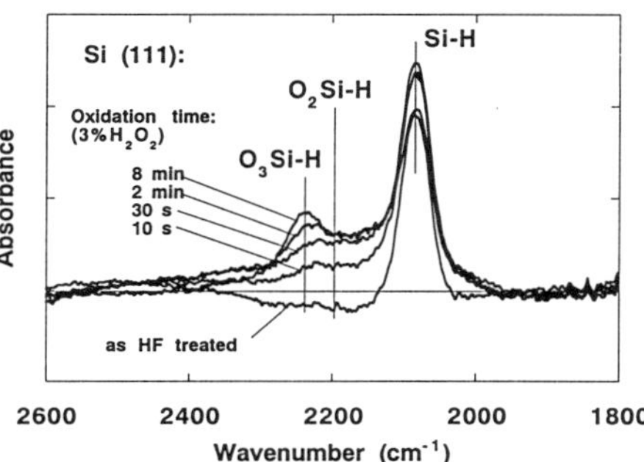

Fig. 2. P-polarized ATR spectra of Si-H stretching on Si (100) in de-ionized water. The reference was a native oxide formed Si (100) prepared in a boiling H_2SO_4 /H_2O_2 solution. It was removed using 0.5% HF then re-oxidized using 3% H_2O_2.

Fig. 3. P-polarized ATR spectra of Si-H stretching on Si (111) in de-ionized water. The reference was a native-oxide formed Si (111) prepared in a boiling H_2SO_4/H_2O_2 solution. It was removed using 0.5% HF then re-oxidized using 3% H_2O_2.

Figures 2, and 3 show the oxidation of (100) and (111) surfaces in 3% H_2O_2 solution. While the Si-H_z (z = 1 - 3) stretching frequencies in figures 2 and 3 were almost the same as these observed by *ex situ* analysis, the peaks are broader. This was already explained in terms of the interaction of Si-H_z (z = 1 - 3) with water [9]. It can be clearly seen from these figures that the increase of $O_xSi_yH_z$ (x = 1 - 3, x + y + z = 5) structures was accompanied by the decrease of SiH_z (z = 1 - 3) structure. In the (100) surface, the O_3SiH structure was dominant even during very early stages of the oxidation process (less than 0.5 monolayer), indicating that the $O_2Si_yH_z$ (y + z = 3, z = 1 - 2) and the OSi_yH_z (y + z = 4, z = 1 - 3) structures were unstable. In contrast, these were clearly observable in the (111) surface, as shown in figure 3. The same observation was made when H_2O_2 was replaced with ozonized water, demonstrating that the instability of $O_2Si_yH_z$ (y + z = 3, z = 1, 2) and OSi_yH_z (y + z = 4, z = 1 - 3) on (100) and the stability of the same on (111) remains independent of the redox potential of the oxidizing solution.

These results suggested that one back bonds oxidation weakened the other back bonds on (100), yet, the partial oxidation on (111) did not weaken neighboring back bonds toward oxidation. In *in situ* IR-ATR experiments recently done by others [12], such intermediate oxides were not observed on (111). This might due to a lack of sensitivity needed to detect these structures.

We found similar spectra between p-polarization and s-polarization conditions on (100) surfaces. On the other hand, polarization experiments on the (111) surface showed meaningful results. S-polarization spectra should detect the oxidation of Si-hydrides at step or kink sites, however, there was no evidence of site specific back bond oxidation in our experiment. It seemed that the back bond oxidation of the topmost Si layers progressed randomly on (111) at least.

When we consider the charge transfer from the back bonds to the chemisorbed oxygen, it may be generally stated that the oxidation of one back bond weakens the other back bonds. In the case of the (111) surface, however, the attachment of oxygen to the back bond of mono-hydride requires a displacement of the topmost Si atoms, thus stressing the remaining two back bonds. This increases the reaction potential for oxidation. On the (100) surface, this effect is expected to be less significant because oxidation of one back bond of di-hydride requires displacement of the

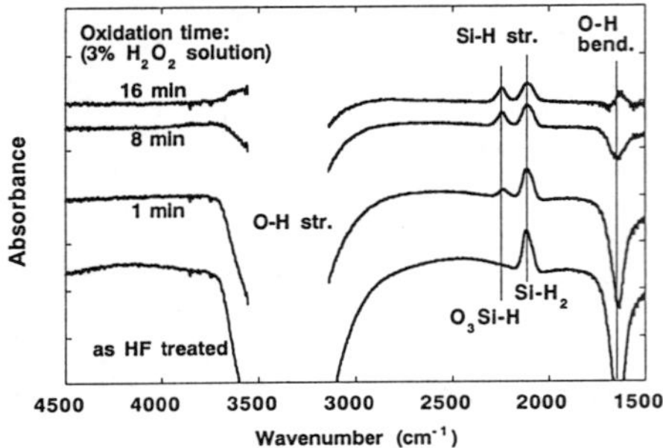

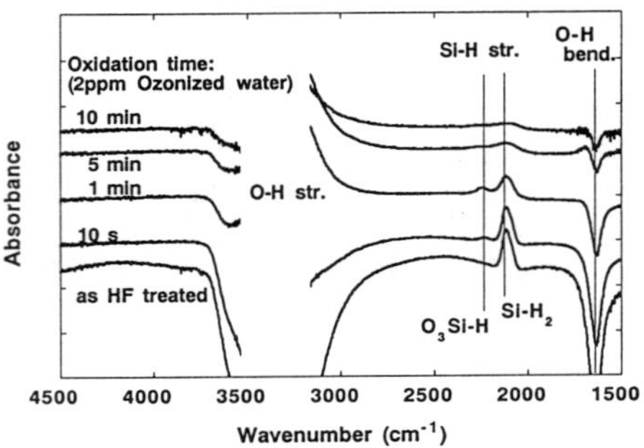

Fig. 4. S-polarized ATR spectra of H/Si (100) in de-ionized water. The reference was a native-oxide formed Si (100) prepared in a boiling H_2SO_4/H_2O_2 solution. It was removed using 0.5% HF then re-oxidized using 3% H_2O_2.

Fig. 5. S-polarized ATR spectra of H/Si (100) in de-ionized water. The reference was a native-oxide formed Si (100) prepared in a boiling H_2SO_4/H_2O_2 solution. It was removed using 0.5% HF then re-oxidized using 2 ppm ozonized water.

topmost Si atoms, thus stressing only one back bond. We think that such stress on (111) basically originates from the crystal structure of Si (111), that is, shorter distance between the top two Si layers. We believe that the stress interrupts the subsequent oxygen attachment to the back bonds on the (111) surface. This model may explain why the back bond oxidation rate on (100) surfaces is higher than on (111) surfaces in solution [13, 14].

2. Interaction of water with substrate surface and the affect of surfactants

The time evolution of the oxidation of hydrogenated Si (100) in a 3% hydrogen peroxide solution is plotted in figure 4. The increase of water related absorption is clearly seen during the oxidation, indicating that oxidation degrades the hydrophobicity of the surface. In other words, the density of water is lower near the hydrophobic surface than it is near the hydrophilic surface. The same observation was also confirmed in the oxidation of Si (100) with 2 ppm ozonized water, as shown in figure 5. However, Si-hydride structures almost disappeared from hydrogenated surfaces after only a few minutes of oxidation in ozonized water. In contrast, most of O_3SiH and SiH_z ($z = 1 - 3$) structures remained stable in the 3% H_2O_2 solution even after long term oxidation. It is noted that the hydrophobicity of hydrogenated Si in ozonized water decreased drastically during 10 s to 1 min oxidation, however, the observed O_3Si-H growth was very weak, as seen in figure 5. This suggests the existence of two oxidation paths, one is back bond oxidation and the other is Si-H itself. The oxidation rate of Si-H was faster than that of Si-Si in ozonized water. We believe that the oxidation of Si-H produces Si-OH in ozonized water which may degrade the hydrophobicity of the surface.

Figure 6 shows ATR spectra of hydrogenated Si (100) measured in the surfactant added 1% HF solution and compared to three different references; native-oxide formed Si in pure water, hydrogenated Si in pure water and hydrogenated Si in a surfactant free 1% HF solution. Broad HF related absorptions are seen in figure 6 with the same wavenumber positions observed in a previous report [15]. Surface adsorption of the surfactant was clearly seen, as was the hydrophobicity degradation of Si in SHF. The changes of Si-hydride structures caused by surfactant adsorption were also observed.

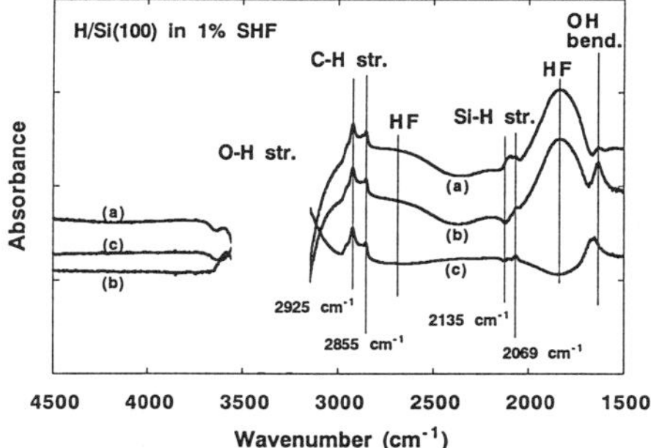

Fig. 6. S-polarized ATR spectra of H/Si (100) in a surfactant added 1% HF solution. Reference spectra were recorded with (a) a native oxide-formed Si (100) in de-ionized water, (b) a hydrogenated Si (100) in de-ionized water, and (c) a hydrogenated Si (100) in no-surfactant added 1% HF solution.

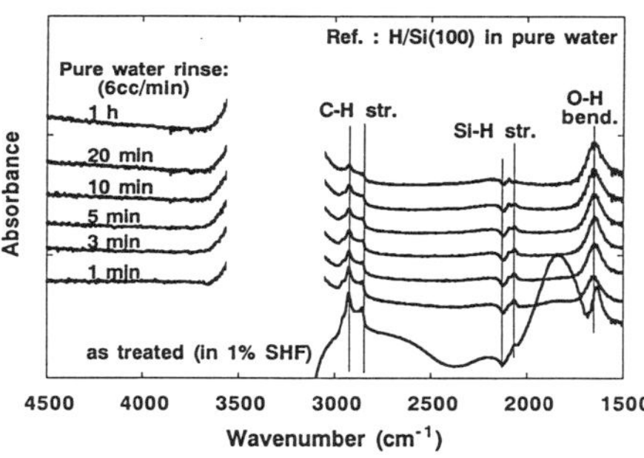

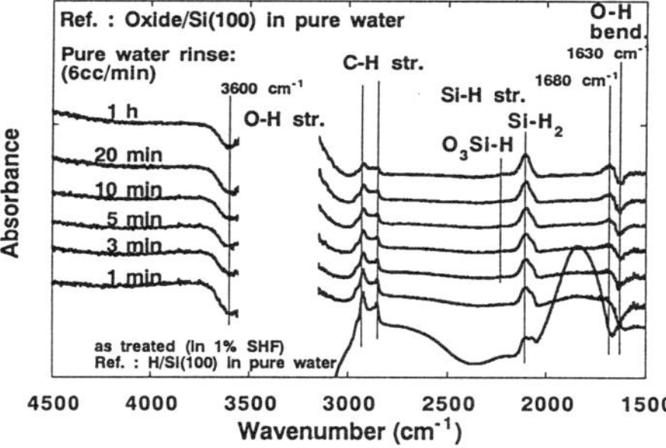

Fig. 7. Time evolution of spectrum changes on H/Si (100) formed in a surfactant added 1% HF solution during de-ionized water rinse. The reference was a native-oxide formed Si (100) in de-ionized water.

Fig. 8. Time evolution of H/Si (100) formed in a surfactant added 1% HF solution during de-ionized water rinse. The reference was a hydrogenated Si (100) in de-ionized water.

Figure 7 shows the affects of a pure water rinse on hydrogenated Si (100) formed with 1% SHF. The reference spectrum in figure 7 was obtained from hydrogenated Si(100) in pure water. The surface is less hydrophobic even after long term rinsing, which may be due to the resident surfactant on the surface, observed in figure 7.

Two shoulder peaks in the C-H stretching region may be attributed to symmetric and asymmetric CH_2 stretching of adsorbed surfactant. The shoulder peaks ratio was almost constant during the water rinse. The details of adsorbed surfactant properties were left for further studies.

In figure 8 we plotted these spectra again with another reference spectrum of a native-oxide formed Si (100) prism in pure water. The O-H related absorption showed some characteristic spectra changes. A less absorption than the typical hydrophilic surface of the reference sample was observed at around 1630 cm⁻¹ and 3600 cm⁻¹. This may be attributed to the lower concentration of non-hydrogen bonded water originating from the surfactant [9]. The increase in the absorption at 1680 cm⁻¹ and 3100 cm⁻¹ during the water rinse accompanied an increase in the intensity of $O_xSi_yH_z$ (x + y + z = 5). We considered that this O-H absorption might relate to some interaction between the water near the surface and the oxide grown on the surface.

It can be stated that the surfactant adsorption on the surface affects both the Si-hydride structure and the hydrogen bonded state of the water near the surface. We suppose that this affect may change the reaction properties between the water and Si surface, such as the oxidation and the etching of Si.

SUMMARY

The insertion of oxygen process to Si-Si is highly dependent on surface orientation. *In situ* IR observation suggests that the oxidation of one back bond of hydrogenated Si on (100) weakens the other bonds against oxidation. It also suggests that insertion oxidation on (111) is interrupted with the stress originating from Si (111) structure.

An instability of the Si-hydrides in ozonized water was detected with *in situ* IR analysis. It is suggested that the chemical stability of the Si-hydrides depends on the redox potential of the solution.

The affect of surfactant adsorption on hydrogenated Si was examined using an electrical industry grade HF based solution. Adsorption of the surfactant caused a degradation in the hydrophobicity of the surface. The existence of surfactant on the Si surface could affect the hydrogen bonded state of water near the surface. The surface concentration of the adsorbed surfactant decreased over water rinsing time, however its complete removal did not occur.

REFERENCES

1. Y. J. Chabal, G. S. Higashi K. Raghavachari and V. A. Burrows, J. Vac. Sci. Technol. **A7** 2104 (1989).
2. M. Morita, T. Ohmi, E. Hasegawa, M. Kawakami and M. Ohwada, J. Appl. Phys. **68** 1272 (1990).
3. T. Ohmi, M. Morita, A. Teramoto, K. Makihara and K. S. Tseng, Appl. Phys. Lett. **60** 2176 (1992).
4. Y. Nagasawa, H. Ishida, T. Takahagi, A. Ishitani and H. Kuroda, Solid State Electronics **33**, supplement 129 (1990).
5. D. Gräf, M. Grundner and R. Schults, J. Vac. Sci. Technol **A7** 808 (1989).
6. E. Iijima, T. Aiba, K. Yamauchi, H. Nohira, T. Tabe, M. Katayama and T. Hattori, <u>Extended Abstract of the 1995 International Conference on Solid State Devices and Materials, Osaka</u> (Business Center for Academic Societies Japan, Tokyo, 1995) p. 497.
7. T. Miura, M. Niwano, D. Shoji, and N. Miyamoto, J. Appl. Phys. **79** 4373 (1996).
8. S. Fujimura, H. Ogawa, K. Ishikawa, C. Inomata and H. Mori, <u>Extended Abstracts of 1993 International Conference on Solid State Devices and Materials, Makuhari</u> (Business Center for Academic Societies Japan, Tokyo, 1993) p. 618.
9. S. Watanabe, Surf. Sci. **341** 304 (1995).
10. M. Miyamot, N. Kita, S. Ishida and T. Tatsuno, J. Electro. Chem. Soc. **141** 2894 (1994).
11. G. Lucovsky, J. Yang, S. S. Chao, J. E. Tyler and W. Czubatyji, Phys. Rev. **B28** 3225 (1983).
12. E. P. Boonekamp, J. J. Kelly, J. Ven and A. H. M. Sondag, J. Appl. Phys. **75** 8121 (1994).
13. M. Hirose, T. Yasaka, M. Takakura and S. Miyazaki, Solid State Technol. Dec. 43 (1991).
14. Y. Sugita and S. Watanabe, <u>Extended Abstracts of 1996 International Conference on Solid State Devices and Materials, Yokohama</u> (Business Center for Academic Societies Japan, Tokyo, 1996) p. 386.
15. P. A. Giguere and S. Turrel: J. Am. Chem. Soc. **102** 5473 (1980).

ANALYSIS OF InP PASSIVATED WITH THIOUREA/AMMONIA SOLUTIONS AND THIN CdS FILMS

H.M. DAUPLAISE, A. DAVIS, K. VACCARO, W.D. WATERS, S.M. SPAZIANI,
E.A. MARTIN, and J.P. LORENZO
USAF Rome Laboratory, Optical Components Branch, Hanscom Air Force Base, MA 01731

ABSTRACT

The use of thiourea/ammonia pre-treatments on (100) InP, followed by chemical bath deposition (CBD) of CdS thin films ($\sim$ 30 Å), with low-temperature, low-pressure chemical vapor deposited SiO_2 has been shown to produce metal-insulator-semiconductor (MIS) samples with near-ideal capacitance-voltage (C-V) response. Here, we report on x-ray photoelectron spectroscopy (XPS) analysis of the near-surface of InP following pre-treatment and CdS deposition. The pre-treatment was shown by XPS to form an indium sulfide layer and effectively remove native oxides from the InP surface. The subsequent deposition of CdS on a sulfur-passivated surface forms a stable layer which protects the substrate from oxidation during SiO_2 chemical vapor deposition. MIS samples prepared using the pre-treatment without CdS deposition showed improved C-V response, while samples prepared with both the pre-treatment and CdS deposition showed a dramatic reduction in the density of interface states.

INTRODUCTION

Sulfur treatments of III-V materials have shown great promise in producing stable, passivated surfaces for MIS device applications. Methods of sulfur passivation include ammonium sulfide solutions[1] and gas-phase polysulfide treatments,[2] both of which form an indium sulfide layer on InP. While these methods have resulted in a reduction in interface-state densities for InP, low-frequency electrical response, drain current drift, and repeatability remain problems for MISFET device fabrication. The efficiency of CBD CdS thin films on InP in improving the MIS C-V response has been recently demonstrated by our group.[3] Since this report, we have found that pre-treating the sample in a solution of thiourea/ammonia prior to CdS deposition results in further improvements in MIS C-V response.[4] In this paper, we report on XPS analysis of InP pre-treated with thiourea/ammonia solutions and with subsequent deposition of thin CdS films. Our hope is to elucidate the surface chemistry of InP following these treatments and better understand their role in improving resulting MIS devices.

EXPERIMENT

Thiourea/ammonia pre-treatment is performed by immersing an unetched n-InP sample in a 20 ml solution of 0.033 M thiourea and 12.3 M ammonia at 85°C for 15 min. The sample is then rinsed with deionized water and dried with nitrogen. CdS thin films were deposited as reported previously.[4] Standard CdS deposition conditions were 0.028 M thiourea ($CS(NH_2)_2$), 0.014 M cadmium sulfate $CdSO_4$, and 11 M NH_3 for 3 min at 85°C. MIS samples were prepared by depositing 300 Å of SiO_2 at 3 Torr and 260°C.[5] A silane overpressure was provided during sample heating from room temperature to the deposition temperature by 50 sccm of 5% SiH_4 in N_2. Gas flows were 33 sccm O_2 and 100 sccm 5% SiH_4 in N_2 during the deposition. After thermal evaporation of Al front contacts and In back contacts, samples were annealed overnight in N_2 at

Mat. Res. Soc. Symp. Proc. Vol. 448 © 1997 Materials Research Society

350°C. *C-V* response was measured at 1 MHz using an HP 4275A multi-frequency LCR meter; quasistatic measurements were made with a Keithley 595 meter. Interface-state densities were calculated by the method of Castagné and Vapaille,[6] using both high-frequency and quasistatic *C-V* data.

In order to determine the chemistry of the passivated InP surface following thiourea/ammonia pre-treatment and CdS thin film deposition, XPS analysis was performed on a Physical Electronics PHI 5100 using non-monochromatic Mg K_α radiation at 1253.6 eV. The anode is operated at 15 KV with an incident power of 400 W. The diameter of the analysis area is 800 μm. A pass energy of 11.75 eV was used for all detail scans. The spectra were corrected for charging effects by referencing the carbon 1*s* peak to 284.8 eV. After Shirley background subtraction, Gaussian-Lorentzian peaks were fitted to the spectra for non-linear least squares optimization. The elemental peak areas were corrected using standard sensitivity factors.[7] For fitting the In $3d_{5/2}$ peak, results obtained on vacuum-cleaved InP were used as a guide, with the FWHM for the substrate component being 1.15 eV, and a Gaussian-to-Lorentzian ratio of 2.2. Layer thicknesses, *t*, were calculated using

$$\ln\left(\frac{C_{layer}}{C_{substrate}}+1\right) = \frac{t}{25 \times \cos\theta} \tag{1}$$

where C_{layer} is the sum of the corrected elemental peak areas in the layer.[8] Carbon and oxygen, if detected, were included in this parameter. $C_{substrate}$ is the sum of the corrected peak areas of In and P. The polar angle of analysis, θ, was 45°. A mean free path of 25 Å was assumed.

Following the thiourea/ammonia pre-treatment and, when applicable, the CdS thin film deposition, samples were immediately loaded into the XPS system. Following analysis of the unannealed surface, samples were heated for one hour at 200°C *in vacuo*, allowed to cool, then re-analyzed. The heating/analysis cycles were repeated at 50°C increments up to 400°C. The base pressure of the system was 1×10^{-10} Torr and rose to ~ 5×10^{-8} Torr during bakeout. An HF-etched InP sample was also analyzed as a reference.

RESULTS AND DISCUSSION

Figures 1(a) and 1(b) show the 1 MHz and quasistatic *C-V* responses for HF-etched InP, pre-treated InP, and InP with a pre-treatment followed by a thin CdS film (~ 30 Å) deposition, respectively. For the HF-etched InP, sample A, it is evident that the sample is prevented from becoming accumulated by a high density of fast traps near the conduction band edge, and both high and low frequency response are impeded by a high density of states throughout the band gap. Following thiourea/ammonia pre-treatment, the *C-V* response is improved, sample B. The sample now goes into accumulation, and the difference in C_{ox} between the 1 MHz and quasistatic curves is significantly reduced. The pre-treated sample still exhibits poor quasistatic response. A higher value of C_{min} at 1 MHz also suggests additional states have been added below midgap. This effect has also been reported by Lau *et al.* for polysulfide-treated InP.[7] The response of the sample prepared with both pre-treatment and CdS film deposition, sample C, is nearly ideal at 1 MHz and exhibits good low frequency response in depletion and inversion. The theoretical *C-V* response is shown for both 1 MHz and quasistatic by the open circles. The parameters for the theoretical fit for both the high frequency and low frequency curves were the same except for the dielectric constant of the oxide, ε_{ox}. For the 1 MHz response curve, ε_{ox} was 4.5, while 4.6 was

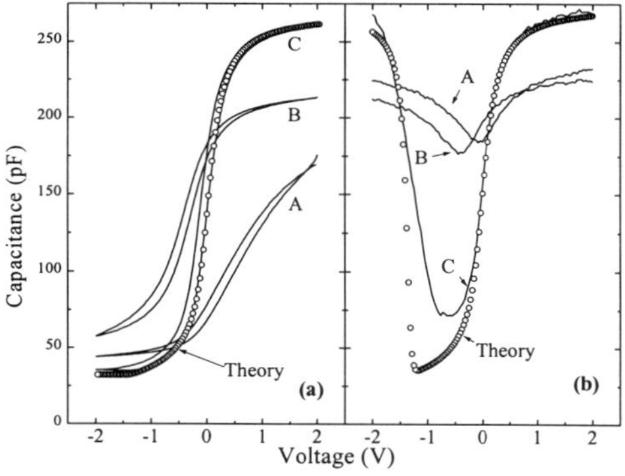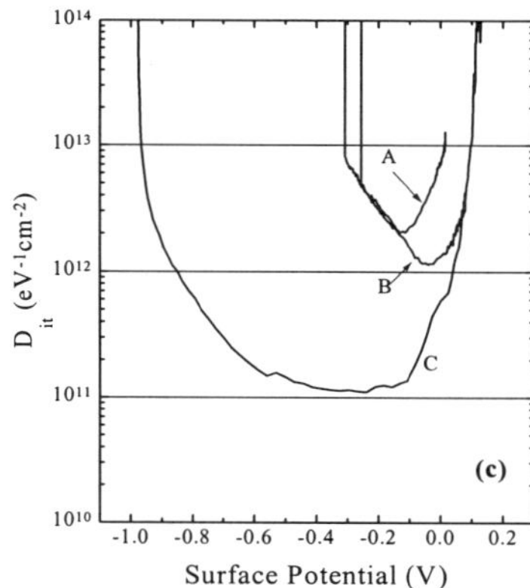

Figure 1. The (a) 1 MHz and (b) quasistatic responses frequency response of *n*-InP MIS samples. Sample A was HF-etched immediately prior to dielectric deposition, while sample B was pre-treated in a thiourea/ammonia solution. Sample C had a CdS thin film deposited after pre-treatment. The open circles show theoretical *C-V* response. (c) D_{it} values calculated for all three samples.

used for the low frequency plot. The variation in C_{ox} with frequency may be due to a low density of fast traps near the conduction band edge, or a dispersion in ε_{ox}. C_{min}/C_{ox} for sample C in Fig. 1(b) is 0.26, compared to the theoretical value of 0.13. The low frequency behavior shows that the InP Fermi level is unpinned, allowing the InP surface to be biased from accumulation to inversion. The D_{it} values for all three samples are shown in Fig. 1(c). We have found that the preparation method for sample C yields the most consistent and repeatable results for MIS capacitor fabrication.

Figure 2(a) shows the S $2p$ detail scan of a pre-treated *n*-InP sample unannealed and after annealing at 200, 300, and 400°C. Sulfur comprises 5.8% of the total detected signal on the unannealed sample; it is reduced to 3.3% following annealing at 400°C. The estimated layer thickness is 11.5 Å, which decreases to ~ 3 Å after the final anneal. The fact that a sulfur signal is still detected after the 400°C anneal indicates the remaining sulfur is strongly bonded to the substrate. The binding energy of the S $2p$ peak is 161.7 eV for the unannealed layer, and the chemical state of the peak can be identified as that of a sulfide. The spin orbit splitting is 1.2 eV. No sulfur oxides (167-169 eV) are observed, even following heating to 400°C.

Figure 2(b) shows the In $3d_{5/2}$ peak for the sample of Fig. 2(a). The peak was fit with a substrate component at 444.5 eV and a component shifted 0.7 eV from the substrate peak. This component may be attributed to In-O or In-S bonding. Due to the observed S $2p$ sulfide signal, we attribute the shifted component to In-S. For pure In_2S_3 samples, the separation between the In $3d_{5/2}$ and S $2p$ peak has been reported as 283.2 eV,[9] which is close to our value (445.2 - 161.7 = 283.5 eV). Therefore, we see that the thiourea/ammonia pre-treatment forms In-S bonds on the InP surface. Free sulfide is formed in this solution according to the equation:[10]

$$CS(NH_2)_2 + 2OH^- \rightarrow S^{2-} + CN_2H_2 + 2H_2O \qquad (2)$$

The high ammonia concentration of the solution is believed to reduce the native oxides present

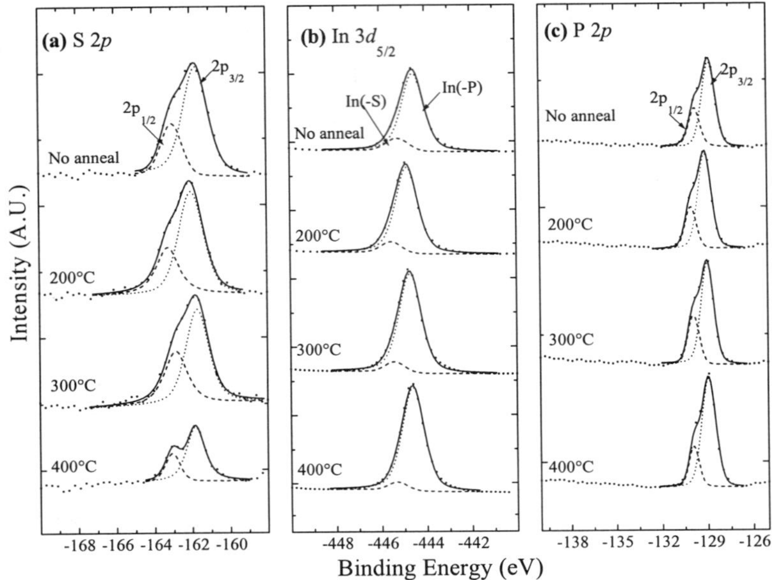

Figure 2. The (a) S $2p$, (b) In $3d_{5/2}$, and (c) P $2p$ detail scans of a pre-treated InP sample unannealed and after anneals at 200, 300, and 400°C. The dotted curves in (a) and (c) are for the $2p_{3/2}$ component, while the dashed curves are for the $2p_{1/2}$ component. The bulk component, In(-P), for the In $3d_{5/2}$ peak in (b) is shown by the dotted curve, and the In(-S) component by the dashed curve.

on the InP surface. Once the thiourea is hydrolyzed and the sulfide ions are released, the sulfide ions react with the InP surface to form In-S and P-S bonds. Since P-S compounds are highly soluble in alkaline solutions, no P-S bonding is evident in the P $2p$ detail scan, Fig. 2(c). The ratio of S/In(-S), where In(-S) indicates the corrected area of the peak attributed to In-S bonding, in the unannealed layer was 1.7, and decreased to 1.3 following annealing at 400°C.

The P $2p$ peak for the sample is shown in Fig. 2(c); it has a bonding energy of 128.8 eV. No evidence of P-O or P-S bonding is seen on either the unannealed sample or following annealing to 400°C. These components have binding energies shifted 4-5 eV from that of the substrate. There is, therefore, no evidence that P is present in the sulfide layer, and the sulfide layer is comprised entirely of indium and sulfur.

Figure 3(a) shows the S $2p$ spectrum from an InP sample that has been pre-treated in a thiourea/ammonia solution, followed by deposition of a thin CdS film for 3 min at 85°C. The S binding energy is 161.8 eV. The peak is attributed to a sulfide and comprises $\sim$ 8.8% of the total detected signal. Sulfate is also present with a concentration of $\sim$ 0.7% at 164.8 eV. This component is removed by annealing at 350°C. It is difficult to determine the amount of In-S and Cd-S bonding for the sample, as the energy shifts for the two components are nearly identical. While the S $2p$ peak occurs at 161.7 eV for the In-S component, published values of the binding

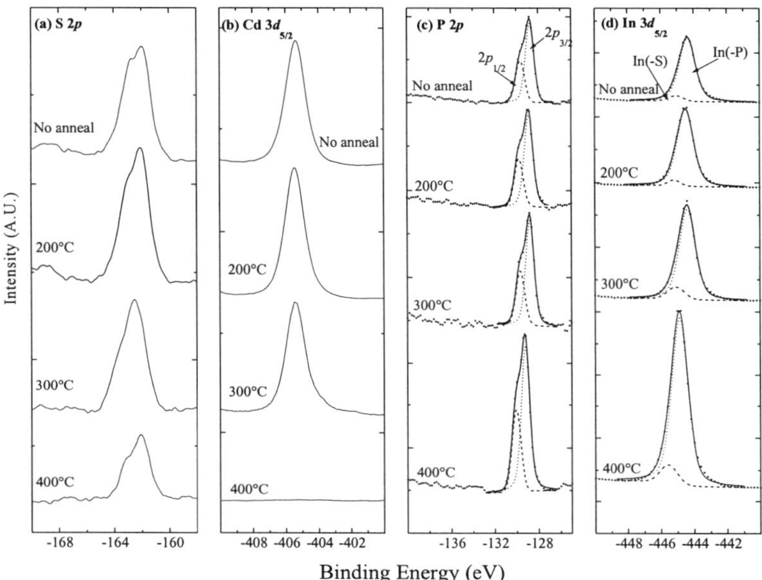

Figure 3. The (a) S $2p$, (b) Cd $3d_{5/2}$, (c) In $3d_{5/2}$, and (d) P $2p$ detail scans of a pre-treated n-InP sample with a thin CdS film unannealed and after anneals at 200, 300, and 400°C. The dotted curves in (c) are for the $2p_{3/2}$ component, while the dashed curves are for the $2p_{1/2}$ component. The bulk component, In(-P), for the In $3d_{5/2}$ peak in (d) is shown by the dotted curve, and the In(-S) component by the dashed curve.

energies for Cd-S have been reported in the range 161.4-161.8 eV.[11] Therefore it is difficult to determine quantitative stoichiometries for the thin layers we are considering here.

Figure 3(b) shows the Cd $3d_{5/2}$ peak of the sample in Fig. 3(a); the binding energy is 405.1 eV. This value agrees well with previously reported data on single-crystal and thin-film CdS samples. The Cd peak has a concentration of 16.3%. It is evident from Fig. 3(b) that, following annealing at 300°C, the Cd signal is no longer detected. There is also a corresponding decrease in the S signal. The evolution of Cd from the sample suggests that Cd is not strongly bonded to the substrate. Previously, we have confirmed, using Auger depth profiling, the presence of CdS thin films at the interface of an MIS device following annealing overnight at 350°C.[4] The deposited oxide of the MIS sample serves as a capping layer, preventing evolution of the CdS thin film. Figure 3(c) shows the P $2p$ peak for the sample. We again see no evidence of P-S or P-O bonding. The In $3d_{5/2}$ peak, Fig. 3(d), is similar to that of the pre-treated sample. Again we see that a sulfide component is present for the In $3d_{5/2}$ peak. Following the 400°C anneal, the S/In(-S) ratio is 1.4. If the same ratio for S/In(-S) is assumed for this case as measured in the pre-treated sample, then the ratio of S/Cd is 0.4 for the unannealed sample. The presence of Cd-P bonding in the sample could not be verified due to lack of a good standard and reference data. The existence of this species cannot be excluded, therefore, and a complete understanding of the chemistry at the CdS-InP interface will require more experimentation.

An important measure of the quality of the InP surface is the degree of P deficiency. P vacancies can act as traps which impair the C-V response of subsequently fabricated devices. The temperature instability of InP is one of the leading causes of poor device performance, as it limits processing temperatures. While the HF-etched and pre-treated samples show similar P/In(-P) ratios (~ 0.8 for the unannealed samples to ~ 0.7 for 400°C-annealed samples), the sample with pre-treatment and a CdS thin film has a higher value (~ 0.9) for temperatures up to 300°C. XPS analysis of annealed MIS samples also gave P/In(-P) ratios near 0.9. This increased value of P/In(-P) is not yet understood.

CONCLUSIONS

Pre-treating n-InP in a thiourea/ammonia solution has been shown to produce an In-S layer at the surface. While the presence of In-S bonding is shown to improve the MIS response, the surface remains slightly P-deficient. Deposition of a thin layer of CdS yields dramatic improvements in MIS C-V response. In-S bonding is also observed, and samples with the deposited CdS layer also exhibit higher P/In(-P) ratios. Pre-treated samples with or without a subsequent thin CdS film showed no evidence of native oxides at the surface. The CdS layer may also serve to protect the In-S layer during dielectric deposition, but it is unstable when unprotected at temperatures between 300 and 350°C. The presence of a 300 Å SiO_2 layer protects the CdS layer and allows for processing at 350°C. Other factors which may impart improved electrical response of MIS samples prepared with CdS thin films are the wide band gap of CdS and its favorable valence band offset on InP. The electrical response also demonstrates that a thin CdS layer is effective in reducing interface states. It is an interesting historical footnote that the first insulated gate thin-film transistor was fabricated on CdS,[12] possibly due to its high degree of ionicity, which results in good interfaces with a low density of states.

ACKNOWLEDGMENTS

The authors wish to thank B.J.H. Stadler for helpful discussion, and G.O. Ramseyer for assistance in Auger analysis. A portion of this research was funded by the Air Force Office of Scientific Research (AFOSR) Entrepreneurial Research Program.

[1] R. Iyer, R.R. Chang, D.L. Lile, *Appl. Phys. Lett.* **53**, 134 (1988).

[2] R.W.M. Kwok, G. Jin, B.K.L. So, K.C. Hui, L. Huang, W.M. Lau, C.C. Hsu, D. Landheer, *J. Vac. Sci. Technol. A* **13** (3), 652 (1995).

[3] K. Vaccaro, H.M. Dauplaise, A. Davis, S.M. Spaziani, J.P. Lorenzo, *Appl. Phys. Lett.* **67** (4), 527 (1995).

[4] H.M. Dauplaise, K. Vaccaro, A. Davis, G.O. Ramseyer, J.P. Lorenzo, *J. Appl. Phys.* **80** (5), 2873 (1996).

[5] B.R. Bennett, J.P. Lorenzo, K. Vaccaro, A. Davis, *J. Electrochem. Soc.* **134** (10), 2517 (1987).

[6] R. Castagné and A. Vapaille, *Surface Sci.* **28**, 157 (1971).

[7] J. Chastain, *Handbook of X-Ray Photoelectron Spectroscopy,* Perkin-Elmer Corporation, (1992).

[8] W.M. Lau, R.W.M. Kwok, S. Ingrey, *Surface Science* **271**, 579 (1992).

[9] Y. Tao, A. Yelon, E. Sacher, Z.H. Lu, M.J. Graham, *Appl. Phys. Lett.* **60** (21), 2669 (1992).

[10] R. Ortega-Borges and D. Lincot, *J. Electrochem. Soc.* **240** (12), 3464 (1993).

[11] NIST Standard Reference Database 20, *NIST X-ray Photoelectron Spectroscopy Database*, Version 1.0, compiled by C.D. Wagner, U.S. Dept of Commerce, Gaithersburg, MD (1989).

[12] P. Weimer, *Proc. IRE* **50**, 1462 (1962).

SURFACE OXIDATION STUDY OF SILICON-DOPED GaAs WAFERS BY FTIR SPECTROSCOPY

R.-H. CHANG *, M. AL-SHEIKHLY *, A. CHRISTOU *, and C. VARMAZIS **
* Department of Materials and Nuclear Engineering, University of Maryland, College Park, MD 20742-2115
** M/A-COM Microelectronic Division, IC Business Unit, 100 Chelmsford St., Lowell, MA 01851-2694

ABSTRACT

A surface study of Si-doped GaAs (100) oriented wafers treated with NH_4OH and HCl following exposure to fluorine containing plasma was conducted using fourier transform infrared spectroscopy (FTIR). These treatments were observed to produce various oxidation products, such as As_2O_5 and GaO. Though inorganic salts, such as $(NH_4)_3GaF_6$, can be formed on the Si-doped GaAs wafers during cleaning with hydrofluoric acid buffered with ammonium fluoride, the applied cleaning method which consisted of NH_4OH and HCl treatments subsequent to exposure to fluorine containing plasma did not induce formation of any inorganic salts. A small amount of hydroxide group was also presented in the samples. Water molecules and ammonium hydroxide can be sources of OH which can then be incorporated interstitially into the wafer surfaces.

INTRODUCTION

It is well known that the presence of chemical contaminants and impurities on wafer surfaces will affect the performance and reliability of microelectronic devices. Effective techniques for cleaning semiconductor wafers before and after oxidation and patterning are now more and more important because of the extreme sensitivity of the semiconductor surface and the submicron scales of device features.

Wet chemical cleaning processes can meet the rigorous demands on wafer smoothness and low metal surface contamination in advanced cleaning processes. Widely used wet cleaning procedures in the semiconductor industry are treatments based on the RCA cleaning sequences [1], containing an HF step for oxide removal, an alkaline NH_4OH/H_2O_2 (SC1) wet chemical oxidation step and an acid HCl/H_2O_2 (SC2) cleaning step. However there are some limitations of wet wafer cleaning processes, such like an incompatibility between wet wafer cleaning operation and process integration, and metallic contaminants plated on silicon surfaces from HF solutions [2]. Also these processes may induce various oxidation reactions on the surface of silicon-doped GaAs wafers [3]. The activation energy of such reactions are relatively low, so that they can proceed at room temperature. These reactions involve oxidation of GaAs, silicon, and oxygen diffusion at various reaction rate constants.

Residues from wet cleaning processes must be removed in order to make sure that the subsequent steps provide accurate electrical and mechanical performance. Dry cleaning procedures followed by wet cleaning treatments could eliminate inherent disadvantages brought by wet cleaning processes. Widely used dry cleaning procedures are thermally enhanced, vapor phase, photochemically-enhanced, and plasma-enhanced treatments. Among these, plasma based processes can effectively remove layers of organic contaminant, residual oxide or metallic residue remaining on the wafer surfaces after wet chemical cleaning processes. Several attempts at plasma-enhanced cleaning have been reported [4][5][6]. In the past few years investigations of plasma cleaning techniques were mainly based on argon, ozone, oxygen, or hydrogen

Mat. Res. Soc. Symp. Proc. Vol. 448 © 1997 Materials Research Society

atmospheres. Most of the applications are focused on silicon wafers; howeve, few of them have been put into GaAs wafers.

In this study, we will identify the products induced by fluorine containing plasma on the GaAs wafer surfaces using FTIR technique in order to understand if the applied cleaning methods could effectively remove oxides that possibly formed after HCl and NH_4OH cleaning.

EXPERIMENT

A surface study was conducted on 8 three-inch (100)-oriented GaAs wafers. Samples were provided by M/A-COM Microelectronic Division. These wafers were implanted with Si and annealed. Both sides of each wafer were optically flat surfaces. Surface preparation of each wafer is described in Table 1.

Table 1. Surface Preparation of 8 (100)-oriented GaAs Wafers

Wafer	Surface Preparation
# 1	None
# 2	NH_4OH clean
# 3	HCl clean
# 4	45 sec. exposure to fluorine containing plasma
# 5	90 sec. exposure to fluorine containing plasma
# 6	HCl clean and 45 sec. exposure to fluorine containing plasma
# 7	HCl clean and 90 sec. exposure to fluorine containing plasma
# 8	NH_4OH clean and 45 sec. exposure to fluorine containing plasma

Measurements were performed by Nicolet Magna-IR^{TM} 550 spectrometer. Experiments were made by the conventional simple transmission geometry at room temperature. Samples were positioned on an optical bench interfaced to spectrometer. Instruments were nitrogen-purged to minimized infrared absorption by atmospheric CO_2 and water vapor. Spectra were acquired at 4 cm^{-1} resolution with 128 scan accumulations. Any data analysis, such like baseline correction, was performed by Windows$^{®}$ compatible OMNIC$^{®}$ software.

RESULTS AND DISCUSSIONS

Figures 1~3 represent the FTIR absorbance spectra of the these wafers at different frequency ranges. Regions of absorption peaks are summarized in Table 2.

Table 2. Regions of Absorption Peaks

wavenumber (cm^{-1})	420	440	490	510	525	575	600~620	768~770
relative strength	*s*	*s,b*	*w, b*	*sh*	*s*	*w*	*w*	*w*

s: strong; *sh*: shoulder; *b*: broad; *w*: weak

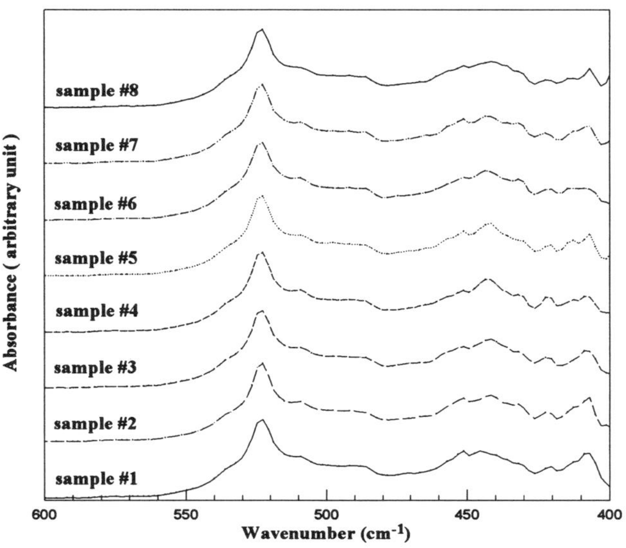

Figure 1. FTIR spectra of 8 (100)-oriented GaAs wafers showing the range between 400 to 600 cm^{-1}. Surface treatments of each sample were described in Table 1.

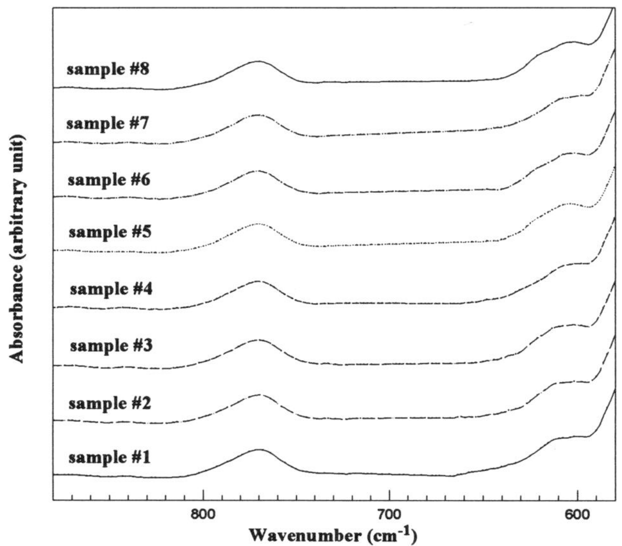

Figure 2. FTIR spectra of 8 (100)-oriented GaAs wafers showing the range between 580 to 880 cm^{-1}. Surface treatments of each sample were described in Table 1.

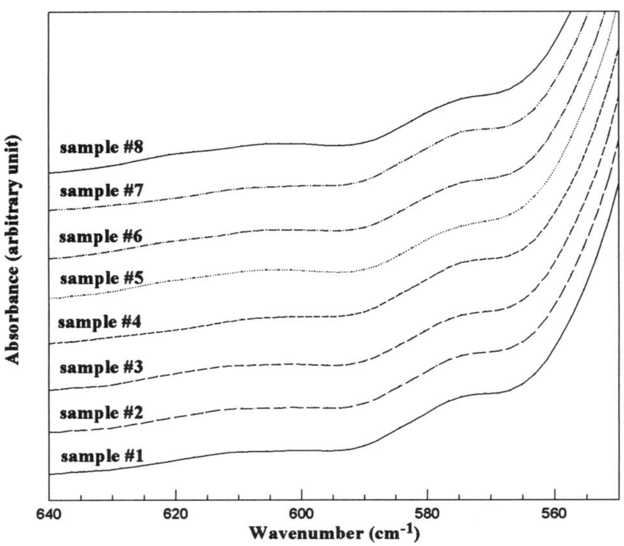

Figure 3. FTIR spectra of 8 (100)-oriented GaAs wafers showing the weak absorption at 575 cm^{-1} and 600~620 cm^{-1} . Surface treatments of each sample were described in Table 1.

Figure 1 shows spectra in the range between 400 to 600 cm^{-1}. Spectra between 400 and 450 cm^{-1} are due to the AlAs-like TO phonon modes from the AlGaAs barrier layers [7]. Thus we doubt there are Al impurities existing in these samples. Actually these vibration modes (AlAs-like TO phonon modes) also appear in the far infrared range at 350~400 cm^{-1}. The presence of the peaks at 510 and 525 cm^{-1} suggest the presence of various types of arsenic oxides or arsenic hydroxides. For example, As_2O_5 exhibits absorption band at 532 cm^{-1} and p-hydrobenzenearsonic acid (arsenic atom bounded to oxygen, two hydroxyl groups, and organic phenol group) exhibits absorption peak at 525 cm^{-1} [8]. The presence of absorption band at 585 cm^{-1} also suggests the presence of As_2O_5.

Figure 2 shows spectra in the range between 580 to 880 cm^{-1}. The peak at 768 cm^{-1} suggests the presence of GaO (the vibration frequency of diatomic molecules of GaO) [9]. A. vom Felde et al. have assigned this peak as O atoms bonded to Ga atoms on the (110) terraces [3]. It should be mentioned that the vibrational frequency of the absorbed oxygen atom overlaps the frequency range expected for the third harmonic of the TO phonon in the bulk of GaAs [10], which located at 760~780 cm^{-1}. Relative broad small peak at 600~620 cm^{-1}, as shown in Figure 3, indicates the presence of OH. The presence of OH can be explained by the fact that these samples were washed by ammonium hydroxide aqueous solutions. Water molecules and ammonium hydroxide can be sources of OH which can then be incorporated interstitially into the surfaces [11]. The presence of this peak (600~620 cm^{-1}) in the reference sample (sample #1) can be explained by the fact that the water molecules in the air were absorbed on the surfaces of the wafers.

When we closely look at the frequency region of 580~880 cm^{-1}, as shown in Figure 2, we do not find peaks at 690 and 800 cm^{-1} for Ga_2O_3 and As_2O_3, respectively [12,13]. There is also no peak showing at 843 cm^{-1} for both the Ga-O-As and As-O-As bending modes of the chemically grown oxide structure [7]. It should be mentioned that both Ga and As form stable oxides, and that extensive oxidation of both GaAs produces a mixture of both [14]. Absence of peak at 715 cm^{-1} for the vibration frequency of oxygen atoms occupying arsenic vacancies suggests there are no substitutional oxygen atoms in the GaAs bulks [15].

The absence of bending and stretching frequencies for the ammonium ions at 1300~1450 cm^{-1}, and 2800~3350 cm^{-1} regions respectively suggest that there is no inorganic salt residue after wet cleaning processes [16~20]. Absences of the $(NH_4)_3GaF_6$ at 1425, 2870, 3040, 3220 cm^{-1} show that fluorine containing plasma did not induce formation of inorganic salts [21].

It should be mentioned that the FTIR spectrum pattern of each sample looks like the same. This could be explained as that these samples have been exposed to the air for a couple of days before the measurements performed. These existing oxides such like As_2O_5 and GaO might be the products of GaAs bulks reacting with atmospheric water or oxygen molecules rather than with the applied cleaning steps. Therefore, in order to realize the reason for these oxides formation, further investigations combining *in situ* observation are needed

CONCLUSIONS

The results suggest the absence of inorganic salts that consist of ammonium fluoride and ammonium bifluoride. These salts could be formed on the Si-doped GaAs wafers during cleaning with hydrofluorid acid buffered with ammonium fluoride. Thus, it is concluded that the applied cleaning methods for these wafers do not induced formation of any inorganic salt under these conditions of temperature and exposure time to fluorine containing plasma. Also there are Al impurities existed in these wafers.

Finally, it is concluded that the applied chemical cleaning methods and fluorine containing plasma (at room temperature and exposure time of 45~90 seconds) have no negative effects on these wafers.

ACKNOWLEDGMENT

The authors would especially like to thank the M/A-COM Microelectronic Division for providing samples for investigation.

REFERENCES

1. W. Kern, in Handbook of semiconductor Wafer Cleaning Technology, edited by W. Kern, Noyes Publications, pp. 19 (1993)
2. E. Hsu, H. G. Parks, R. Craigin, S. Tomaoka, L. S. Ramberg, and R. K. Lowry, in Proceedings of the Second International Symposium on Cleaning Technology in Semiconductor Device Manufacturing, edited by J. Ruzyllo, and R. Novak, The Electrochem. Soc., Pennington, N. J., pp. 170 (1992)
3. A. vom Felde, K. Kern, G. S. Higashi, Y. J. Chabal, S. B. Christman, C. C. Bahr, and M. J. Cardillo, Physical Review B, Vol. 42, No. 8, pp. 5240 (1990)
4. B. Anthony, L. Breaux, T. Hsu, S. Banerjee, and A. Tasch, J. Vacuum Science and Technology, Vol. B7, pp. 621 (1989)
5. T. P. Schneider, J. Cho, D. A. Aldrich, Y. L. Chen, D. Maher, and R. J. Nemanich, in Proceedings of the Second International Symposium on Cleaning Technology in Semiconductor Device Manufacturing, edited by J. Ruzyllo, and R. Novak, The Electrochem. Soc., Pennington, N. J., pp. 170 (1992)

6. R. A. Rudder, G. G. Fountain, and R. J. Markunas, J. Appl. Phys., Vol. 60, pp. 3519 (1986)

7. Z. C. Feng, S. Perkowitz, J. Chen, K. K. Bajaj, D. K. Kinell, and R. L. Whitney, in Semiconductor Characterization Present Status and Future Needs, edited by W. M. Bullis, D. G. Seiler, and A. C. Diebold, AIP, New York, pp. 644 (1996)

8. F. F. Bentley, L. D. Smithson, and A. L. Rozek, Infrared Spectra and Characteristic Frequencies ~700-300 cm^{-1}, Interscience Publishers, New York, pp. 95, 1498 (1968)

9. G. Herzberg, Molecular Spectra and Molecular Structure, I. Spectra of Diatomic Molecules, Van Nostrand Reinhold, New York (1950)

10. J. S. Blakemore, in Key Papers in Physics, GaAs, edited by J.S. Blakemore, AIP, New York, pp. 3 (1987)

11. C. T. Lenczycki and V. A. Burrows, Thin Solid, Films, Vol. 193/194, pp. 610 (1990)

12. N. T. McDevitt and W. L. Baum, Spectrochimica Acta, Vol. 20, pp. 799 (1964)

13. F. A. Miller, G. L. Carlson, F. F. Bentley, and W. H. Jones, Spectrochimica Acta, Vol. 16, pp. 195 (1960)

14. C. D. Thurmond, G. P. Schwartze, G. W. Kammlott, and B. Schwartz, J. Electrochem. Soc., Vol. 127, pp. 1366 (1980)

15. J. Schneider, B. Dischler, H. Seelewind, P. M. Mooney, J. Lagowski, M. Matsui, D. R. Beard, and R. C. Newman, Appl. Phys. Lett., Vol. 54, pp. 1442 (1989)

16. L. F. H. Bovey, J. Opt. Soc. Am., Vol. 41, pp. 836 (1951)

17. K. Nakamoto, Infrared and Raman Spectra of Inorganic and Coordinating Compounds, Wiley, New York, pp. 132 (1978)

18. S. D. Hamann, Aust. J. Chem., Vol 31, pp. 11 (1978)

19. R. C. Plumb and D. F. Horning, J. Chem. Phys., Vol. 23, pp. 947 (1955)

20. G. Herzberg, Infrared and Raman Spectra of Polyatomic Molecules, Van Nostrand Reinhold, New York, pp. 167 (1945)

21. V. A. Burrows and J. Yota, Thin Solid Films, Vol. 193/194, pp. 371 (1990)

CHEMICAL CHARACTERIZATION BY FT-IR SPECTROMETRY AND MODIFICATION OF THE VERY FIRST ATOMIC LAYER OF A TiO$_2$ NANOSIZED POWDER

M.-I. BARATON *, L. MERHARI **, F. CHANCEL * and J. TRIBOUT *
*LMCTS, Faculté des Sciences, 123 avenue Albert Thomas, F-87060 Limoges cedex, (France), baraton@unilim.fr
**CERAMEC R&D, 64 avenue de la Libération, Limoges, F-87000, (France)

ABSTRACT

Nanosized powders exhibit high specific surface areas resulting in enhanced reactivities. Surface tailoring by controlled adsorption of molecules can thus be conveniently performed and more easily monitored by surface-sensitive techniques. *In situ* and *ex situ* grafting procedures of hexamethyldisilazane (HMDS) on nanosized titania (n-TiO$_2$) powder were carried out and studied by Fourier transform infrared spectrometry (FT-IR). In addition to a decrease of the hydrophilic OH groups, the vibration analysis revealed hydrophobic CH$_3$ groups on the grafted samples. Co-adsorption of CO and H$_2$O on the differently grafted samples showed a large reduction of water effect compared to the as-received n-TiO$_2$ powder. Modulation of infrared transmitted energy by controlled adsorption of O$_2$ and CO made it possible to qualitatively compare the electronic properties of the surface-tailored samples.

INTRODUCTION

Most of the experimental techniques providing chemical surface analyses of materials actually probe a depth ranging from a few nanometers to ten nanometers. Under those conditions, it is obvious that the so-determined chemical composition is only an average over several atomic layers and cannot resolve the very first atomic layer. This latter precisely may have a quite specific chemical structure which in fact controls the surface and interface properties of materials including electronic properties of semiconductors. This becomes even more dramatic for nanosized materials for which the surface upon bulk ratio is very high, making them often very reactive. Fourier transform infrared (FT-IR) spectrometry is a powerful tool to characterize their first atomic layer because the surface chemical species generated during the nanostructured material synthesis process and/or generated by surrounding contaminants can be identified as well as reactive sites in a non destructive manner. These surface groups and the reactive sites as well can be modified *in situ* by controlled adsorption of molecules (referred to as *grafting* in the literature) while the resulting modifications in the surface properties can be investigated by the same FT-IR technique. On the other hand, independently of fundamental vibration studies, some semiconducting properties can be deduced from the infrared spectra without requiring electrical contacts [1]. Our choice of n-TiO$_2$ as a candidate for grafting experiments is motivated by the fact that this material is one of the most studied transition metal oxide for numerous applications (e.g. catalysis, gas sensing ...) and thus can be considered as a prototypical metal oxide in surface studies [2]. In this preliminary work, we have investigated the *in situ* and *ex situ* grafting procedures of HMDS on n-TiO$_2$ and characterized the resulting surface species by FT-IR spectrometry. Simultaneously, the changes in n-TiO$_2$ and HMDS-grafted n-TiO$_2$ electronic population upon CO, O$_2$ adsorption and co-adsorption of CO and H$_2$O were accessed to by the same FT-IR technique and qualitatively correlated to existing solid state theories.

Mat. Res. Soc. Symp. Proc. Vol. 448 © 1997 Materials Research Society

EXPERIMENTAL

All the spectra were recorded in transmission mode by means of a Perkin-Elmer Spectrum 2000 FT-IR spectrometer equipped with an MCT cryodetector. The analyzed spectral range extended from 500 to 6500 cm^{-1} with a 4 cm^{-1} resolution. The FT-IR experiments were run *in situ* by using a specially designed heatable vacuum cell [3] placed inside the spectrometer sample compartment. Controlled pressures of gases were adjusted through a precise valve system. The titania powder (P25, Degussa-France) was mainly in the anatase crystalline phase (~70%). The specific surface area measured by the supplier was 50 m^2g^{-1} and the estimated average particle size was 21 nm. For the infrared analyses, the n-TiO_2 powder was slightly pressed into thin pellets (~50 mg) on a stainless grid (Gantois, France) ensuring a homogeneous thermal distribution. These pellets were systematically kept under vacuum at 673 K prior to be subjected to different gases while keeping the temperature constant. All the gases (Alphagaz, France) were 99% pure, hexamethyldisilazane (HMDS) (Fluka, Germany) was 99.5 % pure and water was bi-distilled and de-ionized. The *in situ* and *ex situ* grafting procedures are described below.

INFRARED SURFACE ANALYSIS OF HMDS-GRAFTED TITANIA NANOSIZED POWDER

<u>As-received n-TiO_2 powder</u>

As it is the case for all oxides, the as-received n-TiO_2 powder is covered with molecular water adsorbed on its surface [4]. A thermal treatment under dynamic vacuum (referred to as *activation* throughout the text) leads to a surface freed of adsorbed species according to the temperature. The surface is then no longer in an equilibrium state, and as soon as molecules impinge on this activated surface, they adsorb on reactive sites. This activation not only allows one to obtain a good knowledge of the surface species but also to follow the reactions which will eventually take place during the gas-surface interactions. We must underline here that we are not considering the n-TiO_2 pellet as a gas sensor under its standard working conditions. Instead, we are trying to highlight the surface reactions and to understand their mechanism so as to extend this knowledge to the very complex case of a real gas sensor in an atmosphere whose exact composition is unknown.

The spectrum of n-TiO_2 surface activated at 673 K is given in Fig. 1a. The displayed spectral range only extends from 2000 to 4000 cm^{-1} where the absorption bands of the surface species which are minority by far even for nanosized powders, are clearly visible. The complex band centered at 3600 cm^{-1} corresponds to the $\nu(OH)$ stretching vibrations of different types of surface hydroxyl groups responsible for hydrophilicity. Indeed, according to the cation coordination or/and the number of cations linked to the OH groups, the $\nu(OH)$ frequencies vary over a large range [5].

When the n-TiO_2 pellet is heated at 673 K under air, a broad feature appears around 3400 cm^{-1} (Fig. 2a) and is assigned to hydrogen-bonded hydroxyl groups. Simultaneously a lowering of the baseline is observed corresponding to the IR absorption decrease due to oxygen adsorption (cf following section). Carbon dioxide from atmosphere is also detected. Adsorption of 9 mbar CO at 673 K reduces the surface (Fig. 3a) and consequently increases the baseline. Generation of CO_2 (2344 cm^{-1}) is hardly visible. Upon co-adsorption of water and CO (Fig. 4a) (total pressure=10 mbar and p_{H2O}/p_{CO} pressure ratio=1/4) generation of CO_2 is detected (reduction of the surface) while modification of the OH absorption range is due to dissociation of water [4].

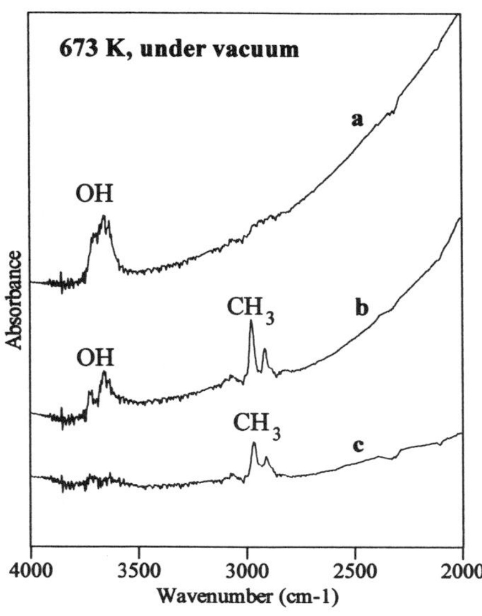

Fig. 1. FT-IR spectra of n-TiO$_2$:(a) pure; and HMDS-grafted n-TiO$_2$: (b) *ex situ*; (c) *in situ*.

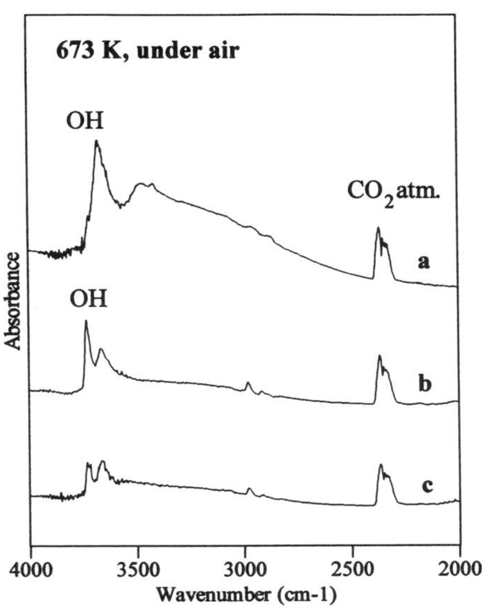

Fig. 2. FT-IR spectra of n-TiO$_2$:(a) pure; and HMDS-grafted n-TiO$_2$: (b) *ex situ*; (c) *in situ*.

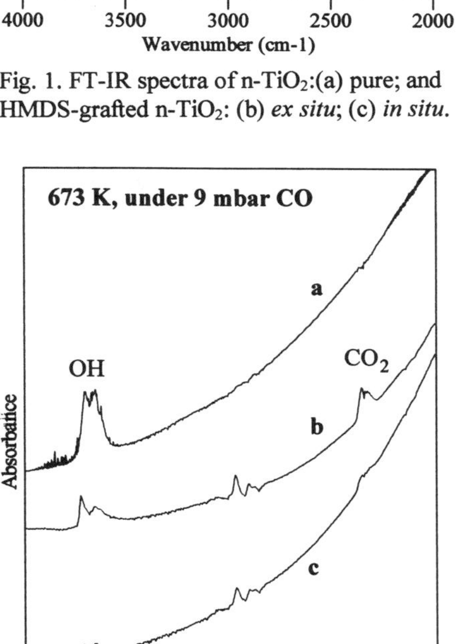

Fig. 3. FT-IR spectra of n-TiO$_2$:(a) pure; and HMDS-grafted n-TiO$_2$: (b) *ex situ*; (c) *in situ*.

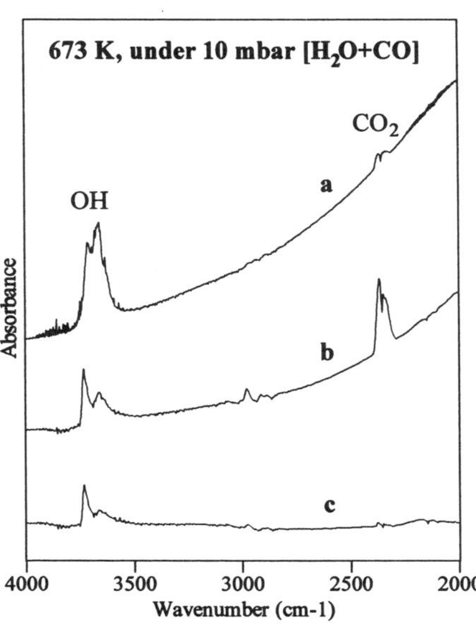

Fig. 4. FT-IR spectra of n-TiO$_2$:(a) pure; and HMDS-grafted n-TiO$_2$: (b) *ex situ*; (c) *in situ*.

Ex situ HMDS-grafted n-TiO$_2$ powder

The *ex situ* grafting procedure consisted in heating the n-TiO$_2$ powder overnight into a standard oven at 473 K. Then, a few drops of liquid HMDS were mixed with the partly dehydrated powder under atmosphere. The resulting mixture was used to make pellets in the same way as for the as-received powder. The most striking features observed in Fig. 1b are the appearance of hydrophobic CH$_3$ groups (2800-3000 cm^{-1} range) and the concurrent decrease of the OH groups. It is worth noting that a very intense band at 1270 cm^{-1} (not shown here) appears in the spectra of both the *ex situ* and *in situ* HMDS-grafted n-TiO$_2$ powders. It is attributed either to ν(Si-C) or δ(CH$_3$) vibrations and is quite characteristic of HMDS grafting. We checked that the grafting was irreversible by activating the samples at 673 K and still observing the same HMDS absorption bands.

When heating the *ex situ* HMDS-grafted n-TiO$_2$ powder at 673 K under atmosphere, the hydroxyl groups partly recover to the detriment of the CH$_3$ groups (Fig. 2b). While SiOH groups (3738 cm^{-1}) [5] are generated on the surface, no hydrogen-bonded hydroxyl groups (around 3400 cm^{-1}) are observed. As in the case of n-TiO$_2$, a lowering of the baseline is observed and similarly explained by oxygen adsorption.

Upon adsorption of 9 mbar CO at 673 K the surface is reduced (Fig. 3b), the baseline increases accordingly and CO$_2$ (2344 cm^{-1}) is generated. Co-adsorption of water and CO (total pressure=10 mbar and p_{H2O}/p_{CO} pressure ratio=1/4) produces CO$_2$ (reduction of the surface) while SiOH groups appear as in the case of air adsorption (Fig. 4b).

In situ HMDS-grafted n-TiO$_2$ powder

A second grafting procedure was concurrently used. This time the n-TiO$_2$ surface was activated at 673 K. Then, the sample kept at room temperature under vacuum was subjected to 7 mbar of HMDS vapor. The grafting mechanism, tentatively described elsewhere [5] is confirmed by the present results. Indeed, a dramatic decrease of most of the OH groups with a concomitant appearance of hydrophobic CH$_3$ groups can be clearly observed in Fig. 1c. After evacuation at 673 K, very few OH groups are restored and the surface remains almost dehydroxylated. In this case again, the sample was heated at 673 K under vacuum, and we checked that HMDS was still grafted on the n-TiO$_2$ surface. Note that, for both *in situ* and *ex situ* grafting procedures, even after the grafted samples were brought back to ambient atmosphere at room temperature, the n-TiO$_2$ original surface was not restored.

The behavior of the *in situ* HMDS-grafted sample at 673 K during adsorption of air (Fig. 2c), CO (Fig. 3c) and co-adsorption of water and CO (Fig. 4c) is qualitatively similar to that of the *ex situ* HMDS-grafted sample. The reduction level of hydrophilic OH groups could differ, though.

VARIATIONS OF THE ELECTRONIC PROPERTIES OF HMDS-GRAFTED TITANIA NANOSIZED POWDER

Point defects (predominantly O ion vacancies), readily created on TiO$_2$ surface by heating under vacuum cause a dramatic change in electronic structure. These surface defects are associated with an increase in the conduction electron density at the surface. In transition metal oxides, reduction (or oxidation) of the surface by an adsorbate is expected to be clearly correlated to a change in population of the d levels [2]. In the following experiments, the IR energy transmitted through the n-TiO$_2$ and HMDS-grafted n-TiO$_2$ pellets is recorded versus adsorbed gases. The absorption of the IR energy is due in part to surface states and to free carriers.

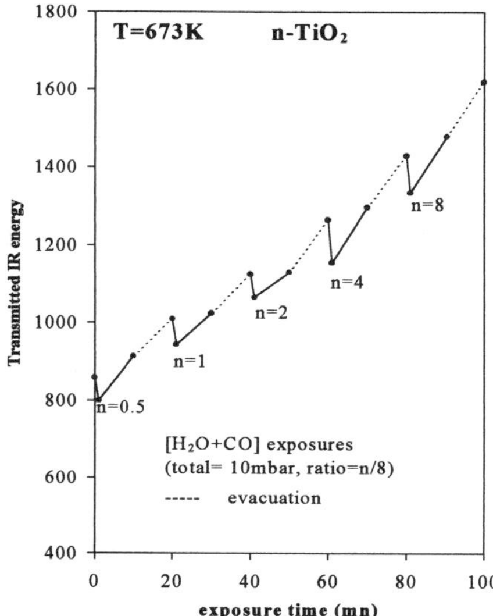

Fig. 5. Transmitted IR energy versus gas exposure for n-TiO₂.

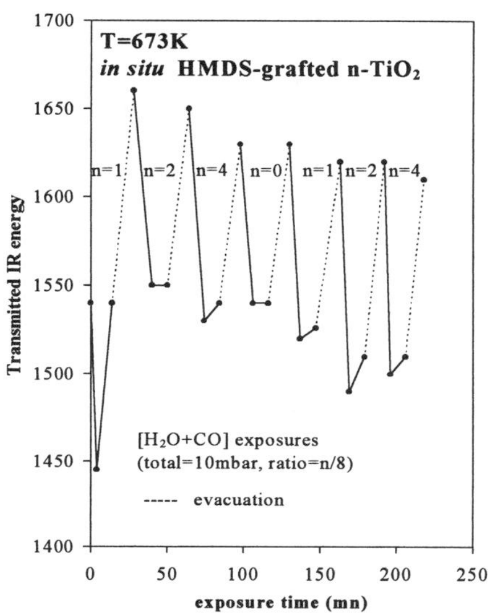

Fig. 6. Transmitted IR energy versus gas exposure for HMDS-grafted n-TiO₂.

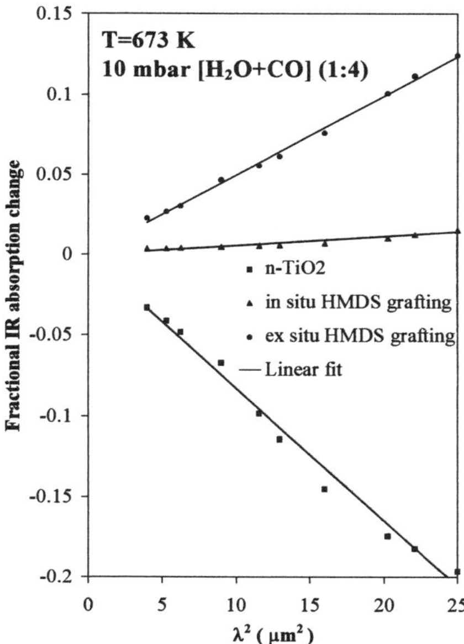

Fig. 7. Fractional IR absorption change versus square wavelength upon H₂O+CO exposure.

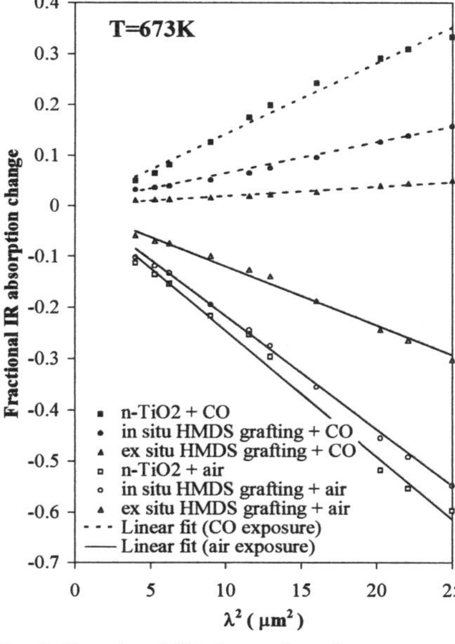

Fig. 8. Fractional IR absorption change versus square wavelength upon CO and air exposure.

Whereas oxidizing adsorbates are supposed to lower the IR absorption, reducing adsorbates should increase it.

Fig. 5 shows sequences of evacuation (10 mn) and co-adsorption of H_2O and CO (1 and 10 mn) performed on n-TiO_2 surface at 673 K. As expected, CO increases the IR absorption but the adverse effect of H_2O appears for longer exposure time. A cumulative effect is observed (steady increase of transmitted IR energy) indicating an evolving hygroscopic surface.

Similar sequences of evacuation and co-adsorption of H_2O and CO are performed on *in situ* HMDS grafted n-TiO_2 surface at 673 K. The most striking phenomenon (Fig. 6) is the stabilization of the surface (no major drift of the transmitted IR energy after the first sequence) which still behaves as expected under reducing gases. Clearly, the water effect is significantly reduced. The results obtained for the *ex situ* HMDS-grafted sample are qualitatively identical.

Modulation of the transmitted IR energy through the n-TiO_2 particles by variation of the gaseous environment leads to a modulation of the electronic population of surface states and of free carriers in the space charge right at the surface of the material [1]. The free carriers are supposed to give rise to a broad square wavelength (λ^2) absorption (intraband absorption) according to the well-known Drude-Zener theory [6]. The fractional IR absorption changes versus λ^2 plotted in Fig. 7 and Fig. 8, in the 2-5 µm range (5000-2000 cm^{-1}) for several gaseous environments show a good agreement with the theoretical λ^2 dependence, thus indicating a free carrier absorption. The curves corresponding to adsorption of air and of 9 mbar CO (Fig. 8) on n-TiO_2, *in situ* and on *ex situ* HMDS-grafted n-TiO_2 at 673 K behave as expected since positive changes correspond to reducing adsorbates whereas negative curves correspond to air adsorption. It is worth noting that Fig. 7 confirms the hydrophobic behavior of the HMDS-grafted samples upon co-adsorption of H_2O and CO (total pressure=10 mbar and p_{H2O}/p_{CO} pressure ratio=1/4). Indeed, the positive curves correspond to these samples only, and indicate a response to CO adsorption qualitatively not affected by water adsorption.

CONCLUSION

Surface FT-IR spectrometry proves to be a valuable technique for fundamental studies of nanosized metal oxides where carriers are present as it makes it possible to investigate the chemical reactions occuring right at the real surface of the material and, <u>simultaneously</u>, access <u>without contact perturbation</u> to the electronic phenomena taking place on the surface and inside the bulk. Our preliminary results show that HMDS-grafted n-TiO_2 is less sensitive to water adsorption while still sensitive to reducing and oxidizing adsorbates. The underlying mechanism is believed to be partly due to synergistic effects including hydrophobicity of the CH_3 groups, shielding of the surface by the CH_3 groups and modification of the electronic surface charge distribution.

REFERENCES

1. N.J. Harrick in <u>Internal Reflection Spectroscopy</u>, Interscience, Wiley, New York, 1967. Second printing by Harrick Scientific Corporation, Ossining, N.Y. 1979, and references therein.
2. V.E. Heinrich and P.A. Cox in <u>The Surface Science of Metal Oxides</u>, Cambridge University Press, 1994, and references therein.
3. M.-I. Baraton, J. High. Temp. Chem. Processes **3**, 545 (1994).
4. M.-I. Baraton, Sensors Actuators B **31**(1-2), 33-38 (1996).
5. M.-I. Baraton, F. Chancel and L. Merhari, Nanostruct. Mat. (1996), in press.
6. Y.J. Chabal, Surface Science Reports **8**, 211 (1988).

IN-SITU ETCH TO IMPROVE CHEMICAL BEAM EPITAXY REGROWN AlGaAs/GaAs INTERFACES FOR HBT APPLICATIONS

Y.M. Hsin, N. Y. Li, C. W. Tu, and P. M. Asbeck
Department of Electrical and Computer Engineering
University of California, San Diego, CA 92093-0407

ABSTRACT

We have studied the etching effect of $Al_xGa_{1-x}As$ ($0 \leq x \leq 0.5$) by tris-dimethylaminoarsenic (TDMAAs) at different substrate temperatures, and the quality of the resulting etched/regrown GaAs interface. We find that the etching rate of $Al_xGa_{1-x}As$ decreases with increasing Al composition, and the interface trap density of the TDMAAs etched/regrown interface can be reduced by about a factor of 10 as deduced from capacitance-voltage carrier profiles. A smooth surface morphology of GaAs with an interface state density of 1.4×10^{11} cm^{-2} can be obtained at a lower in-situ etching temperature of 550°C. Moreover, by using this in-situ etching the I-V characteristics of regrown p-n junctions of $Al_{0.35}Ga_{0.65}As/Al_{0.25}Ga_{0.75}As$ and $Al_{0.35}Ga_{0.65}As/GaAs$ can be improved.

INTRODUCTION

Regrowth techniques can significantly improve performance of heterojunction bipolar transistors (HBTs) by controlling composition and doping in 3 dimensions. Regrown external base layers can reduce base resistance thereby improving r.f. performance and noise characteristics [1]. A thick regrown collector/sub-collector can reduce collector resistance [2], provide for device planarization and can be used to decrease C_{bc}. However the regrown interface usually increases recombination current and thus decreases HBT current gain. With only ex-situ etching before regrowth, it is not easy to obtain a clean regrown interface due to possible contamination in the atmosphere. Recently, many research groups have reported studies of in-situ etching of III-V compounds using different gaseous sources. Mui et al. [3] reported high-quality etched/regrown GaAs interfaces using an in-situ Cl_2 etching process, but a complicated interlocking system of separate growth and etching chambers was used. Also Tappura et al. [4] reported that in-situ cleaning of GaAs surface in MBE using an atomic hydrogen plasma can obtain high quality interfaces. Tsang et al. [5] reported in-situ etching of GaAs and InP using $AsCl_3$ and PCl_3 prior to regrowth in the same chemical beam epitaxy (CBE) chamber; therefore, possible contaminations in the etched/regrown interface can be minimized. However, Cl_2 decomposed from $AsCl_3$ or PCl_3 is corrosive and may etch filaments in the growth chamber, so the long term use of $AsCl_3$ or PCl_3 could be a concern. Tateno et al. [6] and Hou et al. [7] found that gaseous carbon doping sources, such as CCl_4 and CBr_4, can etch GaAs and AlAs in organometallic vapor phase epitaxy (OMVPE), but carbon impurities dissociated from these doping sources during in-situ etching process would contaminate the etched/regrown interface. Therefore, an etching source that has no direct bonds to carbon and halogen (Cl and Br) would be highly desirable.

Villaflor et al. found that tris-dimethylaminoarsenic (TDMAAs) has an etching effect on GaAs in a CBE system [8-9]. This TDMAAs with As directly bonded to N is a promising As source. Because there are no As-H bonds, TDMAAs is expected to be less toxic than arsine

Mat. Res. Soc. Symp. Proc. Vol. 448 © 1997 Materials Research Society

(AsH3) [10]. Also due to the lack of direct As-C bonds, TDMAAs has been successfully used in metalorganic molecular beam epitaxy (MOMBE)/CBE of (Al,Ga)As with a lower carbon incorporation [11-14]. The objective of this work is to investigate the TDMAAs etched/regrown interface of GaAs and $Al_xGa_{1-x}As$ for device application. Our results show that improved etched/regrown interfaces and pn junctions can be obtained. These characteristics are important to realize emitter-up HBTs with low emitter-edge recombination, as well as collector-up HBTs.

EXPERIMENT

The in-situ etching and regrowth experiments were performed in a modified Perkin-Elmer 425B CBE system. Uncracked TDMAAs was carried by hydrogen and transported into the CBE chamber through an ultrahigh vacuum leak valve. The H_2 flow rate was varied from 2 to 4 sccm to adjust the TDMAAs flux, corresponding to an As incorporation rate of 1.2 to 2.0 monolayer per second (ML/s), as determined from As-induced intensity oscillations of reflection high-energy electron diffraction (RHEED) on a Ga-rich GaAs surface.

For the study of etched/regrown interfaces, Si-doped GaAs were regrown twice on n^+ Si-doped (100) GaAs substrates by using triethylgallium (TEGa), tertiarybutylarsine (TBA) or arsenic (As_4), and silicon tetrabromide ($SiBr_4$) . The regrowth temperature and V/III ratio were 510°C and 1.3, respectively. Capacitance-voltage (C-V) measurements were performed for evaluating the quality of etched/regrown GaAs interfaces.

The $Al_xGa_{1-x}As$ ($0 \leq x \leq 0.50$) samples with a 50Å GaAs cap layer were grown on semi-insulating (100) GaAs substrates. 2500 Å of SiO_2 films by plasma enhanced chemical vapor depoition (PECVD) were used to be masks on these samples for determining the TDMAAs etch rate at different substrate temperatures from 550 to 700°C. We use scanning electron microscopy (SEM) and the Dektak stylus profiler to study the etch rate of $Al_xGa_{1-x}As$. The surface topography and roughness were examined using atomic force microscopy (AFM).

For emitter-up HBT applications, our objective is to regrow p-type wide bandgap material on external base region to reduce base resistance and increase f_{max}. So we studied the junction of regrown $p-Al_{0.35}Ga_{0.65}As/GaAs$ by using the optimal TDMAAs in-situ etch and regrowing $p-Al_{0.35}Ga_{0.65}As$ on n-GaAs collector. The idea for collector-up HBT applications is to reduce recombination current in the external base region by using the same regwon technique for emitter-up HBTs. Therefore, we regrew $p- Al_{0.35}Ga_{0.65}As$ on AlGaAs emitter on collector-up HBTs.

RESULTS

Fig.1 shows the TDMAAS etch rate of AlGaAs under the different substate temperature. The lower etch rate on AlGaAs is attributed to the Al-As bonds which are more difficult to break than the Ga-As bonds. Once they are broken, however, the Ga-amine species are more volatile than Al-amine species on the etched surface, resulting in a lower etch rate with increasing Al composition. It should be pointed out here that the etch rate of GaAs at 550°C is 400 Å/hr, but no etching effect is observed for patterned $Al_xGa_{1-x}As$ ($x \geq 0.35$) samples (below the sensitivity of our instrument ~ 20 Å). Therefore, the etching selectivity of GaAs from $Al_xGa_{1-x}As$ ($x \geq 0.35$) at 550°C is greater than 20, which is much higher than samples etched at 650°C with an etching selectivity of 5.

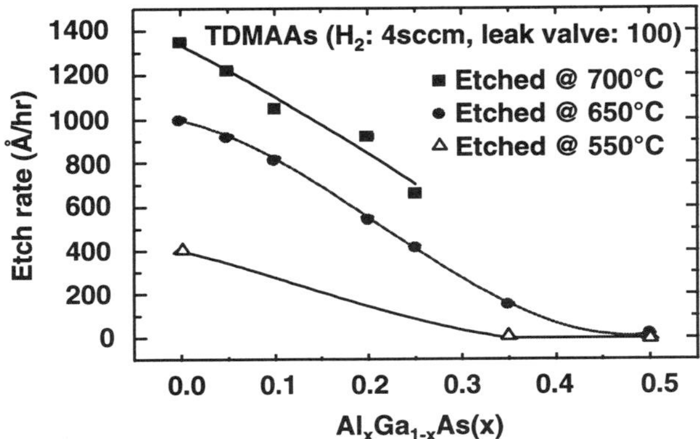

Fig.1 The TDMAAs etch rate of AlGaAs as a function of substrate temperature

The surface roughness is also done by AFM and shown in Fig. 2. The lower etching temperature (rate), the smoother etched surface. At 550°C etching temperature, the RMS roughness of GaAs is 22 Å which is expected to be improved by lowing TDMAAs/TEGa (V/III) ratio [10].

C-V measurement were done on Au/n-GaAs Schottky diodes to study the electrical property of etched/regrown interfaces of GaAs with different samples preparations. The C-V results are shown in Fig. 3. The minimal D_{int} we got is by lowering in-situ etching temperature to 550°C for 20 minutes. So sample A has the lowest D_{int} by integrating the carrier profiles, which means that a clean interface could be achieved by the TDMAAs in-situ etching process compared to the conventional oxide desorption by As_4.

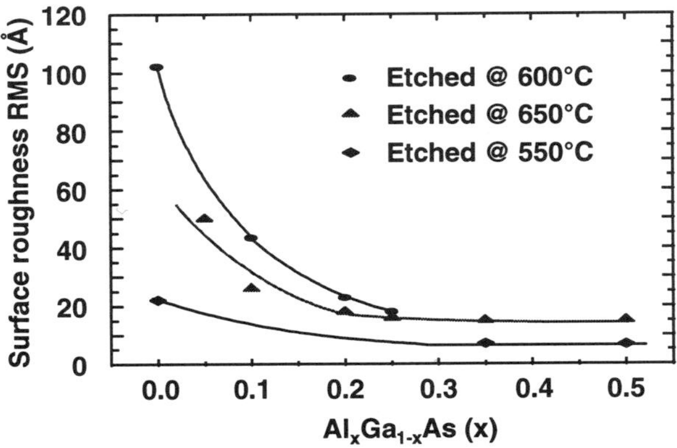

Fig. 2 Surface roughness of the TDMAAs etched AlGaAs surface

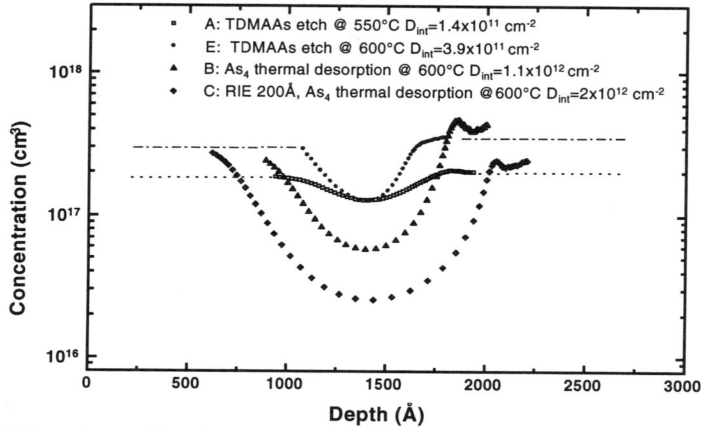

Fig. 3 C-V carrier profiles of etched/regrown samples with different preparations

To evaluate the regrown pn junction for emitter-up HBT applications, we measure pn junctions of regrown p-$Al_{0.35}Ga_{0.65}As$ onto original extrinsic n-GaAs collector region. By using optimal TDMAAs in-situ etch (etch temperature of 550°C for 20 minutes) the regrown pn junction (sample EU_A) didn't show much difference from sample EU_D prepared by conventional TBA thermal desorption as shown in Fig. 4.

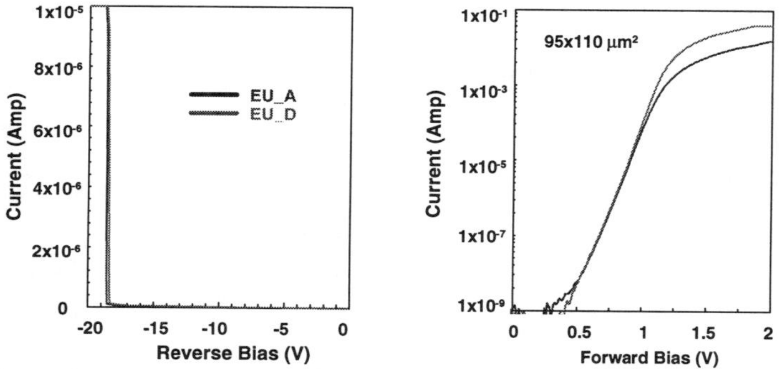

Fig. 4 Regrown p-AlGaAs/n-GaAs junctions by TDMAAs in-situ etch/regrowth (EU_A) and TBA thermal desorption preparation (EU_D).

Another evaluation on regrown p-n junctions as shown in Fig. 5 was done by regrowing p-type wide bandgap $Al_{0.35}Ga_{0.65}As$ on to n-type emitter $Al_{0.25}Ga_{0.75}As$ layers of collector-up HBTs. The layers were prepared by TBA thermal desorption (sample CU_D) and TDMAAs etch (sample CU_A1, CU_A2) preparations. Unlike samples CU_D and CU_A1, sample CU_A2 has 200 Å of p-type GaAs base protection layer on n- $Al_{0.25}Ga_{0.75}As$ emitter to prevent AlGaAs oxidation while loading these three samples into CBE chamber. From Fig. 5, the sample CU_A2 with GaAs protection layer and by using TDMAAs in-situ etch (etch temperature of 550°C for 20 minutes) before regrowth has highest breakdown voltage, while sample CU_D without GaAs protection layer and only by TBA thermal desorption before regrowth showed the lowest breakdown voltage. The diode trun-on voltage also varied among samples. The lowest value was shown by sample CU_D from exposed AlGaAs layer (as illustrated in Fig. 5.) This reduced turn-on voltage may be attributed to increased minority carrier recombination.

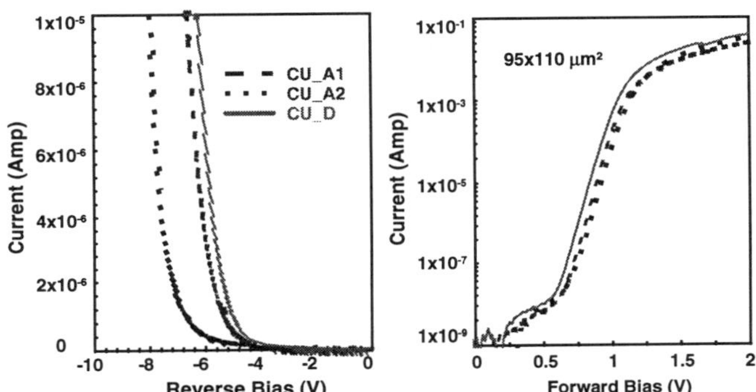

Fig. 5 Regrown p-AlGaAs/n-AlGaAs junctions by TDMAAs in-situ etch/regrowth (CU_A1 & 2) and TBA thermal desorption preparation (CU_D).

CONCLUSIONS

TDMAAs etched/regrown interfaces has been investigated and optimized. We found that by lowering the in-situ TDMAAs etching temperature of GaAs, the etched/regrown GaAs and $Al_xGa_{1-x}As$ interfaces can be improved. The lowest D_{int} of $1.4x10^{11}$ cm^{-2} reported in this study using TDMAAs is comparable to using other etching sources. The etching selectivity over 20 can also be achieved on GaAs from AlGaAs (x>0.35) at 550°C. Finally by applying these optimal TDMAAs etching/regroth conditions to HBT applications, the regrown pn junctions show the improved breakdown voltage and higher turn-on voltage.

ACKNOWLEDGMENTS

This work was partially supported by U.S. Air Force Wright Laboratory and by Rockwell International. Authors also would like to thank Kopin Coroperation for providing HBT epilayer wafers.

REFERENCES

1. H. shimawaki, Y. Amamiya, N. Furuhata, and K. Honjo, Device Research Conference, 1993.

2. H. J. Gregory, J. M. Bonar, P. Ashburn, and G. J. Parker, Electronics Letters, V32 N9, p 850-851, 1996

3. D. S. L. Mui, T. A. Strand, B. J. Thibeault, L. A. Coldren, P. M. Petroff and E. L. Hu, Inst. Phys. Conf. Ser.141 (1995) 69.

4. K. Tappura, A. Salokatve, K. Rakennus, H. Asonen, and M. Pessa, Appl. Phys. Lett. 57 (1990) 2313.

5. W.T. Tsang, R. kapre, and P.F. Sciortino, Jr., J. Crystal Growth 136 (1994) 42.

6. K. Tateno and Y. Kohama, Electronic Materials Conference, 1996.

7. H.Q. Hou, B.E. Hammons, and H.C. Chui, Electronic Materials Conference, 1996.

8. A. B. Villaflor, H. Asahi, D. Marx, K. Miki, K. Yamamoto and S. Gonda, J. Crystal Growth 150 (1995) 638.

9. D. Marx, H. Asahi, X. F. Liu, M. Higashiwaki, A. B. Villaflor, K. Miki, K. Yamamoto, S. Gonda, S. Shimomura and S. Hiyamizu, J. Crystal Growth 150 (1995) 551.

10. P. Gimmnich, A. Greiling and J. L. Lorberth, C. Thalmann, K. Rademann, G. Zimmermann, H. Protzmann, W. Stolz, and E. O. Göbel, Mater. Sci. Eng. B17 (1993) 21.

11. C. R. Abernathy, P. W. Wisk, D. A. Bohling and G. T. Muhr, Appl. Phys. Lett. 60 (1992) 2421.

12. S. Salim, J. P. Lu, K. F. Jensen and D. A. Bohling, J. Crystal Growth 124 (1992) 126.

13. K. Ishikura, A. Takeuchi, M. Kurihara, H. Machida and F. Hasegawa, Jap. J. Appl. Phys. 33 (1994) L494.

14. Yoshida and M. Sasaki, J. Crystal Growth 150 (1995) 557.

Part II

Control of Growth I: Surfaces and Interfaces

Ultra High Vacuum Scanning Tunneling Microscopy Observation of Multilayer Step Structure on GaAs and AlAs Vicinal Surface Grown by Metalorganic Vapor Phase Epitaxy

Jun-ya ISHIZAKI, Yasuhiko ISHIKAWA and Takashi FUKUI
Research Center for Interface Quantum Electronics, Hokkaido University, Sapporo 060, Japan,
ishizaki@ryouko.rciqe.hokudai.ac.jp / fukui@ryouko.rciqe.hokudai.ac.jp

Abstract

We observe the atomic structures at the multilayer step region on MOVPE-grown GaAs (001) vicinal surface using ultra high vacuum scanning tunneling microscopy (UHV-STM), and clarify that (4x2) or (4x3) like reconstruction units are dominant. Oxide free AlAs surfaces grown on GaAs vicinal surface are also successfully observed by UHV-STM. The reconstruction units at the multilayer step region on AlAs surface have the same units on GaAs vicinal surface. GaAs surface has the lack of dimmer rows on the terrace region just below the multilayer step region, while AlAs surface has dimmer rows even on the terrace just below the multilayer step region. GaAs layer growth leads tothe step bunching phenomenon and AlAs surface leads to the step debunching phenomenon.

1.Introduction

Multilayer step formations i.e. step bunching phenomena have been widely observed on GaAs vicinal surface grown by metalorganic vapor phase epitaxy (MOVPE) and/or after thermal treatment. Step bunching phenomena is a long standing issue of crystal growth, but it is not clear whether the shape of multilayer steps is determined by the thermal equilibrium or by the kinematical motion of adatom on the terrace. Schwoebel and Shipsey pointed out that step bunching phenomena are caused by the anisotropic migration barrier for adatom to up- and down-side step sites, and statistically analyzed the step motion [1]. However, the size of multilayer steps increase monotonously with increasing the growth thickness. Although, their model can explain the classical step bunching phenomena which include the facetting of metal surfaces [2], for the step bunching phenomena on GaAs vicinal surface the size of multilayer steps saturates as the growth thickness increases.

In order to explain the step bunching phenomena, we performed Monte Carlo simulation assuming the anisotropic migration barrier for adatom to up- and down-side step sites. [3] From the comparison of the simulation to the experimental data, we estimated the barrier height near the step sites for migrating adatom to up-side step site, and also clarified the saturation mechanism of the size of multilayer steps. Moreover, we clarified that the atomic structure of multilayer step region is determined by the reconstruction unit between monolayer steps. [4] However, it is not clear why step site has anisotropic migration barrier.

The purpose of this work is to clarify the detail atomic structures at the multilayer step region on both GaAs and AlAs vicinal surfaces by ultra high vacuum scanning tunneling microscopy (UHV-STM) observation.

2.Experimental

GaAs and AlAs epitaxial growth was done using low pressure MOVPE system with UHV chamber for reflection high energy electron diffraction (RHEED) observation and surface passivation by amorphous As. Detail of the MOVPE system was described in previous work. [4] The substrates were n-type (001) GaAs misoriented toward [110] direction by 2°. The carrier concentration was $1 \sim 3 \times 10^{18} cm^3$. Three kinds of samples were prepared for UHV-STM observation. After removal of surface oxide in arsine atmosphere at $650°C$, 100nm-thick GaAs buffer layer, 40cycle $(AlAs)_3(GaAs)_3$

Mat. Res. Soc. Symp. Proc. Vol. 448 © 1997 Materials Research Society

period superlattice, 60nm-thick GaAs layer, 15nm-thick Si doped GaAs conductive layer and 5nm-thick non-doped GaAs top layer were grown at 600°C for GaAs surface observation, schematically shown in Fig.1(a). In order to observe non-doped GaAs vicinal surface by UHV-STM, Si doped GaAs layer was inserted near the surface. For AlAs surface observation, 1 monolayer (ML) AlAs was also grown on GaAs top layer at 600°C. (Fig.1(b)). For AlAs sample grown at 700°C, the growth interruption was introduced at AlAs/GaAs interface during increasing the temperature.

After the growth, the samples were transferred from the reactor to the RHEED chamber. For both GaAs and AlAs surfaces, c(4x4) reconstruction were observed by RHEED. Next, sample surfaces were passivated by amorphous As using Knudsen cell within RHEED chamber. The samples were loaded out from MOVPE system, and were loaded into UHV analysis system having UHV-STM chamber and X-ray photoelectron spectroscopy (XPS) chamber. Amorphous As on the sample surface was removed by heating in UHV analysis system. Oxidation free surfaces were confirmed by XPS measurement. UHV-STM images were observed under the negatively biased condition for sample to tungsten tip which was held at ground potential (filled images). The bias voltages were between -2 and -3 V, at constant current mode of 0.15 and 0.30 nA.

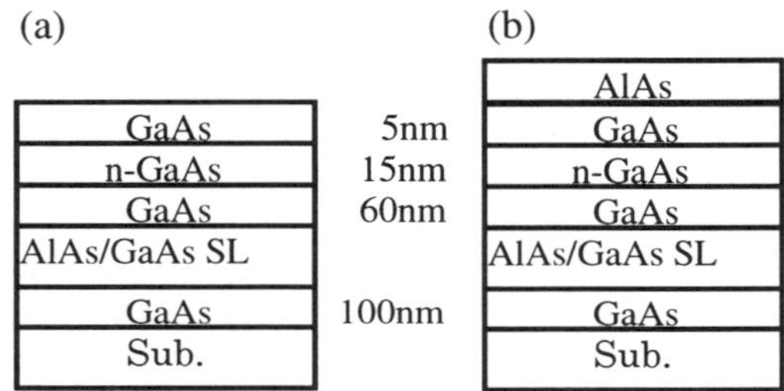

Fig.1 Schematic illustration of sample structures for (a) GaAs surface observation and (b) AlAs surface observation.

3.Results and discussion

First, UHV-STM image for GaAs vicinal surface is shown in Fig.2 (a). GaAs surface grown at 600°C has multilayer steps and atomically flat terraces. Average distance between each multilayer step is about 70nm, and the terrace region has no island. At the multilayer step region, $6 \sim 10$ ML steps were bunched together, and (4x3) or (4x2) reconstruction units were observed between neighboring monolayer steps. [4] On this GaAs surface with multilayer steps, 1 ML -thick AlAs layer was grown at 600 or 700°C. AlAs surface grown at the temperature of 600°C has two dimensional nucleation on the terrace region, shown in Fig.2 (b). Island density on the terrace region is about 3×10^{10} cm^{-2}, and the average island separation is about 30nm. Because the two dimensional nucleation is only formed when AlAs layer grows on GaAs vicinal surface, these nucleation are composed of AlAs. The reason that the two dimensional nucleation is formed on terrace is that the distance between each multilayer step is larger than the migration distance of Al adatom. In previous works for GaAs growth on vicinal surfaces, we clarified from experiment [5] and simulation [3] that the distance between each multilayer step is limited by the migration distance of adatom. On the other hand, AlAs surface grown on GaAs vicinal surface at the temperature of 700°C has no island on the terrace region, shown in Fig.2 (c). These results suggest that the surface migration length of Al adatom increases with increasing growth temperature, and Al migration distance becomes larger than the terrace width. AlAs surface grown at 700°C has clearer reconstruction unit than that at 600°C, shown in Fig.2(b) and (c).

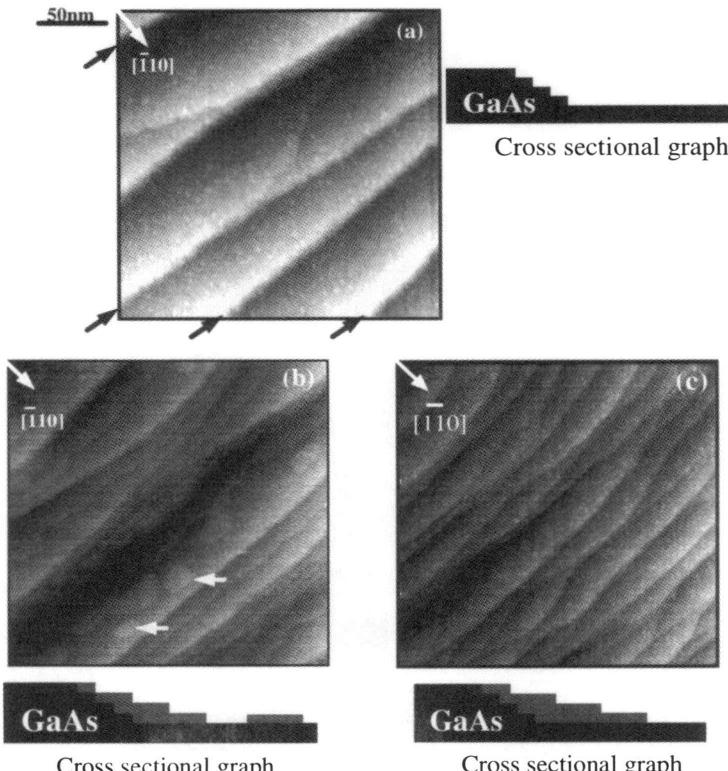

Fig.2 (a) GaAs surface morphology with multilayer steps observed by UHV-STM. Distance between each multilayer step is about 70nm, and average height of multilayer steps is about 9ML. (b) AlAs surface grown on GaAs with multilayer step at the temperature of $600\,^{\circ}C$. Terrace region has two dimensional nucleation which density is $3 \times 10^{10}\,cm^{-2}$. (c) AlAs surface grown at $700\,^{\circ}C$ which has no nucleation on the terrace.

Next, the distributions of inter-step distances at the multilayer step region on GaAs and AlAs surface are shown in Fig.3. When the single domain of (4x2) or (4x3) reconstruction units occur between two neighboring monolayer step at multilayer step region, the mean inter-step separation should be 1.8nm, which correspond to (119)B surface [4]. For GaAs vicinal surface, mean inter-step separation is 1.8nm, shown in Fig.3(a). The result show that the single domain of (4x2) or (4x3) reconstruction unit determines the inter-step separation distance.

However, the inter-step distances at the multilayer step region on AlAs surface tends to be larger than that on GaAs surface. Moreover, for AlAs surface, several monolayer steps were also observed even on the terrace region. Distributions of inter-step distances on AlAs surface grown at both $600\,^{\circ}C$ and $700\,^{\circ}C$ are broader than underlying GaAs (001) vicinal surface. The results show that the multilayer steps tend to debunch during AlAs layer growth.

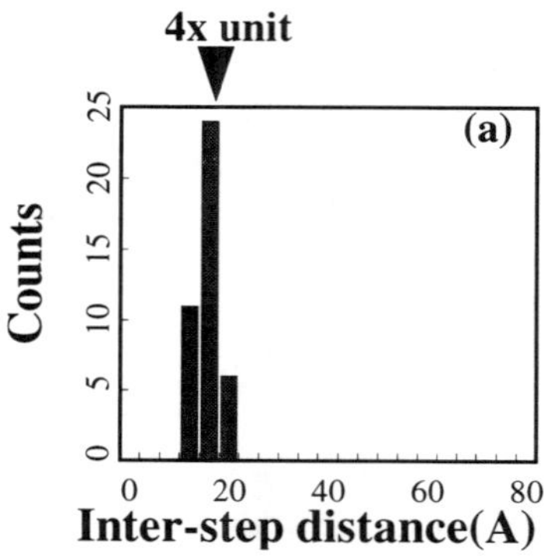

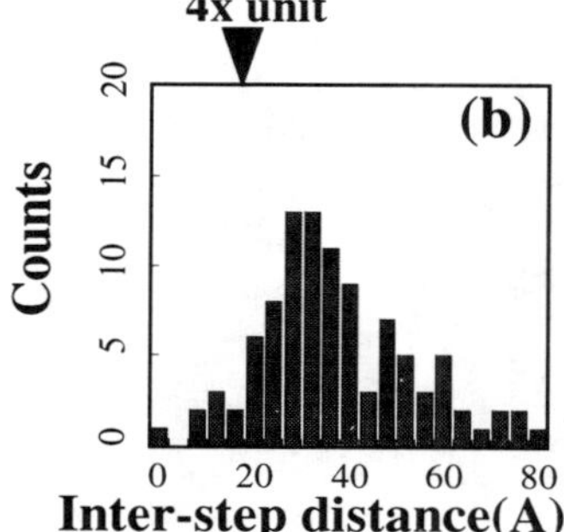

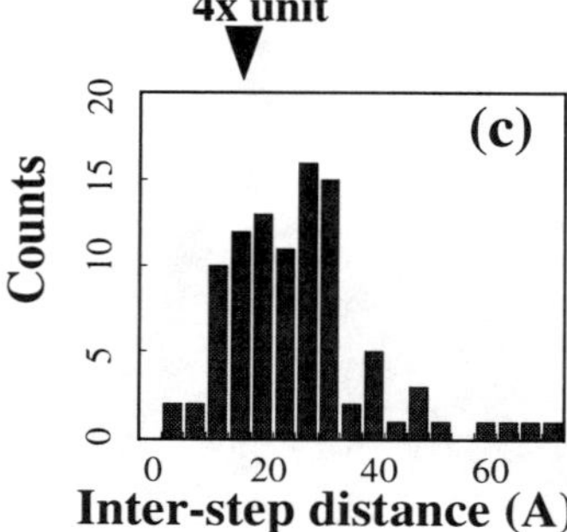

Fig.3 The distributions of inter-step distances at the multilayer step region for (a) GaAs surface grown at 600°C, (b) AlAs surface grown at 600°C and (c) AlAs surface grown at 700°C

Now, we can discuss the mechanism for debunching of the multilayer step during AlAs growth. First, we consider the migration length of adatom. The surface migration length of Al adatom on the terrace is shorter than that of Ga adatom [6]. For thick AlAs growth on vicinal surface, the terrace widths should be narrower than that for GaAs, because average terrace widths are determined by the migration length of adatom on the terrace [3,5]. However, in this experiment, average thickness of AlAs layer was only 1ML, and the mean distance between the multilayer step region was almost unchanged from UHV-STM observation. Therefore, the tendency of debunching during AlAs growth at the multilayer step region can not be explained simply by the surface migration length of adatom.

Next, we discuss the Schwoebel barrier. Schwoebel and Sphisy suggest that the step bunching phenomena can arise only when the anisotropic barriers exist for surface migration adatom to the up- and down-side step sites. [1] STM images suggest that the reconstruction units at the multilayer step region is almost the same for GaAs and AlAs surfaces with (4x3) like structures.

However, GaAs vicinal surface lacks dimer rows on terrace region just below the multilayer steps, while AlAs grown surface has always dimer rows even on the terrace region just below the multilayer steps, shown in Fig.4. These results suggest that the lack of dimer rows might be the reason which causes Schwoebel barrier for migration adatoms to up-side step site and, so that, the step bunching phenomena might occur only for GaAs growth on vicinal surface.

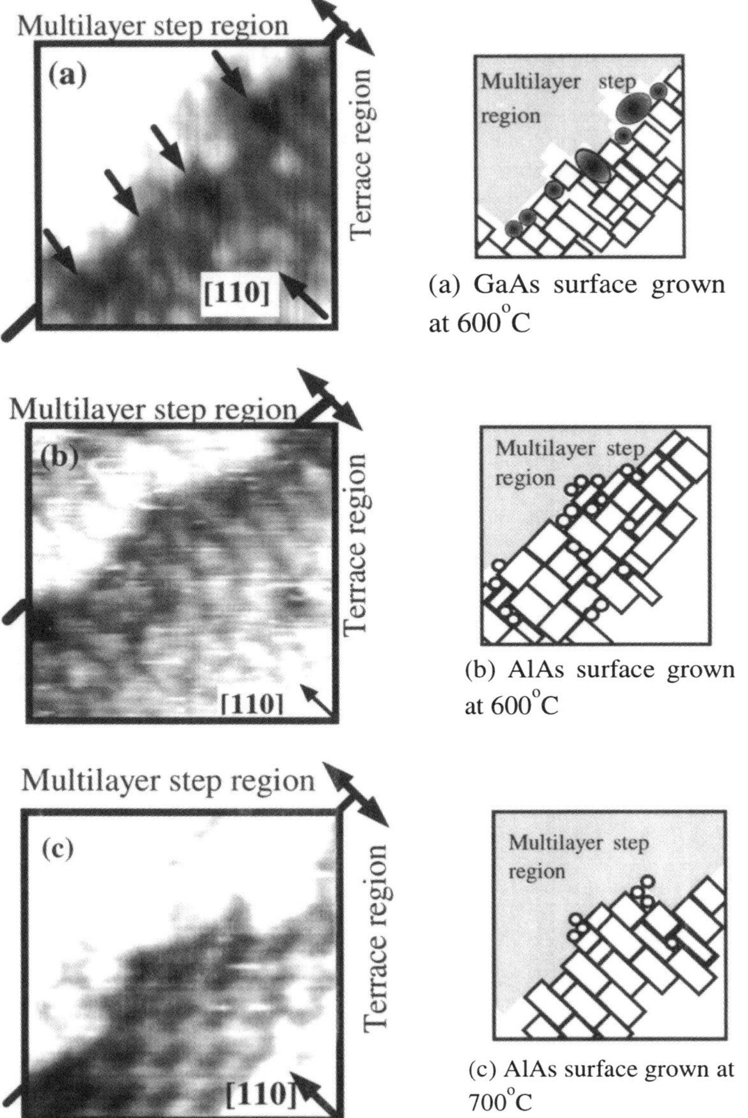

Fig.4 STM images and their schematic illustrations of (a) GaAs and (b) (c) AlAs surfaces just below the multilayer step regions. (a) GaAs surface has the lack of dimmer rows on the terrace region just below multilayer step, and (b) AlAs surface has dimmer rows on the terrace region just below multilayer step. The arrow in Fig (a) show the empty site on GaAs surface near the terrace region.

4.Conclusions

We observed the reconstruction unit at the multilayer step region on both AlAs and GaAs surface grown on GaAs (001) vicinal surface using UHV-STM. Oxide free AlAs surfaces grown on GaAs vicinal surface are successfully observed by UHV-STM for the first time. The reconstruction unit at the multilayer step region on GaAs surface has (4x2) or (4x3) like unit, and the reconstruction unit at the multilayer step region on AlAs surface has the same unit on GaAs vicinal surface. AlAs surface grown at 600°C has many islands on the terrace region. The step distribution on AlAs surfaces at multilayer step regions have broad distribution compared with that on GaAs surface. These results suggest that during GaAs layer growth monolayer steps bunch each other to form multilayer step regions, while these bunching steps tend to release during AlAs growth. Moreover, GaAs surface has the lack of dimer rows on the terrace region just below the multilayer step region, and AlAs surface has dimer rows even on the terrace just below the multilayer step. Schwoebel barrier must exist just below the step site in order to arise step bunching phenomenon, and these results suggest that the step bunching phenomena might occur by the lack of dimmer lows on the terrace region just below the multilayer step region for GaAs growth on vicinal surface.

Acknowledgment

Authors would like to thank Profs. H.Hasegawa and T.Hasizume for their stimulating suggestions and also Prof. J.Motohisa for his valuable discussion and also Mr. M.Akabori and Mr. S.Kasai for their technical supports. In this study, one of us (J.I.) is supported by JSPS Research Fellowship for Young Scientists.

References

[1] R.L.Shwoebel and E.J.Sphisy, J.Appl.Phys. **37,** 3682, (1966); R.L.Schwoebel, J.Appl.Phys. **40,** 614, (1969).
[2]W.W.Mullins, Philos. Mag. **48,** 1313, (1961).
[3]J.Ishizaki, K.Ohkuri and T.Fukui, Jpn.J.Appl.Phys. **35,** 1280, (1996).
[4]J.Ishizaki, Y.Ishikawa, K.Ohkuri, M.Kawase and T.Fukui, to be published in Appl.Surf.Sci.
[5]J.Ishizaki, S.Goto, M.Kishida, T.Fukui and H.Hasegawa, Jpn.J.Appl.Phys. **33,** 721, (1994).
[6] M.Kasu and N.Kobayashi, J.Appl.Phys. **78,** 3026, (1995).

IN SITU OPTICAL OBSERVATION AND CONTROL OF INITIAL STAGES OF GaAs GROWTH ON CaF₂ SURFACE MODIFIED BY ELECTRON BEAM IRRADIATION

K. KAWASAKI and K. TSUTSUI
Department of Applied Electronics, Interdisciplinary Graduate School of Science and Engineering, Tokyo Institute of Technology, 4259 Nagatsuta, Midori-ku, Yokohama, 226, Japan, koji@ae.titech.ac.jp

ABSTRACT

We investigated the electron beam induced surface modification of $CaF_2(111)$ and initial stage of GaAs growth on the modified CaF_2 surface by means of the surface photoabsorption technique and atomic force microscopy (AFM). The CaF_2 surface was modified by 300 eV electron beam irradiation at 200°C in a As_4 molecular beam. The amount of adsorbed As atoms increased with electron dose and it follows Langmuir adsorption principle and saturated at a value equivalent to 1 monolayer adsorption. In situ observation of GaAs growth on this modified surface clarified that the sticking coefficient of GaAs on CaF_2 surface was drastically improved by the surface modification. AFM observation revealed that the surface roughness of initial growth of GaAs on modified CaF_2 was improved at the growth temperature of 550°C.

INTRODUCTION

Heteroepitaxy of III-V compound semiconductors on crystalline insulator such as alkaline earth fluorides [1-10] (CaF_2, SrF_2, BaF_2 and their alloy) is considered to be attractive for future high performance devices. Especially GaAs/fluoride/Si [5-10] is interesting because it can be integrated with Si based devices. We have proposed the electron beam surface modification technique on the fluoride surface in order to overcome poor wettability of semiconductors on fluorides [6]. Although W. Li *et al.* reported direct growth of GaAs with mirror-like surface on CaF_2/Si(111) substrate, its growth condition was very critical [7]. The electron beam surface modification technique showed drastic improvement in surface morphology [6], dislocation density [9], electron mobility [8,9] and photoluminescence intensity [10]. The mechanism of the modification process has been considered where F ions desorb from the CaF_2 surface by electron beam assist and group V atoms, As or P, supplied simultaneously occupied the surface F vacancy sites so as to improve wettability of over grown GaAs. Although the improvement of crystallinity of the GaAs layer grown by this technique was remarkable, the electrical properties of the layer have not been satisfactory compared to that of bulk crystal so far. In order to improve the film property, in situ observation of surface modification process and control of growth process are expected to be useful. However, it is difficult to observe the fluoride surface using electron diffraction method because the fluoride surface is very sensitive to electron beam irradiation and the beam effect can not be neglected. So we applied the surface photoabsorption (SPA) method [11] using a visible light which is expected not to affect the fluoride surface. Recently, we observed the surface modification process by the SPA method in which the sensitivity of the surface condition changes was increased by utilizing interference effects [12]. As a result, it was found that Ca colloids were generated on the CaF_2 surface by low energy electron beam irradiation and that it could be suppressed when the surface modification was carried out using 300 eV electron beam at 200°C.

In this work, the progress of the modification process and initial growth stage of GaAs were observed by means of SPA technique. And surface morphologies of GaAs grown on the modified CaF_2 surfaces were observed by atomic force microscopy (AFM) in order to comprehend the basis of surface modification effect.

EXPERIMENT

The experiment was carried out using a 3 chambers MBE system composed of a load lock and of two growth chambers for fluorides and for GaAs. After thermal cleaning of Si(111)

Mat. Res. Soc. Symp. Proc. Vol. 448 © 1997 Materials Research Society

substrate at 900°C to obtain 7×7 superstructure in reflection electron diffraction pattern, 20 nm thick CaF_2 films were grown on the substrate. Base pressure of fluoride growth chamber was less than 6×10^{-9} torr and pressure during the growth was 1×10^{-8} torr. After the substrate was transferred to the GaAs MBE growth chamber in which base pressure was less than 2×10^{-9} torr, its surface was irradiated by an 300 eV electron beam at 200°C using an electron gun facing to the substrate under As_4 molecular beam impingement. After the surface modification process, GaAs growth was followed by opening a Ga cell shutter. The As_4 pressure during the electron beam irradiation and the GaAs growth was estimated as the equivalent pressure of the order of 10^{-5} torr at the sample surface. The surface modification and succeeding growth process was observed in situ by the SPA method as shown in Fig. 1. A p-polarized Ar ion laser light (488 nm) of which incident angle was set to 83° was employed as the probing light. The reflected light was detected by a Si photodiode and the output was lock-in amplified. The SPA signal was taken as $\Delta R/R$, where R is initial reflected intensity and ΔR is the variance from the observed intensity to the R. Finally, the surface morphology of the samples was investigated by an AFM, NanoScope III, in the atmosphere.

RESULTS AND DISCUSSION

Figure 2 shows a SPA signal during the surface modification process. The previous work shows that signal at the wavelength employed here is insensitive to defects in CaF_2 but increases in proportional to the amount of adsorbed material on the CaF_2 surface as far as $1 >> d/\lambda$ (d: thickness of adsorbed material, λ :wavelength) [12]. The observed signal was found to increase when the CaF_2 surface was irradiated by electrons in the As_4 molecular beam and to saturate even though electron beam was continued to irradiate on the CaF_2 surface. This dependence on electron dose could be fitted to an expression related to Langmuir's adsorption. In addition, the saturated value of 4.4 % in the formula was confirmed to correspond to approximately one monolayer adsorption of As to the CaF_2 surface by means of secondary ions mass spectroscopy. In addition, Lee et al. reported through the electron energy loss spectroscopy that a loss energy peak of 1.8 eV assigned to the surface F-vacancy state was observed when a $CaF_2(111)$ surface was exposed to an electron beam with energy of 3 keV and it vanished after supply of As_4 molecular beam to the CaF_2 surface [13]. This result implies that F-vacancies were replaced by As atoms. Although the electron energy used here was lower than that of the previous work, it is considered that our SPA results indicates a sequential process that electrons desorbs F ions only from the surface sites and then As

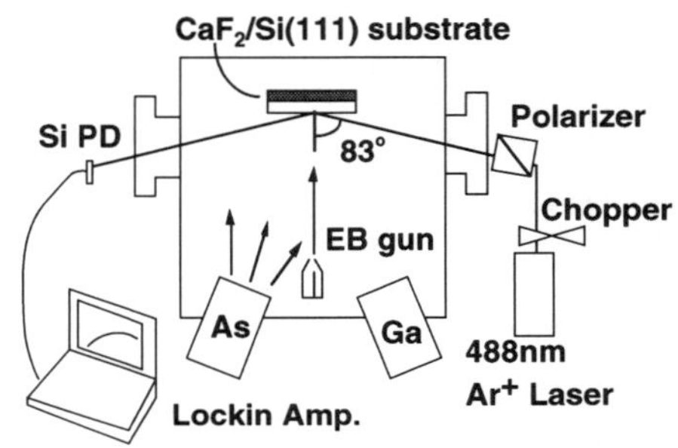

Fig.1 Experimental system for the *in situ* observation of the surface modification and GaAs growth process.

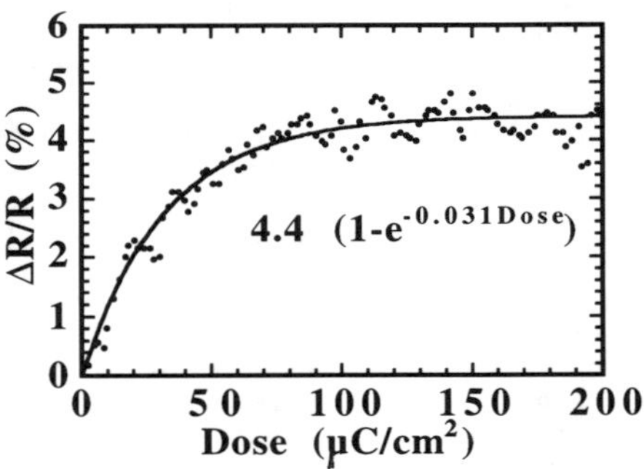

Fig.2 The SPA signal during the surface modification process under As_4 molecular beam impingement at a substrate temperature of 200°C. The electron energy was 300 eV. The signal curve during electron beam irradiation was fitted to $4.4 \{ 1 - \exp(-0.031 \cdot \text{dose } [\mu C/cm^2]) \}$.

atoms are adsorbed in the vacancy sites so as to saturate at one monolayer adsorption.

Figure 3 shows AFM images of the initial stage of GaAs growth on the modified CaF_2 surface as a function of the coverage of the As estimated from the value of $\Delta R/R$ before growth. The changes of $\Delta R/R$ during GaAs growth were also shown in Fig. 4, in which the notations are corresponding to those shown in Fig. 3. The substrate temperature during the growth was 450°C, the Ga beam flux was equivalent to that for homoepitaxial GaAs growth with 6 nm/min rate and the growth time was 10 seconds. Figure 3(a) shows the case that the surface coverage of As is equal to zero, *i.e.* no electron beam was irradiated on the CaF_2 surface. In this case, it can be said that GaAs was not deposited since the SPA signal was not changed as shown in Fig. 4(a). In contrast, GaAs was grown along the several step edges of the CaF_2 in the case of a half surface coverage by As as shown in Fig. 3(b). The SPA signal shown in Fig. 4(b) was also increased during the growth. In the case of the full surface coverage by As, GaAs was found to be grown along the almost all of step edges as shown in Fig. 3(c). The increase of the SPA signal shown in Fig. 4(c) is approximately twice compared to that of Fig. 4(b). It is considered that approximately twice amount of GaAs was deposited on the surface. It is quite interesting that the adsorbed amount of GaAs was proportional to the coverage of As. In addition, the almost part of surface was replaced by As in the case of Fig. 3(c), nevertheless the GaAs islands were formed only on the step edges. It is considered that the nuclei of GaAs were selectively formed along the step edges and then GaAs islands were formed at the sites by collecting the migrating atoms on the modified surface.

Figure 5 shows the growth temperature dependence of GaAs grown on the modified CaF_2 surface. The growth rate of GaAs was 1 µm/h and the growth time was 70 sec. The dose of the electrons before the GaAs growth was controlled so that the SPA signal intensity was saturated. Figure 5(a), 5(b) and 5(c) are corresponding to the growth temperature of 500, 550 and 600°C, respectively, and the RMS of their surface roughness are 2.0, 0.9 and 1.1 nm, respectively. Figure 5(d) shows surface morphology of CaF_2 before electron beam irradiation as a reference of the surface roughness. The morphology shows well-ordered multi-steps and the value of RMS was 0.84 nm. Comparing to the roughness of the CaF_2 surface, it can be said that smooth surface of GaAs at 550°C was obtained as same level as the substrate. However, in microscopic, the initial growth mode of GaAs on the CaF_2 was said to be 3-dimensional growth mode rather than 2-dimensional one since the inheritance of CaF_2 surface morphology can not be seen on the GaAs surface morphology.

The changes of the SPA signal intensity during growth were shown in Fig. 6. The lines (a), (b) and (c) in Fig. 6 are corresponding to growth of the sample shown in Fig. 5(a), 5(b) and 5(c),

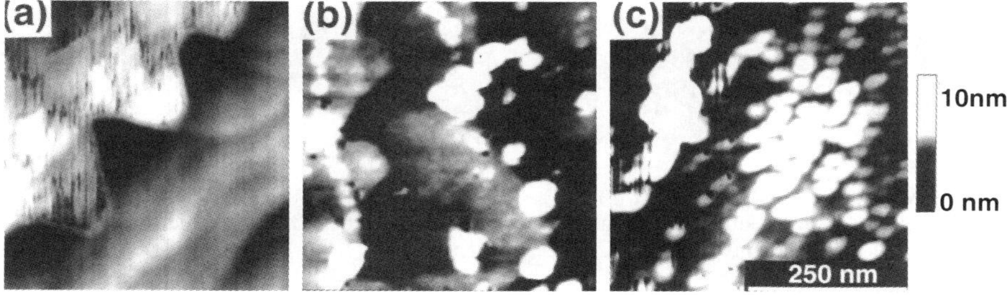

Fig. 3 AFM images of GaAs grown on modified CaF_2 surfaces. The coverage of As are (a) 0 (no electron beam irradiation), (b) 0.5 and (c) 1, respectively. Substrate temperature during surface modification by electron beam was 200°C and GaAs growth temperature was 450°C. The amount of GaAs supplied was equivalent to 1 nm thick in homoepitaxial growth.

respectively. Just after started the GaAs growth, the signal intensity increased linearly and then it was rolled off as shown in Fig. 6. This roll-off is due to an optical interference effect in the GaAs layer. The absolute value of $\Delta R/R$ is changed by the fluctuation in the probing light incidence and light scattering due to 3-dimensional growth, however, this interference period for the GaAs thickness is not changed. It can be seen that it takes longer time to reach the maximum intensity point as the growth temperature becomes higher. It means that growth rate was decreased as the growth temperature become higher. This is probably due to enhancement of the Ga evaporation from the growth surface. In the case of 500°C, as shown in Fig. 6(a), the signal curve was suddenly changed at the point of 50 sec past. It is considered that 3-dimensional growth was enhanced in this point since the theoretical calculation under the condition of layer-by-layer growth mode shows that the periodical intensity change should be maintained to the end of thick (1 μm) GaAs growth (as shown by broken line). The AFM image in Fig. 5(a) also shows higher value of RMS than other samples. In the other

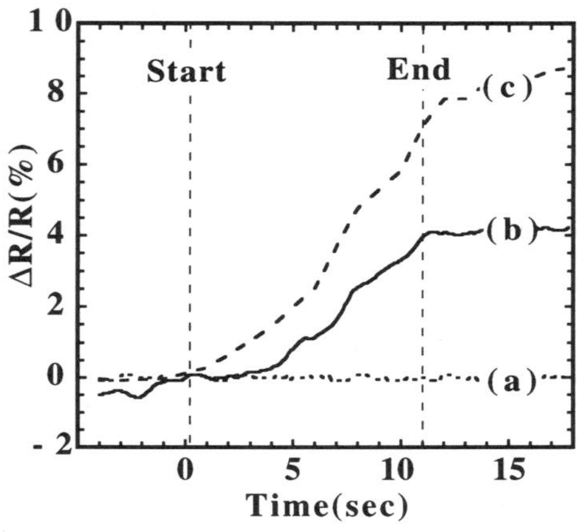

Fig.4 SPA signal intensity changes for GaAs growth on (a) as-grown CaF_2 surface and (b) modified CaF_2 surface with a half coverage of As and (c) that with full coverage of As. Samples are the same those shown in Fig. 3. The shutter of a Ga effusion cell was opened at "Start" point and was closed at "End" point.

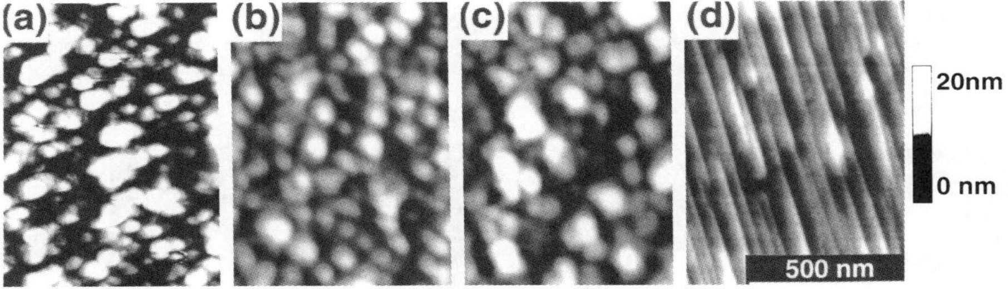

Fig.5 AFM images of GaAs grown on CaF_2. The growth temperature are (a) 500°C, (b) 550°C, (c) 600°C and (d) shows surface of as-grown CaF_2. The amount of GaAs supplied was equivalent to 20 nm thick in homoepitaxial growth.

samples, oscillation breakdown in the signal was not observed in these growth thickness. However, a kink at which the gradient of intensity change become large can be found in the each signal change. Furthermore, it is found that the kink occurred at 10 sec later from the start point of the growth for the at 550°C growth, while it occurred at 20 sec later for the 600°C growth. It can be considered that growth rate was rather slow in the initial stage because the surface was not completely covered by GaAs due to 3-dimensional growth mode and reevapolation rate of Ga atom from the CaF_2 surface would be rather high. After the surface was completely covered by

GaAs, however, sticking coefficient was increased and the gradient of intensity change became large. In the high temperature growth at 600°C, it is considered that evaporation was more enhanced and it took twice time compared to the case of 550°C to cover whole CaF_2 surface by GaAs. So, the surface morphology become rough (RMS of 1.1 nm) due to reduction of sticking coefficient and the large island formation at 600°C. It is considered to be desirable that growth temperature is decreased in the initial stage of GaAs growth in order to increase the sticking coefficient and/or to reduce the large island formation. However, due to difference of thermal expansion coefficients between GaAs and CaF_2 [14], lowering the temperature enlarges the lattice mismatch. For example, the mismatch of 2.5 % at 550°C is increased to 2.7 % at 500°C. Although we are not sure whether such a small difference causes the apparent difference in the growth mode, it might be considered that island growth of GaAs was enhanced by the interface strain due to large mismatch of the lattice parameters at lower temperature.

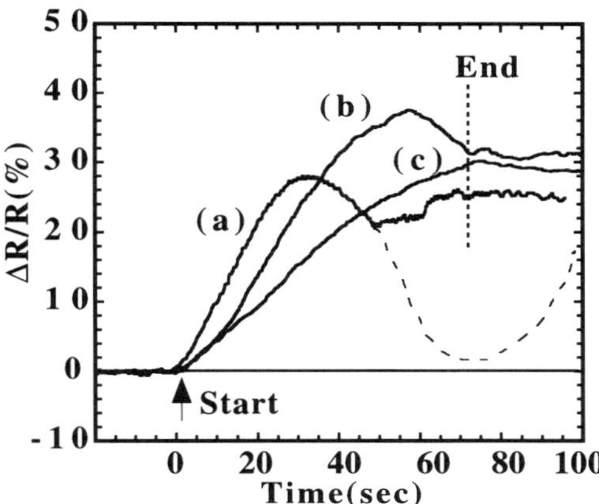

Fig.6 SPA signal intensity changes during growth. The growth temperature are (a) 500°C, (b) 550°C and (c) 600°C. Samples are the same those shown in Fig.5. The shutter of a Ga effusion cell was opened at "Start" point and was closed at "End" point.

CONCLUSIONS

We investigated the electron beam induced surface modification of CaF_2(111) and initial stage of GaAs growth on the modified CaF_2 surface using SPA method and AFM. The amount of adsorbed As atoms was found to follow Langmuir adsorption principle during the electron beam irradiation. It increased with electron dose and saturated at a value equivalent to 1 ML adsorption. It was also found that the surface modification process increases sticking coefficient of GaAs on the CaF_2 surface. It was shown that rather flat initial surface of GaAs was obtained at the growth temperature of 550°C. It is expected that high quality GaAs on CaF_2 will be realized by future optimization of modification and initial growth conditions.

REFERENCES

1.   S. Siscos, C. Fontain and A. M-Yague, Appl. Phys. Lett. **44** (1984) 1146.

2.   P. W. Sullivan, G. M. Metze and J. E. Bower, J. Vac. Sci. & Technol. **B3** (1985) 500.

3.   K. Tsutsui, H. C. Lee, H. Ishiwara, T. Asano and S. Furukawa, GaAs and Related Compounds 1985 (Inst. Phys. Ser. 79), (1986) 109.

4.   C. W. Tu, S. R. Forrest and W. D. Jonston Jr., Appl. Phys. Lett. **43** (1983) 569.

5.   T. Asano, H. Ishiwara, H. C. Lee, K. Tsutsui and S. Furukawa, Jpn. J. Appl. Phys. **25** (1986) L139.

6.   H. C. Lee, T. Asano, H. Ishiwara and S. Furukawa, Jpn. J. Appl. Phys. **27** (1988) 1616.

7.   W. Li, T. Anan and L. J. Schowalter, J. Cryst. Growth **135** (1994) 78.

8.   A. Ono, K. Tsutsui and S. Furukawa, Jpn. J. Appl. Phys. **30** (1991) 454.

9.   S. M. Hwang, K. Miyasato, K. Kawasaki and K. Tsutsui, Jpn. J. Appl. Phys. **35** (1996) 1701.

10. M. Uchigoshi, K. Tsutsui and S. Furukawa, Jpn. J. Appl. Phys. **30** (1991) L444.

11. Y. Kobayashi and N. Kobayashi, Jpn. J. Appl. Phys. **31** (1992) L71.

12. K. Kawasaki and K. Tsutsui, submitted to Jpn. J. Appl. Phys.

13. H. C. Lee, H. Ishihara, S. Furulawa, K. Saiki and A. Koma, Appl. Surf. Sci. **41/42** (1989) 553.

14. S. Furukawa, H. Ishiwara and K. Tsutsui, <u>Proc. of the 1988 Electron. Devices and Mat. Symp.</u> (1988) 266.

FORMATION OF ZnSe/GaAs HETEROVALENT HETEROSTRUCTURES BY MOVPE

Mitsuru Funato *, Satoshi Aoki, Shizuo Fujita and Shigeo Fujita
Department of Electronic Science and Engineering,
Kyoto University, Kyoto 606-01, Japan
* funato@kuee.kyoto-u.ac.jp

ABSTRACT

ZnSe/GaAs (001) heterovalent heterostructures are fabricated by metalorganic vapor phase epitaxy. During the growth, both GaAs and ZnSe surfaces are kept atomically flat to achieve precise control of the interface formation. Interface composition, Ga/As, are controlled by means of either Zn or Se treatment of a GaAs surface, and then ZnSe growth follows. Consequently, it is revealed by X-ray photoemission spectroscopy (XPS) that artificial control of Ga/As from 1.0 to 2.8 leads to the variation of valence band offsets from 0.6 to 1.1 eV. Based on the electron counting model and layer-attenuation model, it is proposed that the As plane just below the interface consists of As, anti-site Ga and As vacancy.

INTRODUCTION

Heterovalent heterostructures, which are characterized by chemical valence mismatch at the interfaces, possess attractive feature of variable band offset. The band offset at a heterojunction interface plays an important role to determine carrier transport and confinement properties, and it has been recognized to be constant for a given heterostructure. However, recent considerable works reveal that it may be changed by an interfacial atomic structure[1-6]. Theoretically calculated changes of the offsets in several heterovalent heterostructures are, for example, 1.0, 0.6 and 0.8 eV for ZnSe/GaAs[1,2], Ge/GaAs[3,4] and Si/GaP[3], respectively. The large variation in band offsets at the heterovalent interfaces is originated from the existence of nonoctet bonds which form donor and acceptor states. Due to the charge transfer from the donor to the acceptor states electronic dipoles which determine the offsets are induced. The strength and the direction of the dipoles depend sensitively on the microscopic interface atomic configuration and thus, so do the band offsets. Therefore, to achieve tunable band offsets, it is crucially important to control formation processes of an interface in an atomic order.

Experimentally, X-ray photoemission spectroscopy (XPS) has chiefly been utilized to study contributions of interface structures to the band offsets[1,5,6]. It has been reported that interface composition can be controlled by the Zn/Se flux ratio employed during molecular beam epitaxy (MBE) of ZnSe on GaAs (001), resulting in the variation of valence band offsets from 0.58 eV (Se-rich condition) to 1.20 eV (Zn-rich condition)[1]. In the study, however, since the interface composition is modified by changing the epitaxial growth condition of beam pressure ratio, bulk properties as well as interface properties would be influenced. Further, the interface formation and the epitaxial growth occur at the same time, and thus, the strict control of the formation processes of the interfaces has not been achieved yet.

In this study, we demonstrate the tunability of the band offset in the ZnSe/GaAs (001) heterostructure by changing compositional ratio, Ga/As, at the interface. The local composition is modified by treatments of GaAs surfaces by Zn or Se precursors. The conditions of the following ZnSe growth are unchanged in order to concentrate the discussion on the

107

interface chemistry. Surface structures during the formation of the heterointerface are investigated by means of atomic force microscopy (AFM) for the purpose to keep an atomically flat surface which is an implication of an atomic order control of the interface structure. The valence band offsets are estimated by XPS.

FORMATION OF ZnSe/GaAs HETEROSTRUCTURE

Samples were grown on (001) nominally singular GaAs substrates by atmospheric-pressure metalorganic vapor phase epitaxy (MOVPE). The substrates were chemically and thermally etched in the same manner reported previously[7]. A GaAs buffer layer 1500 Å thick was grown on the GaAs substrate at the growth temperature of 700 °C. Source precursors were triethylgallium (TEGa) and tertiarybutylarsine (TBAs) and their flow rates were 10 and 100 μmol/min, respectively. To obtain an atomically flat surface, post-growth annealing was performed at 700 °C for 10 min and the substrate temperature was cooled down with TBAs flow. Therefore, GaAs surface must be terminated by As at this time[8,9]. Since both ZnSe and GaAs layers were grown in the same reactor, the GaAs buffer layers were kept in the reactor under hydrogen flow at 200 °C for typically 1 hour to purge the source precursors for GaAs growth.

In order to relax the chemical valence mismatch, there should exist the same number of donor and acceptor states at the interface. As one of the most simplified examples, a mixed Ga-Zn plane with 50-50 compositions or a As-Se plane with the similar compositions at the interface may satisfy the above situation. The resulting valence band offsets for the former and the latter interfaces have been calculated to be 1.59 and 0.62 eV, respectively[1]. To realize such interfaces experimentally, prior to the growth of ZnSe, GaAs surfaces were exposed to Zn or Se precursors, that is, diethylzinc (DEZn) or dimethylselenium (DMSe), at 450 °C. In this paper, we will denote these treatments as Zn or Se treatment. Then, ZnSe layers were grown at 450 °C. Molar flow rate of DEZn was 1.2 μmol/min and the molar flow ratio of DMSe/DEZn was 10.

A GaAs surface after the purge at 200 °C, was observed by AFM as shown in Fig.1(a). The surface consists of atomically flat terraces and monolayer (ML) steps. No islands are detected on the terraces. GaAs surfaces after Zn or Se treatment were also investigated,

(a)   (b)

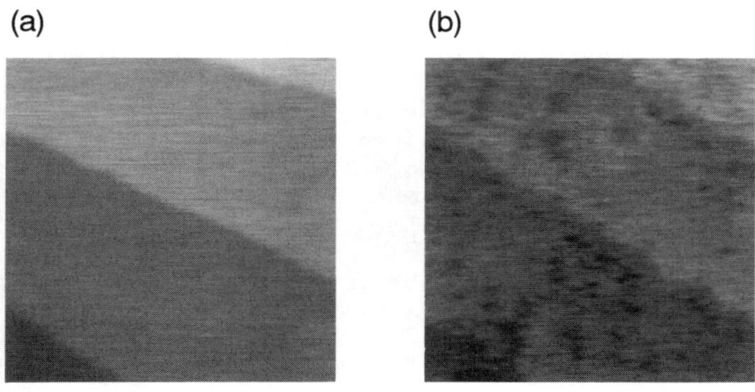

200 nm

Figure 1: AFM images of (a) GaAs epitaxial layer grown and annealed at 700 °C and (b) ZnSe layer 30 Å thick grown at 450 °C.

and it was confirmed that those treatments did not affect the surface flatness. The growth mode of ZnSe depends strongly on the growth temperature. As is discussed in detail in ref.[10], 450 °C is suitable temperature to obtain a smooth surface. Fig.1(b) illustrates the surface of ZnSe 30 Å thick grown at 450 °C. Atomically flat terraces and ML steps can be clearly found, indicating that the growth mode of ZnSe at 450 °C is layer-by-layer mode. Regardless of the treatment of GaAs surfaces, the ZnSe surfaces are atomically flat like Fig.1(b). The surface structure was preserved at least to the thickness of 900 Å. From the results, we can conclude that the ZnSe/GaAs interfaces are relatively abrupt.

XPS MEASUREMENTS

ZnSe layers 5–50 Å thick were prepared for XPS. The XPS measurements were performed *ex situ* using Al $K\alpha$ (1486.6 eV) as X-ray source and signals from $3d$ core levels of Zn, Se, Ga and As, were detected. Fig.2 shows schematic views of XPS with different thickness of ZnSe. Note that XPS signal includes the information within about 50 Å from a surface. Therefore, when ZnSe is thick (~ 40Å, Fig.2(a)), Ga and As $3d$ signals represent the information in the vicinity of the interface and Zn and Se $3d$ signals show bulk properties of ZnSe, whereas as ZnSe becomes thinner (Fig.2(b)), Ga and As $3d$ signals more include the GaAs bulk properties and Zn and Se $3d$ signals do the interface properties on ZnSe side. Therefore, depth profiles were obtained from the heterostructures with different ZnSe thicknesses.

Several heterostructures were fabricated also varying the supplying amount of DEZn or DMSe for the GaAs surface treatment. The integrated intensity ratio of $3d$ core levels observed in XPS, I_{Zn}/I_{Se} and I_{Ga}/I_{As}, as a function of ZnSe thickness provides depth profile. Both I_{Zn}/I_{Se} and I_{Ga}/I_{As} were normalized so as to exhibit unity in bulk standards. If we look at I_{Zn}/I_{Se}, regardless of GaAs surface treatments and ZnSe layer thicknesses, I_{Zn}/I_{Se} is equal to unity, indicating that the Zn/Se composition at ZnSe/GaAs interface is stoichiometric, i.e. unity. As for I_{Ga}/I_{As}, it depends on the GaAs surface treatments. Se treatment does not provide an influence on the composition of GaAs; I_{Ga}/I_{As} remains unity at any ZnSe thickness. If GaAs is exposed to DEZn, on the other hand, I_{Ga}/I_{As} increases with increase of the ZnSe layer thickness. Further, this characteristics becomes

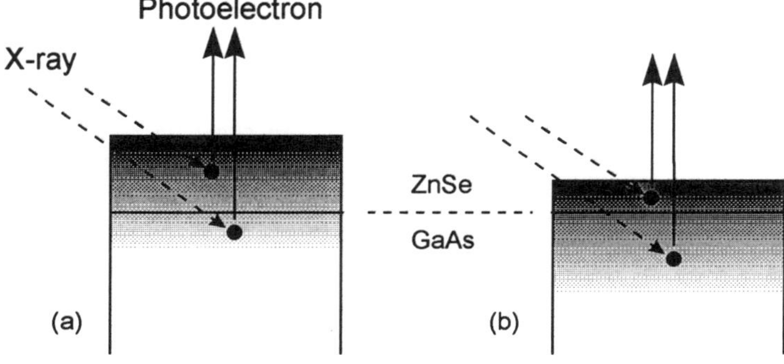

Figure 2: Schematic views of XPS of ZnSe/GaAs heterostructures with (a) thick and (b) thin ZnSe. Shade indicates detectable region.

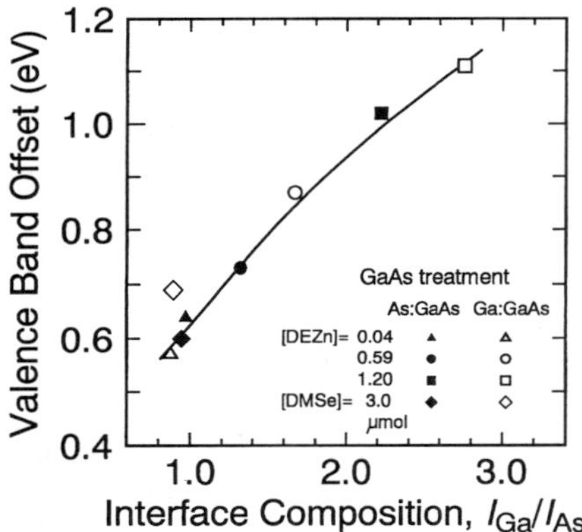

Figure 3: Estimated valence band offset as a function of interface composition, I_{Ga}/I_{As}. Valence band offsets were determined by the ZnSe thickness of 20–30 Å, while I_{Ga}/I_{As} was observed value in the interface region where the ZnSe thickness was above 30 Å(see text in detail). In addition to Zn or Se treatment of As terminated GaAs (As:GaAs), Zn or Se treatment of Ga terminated GaAs (Ga:GaAs) was examined and the results were summarized together.

pronounced as the amount of supplying DEZn increases. Since I_{Ga}/I_{As} is obtained very near the interface when ZnSe is thick as shown in Fig.2(a), this behavior of I_{Ga}/I_{As} indicates that the local relative solid composition at the interface is Ga-rich. DEZn reached at the GaAs surface seems to bring about the formation and evaporation of a volatile compound, Zn_3As_2[11], leaving the surface Ga-rich. It should be noted that the supplying amount of Zn was adjusted by changing duration time, not flow rate, so that the formation of Zn_3As_2 was limited by supply of DEZn. If not, deposition of Zn takes place and morphology of ZnSe becomes poor, suggesting the break of the interface control in an atomic order.

Valence band offsets were extracted by the standard approach in which core level to core level separations in heterostructures are measured[12]. Variation of valence band offsets is shown in Fig.3 as a function of interface composition of I_{Ga}/I_{As}. The I_{Ga}/I_{As} value in Fig.3 is that observed in ZnSe thickness of above 30 Å, so as to reflect well the compositional ratio at very close to the interface, while the valence band offsets were determined by the ZnSe of thickness 20–30 Å, because this thickness can minimize the influence on the XPS signal by the chemical shift due to interfacial structures. In addition to Zn or Se treatment of As terminated GaAs (denoted by As:GaAs), Zn or Se treatment of Ga terminated GaAs (Ga:GaAs) which was completed by supplying TEGa by the amount corresponding to 1 atomic layer, was examined and the results were summarized all together. Fig.3 clearly shows that valence band offsets increase from 0.6 to 1.1 eV as I_{Ga}/I_{As} increases, that is, interface composition becomes Ga-rich. It is worth noting that the correlation between interface composition and valence band offsets can be expressed by a unique curve regardless of the procedures to prepare interfaces. In this sense, it is essential to modify the interface composition, I_{Ga}/I_{As}, to control band offsets, and the method of surface treatment itself is not important.

STRUCTURAL ANALYSIS

Valence band offset of 0.6 eV is obtained mainly by the Se treatment of a GaAs surface. In this heterostructure, both I_{Zn}/I_{Se} and I_{Ga}/I_{As} are equal to 1, indicating that the interface is relatively abrupt without compositional grading. Therefore, we can compare the experimental result with the theoretical predictions on the correlation between interface

structures and band offsets[1,2], because in the theoretical analyses, interfaces without remarkable interdiffusion are selected as test structures. Consequently, this interface is likely
to have a mixed anion plane formed by 50 % As and 50 % Se atoms. The atomic configuration provides dipole moments pointing toward ZnSe and the resulting valence band offset
becomes smaller than that without interface dipoles.

On the other hand, the interfaces with I_{Ga}/I_{As} larger than 1 cannot be compared with
the theoretical analyses. In order to provide a possible structural model, the measured
intensity ratios of core level photoemission are analyzed through the standard electron
counting model and layer-attenuation model. Taking into account the experimental results
that I_{Zn}/I_{Se} is unity and I_{Ga}/I_{As} becomes large at the interface, we assume as follows;
(1) ZnSe is stoichiometric at/near the ZnSe/GaAs interface, (2) the plane which forms
a junction with ZnSe is a complete Ga cation plane and the Ga atoms bond to Se, and
(3) GaAs is also stoichiometric except for the anion plane just below the Ga cation plane;
the anion plane consists of As, anti-site Ga (Ga_{As}) and As vacancy (V_{As}). Here, the latter
two assumptions attribute to the Ga rich interface. We ruled out the possibility of the
formation of Ga_2Se_3 which has been often reported as an interfacial layer of ZnSe grown
on Ga-rich GaAs surface[13,14]. This is because as shown in Fig.4, transmission electron
microscopy (TEM) 002 dark field image of the interface between ZnSe and Zn treated
(Ga-rich) GaAs does not show the bright contrast which distinguishes the existence of the
compound, Ga_2Se_3, indicating the absence of formation of Ga_2Se_3. This assumption is also
supported by the finding that shape of Se $3d$ spectrum in XPS is not changed for all samples
in this study.

The detailed procedure of the calculation will be given elsewhere, but the preliminary
results show that both Ga_{As} and V_{As} are necessary to satisfy the charge neutrality across
the interface and I_{Ga}/I_{As} larger than unity simultaneously. For example, to realize the
observed I_{Ga}/I_{As} of 1.5 in XPS for the ZnSe (5 Å) on GaAs heterostructure, where the
valence band offset is 1.0 eV, the fractional occupancies of Ga_{As} and V_{As} are calculated
to be 0.52 and 0.18, respectively. It should be noted that the atomic structure in the
structural model induces Ga-Se donor bonds at the ZnSe side of the interface, which is the
completely opposite atomic configuration compared with stoichiometric (50 % As and 50 %
Se mixed) interface where the valence band offset is 0.6 eV, as described in the beginning
of this subsection. This change of atomic geometry is responsible for the larger band offset.

Figure 4: TEM 002 dark field image of (110) cross section of
ZnSe/GaAs heterostructure. GaAs surface was treated by DEZn of
1.2 μmol at 450 °C and the resulted I_{Ga}/I_{As} from XPS is about 2.

CONCLUSIONS

ZnSe/GaAs (001) heterovalent heterostructures were fabricated by MOVPE, keeping both ZnSe and GaAs surface atomically flat to realize strict controll of formation processes of the interfaces. XPS showed that tunability of band offsets was achieved by changing the local interface composition of I_{Ga}/I_{As} by means of either Zn or Se treatment of a GaAs surface followed by ZnSe growth. The proposed interface atomic structure supported the hypothesis that a dipole due to interface geometry played a role to determine the band offset.

ACKNOWLEDGEMENTS

The authors wish to acknowledge the Mazda foundation's research grant and the Hoso Bunka Foundation for the financial assistance. XPS measurements were carried out at Micro Photonics Materials Laboratory at Kyoto University. This work was partially supported by KU-VBL (Kyoto University Venture Business Laboratory) Project.

REFERENCES

1. R. Nicolini, L. Vanzetti, G. Mula, G. Bratina, L. Sorba, A. Franciosi, M. Peressi, S. Baroni, R. Resta, A. Baldereschi, J. E. Angelo and W. W. Gerrich, Phys. Rev. Lett. **72**, 294 (1994)

2. A. Kley and J. Neugebauer, Phys. Rev. B **50**, 8616 (1994)

3. R. G. Dandrea, S. Froyen and A. Zunger, Phys. Rev. B **42**, 3213 (1990)

4. K. Kunc and R. M. Martin, Phys. Rev. B **24**, 3445 (1981)

5. G. Bratina, L. Vanzetti, A. Bonanni, L. Sorba, J. J. Paggel, A. Franciosi, T. Peluso and L. Tapfer, J. Crystal Growth, **159**, 703 (1996)

6. R. W. Grant, J. R. Waldrop, S. P. Kowalczyk and E. A. Kraut, J. Vac. Sci. Technol. B **3**, 1295 (1985)

7. M. Funato, Sz. Fujita and Sg. Fujita, Jpn. J. Appl. Phys. **33**, 4851 (1994)

8. I. Kamiya, D. E. Aspnes, H. Tanaka, L. T. Florez, J. P. Harbison and R. Bhat, Phys. Rev. Lett., **68**, 627 (1992)

9. F. Reinhardt, W. Richter, A. B. Müller, D. Gutsche, P. Kurpas, L. Ploska, K. C. Rose and M. Zorn, J. Vac. Sci. Technol. B11, 1427 (1993)

10. M. Funato, S. Aoki, Sz. Fujita and Sg. Fujita, Jpn. J. Appl. Phys. (submitted)

11. B. Chelluri, T. Y. Chang, A. Ourmazd, A. H. Dayem, J. L. Zyskind and A. Srivastava, Appl. Phys. Lett., **49**, 1665 (1986)

12. J. R. Waldrop, R. W. Grant, S. P. Kowalczyk and E. A. Kraut, J. Vac. Sci. Technol. A3, 835 (1985)

13. D. Li, J. M. Gonsalves, N. Otsuka, J. Qiu, M. Kobayashi and R. L. Gunshor, Appl. Phys. Lett. **57**, 449 (1990)

14. L. H. Kuo, K. Kimura, T. Yasuda, S. Miwa C. G. Jin, K. Tanaka and M. Yao, Appl. Phys. Lett. **68**, 2413 (1996)

ATOMIC HYDROGEN ASSISTED GROWTH OF Si-Ge HETEROSTRUCTURES ON (001) Si

J.-M. BARIBEAU, D.J. LOCKWOOD, S.J. ROLFE, R.W.G. SYME AND H.J. LABBÉ
Institute for Microstructural Sciences, National Research Council Canada,
Ottawa, K1A 0R6, CANADA

ABSTRACT

A study of the interface chemical and physical abruptness of Si-Ge heterostructures grown on (001) Si by molecular beam epitaxy under atomic hydrogen exposure is reported. Atomic hydrogen (AH) was produced by the dissociation of molecular hydrogen interacting with a hot tungsten filament. Secondary-ion mass spectroscopy (SIMS) of structures made of alternating Ge (0.5 nm)/Si (40 nm) layers demonstrated that AH can effectively suppress Ge surface segregation. The segregation length was reduced from 1.5 nm to about 0.5 nm in films grown at a hydrogen partial pressure of ~5 × 10^{-3} Pa and cell temperature of 2140 °C with an estimated cracking efficiency of ~5%. However, the high hydrogen background pressure had detrimental effects on the physical sharpness of the interfaces. This was evidenced by comparing the interface quality of Si/Ge atomic layer superlattices grown with and without AH exposure. X-ray reflectivity and Raman spectroscopy revealed a significant increase of the interface roughness, although the periodic character and the good crystallinity of the structures were preserved.

INTRODUCTION

There is currently enormous interest in the growth of strained Si/Si$_{1-x}$Ge$_x$ heterostructures. These offer enhanced electrical characteristics in bipolar devices[1] and are promising structures in optoelectronics.[2] In all these applications it is crucial that the interfaces between Si and Si$_{1-x}$Ge$_x$ be of high quality. For example, interface roughness at the atomic scale or chemical intermixing may affect the carrier transport properties at the hetero-junctions. Surface segregation is another phenomenon that may have detrimental effects on the electronic properties of a hetero-junction. This phenomenon is common in molecular beam epitaxy (MBE) where impurities that lower the surface free energy have a tendency to float at the surface rather than incorporate into the growing film. In Si-based MBE this is the case for most n-type dopants. In Si/Si$_{1-x}$Ge$_x$ hetero-epitaxy, the Ge atoms also tend to segregate at the surface resulting in a trailing edge in the Ge concentration profile after Ge deposition has ceased. In a first approximation, the Ge concentration profile N(z) at a Si$_{1-x}$Ge$_x$/Si interface may be written N(z) = (N(0)/ℓ) exp-(z/ℓ) where ℓ is the segregation length and N(0) is the initial Ge surface coverage (per unit area). The value of ℓ is a function of the growth conditions and is typically of the order of 1-2 nm.[3]

It is possible to reduce Ge surface segregation by exposing the surface during growth to another atomic species that has a stronger tendency to segregate. Impurities such has Sb, Ga, Sn and Bi act as surfactants and can suppress Ge surface segregation.[3] However, the introduction of such impurities in a high vacuum system may be difficult or even undesirable, as it may adversely affect other processes. Ge surface segregation is significantly inhibited in techniques that use gases as precursors, such as gas source MBE or ultra-high vacuum chemical vapor deposition. The Si surface is then passivated by a hydrogen adlayer acting as a surfactant.[4,5] Suppression of Ge segregation has been observed in solid source MBE by exposing the surface to an AH flux during deposition.[6-8] Hydrogen-assisted growth of Si$_{1-x}$Ge$_x$ is thus a promising approach to reducing segregation with minimum effects on the electrical or optical properties of the epitaxial film. AH has also been found to favor planar growth of Ge on (001) Si by reducing Ge adatom mobility.[6] High hydrogen partial pressure during Si MBE, however, has detrimental effects on the interface/surface roughness and promotes the transition to poly-crystalline or amorphous growth at higher temperatures.[9]

In this work we report the growth and characterization of Si/Ge heterostructures grown under AH exposure. The operating conditions of the AH source required to inhibit Ge surface

Mat. Res. Soc. Symp. Proc. Vol. 448 © 1997 Materials Research Society

segregation are determined. The quality of the interfaces in 50 period Si/Ge atomic layer superlattices grown with or without the presence of a hydrogen flux are then compared to assess the effect of a high hydrogen partial pressure on the interface morphology.

EXPERIMENT

The epitaxial layers were deposited on lightly doped p(B)-type (001) Si wafers by MBE in a VG Semicon V80 system.[10,11] Epitaxy was performed at a growth rate of 0.1-0.2 nm/s in the temperature range 300-550 °C. AH was produced by thermal dissociation of molecular hydrogen with a hot tungsten filament using a commercial cracker cell (EPI MBE Products Group). The source was located ~30 cm from the substrate and hydrogen gas was admitted by a leak valve. The filament temperature was measured by optical pyrometry. The pressure during the cell operation was monitored by an ion gauge located in vicinity of the wafer and by quadrupole mass spectrometry. Test structures consisting of ten Ge (0.5 nm)/Si (40 nm) layers were prepared under different cell operating conditions (1700 °C < TH < 2200 °C, 1×10^{-4} Pa < P_{H2} < 5×10^{-3} Pa). Atomic layer superlattices made of alternating Si(2.5 nm) and Ge(0.5 nm) layers were grown under standard growth conditions[11] or under optimized AH flux.

The Ge concentration profiles were obtained by SIMS using a Cameca 4f and a O_2^+ primary beam. The chemical modulation and interfacial perfection of the multilayers were investigated by x-ray reflectivity in the specular and transverse geometry using a Philips 1820 θ-2θ vertical goniometer.[12] Specular reflectivity was analyzed using a recursive formalism[13] that includes a Debye-Waller model[14] of the interface roughness and transverse scans were simulated using the distorted wave Born approximation.[15] The Raman spectra of the atomic layer superlattices were excited with 300 mW of 457.9 nm laser light, recorded at 295 K and analyzed following a method outlined elsewhere.[16]

RESULTS AND DISCUSSION

<u>Secondary ion mass spectrometry of test structures</u>

There have been conflicting reports in the literature as to how the operating conditions of a hydrogen cracking cell (temperature and pressure) influence the surface chemistry during Si-Ge heteroepitaxy. Cracker cell temperatures in the range 1800-2000 °C are generally used[7-9] although there have been studies performed at lower temperatures (1500-1600 °C).[17] The source geometry and distance from the surface and the wafer temperature, among other factors, may explain these variations. We have used SIMS to compare the Ge distribution in test structures comprising ten 4-monolayer thick Ge layers separated by 40 nm Si spacer layers. Figure 1(a) compares the Ge profiles of two similar structures prepared at different AH cell temperatures. The first four Ge spikes (zone 1) were deposited without AH while the last six (zones 2, 3 and 4) were grown at *increasing* hydrogen partial pressures. For a cell temperature of 1840 °C, the Ge profile exhibits only marginal variations when the surface is exposed to high doses of hydrogen. In contrast, the Ge profile becomes sharper when the surface is exposed at the same doses, but at a higher hydrogen cell temperature of 2140 °C. The dependence of the Ge profile on the hydrogen dose is, however, difficult to evaluate in this profile because loss of depth resolution away from the surface region (due to sputtering intermixing) may also contribute to the broadening of the Ge profile in the deeper Ge spikes. Figure 1(b) shows the SIMS profile of a similar structure, but where the last six Ge spikes were grown at *decreasing* hydrogen pressure. Again, sharper Ge profiles are observed when the surface is exposed to AH. The effect is more pronounced at higher doses as evidenced by the sharpening of the Ge profiles in zones 3 and 2 as compared to zone 4. A segregation length ℓ can be estimated from the derivative of the SIMS profile (ℓ^{-1} = 1/N × dN/dz) as illustrated in the inset of Fig 1(b). Under standard growth conditions $\ell \sim 1.5$ nm, while $\ell \sim 0.5$ nm for growth under 4×10^{-3} Pa hydrogen partial pressure. Smaller gas exposures (zones 3 and 4) have a much reduced effect on the segregation length.

The results reported here are in good agreement with those of Ohta et al.[7] in a similar study. It is interesting to estimate the AH flux on the substrate needed to inhibit the Ge segregation. Assuming a cracking efficiency of 5% at 2140 °C and a hydrogen gas pressure of 10^{-3} Pa[18] one

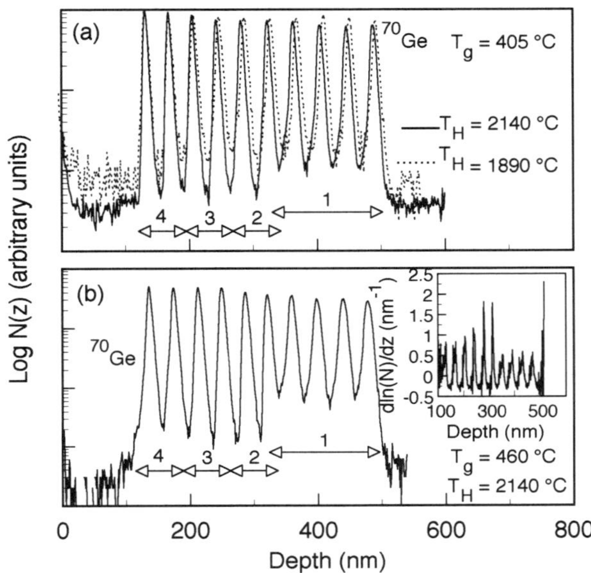

Fig. 1: ^{70}Ge SIMS profile of structures comprising ten repeats of Ge(0.5 nm)/Si(40 nm). (a) Growths at two different cell temperatures. Zones 1, 2, 3 and 4 were grown at hydrogen pressures of ~0, 0.1, 0.5 and 2 × 10^{-3} Pa, respectively. (b) Same as (a), but zones 1, 2, 3 and 4 grown at hydrogen pressures of ~0, 4, 1 and 0.2 × 10^{-3} Pa, respectively. The inset shows the logarithmic derivative of the ^{70}Ge signal with respect to depth.

finds an impingement rate of ~5 × 10^{14} hydrogen atoms cm^{-2}s^{-1}. This is comparable to the impingement rate of Ge atoms during epitaxy at a rate of 0.1 nms^{-1}. This suggests that in order to be incorporated in the growing crystal, the incoming Ge atoms have to displace surface hydrogen atoms. To suppress segregation, AH should be supplied at the same rate as Ge to saturate the surface bonds. This is in contrast with other common surfactants (Sb, Bi, As) which do not desorb but rather float on the surface during growth thus not requiring continuous supply.

Hydrogen-assisted growth of Si/Ge multilayers

We have prepared a series of Si/Ge short-period superlattices under an AH flux in the hope of improving the interfacial sharpness of these structures. Figure 2 compares the x-ray specular reflectivity for two 50-period superlattices prepared with and without AH exposure. The structure grown under standard vacuum conditions exhibits much more intense satellite peaks indicative of higher quality interfaces. Calculated reflectivities are also displayed in Fig. 2 and they show that the interfacial widths σ_{Si} and σ_{Ge} for the structure grown under AH flux have values about twice as large as those found in the structure grown under standard conditions.

Specular reflectivity can be used to determine interfacial widths but cannot discriminate between interfacial broadening caused by intermixing or physical roughness. The lineshape of the diffuse scattering in a transverse scan can be used to assess the lateral characteristics of the interface roughness. Transverse scans of the first satellite peak for the two same superlattices are presented in Fig. 3. The scans exhibit a resolution limited specular peak and a diffuse component that extends to large values of parallel wave vector transfer $q_{//}$. On the sample grown with AH exposure, the specular component is less intense and a much stronger diffuse scattering is observed. The diffuse scattering at large $q_{//}$ has been modeled using a correlation function of the

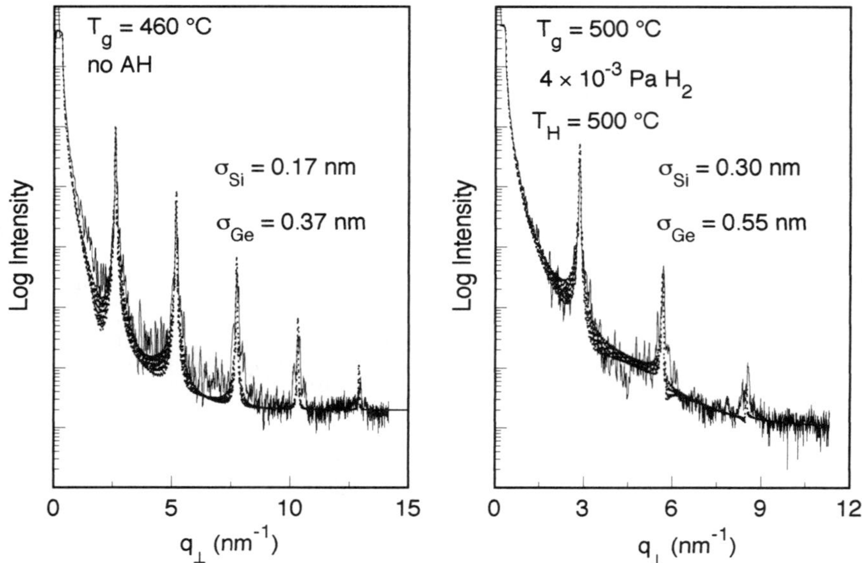

Fig. 2. Measured (full line) and calculated (dotted line) reflectivity from two 50-period Si/Ge superlattices grown without or with an AH flux. The interface widths σ_{Si} and σ_{Ge} used in the calculation are indicated.

type[19] $C(X) = \sigma^2 \exp\text{-}(X/\xi)^{2h}$ where ξ is the cutoff length (the scale at which the interfaces begin to look rough) and the exponent $0 < h < 1$ determines how jagged the interface is (smoother interfaces for larger values of h). Using the values of σ determined from specular measurements, the diffuse scattering on the sample grown under nominal conditions is qualitatively reproduced using $\xi = 200$ nm and h = 0.35. The model is adequate at large values of $q_{//}$ but fails to reproduce the shoulders of the specular peak (arrows in Fig. 3). These originate from a long wavelength undulation (~ 1 μm) of the interfaces associated with the wafer residual miscut and have been discussed elsewhere.[12] The calculation for the structure grown with AH exposure was obtained using $\xi = 20$ nm and h = 0.9 which indicates that the interfaces are rougher at a shorter length scale, although less jagged on that scale. The agreement between experiment and theory is only approximate. The fit could presumably be improved by using a more realistic correlation function, by varying the interface widths from layer to layer and by including the long range modulation of the interfaces in the model, but the analysis captures the essential physics.

Raman spectra of the two superlattices discussed above are shown in Fig. 4. The spectrum of the reference superlattice grown in vacuum at 460 °C exhibits peaks at 294, 416 509.8 (shoulder) and 519.3 cm^{-1} that are associated with superlattice longitudinal optic (LO) modes.[20] The lower frequency peaks at 104 and 203 cm^{-1} are due to superlattice folded longitudinal acoustic (FLA) modes[12] and their intensities are sensitive to interface atomic abruptness.[21] Significant changes in these FLA peaks are evident in the Raman spectrum of the equivalent superlattice grown with AH exposure. Apart from a small shift to higher frequency (111 cm^{-1}) due to a slight decrease in the superlattice period, the first FLA mode is noticeably weaker in intensity, while the second FLA peak at 210 cm^{-1} is now barely discernible. This sharp drop in intensity is the consequence of a loss of atomic abruptness at the Si/Ge interfaces that is consistent with the doubling of the interfacial width found by x-ray reflectivity. The LO peaks in the AH-exposed superlattice occur at frequencies of 293, 413 and 519.4 cm^{-1} and are not too different in intensity and position from the reference superlattice case. This would be expected for superlattices with similar respective Si and Ge layer thickness. In general, the FLA peaks in

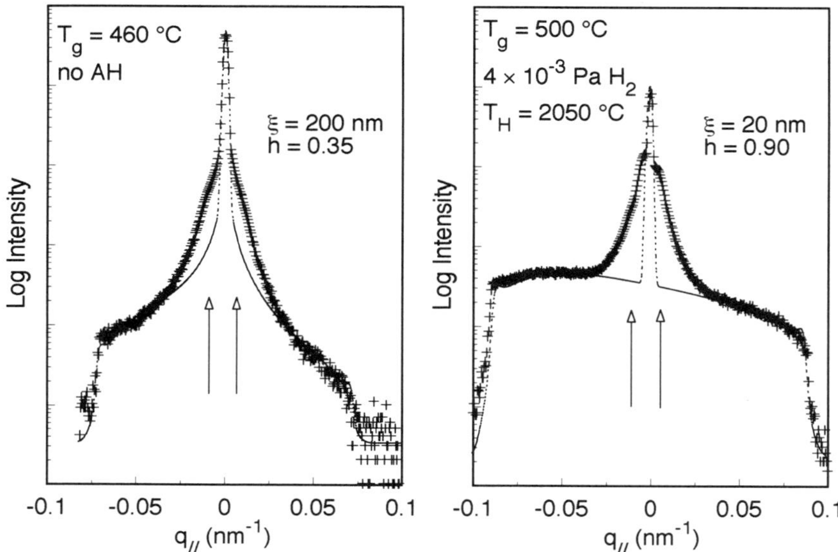

Fig. 3: Transverse scans at the first satellite reflection for two 50-period Si/Ge superlattices grown without and with AH exposure. The values of ξ and h used in the calculation are indicated.

superlattices grown under a variety of conditions with AH surfactant were always weaker than in reference superlattices grown in vacuum, showing a consistent loss of interface sharpness.

The above results are not necessarily incompatible with the suppression of Ge surface segregation, but show that growth under high hydrogen pressure leads to a deterioration of the interface morphology. The data is consistent with interfaces having characteristic roughness on a shorter length scale in hydrogen-assisted epitaxy. Saturation of the chemical bond at the surface by hydrogen may reduce surface diffusion thereby precluding layer-by-layer growth and causing a short scale corrugation. It is noteworthy that in these experiments AH was supplied continuously during growth. It is likely that Ge segregation can be suppressed while maintaining good interface morphology if AH is used only during short intervals corresponding to the interface formation.

CONCLUSION

In this work we have investigated atomic hydrogen assisted growth as a means to reduce Ge surface segregation in SiGe heteroepitaxy on (001) Si. Thin Ge layers imbedded in Si were found to have sharper chemical profiles when deposited under a flux of atomic hydrogen. The atomic hydrogen flux required to inhibit segregation is comparable to the impingement rate of Ge atoms on the surface. This suggests that the incorporation of Ge in the growing films involves the desorption of hydrogen surface atoms. Due to the small hydrogen cracking efficiency of the thermal cell, efficient use of atomic hydrogen requires conducting the growth under a high hydrogen partial pressure (10^{-3} Pa). This was found to have detrimental effects on the interface perfection in Si/Ge multilayer structures. A larger interface width and shorter length scale were measured in structures grown under high hydrogen exposure, presumably because of a reduced surface mobility. This problem could possibly be overcome by exposing the surface to large doses of atomic hydrogen only during interface formation. Work in that direction is currently in progress.

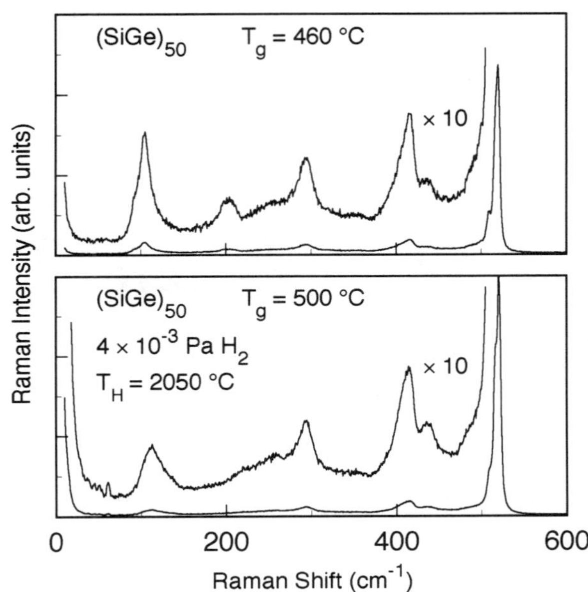

Fig. 4: Raman spectra of two 50-period Si/Ge superlattices grown without and with AH exposure.

REFERENCES

1. J.C. Sturm in <u>Properties of Strained and Relaxed Germanium</u>, edited by E. Kasper, INSPEC, The Institution of Electrical Engineers, 1995, pp 193-204.
2. R. Soref, J. Vac. Sci. Technol. **A14**, 913 (1996).
3. H. Jorke in <u>Properties of Strained and Relaxed Germanium</u>, edited by E. Kasper, INSPEC, The Institution of Electrical Engineers, 1995, pp 180-189.
4. M. Copel and R.M. Tromp, Appl. Phys. Lett. **58**, 2468 (1991).
5. K. Fujinaga, Jpn. J. Appl. Phys. **34**, 4004 (1995).
5. N. Ohtani N., S.M. Mokler, H.H. Xie, J. Zhang, and B.A. Joyce, Int. Conf. Solid State Devices and Materials, Chiba, 1994, pp 249-251.
6. A. Sakai and T. Tatsumi, Appl. Phys. Lett. **64**, 52 (1994).
7. G. Ohta, S. Fukatsu, Y. Ebuchi, T. Hattori, N. Usami and Y. Shiraki, Appl. Phys. Lett. **65**, 2975 (1994).
8. K. Nakagawa, A. Nishida, Y. Kimura and T. Shimada, Jpn. J. Appl. Phys. 33, L1331 (1995).
9. P.D. Adams, S.M. Yasilove, D.J. Eaglesham, Appl. Phys. Lett. **63**, 3571 (1993).
10. J.-M. Baribeau, T.E. Jackman, D.C. Houghton, P. Maigné, Denhoff M.W., J. Appl. Phys. **63**, 5738 (1988).
11. J.-M. Baribeau, D.J. Lockwood, M.W.C. Dharma-wardana, N.L. Rowell and J.P. McCaffrey, Thin Solid Films **183**, 17 (1989).
12. J.-M. Baribeau, D.J. Lockwood and R.W.G. Syme, J. Appl. Phys. **80**, 1450 (1996).
13. L.G. Parratt, Phys. Rev. **95**, 359 (1954).
14. P. Croce and L. Névot, Rev. Phys. Appl. **11**, 113 (1976).
15. Calculations performed using GIXS simulation package from BEDE Scientific.
16. D.J. Lockwood, M.W.C. Dharma-wardana, J.-M. Baribeau and D.C. Houghton, Phys. Rev. B **35**, 2243 (1987).
17. K. Sakamoto, H. Matsuhata, K. Miki and T. Sakamoto, J. Cryst. Growth **157**, 295(1995).
18. G.W. Wicks, E.R. Rueckwald and M.W. Koch, J. Vac. Sci. Technol. **B 14**, 2184(1996).
19. S.K. Sinha, E.B. Sirota , S. Garoff S. and H.B. Stanley, Phys. Rev. B **38**, 2297 (1988) .
20. M.W.C. Dharma-wardana, G.C. Aers, D.J. Lockwood and J.-M. Baribeau, Phys. Rev B **41**, 5319 (1990).
21. C. Colvard, T.A. Gant, M.V. Klein, R. Merlin, R. Fischer, H. Morkoç and A.C. Gossard, Phys. Rev. B **31**, 2080 (1985).

SURFACE REACTIONS DURING THE DEPOSITION OF Ge FROM CHEMICAL SOURCES ON Ge(100)-(2×1)

C. MICHAEL GREENLIEF * and JIHONG CHEN **
* University of Missouri-Columbia, Department of Chemistry, Columbia, MO 65211, chemcmg@showme.missouri.edu
** University of Cincinnati, Department of Electrical and Computer Engineering, Cincinnati, OH 45221-0030

ABSTRACT

The adsorption and decomposition of diethylgermane, triethylgermane, and digermane on the Ge(100) surface are investigated with the intent of elucidating the surface processes leading to the deposition of epitaxial Ge films. Room temperature adsorption of diethylgermane or triethylgermane leads to the formation of surface germanium hydrides and ethyl groups. The ethyl groups decompose at higher temperatures and form ethylene via a β-hydride elimination reaction. Isotopic labeling experiments are used to confirm this reaction step. This is in contrast to the Si(100) surface where both α- and β-hydride elimination is observed for the decomposition of surface ethyl groups. The adsorption and reaction of digermane with the Ge surface is also determined to help provide a comparison with the ethylgermanes. Low energy electron diffraction is used to evaluate the quality of the deposited germanium films.

INTRODUCTION

As the development of the semiconductor industry continues, the size of microelectronic and optical devices are smaller and the structure of these devices is more complex. The atomic level understanding of these structures is crucial for continued development of methods for fabrication of these devices. The silicon-germanium system has become increasingly important because of its potential applications in electronic devices [1-7]. Digermane, Ge_2H_6, and germane, GeH_4, are the two most common germanium containing molecular precursors that are used in the chemical vapor deposition (CVD) of Ge [8-20]. However, these two gases are pyrophoric and their handling and disposal are important safety concerns. With these concerns in mind, there is interest in the development of alternative precursors for germanium. Alkylgermanes, such as diethylgermane (GeH_2Et_2) and triethylgermane ($GeHEt_3$), are possible substitutes because they are liquids at room temperature with suitable vapor pressures for CVD, are less reactive with air, and are easier to handle than conventional hydride sources.

The thermal decomposition of mono–, di–, and, tri–ethylgermane on silicon have been studied to explore the possible use of these molecules as CVD and atomic layer epitaxy (ALE) precursors [21-29]. Each of these molecules was able to deposit Ge in an ALE reaction cycle. H_2 and C_2H_4 were observed as the gas phase. A similar study examining the surface chemistry of GeH_2Et_2 on Ge(100) was recently reported [30]. Again, as on silicon, surface ethyl groups were detected. These ethyl groups decomposed to form ethylene. However little insight as to the formation of ethylene

Mat. Res. Soc. Symp. Proc. Vol. 448 © 1997 Materials Research Society

was given. The pathway(s) for ethylene desorption from the decomposition of surface ethyl groups on Si(100) and Si(111) were examined previously [31, 32]. Two possible pathways for ethyl group decomposition were determined. One reaction sequence was α-hydride elimination followed by a hydrogen shift to generate ethylene. The other pathway was β-hydride elimination. On silicon surfaces, these previously obtained results showed that both α-hydride and β-hydride elimination are involved in ethylene formation.

In this paper we report the reactions of diethylgermane (GeH_2Et_2), triethylgermane ($GeHEt_3$), and digermane (Ge_2H_6) on the Ge(100) surface. These studies provide a direct comparison to those previously completed on silicon surfaces. Room temperature exposure of GeH_2Et_2 or $GeHEt_3$ to Ge(100) results in dissociative adsorption of each molecule and the formation of surface ethyl groups. These ethyl groups decompose at higher temperatures to form C_2H_4. H_2 is also observed as the other desorption product.

EXPERIMENT

The experiments are performed in a custom–built stainless steel UHV chamber and the chamber is pumped by a 500 ℓ/s turbomolecular pump (Balzers TPU 520). The main chamber is equipped with a quadrupole mass analyzer (UTI model 100C) for residual gas analysis and TPD studies, a set of reverse–view low energy electron diffraction (LEED) optics (Princeton Research Instruments), and an ion gun for sputtering. The base pressure of the system is 1.5×10^{-10} torr with a typical working pressure of 2.0×10^{-10} torr.

The samples are cleaved into $10 \times 25 \times 0.2$ mm rectangles from n-type Ge(100) substrates (Eagle–Picher, $\pm 0.5°$ of the (100) plane, Sb doped, 0.04–0.4 $\Omega \cdot$cm resistivity). Sample preparation and mounting procedures are discussed in detail elsewhere [33].

Digermane (Voltaix, ultrahigh purity grade, minimum purity 99.999%), GeH_2Et_2 (Gelest, purity >98%), and $GeHEt_3$ (Gelest, purity >98%) are further purified by several freeze-pump-thaw degassing cycles and the purity of the gases is checked *in–situ* by mass spectrometry. The exposures are reported as Langmuirs ($1L=10^{-6}$ torr·sec) as measured directly by the ion gauge.

Temperature programmed experiments are conducted with a linear temperature ramp of 5 K s^{-1} with the crystal in line-of-sight of the quadrupole mass spectrometer. The coverage at saturation for hydrogen on Ge(100) at room temperature has been determined as 1 ML (1 ML= one adsorbate per surface Ge atom) [30, 34-37]. The temperature programmed desorption area from a saturation coverage of H atoms is then used as a internal standard for H_2 thermal desorption.

RESULTS AND DISCUSSION

Figure 1 presents the results of a series of thermal desorption experiments using digermane as the adsorbate. In these experiments, Ge_2H_6 is exposed to Ge(100) at a surface temperature of 375 K. The Ge(100) sample is then heated and the desorbing gases are monitored by mass spectrometry. Hydrogen is the only desorbing species observed. At the lowest Ge_2H_6 exposures (Fig. 1a), a single desorption state is detected at 570 K. The intensity of this state increases with increasing Ge_2H_6 exposure.

The peak area saturates for exposures greater than 900 L. The hydrogen desorption resulting from the decomposition of Ge_2H_6 is at the same temperature as that observed for atomic hydrogen exposures to Ge(100) [38]; meaning that the decomposition of Ge_2H_6 leads to the same surface hydrides that are created upon atomic hydrogen exposure to Ge(100). The LEED pattern after H_2 desorption is a sharp (2×1) pattern, indicating a high quality crystalline surface remains after Ge deposition.

Figure 2 presents a series of thermal desorption spectra resulting from the decomposition of GeH_2Et_2 on Ge(100). Only two desorbing species, hydrogen and ethylene, are observed as desorption products. The C_2H_4 desorption spectra (dots, lower portion of Fig. 2) contain a single peak at 650 K. This feature increases with increasing GeH_2Et_2 exposures and saturates for exposures greater than 600 L. The H_2 desorption states (squares and upper portion of Fig. 2) are wider than the C_2H_4 desorption states and are also wider than those observed for H_2 from Ge_2H_6 decomposition as shown in Fig. 1. At the highest exposures (Fig. 2f), the H_2 desorption peak maxima is coincident with the C_2H_4 desorption at 650 K. This is 80 K higher in temperature than normal H_2 desorption from Ge. There is also a shoulder to the low temperature side of the H_2 desorption peak. At lower GeH_2Et_2 exposures (Fig. 2b), two H_2 peaks can be resolved, with one peak maxima near 580 K and the second

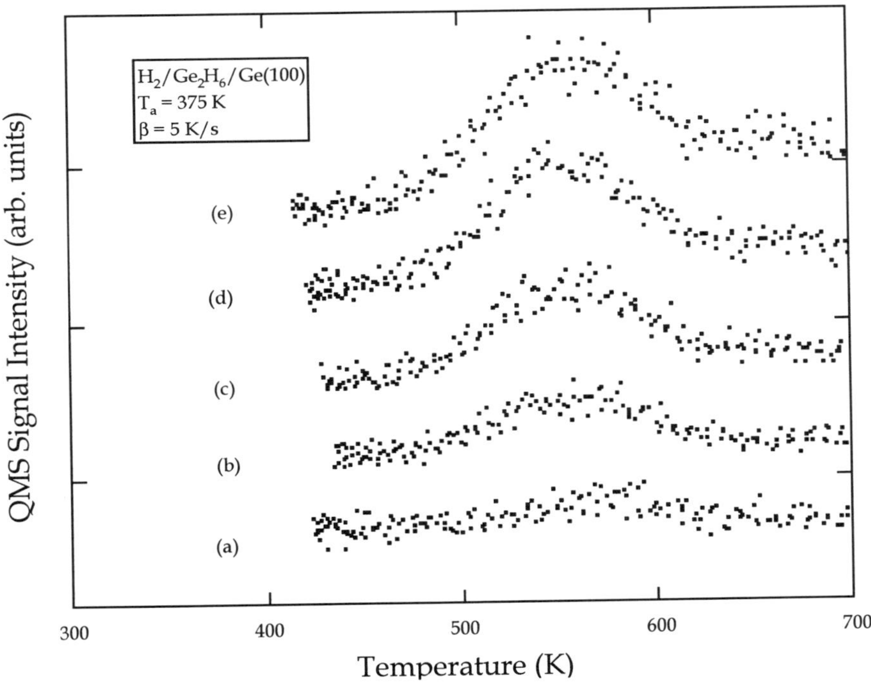

Figure 1: TPD spectra of H_2 following different exposures of Ge_2H_6 at a surface temperature of 375 K. The exposures shown are (a) 0.09, (b) 9, (c) 90, (d) 390, and (e) 900 L.

peak maxima at 650 K.

A series of TPD experiments using GeHEt$_3$ were also completed. Qualitatively, the H$_2$ and C$_2$H$_4$ desorption spectra are the same as those obtained for GeH$_2$Et$_2$. The main difference between the two sets of experiments is that the C$_2$H$_4$ desorption peak area is greater with GeHEt$_3$ for a given exposure compared to GeH$_2$Et$_2$. The H$_2$ peak areas show the same trend, but in the opposite direction.

In these thermal desorption studies of GeH$_2$Et$_2$ and GeHEt$_3$ on Ge(100), no parent fragments were detected in the TPD experiments. This result indicates that the incident precursor molecules dissociatively chemisorb on the surface and is consistent with the dissociative adsorption of GeH$_2$Et$_2$ on Ge(100) observed previously by high resolution electron energy loss spectroscopy (HREELS) [30]. One peak is observed at 570 K and is due to the recombinative desorption of H atoms. The second peak at

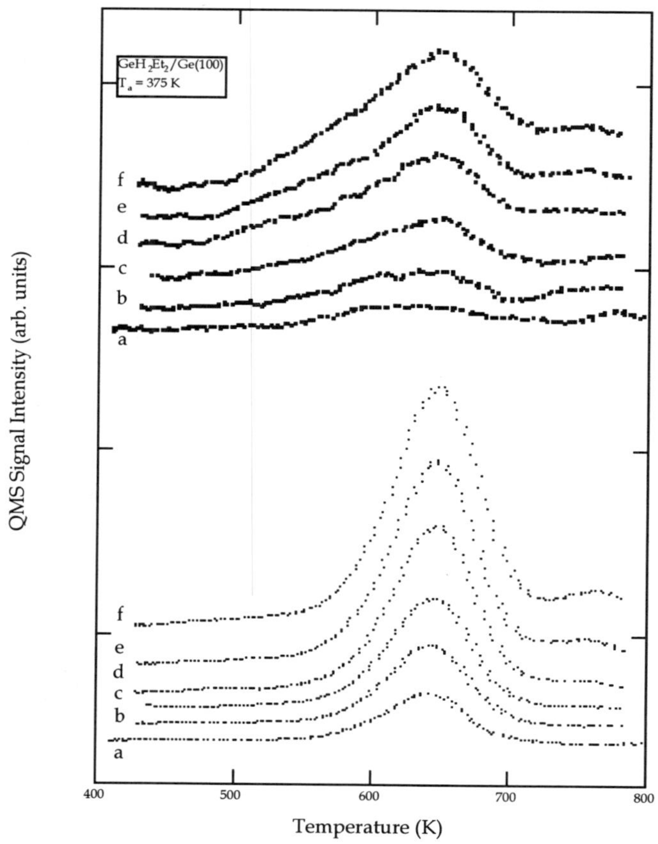

Figure 2: TPD spectra of C$_2$H$_4$ (small dots, lower portion) and H$_2$ (squares, upper portion) following different exposures of GeH$_2$Et$_2$ at a surface temperature of 375 K. The exposures shown are (a) 0.09, (b) 0.9, (c) 9, (d) 90, (e) 300, and (f) 600 L.

650 K is due to the desorption of H atoms generated by hydride elimination of surface ethyl groups [33]. Isotopic labeling experiments [33], have shown that the ethyl group decomposition occurs by β–hydride elimination. If the two H atoms attached to Ge in GeH_2Et_2 are transferred to Ge(100) upon adsorption, two peaks should be observed in TPD for H_2 with the same area because the hydride elimination of the two ethyl groups will generate two H atoms. In the H_2 TPD results, there is only one peak at 650 K with a low temperature shoulder. This peak was fit by two Gaussian peaks at 575 K and 645K. The area for the peak at 570 K is only 26% of the peak at 645 K. In the $GeHEt_3$ TPD experiments, the low temperature shoulder in the H_2 desorption peak at 650 K is smaller than that observed for GeH_2Et_2. Fitting the H_2 desorption spectrum in this case revealed that the peak area at 590 K is 9% of that at 655 K. These results suggest that GeH_2Et_2 and $GeHEt_3$ dissociatively adsorb as more than one organometallic species; such as $GeH(C_2H_5)_x$ ($x = 0 - 3$). If GeH_2Et_2 and $GeHEt_3$ transfer small amounts of H to the surface upon adsorption to the Ge(100) surface, and the adsorbed organometallic species decompose at 650 K, the above H_2 TPD spectra can be obtained.

CONCLUSIONS

We have studied the surface chemistry of GeH_2Et_2 and $GeHEt_3$ on the Ge(100) surface. The compounds dissociatively adsorb producing ethyl groups and adsorbed hydrogen. As the surface is heated, the hydrogen produced during the adsorption step desorbs at 570 K, the temperature for the desorption of hydrogen from a monohydride covered Ge(100) surface. The ethyl groups undergo β–hydride elimination at 650 K, leave the surface as ethylene, and deposit atomic hydrogen on the surface. The hydrogen immediately recombines to form molecular H_2 and desorbs producing simultaneous C_2H_4 and H_2 desorption peaks.

ACKNOWLEDGMENTS

The authors acknowledge the National Science Foundation (Grant CHE-9100429) and the University of Missouri Research Board for support of this research. CMG also acknowledges the National Science Foundation for a Young Investigator Award.

REFERENCES

1. M. L. Green, D. Brasen, H. Temkin, R. D. Yadvish, T. Boone, L. C. Feldman, M. Geva, and B. E. Spear, *Thin Solid Films* **184**, 107 (1990).
2. S. S. Iyer, G. L. Patton, J. M. C. Stork, B. S. Meyerson, and D. L. Harame, *IEEE Trans. Electron. Dev.* **36**, 2043 (1989).
3. S. S. Iyer, G. L. Patton, D. L. Harame, J. M. C. Stork, E. F. Crabbé, and B. S. Meyerson, *Thin Solid Films* **184**, 153 (1990).
4. T. P. Pearsall, *CRC Crit. Rev. Solid State Mater. Sci.* **15**, 551 (1989).
5. J. Welser, J. L. Hoyt, and J. F. Gibbons, *Jap. J. Appl. Phys. Pt. 1* **33**, 2419 (1994).
6. K. Werner, *et al., J. Crys. Growth* **164**, 223 (1996).
7. G. L. Zhou, and H. Morkoç, *Thin Solid Films* **231**, 125 (1993).
8. P. D. Agnello, T. O. Sedgwick, M. S. Goorsky, J. Ott, T. S. Kuan, and G. Scilla, *Appl. Phys. Lett.* **59**, 1479 (1991).

9. M. Cao, A. W. Wang, and K. C. Saraswat, *Proc. Electrochem. Soc.* **93-6**, 350 (1993).
10. D. Dutartre, P. Warren, I. Berbezier, and P. Perret, *Thin Solid Films* **222**, 52 (1992).
11. P. M. Garone, J. C. Strum, and P. V. Schwartz, *Appl. Phys. Lett.* **56**, 1275 (1990).
12. S. M. Jang, and R. Reif, *Appl. Phys. Lett.* **59**, 3162 (1991).
13. T. I. Kamins, and D. J. Meyer, *Appl. Phys. Lett.* **59**, 178 (1991).
14. H. Kühne, T. Morgenstern, P. Zaumseil, D. Krüger, E. Bugiel, and G. Ritter, *Thin Solid Films* **222**, 34 (1992).
15. B. S. Meyerson, K. J. Uram, and F. K. LeGoues, *Appl. Phys. Lett.* **53**, 2555 (1988).
16. M. Racanelli, and D. W. Greve, *Appl. Phys. Lett.* **56**, 2524 (1990).
17. D. J. Robbins, J. L. Glasper, A. G. Cullis, and W. Y. Leong, *J. Appl. Phys.* **69**, 3729 (1991).
18. T. O. Sedgwick, and P. D. Agnello, *J. Vac. Sci. Technol. A* **10**, 1913 (1992).
19. M. Suemitsu, F. Hirose, and N. Miyamoto, *J. Crys. Growth* **107**, 1015 (1991).
20. Y. Zhong, M. C. Öztürk, D. T. Grider, J. J. Wortman, and M. A. Littlejohn, *Appl. Phys. Lett.* **57**, 2092 (1990).
21. A. Mahajan, B. K. Kellerman, N. M. Russell, S. Banerjee, A. Campion, J. G. Ekerdt, A. Tasch, J. M. White, and D. J. Bonser, *J. Vac. Sci. Technol. A* **12**, 2265 (1994).
22. P. A. Coon, M. L. Wise, A. C. Dillon, and S. M. George, *Mater. Res. Soc. Symp. Proc.* **282**, 413 (1993).
23. A. C. Dillon, M. B. Robinson, and S. M. George, *Surf. Sci.* **286**, L535 (1993).
24. P. A. Coon, M. L. Wise, and S. M. George, *J. Chem. Phys.* **98**, 7485 (1993).
25. P. A. Coon, M. L. Wise, Z. H. Walker, S. M. George, and D. A. Roberts, *Appl. Phys. Lett.* **60**, 2002 (1992).
26. C. M. Greenlief, D.-A. Klug, and L. A. Keeling, *Mater. Res. Soc. Symp. Proc.* **282**, 427 (1993).
27. Y. Takahashi, H. Ishii, and K. Fujinaga, *J. Electrochem. Soc.* **136**, 1826 (1989).
28. W. Du, L. A. Keeling, and C. M. Greenlief, *J. Vac. Sci. Technol. A* **12**, 2281 (1994).
29. L. A. Keeling, Ph.D., University of Missouri (1995).
30. A. Mahajan, B. K. Kellerman, J. M. Heitzinger, S. Banerjee, A. Tasch, J. M. White, and J. G. Ekerdt, *J. Vac. Sci. Technol. A* **13**, 1461 (1995).
31. D.-A. Klug, and C. M. Greenlief, *J. Vac. Sci. Technol. A* **14**, 1826 (1996).
32. L. A. Keeling, L. Chen, C. M. Greenlief, A. Mahajan, and D. Bonser, *Chem. Phys. Lett.* **217**, 136 (1994).
33. J. Chen, and C. M. Greenlief, *J. Vac. Sci. Technol. A*, submitted.
34. S. M. Cohen, Y. L. Yang, E. Rouchouze, T. Jin, and M. P. D'Evelyn, *J. Vac. Sci. Technol. A* **10**, 2166 (1992).
35. M. P. D'Evelyn, S. M. Cohen, E. Rouchouze, and Y. L. Yang, *J. Chem. Phys.* **98**, 3560 (1993).
36. M. P. D'Evelyn, Y. L. Yang, and S. M. Cohen, *J. Chem. Phys.* **101**, 2463 (1994).
37. L. B. Lewis, J. Segall, and K. C. Janda, *J. Chem. Phys.* **102**, 7222 (1995).
38. L. Surnev, and M. Tikhov, *Surf. Sci.* **138**, 40 (1984).

PHYSICS AND CONTROL OF Si/Ge HETEROINTERFACES

S.Fukatsu*, N.Usami**, H.Sunamura**, Y.Shiraki**, and R.Ito***

Department of Pure and Applied Sciences, The University of Tokyo,
3-8-1 Komaba, Meguro-ku, Tokyo 153, Japan, fkatz@srv.bme.rcast.u-tokyo.ac.jp
**RCAST, The University of Tokyo, 4-6-1 Komaba, Meguro-ku, Tokyo 153, Japan
***Department of Applied Physics, Faculty of Engineering, The University of Tokyo,
7-3-1 Hongo, Bunkyo-ku, Tokyo 113, Japan

ABSTRACT

We describe physics and control of Si/SiGe heterointerfaces. A clear distinction will be made between the vertical and lateral effects of the Si/SiGe interface from the viewpoint of interface engineering. Ge surface segregation during nonequilibrium MBE growth and surfactant-mediated-growth are highlighted as prominent examples for the vertical effects while interface microroughness is addressed for the lateral effects. The influence of the interface effects on radiative recombination of indirect excitons is described in the context of SiGe-based optoelectronic applications.

I.Introduction

Creating a sharp Si/Ge interface has been the target of extensive research[1-14] presumably directed toward future device applications where band engineering is expected to play the central role. A kinetic limitation placed on the motions of atoms is one way to defeat thermodynamic obstacles in growth whereas, in view of crystallinity degradation, a growth mode close to the equilibrium condition is desirable. Most of previous studies have dealt only with the vertical directions so far, it has gained a general recognition that the lateral interface is of significance when discussing electronic and optical properties of heterointerfaces. A better understanding of kinetics and energetics of interface formation, and a good command of growth techniques are necessary to further advance the potential of interface engineering to a level for interface-based band engineering of heterostructures.

In this paper, physics and control of Si/SiGe heterointerfaces are described in reference to the vertical and lateral effects from the perspective of interface engineering. The influence of the both interface effects on radiative recombination of indirect excitons is studied in the context of SiGe-based optoelectronic applications.

II.Experimental

The samples were grown either by solid-source molecular beam epitaxy (MBE) [VG Semicon V80M] [5,6]or by gas-source molecular beam epitaxy (Daido Hoxan VCE S-2020)[15] on nominally on-axis Si(100) wafers. Elemental Si was evaporated from an electron beam evaporator while effusion cells were used for deposition of Ge, Sb, Ga, Pr, and Er. The growth rate was controlled by timed-exposure. Atomic hydrogen (deuterium) was produced by pyrolysis on a heated tungsten filament. The sample temperature (T_S) was monitored by a thermocouple to an accuracy ± 20 °C. X-ray photoemission spectroscopy (XPS) was performed in vacuo. SIMS depth profiling was done ex situ using either oxygen or cesium primary ions. Photoluminescence (PL) was recorded by a standard lock-in technique using a visible laser.

Mat. Res. Soc. Symp. Proc. Vol. 448 © 1997 Materials Research Society

III. Results and discussion

III-1. Vertical interface effects

III-1-1. Revisiting the Ge surface segregation during MBE

As is well documented in the literature, one observes the Ge surface segregation when growing Si/Ge heterointerfaces by solid-source MBE. The Ge segregation has turned out to be describable reasonably well by the two-state exchange model or surface bilayer model analogous to those established for dopant segregation [1-12]. The problem of these simple-minded schemes is the assumption of ballistic or instantaneous deposition of the Si cap. Recently, further sophistication of calculation has met success to some extent[9,16]. However, the essence is almost fully contained in the surface bilayer model and hence a more advanced approach will be left over as the subject of future study. Instead we will focus on the surface bilayer model and draw important results in the perspective of thermodynamics which are directly derived from that model.

III-1-2. Thermodynamic interpretation of "self-limiting" site exchange

A cut-away view along <110> of the surface and subsurface sites is illustrated in Fig.1. As was already pointed out by the authors, the key feature of the Ge surface segregation within the surface bilayer model is the "self-limiting" site exchange [5,7] that distinguishes the segregation of Ge from those of dopants. The point is that only a Si-Ge exchange contributes to the Ge segregation. Conversely, a Ge-Ge exchange event has only the effect equivalent to an enhanced kinetic limitation on the Ge segregation. Standard energy diagram is assumed for simulation and the relevant energies are the kinetic barrier E_a and the segregation energy E_b. Let n_1 and n_2 denote the population of surface and subsurface Ge, respectively, and the rate equation for the kinetics controlling the Ge site exchange becomes of the form,

$$\frac{\partial n_1}{\partial t} = -2pn_1 (1-n_2) + 2qn_2 (1-n_1) , \tag{1}$$

with n_1+n_2=const. (mass balance) unless the loss due to Ge thermal desorption is pronounced. Note that $p=f_0 \exp(-E_a/kT)$ and $q=p \exp(-E_b/kT)$ where f_0 is an exchange frequency.

As the temperature is raised, the thermal equilibrium is reached early. Then the Ge segregation becomes insensitive to the kinetic barrier, E_a. This is true for $T_s>450°C$ under normal MBE growth conditions. The steady-state solution rather than that at thermal equilibrium is obtained by putting l.h.s. of (1) to zero, or $\partial/\partial t \equiv 0$. One finds after a simple calculus,

$$\frac{n_2}{1-n_2} = \frac{n_1}{1-n_1} \exp(\frac{E_b}{kT}) , \tag{2}$$

i.e., the formula for the equilibrium segregation of a binary alloy. On the other hand, this is readily obtained from thermodynamic arguments of free energy F containing the entropy of mixing,

$$F = n_1 E_b - kT \{-n_2 \ln n_2 -(1-n_2) \ln(1-n_2) - n_1 \ln n_1 -(1-n_1) \ln(1-n_1)\}. \tag{3}$$

The bracket contains the configurational (mixing) entropy terms. Differentiating Eq.(3) with respect to n_1 and putting $\partial F/\partial n_1 =0$ with $\partial n_2 /\partial n_1 =-1$ (normalization) yields Eq.(2). Therefore, it is seen that the "self-limitation" drawn from the purely kinetic argument is of physical significance and the

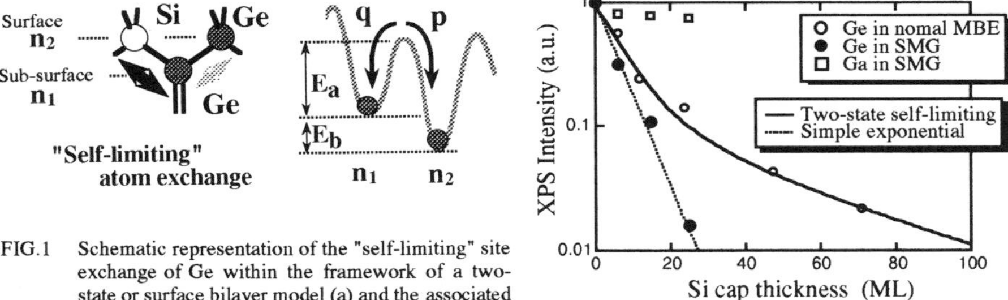

FIG.1   Schematic representation of the "self-limiting" site exchange of Ge within the framework of a two-state or surface bilayer model (a) and the associated energy diagram relevant to Ge surface segregation.

FIG.2   Depth profiles of Ge and Ga measured by XPS. Solid line has been drawn after simulation using the surface-bilayer model with self-limiting site swap of Ge. Broken line is a single exponential fit. SMG stands for surfactant-mediated-growth.

roles of the corresponding terms are taken over by the configurational entropy in the thermodynamic representation. Importantly, these terms are *not omissible* in the kinetic formalism within the surface bilayer model in general.

The relationship of the atom site exchange for dopants has been shown to be of the form [12],

$$\frac{n_2}{n_1} = \exp\left(\frac{E_b}{kT}\right). \tag{4}$$

Essentially, Eq.(4) is a reduced form of Eq.(2) in the dilute limit of $n_{1(2)}$.

III-1-3. Surfactant-mediated-growth

The conventional wisdom to create abrupt Si/SiGe and Si/Ge interfaces is to deposit a functional adlayer of a third element known as "surfactant" [2]. The corresponding MBE growth using surfactants has been referred to as either surfactant-mediated-growth (SMG) or segregant-assisted-growth [17] depending on their functions. In Si/Ge growth, SMG is performed by depositing a surfactant adlayer of appropriate amount prior to growth. Since most of surfactants are of strong segregation tendency, e.g., Sb[6], Ga[3], Bi[10], a single deposition rather than continuous supply is sufficient to warrant the continuous operation of SMG unless the incorporation loss of surfactants into the overgrown layer is pronounced.

A. Ga

The compositional profiles of Ge can be monitored in situ or ex situ. A typical result of in situ XPS profiling is shown in Fig.2. The Si cap thickness is expressed in units of monolayers (MLs). The growth temperature was 400°C and the Si growth rate was 1.0Å/s. The solid line has been drawn by simulating cumulative XPS counts with standard layer-by-layer deposition model using Eq.(2). The Ge profile (open circles) corresponding to the surface segregation is well explained. Another Ge trace under SMG using Ga shown by filled circles is describable by a single exponential decay with a 1/e decay length of 8Å, indicating a sharp Si/Ge interface. The almost leveled Ga profile indicates that Ga atoms stay on the surface during SMG. Figure 3 shows SIMS depth profiles of Ga and Ge in a 8-period Si/Ge superlattice. Si is 300Å while Ge is 1 ML. The Ga was supplied at the layer with the arrow. The higher magnitudes of profile slopes for the last 4-layers shown in the

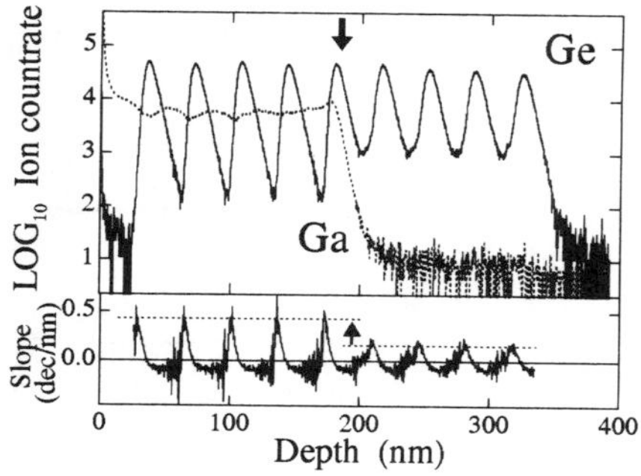

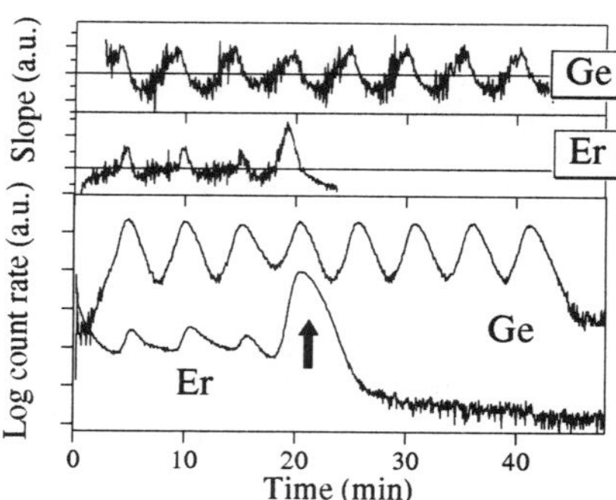

FIG.3 (a) SIMS profiles of Ge, Si and Ga in a 8-period Si(300Å)/Ge(3 ML) strained-layer superlattice grown on Si(100). The last four Ge layers were grown by SMG using Ga of 0.5ML. The arrow indicates the position of the Ga adlayer deposition. (b) Logarithmic derivative of the Ge profile in (a). The arrow indicates the profile sharpening of 2.5 times for the Ge layers grown by SMG using Ga.

FIG. 5 Ge and Er SIMS profiles in a Ge(1ML)/Si(300Å) superlattice grown on Si(100). The absence of changes in slope following the Er deposition indicates that Er does not function as a surfactant though Er is one of strong segregants in Si.

lower panel are evidence of the interface sharpening following the Ga-surfactant deposition. Clearly, the SMG is continuously operable for more than 150nm in thickness without appreciable damping. However, the Ga doping is evident as well. The level of the unintentional Ga doping was measured to amount to 10^{17}-10^{19}/cm^3. These characteristics are analogous to the SMG using Sb.

B. Atomic hydrogen (deuterium)

Recently, the authors and co-workers have reported the SMG using atomic hydrogen (AH) and deuterium (AD)[18]. The latter was used to track down the H(D) incorporation. The choice of hydrogenic surfactant is appropriate in view of meager electrical activity unlike Sb nor Ga when

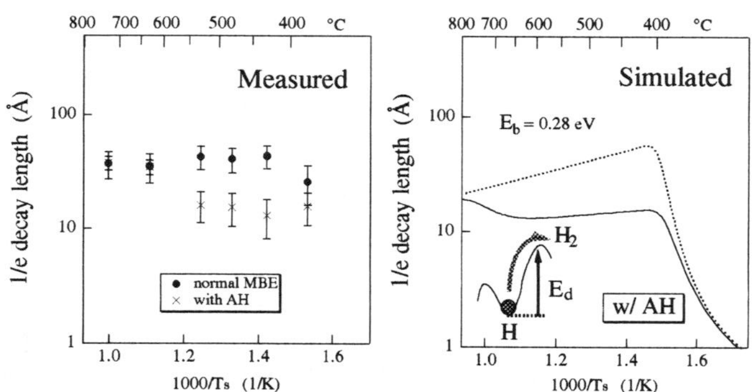

FIG.4 (a) Experimental Ge decay lengths vs substrate temperature as logarithmic derivatives of SIMS profiles. The sample was a Ge(1ML)/Si(300Å) superlattice grown on Si(100) using AH as surfactant. The surfactant] role of AH is weakened at high temperature due to thermal recombinative desorption.
(b) Theoretical Ge decay lengths simulated by using the relevant parameters E_a=1.63eV, E_b=0.28eV and the desorption energy E_d of 1.84eV with trial frequency of 10^9s^{-1}. Agreement with the experiment is reasonable.

128

incorporated[11,19,20]. However, the bottom line is that AH(AD) is not of segregation tendency and therefore continuous supply is necessary. Moreover, SMG using AH(D) suffers from thermal desorption at high temperature while a degradation of crystallinity is inevitable below 400˚C. Nevertheless, the SMG using AH(D) is effective since the intrinsic nature of the Ge segregation makes it pronounced for the range of T_s=400-450˚C in the normal MBE growth[5].

The $1/e$ decay lengths of Ge atoms are summarized as a function of temperature in Fig.4. The decay lengths were obtained by taking the logarithmic slope. Theoretical values were obtained by simulating Ge compositional profiles using Eq.(2). The model used here is that an adsorbed AH(D) atom hinders the Ge site exchange by terminating the surface. The energy values used are E_a=1.63eV, E_b=0.28eV and the desorption energy E_d of 1.84eV with a trial frequency of $10^9 s^{-1}$. In the experiments, H_2 pressure was 5×10^{-5}Torr with the corresponding hydrogen coverage of 1ML in 10s [16]. The overall agreement between the experiment and theory is reasonable. The increase of decay lengths for Ts>600˚C is readily attributed to the AH desorption.

C. Limitation of segregant-assisted-growth

The terminology of segregant-assisted-growth is based on the recognition of the driving force. Figure 5 shows SIMS depth profiles of Er and Ge deposited in a similar way to the SLS shown in Fig.3. The observation of unchanged Ge slopes before and after Er deposition at the 5th Ge layer is clearly in conflict with the picture that the weaker segregant is defeated by the second strong segregant. Clearly, Er does not function as a surfactant though Er is a segregant of strong tendency in Si compared to Ge. Therefore, it is inferred that the SMG operation is not a competitive kinetic processes of two segregants with differential segregation tendencies, but rather an argument taking account of appropriate energetics is necessary to understand the underlying mechanism.

III-1-4. Impact of the vertical effects on optoelectronics

A perfect Si/Ge interface morphology in the growth direction is the starting point of discussion of optoelectronic applications. The so-called Si/Ge strain-symmetrized type-II short-period superlattices (s-SLSs) represent a prominent example that requires such properties as atomically abrupt interface transient as well as perfect periodicity [21,22]. Conversely, s-SLSs are vulnerable to vertical interface disorder such as layer-width fluctuation and loss of periodicity.

In general, the monolayer fluctuation is controlled by thermodynamics and therefore should more or less accompany any type of interfaces. Consider a s-SLS consisting of a few monolayers of Si and Ge. If there is a monolayer fluctuation of ±1ML, it is enough to give a large inhomogeneous broadening of the ground state energy, ΔE. On the other hand, the miniband width U is usually small due to the heavy electron mass of $0.916 m_0$ in the growth direction where m_0 is the free-electron mass. Therefore the criterion of Anderson localization $\Delta E \gg U$ is readily fulfilled. This results in disruption of the superlattice effect, i.e., zone-folding, and s-SLS changes into a disordered SLS (d-SLS). Thus, what is observed in experiments is nothing but the PL due to localized excitons.

The envelopes of localized wavefunctions in a model calculation are schematically shown in Fig.6. Clearly, the conduction and valence band wavefunctions are spatially separated from each other in the presence of disorder. As a matter of fact, such split envelope functions are an analogue of disordered superlattices. The resultant small wavefunction overlap proportional to the oscillator

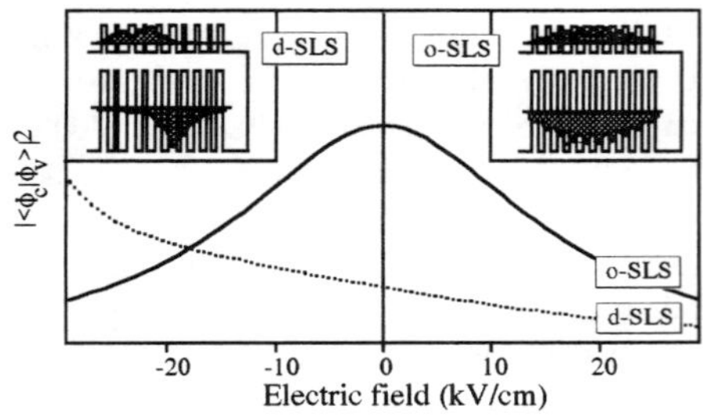

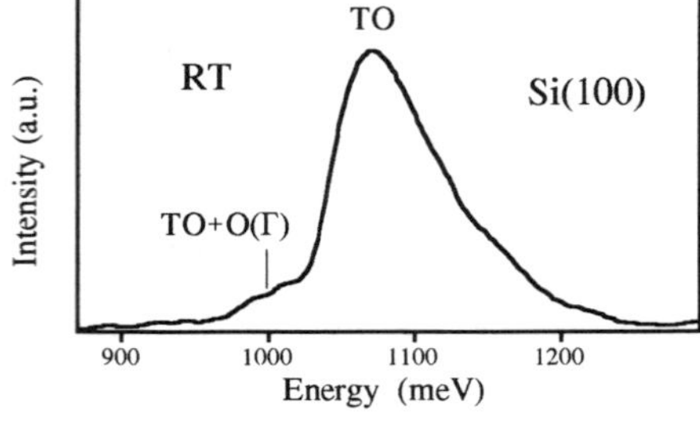

FIG.6   Theoretical variations of wavefunction overlaps under a vertical electric field in a disordered (d-) and ordered (o-) short-period Si/Ge superlattices (SLS). The insets show schematic representations of wavefunction envelopes.

FIG.8   Room temperature PL spectrum of Si. No-phonon transition is absent.

strength may account for rather weak intensities of PL of localized excitons as reported previously although the exciton localization is expected to significantly reduce nonradiative contributions. This could be verified by observing a monotonic increase of PL by applying electric fields vertically. The simulated results are shown in Fig.6. Such variations of PL have also been recently observed in type-I strained Si/SiGe alloy disordered superlattices by the authors.

III-2. Lateral effects

The well-known example of the lateral effects of the Si/Ge interface could be Ge islanding or lateral clustering in the context of the original motivation of the SMG[23,24]. In Ge islanding, the vertical effects also come into play due to its cross-sectional morphology. In this sense, lateral "morphological" effects should be discussed along with appropriate vertical effects. The Ge islanding is known to set in when the Ge thickness exceeds the critical thickness. The kinetic critical thickness of 3.7ML [24] was observed for gas-source grown Ge/Si while the thermodynamic value has been found to be 3.0ML after prolonged growth interruption experiments [25].

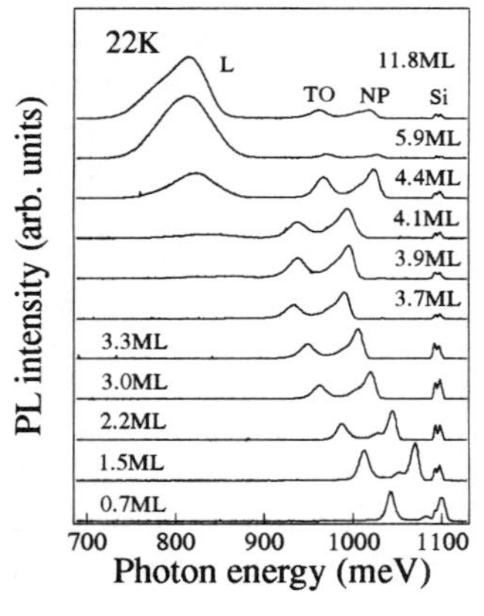

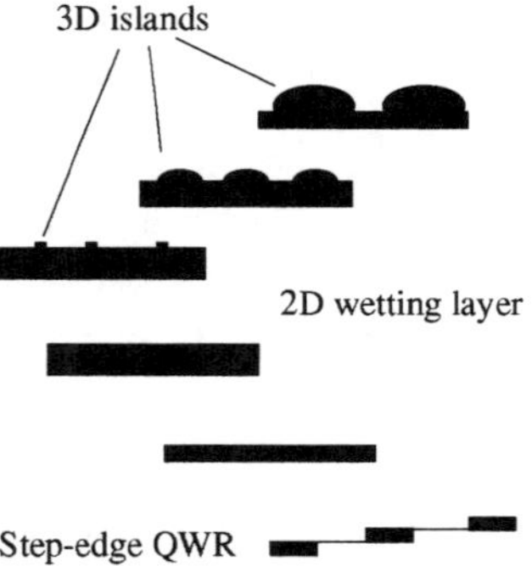

FIG.7   22-K PL spectra and the corresponding cross-sectional morphologies of Si/Ge/Si type-II quantum wells with various Ge coverages. The doublet peaks observed for all Ge coverages are no-phonon (NP) and phonon-aided (TO) transitions of two-dimensional wetting layers while the broad bands at lower energies are PL due to 3-dimensional Ge islands or quantum dots.

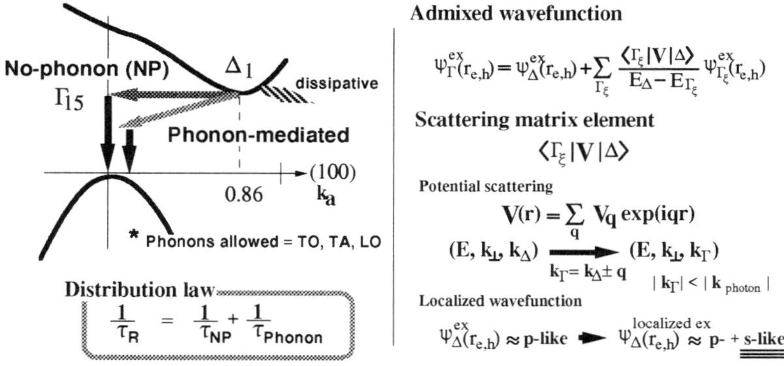

FIG.9   Schematic diagram of indirect-gap interband transitions. No-phonon transition requires an appropriate scattering momentum for the electron wavefunction to have an s-like parity. The zone center electron (exciton) wavefunction is constructed from the orthonormal set of Bloch functions at the Γ point.

From the optoelectronic point of view, especially from the perspective of control over interband transitions, the Ge islands are of particular interest for use as quantum dots since the valence band offset amounts to 840meV. PL spectra and the corresponding cross-sectional morphologies are summarized as a function of Ge coverage in Fig.7. Note that Si-lattice-matched Si/Ge QWs are of type-II band lineup where the electrons are rejected by the Ge layer. The pairwise double peaks that change emission energies systematically with changing Ge coverage are the signatures of 2-dimensional wetting layers. These peaks are attributed to the no-phonon (NP) transition and its TO-phonon-replica. The broad peaks in the low energy regime are the PL from Ge quantum dots or islands. The PL from these islands is fairly robust and survives up to room temperature, reflecting exciton localization due to a deep trapping potential.

Aside from the interest in the quantum confinement shift, the puzzling observation of Fig.7 is the development of the NP peaks [26]. In Si where NP peaks are absent particularly at high temperature, the momentum necessary to establish a dipole-allowed interband transition is mostly taken up by TO-phonons and then the PL is due to the TO-phonon-aided transition as shown in Fig.8. Following the discussions of the schematic diagram and captions Fig.9, the origin of the NP peak should be associated with admixture of the electron (exciton) wavefunction at Δ valleys along a <100>-equivalent axis and the zone center wavefunction. In other words, appropriate scattering should take place or localized wavefunction should contain an s-like character in the longitudinal direction in addition to otherwise p-like character of the electron (exciton) wavefunction.

Since the k-dispersion in the vertical direction is totally missing except for very thin or thick Ge layers because of the quantum confinement, a Δ–Γ scattering event must occur along the lateral direction. As illustrated in Fig.10, the observation of NP peaks for Si/Ge quantum wells indicates the presence of microscopic roughness which is distributed laterally while sharply localized at the heterointerface. The roughness is expected to be short-ranged so that the wavelength, i.e., correlation, is comparable to the unit-cell size. Here the type-II band alignment plays another important role for a Δ–Γ scattering to occur since the Hartree potential arising from the holes

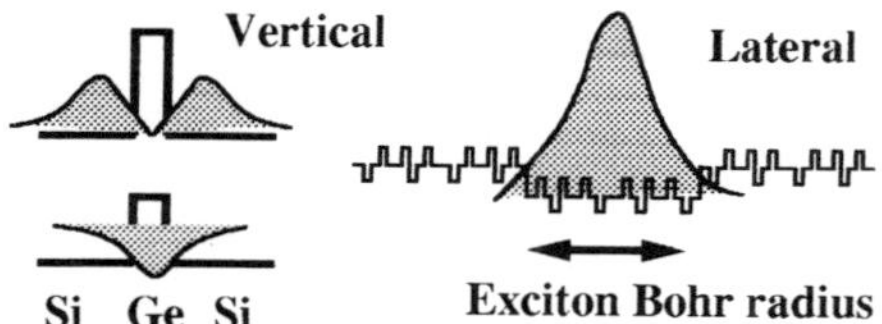

FIG.10 Schematic diagram of microroughness at the heterointerface sensed by indirect excitons. Note the k-dispersion is absent in the vertical direction for a quantum confinement system.

confined in the Ge layer forces the electrons to sense the heterointerface. Within this picture, excitons are allowed to move along the interface plane almost freely thereby giving a normal PL line width while NP transitions develop due to the electron $\Delta-\Gamma$ scattering along one of <100>-equivalent axes, which occur within the exciton Bohr radius.

It is further noted in Fig.7 that the NP peak intensity for the 0.7-ML quantum wires (bottom trace) is diminished although the hole confinement at Ge and the type-II band lineup are still maintained. This can also be understood from the above scenario. Since the Ge wires are aligned only to <110> step-edge directions [27], a $\Delta-\Gamma$ scattering event along <100>-equivalent directions is less likely, and therefore only a weak contribution of NP transition is observed experimentally.

From the foregoing discussion, the interface-based engineering of indirect optical transitions seems to be promising if we take advantage of the type-II lineup and the lateral $\Delta-\Gamma$ scattering of the electrons. The extreme example of this is shown in Fig.11 where a Si quantum well of the electrons is placed adjacent to the SiGe quantum well for the holes with a type-II lineup (ACP/NCS) using a graded SiGe relaxed buffer [28,29]. A 5-fold increase of the NP peak for is apparent as compared to the symmetric type-II Si quantum well where the interface effect is less pronounced. Although the detailed mechanism is not identified for the NP enhancement, previous studies indicate strongly that it is closely related with the interface properties and the PL is due to radiative recombination of localized excitons. Line width characteristics, prolonged decay lifetimes and their temperature dependence as well as the reduction of the NP intensity after interface grading (diffusion) and a further enhanced NP peak by insertion of an islanded Ge layer have clearly supported the argument of the relevance of the interface-localized excitons and associated $\Delta-\Gamma$ scattering.

Once the interface localization and hence an enhancement of the NP transition is established, a technique based on quantum electrodynamics will be ready to be applied. The microcavity effects and the control of spontaneous emission process may be exploited for optoelectronic applications [31].

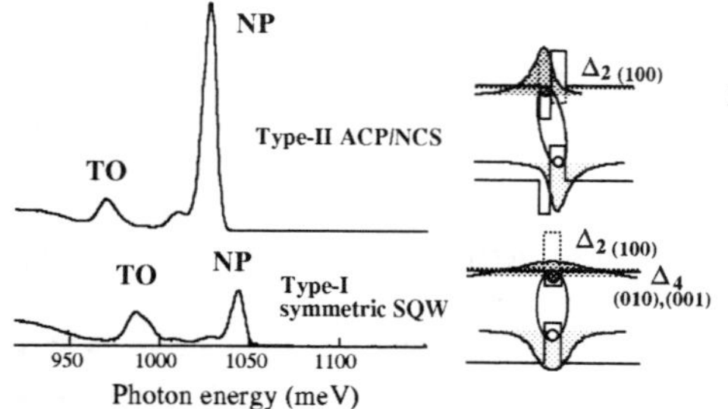

FIG.11 PL enhancement with interface-localized excitons which sense microroughness of an asymmetric confinement potential established by two neighboring confinement layers for the electrons and holes. The quantum wells were grown on a step-graded relaxed SiGe(x=0.18) buffer. A 5-fold increase of NP intensity is clearly observed as compared to the symmetric single type-I quantum well.

III-3. What is a realistic Si/Ge interface?

A. Thermodynamic requirement

This is not a simple question to ask. As we have seen, it is important to discriminate otherwise conflicting concepts centered around the "interface". From thermodynamic considerations, it is obvious that we have to take into account the free-energy *even if the system is far from equilibrium.* This is plausible since the system tends to reach a (local) free-energy minimum which provides the driving force for the evolution of mass transport and morphological transformation along allowable paths. In other words, the configurational (mixing) entropy cannot be neglected since the MBE growth is performed at elevated temperatures which drive the system closer to thermal equilibrium.

For many applications using band engineering, an "abrupt" compositional transient is desired in the vertical direction whereas a "flat" interface morphology is favored along the lateral direction. However, thermodynamics place a limitation on the realization of "abrupt" and "flat" interface morphology due to the configurational entropy. In some extreme cases, e.g., quantum well interfaces prepared by migration-enhanced-epitaxy, the interface morphology appears to fulfill the desired conditions. As such, the interface fluctuations on an atomic scale do occur while the exciton probe is insensitive to these small interface perturbations. In contrast, as shown clearly, the interface-localized indirect excitons are quite sensitive to such small roughness. As a matter of fact, the indirect exciton probe has demonstrated the intrinsically "rough" nature of "abrupt" Si/SiGe interfaces.

B. How to specify the interface

Summarizing the foregoing scenario, fundamental terms relevant to the interface come in a pair that a vertically "abrupt" interface is "rough" laterally while a "smooth" or "graded" vertical transient indicates averaged characteristics in the lateral direction and therefore a "flat" interface morphology. The latter has been experimentally confirmed by the authors. We observed that PL linewidths of strained InGaAs/GaAs allow quantum wells are small when the interface is compositionally graded due to In segregation. On the contrary, the linewidths were broader for rectangle QWs with abrupt interfaces which were created by suppressing In segregation [32].

IV. Summary

In summary, we have studied physics and control of Si/SiGe heterointerfaces. The vertical and lateral morphologies and their influence on optical properties were discussed in the context of SiGe-based optoelectronics. A guideline was established to differentiate otherwise confusing terminologies related to the heterointerface morphology.

Acknowledgments The authors gratefully acknowledge K.Fujita, K.Miyashita, G.Ohta, T.Hattori, Y.Miyake for their assistance. One of the authors (S.F.) would like to acknowledge the support from Toray Science Foundation, Asahi Glass Foundation, Murata Science Foundation, Kawasaki Steel 21st Century Foundation, and Casio Science Promotion Foundation. This work was in part supported by a Grant-in-Aid from the Ministry of Education, Science, Sports, and Culture.

References

1. S.S.Iyer, J.C.Tsang, M.R.Copel, P.R.Pukite and R.M.Tromp, Appl.Phys.Lett. **54**,219(1989).
2. M.Copel, M.C.Reuter, E.Kaxiras and R.M.Tromp, Phys.Rev.Lett. **62**,632(1990).
3. P.C.Zalm, G.F.A.van de Walle, D.J.Gravesteijn, and A.A.van Gorkum, Appl.Phys.Lett. **55**, 2520 (1990).
4. K. Nakagawa and M. Miyao, J.Appl.Phys. **69**, 3058 (1991).
5. S.Fukatsu, K.Fujita, H,Yaguchi, Y.Shiraki, and R.Ito, Surf.Sci. **267**,79 (1992); Appl.Phys.Lett. **59**(1991)2103; Mat.Res.Soc.Sym.Proc. **220**,217(1991).
6. K.Fujita, S.Fukatsu, T.Igarashi, H,Yaguchi, Y.Shiraki, and R.Ito, Jpn.J.Appl.Phys. **29**, L1981 (1990); Mat.Res.Soc.Sym.Proc. **220**,193(1991).
7. D.Godbey and M.Ancona, Appl.Phys.Lett. **61**,2217 (1992); J.Vac.Sci.Technol. **B11**, 1120 (1193); ibid **B11**,1392 (1993).
8. U.Mencziger, G.Absteiter, J.Olajos, H.G.Grimmeiss, H.Kibbel, H.Presting and E.Kasper, Phys.Rev.**B47**,4009(1993).
9. W.-X.Ni, J.Knall, M.A.Hasan, G.V.Hansson, J.-E.Sundgren, S.A.Barnett, L.C.Markert, and J.E.Greene, Phys.Rev.**B40**,10449 (1989).
10. K.Sakamoto, K.Kyoya, K.Miki, H.Matsuhat, and T.Sakamoto, Jpn.J.Appl.Phys.**32**,L204(1993).
11. N.Ohtani, S.Mokler, M.H.Xie, J.Zhang, and B.A.Joyce, Jpn.J.Appl.Phys.**33**,2311 (1994).
12 Y.Li, G.G.Hembree, and J.A.Veanables, Appl.Phys.Lett **67**, 276 (1995).
13. J.Brunner, R.Menczigar, and G.Absteiter, J.Vac.Sci.Technol **B11**, 1097 (1993).
14. S.Fukatsu, N.Usami, and Y.Shiraki, Jpn.J.Appl.Phys. **31**,1502(1992).
15. S.Fukatsu, N.Usami, Y.Kato, H.Sunamura, Y.Shiraki, H.Oku, T.Ohnishi, Y.Ohmori and K.Okumura, J. Crystal Growth **136** (1994) 315.
16. G.G.Jernigan, P.E.Thompson, and C.L.Silverstre, Appl.Phys.Lett **69**, 1894 (1996).
17. N.Usami, S.Fukatsu, and Y.Shraki, Appl.Phys.Lett **63**, 388 (1993).
18. G.Ohta, S.Fukatsu, Y.Ebuchi, T.Hattori, N.Usami, and Y.Shiraki, Appl.Phys.Lett **65**, 2975 (1994).
19. K.Sakamoto , H.Matsuhata, K.Miki, and T.Sakamoto, J.Cryst. Growth **157**,295 (1995).
20. K.Nakagawa, A.Nishida, Y.Kimura, and T.Shimada, J.Cryst. Growth **150**, 939 (1995).
21. J.Olajos, J.Engvall, H.G.Grimmeiss, U.Mencziger, G.Absteiter, H.Kibbel, E.Kasper and H.Presting, Appl.Phys.Lett.**63**,493 (1993).
22. U.Mencziger, G.Absteiter, J.Olajos, H.G.Grimmeiss, H.Kibbel, H.Presting and E.Kasper, Phys.Rev.**B47**,4009(1993).
23. P.Schittenhelm, M.Gail, and G.Absteiter, J.Cryst.Growth **157**, 260 (1995).
24. H.Sunamura, Y.Shiraki, and S.Fukatsu, Appl.Phys.Lett **66**, 953 (1995); ibid **66**,3024 (1995).
25. H.Sunamura, S.Fukatsu, N.Usami, and Y.Shiraki, J.Cryst.Growth **157**, 265 (1995).
26. J.Olajos, J.Engvall, H.G.Grimmeiss, M.Gail, G.Absteiter, H.Presting, and H.Kibbel, Phys.Rev. **B54**, 1922 (1996).
27. H.Sunamura, N.Usami, Y.Shiraki, and S.Fukatsu, Appl.Phys.Lett. **68**,1847 (1996).
28. N.Usami, F.Issiki, D.K.Nayak, Y.Shraki, and S.Fukatsu Appl.Phys.Lett. **67**, 524 (1995)
29. N.Usami, Y.Shiraki, and S.Fukatsu, Appl.Phys.Lett. **68**, 2340 (1996).
30. S.Fukatsu, N.Usami, and Y.Shiraki, J. Vac. Sci. and Technol. **B14**, 2387 (1996).
31. S.Fukatsu, J.Mat.Sci., Electronic Materials 6,341-349 (1995).
32. K.Muraki, S.Fukatsu, Y.Shiraki, and R.Ito, Surf.Sci. **267**, 107 (1992).

SURFACTANT-MEDIATED Si/Ge EPITAXIAL CRYSTAL GROWTH

EUNJA KIM[+], CHAN WUK OH, and YOUNG HEE LEE

Department of Physics and Semiconductor Physics Research Center, Jeonbuk National University, Jeonju, 561-756, Korea

1 ABSTRACT

We investigate the kinetic role of a surfactant in the epitaxial Si/Ge crystal growth using *ab initio* molecular dynamics approach. We examine the previously suggested dimer-exchange mechanisms and find that kinetics plays a crucial role in determining the exchange process. We further find that the diffusion of adatoms on an island in the presence of a surfactant is quite different from the dimer-exchange process on a flat surface.

2 INTRODUCTION

Despite the fact that the current sophisticated growth techniques provide various novel systems of quantum structures, there exists an intrinsic problem in an heterostructure formation that A material can grow easily on B material whereas the opposite is always difficult to achieve due to the surface free energy difference between two materials. For instance, the Ge on a Si layer can grow layer-by-layer at least a few monolayers whereas the followed Si deposition on the Ge overlayered structure does not always leave an abrupt interface due to the Ge segregation to the front growing surface, which is induced by the surface free energy difference. One way to overcome this difficulty is to introduce a surfactant prior to the Ge deposition. This lowers the surface free energy drastically and therefore the Ge segregation is suppressed by surfactants being at the front growing surface and thus the interface becomes abrupt.

Several experiments have been done to investigate the role of surfactants [1-4]. Since Tromp and Reuter suggested a needlelike dimer-exchange process based on low energy electron microscope measurement that the As dimer grows along the direction of the dimer row [3], several calculations based on the energetics have been performed to explain surfactant-mediated growth [5,6]. Eaglesham *et al.* have reported that the post annealing of Ge islands on the Si(001) surface in the presence of Sb surfactants reduces the islands drastically [4]. These observations suggest that the role of surfactants on islands and on a flat surface are quite different and furthermore under-

135

Mat. Res. Soc. Symp. Proc. Vol. 448 © 1997 Materials Research Society

standing these phenomena requires the evaluattion of the activation energies to assure the related atomic motion. In this report we emphasize that it is not the energetics but the kinetics that governs the dimer-exchange process and the annealing of islands. Introducing *ab initio* molecular dynamics (MD) method, we evaluate the previously suggested dimer-exchange mechanisms by constructing appropriate pathways and examine the validity of these models. We also evaluate kinetics on islands in the presence of surfactants and prove that the islands are annealed out by enhancing the diffusion of adatoms on the island in the presence of Sb at the edge of islands.

3 THEORETICAL APPROACHES

For the dimer-exchange process on a flat surface, we construct a periodic supercell which contains four layers of Si (4×2), one monolayer of Ge, and the surfactant (Sb) along the Si(001) direction. The dangling bonds of the bottom Si layer are saturated by 16 H atoms and are not allowed to relax. The surface is simulated by a vacuum region of 7 Å. For the island calculation, we introduce a rebonded double layer step (D_B) on the Si(001) surface since islands often possess {311} facets which are composed of consecutively rebonded D_B steps. Therefore we expect that the diffusion phenomenon on the D_B step is similar to that on the islands with experimentally observed {311} facets. The D_B stepped Si(001) surface is simulated by the periodically repeated triclinic supercell of 54 Si atoms with H atoms saturating the dangling bonds of the bottom Si layer, similar to the flat surface. The number of layers at upper and lower terraces is kept equivalent in order to prevent any unnecessary effects on both terraces. This unusual periodic boundary condition is necessary to simulate the D_B with the limited number of atoms. The cell dimension is $l = 17.278$ Å in the [110] direction. Given that the step height is $h = 2.715$ Å we get a possible misorientation angle of $\gamma = \tan^{-1}(h/l) = 9°$, which is approximately the maximum angle for a stable vicinal surface. Our calculations have been performed using Car-Parrinello approach [8]. The interaction between ionic cores and valence electrons was described by a fully nonlocal pseudopotential with s-only (for without Sb), s,p (for with Sb) nonlocality. We used a plane-wave basis set with a typical cutoff kinetic energy of 8 Ry and Bloch functions only at the Γ point of the supercell surface Brillouin zone.

4 RESULTS AND DISCUSSION

We first construct the dimerized flat surfaces of Sb/Ge/Si and Ge/Sb/Si system which are fully relaxed by *ab initio* MD method. The surface free energy is lowered by 3.5 eV/dimer for Sb/Ge/Si

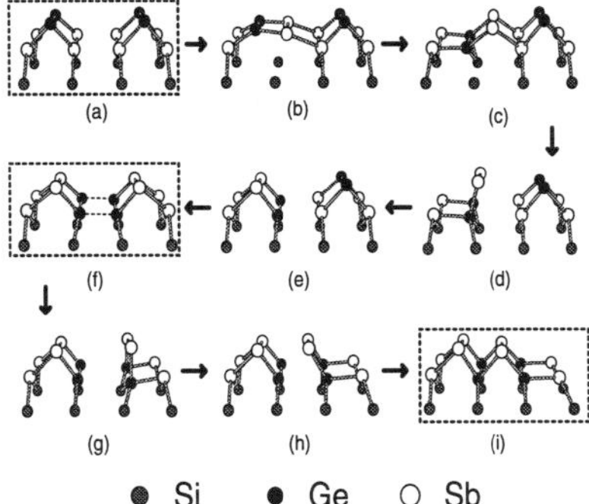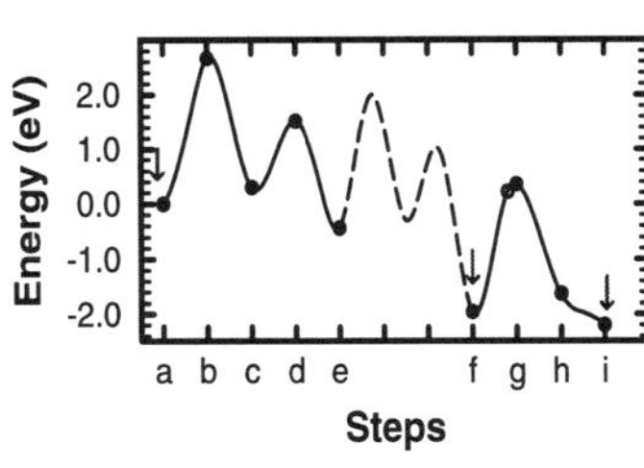

Figure 1: Constructed pathway and the corresponding potential profile for two Ge ad-dimers located on adjacent two Sb-dimer rows. The pathway in the dotted boxes are previously suggested configurations which are energetically favorable by Ohno (Ref. [5]).

system than for Ge/Sb/Si system, clearly indicating a tendency that Sb prefers to float to the front growing surface. In order for the surfactant to float to the top surface, the exchange with the adatoms should occur. The exchange mechanism is possible only when surface diffusion barrier (E_s) for the adatom to diffuse on the surfactant-covered surface is comparable to or higher than the interdiffusion barrier (E_b). Otherwise the surface diffusion will be dominant over the exchange process. We first construct an appropriate pathway and calculate the surface diffusion barrier for the Ge ad-dimer on the Sb-covered surface. We find the E_s to be 1.15 eV [9] when the Ge dimer moves perpendicular to the subsurface Sb dimer rows. This will be a reference energy to examine the validity of the dimer-exchange mechanism of previously suggested models.

Two models have been suggested based on the energetics [5,6]. We first construct an appropriate pathway for Ohno's model [6], as shown in Fig. 1 and calculate the activation energy. This model starts from two Ge ad-dimers located on two adjacent dimer rows, which is a local minimum. This pathway involves the pushing-out and rolling-over processes that give the large diffusion barrier of 2.67 eV. This value is too large compared to the surface diffusion barrier of 1.15 eV, despite the fact that the model eventually leads to the experimentally suggested needlelike dimer growth with a significant energy gain of 2.2 eV. Our kinetics suggests this model, which is based on the energetics, to be improbable due to too large activation energy.

We next examine another model suggested by Yu and Oshiyama [5]. This model starts from

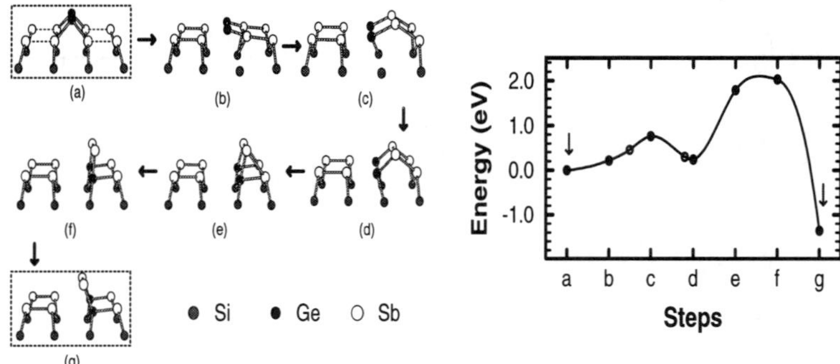

Figure 2: Constructed pathway and the corresponding potential profile for the Ge ad-dimer located in the trough between two dimer rows. The pathway in the dotted boxes are previously suggested configurations which are energetically favorable by Yu and Oshiyama (Ref.[4]).

the Ge ad-dimer in the trough between dimer rows, as shown in Fig. 2. Pushing down the Ge ad-dimer to the subsurface direction requires an energy barrier of 0.76 eV only, as shown in the potential profile. However, rolling-over the Sb dimer across the Ge atoms in the subsurface as shown in step (f) requires a large activation energy of 2 eV due to the complete bond-breaking of the Sb-dimer with surface Sb atoms. Although the complete rolling-over process (step (f)) which achieves the exchange process, gains the energy by 1.4 eV compared to the original step (a), the activation energy is still too large compared to the surface diffusion barrier. Although the dimer-exchange process seems to occur in real experimental situations, the previously suggested models for the exchange process should be reevaluated so as to give low activation energies in the future works [9].

In the following we describe shortly the diffusion process on the island. The detail has been published elsewhere [7]. We first search the preferable site of the Sb dimer on D_B step edge. We find that the Sb dimer prefers to stay in the dimer site near the edge of the island. We then evaluate the diffusion barrier of the adatom on island and find that the Schwoebel barrier for a single adatom in the presence of the Sb dimer at the step edge requires the large Schwoebel barrier of 2.1 eV compared to 0.93 eV without the presence of Sb at the edge. This strongly suggests that the diffusion on a island is governed by the dimer diffusion rather than the single-adatom diffusion. We further examine the dimer diffusion with the various exchange processes,

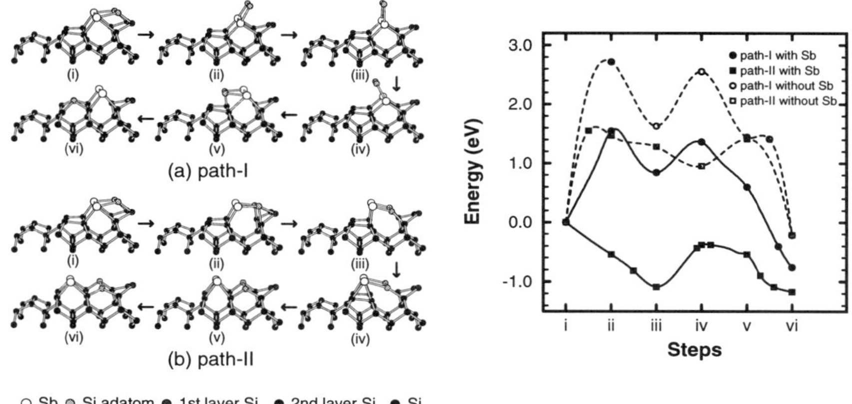

Figure 3: Constructed pathway and the corresponding potential profile for a Si dimer that moves across the step edge.

as shown in Fig. 3. We find that the pushing-out process is preferable to the rolling-over process by an energy difference of 0.84 eV [7]. Unlike the diffusion on a flat surface that the diffusion is suppressed by the surfactant-mediated exchange process, the diffusion on a island is greatly enchanced by the surfactant. This excellently explains why the islands anneal out in the presence of Sb surfactant, as suggested in the experiment [4]. In real experimental situation both diffusion phenomena will play an important role in the epitaxial crystal growth. Islands may be formed at any instant but those will soon be annealed out to a flat surface so as to promote the epitaxial growth by the dimer-exchange process.

ACKNOWLEDGMENTS

This work is supported by the Korea Science and Engineering Foundation through the Semiconductor Physics Research Center at Jeonbuk National University. One (YHL) of us acknowledges financial support by the LG Yonam Foundation.

REFERENCES

[+] Present Address: Department of Physics, University of Nevada, Las Vegas, NV 89154, USA

1. K. Sakamoto *et al.*, Jpn. J. Appl. Phys. **32**, L204 (1993).

2. M. Copel *et al.*, Phys. Rev. Lett. **63**, 632 (1989).

3. R. M. Tromp and M. C. Reuter, Phys. Rev. Lett. **68**, 954 (1992).

4. D. J. Eaglesham *et al.*, Phys. Rev. Lett. **70**, 966 (1993).

5. B. D. Yu and A. Oshiyama, Phys. Rev. Lett. **71**, 585 (1993); **72**, 3190 (1994).

6. T. Ohno, Phys. Rev. Lett. **73**, 460 (1994).

7. C. W. Oh, E. Kim, and Y. H. Lee, Phys. Rev. Lett. **76**, 776 (1996).

8. R. Car and M. Parrinello, Phys. Rev. Lett. **55**, 2471 (1985).

9. E. Kim, C. W. Oh, J. Y. Kim, and Y. H. Lee, unpublished.

GROWTH OF ZnSe-BASED COMPOUNDS
ON Ge-TERMINATED GaAs SURFACE

T. SAITOH, A. TSUJIMURA, T. NISHIKAWA, A. WATAKABE and Y. SASAI
Semiconductor Research Center, Matsushita Electric Industrial Co., Ltd.

ABSTRACT

Growth of ZnSe-based compounds on the GaAs surface terminated by ultra-thin Ge epitaxial layer was carried out by molecular beam epitaxy and the influence of Ge layer on the growth of ZnSe was investigated. When the thickness of Ge layer was 1 atomic layer (AL), 2-dimensional growth occurred in the initial stage of ZnSe layer growth and anti-phase boundary (APB) free ZnSe layer was obtained. For Ge layer thickness of 10 AL, ZnSe grew 3-dimensionally and APBs were generated in the ZnSe layer. The crystalline quality of ZnMgSSe layer was also strongly influenced by the thickness of Ge layer. These phenomena were identified to be due to the transition of Ge surface structure from single domain to double domain with increasing Ge layer thickness.

1. INTRODUCTION

ZnSe-based lasers have attracted much interest as light sources in the blue and green regions for use in high density optical disks, full color displays and many other applications. Heteroepitaxial growth of ZnSe-based compounds on GaAs substrates is commonly used for fabricating ZnSe-based lasers. It has been revealed that stacking faults are generated at the ZnSe/GaAs interface and the laser degradation is induced by the presence of such defects[1,2]. Control of the initial growth of ZnSe on GaAs surface is the most important issue for improving the property of ZnSe-based lasers.

Recently, it has been pointed out that the existence of Ga-Se bonds plays an important role in such defect nucleation[3]. Termination of GaAs surface is an attractive method for avoiding the direct absorption of Se atoms on GaAs surface. Zn irradiation onto GaAs surface is known to be an effective method to suppress the formation of Ga-Se bonds [3-5]. However, defect density as low as 10^4 cm^{-2} still exists even using Zn irradiation [5]. One of the reasons is that the GaAs surface is not completely terminated by Zn layer since the sticking coefficient of Zn atoms on the As-stabilized GaAs surface is quite low.

We propose a termination of GaAs surface using ultra thin Ge epitaxial layer. Germanium is one of the most suitable candidates as an interlayer at the ZnSe/GaAs interface since the lattice constant of Ge is very close to those of ZnSe and GaAs. Additionally, Ge interlayer can be used for controlling the charge imbalance at the ZnSe/GaAs interface since Ge is a Column IV element. However, there exists a severe problem that such an interlayer of non-polar materials will become the origin generating anti-phase boundaries in the ZnSe layer. Fundamental growth process of the ZnSe/Ge/GaAs system must be clarified to use Ge layer as an interlayer for controlling the ZnSe/GaAs interface.

In this study, growth of ZnSe-based compounds on the GaAs surface terminated by ultra-thin Ge epitaxial layer was carried out and the influence of Ge layer on the growth of ZnSe was investigated. The relationship between ZnSe initial growth and the domain structure of Ge surface is discussed.

Mat. Res. Soc. Symp. Proc. Vol. 448 © 1997 Materials Research Society

2. EXPERIMENTAL PROCEDURES

The crystal growths were carried out by molecular beam epitaxy (MBE) in a system having two growth chambers for III-V and II-VI compounds, which are connected through ultra-high vacuum transfer chamber. Si-doped n-type GaAs(001) substrates were used. The substrates were first introduced into the III-V chamber. After the thermal deoxidization, 500 nm thick Si-doped GaAs buffer layers were grown at 600 °C. Then, Ge layers with the thickness from 1 to 10 atomic layer (AL) were grown onto the GaAs buffer layer at 400 °C. The Growth rate of Ge layer was 0.1 AL/sec, which was calibrated by the thickness measured by cross-sectional TEM image. Metal Ga, As and Ge were used as source materials. After the growth of Ge layers, the samples were transferred into the II-VI chamber through the transfer chamber. ZnSe and ZnMgSSe layers were grown on the Ge-terminated GaAs substrates at 280 °C, where poly-crystalline ZnSe, ZnS and metal Mg were used as source materials[6]. $ZnCl_2$ was used as the n-type dopant.

Reflection high-energy electron diffraction (RHEED) was used for real time observations of the surface reconstruction structures during MBE growth. The structures of the ZnSe/Ge/GaAs interface were observed by cross-sectional transmission electron microscopy (TEM). Photoluminescence (PL) was used for characterizing crystalline quality of the epitaxial layers. The 325 nm line of the He-Cd laser was used as an excitation source.

3. RESULTS

3.1. *TEM observation of the ZnSe/Ge/GaAs structures*

ZnSe layer was grown on the Ge-terminated GaAs surface and the cross-sectional structure was investigated by TEM. Figure 1 shows a cross-sectional TEM image of the ZnSe/Ge/GaAs structure where the thickness of Ge layer was about 4 AL. Flat and continuous Ge layer is observed between the ZnSe layer and the GaAs substrate. The lattice spots are arranged in the straight line in the ZnSe layer, Ge layer and GaAs substrates. This means that the Ge layer and the following ZnSe layer were epitaxially grown on the GaAs substrate without evident formation of dislocations. These results show that ZnSe layers can be grown without interaction with GaAs substrates by using Ge-termination of GaAs surfaces.

Fig. 1
Cross-sectional TEM image of the ZnSe/Ge/GaAs structure where 4 AL thick Ge layer was inserted between ZnSe and GaAs.

3.2. *Growth of ZnSe on Ge-terminated GaAs surfaces studied by RHEED observations*

First, the initial stages of the growth of Ge layers on As-stabilized GaAs surfaces were studied by RHEED observations. Figure 2 shows the RHEED patterns of the surfaces before and after Ge growth. Before Ge growth, the RHEED pattern showed a clear (2x4) streaky pattern (Fig. 2 (a)), which indicates formation of As-stabilized GaAs surface. When the Ge layer with a thickness of 1 AL was grown on the GaAs surface, the RHEED pattern changed from (2x4) to (1x2) streaky pattern (Fig. 2 (b)). This transition of the RHEED patterns shows that the two-fold periodic structure was constructed by formation of the Ge dimers along the [1-10] direction.

When the thickness of Ge layer increased to 10 AL, the RHEED pattern changed to the (2x2) pattern (Fig. 2 (c)). We identify that the (2x2) pattern observed here is the (2x1)-(1x2) mixed pattern, which is commonly observed from the surfaces of non-polar substrates such as Si or Ge. The transition of the RHEED pattern from (1x2) to (2x1)-(1x2) mixed pattern shows that the Ge dimers were formed along both [110] and [1-10] directions. This means that the surface structure of Ge layer changed from single domain to double domain with increasing Ge layer thickness. The details of the transition of the surface structures are discussed in Section 4.

Although the reconstruction pattern depended on the layer thickness, the RHEED patterns of the Ge surfaces were always streaky, which indicates that atomically flat Ge layer was formed on the GaAs surface. This is in good agreement with the results of the TEM observation as shown in the Fig. 1 .

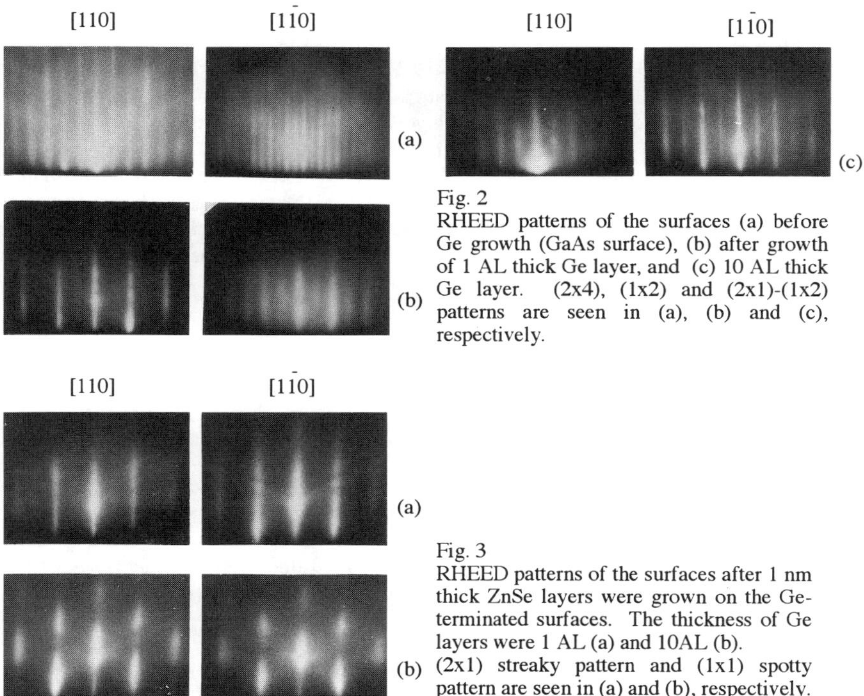

Fig. 2
RHEED patterns of the surfaces (a) before Ge growth (GaAs surface), (b) after growth of 1 AL thick Ge layer, and (c) 10 AL thick Ge layer. (2x4), (1x2) and (2x1)-(1x2) patterns are seen in (a), (b) and (c), respectively.

Fig. 3
RHEED patterns of the surfaces after 1 nm thick ZnSe layers were grown on the Ge-terminated surfaces. The thickness of Ge layers were 1 AL (a) and 10AL (b). (2x1) streaky pattern and (1x1) spotty pattern are seen in (a) and (b), respectively.

Next, the initial stages of ZnSe growth on the Ge-terminated GaAs substrates were studied by RHEED. ZnSe layers were grown on the two different domain structure of Ge surfaces. Figure 3 (a) and (b) show the RHEED patterns of the surfaces after 1 nm thick ZnSe layers were grown on the Ge-terminated GaAs substrates, where the thickness of Ge layer was 1 AL and 10 AL, respectively.

When ZnSe layer was grown on the 1 AL thick Ge layer, the RHEED pattern of the ZnSe surface showed a streaky (2x1) pattern (Fig. 3 (a)). The streaky (2x1) pattern indicates a formation of smooth Se-stabilized ZnSe surface. This means that the 2-dimensional growth occurred even during the initial stage of ZnSe growth. Since the two-fold pattern was observed in the [110] direction only, it was confirmed that the APBs were not generated in the ZnSe layer in spite of the growth on non-polar material.

On the other hand, when ZnSe was grown on the 10 AL thick Ge layer, the RHEED pattern showed a spotty (1x1) pattern (Fig. 2 (b)). This means that the 3-dimensional island growth occurred in the initial stage and the surface of the ZnSe layer was rough. When the thickness of ZnSe layer increased to more than about 20 nm, the RHEED pattern changed to the (2x1)-(1x2) mixed pattern, which indicates a generation of APBs in the ZnSe layer. The change in the RHEED patterns of the ZnSe layer grown on the Ge layers with different thickness indicates that the initial growth of ZnSe is strongly influenced by the domain structure of the Ge surfaces. The APB free ZnSe layer can be grown on the Ge-terminated GaAs substrate by controlling the thickness of Ge layer to 1 AL.

3.3. Crystalline quality of ZnMgSSe layers grown on the Ge-terminated GaAs substrates

Cl-doped $Zn_{0.92}Mg_{0.08}S_{0.2}Se_{0.8}$ layers ($n=4\times10^{17}$ cm^{-3}) with a thickness of 1 μm were grown on the Ge-terminated GaAs substrates. 25 nm thick ZnSe buffer layers were inserted before the growth of ZnMgSSe layers. Crystalline quality of the ZnMgSSe layers grown on the Ge layers with different thickness were investigated by photoluminescence (PL) measurements. Figure 4 shows the PL spectra of the Cl-doped ZnMgSSe layers measured at 8 K. The solid and dotted lines represent the cases where the thickness of Ge layer was 1 and 10 AL, respectively. When the thickness of Ge layer was 1 AL, the bound-exiton emission (I_2 line) was dominantly observed at 419nm in wavelength and deep level emissions were negligible.

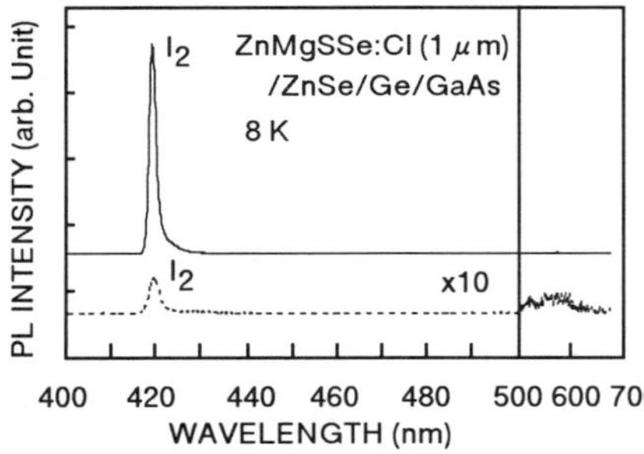

Fig. 4
Photoluminescence spectra of Cl-doped ZnMgSSe layers grown on Ge-terminated GaAs. 25 nm thick ZnSe layers were inserted before ZnMgSSe growth. Solid and dotted lines represent the cases where the thickness of Ge layer was 1 AL and 10 AL, respectively.

When the thickness of the Ge layer was 10 AL, I_2 line at 419 nm was observed to a much lesser extent. The intensity of the I_2 line was about 60 times lower than that of the sample where the Ge thickness was 1 AL. The deep level emission was detected in the range from 500 nm to 700 nm. This shows that the crystalline quality of the ZnMgSSe layer grown on the 10 AL thick Ge layer was worse than that grown on the 1 AL thick Ge layer. We consider that the change in the crystalline quality was caused by the difference in the initial growth stage of ZnSe layer which is discussed in Section 3.2.

4. DISCUSSION

From the results of RHEED observation and PL measurements, it was revealed that the ZnSe growth mode during the initial stage and crystalline quality of ZnMgSSe layers were strongly influenced by the thickness of the Ge layer. We discuss the relationship between ZnSe growth mode and the initial surface structure of Ge layer which is dependent on the thickness of Ge layer.

Figure 5 shows a schematic of our model. When 1 AL thick Ge layer was grown on the (2x4) GaAs surface, the RHEED pattern showed the (1x2) reconstruction due to the formation of Ge dimer rows along the [1-10] direction, as shown in the Fig. 2 (b). The fact that the direction of Ge dimer rows was rotated by 90 degrees against the As dimer rows ([110] direction) indicates that the Ge atoms in the 1 AL thick Ge layer occupy the Ga sites only (Fig. 5 (a)). As a result, the step height of the Ge surface was double atomic height, reflecting the step structure of the GaAs surface. After the growth of ZnSe on this surface, the RHEED pattern changed to the Se-stabilized (2x1) pattern as shown in the Fig. 3 (a). Considering that the direction of Se dimer rows ([110] direction) agrees with that of As dimer rows, the Ge atoms make bonds with Se atoms at the interface (Fig. 5 (b)). ZnSe layer is grown without APBs due to double atomic step structure of the Ge surface.

When 10 AL thick Ge layer was grown on the GaAs surface, the RHEED pattern showed the (2x1)-(1x2) mixed pattern. This indicates that the Ge atoms begin to occupy both Ga and As sites with increasing thickness of the Ge layer. As a result, step structure changes from double atomic height to single atomic height (Fig. 5 (c)). When the ZnSe layer is grown on this surface, APBs are formed at the step sites (Fig. 5 (d)). It is supposed that the 3-dimensional island growth in the initial stage of ZnSe growth is originated from the nucleation of APBs.

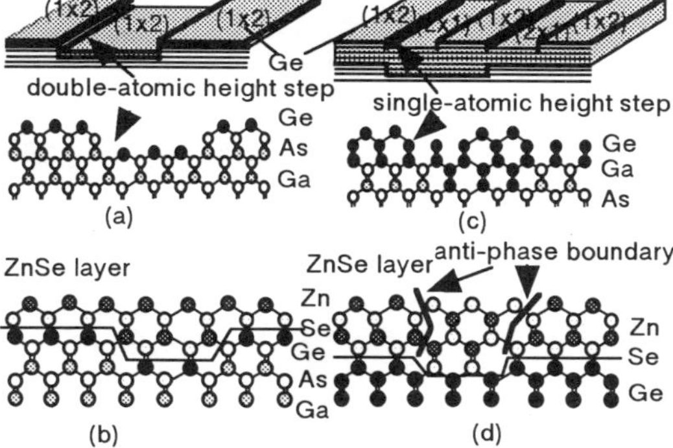

Fig. 5
Schematic of the model of ZnSe growth on Ge-terminated GaAs. The atom arrangement is shown (a) after 1 AL thick Ge layer was grown on GaAs,(b)after ZnSe was grown on 1AL thick Ge layer, (c)after more than 1 AL thick Ge layer was grown on GaAs surface , and (d)after ZnSe was grown on more than 1 AL thick Ge .

5. CONCLUSION

We have investigated the growth process of ZnSe-based compounds on the Ge-terminated GaAs surfaces. The cross-sectional TEM image showed that the ZnSe layer was epitaxially grown on the Ge layer without interaction with the GaAs substrates. From the results of RHEED observations, it was clarified that 2-dimensional growth occurred in the initial stage of ZnSe growth and APB free ZnSe layer was obtained when the thickness of Ge layer was 1 AL. On the other hand, in case that the thickness of Ge layer was 10 AL, ZnSe grew 3-dimensionally and APBs were generated in the ZnSe layer. The crystalline quality of the ZnMgSSe layers characterized by PL measurement was also influenced by the thickness of Ge layer. The dependence in the growth of ZnSe-based compounds on Ge layer thickness is caused by the transition of Ge surface structure from single domain to double domain with increasing Ge layer thickness.

In conclusion, the formation of the single domain structure of the Ge surface by controlling the thickness of Ge layer is important for obtaining APB free epitaxial layers of ZnSe-based compounds.

ACKNOWLEDGMENTS

The authors would like to thank Dr. K. Ohnaka for his encouragement and advises. They also thank Dr. Y. Yabuuchi for TEM observations, and Drs. T. Yokogawa and G. Tohmon for useful discussions and advises.

REFERENCES

1. G. C. Hua, N. Otsuka, D. C. Grillo, Y. Fan, J. Han, M. D. Ringle, R. L. Gunshor, M. Hovinen and V. Nurmikko, Appl. Phys. Lett. 65, p1331 (1994).
2. S. Tomiya, E. Morita, M. Ukita, H. Okuyama, S. Itoh, K. Nakano and A. Ishibashi, Appl. Phys. Lett. 66, p1208 (1995).
3. L. H. Kuo, K. Kimura, T. Yasuda, S. Miwa, C. G. Jin, K. Yanaka and T. Yao, Appl. Phys. Lett. 68, p2413 (1996).
4. L. H. Kuo, L. Salamanca-Riba, B. J. Wu, G. Hofler, J. M. Depuydt and H. Cheng, Appl. Phys. Lett. 67, p3298 (1995).
5. S. Saito, P. J. Parbrook, S. Nakamura, Y. Nishikawa and G. Hatakoshi, Proc. Int. Symp. Blue Laser and Light Emitting Diodes, p66 (Chiba, 1996).
6. T. Karasawa, K. Ohkawa and T. Mitsuyu, J. Appl. Phys. 69, p3226 (1991).

ANISOTROPY IN ATOMIC-SCALE INTERFACE STRUCTURE AND MOBILITY IN INAS/GA$_{1-x}$IN$_x$SB SUPERLATTICES

A. Y. Lew*, S. L. Zuo*, E. T. Yu* , R. H. Miles**
*ECE Department, University of California at San Diego, La Jolla, CA 92093-0407
**Hughes Research Laboratories, Malibu, CA 90265

ABSTRACT

We have used cross-sectional scanning tunneling microscopy to study the atomic-scale interface structure of InAs/Ga$_{1-x}$In$_x$Sb superlattices grown by molecular-beam epitaxy. Detailed, quantitative analysis of interface profiles obtained from constant-current images of both (110) and $(1\bar{1}0)$ cross-sectional planes of the superlattice indicates that interfaces in the $(1\bar{1}0)$ plane exhibit a higher degree of interface roughness than those in the (110) plane, and that the Ga$_{1-x}$In$_x$Sb-on-InAs interfaces are rougher than the InAs-on-Ga$_{1-x}$In$_x$Sb interfaces. The roughness data are consistent with anisotropy in interface structure arising from anisotropic island formation during growth, and in addition with a growth-sequence-dependent interface asymmetry resulting from differences in interfacial bond structure between the superlattice layers. Roughness data are compared with measurements of anisotropy in low-temperature Hall mobilities of the samples.

INTRODUCTION

InAs/Ga$_{1-x}$In$_x$Sb strained-layer superlattices have shown great promise for application in mid-to-long wavelength infrared imaging applications.[1-4] However, the atomic-scale interfacial properties of the superlattice structures have been found to be of crucial importance in determining material and device properties. Because both Group III and Group V constituents change from one superlattice layer to the next, two distinct bond configurations – InSb-like and Ga$_{1-x}$In$_x$As-like – can be present at each interface. Ga$_{1-x}$In$_x$Sb/InAs superlattices grown with InSb-like interfacial bonds have been demonstrated to possess superior device characteristics compared to those grown with Ga$_{1-x}$In$_x$As-like bonds.[5,6] Previous studies provided evidence of atomic-scale interface roughness and asymmetry in InAs/GaSb[7-10] and InAs/Ga$_{1-x}$In$_x$Sb[11] superlattices. A detailed understanding of the atomic-scale structural and compositional properties of these interfaces is therefore essential to the optimization of electrical and optical properties of device structures based on the InAs/Ga$_{1-x}$In$_x$Sb and related material systems.

In this paper we describe a detailed, quantitative analysis of interface structure in InAs/Ga$_{1-x}$In$_x$Sb superlattices using cross-sectional scanning tunneling microscopy (STM). STM data are used to investigate directly and quantitatively the degree of directional anisotropy and growth-sequence dependence in interface structure. To quantify the interface structure observed by STM, individual interface profiles are extracted from the images and their roughness spectra calculated. The roughness spectra show that interfaces in the $(1\bar{1}0)$ plane of the superlattice exhibit larger roughness amplitudes and correlation lengths than interfaces in the (110) plane, and that interfaces in which Ga$_{1-x}$In$_x$Sb has been grown on InAs exhibit larger roughness amplitudes and correlation lengths than interfaces in which InAs has been grown on Ga$_{1-x}$In$_x$Sb.

Mat. Res. Soc. Symp. Proc. Vol. 448 © 1997 Materials Research Society

We attribute these features in interface structure to anisotropy in island formation during sample growth, and to a dependence on growth sequence of bond configurations at the interfaces. Results obtained from STM studies are also compared with low-temperature Hall mobility studies of similar structures grown on GaAs substrates.

EXPERIMENT

The InAs/$Ga_{1-x}In_xSb$ superlattices used in the STM study were grown by solid-source molecular-beam epitaxy (MBE) in a VG V80 MKII MBE system on n-type GaSb (001) substrates. The growth system and substrate preparation techniques have been described elsewhere.[7] The superlattice consisted of 50Å $Ga_{0.75}In_{0.25}Sb$ alternating with 17Å InAs for 150 periods, capped with 500Å GaSb. The epitaxial layers were grown at 380°C on a 1000Å GaSb buffer layer. At each interface in the superlattice layers, a 5-sec. Sb soak was used to induce the formation of InSb-like bonds. The average composition and overall structural quality of the sample were confirmed by high-resolution x-ray diffraction.

STM experiments were conducted in an ultra-high vacuum system at a base pressure of approximately $1-2\times10^{-10}$ Torr. Superlattice samples were cleaved *in situ* to expose either a (110) or a $(1\bar{1}0)$ cross-sectional face on which STM imaging of the epitaxial layers was performed. While atomically flat cross-sectional surfaces were obtained for both the (110) and $(1\bar{1}0)$ planes, we found that cleaving to expose the (110) surface was generally more successful. Both Pt-Ir and W tips cleaned *in situ* by electron bombardment were used for these studies. Because the cleaved surfaces are atomically flat, the contrast seen in constant-current STM images corresponds primarily to variations in the local electronic structure of the sample, rather than to actual physical topography of the surface.

Hall measurements were obtained for samples consisting of 75-period 50Å $Ga_{0.75}In_{0.25}Sb$/17Å InAs superlattices grown on GaSb buffer layers on semi-insulating GaAs substrates. Van der Pauw structures were fabricated on the samples, and temperature-dependent mobility measurements were taken along both the $[110]$ and $[1\bar{1}0]$ directions of the superlattice.

RESULTS

Figure 1 shows a 200Å×500Å high-resolution constant-current cross-sectional image of the superlattice structure, obtained at a sample bias voltage of -1.5 V and a tunneling current of 0.1 nA. Contrast between the darker InAs and brighter $Ga_{0.75}In_{0.25}Sb$ layers can clearly be seen in the image, as can monolayer-level roughness at the interfaces between the InAs and the $Ga_{1-x}In_xSb$ layers. A large number of images of similar quality were obtained over regions up to 1000Å×1000Å in size from both (110) and $(1\bar{1}0)$ cross-sections of the sample.

Individual interface profiles are extracted from the images by enhancing the contrast between the InAs and $Ga_{1-x}In_xSb$ layers, and then using an edge-detection algorithm. Quantitative profiles *a(x)* are obtained by measuring the distance between the interfaces and a baseline corresponding to the profile of the atomic bilayers in the image from which the interfaces have been extracted. Figure 2 shows representative interface profiles extracted from images similar in quality to that of Figure 1. In total, several dozen interface profiles, each 500Å in length, were extracted from multiple images of both (110) and $(1\bar{1}0)$ cross-sectional surfaces.

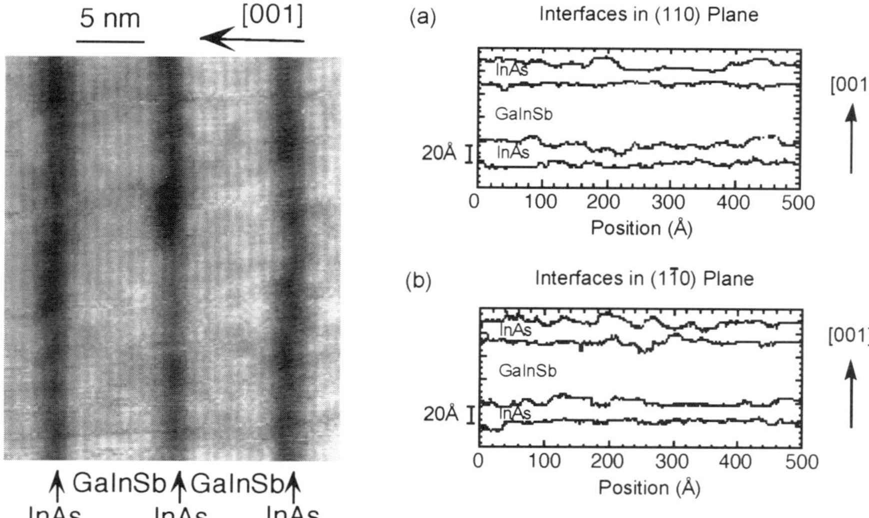

FIG. 1: 200Å×500Å constant-current STM image of the $Ga_{0.75}In_{0.25}Sb$/InAs superlattice, obtained at a sample bias of -1.5 V and a tunneling current of 0.1 nA. The gray-scale range of the image is 2.5Å. Profiles of the interfaces between the InAs layers and the $Ga_{1-x}In_xSb$ layers are extracted from several such images.

FIG. 2: Representative interface profiles obtained from (a) (110) and (b) $(1\bar{1}0)$ cross-sectional STM images of the $Ga_{0.75}In_{0.25}Sb$/InAs superlattice sample. The [001] growth direction is indicated.

The roughness spectra of the interfaces are calculated by taking discrete Fourier transforms of the extracted profiles. The roughness frequency components A_q are given by

$$A_q = \frac{2}{N} \sum_{n=0}^{N-1} a(nd)e^{-iqd} \tag{1}$$

where $q = 2\pi n/L$, L is the length of the interface, and d is the spacing between data points along the interface (typically 0.5Å). The power spectrum of each interface, $|A_q|^2$, is fitted to a Lorentzian function,

$$|A_q|^2 = \frac{1}{L} \cdot \frac{2\Delta^2 \cdot (\Lambda/2\pi)}{1 + (q\Lambda/2\pi)^2} \tag{2}$$

where Δ is the roughness amplitude and Λ the correlation length. We find the Lorentzian to be a better fit to our data than a Gaussian or other functional forms. The Lorentzian spectral distribution corresponds to an exponential correlation in real space, and is expected for a random distribution of steps at the interface.[12] Figure 3 shows a plot of the power spectra and also of

Lorentzian fits to the power spectra for $Ga_{0.75}In_{0.25}Sb$-on-InAs and InAs-on-$Ga_{0.75}In_{0.25}Sb$ interfaces from both (110) and $(1\bar{1}0)$ cross-sections. Table I lists the roughness parameters for each interface type. The spectra shown are averages of individual roughness spectra from at least 6 different profiles for each interface type. Note that interfaces imaged in the $(1\bar{1}0)$ cross-sectional plane are oriented parallel to the [110] direction in the sample.

As indicated in Table I, the interfaces in the $(1\bar{1}0)$ plane exhibit larger roughness amplitudes and correlation lengths than those in the (110) plane. This observation is consistent with anisotropic island formation during growth, a phenomenon that has been observed, using both reflection high-energy electron diffraction (RHEED)[13] and STM,[14,15] in growth of (001) GaAs. These studies showed that islands on the GaAs surface tend to be elongated in the $[1\bar{1}0]$ direction. Similar island formation during growth of the InAs/$Ga_{0.75}In_{0.25}Sb$ layers should lead to anisotropic interface structure: interfaces from the (110) cross-section, which profile the elongated-island cross-sections present along the $[1\bar{1}0]$ direction, would be expected to contain smaller roughness components in their roughness spectra than interfaces from the $(1\bar{1}0)$ cross-section, which profile the shorter-island cross-sections found along the [110] direction. The quantitative results obtained from our interface roughness analysis are consistent with this interpretation.

The parameters in Table I also show a substantial dependence of interface roughness on growth sequence, with the $Ga_{1-x}In_xSb$-on-InAs interfaces being substantially rougher than the InAs-on-$Ga_{1-x}In_xSb$ interfaces. In earlier studies of InAs/GaSb superlattices using STM, a substantial growth-sequence dependence in interface structure was observed, with the degree of asymmetry depending upon growth conditions.[8-10] X-ray diffraction studies of InAs/$Ga_{1-x}In_xSb$ superlattice samples have suggested that the $Ga_{1-x}In_xSb$ layers are terminated with InSb-like bonds, while the InAs layers are terminated by roughly equal numbers of InSb-like and $Ga_{1-x}In_xAs$-like bonds.[7] The $Ga_{1-x}In_xSb$-on-InAs interfaces would then be of mixed InSb-like and $Ga_{1-x}In_xAs$-like character, while the InAs-on-$Ga_{1-x}In_xSb$ interfaces would be of a more homogenous InSb-like character. Since STM is sensitive to nanometer-scale changes in electronic structure, differences

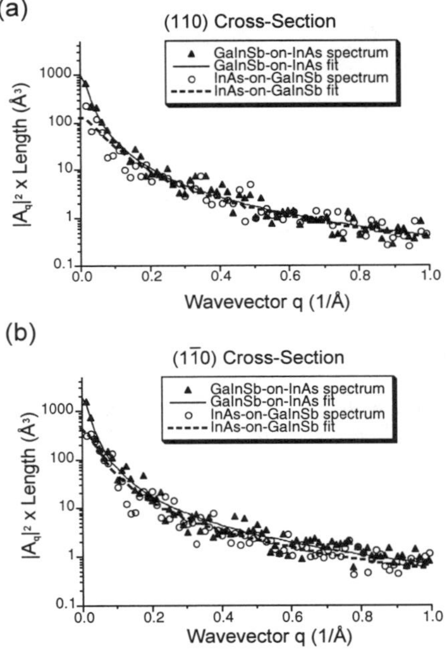

FIG. 3: Interface roughness power spectra (symbols) and Lorentzian fits to the spectra (lines) for $Ga_{0.75}In_{0.25}Sb$-on-InAs and InAs-on-$Ga_{0.75}In_{0.25}Sb$ interfaces from (a) (110) and (b) $(1\bar{1}0)$ cross-sections of the superlattice. Roughness amplitudes and correlation lengths for each interface type are listed in Table I.

TABLE I: Roughness amplitudes and correlation lengths obtained from fitting a Lorentzian to the interface roughness power spectra calculated for $Ga_{0.75}In_{0.25}Sb$-on-InAs and InAs-on-$Ga_{0.75}In_{0.25}Sb$ interfaces from (110) and $(1\bar{1}0)$ cross-sectional images.

Cross-Section	Interface	Amplitude Δ (Å)	Correlation Length Λ (Å)
$(1\bar{1}0)$	$Ga_{1-x}In_xSb$-on-InAs	4.3±0.2	327±38
	InAs-on-$Ga_{1-x}In_xSb$	2.8±0.2	174±21
(110)	$Ga_{1-x}In_xSb$-on-InAs	3.2±0.2	301±39
	InAs-on-$Ga_{1-x}In_xSb$	1.9±0.1	112±16

such as these in interfacial bond structure would be reflected in the roughness as measured by STM, and could account for the observed growth-sequence dependence of the superlattice interface structure.

The results of the interface roughness analysis described previously have been compared with results obtained from low-temperature Hall measurements of mobility. Interface roughness scattering has been found to dominate the electron mobility of InAs/$Ga_{1-x}In_xSb$ superlattices under almost all conditions of interest.[16] Furthermore, anisotropic Hall mobilities in high-mobility modulation-doped $Al_xGa_{1-x}As$/GaAs heterostructures have been attributed to an anisotropy of interface islands.[17] Temperature-dependent measurements of Hall mobility were taken along both the $[110]$ and $[1\bar{1}0]$ directions of the superlattice on samples consisting of 75 periods of superlattice structure nominally identical to that in the STM samples (50Å $Ga_{0.75}In_{0.25}Sb$ alternating with 17Å InAs), but grown on GaSb buffer layers on semi-insulating GaAs substrates. These measurements demonstrated that a substantial anisotropy in Hall mobility exists in the superlattice: observed mobility was larger in the $[1\bar{1}0]$ direction than in the $[110]$ direction, and the degree of anisotropy increased at lower temperatures. This correlates with the directional anisotropy in interface roughness seen in the STM roughness analysis: images of the $(1\bar{1}0)$ cross-section, which show a greater degree of roughness, correspond to interfaces along the $[110]$ direction. The greater roughness along this $[110]$ direction would be expected to lead to more interface roughness scattering for transport in the $[110]$ direction than in the $[1\bar{1}0]$ direction, and will contribute to a decrease in mobility along the $[110]$ superlattice direction. Continuing work is under way to measure Hall mobilities of superlattice samples grown on GaSb substrates, as well as to correlate the measured mobilities with results of theoretical modeling incorporating detailed band structure calculations and the interface structure observed by STM.

CONCLUSIONS

In summary, we have used cross-sectional STM to investigate directly the atomic-scale interface morphology of InAs/$Ga_{1-x}In_xSb$ superlattices grown by MBE. Spectral analysis of interface roughness observed by STM shows that interfaces in the $(1\bar{1}0)$ cross-sectional plane have larger roughness amplitudes and correlation lengths than those in the (110) cross-sectional plane; this directional anisotropy in the interface structure of the superlattice layers is interpreted as arising from anisotropic island formation during growth. The roughness spectra also indicate the presence of a growth-sequence dependent interface asymmetry: interfaces in which the $Ga_{1-x}In_xSb$ has been grown on InAs are rougher than interfaces in which InAs has been grown on $Ga_{1-x}In_xSb$. This directly supports earlier studies suggesting a more homogenous InSb-bond character at the InAs-on-$Ga_{1-x}In_xSb$ interfaces than at the $Ga_{1-x}In_xSb$-on-InAs interfaces. Our STM results have also been found to be consistent with measured anisotropic low-temperature Hall mobilities of the superlattice structures.

ACKNOWLEDGEMENTS

AYL, SLZ and ETY would like to acknowledge support from GM Hughes Electronics via the UC MICRO program, the National Science Foundation (Award ECS 93-07986), and the DARPA Optoelectronics Technology Center.

REFERENCES

1. D. L. Smith and C. Mailhiot, J. Appl. Phys. **62**, 2545 (1987).
2. D. H. Chow, R. H. Miles, J R. Söderström, and T. C. McGill, Appl. Phys. Lett. **56**, 1418 (1990).
3. T.C. Hasenberg, D. H. Chow, A. R. Kost, R. H. Miles, and L. West, Elect. Lett. **31**, 275 (1995).
4. J. L. Johnson, L.A. Samoska, A. C. Gossard, J. L. Merz, M. D. Jack, G. R. Chapman, B. A. Baumgratz, K. Kosai, and S. M. Johnson, J. Appl. Phys. **80**, 1116 (1996).
5. D. H. Chow, R. H. Miles, and A. T. Hunter, J. Vac. Sci. Technol. B **10**, 888 (1992).
6. R. H. Miles, J. N. Schulman, D. H. Chow, and T. C. McGill, Semicond. Sci. Technol. **8**, S102 (1993).
7. R. H. Miles, D. H. Chow, and W. J. Hamilton, J. Appl. Phys. **71**, 211 (1992).
8. R. M. Feenstra, D. A. Collins, D. Z.-Y. Ting, M. W. Wang, and T. C. McGill, J. Vac. Sci. Technol. B **12**, 2592 (1994).
9. R. M. Feenstra, D. A. Collins, D. Z.-Y. Ting, M. W. Wang, and T. C. McGill, Phys. Rev. Lett. **72**, 2749 (1994).
10. P. M. Thibado, B. R. Bennett, M. E. Twigg, B. V. Shanabrook, and L. J. Whitman, Appl. Phys. Lett. **67**, 3578 (1995).
11. A. Y. Lew E. T. Yu, D. H. Chow, and R. H. Miles, Appl. Phys. Lett. **65**, 201 (1994).
12. S. M. Goodnick, D. K. Ferry, C. W. Wilmsen, Z. Liliental, D. Fathy, and O. L. Krivanek, Phys. Rev. B **32**, 8171 (1985).
13. P. R. Pukite, G. S. Petrich, S. Batra, and P. I. Cohen, J. Cryst. Growth **95**, 269 (1989).
14. E. J. Heller and M. G. Lagally, Appl. Phys. Lett. **60**, 2675 (1992).
15. V. Bressler-Hill, R. Maboudian, M. Wassermeier, X.-S. Wang, K. Pond, P. M. Petroff, and W. H. Weinberg, Surf. Sci. **287/288**, 514 (1993).
16. C. A. Hoffman, J. R. Meyer, E. R. Youngdale, F. J. Bartoli, and R. H. Miles, Appl. Phys. Lett. **63**, 2210 (1993).
17. Y. Tokura, T. Saku, S. Tarucha, and Y. Horikoshi, Phys. Rev. B **46**, 15558 (1992).

INTERFACE ROUGHNESS IN STRAINED Si/SiGe MULTILAYERS

A. A. DARHUBER[1], V. HOLY[1]*, J. STANGL[1], G. BAUER[1], J. NÜTZEL[2] and G. ABSTREITER[2]
[1]Institut für Halbleiterphysik, Johannes Kepler Universität, A-4040 Linz, Austria
[2]Walter Schottky Institut, TU München, Am Coulombwall 2, D-85748 Garching, Germany

ABSTRACT

Diffuse x-ray reflection from a SiGe/Si multilayer grown pseudomorphically on slightly miscut Si(001) substrates has been studied theoretically and experimentally. In the framework of the Distorted-Wave Born Approximation (DWBA), we demonstrated that the distribution of the diffusely scattered intensity gives conclusive information on both the amount and the in-plane and inter-plane correlation properties of the interface roughness. The best model for the description of the interface-morphology was found to be a combination of a two-level model and a staircase model.

INTRODUCTION

The assessment and control of the interface roughness is important for the electrical [1,2] and optical properties of semiconductor epi-layers. In multilayers, the evolution of the roughness with increasing number of periods is of interest. For thin layers, uncorrelated roughness of the two hetero-interfaces leads to substantial well width fluctuations, whereas correlated roughness does not. Therefore a comprehensive picture of the interface morphology and its correlation parallel and perpendicular to the growth direction is desirable.

The only non-destructive method for the characterization of the roughness of the surface and buried interfaces is x-ray reflectivity, in particular two-dimensionally resolved measurements. X-ray scattering from rough interfaces in SiGe multilayers has been studied by several authors [3-6]. Their experimental results were that the interface profiles of different interfaces are strongly correlated, and, moreover, the direction of the highest correlation does not coincide with the growth direction. In the case of multilayers grown on miscut [001]-substrates, the distribution of the diffusely scattered intensity in reciprocal space reflects the terraced structure of the interfaces. The mean width and height of the terraces are correlated via the miscut angle (between the growth direction **n** and [001]) and the direction along the steps is always perpendicular to the plane determined by **n** and [001]. In [5], the shape of the terraces has been studied. The authors found that steps are created by bunching of existing substrate steps which results in a wavy shape of the interfaces. However, this finding was supported only by qualitative considerations.

In this paper, we use the concepts introduced in [10] to simulate the reciprocal space maps of diffuse x-ray reflection from a SiGe multilayer. The simulation of the scattering is based on the Distorted-Wave Born Approximation (DWBA, [7]) and the roughness replication model described in [13]. From the analysis the most suitable model for the interface-morphology is found.

THEORETICAL CONSIDERATIONS

Theoretical works on diffuse x-ray reflectivity from multilayers [7-9] allow the simulation the reciprocal space maps, if the correlation function of the roughness profiles of

* permanent address: Dept. of Solid State Physics, Masaryk University, Kotlarska 2, 61137 Brno, Czech Republic

Mat. Res. Soc. Symp. Proc. Vol. 448 © 1997 Materials Research Society

different interfaces is known. The correlation properties of a single terraced surface can be formulated using the concept of Markov random processes [10]. This procedure yields a formula for the conditional probability $w(\Delta x, \Delta z)$ of finding a surface scatterer in the point $(\Delta x, \Delta z)$ provided that the point $(0,0)$ is occupied by another surface atom. In the cited works, surfaces with both infinite and finite number of levels have been treated. The former are represented by an infinite number of terrace steps, the latter correspond to surfaces with a "castellated" roughness profile (see insets in Fig. 1). The correlation of the roughness profiles of different interfaces (roughness replication) has been studied thoroughly in [11,12].

A reciprocal space map represents the two-dimensional dependence of the diffusely scattered intensity I on the x- and z-components of the wave vector transfer $\mathbf{Q} = \mathbf{K}_2 - \mathbf{K}_1$, where $\mathbf{K}_{1,2}$ are the wave vectors of the incident and scattered radiation, respectively. x and z are the coordinates parallel and perpendicular to the mean sample surface. The intensity scattered by a sample with rough interfaces consists of coherent and incoherent components. The coherent component is concentrated along the Q_z-axis $(Q_x = 0)$ of the reciprocal plane (the truncation rod) and it represents the specularly scattered intensity, the incoherent component is the diffuse scattering.

In the following, we will only deal with the diffuse scattering. The DWBA method yields the following general formula for the diffusely scattered intensity

$$I(Q_x,Q_z) = \text{const} \frac{|T_1 T_2|^2}{|q_z|^2} \sum_{j,k} \delta_j \, \delta_k{}^* \, \exp(iq_z \, (z_j - z_k)) \int d\Delta x \, \exp(i \, Q_x \, \Delta x) \, C_{jk}(\Delta x, q_z) \quad (1)$$

where $T_{1,2}$ are the transmittivities of the interface between vacuum and an effective layer having the average refractive index of SiGe and Si layers, these transmittivities are calculated for the direction of the primary and scattered beams, respectively. q_z is the vertical component of $\mathbf{Q}$ corrected for refraction and absorption in the effective layer. The double summation is performed over all the interfaces, z_j is the depth of the j-th interface below the mean free surface. δ_j denotes the difference of the refractive indices above and below the j-th interface.

The essential quantity describing the interface roughness is the correlation function

$$C_{jk}(x-x', q_z) = \langle \, \exp(i \, q_z \, [U_j(x) - U_k(x')]) \, \rangle \qquad (2)$$

of the roughness profiles $U_{j,k}(x)$ of the j-th and k-th interfaces. Let us assume that the roughness profiles of different interfaces are partially correlated in the direction making an angle χ with the surface normal $\mathbf{n}$. According to [13] we assume an exponential dependence of this correlation on $z_j - z_k$, thus,

$$C_{jk}(\Delta x, q_z) = \text{const} + C(\Delta x + (z_j - z_k)\tan\chi, q_z) \exp(-|z_j - z_k| / \Lambda_\perp), \qquad (3)$$

where $\Lambda_\perp$ is the effective length of the vertical replication. The constant term plays no role if $Q_x \neq 0$ and it will be omitted subsequently.

The correlation function $C(\Delta x, q_z)$ of an interface is simply connected with the above mentioned conditional probability $w(\Delta x, \Delta z)$ by the Fourier transformation

$$C(\Delta x, q_z) = \int d\Delta z \, w(\Delta x, \Delta z) \, \exp(i \, q_z \, \Delta z) \qquad (4)$$

From eq. (1) it is obvious that the intensity distribution in reciprocal space is substantially influenced by the Fourier transformation of the conditional probability

$$w^{FT}(Q_x,q_z) = \int dQ_x \, \exp(i \, Q_x \, x\} \, C(\Delta x,q_z) \equiv$$

$$\int d\Delta x \int d\Delta z \, w(\Delta x, \Delta z) \exp(i(Q_x \Delta x + q_z \Delta z)) \qquad (5)$$

which depends on the model of the interface morphology. On the basis of the theory in [10], we have simulated the reciprocal plane distribution of $w^{FT}(Q_x, q_z)$ for four different models (see Fig. 1). In all these models, the x-axis is perpendicular to the terraces, i.e. it lies in the plane of **n** and [001].

- two level model - this model represents a "castellated" interface having two levels separated by steps with heights $\pm$ d, The level widths are random and they are distributed according to the Gamma distribution of the m-th order with the mean values $L_{1,2}$. The averaged interface is parallel to the levels.

- staircase model - the model has an infinite number of levels creating a staircase structure. The step widths are distributed according to the Gamma distribution of the m-th order and the mean value L. The step height is d. The surfaces of the steps are parallel and they make an angle $\Theta = \arctan(d/L)$ with the mean surface. This model seems to be the most appropriate one for the description of a vicinal surface, Θ corresponds to the miscut angle and the step surfaces are (001).

- roof model - is a symmetric modification of the previous model, both the step (001) surfaces and the slopes have the same angle Θ with the mean surface. The surfaces and the slopes have the same mean width L, the order of the Gamma distribution of their widths is m.

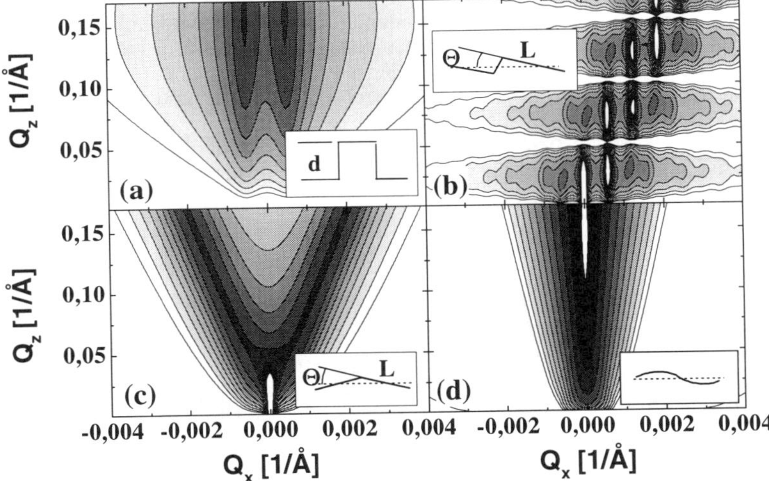

Fig. 1: Simulations of $w^{FT}(Q_x, q_z)$ in the reciprocal plane. (a) Two-level model (L = 1 μm, m = 3, d = 20 Å), (b) many-level staircase model (L = 1 μm, m = 3), (c) many-level roof model (L = 1 μm, m = 3) and (d) Gaussian surface (Λ = 1 μm). In models (b) and (c) the heights are chosen to correspond to the miscut angle $\Theta = 0.25°$.

w^{FT} of the two-level model is symmetrical with respect to the q_z-axis. Distinct subsidiary maxima can be seen being parallel to this axis, the sharpness of this maxima depends on the order m of the width distribution. For m $\longrightarrow \infty$ the subsidiary maxima create a periodical sequence with the period $2\pi/(L_1+L_2)$.

The staircase model yields an asymmetrical distribution. The main maximum of w^{FT} is elongated in the direction perpendicular to the (001) step surfaces. In the case of larger m_L, this maximum is modulated by a periodical sequence of further maxima, whose separation is $2\pi/L$. These maxima are parallel to the *mean* surface. The distribution of w^{FT} for the roof model is symmetrical. Two main maxima are perpendicular to the (001) surfaces and the slopes, respectively, thus they make an angle 2Θ. Similarly to the previous models, the Gamma distribution of the lengths of the surfaces and the slopes gives rise to a periodical series of the maxima perpendicular to the mean surface. Their distance is $2\pi/L$ if the order m is large enough. The distribution of w^{FT} for a Gaussian surface is shown in Fig. 1 as well. This distribution contains only one central maximum along the q_z-axis, no subsidiary maxima can be seen.

The reciprocal plane distribution of the intensity diffusely scattered from a rough multilayer equals the product of w^{FT} with the term

$$Z(Q_x,Q_z) = \text{const} \frac{|T_1 T_2|^2}{|q_z|^2} \sum_{j,k} \delta_j \, \delta_k{}^* \exp(i(q_z - Q_x \tan\chi)(z_j-z_k)) \exp(-|z_j - z_k| / \Lambda_\perp) \quad (6)$$

This term depends on the correlation of the roughness profiles of different interfaces. If these profiles are at least partially correlated, $Z(Q_x,Q_z)$ has a series of maxima stretched in the direction perpendicular to the direction of maximum correlation, i.e. they make angle χ with the Q_x-axis (so called resonant diffuse scattering - RDS - see [8]). If the multilayer is periodical, the pattern of these maxima is periodical as well, their distance is $2\pi/D$, where D is the multilayer period. Due to x-ray refraction, these maxima are slightly curved which gives them a banana-like shape. The lateral correlation is described by w^{FT} and, therefore, it influences mainly the intensity distribution along the RDS maximum. The correlations of different interfaces are characterized by the parameters χ and $\Lambda_\perp$ and they contribute to $Z(Q_x,Q_z)$. Therefore, this correlation is responsible for the intensity profile across the RDS maximum.

Further details of the theoretical treatment will be published elsewhere [15] along with the discussion of the experimental results of a series of samples prepared at different growth conditions.

DISCUSSION

The investigated SiGe/Si multilayer sample (10 periods) has been grown by molecular beam epitaxy (MBE) on an unintentionally miscut (miscut angle $\Theta = 0.25°$) Si (001) substrate. The x-ray scattering measurements have been performed at the OPTICS beamline of the ESRF, Grenoble. A wavelength of $\lambda = 1.05$ Å just below the Ge absorption edge has been used.

From the measurement of the specular reflectivity, the individual layer thicknesses, the Ge content of the SiGe layers as well as the root mean square (r.m.s.) roughness of the interfaces have been determined : the multilayer period is 209 ± 1 Å, the ratio of the layer thicknesses is $T_{Si}/T_{SiGe}=7.6 \pm 0.2$

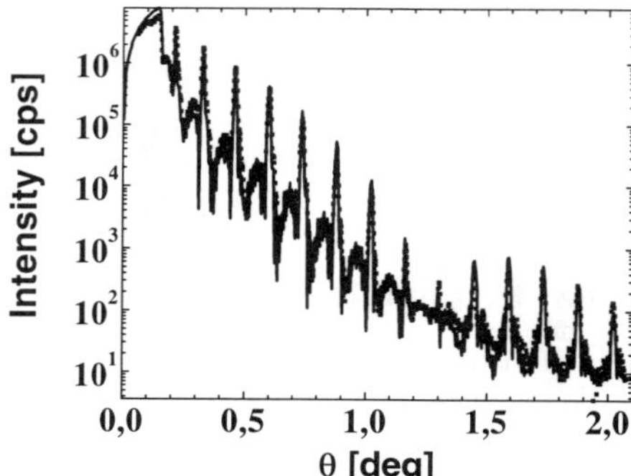

Fig. 2: Measured (points) and simulated (line) specular reflectivity.

and the Ge-concentration in the SiGe layers is 40 ± 10 %. The r.m.s. roughness of the interfaces are 5.2 ± 0.2 Å, the surface-roughness is 6 ± 0.2 Å. The thickness of the Si-cap layer is 225 ± 10 Å. The specular reflection curve is shown in Fig. 2. The good coincidence of the measured and the simulated curves is an evidence of a excellent homogeneity and structural quality of the sample.

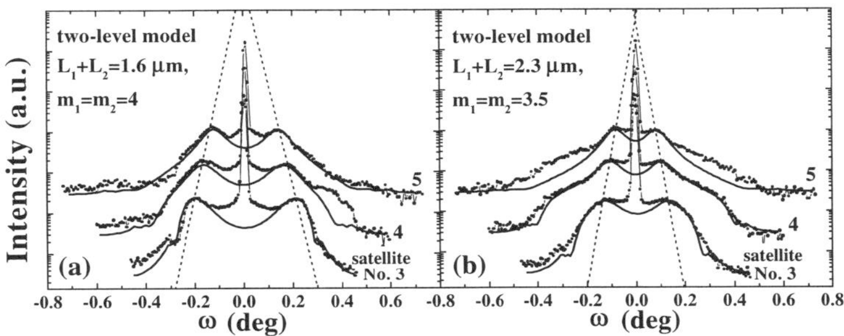

Fig. 3: Measured (points) and simulated (lines) ω–scans across the 3rd, the 4th and the 5th superlattice satellites. The azimuthal angle was φ = 45° (a) and 0° (b).

The inter-plane correlation of the roughness profiles follows from measured 2θ–scans, where the angle of incidence ω has been kept constant. Thus, the measured intensity is a function of the exit angle. The 2θ-scans have been measured for two azimuthal directions - φ = 0° and φ = 45°. Besides the coherent (specular) maximum, peaks corresponding to the cross-sections of the measurement trajectory with the RDS bananas can be seen. From the comparison of the width and shape of those maxima with the parameters of the coherent peaks, the vertical correlation length can be derived. Since the widths are nearly equal, the roughness profiles of all the interfaces in the stack are almost entirely correlated.

In Fig. 3, the measured and simulated ω scans through the 3rd, the 4th and the 5th satellite maximum of the specular reflectivity are plotted. During the ω–scan, the angle between the primary beam and the detector is constant, only the sample is rotated. The trajectory of the ω–scan in reciprocal space is therefore nearly parallel to the RDS banana.

The ω–scans across the terraces exhibit two distinct maxima. Their intensities are slightly asymmetric so that the two-level model does not fully describe the actual structure. Unlike to the 2θ-scans, the ω–scans substantially depend on the azimuth angle φ. Across the terraces (φ = 45°), the peak separation is larger than at φ = 0° due to the smaller projected terrace step length. Using the two-level model, we determined the mean widths of the terraces L_1+L_2 = 1.6 ± 0.2 µm (φ = 45°) and L_1+L_2 = 2.3 ± 0.2 µm (φ = 0°), their heights d = 15 Å. The attempts to compare the measured ω–scans

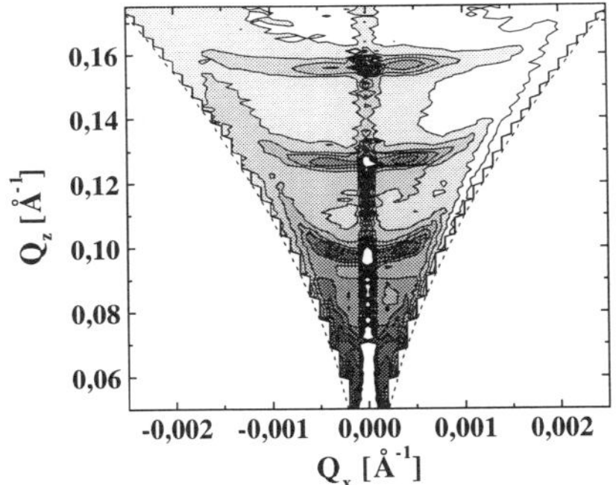

Fig. 4: Two-dimensional distributions of the diffusely scattered intensity for φ=0°. The increment between the (logarithmic) iso-intensity contours is $10^{0.2}$.

with those following from the other models were not successful. Most likely, the true structure can be represented by some "mixture" of the two-level and staircase models. This will be the subject of further investigations.

We have also measured the two dimensional distribution of the scattered intensity (reciprocal space map, Fig. 4) for a similar sample. We found distinct banana-shaped maxima having the angle $\alpha = 60° \pm 5°$ with the vertical axis. In the map measured with the sample rotated by 180° around the surface normal (not shown), the sense of rotation of the RDS maxima is opposite, thus this inclination is not an experimental artefact. Similarly to the ω−scans, the intensity distribution in a RDS maximum is slightly asymmetric which cannot be explained by the two-level model alone.

CONCLUSIONS

The distribution of the diffusely scattered intensity in reciprocal space yields information on both in-plane and inter-plane correlations of the roughness profiles in multilayers. In the direction parallel to the miscut, a well pronounced spatial frequency maximum has been found and the distribution of the scattered intensity can be explained by a two-level model. However, a slight asymmetry of distribution indicates that the actual morphology could be described by a combination of the castellation and staircase models.

ACKNOWLEDGEMENTS

The experiments were performed at the ESRF, Grenoble. We thank A. Souvorov for experimental assistance. Work partially supported by FWF (project 10083 PHY), Vienna.

REFERENCES

1. Y. H. Xie, D. Monroe, E. Fitzgerald, P. J. Silverman, F. A. Thiel, G. P. Watson, Appl. Phys. Lett. **63**, 2263 (1993)
2. C. R. Bolognesi, H. Kroemer, J. H. English, Appl. Phys. Lett. **61**, 213 (1992)
3. R. L. Headrick, J.-M. Baribeau, Phys. Rev. B **48**, 9174 (1993).
4. R. L. Headrick, J.-M. Baribeau, J. Vac. Sci. Technol. B 11, 1514 (1993)
5. Y. H. Phang, C. Teichert, M. G. Lagally, L. J. Peticolas, J. C. Bean, E. Kasper, Phys. Rev. B **50**, 14435 (1994)
6. R. L. Headrick, J.-M. Baribeau, Y. E. Strausser, Appl. Phys. Lett. **66**, 96 (1995)
7. S. K. Sinha, E. B. Sirota, S. Garoff, H. B. Stanley, Phys. Rev. B **38**, 2297 (1988)
8. V. Holy, T. Baumbach, Phys. Rev. B **51**, 10668 (1994)
9. D. K. G. de Boer, Phys. Rev. B **51**, 5297 (1995)
10. C. S. Lent, P. I. Cohen, Surf. Sci. **121**, 121 (1984) and Surf. Sci. **161**, 39 (1985).
11. E. Spiller, D. Stearns, and M. Krumrey, J. Appl. Phys. **74,** 107 (1993).
12. V. M. Kaganer, S. A. Stepanov, R. Köhler, Phys. Rev. B **52**, 16369 (1995)
13. Z. H. Ming, A. Krol, Y. L. Soo, Y. H. Kao, J. S. Park, K. L. Wang, Phys. Rev. B **47**, 16373 (1993)
14. T. Salditt, T. H. Metzger, J. Peisl, Phys. Rev. Lett. **73**, 2228 (1994)
15. V. Holy et al., to be published

Surface Roughening and Composition Modulation of ZnSe-related II-VI epitaxial films

Shigetaka Tomiya, Hironori Tsukamoto, Satoshi Itoh, Kazushi Nakano, Etsuo Morita and Akira Ishibashi
Sony Corporation Research Center, 174, Fujitsuka Hodogaya, Yokohama, 240 Japan

Abstract
We have investigated ZnSSe and ZnMgSSe epitaxial layers lattice-matched to GaAs (001) substrates grown by molecular beam epitaxy using atomic force microscopy and transmission electron microscopy. Under II-rich conditions with c(2x2) surface reconstruction, surface morphology exhibited corrugation aligned in the [1$\bar{1}$0] direction and composition modulation was observed in the same [1$\bar{1}$0] direction. Under VI-rich condition with (2x1) surface reconstruction, the surface morphology becomes rounded grain-like and composition modulation was not observed. The formation of composition modulation is associated with the surface corrugated structures.

1. Introduction
Ternary ZnSSe and quaternary ZnMgSSe alloy materials lattice matched to GaAs substrates have been playing significant role as guiding and cladding layers in ZnSe based blue/green II-VI laser diodes [1]. Composition modulation has been observed in many III-V alloy epilayers such as InGaAsP. However, to date, there have been only a few reports on TEM investigations of composition modulation of quaternary ZnMgSSe alloys [2,3]. In this paper, we report for the first time the relationship between the surface morphology and composition modulation. We will show that the surface roughening is responsible for the composition modulated structures.

2. Experimental
The samples investigated in this study were ZnSe, ternary ZnSSe alloy and quaternary ZnMgSSe alloy epitaxial layers and laser structures containing these alloy layers grown on semi-insulating GaAs (001) substrates using MBE using elemental Zn, Se, Mg and compound ZnS as source materials. The growth temperatures ranged from 250 ℃ to 350 ℃. Growth was performed under group II-rich conditions with (2x1) surface reconstruction and under group VI-rich conditions with c(2x2) surface reconstruction. The detailed II-VI laser structures have been reported elsewhere [1].

Atomic Force Microscope (AFM) measurements of the surface morphology were performed with a Digital Instrument Nanoscope II, in contact mode. AFM images were obtained for a 5 μm^2 area in the middle of each sample. The average compositions of the epilayers was determined using the double crystal X-ray rocking curve and electron-probe microanalysis (EPMA). Transmission Electron Microscope (TEM) experiments were performed using

Mat. Res. Soc. Symp. Proc. Vol. 448 © 1997 Materials Research Society

Hitachi H-700H, HF-2000 and JEOL 4000FXS electron microscopes operating at 200 kV and 400kV. Nano-probe X-ray energy dispersive spectrometer (EDS) analysis was also carried out with 1 nm probe size in the HF-2000. Cross-sectional and plan-view TEM samples were prepared by mechanical polishing followed by ion-milling and chemical etching.

3. Results and Discussion

Figure 1(a) and (b) shows AFM images of ZnSe and $ZnS_{0.07}Se_{0.93}$ alloy layers grown on GaAs (001) at 280 ℃ under II-rich conditions. A c(2x2) surface reconstruction was observed in the reflection high-energy electron diffraction during growth. The image shows the formation of elongated corrugations oriented in the [1$\bar{1}$0] direction. The average periods and amplitude of the corrugations in Fig. 1(a) and (b) was about 10 nm and about 60 nm, respectively. A similar morphology was also observed in $Zn_{0.93}Mg_{0.07}S_{0.18}Se_{0.82}$ alloy layers grown at 280 ℃ under a II rich condition with c(2x2) reconstruction (Figs. 1(c)). The average amplitude and the average period of the corrugations was about 6 nm and about 45 nm. The smaller amplitude and period in Fig.1 (c) may be due to the shorter diffusion length of Mg atoms, which adhere more easily to the surface

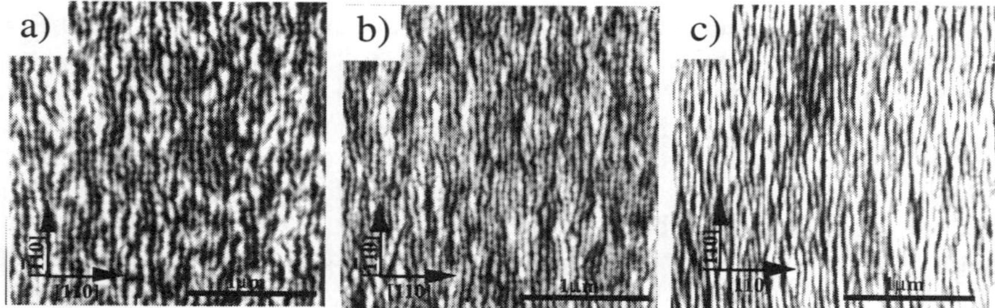

Figure 1. AFM image of a) ZnSe, b) ZnSSe, c) ZnMgSSe epilayers grown on GaAs (001) at a temperature of 280 ℃ under group II-rich conditions.

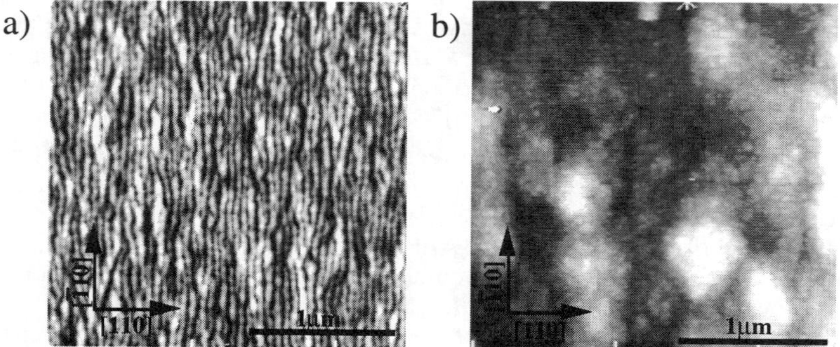

Figure 2. AFM image of $ZnS_{0.07}Se_{0.93}$ epilayers grown at a temperature of 280 ℃ under two different VI/II ratio of a) 0.34 and c) 1.5. Corrugated structures are not seen in b).

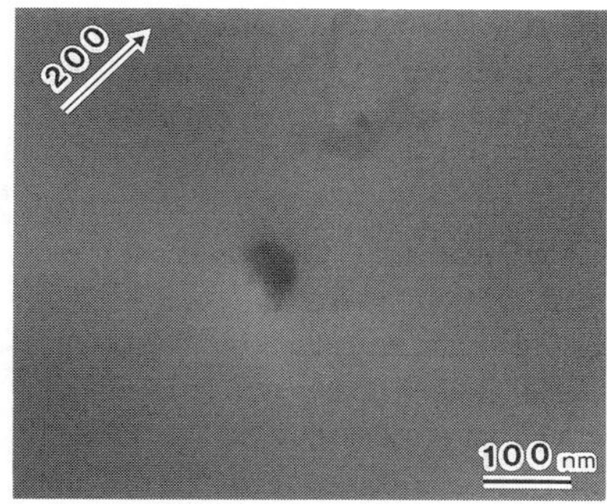

Figure 3. Plan view TEM image of ZnS$_{0.07}$Se$_{0.93}$ epilayers grown at a temperature of 280 ℃ under two different VI/II ratio of a) 0.34 and b) 1.5.

than Zn atoms [4]. Cross-sectional high resolution TEM observations reveals that the corrugations have a sinusoidal shape [5].

Figure 2 shows AFM images of ternary ZnS$_{0.07}$Se$_{0.93}$ alloy layers grown at 280 C under two different VI/II ratios. Corrugated structures was formed with small VI/II ratios with c(2x2) reconstruction. At higher VI/II ratios with (2x1) reconstruction, rounded grains form instead of corrugated structures. Our detailed AFM observation reveals that surface roughening is mainly due to growth kinetics and is not misfit stress-driven [5].

Figures 3(a) and (b) show 002 dark field plan-view TEM images of ternary ZnS$_{0.07}$Se$_{0.93}$ alloy layers grown at 280 ℃ under two different VI/II ratios. The corresponding surface morphology is shown in figures 2(a) and (b). In figure 3(a), regular periodic stripes of bright and dark bands approximately aligned along the [1$\bar{1}$0] direction are seen. The average period of the strip contrast is

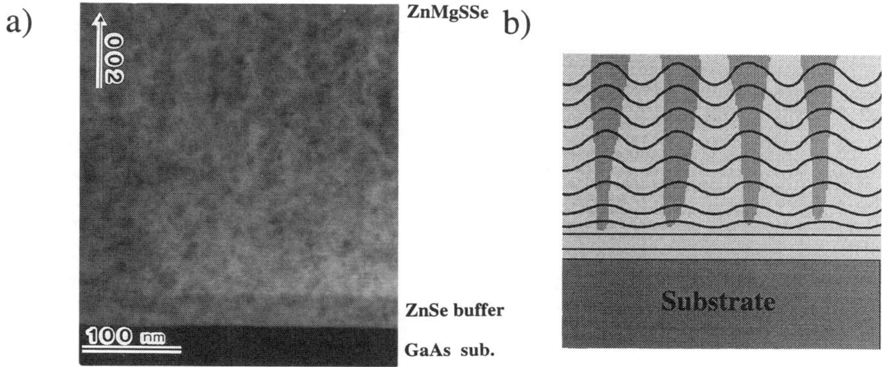

Figure 4. a) Cross-sectional dark filed TEM image of Zn$_{0.93}$Mg$_{0.07}$S$_{0.18}$Se$_{0.82}$ epilayers grown at a temperature of 280 ℃ under group II-rich conditions with substrate rotation. b) Schematics of Figure of a).

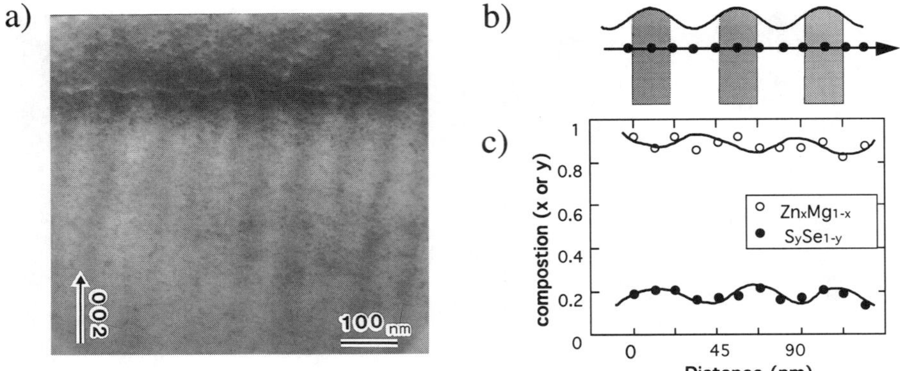

Figure 5. a) Cross-sectional dark filed TEM image of $Zn_{0.93}Mg_{0.07}S_{0.18}Se_{0.82}$ epilayers grown at a temperature of 280 ℃ under group II-rich conditions. b) Nano-probe EDS analysis were performed along arrow-line, which is 10 nm below the valleys of the corrugations. c) Result of EDS analysis

about 60 nm and corresponds to the period of the surface corrugations. On the other hand, no such contrast was observed in Fig. 3 (b). This contrast is due to the modulated composition of S and Se.

Figure 4 (a) shows a cross-sectional TEM image of quaternary $Zn_{0.93}Mg_{0.07}S_{0.18}Se_{0.82}$ alloy viewed from the $[1\bar{1}0]$ direction. Figure 4(b) shows a schematic representation of fig. 4(a). The sample was rotated during epitaxial growth. Lateral stripes were observed in the ZnMgSSe alloy and these stripes were caused by the compositional oscillation due to the substrate rotation. We can use this lateral stripes as the marker of the growth front. The growth front is flat at the beginning and then becomes corrugated. Vertical stripes were also observed with a period of 45 nm. When the growth front becomes corrugated, the vertical stripes appear. Vertical stripes were also caused by the composition modulation. This indicates that vertical composition modulation was associated with the corrugated structures.

To clarify the modulated contrast in the TEM image, we performed nano-probe EDS analysis of the ZnMgSSe alloy. Figure 5(a) shows a 002 dark field cross-sectional TEM image of the ZnMgSSe alloy grown on GaAs without sample rotation. A sinusoidal corrugated interface was observed. EDS measurement was performed 10 nm below the interface as indicated in Fig.5(b). The result is shown in Fig.5(c). The top of the corrugation is Mg and Se-rich and Zn and S-deficient, while the valley of the corrugation is Zn and S-rich and Mg and Se-deficient. We note that in Fig.5 (a), the interface contrast is somewhat darker. This contrast is due to additional strain created by the corrugations.

Next, we will discuss the mechanism of formation of the composition modulation. The corrugated structures observed here were formed mainly due to

162

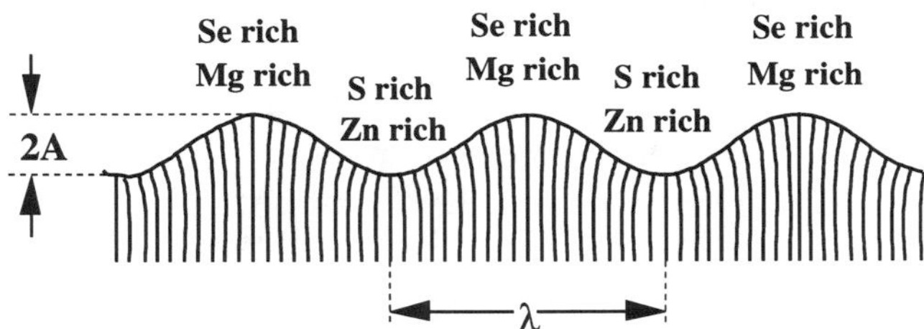

Figure 6. Schematic illustration showing the stress relaxation at the regions of peaks and stress concenteration at the regions of valleys for a sinusoidally corrugated surface.

growth kinetics [5]. In addition, the corrugated structures were formed not only in ternary ZnSSe and quaternary ZnMgSSe epilayer but also in binary ZnSe epilayer as was observed in Fig.1. Thus the formation of the one-dimensional composition modulation is not necessary for the driving force of the formation of corrugated structures. On the contrary, the formation of corrugated structures creates the composition modulation, as we will discuss below. When the surface becomes corrugated, an additional stress field is caused. The lattice spacing in the valleys and at the peaks is affected by compressive and tensile stress, respectively as illustrated in Fig.6. The stress $\sigma_\tau(x)$ tangential to the surface of a sinusoidally corrugated surface has been considered by Gao [6] as,

$$\sigma_\tau(x) = \sigma_B\left[1 + 4\pi\frac{A}{\lambda}\cos\left(\frac{2\pi}{\lambda}x\right)\right] \tag{1}$$

Here, σ_B is the normal bulk stress in epitaxial layer, 2A is the corrugation height, λ is the period of the corrugation, and x-axies is defined along the [110] direction in this case. For ZnSSe in Fig.1(b), λ is 60 nm and 2A is 10 nm, therefore, σ_τ/σ_B=2.047 at the valley. The stress at the valleys is higher than the stress at the peaks. Thus, the strain energy density at the peaks is lower than the strain energy density at the valleys. This difference in strain energy densities creates a gradient in surface chemical potential given by [7],

$$\mu(x) = \mu^* + \gamma\Omega\kappa(x) + \left\{\left[\sigma_\tau(x)\right]^2 - \sigma_B^2\right\}\frac{\Omega}{2E} \tag{2}$$

where μ^* is the chemical potential of the unstressed flat surface, Ω is the atomic volume, and γ is the surface free energy per unit area. The second term, involving the surface free γ energy per unit area and the surface curvature κ, reflects the fact that the surface would prefer to become flat in order to minimize the surface energy. The third term, involving the atomic volume Ω and the Young's modulus E, represents local elastic energy correction for a stressed surface. Since the lateral diffusion of adatoms is driven by the gradient in surface chemical potential, adatoms tend to move from the regions of high strain energy (valleys) to the regions of low strain energy (peaks).

Now, let us consider when epilayer is composed of several species. We have several species adatoms such as Zn, Mg, S and Se. Each adatom feels a different stress field depending on the species when they landed on the surface. To take into account of this effect, we simply modify $\mu(x)$ and σ_B of eq.(2) into $\mu^i(x)$ and $\sigma^i_B(x,)$, respectively, where the superscript i denotes the adatom species, and $\sigma^i_B(x)$ is the modified local bulk stress of species i., which is taken account of adatom size. We observed that the valleys contained more Zn and S and the peaks contained more Mg and Se. The lattice constants of the binary compounds of Zn, Mg, S and Se have the following size order [8]:

$$ZnS(0.541) < MgS(0.562) < GaAs(0.565) < ZnSe(0.567) < MgSe(0.586),$$
where the number in the bracket is the lattice constant in nm. The compounds containing Zn and S tend to have smaller lattice constant, while the compounds containing Mg and Se tend to have larger lattice constant. The average lattice constants of ternary alloy ZnSSe and quaternary alloy ZnMgSSe in this study are close to that of GaAs. Thus, Zn adatoms and S adatoms migrate towards the valleys and Mg and Se adatoms migrate towards the peaks to minimize strain energy.

4. Summary

We have investigated surface roughening and composition modulation of ZnSe-related II-VI epitaxial films grown by MBE. Under II-rich conditions with a c(2x2) surface, corrugations aligned in the [110] direction are observed. Under VI-rich conditions with a (2x1) surface, rounded grains form. Composition modulation along the [110] direction was observed in both ZnSSe and ZnMgSSe alloy epitaxial layers and corresponds to surface roughness. Since the corrugated surface produces additional strain, composition modulation occurs in order to minimize strain at the surface of the film.

Acknowledgments

We would like to acknowledge Y. Yamada and Y. Hayafuji for their encouragement of this study. We also thank H. Okuyama for helpful discussions.

References

1. N. Nakayama, H. Okuyama, E. Kato, S. Itoh, M. Ozawa, T. Ohata, K. Nakano, M. Ikeda, A. Ishibashi and Y. Mori, Electron Lett. 30, 568 (1994).
2. G. C. Hua, N. Otsuka, D.C. Grillo, J. Han, L. He, and R.L. Gunshor, J. Crystal Growth **138**, 367 (1994).
3. L. H. Kuo, L. Salamanca-Riba, B. J. Wu, J. M. DePuydt, G.M. Haugen, H. CHeng, S. Guha, and M.A. Hasse, Appl. Phys. Lett. **65**, 1230 (1994)
4. H. Okuyama, T. Kawasumi, A.Ishibashi and M. Ikeda (to be published)
5. S. Tomiya. R. Minatoya, H. Tsukamoto, S. Itoh, K. Nakano, E. Morita, A. Ishibashi, ICPS proceedings 1079, **3** (1996); (submitted to J. Appl. Phys.)
6. H. Guo, J. Mech. Phys. Solids **39**, 443 (1991) .
7. D. J. Srolovitz, Acta metall. **37**, 621 (1989)
8. H. Okuyama, K. Nakano, T. Miyajima, and K. Akimoto, Jpn. J. Appl. Phys., **30**, L1620 (1991)

DIRECT BANDGAP QUANTUM WELLS ON GaP

Jong-Won Lee[a], Alfred T. Schremer, Dan Fekete[b], James R. Shealy, and Joseph M. Ballantyne
School of Electrical Engineering, Cornell University, Ithaca, NY 14853

ABSTRACT

Currently, there are no direct-bandgap alloy semiconductors that can be grown lattice-matched to GaP substrates. A strained layer of GaInP can be grown on GaP, however, with difficulties. First, GaInP is an indirect-bandgap material for In concentrations up to ~30%. Second, the band alignment between GaInP and GaP is type-II for In concentrations up to ~60%. The Mathews-Blakeslee critical thickness of GaInP layer on GaP is prohibitively small in the useful In concentration range. GaInP is known to grow in an ordered phase in certain growth conditions. By changing the growth conditions, a heterojunction of ordered GaInP and disordered GaInP can be grown. The conduction band offset going from a disordered GaInP phase to an ordered GaInP phase has been reported to be about 150 meV. Using a layer of ordered GaInP, a QW with type-I band alignment may be grown on GaP for a wider range of composition.

We have grown a series of approximately 60 Å thick GaP/GaInP/GaP strained quantum wells of various compositions using OMVPE. Strong photoluminescence, which exhibited an unusual temperature dependence, has been observed on many samples. A study of the QW's using X-ray diffraction, TEM, and variable temperature PL reveals behaviors consistent with direct bandgap GaInP quantum wells containing ordered and disordered domains.

INTRODUCTION

One of the hurdles in achieving integration of optoelectronics into Si-based electronics is the lack of light emitting materials that can be grown on Si with few defects. III-V compounds such as GaP and AlP have lattice constants close to that of Si. GaP can be grown on Si with a low defect density [2,3]. GaP and AlP have indirect bandgaps, and therefore they cannot be used as the active regions of light-emitting diodes or laser devices. Strained layers of other direct bandgap III-V compounds such as GaInP can be used as the active region. Thus, if one can grow a light-emitting structure with a strained GaInP layer on GaP, in conjunction with the ability to grow GaP on Si, one can achieve the goal of monolithic integration of optoelectronics into Si-based electronics.

Recently, growth of direct-bandgap GaInP/GaP strained quantum well has been reported. [3,4] It was suggested that GaInP contained both ordered and disordered phases, based on photoluimnescence observations. [4] In this paper, we present further evidences to substantiate the claim, and determination of the energy levels and the ordering parameter of the quantum wells.

EXPERIMENTAL

A series of GaInP strained layers of various compositions were grown on vicinal [001] GaP substrates by low-pressure metalorganic chemical vapor deposition (MOCVD). According to TEM, the quantum wells consisted of a 1000Å GaP buffer layer, a 80Å - 90Å thick InGaP layer, and a 1000Å GaP cap layer. TEGa, TMIn, and PH_3 were used as source materials. The quantum wells were grown at 650°C in a 76 Torr H_2 ambient. The growth rate was ~ 4800Å/hr. A 363 nm line from an Ar ion laser was used as the pump for the photoluminescence measurement. The photoluminescence spectra were taken after the samples were cooled down to 79K. A thick layer (5000Å) of GaInP was grown on GaP and was analyzed with high-resolution x-ray diffraction. Using the composition of the thick layer determined with x-ray diffraction as a

---

[a] Corresponding author: jongl@msc.cornell.edu
[b] On leave from Technion-Israel Institute of Technology, Haifa, Israel.

Mat. Res. Soc. Symp. Proc. Vol. 448 © 1997 Materials Research Society

calibration point, the compositions of the quantum wells were estimated from the group III precursor flow rates. High-resolution cross-section TEM was employed to measure the thicknesses of the quantum wells.

RESULTS AND DISCUSSIONS

According to the model-solid calculation (reviewed in Ref. 5), the band line-up between GaInP and GaP is type-II for In concentrations up to 58%; the Γ-point GaInP band edge does not fall below the X-point GaP band edge until the In concentration goes up to 58%. Therefore, an electronic confined state does not exist for much of In concentrations. Without both electrons and holes confined, low emission efficiency is expected for low In concentrations (<58%). However, quantum well photoluminescence was observed for In concentrations as low as 45%. The quantum well emission was at around 634 nm as soon as it became observable. The emission wavelength didn't vary much with increasing In concentration. When it did, the shift in emission wavelength was abrupt; at In concentration of ~ 62%, a yellow-green 580 nm emission appeared, (Fig. 1) and then at higher In concentrations the red 634 nm emission vanished. Another peculiar feature about the quantum well photoluminescence was its temperature dependence. (Fig. 2) At low temperatures (<100K), photoluminescence became significantly weaker, as if there were another electronic state in which charge carriers did not contribute to radiative transition, with comparable density of states as the quantum well electronic state. The

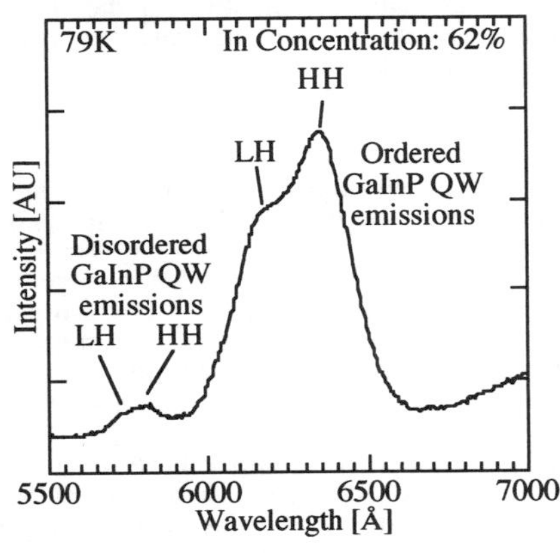

Fig. 1. 79K photoluminescence from GaInP/GaP quantum well with In concentration of 62%. GaInP quantum well contains both ordered and disordered domains. Light and heavy hole degeneracy is lifted by presence of elastic strain and ordering.

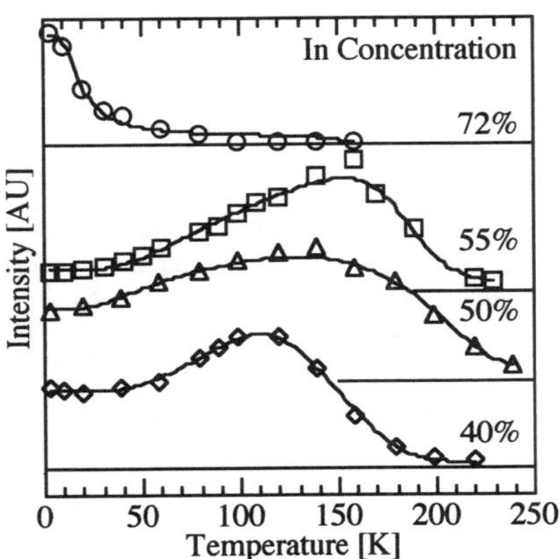

Fig. 2. Photoluminescence peak intensity vs. sample temperature for various In concentrations. Circles, squares, triangles, and rhombi are experimental data. Solid curves are calculated curvefits. For 40%, 50%, and 55%, the points are the peak intensities of 640 nm peak. For 72%, the points are the peak intensities of 580nm peak.

temperature at which the emission intensity was the maximum depended on the GaInP composition. Note that the 580nm peak exhibits behavior expected from a regular quantum well.

One possible explanation is that GaInP in the quantum well undergoes spontaneous long-range ordering during growth, and the quantum well contains both ordered and disordered phases of GaInP. Spontaneous ordering of GaInP has been reported to shrink the bandgap. [6] In the disordered domains, the band lineup between GaInP and GaP is type-II, and no electronic confined states exist. In the ordered domains, the band lineup is type-I, and an electronic confined state exists. According to ref. 7, the band lineup between ordered GaInP and disordered GaInP is type-II with valence band of the ordered phase lower than that of the disordered phase. Therefore, the boundary between the ordered and disordered domains of the quantum well form an interface with type-II band line-up with the hole energy level in the ordered domain lower than the hole energy level in the disordered domain. (Fig. 3a.)

At low temperatures, electrons are confined to the ordered domains, and holes are confined to the disordered domains; electrons and holes are spatially separated, thus reducing radiative recombination efficiency. At high temperatures, holes excite into the ordered GaInP domains; electrons and holes are confined together thus boosting radiative recombination efficiency. Based on these assumptions, an expression for emission intensity as a function of temperature can be derived. [8,9]

$$I = \frac{P\left(\dfrac{1}{1 + e^{\Delta E_V/k_B T}} + \dfrac{R_i/R_d}{1 + e^{-\Delta E_V/k_B T}}\right)}{\left(1 + \dfrac{R'}{U}\right)\left(1 + \dfrac{U_1}{R_d} e^{-E_1/k_B T}\right) + \dfrac{R'}{R_d} e^{-\Delta E_a/k_B T}} \tag{1}$$

P is the photo-excitation rate in the barriers. R_d and R_i are the radiative recombination rate constants for the direct and the spatially-separated transitions. U and U $e^{-\Delta E_a/kT}$ are the capture rate into and the escape rate from the quantum well of electron-hole pair complexes, respectively. U_1 is the thermal escape rate from the quantum well of individual carriers. R' is the rate of non-radiative recombination in the barriers. ΔE_v is the energy difference between the hole energies in the ordered GaInP domain and the disordered GaInP domain. ΔE_a is the energy difference between the GaP indirect band gap and quantum well luminescence energy. E_1 is the lesser of the electron and hole confinement energies. The curvefits are plotted in Fig. 2. The excellent agreement between the experimental data and the calculated curves supports the validity of our assumptions.

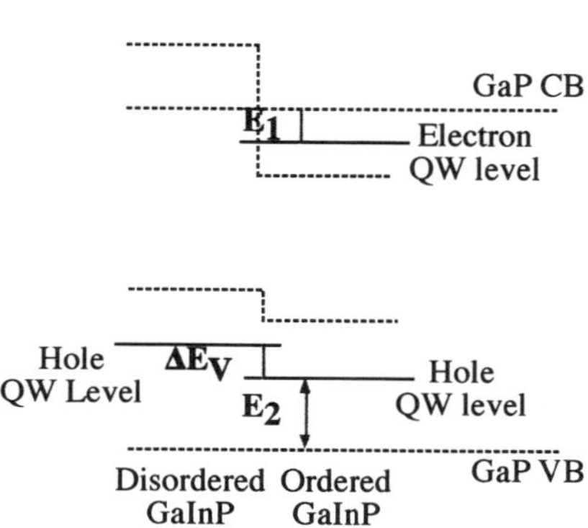

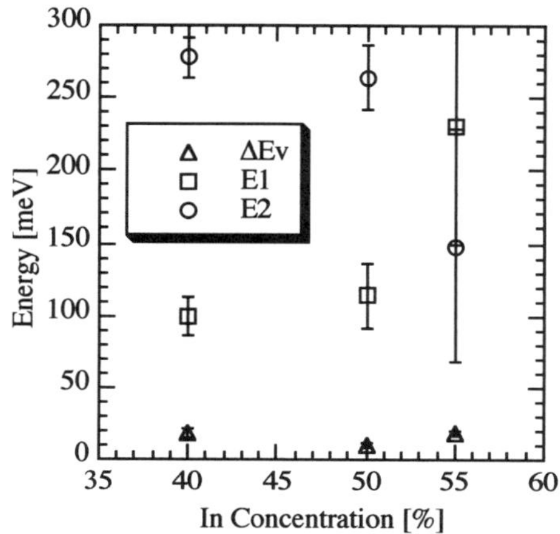

Fig. 3a. A schematic of band structure of a GaInP/GaP ordered/disordered heterogeneous quantum well.

Fig. 3b. Various energy levels in quantum well extracted from the curvefit. Energies are indicated in Fig. 3a.

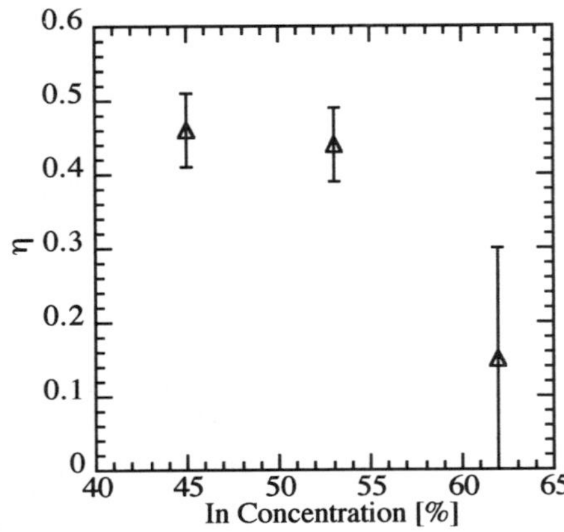

Fig. 4. Long-range ordering parameter of ordered GaInP domains versus In concentration. The extent of ordering reduces with increasing In concentration.

As the In concentration is increased, even the disordered domains eventually become type-I quantum well. Thus ordered domains produce the red 634 nm emission, and the disordered domains produce the yellow-green 580nm emission. At high In concentrations, excessive strain prevents spontaneous ordering from occurring, and only ordered domains remain; only the yellow-green emission is observed. The intensity of the yellow-green emission versus the temperature shows the usual behavior. (Fig. 2)

From the curvefit, all the energy levels in the quantum well can be extracted. The electron confinement energy, the hole confinement energy, and the energy barrier between the disordered and the ordered domains as functions of the In concentration are plotted in Fig. 3b.

Another quantity of interest is the degree of ordering in GaInP. The degree of ordering is described in terms of the long-range ordering parameter, η. η is defined as follows: GaInP crystal consists of alternating (111) planes of $Ga_{x-\eta/2}In_{x+\eta/2}P$ and $Ga_{x+\eta/2}In_{x-\eta/2}P$. The effect of spontaneous ordering to the band structure of GaInP has been previously reported. [10] The ordering produces shift of band edges in addition to any shift produced by elastic strain. In particular, the splitting between the heavy-hole band and light-hole band can be used determine the ordering parameter. Following the method described in Ref. 10, the band-edges of GaInP versus composition taking into account the elastic strain and the ordering have been calculated. Then from these energies, emission energies of transitions involving the heavy-hole and the light-hole were calculated. The calculated values were compared to the experimentally measured differences between the heavy-hole and the light-hole transitions to determine the ordering parameter. Thus determined ordering parameter is plotted in Fig. 4. As suggested earlier, the extent of ordering reduces as the In concentration is increased.

CONCLUSIONS

Strained GaInP layers were grown on GaP by MOCVD. Even though the quantum wells were thicker than the critical layer thickness, strong photoluminescence was observed. Careful study of photoluminescence revealed that the quantum well consisted of both ordered and disordered GaInP. Electronic energy levels within the quantum well structure were determined. The ordering parameter of GaInP quantum were was also determined from photoluminescence. Some quantum wells had photoluminescence persisting at room temperature suggesting this material structure could be used as the active region of light-emitting devices on GaP as well as on Si substrates.

168

ACKNOWLEDGMENTS

This work was done at Cornell OMVPE Facility and Cornell Materials Science Center (funded by NSF). This work was supported by the Semiconductor Research Corporation under the contract number 96-SC-419. Supplementary support came from the USAF and ARPA under contracts F3063-92-C0082, F3062-95-I0041, and MDA 972-94-1-0002.

REFERENCES

1. T. Suzuki, T. Soga, T. Jimbo, M. Umeno, J. Crystal Growth **115**, 158 (1995).

2. J.S. Lee, J. Salzman, D. Emerson, J.R. Shealy, J.M. Ballantyne, Materials Research Symposium Proceedings Vol. 417, p301 (1996).

3. J.G. Neff, M.R. Islam, R.V. Chelakara, K.G. Fertitta, F.J. Ciuba, R.D. Dupuis, Journal of Crystal Growth **145**, 746 (1995).

4. J.-W. Lee, A. Schremer, D. Fekete, J. Ballantyne, 9th annual meeting of the IEEE Lasers and Electro-Optics Society (IEEE/LEOS), Conference Proceedings, WQ4, Boston, MA, Nov. 18-Nov. 21 (1996)

5. C.G. Van de Walle, Phys. Rev. B, **39**, 1871 (1989)

6. L.C. Su, I.H. Ho, and G.B. Stringfellow, Appl. Phys. Lett., **65**, 749 (1994).

7. R. P. Schneider, Jr., E. D. Jones, and D. M. Follstaedt,. Appl. Phys. Lett. **65**, 587 (1994).

8. J.R. Botha and A.W.R. Leitch, Physical Review B, **50**, 18147 (1994)

9. J.-W. Lee, A.T. Schremer, D. Fekete, and J.M. Ballantyne, Applied Physics Letters, in press

10. Su-Huai Wei, A. Franceschetti, and A. Zunger MRS symposium proceedings edited by E.D. Jones, A. Mascarenhas, and P. Petroff (1995, Boston) **417**, 3. Su-Huai Wei and A. Zunger, Applied Physics Letters, **64**, 757 (1994).

(1971)]

3. T. Suemasu, M. Watanabe, J. Suzuki, Y. Kohno, M. Asada and N. Suzuki, Jpn. J. Appl. Phys. **33,** 57 (1994).

4. M. Hirose, M. Morita and Y. Osaka, Jpn. J. Appl. Phys. **16,** suppl.16-1, 561 (1977).

5. Q. Ye, R. Tsu and E. H. Nicollian, Phys. Rev. B **44,** 1806 (1991).

6. S. Y. Chou and A. E. Gordon, Appl. Phys. Lett. **60,** 1827 (1992).

7. Y. Yuki, Y. Hirai, K. Morimoto, K. Inoue, M. Niwa and J. Yasui, Jpn. J. appl. Phys. **34,** 860 (1995).

8. M. Fukuda, K. Nakagawa, S. Miyazaki and M. Hirose, Ext. Abs. 1996 Int. Conf. on Solid State Devices and Materials, (1996, Yokohama) 175.

9. X. Zhou and W. P. Kirk, Mat. Res. Soc. Symp. Proc. **318,** 207 (1994).

10. J. Kolodzey, S. Zhang, P. O'Neil, E. Hall, R. McAnnally and C. P. Swann, Proc. Int. Conf. on SiC and Rel. Mat., Inst. Phys. Conf. Ser. No. 137, 357 (1993).

11. S. C. Kan, H. Mrkoc and A. Yariv, Appl. Phys. Lett. **52,** 2250 (1988).

12. in <u>Crystals with the fluorine structure</u>, edited by W. Hayes (Oxford univ. press, 1974).

13. R.T. Poole and D. R. William, Chem. Phys. Lett. **36,** 401 (1975).

14. D. Shanarabny, M. Wolf and D. Gerlich, J. Phys. Chem. Solids, **37,** 577 (1976).

15 S. M. Sze in <u>Physics of Semiconductor Devices 2nd ed.</u>, (Wiley Interscience, 1981).

16. N. S. Sokolov, S. V. Gastev, S. V. Novikov, N. L. Yakovlev, A. Izumi and S. Furukawa, Appl. Phys. Lett. **64,** 2964 (1994).

17. N. S. Sokolov, N. N. Faleev, S. V. Gastev, N. L. Yakovlev, A. Izumi and K. Tsutsui, J. Vac. Sci. Technol. A **13,** 2703 (1995).

18. A. Izumi, K. Tsutsui, N. S. Sokolov, N. N. Faleev, S. V. Gastev, S. V. Novikov, N. L. Yakovlev, J. Crystal Growth **150,** 1115 (1995).

19. A. Izumi, K. Kawabata, K. Tsutsui, N. S. Sokolov, S. V. Novikov, A. Yu. Khilko, Appl. Surf. Sci. (1996) to be published.

20. A. Izumi, T. Hirai, K. Tsutsui and N. S. Sokolov, Appl. Phys. Lett. **67,** 2792 (1995).

21. L. J. Schowalter and R. W. Fathauer, CRC Critical Rev. **15,** 367 (1989).

22. T. Asano, H. Ishiwara and N. Kaifu, Jpn. J. Appl. Phys. **22,** 1474 (1983).

23. N. S. Sokolov, T. Hirai, K. Kawasaki, S. Ohmi, K. Tsutsui, S. Furukawa, I. Takahashi, Y. Itoh and J. Harada, Jpn. J. Appl. Phys. **4B,** 2395 (1994).

24. A. Ishizaka and Y. Shiraki, J. Electrochem. Soc. **133,** 666 (1986).

25. R. P. Khosla and D. Matz, Solid State Commun. **6,** 859 (1968).

FORMATION OF LARGE CONDUCTION BAND DISCONTINUITIES OF HETEROINTERFACES USING CdF_2 AND CaF_2 ON Si(111)

Akira Izumi, Noriyuki Matsubara[*‡], Yusuke Kushida[*], Kazuo Tsutsui[*] and Nikolai S. Sokolov[**]

School of Materials Science, JAIST (Japan Advanced Institute of Science and Technology)
1-1 Asahidai Tatsunokuchi, Ishikawa 923-12, Japan

*Interdisciplinary Graduate School of Science and Engineering, Tokyo Institute of Technology
4259 Nagatsuta Midori-ku, Yokohama 226, Japan

**Ioffe Physico-Technical Institute, Russian Academy of Sciences
St. Petersburg 194021, Russia

ABSTRACT

We proposed use of a new CdF_2/CaF_2 heterointerface for the formation of large conduction band discontinuities to apply quantum effect devices fabricated on Si substrates. Resonant tunneling diodes using this heterointerface on Si were fabricated and negative differential resistance whose P/V current ratio of 24 at highest was observed at room temperature.

INTRODUCTION

Quantum effect devices fabricated on Si substrate will be very attractive in future [1]. Abrupt heterointerfaces which have the large conduction band discontinuity (ΔE_c) are very desirable for them. Such heterointerfaces are very useful not only for the operation of electrical devices such as resonant tunneling devices but also for short wave length emission of intersubband laser [2]. A lot of different materials such as CaF_2 [3], SiO_2 [4-8], ZnS [9], ternaries based on Mg and Se, and SiGeC [10] were proposed for application in quantum effect devices.To date, most successful resonant tunneling diode (RTD) fabricated on Si substrates has been obtained by the III-V heterojunctions on Si [11]. However, in the III-V on Si case, a fairly thick buffer layer of III-V is necessary in order to obtain good crystallinity. Moreover, autodoping between III-V and Si becomes problem in device fabrication process. Thus, materials and structures which do not have such problems and make it possible to realize quantum structures directly formed on Si surface are very desirable.

We suggested a use of new CdF_2/CaF_2 interface for the formation of large ΔE_cs which is fabricated on Si. Some properties of CdF_2 and CaF_2 are summarized in Table 1 comparing with that of Si [12-15]. CdF_2 and CaF_2 have the fluorite lattice structure nearly lattice matched to Si, so one can expect that they can be grown on Si epitaxially. Recently we realized epitaxial growth

---

‡ Present address YAMAHA Corp., Iwata, Shizuoka 438-01, Japan

Mat. Res. Soc. Symp. Proc. Vol. 448 © 1997 Materials Research Society

Table 1 Material constants of CdF_2, CaF_2 and Si at room temperature.

Materials	CdF_2	CaF_2	Si
Structure	Fluorite	Fluorite	Diamond
Lattice constant (A)	5.388	5.463	5.431
Mismatch with Si(%)	-0.8	0.6	–
Melting point (°C)	1100	1360	1414
Thermal expansion coef.at RT (deg⁻¹)	2.1×10^{-5}	1.82×10^{-5}	2.5×10^{-6}
Energy bandgap (eV)	8.0	12.11	1.1
Electron affinity (eV)	4.1	0<	

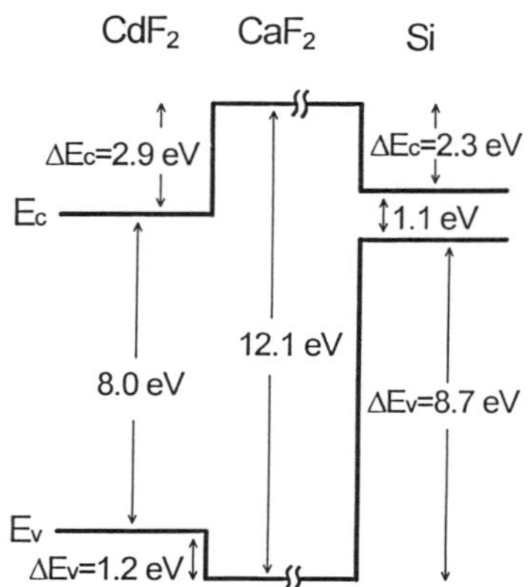

Fig.1 Energy band diagram for the CdF_2/CaF_2 /Si(111) heterostructure obtained by XPS.

CdF_2 on CaF_2/Si(111) for the first time [16].

In addition, it was found that an epitaxial CdF_2 film can be grown with high crystalline quality on a Si substrate by using a CaF_2 buffer layer which is coherently grown on the Si substrate (pseudomorphic CaF_2) [17-18], and abrupt interfaces of CdF_2/CaF_2 were confirmed by double crystal x-ray diffractometory (XRD) observation of a short-period CdF_2/CaF_2 superlattice (SL) which was grown on CaF_2/Si(111) [19]. Moreover, an important property of the CdF_2/CaF_2 heterointerface, large ΔE_c can be expected from the difference of the electron affinities of them. Indeed, we have revealed through the x-ray photoelectron spectroscopy analysis that CaF_2 provides large ΔE_c of CdF_2/CaF_2 and CaF_2/Si, these values are 2.9eV and 2.3eV, respectively and conduction band edge (E_c) of CdF_2 is located 0.6eV below that of Si as shown in Fig.1 [20]. Based on these fundamental results, CdF_2/CaF_2/Si(111) heterostructure is considered to be an attractive candidate for the applications mentioned above.

In this paper, we propose a use of new CdF_2/CaF_2 heterointerface for the formation of large ΔE_cs to apply quantum effect devices fabricated on Si substrates and demonstrate CdF_2/CaF_2 RTD grown directly on Si(111) substrate to confirm these large ΔE_cs.

EPITAXIAL GROWTH OF CdF₂

Epitaxial growth of CaF_2 is established technology and there are a lot of reports concerning this technology [21-22]. But in the case of CdF_2, it was no report before our success of epitaxial growth of CdF_2 by molecular epitaxy method. In order to obtain epitaxial CdF_2, we investigated the growth conditions for it. It was shown that use of a thin CaF_2 buffer layer was very effective to obtain good CdF_2 epitaxial film on Si(111) [19]. Figure 2 shows θ scanning of XRD rocking curve of a CdF_2(50nm)/CaF_2(10nm)/Si(111) structure, grown in Tokyo Institute of Technology. Growth temperature of CdF_2 was 50°C. CaF_2 buffer layer was grown by two step growth technique where several monolayers were grown at 600°C followed by succeeding growth at 100°C until the total thickness reached 10nm [23]. Minimum value of full width of half maximum (FWHM) of 67 arcsec was obtained. In the case of growth at 300°C the rocking curve became too broad to evaluate FWHM. Although an origin of the degradation at the higher growth temperatures is not clear at moment, it was found that CdF_2 layer should be grown at fairly lower temperature.

Figure 3 shows θ-2θ, double crystal XRD patterns obtained from a short-period SL of $[CdF_2$-0.9nm/CaF_2-0.9nm$]_{10}$ which was grown at Ioffe Physico-Technical Institute on the pseudomorphic CaF_2 layer in previous work [17-18]. The growth temperature of SL was 100°C. These satellite peaks show that the layered heterostructure has remained without being smeared at all by interdiffussion. In addition, the pronounced oscillating structure shows well flatness of the SL structure. Because each layer in this SL consisted only of three fluoride monolayer, one can conclude that the CaF_2/CdF_2 interface is abrupt and its average roughness is less than one monolayer.

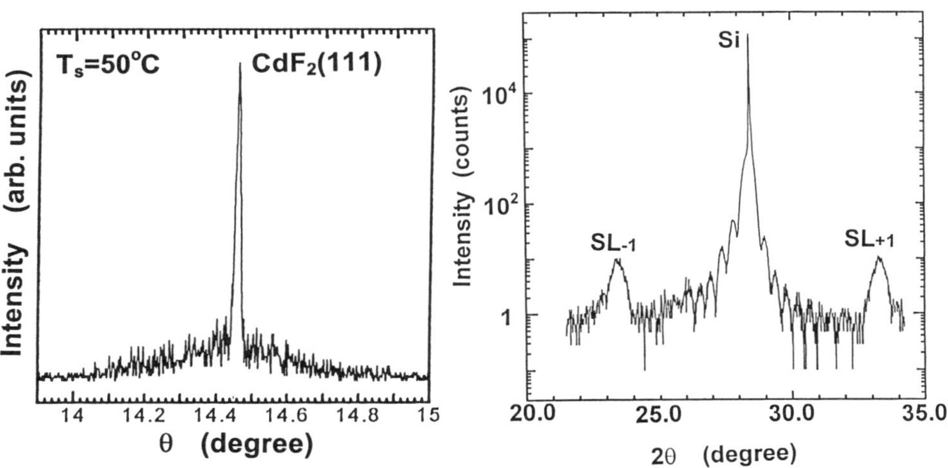

Fig.2 θ scanning of XRD rocking curve of CdF_2(50nm)/CaF_2(10nm)/Si(111) structure.

Fig.3 Double crystal XRD patterns obtained from a short-period SL of $[CdF_2$-0.9nm/CaF_2-0.9nm$]_{10}$ which was grown on the psudomorphic CaF_2 layer.

RESONANT TUNNELING DIODE

Process of fabrication

Figure 4 shows the cross-sectional structure of a RTD fabricated in this work and a corresponding conduction edge energy diagram of the structure. This diagram was drawn based on the energy band diagram of $CdF_2/CaF_2/Si(111)$ which was shown in Fig.1 [19]. The diode structure is asymmetric in which the double barrier structure of $CaF_2/CdF_2/CaF_2$ is sandwiched between n^+-Si in one side and Al in the other side. Thus, negative differential resistance (NDR) characteristic is expected to be observed only when the Si side is negatively biased. The previous work showed that thin but more than 0.9nm CaF_2 buffer layer was necessary to obtain good CdF_2 epilayer on Si(111) [19]. This buffer layer served as the first barrier layer of the double barrier structure in this RTD.

The RTD structure was grown by MBE method as follows on the clean Si(111) substrate referred to the results as we mentioned above. An n-type ($N_D \sim 2x10^{18}cm^{-3}$) Si(111) substrate was cleaned by the Shiraki-Ishizawa method [24], then its surface was terminated with hydrogen by 2% diluted HF dip cleaning. After that, it was loaded into the high vacuum chamber ($\sim 10^{-8}$Torr), followed by thermal flashing at 600°C for 15min. The first CaF_2 layer (0.9nm) combining a buffer and a barrier layer was grown at 600°C. Then, 3.1nm-thick CdF_2 layer was grown at 50°C as the well layer referring to the result shown in Fig.1. After that, the second CaF_2 layer was grown at 200°C as the other barrier layer. Finally, back contact and top electrodes were formed by vacuum evaporation of Al. The top electrodes were patterned with circular pattern (ϕ100-800μm) using a hard mask. No photolithography technique was used for fabrication of the RTD.

Current-voltage characteristics

The current-voltage (I-V) characteristics were measured at room temperature (RT) using a HP-4145B parameter analyzer. Figure 5 shows a I-V curve with the P/V current ratio of 24 which was the largest value observedto date. The diameter of Al top electrode was 800μm. However, correlation between the electrode size and current was weak. One origin of the scattered characteristics and the nonuniform current in plane was considered to be localized leakage current through some defects or nonuniform tunnel current due to fluctuation of thickness of each layer. Especially, since very large barrier height and very thin layers are used in this structure, very small fluctuation of structure can cause tunnel current concentrated in a few limited small regions. The bias voltage at which the first NDR should be observed is 0.84 V from a simple theoretical calculation using the effective mass of CdF_2, m^*_{CdF2}=0.45 m_0 [25], which corresponds to resonant tunneling through third quantum level schematically shown in Fig.4(b). This voltage was observed from 0.8V to 2.0V for most samples. The fairly large deviation from the calculated value can be partly due to the voltage drop by the serial resistances and partly due to fluctuation of thickness of the layers.

Although further study of growth method and control of quality of the structure is necessary, the large P/V current ratio at RT indicates a possibility of quantum well structure with large ΔE_c which can be directly coupled with Si.

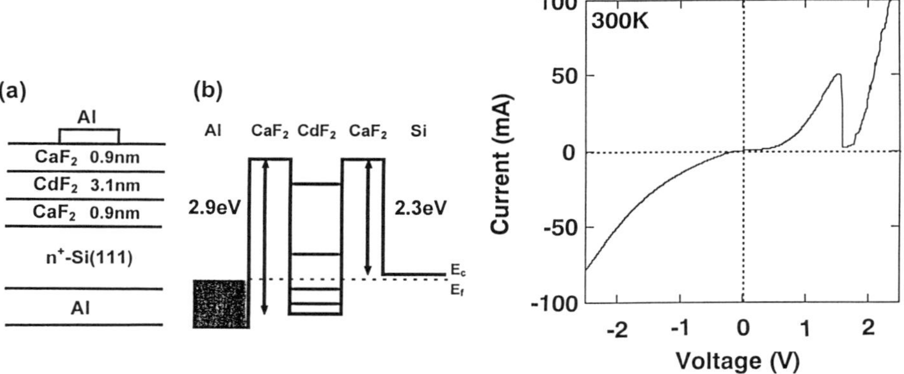

Fig.4 (a) RTD structure, (b) conduction band edge energy diagram of the structure.

Fig.5 The best I-V curve with large P/V current ratio.

CONCLUSIONS

We proposed use of a new CdF$_2$/CaF$_2$ heterointerface for the formation of large ΔE_cs to apply quantum effect devices fabricated on Si substrates. RTDs using this heterointerface were fabricated on Si and NDR whose P/V current ratio of 24 at highest was observed at RT. CdF$_2$ and fluoride heterointerfaces including it will be one of promising material for future devices based on Si.

ACKNOWLEDGMENTS

This work was supported by Kanagawa Academy of Science and Technology Research Grants and the Nippon Sheet Glass Foundation for Materials Science and Engineering. The authors acknowledge Associate Professors M. Asada and M. Watanabe of Tokyo Institute of Technology for fruitful discussions. They appreciate encouragement of this work by Prof. H. Matsumura of JAIST. They are thankful to Mr. A. Yu. Khilko for discussion of CdF$_2$ growth.

REFERENCES

1. A. C. Seabaugh, C. -C. Cho, R. M. Steinhoff, T. S. Moise, S. Tang, R. M. Wallance, E. A. Beam and Y. -C. Kao, Proc. 2nd Int. Workshop on Quantum Functional Devices (1995, Matsue, Japan) 32.
2. R. F. Kazarinov, R. A. Suris, Fiz. Tekh. Poluprov. **5,** 797 (1971) [Sov. Phys. Semicond. **5,** 707

MODIFICATION OF THE SURFACE BAND-BENDING OF A SILICON CCD FOR LOW-ENERGY ELECTRON DETECTION

Aimée L. Smith*, Qiuming Yu, S. T. Elliott, T.A. Tombrello**, and Shouleh Nikzad
Center for Space Microelectronic Technology, Jet Propulsion laboratory
California Institute of Technology, Pasadena, CA 91109
*currently at Massachusetts Institute of Technology, Cambridge, MA 02139
**California Institute of Technology, Pasadena, CA 91109

ABSTRACT

Silicon CCDs have limited sensitivity to particles and photons with short penetration depth, due to the surface depletion caused by the inherent positive charge in the native oxide. Because of surface depletion, internally-generated electrons are trapped near the irradiated surface and therefore cannot be transported to the detection circuitry. This deleterious surface potential can be eliminated by low-temperature molecular beam epitaxial (MBE) growth of a delta-doped layer on the Si surface. This effect has been demonstrated through achievement of 100% internal quantum efficiency for UV photons detected with delta-doped CCDs.

In this paper, we will discuss the modification of the band bending near the CCD surface by low-temperature MBE and report the application of delta-doped CCDs to low-energy electron detection. We show that modification of the surface can greatly improve sensitivity to low-energy electrons. Measurements comparing the response of delta-doped CCDs with untreated CCDs were made in the 50 eV-1.5 keV energy range. For electrons with energies below 300 eV, the signal from untreated CCDs was below the detection limit for our apparatus, and data are presented only for the response of delta-doped CCDs at these energies. The effects of multiple electron hole pair (EHP) production and backscattering on the observed signals are discussed.

INTRODUCTION

Imaging systems for low energy particles generally involve the use of microchannel plate electron multipliers followed by position sensitive solid state detectors, or phosphors and position sensitive photon detectors. These systems work well and can process up to 10^6 electrons/sec., however, the spatial resolution of these compound systems is considerably less than that of a directly imaged CCD. Also, these systems have difficulties with gain stability and they require high voltages. The present large format of CCDs, up to 4000x4000 pixels, could represent a major advance for the imaging of low energy particles. CCDs exhibit a highly linear response which is advantageous for quantitative detection applications. The full well capacity of buried channel CCDs corresponds to a collected electron density of about 10^{11} electrons/cm^2, which together with the low readout noise, gives CCDs a large dynamic range.

Charge coupled devices (CCDs) are high resolution imaging devices which are typically n-channel fabricated in a p-type substrate and frontside, or processed-side, illuminated. Incident radiation is required to penetrate the CCD polycrystalline silicon gates (typically ~5000 Å) before being able to generate electron-hole pairs (EHP) in the pixel. This configuration makes radiation of low penetration depth undetectable. One attempt to eliminate this problem involves turning the chip around in order to illuminate from the back side, thus eliminating attenuation due to the CCD processed layers. Backside illumination requires removal of the thick p$^+$ substrate in order to bring the exposed back surface in close proximity to the intended frontside potential well. However, thinning the CCD by chemically removing the substrate is not sufficient to obtain high quantum efficiency, because positive charge in the native oxide traps electrons generated near the back surface of the CCD. Termination of a Si surface with SiO$_2$ leads to depletion of carriers at the surface, and in p-type Si the band bending due to surface depletion serves to create a surface potential well for electrons. This potential well can extend approximately 0.5 μm into the p-

Mat. Res. Soc. Symp. Proc. Vol. 448 © 1997 Materials Research Society

doped epilayer which comprises the back surface of the thinned CCD, making the CCD insensitive to radiation which generates electrons near the surface. Moreover, the width of the potential well is sensitive to illumination, leading to hysteresis in the response of the thinned CCD. Electrons generated in this surface potential region, or diffusing to this region, recombine and are never detected. Hoenk et al have successfully eliminated this effect for detection of UV light by MBE modifying the back surface with a p++ delta layer. Internal quantum efficiencies of unity were achieved[1] in the UV as well as visible wavelength regimes, and stability over years has been demonstrated.[2] The 100% internal quantum efficiency implies the detection of every electron generated by UV photons that have penetration depths of 40-100 Å.

Low-energy electrons also have short penetration depths in Si and transfer a fraction of their energy to the crystal through electron-hole pair (EHP) production, motivating the attempt to extend application of the delta-doped CCD to direct electron imaging. Previous work on electron detection with CCDs modified by ion implantation[3] and flash gate treatment demonstrated sensitivity down to electron energies of 0.9 keV.[4] Using delta-doped CCDs, we have successfully detected electrons down to 50 eV with high efficiency. This paper will briefly discuss the MBE modifications made to fully-processed CCDs and discuss the experimental results of application of the CCD to low energy electron detection.

Electrons with energies above 1.8 keV are capable of generating x-rays in silicon that can damage the gate oxide on the process-side of the device. While backside illumination provides some protection due to the 10-15 μm membrane of material between the region where incident electrons are likely to deposit their energy and the frontside gate oxide, low dark current for the device requires minimizing exposure to electrons of energy above 1.8 keV.

Delta-Doped CCDs

Delta-doped CCD processing is a recent development at JPL which uses MBE to enhance the UV response of back-illuminated CCDs by removing the dead layer associated with these devices. The general processing procedure is as described by Hoenk *et al.*[1] MBE modifications are made to the back surface of thinned, fully-processed CCDs by growing at low-temperature, 10 Å of boron-doped Si followed by deposition of 2×10^{14} B/cm^2, and a final 15 Å layer of undoped silicon. The delta-doping process is possible due to the development of low-temperature MBE technology. MBE allows for the growth of atomically sharp, high concentration doping profiles and low-temperature growth ensures that the processing temperatures do not approach 500°C, thereby avoiding dissolution of the silicon beneath the Al metallization, or spiking, of fully processed devices. During the *in-situ* preparation and subsequent MBE modification of the surface, the maximum temperature of the device is 450°C for a duration of four minutes. Boron diffusion is extremely slow at this temperature and therefore allows for an extremely thin layer of charge to be produced 5 Å from the Si/ SiO$_2$ interface. TEM analysis has demonstrated that this low-temperature MBE modification is defect free and unlike ion implantation, will not require annealing to remove damage or to incorporate boron onto lattice sites.[5]

Delta-doped CCDs have been extensively tested and have shown 100% internal quantum efficiency in the ultraviolet and visible part of the spectrum indicating that the deleterious backside potential well responsible for the detector dead layer has been effectively eliminated.

EXPERIMENT

To gain an understanding of different aspects of low-energy electron response of delta-doped CCDs, we performed measurements using various electron sources and different device configurations. The various setups, electron sources, device configurations, and the specific points that can be gleaned from each measurement are described below. The CCDs used in these experiments were thinned, back-illuminated EG&G Reticon CCDs. All measurements were repeated with both delta-doped and untreated CCDs. In some of the measurements, direct comparisons of delta-doped CCDs with untreated CCDs were made on the same device, using a delta-doped CCD which included a controlled (untreated) region. The controlled region was provided on the back surface of the array by masking off a portion of the surface during the MBE growth. All devices were fully-characterized prior to the electron measurements using UV illumination. Due to enhancement of quantum efficiency (QE) in the UV by the delta-doping process, the untreated region of the partially delta-doped device were readily apparent as dark regions in the image made with uniform exposure to incident light radiation, i.e. flat-field exposure, using 250 nm photons. For 250 nm light, with absorption length of approximately 70 Å in silicon,[6] the untreated region exhibited zero quantum efficiency whereas the delta-doped region exhibited reflection-limited response.

One set of measurements was performed in an SEM to take advantage of its highly-focused electron beam. The SEM apparatus was a JEOL, model JSM 6400, and the measurements were made with beam energies ranging between 200 eV and 1 keV. While it was not possible for modifications to be made to the SEM in order to accommodate the electronics necessary for collecting CCD images, performing photo-diode mode measurements was quite straightforward and informative. A CCD can be operated in such a way as to integrate the entire signal collected over the surface of the device, photo-diode mode, by grounding all pins except for the output amplifiers. The signal is then read from the pin of one of the output amplifiers, giving the compounded response of each of the pixels in the irradiated region of the device. Photo-diode mode measurements indicate the integrated response of the CCD to incident radiation and demonstrate the effect of the delta-doping treatment on overall collection efficiency. The fact that these measurements compound the response of all irradiated pixels into one measurement effectively averages out much of the error that would result in a pixel by pixel measurement. With the highly-focused beam of the SEM, we were able to make measurements in the untreated region as well as delta-doped regions and therefore directly observe the effect of the delta-doping process on collection efficiency. For each position measured on the surface of the device and for each energy, beam currents were first measured with a Faraday cup. CCD response to the electron beam at each position was measured in photo-diode mode, and finally, the beam current was again measured with the faraday cup to insure the stability of the beam current. Since the CCD is very sensitive to background light, response of the CCD was measured while deflecting the electron beam and it was found to be negligible.

Another set of measurements was made in a UHV system in photo-diode mode. For this mode of measurement, each CCD in turn was mounted in plane with a Faraday cup and a phosphor screen onto a manipulator. Using the custom UHV system afforded the use of two different electron sources, one of very low energy and one of similar energies as used in the SEM measurements. The low-energy electron gun is a hot-filament cathode that produces electron energies of several 10 eV while generating a strong light background. Comparison was made between the observed response of the CCD and the response of the CCD with the electron beam magnetically deflected. Because of the strong CCD response to the background light measurements with this electron gun beam are reported only qualitatively. The higher energy electron source which is a modified cathode ray tube (CRT) has reasonably stable beam energies varying from 300 eV to several keV. Photo-diode mode measurements were made with beam energies ranging from 300eV to 1000eV. Because it is an indirectly-heated cathode, this gun has very small background light, as was verified with our measurements. This background illumination was quantified by magnetically deflecting the electron beam. Repeated measurements were made on each CCD with calibration of the beam current in the Faraday cup both before and after each CCD measurement to insure beam stability. In this chamber geometry the beam spot was about one centimeter in diameter at the CCD. A circular aperture of 0.64 cm diameter (the same as the Faraday cup opening) was defined by a grounded aluminum sheet in front of the CCD to allow the exposure for the Faraday cup and the CCD to the same part of the electron beam.

The UHV system set-up further allowed for the later attachment of the electronics necessary for operating the CCD in imaging mode. This mode of operation allows for observation of electron irradiation on operating parameters only apparent in imaging mode such as charge transfer efficiency (CTE), individual pixel response, and surface charging. For using the CCD in the imaging mode, we mounted a camera directly onto the UHV chamber. The electron source used for these measurements was the indirectly-heated cathode gun. Because of the highly-sensitive imaging mode of operation, the incoming flux of electrons was controlled by using a mechanical shutter thereby taking snap shots of the beam in 10 msec to 2 second exposures. Preliminary measurements have been made at 500 eV and more measurements are underway.

At electron beam energies lower than the silicon K_α edge, there is no risk of damage to the silicon CCD due to the low absorption length of x-rays in silicon for this energy range. Electrons at energies higher than approximately 1.8 keV are capable of producing silicon K_α x-rays, which can penetrate the ~10-15 μm silicon membrane and damage the sensitive gate oxide on the front surface of the CCD. We verified the CCD's high tolerance to electrons at energies below the silicon K_α edge by exposing the delta-doped CCD to 1.5 keV electrons for several hours. Extensive UV testing was performed after this exposure as a test of effect of electron beam on the delta-doping treatment. No degradation of device performance was observed to result from exposure to electrons.

RESULTS

The response of a delta-doped CCD and an untreated backside-thinned CCD to electrons were repeatedly measured in the range of 200 eV through 1000 eV using the modified CRT and the SEM as sources. In figure 1, the electron quantum efficiency is plotted as a function of incident energy. Quantum efficiency was calculated by dividing the measured current from the CCD configured in photodiode mode to the measured electron beam current (measured by a Faraday cup), which is equivalent to the number of electron-hole pairs detected divided by the number of incident electrons. Because portions of the delta-doped CCD were masked during processing to serve as control regions, data taken in the UHV system were corrected to account for the fraction of untreated exposed CCD area. Due to the negligible response of the untreated back-illuminated CCD at these energies, it was assumed that the control region of the delta-doped CCD does not contribute to the signal. The measured quantum efficiency of the delta-doped CCD increases with increasing energy of the incident beam. The dependence of quantum efficiency on incident energy is due to the complicated interaction of electrons with silicon which results in the generation of multiple electron-hole pairs in the cascade initiated by each incident electron. A significant fraction of the incident energy is undetected, due to backscattering of incident electrons and other energy dissipation mechanisms (e.g., secondary and Auger electron emission), as discussed in the next section. Multiple electron-hole pair production, also known in the literature as quantum yield, is also observed in the measured UV and x-ray response of delta-doped CCDs and other devices. Quantum yield greater than unity has been previously observed in backside-illuminated CCDs modified using the flashgate[4] and ion implantation[7] at electron energies greater than 1 keV. Further discussion follows in the next section.

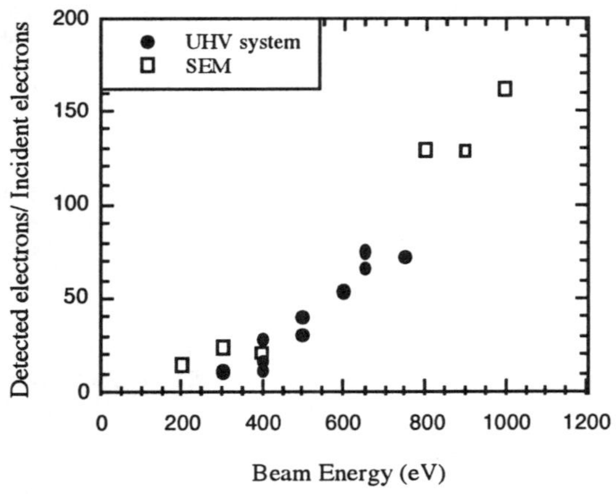

Figure 1 Ratio of detected electrons to incident electrons as a function of energy. The response of the CCD increases with increasing energy as result of multiple electron-hole pair generation.

The delta-doped CCD is the first CCD shown to respond to electrons with energies lower than 0.3 keV. At the previously reported lower limit of 900 eV and 1 keV for the flashgate CCD, the quantum efficiency of the delta-doped CCD is approximately twice as great. In the UHV chamber, the untreated backside-thinned CCD showed a dramatically lower quantum efficiency than the delta-doped CCD. The response of the untreated CCD to electrons was unstable, decaying with a time constant on the order of 20 minutes at an incident electron energy of 1 keV. This decay was not reversible by a thermal anneal at a temperature of 90°C. In the SEM, the control regions of the delta-doped CCD showed no response to electrons at energies less than 300 eV. Even at 1 keV, the response was very low and unstable in these control regions. The delta-doped CCD exhibited a response above the noise at energies as low as 50 eV, using electrons from a directly heated filament source. In measurements with the hot filament, the electron signal was distinguished from the background light signal by measuring the CCD response before and after magnetically deflecting the electron beam.

In preliminary measurements conducted in our laboratory, we report the first use of CCDs to image electrons. Flat-field images of 500 eV electrons with the delta-doped CCD show excellent qualitative similarity to UV images at 250 nm, with nearly identical contrast between the delta-doped and control regions of the CCD. Some small dark blemishes are apparent in one corner of the electron flat-field image that are not seen on the UV flat-field, but this could be due to dust or debris that has been introduced to the membrane surface in the course of handling, transporting, and storing the device in the months following the date when the UV flat-field image was taken. Additional studies of electron imaging with the delta-doped CCD are under way.

DISCUSSION

In the ultraviolet, the measured quantum efficiency of a CCD is the product of three important quantities: the transmission coefficient, the quantum yield, and the internal quantum efficiency of the CCD.[2] The transmission coefficient accounts for reflection from the surface and absorption in the native oxide, the quantum yield accounts for the statistically-averaged number of electron-hole pairs produced at the energy of the incident photon, and the internal quantum efficiency accounts for internal losses in the CCD, such as recombination of electron-hole pairs at the back surface of the CCD. Ultraviolet measurements of the delta-doped CCD indicate that the internal quantum efficiency is very nearly 100%, even at 270 nm where the absorption length in silicon is only 4 nm. The UV data suggest that the internal quantum efficiency of the delta-doped CCD is approximately 100% for electrons-provided the CCD is not damaged during the measurements. As discussed in the experimental section, we verified that the electron exposure did not degrade the performance of the CCD.

Incident electron radiation deposits energy in semiconductors through low-energy processes. Some of these mechanisms include secondary electron generation, Auger processes, Compton scattering, and backscattering. Part of the incident electron energy is transferred to the semiconductor through generation of EHPs. The average fraction of energy dissipated through these processes, EHP generation and all other losses, is a characteristic of the material.[8] For silicon, the statistical average number of EHPs generated by high-energy electrons or photons, also known as quantum yield, can be estimated by dividing the incident energy by 3.63 eV over a wide range of incident energies.[9] The quantum yield has been measured for silicon using x-ray and ultraviolet radiation. The quantum yield for low-energy electrons has never been measured.

Among the important factors that influence the observed response to incident electron irradiation is backscattering of electrons. A large fraction of electrons are lost in backscattering as energetic electrons impinge upon the surface of the material. It is therefore necessary to have a good estimate of the backscattering coefficient in order to interpret the measured CCD quantum efficiency. Theoretical and experimental studies, alike, have concentrated on the backscattering coefficient of higher energy electrons (generally for energies greater than 5 or 10 keV). Drescher *et al.* have measured backscattering of 10-25 keV electrons from silicon and aluminum targets[10] and Darlington *et al.* have measured backscattering from aluminum of electrons of energies down to 0.5 keV. [11] These are shown in figure 2 along with theoretical estimates from Staub *et al.* [12]

The theory does not correlate well with the low-energy Al measurements. An estimate for the low energy backscattering coefficients of Si can be obtained by using a fit to Darlington's experimental Al data and then extrapolating the fit to 200 eV. While using this model gives some qualitative indication of the effect of back-scattering on quantum efficiency of the delta-doped CCD for low-energy electron irradiation, the backscattering coefficient of low-energy electrons from silicon has not yet been measured. Using the measured backscattering coefficient of Al as an estimate for silicon, we have estimated that the backscattering coefficient for silicon is approximately 40-50% in the 200-1500 eV energy range. Even after taking backscattering into account, we are not detecting enough electrons to give us one electron for every 3.63 eV of incident energy. This means that either the actual quantum yield is lower for electrons in this energy range, (or 3.63 eV does not apply in this range) or other electron interactions contribute significantly to the transmission factor for low-energy electrons.

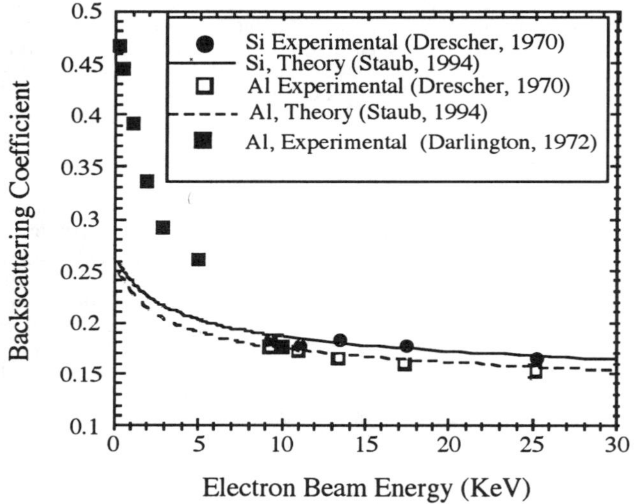

Figure 2. Experimental and theoretical backscattering results with extrapolation down to 200eV.

Analogous to the UV quantum efficiency discussed above, our electron response measurements represent the product of the effective quantum yield, the transmission factor (a factor representing the fraction of incident beam absorbed in the device which includes backscattering coefficient), and the quantum efficiency of the device. Assuming that all the generated electrons are detected by the delta-doped CCD (internal QE~100%), our measurements will represent the product of the effective quantum yield of silicon and the transmission factor for low-energy electrons. If the transmission factor is dominated by the backscattering coefficient, i.e., 40-50% for 200-1500 eV electrons, we have measured the effective quantum yield.

While separating the effects of transmission and quantum yield is interesting from a theoretical standpoint, the convolution of the two, as measured in these experiments, is the quantity of interest for solid-state electron detectors. It is significant that no other solid-state devices detect low-energy electrons as efficiently as the delta-doped CCD, due to the presence of a dead layer near their surfaces. In addition to its high efficiency, the delta-doped CCD also has the capability to image low-energy particles, which may prove valuable in energy-selective particle detector applications.

CONCLUSIONS

Because of their high resolution, linearity, and large dynamic range, CCDs could make major advances in particle detection. Delta-doped CCDs have been used for low-energy electron detection in the 50-1500 eV energy range, this represents the first measurements using CCDs to detect electrons in this energy range. Using delta-doped CCDs, we have extended the energy threshold for detection of electrons by approximately two orders of magnitude. We have also demonstrated the highest gain achieved to date by back-illuminated CCDs in response to low-energy electrons. Surface modification by delta-doping using MBE has demonstrated the highest quantum yield yet achieved for a backside electron-irradiated CCD. For the first time, electrons have been imaged with a CCD for the case of 500 eV electrons with a delta-doped CCD.

ACKNOWLEDGMENTS

The authors gratefully acknowledge the invaluable assistance of Drs., L. Douglas Bell, Michael Hoenk, Steve Manion, Tom Van Zandt, Mr. Walter Proniawicz, and Professor L.C. Kimerling. The work presented in this paper was performed by the Center for Space Microelectronics Technology, Jet Propulsion Laboratory, California Institute of Technology, and was jointly funded by the Caltech President's fund and the NASA Office of Space Science.

REFERENCES

1. M.E. Hoenk, P.J. Grunthaner, F.J. Grunthaner, M. Fattahi, H.-F. Tseng and R.W. Terhune, Appl. Phys. Lett., **61** (9) 1084 (1992).
2. S. Nikzad, M.E. Hoenk, P.J. Grunthaner, R.W. Terhune, R. Wizenread, M. Fattahi, H-.F. Tseng, and F.J. Grunthaner, Proc. of SPIE, **2217**, *Surveillance Technologies III*, April 4-8, Orlando, Fl. (1994).
3. K.L. Luke and L.-J. Cheng, , J. Appl. Phys. **60**, 589 (1986).
4. T. Daud, J.R. Janesick, K. Evans, and T. Elliot, Opt. Eng., **26** (8) 686 (1987).
5. S. Nikzad, M.E. Hoenk, P.J. Grunthaner, R.W. Terhune and F.J. Grunthaner, Proc. SPIE, **2198**, *Astronomical Telescopes & Instrumentation for the 21st Centruy*, March 13-18, Kona, Hawaii (1994).
6. S.M. Sze, Physics of Semiconductor Devices, 2nd Ed, Wiley & Sons, New York (1981) p.42
7. D.G. Stearns and J.D. Wiedwald, Rev. Sci. Instrum. **60** (6) 1095 (1989).
8. C.A. Klein, , J. Appl. Phys. **39** (4) 2029 (1968).
9. L.R. Canfield, J. Kerner, and R. Korde, Appl. Opt. **28** 3940 (1989).
10. H. Drescher, L. Reiner, and H. Seidel, Zeitschrift fur Angewandte Physik, **29** (6) 331(1970).
11. E.H. Darlington and V.E. Cosslett, J. Phys. D: Appl. Phys., **5** (11) 1969 (1972).
12. P. F. Staub, J. Phys D: Appl. Phys, **27** (7) 1533 (1994).

Comparison of the Morphological and Optical Characteristics of InP Islands on GaInP/GaAs (311)A and (100)

R. I. Pelzel [*], C. M. Reaves [**], S. P. DenBaars [**], and W. H. Weinberg [*][†]
 Center for Quantized Electronic Structures, University of California, Santa Barbara, CA 93106
[*]Department of Chemical Engineering, University of California, Santa Barbara, CA 93106
[**] Department of Materials, University of California, Santa Barbara, CA 93106
[†] e-mail: chari@engineering.ucsb.edu

ABSTRACT

We have studied the effect of surface orientation on the optical and morphological characteristics of coherently-strained InP islands grown on GaInP/GaAs. The differences between islands grown on the (100) orientation and the (311)A orientation are studied. Islands grown on the (311)A orientation are more dense than the islands grown on the (100) orientation. For the (100) orientation, the island height distribution is bimodal peaked at 20 Å and 220 Å. For the (311)A orientation, the island height distribution is also bimodal peaked at 15 Å and 60 Å. Photoluminescence measurements for the (311)A orientation show a peak at 1.9 eV attributed to small islands. This peak is shifted to higher energies in comparison to the corresponding peak for the (100) orientation which is at 1.77 eV. This peak shift is due to the fact that the small islands on the (311)A orientation are smaller than the corresponding islands on the (100) orientation.

INTRODUCTION

The ability to fabricate a self-assembling quantum structure without using ex-situ techniques is a desirable alternative to some of the lithography and etching techniques presently used in device fabrication [1]. Yet, self-assembling growth techniques still lack the control achievable with ex-situ techniques. In order to more precisely control growth and, in turn, produce high quality materials, it is helpful to have an understanding on a fundamental level. Specifically, it is desirable to have a fine level of control over the density and size of the quantum structures as these two parameters determine the optoelectronic characteristics of the material.

Recently, the utilization of self-assembling islands resulting from the coherent Stranski-Krastanov [2] growth mode as quantum dots has been attracting interest [3]. The island size and density and, in turn, the optical characteristics of the islanded surface can be altered by varying such growth parameters such as temperature and growth rate [4-8]. One particular growth parameter that has been a focus is the crystallographic orientation of the growth surface [9-10]. In this work, we have grown self-assembled InP islands on GaInP/GaAs(311)A and GaInP/GaAs(100) and compared the surface morphology and photoluminescence (PL) of the two orientations.

METHODOLOGY

The samples used in this study were grown by metalorganic chemical vapor deposition under conditions similar to those reported in detail elsewhere [11]. GaAs(100) and GaAs(311)A substrates were loaded into the growth chamber to undergo simultaneous deposition. After an anneal in an arsenic-containing environment, a 1000 Å GaAs buffer layer was grown, followed by a 300 Å $Ga_{0.51}In_{0.49}P$ layer which is lattice matched to GaAs. The InP was then deposited. For PL samples, a 500 Å GaInP capping layer was

Mat. Res. Soc. Symp. Proc. Vol. 448 © 1997 Materials Research Society

grown. Growth was stopped and the sample was cooled in a phosphorus-containing environment. The estimated InP deposition rate is one monolayer per second for lattice matched growth on (100).

Morphological characterization was done with tapping-mode atomic force microscopy performed in air. Island sizes were measured using the cross-section profile routine in the microscope software. PL was performed at 2 K using the 488 nm emission of an Ar-ion laser, dispersed by a one meter monochromator. The luminescence was detected using a GaAs photomultiplier.

RESULTS AND DISCUSSION

Different amounts of InP were deposited on GaInP/GaAs surfaces of both orientations. An example of such surfaces are shown in Fig. 1. These atomic force micrographs are top views of the InP/GaInP surfaces after 4 s of InP deposition on both the (311)A (cf., Fig. 1(a)) and the (100) orientation (cf., Fig. 1(b)). The gray features are short InP islands, and the white features are taller InP islands. Qualitatively, we see that the islands are laterally smaller and more dense on the (311)A orientation than on the (100) orientation.

The difference between types of islands, both for a given orientation or between different orientations, is size. In general, small islands are formed in the first stage of three-dimensional growth. With additional deposition, medium-sized islands evolve and co-exist with the small islands. If any of the islands develop strain-relieving defects they become larger than the other islands growing linearly with time [11]. For this study, we

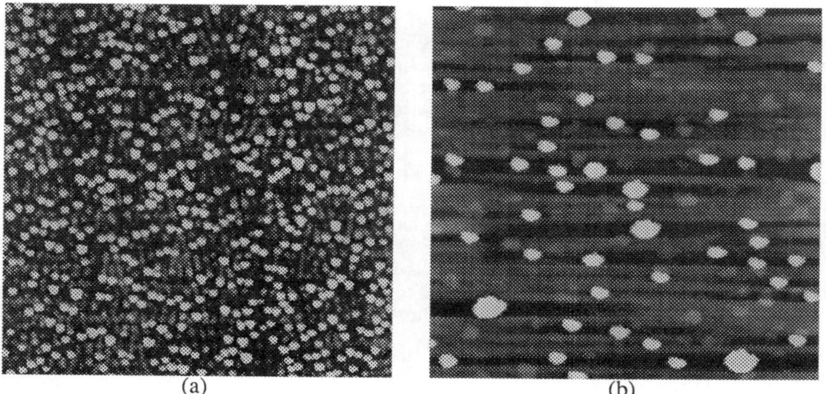

(a)   (b)

Fig. 1 Atomic force micrographs of InP/GaInP/GaAs after 4 s of InP deposition. Deposition of InP on GaInP/GaAs(311)A is shown in (a), and deposition of InP on GaInP/GaAs(100) is shown in (b). The micrographs represent top views of the two orientations. The gray features are small islands, and the white features are tall islands. The dark streaks in (b) are artifacts of the imaging. Both micrographs are 3 μm by 3 μm. The height ranges are 150 Å and 300 Å for (a) and (b), respectively. Both images were plane-fitted and flattened using an third order polynomial.

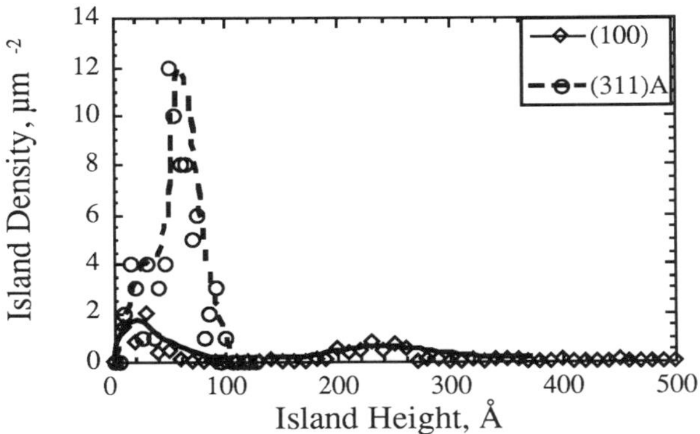

Fig. 2. Island height distribution for InP/GaInP/GaAs(311)A and (100) after 4 s of InP deposition. The lines serve as guides to the eye. Notice that even though the distributions for both orientations are bimodal the islands for the (311)A orientation are shorter.

have interrupted growth prior to forming a large population of defected islands; we are studying a population of primarily coherent islands. This is seen by the bimodal distributions of island heights shown in the histograms in Fig. 2. For the (311)A orientation, small islands have a height distribution centered around 15 Å and the medium islands have a height distribution centered around 60 Å. Moving to the (100) orientation, the island height distribution is peaked at approximately 30 Å and approximately 240 Å. Also, notice that the height of the peaks for the (311)A distribution are larger than the corresponding peaks for (100) indicating that the island density on the (311)A orientation is higher than the island density on the (100) orientation.

Table I gives a summary of the size measurements for both types of islands on the two surfaces. As previously discussed, there are two different heights of islands for both surfaces. The smallest islands formed on both orientations are comparable in height. Yet, for the (311)A orientation, the medium islands are smaller than the equivalent islands on the (100) orientation. Furthermore, the small and medium islands are closer in height for the (311)A orientation than are the small and medium islands on the (100) orientation. For each surface studied, the small and medium islands have the same base width. The base width for the islands on the (311)A orientation is smaller (750 Å) than the base width of the islands on the (100) orientation (1200 Å). Thus, for a complete ensemble of islands, the (311)A orientation gives islands that are smaller in dimension.

Table I. Peak values of island size distributions for InP/GaInP/GaAs.

	(311)A	(100)
height of small islands	15 Å	20 Å
height of medium islands	60 Å	240 Å
base width of small islands	750 Å	1200 Å
base width of medium islands	750 Å	1200 Å

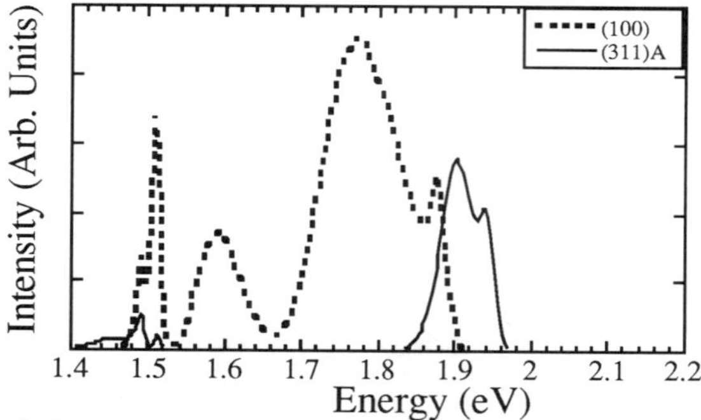

Fig. 3. PL spectra for InP/GaInP/GaAs(311)A and (100) after 2 s of InP deposition. Notice that the small island peak for the (311)A orientation at 1.9 eV is shifted to higher energy in comparison to the small island peak for the (100) orientation which is at 1.77 eV.

The differences in island dimensions between the two surface orientations give rise to differences in optical properties. The PL spectra for InP island samples grown on both the (311)A and the (100) orientations of GaInP are shown in Fig. 3. We see several features for the (100) orientation. The peaks attributed to GaAs are around 1.5 eV and the peak attributed to GaInP peak is at 1.88 eV. The peak at 1.6 eV is attributed to the medium islands, and the peak at 1.77 eV is attributed to the small islands. This spectrum is typical of the (100) orientation over a range of deposition times [12,13]. One effect observed for the (311)A orientation is that at short deposition times, only small islands are observed [14]. Examining the spectrum for a short deposition on (311)A (cf., Fig. 3), we again see the peaks attributed to GaAs at around 1.5 eV. The peak attributed to GaInP for (311)A is at 1.94 eV. The shift in the GaInP peak to higher energies for the (311)A orientation is attributed to less atomic ordering of the GaInP. For (311)A, we only see one peak at 1.9 eV which is attributed to the InP islands.

In comparing the spectra from the two surfaces, note that the InP island PL peak for the (311)A orientation is shifted to higher energies in comparison to the (100) small island peak. The primary reason for this shift is that the small islands on the (311)A orientation are smaller than the equivalent islands on the (100) orientation, (cf., Table I). There are also some other reasons that may lead to a shift in the luminescence emission energy such as an increase in the band gap of the barrier materials (i.e., GaInP) or the different orientation. As seen by other studies of self-assembled quantum dots on different orientations [15], the effects of different orientations often have nominal effects on emission energy. We believe that the primary cause of the energy shift is the smaller islands.

SUMMARY

We have studied the effect of surface orientation on the morphological and optical characteristics of InP islands on GaInP/GaAs(311)A and (100). The island density on the (311)A orientation is higher than for the (100) orientation. The islands on the (311)A orientation are smaller in height and base width than for the corresponding islands on the (100) orientation. The PL peak for the smaller islands on the (311)A orientation is shifted to higher energies in comparison to the corresponding island peak for the (100) orientation. With this work we have shown it is possible to adjust the dimensions and optical characteristics of self-assembled islands by altering the growth surface orientation.

ACKNOWLEDGMENTS

The authors would like to thank B. Nosho, H. Marchand, and J. Owen for useful comments on the manuscript. This work is supported by the National Science Foundation Science and Technology Center for Quantized Electronic Structures (Grant no. DMR 91-20007), by the National Science Foundation Materials Research Laboratory Central Facilities Program (Grant no. DMR 91-23048), by NSF Grant DMR-9504400 (W.H.W.), and by the W.M. Keck Foundation (W.H.W.).

REFERENCES

1. C. Weisbuch and B. Vinter, <u>Quantum Semiconductor Structures</u>, Academic Press, Inc., San Diego, CA, 1991, pp. 191-192 and references therein.

2. D. Eaglesham and M. Cerullo, Phys. Rev. Lett. **64**, p. 1943 (1990).

3. R. Nötzel, Semiconductor Sci. and Technol. **11**, p. 1365 (1996) and references therein.

4. G.S. Solomon, J.A. Trezza, and J.S. Harris, Jr., Appl. Phys. Lett. **66**, p. 991 (1995).

5. D. Leonard, M. Krishnamurthy, S. Fafard, J.L. Merz, and P.M. Petroff, J. Vac. Sci. and Technol. B **12**, p. 1063 (1994).

6. M. Sopanen, H. Lipsanen, and J. Ahopelto, Appl. Phys. Lett. **67**, p. 3768 (1995).

7. V. Bressler-Hill, C.M. Reaves, S. Varma, S.P. DenBaars, and W.H. Weinberg, Surf. Sci. **341**, p. 29 (1995).

8. C.M. Reaves, V. Bressler-Hill, W.H. Weinberg, and S.P. DenBaars, J. Electron. Mater. **24**, p. 1605, (1995).

9. J. Temmyo, A. Kozen, T. Tamamura, R. Nötzel, T. Fukui, and H. Haasegawa, J. Electron. Mater. **25**, p. 431 (1996).

10. R. Nötzel, J. Temmyo, and T. Tamamura, Nature **369**, p. 131 (1994).

11. C.M. Reaves, V. Bressler-Hill, S. Varma, W.H. Weinberg, and S.P. DenBaars, Surf. Sci. **326**, p. 209 (1995).

12. C.M. Reaves, N.A. Cevallos, G.C. Hsueh, Y.M. Cheng, W.H. Weinberg, P.M. Petroff, and S.P. DenBaars, Seventh International Conference on Indium Phosphide and Related Materials (New York, IEEE, 1995) pp. 307-10.

13. N. Carlsson, W. Seifert, A. Petersson, P. Castrillo, M.E. Pistol, and L. Samuelson, Appl. Phys. Lett. **65**, p. 3093 (1994).

14. C.M. Reaves, R.I. Pelzel, G.C. Hsueh, W.H. Weinberg, and S.P. DenBaars, Appl. Phys. Lett., in press.

15. D.I. Libyshev, P.P. Gonzalez-Borrero, E. Marega, Jr., E. Petitprez, and P. Basmaji, J. Vac. Sci. Tech. B **14**, p. 2212 (1996).

A SELF-ORGANIZED MOLECULAR BEAM EPITAXIAL GROWTH OF THE InSb/AlGaSb QUANTUM DOTS ON HIGH-INDEX GaAs SUBSTRATES

Mitsuaki Yano, Kazuto Koike, and Masataka Inoue
New Materials Research Center, Osaka Institute of Technology, Asahi-ku Ohmiya, Osaka 535, Japan
Toshiya Saitoh, and Kanji Yoh
RCIQE Hokkaido University, Kita-ku N13W8, Sapporo 060, Japan

ABSTRACT

In-situ organization of InSb quantum dots on $Al_{0.5}Ga_{0.5}Sb$ is reported. Samples were grown on just (100), 5°off (100) towards [0-1-1], (311)A, and (311)B surfaces of GaAs by molecular beam epitaxy. The growth mechanism and characteristics of quantum dots were analyzed using reflection high-energy electron diffraction, atomic force microscopy, and photoluminescence. Observed photoluminescence peak shift towards lower energy side with InSb thickness was interpreted by the development of quantum dots. Substrate orientation effect was examined and found to be useful to increase the dot density. As a result, as high as $\sim 3 \times 10^9$ cm^{-2} dot density was achieved on the (311)B substrate whereas the typical value on the (100) was $\sim 2 \times 10^8$ cm^{-2}.

1. INTRODUCTION

Strain induced three-dimensional growth has stimulated many research groups to realize novel heterostructures containing nanoscale islands acting as quantum dots (QDs) with relative ease. In the last several years, many previous studies using molecular beam epitaxy (MBE) have focused on InAs/GaAs [1,2] and InGaAs/GaAs [3] systems to understand the island growth, Stranski-Krastanov (S-K) mode growth. More recently, this in-situ approach of fabricating QDs has been extended to other material systems [4-6]. According to the report by Bennett et al. [6], S-K mode MBE growth is also effective to yield QDs of InSb on (100) GaAs substrates. They observed a photoluminescence (PL) emission at around 1.1 eV and related it to the radiative recombination between the electrons in the GaAs and the holes in the InSb QDs [7]. They explained the transition energy assuming a type II band alignment combined with a Hartree potential at the interface.

In this work, the S-K mode MBE growth of InSb QDs on $Al_{0.5}Ga_{0.5}Sb$ is reported for the first time. Since InSb/$Al_{0.5}Ga_{0.5}Sb$ is expected to form a type I band alignment, this experiment can provide simpler examples to understand the electron confinement in the InSb QDs. This material system is also promising for device applications such as long wavelength QD lasers. We have grown the InSb QDs/$Al_{0.5}Ga_{0.5}Sb$ using (100) and high-index GaAs substrates to study the S-K mode growth on off-angle surfaces. Their structural and optical characteristics are discussed.

2. EXPERIMENTAL

Samples were grown from elemental sources using an MBE machine equipped with non clacking K-cells. Substrates of (311)A, (311)B, 5°off (100) towards [0-1-1], and just (100) GaAs were mounted side by side on a same substrate holder. Due to 8 % larger lattice constant than that of GaAs, the $Al_{0.5}Ga_{0.5}Sb$ epilayer is expected to be strained compressively. In order to accommodate the strain, we employed a GaSb/AlSb superlattice followed by a thick GaSb buffer layer on the GaAs substrates. A schematic sample structure is shown in Fig.1. After oxide desorption at 580 °C, smoothing layers of a 300 nm GaAs and a 10 nm AlAs were grown at 600 °C. Then, the antimonide layers consisted of a 100 nm AlSb, a 15 period

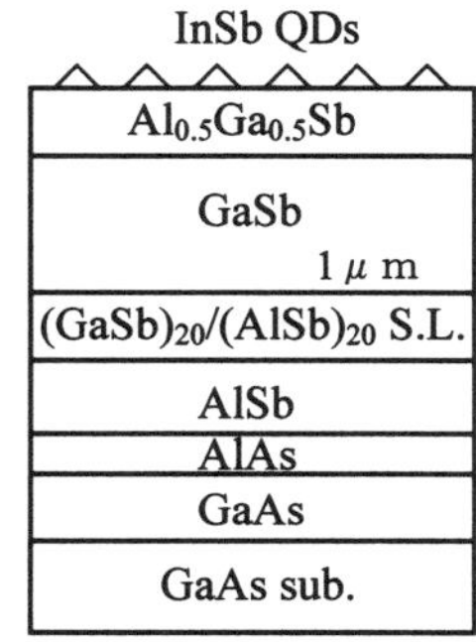

Fig.1   Schematic sample structure.

of (GaSb)$_{20}$/(AlSb)$_{20}$, a 1 μm GaSb, and a 15 nm Al$_{0.5}$Ga$_{0.5}$Sb were grown at 550 °C. Subsequently, the growth was interrupted until the substrate had cooled down to 400 °C. On top of the Al$_{0.5}$Ga$_{0.5}$Sb whose lattice constant is about 6 % smaller than that of InSb, we grew InSb QDs at a rate of 0.03 ML/s. For samples to measure PL spectra, a 50 nm Al$_{0.5}$Ga$_{0.5}$Sb cap layer was grown on the InSb QDs. The whole growth process was in-situ monitored by reflection high-energy electron diffraction (RHEED). The surface morphology of the film was ex-situ characterized in air by atomic force microscopy (AFM) at room temperature. The PL from the samples was measured at 17 K using the excitation of the 514.5 nm line of a 50 mW Ar ion laser.

3. GROWTH ON (100) SUBSTRATES

Growth on just (100) GaAs substrates is discussed in this section. During the Al$_{0.5}$Ga$_{0.5}$Sb growth, we observed a RHEED pattern of streaked (1x3) indicating an Sb-stabilized surface. The succeeding growth interruption made the reconstruction much sharper due to the flattening of the surface in an atomic scale. During the InSb growth, this streaked (1x3) reconstruction continued up to 2.5 ML and then gradually changed to a faint chevronlike pattern. In Fig.2, we show an in-situ measured response of the specular spot intensity in the RHEED pattern. The electron beam was introduced in the [010] direction. After 2.5 cycles of oscillations, the RHEED intensity started a rapid and monotonous decrease, which indicates the change of growth mode from two-dimensional (2D) to 3D at 2.5 ML. These experimental data suggest that the coherent growth to Al$_{0.5}$Ga$_{0.5}$Sb is limited within the initial 2.5 ML of the InSb. The 2D/3D transition, however, depended on the growth condition as is observed generally in other material systems, and the chevronlike pattern also developed after closing the cell shatter even though the InSb growth was interrupted at less than 2.5 ML. This change can be understood by a formation of 3D islands to relax the excess strain in the 2D InSb film via surface transport. Because such accommodation was not observed for films less than 1.4 ML, we can conclude that the critical thickness under thermal equilibrium was about 1.4 ML for the nucleation of 3D islands on the 2D wetting layer.

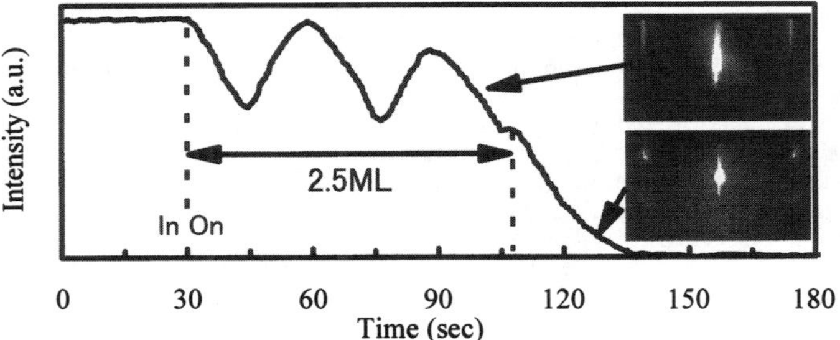

Fig.2   In-situ measured [010] RHEED patterns and corresponding
change of the specular spot intensity.

In agreement with the RHEED experiment, an AFM observation revealed no 3D islands characteristic of S-K mode growth on samples with 1.4 ML InSb. At 1.6 ML, square-cone shaped islands, QDs, developed clearly although their areal density was relatively low, ~1x10^7 cm^{-2}. The QD density increased with InSb thickness and reached to ~1x10^8 cm^{-2} at around 2 ML, and then saturated to ~2x10^8 cm^{-2} at around 6 ML. In this saturation region, from 2 to 6 ML, the increase in the total volume of the QDs roughly corresponded to the total amount of the InSb deposited. More than 6 ML, however, the QD density gradually decreased with increasing the average QD size. This change in more than 6 ML probably due to beginning of coalescence since many QDs have closely coupled each other forming a partly overlapped island shape. For the QD growth on (100) substrates, we obtained nearly the same thickness dependence over the wide range of Sb/In beam equivalent pressure ratio, BEP(Sb/In), between 25 and 100. A typical AFM image from a 4.8 ML InSb sample grown at BEP(Sb/In)=100 is shown in Fig.3. The average values of density, width, and height of the QDs were about 2.6×10^8 cm^{-2}, 180 nm, and 45 nm, respectively. It was indicated by the chevronlike RHEED pattern during growth that the QD side planes consisted of (122) facets.

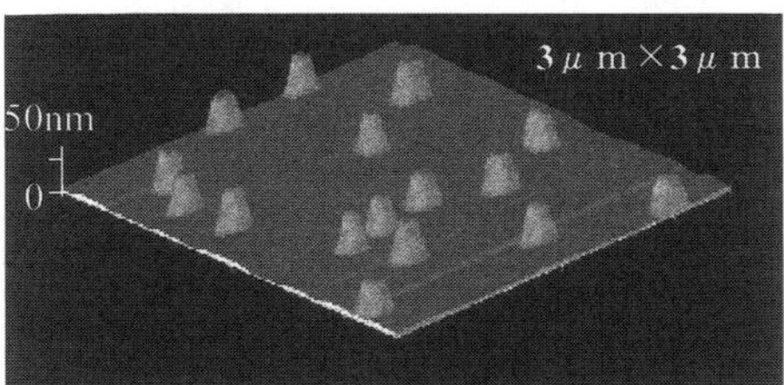

Fig.3   A typical AFM image from a 4.8ML InSb
sample grown at BEP(Sb/In)=100.

Low temperature PL spectra from three different samples covered with a 50 nm thick Al$_{0.5}$Ga$_{0.5}$Sb are shown in Fig.4. The average InSb thicknesses for these spectra (a), (b), and (c) were 1.4, 1.6, and 4.8 ML, respectively. From the 1.4 ML sample, on which no QDs were confirmed by AFM, only a single and intense emission was observed in the PL spectrum being strongly shifted to higher energy side with respect to InSb band gap (0.24 eV). To understand the PL peak of Fig.4(a), we estimate the transition energy of electrons in the InSb wetting layer clad with Al$_{0.5}$Ga$_{0.5}$Sb. In the calculation, a model solid theory [8] combined with deformation potentials and an envelope function approximation are used. We assume that the InSb layer is coherently strained under the biaxial strain due to the lattice mismatch. The effect of nonparabolic InSb conduction band is included in the subband calculation using a Kane model [9]. This calculation leads to the transition energy of 1.11 eV from the lowest electronic subband to the heavy hole subband (from E$_1$ to HH$_1$) in the type I quantum well. Although the energy is 0.07 eV smaller than the experiment, the agreement is fairly good.

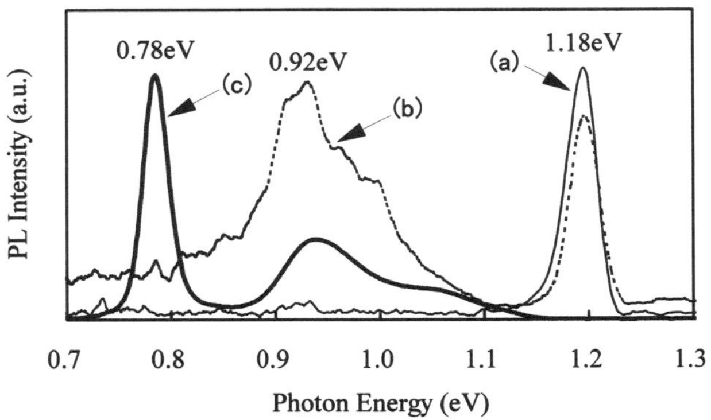

Fig.4   PL spectra from three different covered samples at 17K.
The thicknesses of InSb are (a) 1.4ML, (b) 1.6ML, (c) 4.8ML,
respectively.

The formation of the InSb QDs resulted in a strong modification of the PL spectra. From the 1.6 ML sample with $\sim$1x10^7 cm^{-2} QDs, a new broad emission band appeared at around 0.92 eV, as shown by the spectrum Fig.4(b), in addition to the sharp peak from the wetting layer. This broad band is considered to be the emission from the quntized states in the QDs. We attributed the multiple peaks in the broad band to the widely distributed QD size at the initial islanding stage of the S-K transition. With increasing the InSb deposition, the emission band from the QDs shifted to the lower energy side. In the PL spectrum from the 4.8 ML sample with $\sim$2x10^8 cm^{-2} QDs, Fig.4(c), the signal from the wetting layer disappeared and instead, a new peak at 0.78 eV appeared to dominate the emission. The observed peak shift is reasonably understood by the increased QD size with the InSb thickness. However, these emission peaks in spectra (a), (b), and (c) have larger energies compared with the theoretical calculations on the electron transitions in the respective quantum structures. According to the recent Raman scattering study on AlSb QDs/GaAs [10], the strain due to the lattice mismatch enhances interdiffusion of component elements and leads to an alloyed composition (AlGa)Sb in the QDs. Therefore, although not probed yet in our experiment, this deviation is presumably caused by the alloyed (AlGaIn)Sb composition of the QDs. Its larger band gap with respect to InSb can explain the increased emission energy.

4. GROWTH ON HIGH-INDEX SUBSTRATES

We report, in this section, preliminary experimental results of the substrate orientation effect on QD growth. Since step density and surface chemistry vary with the orientation, different growth kinetics might be possible for high-index substrates. For instance, Lubyshev et al. have reported that PL from QDs depends considerably on the substrate orientation [11]. As well, we can expect an improvement of the low QD density on (100) substrates ($\sim$10^2 lower than typical experimental data on InAs QDs/GaAs) by using high-index substrates since the migration length in general becomes short with increasing the off-angle.

In order to study the orientation effects, we compared four different samples with the same InSb thickness of 4.8 ML simultaneously grown at BEP(Sb/In)=100. The dot densities observed are 2.6x10^8, 4.8x10^8, 1.1x10^9, and 1.6x10^9 cm^{-2} for just (100), 5°off (100), (311)A, and (311)B substrates, respectively. The QD density on the high-index substrates considerably increased, roughly twice on the 5°off (100) and four times on both of the (311)A and B, as compared with that on the just (100) substrate. In Fig.5(a) to (d), we show the height distributions of QDs on these samples. These histograms nearly correspond to the size distribution since the width/height ratio of QDs was about 4, which is independent of the substrate orientation. On the just (100) substrate, as shown by Fig.5(a), most of QDs were higher than 40 nm although another isolated peak presumably due to embryonic QDs appeared at 5 nm. On the 5°off (100), Fig.5(b), the most frequent QDs were around 35 nm in height and only QDs lower than 45 nm were observed. On the contrary, both histograms of Fig.5(c) and (d) for the QDs on (311)A and B did not have clear peak but widely distributed from the smallest size, 5 nm. Note that the percentage of the small size dots became large on these high-index substrates. These results strongly suggest that the QD growth kinetics is closely connected with the surface orientation.

Next, we explored the QD growth process by changing the amount of the InSb deposited. We found that the QD density on the (311)B dramatically increased and reached to $\sim$5x10^9 cm^{-2} within the initial 2.0 ML growth (0.6 ML from the S-K transition at 1.4 ML). Then, the QD density once decreased to $\sim$1x10^9 cm^{-2} with the succeeding 1.0 ML InSb deposition and again increased to $\sim$2x10^9 cm^{-2} at around 6 ML. On the 5°off (100), the initial increase was also dramatic but saturated at $\sim$1x10^9 cm^{-2} and then the density decreased to $\sim$5x10^8 cm^{-2} at around 3 ML. More than 3 ML, the density did not change till 6 ML. Compared with these two cases, the initial rise on the (311)A was a little slow and reached to $\sim$1x10^9 cm^{-2} at around 2.5 ML. Further InSb deposition till 6 ML did not change the density but increased the size. On the just (100), however, much more gradual increase was observed for the QD density as discussed in the previous section.

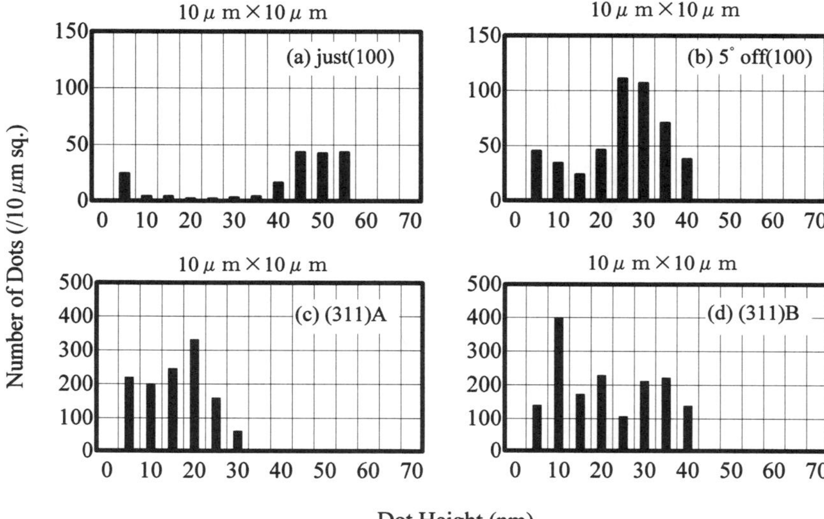

Fig.5   Dot height histograms. The total amount of the deposited InSb is 4.8ML, and
the substrates used are (a) just(100), (b) 5° off(100), (c) (311)A, and (d) (311)B.

The AFM measurement for the samples on (311)B and 5°off (100) also revealed that the decrease of the QD density after peaking was accompanied with an increase of the QD size, in other words, their high density QDs just after the S-K transition mostly consisted of small size dots. In Fig.6, an AFM image and its height distribution obtained from the 1.8 ML InSb sample grown on (311)B are shown. It is seen that 3.0×10^9 cm^{-2} of very small QDs, ~2.5 nm height and ~30 nm width, have developed on the surface. This size distribution is quite different from the 4.8 ML case in Fig.5(d), indicating a substantial change in the growth kinetics. On the other hand, the corresponding QDs on just (100), ~6x10^7 cm^{-2} at 1.8 ML, contained much more large size islands. Their distribution was close to that of the 4.8 ML case in Fig.5(a) although the percentage of the embryonic dots became large, ~50 %, and the most frequent height for the larger size group was a little smaller, ~35 nm. On 5°off (100) and (311)A, their QD size

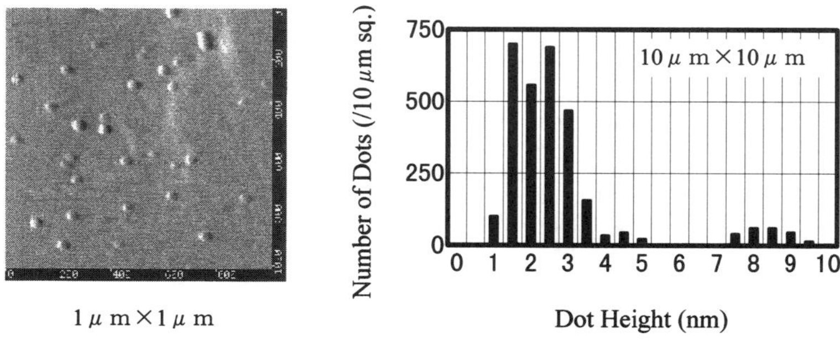

Fig.6   A typical AFM image and its QD height histogram observed from the
1.8ML InSb sample grown on (311)B.

dependencies on thickness were found to be intermediate between the cases on just (100) and (311)B. These results imply that the high-density dot formation is due to the reduced migration length of adatoms on high-index substrates.

Since S-K transition is expected to be governed by local strains depending on steps, defects, thickness fluctuation, and repulsive force between islands [12], it is reasonable to relate the high density QDs with the characteristics of the growth on high-index substrates. After developing initial small size QDs, surface transport will coalesce them since larger size islands should be the most stable. The growth on (100) can be understood by this type of QD formation indicating a sufficient migration length. On the contrary, short migration length may be responsible for the growth on high-index substrates. The critical size for island formation is reached before the substantial surface transport has occurred and more QDs of smaller height are formed [13]. This understanding was supported in part by the terrace-like $Al_{0.5}Ga_{0.5}Sb$ growth on high-index substrates. The terrace-like structure with steplines along [-233] and [0-11] became predominant on (311)B and 5°off (100) substrates, respectively. These terraces on $Al_{0.5}Ga_{0.5}Sb$, ~1.5μm width and ~15 nm height for both substrates, were presumably due to bunching growth and can hinder the surface transport across the stepline. Clearly we must examine much more details to explain these orientation effects. Our findings show the importance of substrate orientation in correlation with the surface transport for the growth of high-density QDs.

5. SUMMARY

We have studied the S-K mode MBE growth of $InSb/Al_{0.5}Ga_{0.5}Sb$ on (100) and high-index GaAs substrates. Development of InSb QDs by the S-K transition was confirmed by using in-situ RHEED and ex-situ AFM measurements. PL spectra from the samples were reasonably understood assuming a type I band alignment for the wetting layer and QDs. Increase of QD density was realized by using the high-index substrates. Characteristic QD growth on these high-index surfaces was examined preliminary and interpreted qualitatively by considering an insufficient migration length.

ACKNOWLEDGEMENTS

The authors wish to thank Dr. S. Sasa for helpful discussion. We also thank Mr. T. Ikeda and Mr. Y. Seki for experimental assistance.

REFERENCES

1. P.Chen, Q.Xie, A.Madhukar, Li Chen, and A.Konkar, J.Vac.Sci.Technol. B12, 2568 (1994).
2. D.Leonard, K.Pond, and P.M.Petroff, Phys.Rev. B5, 11687 (1994).
3. S.Fafard, D.Leonard, J.L.Merz, and P.M.Petroff, Appl.Phys.Lett. 65, 1388 (1994).
4. K.M.Chen, D.E.Jesson, S.J.Pennycook, T.Thundat, and R.J.Warmack, J.Vac.Sci.Technol. B14, 2199 (1996).
5. F.Hatami, N.N.Ledentov, M.Grundmann, F.Heinrichsdorff, M.Beer, D.Bimberg, S.S.Ruvimov, P.Werner, U.Gösele, J.Heydenreich, U.Richter, S.V.Ivanov, B.Ya.Meltser, P.S.Kop'ev, and Zh. I.Alferov, Appl.Phys.Lett. 67, 656 (1995).
6. B.R.Bennett, P.M.Thibado, M.E.Twigg, E.R.Glaser, R.Magno, B.V.Shanabrook, and L.J. Whitman, J.Vac.Sci.Technol. B14, 2195 (1996).
7. E.R.Glaser, B.R.Bennett, B.V.Shanabrook, and R.Magno, Appl.Phys.Lett. 68, 3614 (1996).
8. C.G.Van de Walle, Phys.Rev. B39, 1871 (1989).
9. O.E.Kane, J.Phys.Chem.Solids 1, 249 (1957).
10. B.R.Bennett, B.V.Shanabrook, and R.Magno, Appl.Phys.Lett. 68, 958 (1996).
11. D.I.Lubyshev, P.P.Gonzalez-Borrero, E.Marega,Jr., E.Petitprez, and P.Basmaji, J.Vac.Sci. Technol. B14, 2212 (1996).
12. A.Ponchet, A.Le Corre, H.L'Haridon, B.Lambert, and S.Salaün, Appl.Phys.Lett. 67, 1850 (1995).
13. M.Berti, A.V.Drigo, A.Giuliani, M.Mazzer, A.Camporese, G.Rossetto, and G.Torzo, J.Appl.Phys. 80, 1931 (1996).

STRUCTURAL INVESTIGATIONS OF SELF-ASSEMBLED Ge-DOTS BY X-RAY DIFFRACTION AND REFLECTION

A. A. DARHUBER[1], V. HOLY[1*], J. STANGL[1], G. BAUER[1], P. SCHITTENHELM[2], G. ABSTREITER[2]
[1]Institut für Halbleiterphysik, Johannes Kepler Universität, A-4040 Linz, Austria
[2]Walter Schottky Institut, TU München, Am Coulombwall 2, D-85748 Garching, Germany

ABSTRACT

Self-organized Ge-dots on (001)-oriented Si-substrates have been studied using two-dimensionally resolved high resolution x-ray diffraction and reflectivity. The degree of the vertical correlation of the dot positions ("stacking") has been derived as well as a lateral ordering of the dots in a (disordered) square array with main axes parallel to [100] and [010].

INTRODUCTION

The fabrication of quantum dots by self-organizing processes [1,2] during molecular beam epitaxial or MOCVD-growth has attracted increasing attention in the past years [3,4]. The dots occur when the thickness of the epi-layers exceeds a critical value and the growth mode changes from two-dimensional strained layer growth to a three-dimensional one (Stranski-Krastanow growth). In the Si-Ge system, the transition of this crossover in the growth mode has been studied using photoluminescence (PL) and transmission electron microscopy (TEM) [3,5-7], atomic force microscopy (AFM) [6], and scanning tunneling microscopy [8]. Lateral undulations of Si/SiGe interfaces have been studied by high resolution x-ray diffraction [9].

The remaining two-dimensional wetting layers are thinned due to the formation of the islands. In contrast to reactive ion etched lateral structures, which have been studied previously by x-ray diffraction [10,11], these self-organized nanostructures do not exhibit surface damage. In this article, we report on high resolution x-ray reciprocal space mapping of self-assembled Ge quantum dot multilayer structures and analyse the coherent and diffuse scattering. Information on vertical and lateral ordering is obtained.

EXPERIMENTAL

Multilayer samples [12] have been grown on (001)-oriented Si substrates using solid source molecular beam epitaxy (MBE) in a Riber Siva-32 MBE at the growth temperature of 670°C and a Ge-growth rate of 0.075 Å/s. The samples contain 19 periods of nominally 5.5 ML Ge separated by Si-spacers with thicknesses in the range from 10 to 40 nm. X-ray diffraction measurements have been performed with a laboratory diffractometer with a four crystal Ge-monochromator using the CuKα_1 line and at the OPTICS beamline of the ESRF and the D4 beamline of HASYLAB, Hamburg. For the reciprocal space maps (RSMs) a two crystal Ge(220) analyzer has been used, with an angular resolution of 12 arcsec.

The samples were characterized by PL and AFM [12]. The Ge-layer thickness has been chosen in order to prevent misfit dislocation formation in the multilayer.

In conjunction with the Si spacer layers the wetting layers form a two-dimensional multi quantumwell (MQW). Far from any dot the in-plane lattice constant of these layers is the same as in the Si substrate. This MQW gives rise to narrow ("coherent") diffraction

---

Mat. Res. Soc. Symp. Proc. Vol. 448 © 1997 Materials Research Society

peaks at the same lateral reciprocal coordinate as the Si substrate peak. Due to strain relaxation, the elastic strains above and below the islands are tensile, whereas between the dots, the Si spacer is compressively strained. Hence, the in-plane lattice constant in, above and below the dots is larger than the Si bulk lattice constant, whereas it is smaller in the region between them. Therefore, we expect additional diffracted intensity on both sides of the coherent signal due to the formation of the islands.

In Fig. 1, we present double-crystal (DCD) (004) rocking curves (solid line) of the samples with 30 nm Si spacer thickness as well as a simulation using dynamical diffraction theory (dashed line), which neglects the formation of Ge-islands. MQW-satellites up to the 8th order are visible, which serve as a qualitative measure of the overall crystalline quality of the dot-multilayer. From the fitted curves the period 29.4 nm and the average Ge-content of 2.04 % are obtained. Due to layer width fluctuations and the fact that the height of the Ge-island increases towards the surface, the measured MQW-satellites are broader than in the simulated curves.

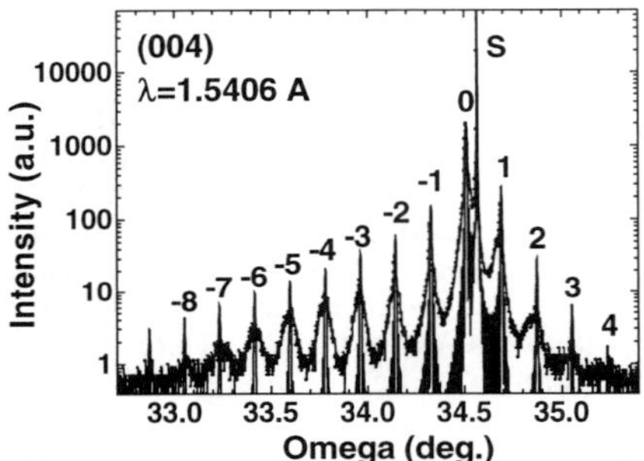

Fig. 1 : (004) rocking curves of the 30 nm spacer sample and dynamical simulation.

In Fig. 2, the measured asymmetric RSMs of the 40 nm spacer sample are shown. The intense peak labelled "S" is the Si substrate peak. Above and below the substrate peak, the MQW satellites of zeroth (SL_0) and first ($SL_{\pm 1}$) orders can be seen. They originate in the remaining two-dimensional MQW, which consists both of wetting and Si spacer layers. Their FWHM in lateral direction is determined by the experimental resolution. The RSMs have been measured in a low incidence geometry. On the left and right hand sides of the coherent peaks in both RSMs additional diffuse maxima are visible, which stem from the dots and the strained regions around them. The RSMs of the 30 nm spacer sample were nearly identical, in the case of 10 and 20 nm spacers, the diffuse maxima were much broader in G_z-direction.

DISCUSSION

X-ray diffraction from an arrangement of self-assembled quantum dots is influenced by the strains inside and outside the dots as well as by the shape of these dots and their relative positions. For an understanding of the diffraction patterns, we have acquired RSMs around several reciprocal lattice points (RELPs) (hkl) (see Fig. 2). The most prominent signatures of the presence of dots are the side peaks on the right hand side of SL_0 (labelled D_0 in Fig. 2) and on the left hand side of SL_1 (labelled D_1) as well as a more or less broad distribution of diffuse scattering around SL_0.

Since x-ray diffraction is a reciprocal method, only small regions in reciprocal space need to be probed for the study of e.g. long-range correlations or large crystallites. On the other hand, for the investigations of such tiny structures as quantum dots, the scattered intensity is distributed over a wide range in reciprocal space. This qualitative argument implies that peaks close to the coherent signal (where $2\pi/|\mathbf{q}|$ is much larger than the size of the dots) originate in regions in "real space" far from the dot center. Consequently, in order to assess detailed information on the strain-fields inside the dots, the peaks with a larger separation from the coherent signal must be investigated. (Unfortunately, the diffracted intensity diminishes rapidly with increasing distance from the Bragg peaks). Far away from the dots, the strain fields in the Si-spacers are compressive and the lattice constant is decreased, therefore a corresponding diffraction-peak should have a larger G_X-coordinate than the unstrained Si-substrate. The opposite is true for peaks originating in the dots themselves or in the regions above and below them.

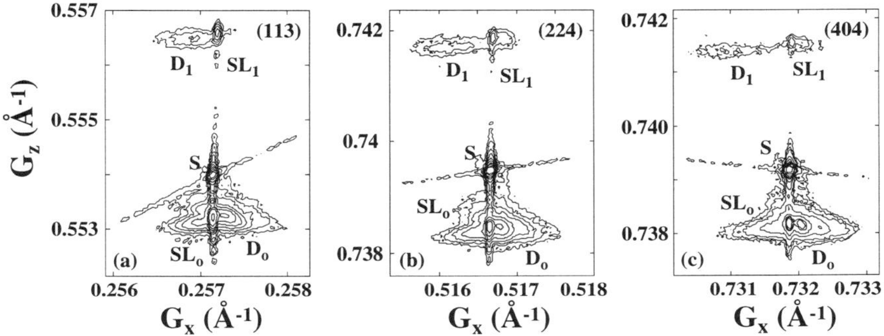

Fig. 2 : Asymmetrical RSMs around the (113) (a), (224) (b) and (404) (c) RELPs.

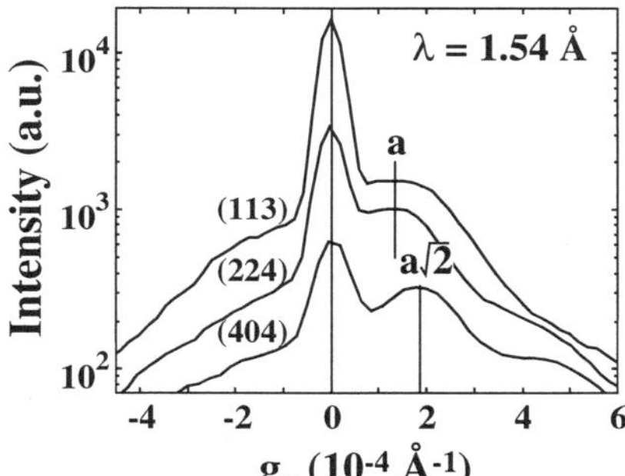

Fig. 3: Integration of the RSMs of the 30 nm spacer sample around the (113), (224) and (404) RELPs in q_z-direction.

In view of the above arguments, we can now identify the peak D_0 as corresponding to the regions around the dots, whereas D_1 corresponds to the regions in and above/below the islands. However, a closer look at the RSMs revealed that the separation between SL_0 and D_0 is identical for (113) and (224) and (-2-24) - the latter is not shown. This means that it's *position* is not due to a decreased lattice constant but due to a lateral periodicity in the dot separation. It is best seen in Fig. 3, where integrations of the reciprocal space maps in G_z-direction are presented.

If the dots were perfectly periodic, a series of satellites with equal spacing would be visible. Since there is at best only a short range ordering in the dot ensemble, the higher satellites are suppressed or concealed in the diffuse scattering. The reason why it appears on the right hand side of the coherent signal *is* due to the compressive strain in the barriers and is discussed in Ref. [10] for a perfectly periodic arrangement of strained scatterers.

Most interestingly, the separation between SL_0 and D_0 does change in the (404) RSM and approximately by a factor of $\sqrt{2}$. The latter was measured in an azimuth of 45° relative to the sample orientation in the (113) and (224) measurements (i.e. the sample was rotated around the surface normal by 45°. This finding is consistent with a short range ordering of the dots in a square array with a lattice constant of 520 ± 70 nm and main axes parallel to [100] and [010]. The same phenomenon has been observed by Tersoff *et al.* [4] in AFM-investigations of SiGe dot multilayers grown at lower temperatures, with an identical orientation of the main axes. This preferred alignment along the <100>-crystal-axes is due to the anisotropy of the elastic properties - the stiffness of Si and Ge has a pronounced minimum along the <100>-axes (see Fig. 4).

In contrast to D_O, the separation of D_1 an SL_1 increases significantly from (113) to (224) and even more so in the (404) RSM. It scales with the in-plane projection $(h^2+k^2)^{1/2}$ of the reciprocal lattice vector (hkl). Some ambiguity is introduced by the asymmetry and low intensity of the D_1-peaks, but the tendency is clearly visible in Fig. 2. Therefore, the position of D_1 reflects the elastic relaxation of the quantum structure. Previously, uncapped single Ge-dot layers have been studied by grazing incidence diffraction by Williams *et al.* [13] and Steinfart *et al.* [14].

So far the vertical correlation (stacking) of the dots has been studied by TEM [3], which is a destructive method. Theoretical models for the non-destructive assessment of the correlation of interface-profiles in multilayers using x-ray diffraction and in particular x-ray reflection have been

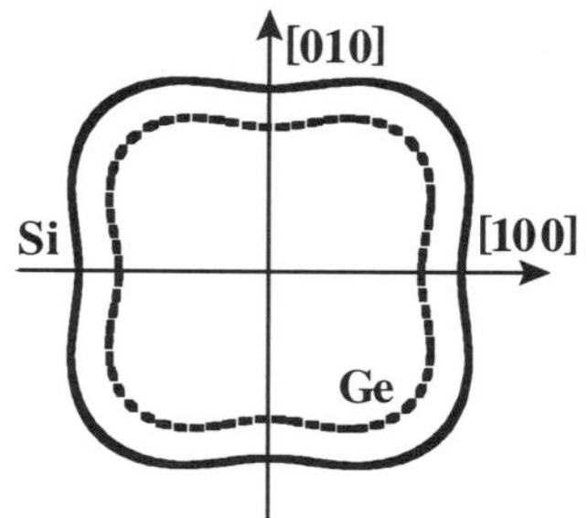

Fig. 4 : Polar diagram of Young's modulus for Si and Ge

developed in the last years [15]. These concepts are also applicable for dot-multilayers. In Fig. 5, we have depicted the measured and simulated specular reflectivity of the 20 nm spacer sample. The specular reflectivity is mainly influenced by the MQW and the r.m.s. values of the interface-"roughness". The multilayer period was determined as $D = 19$ nm and the wetting layer thickness as $T_W = 5$ Å. However, the specular reflectivity is not sensitive to the correlation of the interface profiles throughout the multilayer. In order to determine the latter, one has to measure the diffuse scattering which accompanies the coherent multilayer peaks in the reciprocal plane. As it is discussed in Ref. [15] in the framework of the Distorted Wave Born Approximation (DWBA), in the case of a rough multilayer, the width and height of the so-called Resonant-Diffuse-Scattering (RDS)-maxima is very sensitive to the vertical correlation of the interface-profiles. Since this diffuse signal is weak, the use of synchrotron sources is necessary. In Fig. 6, a 2θ-scan is shown, which corresponds to a curved trajectory in reciprocal space intersecting the coherent signal at one point and many RDS-maxima in the vicinity of this intersection point. From the simulation (full and dashed lines in Fig. 6) we obtained the vertical correlation length as 50 ± 10 % of the total multilayer thickness.

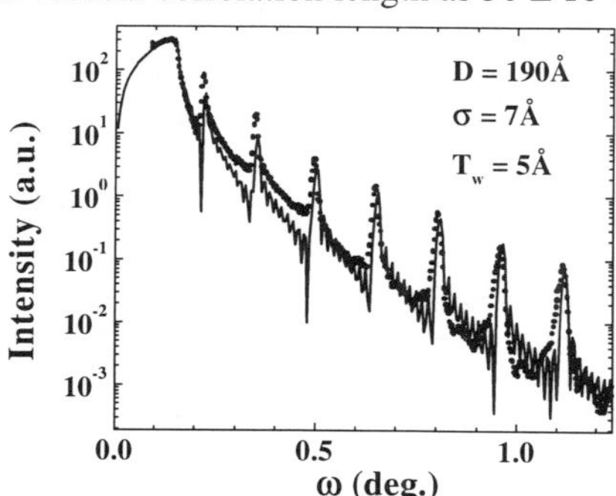

Fig. 5 : Specular reflectivity of the 20 nm spacer sample

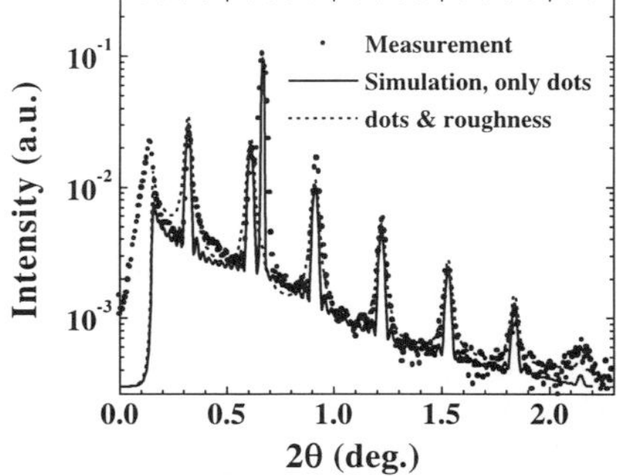

Fig. 6 : 2θ-scan of the 20 nm spacer sample

In our experiments for the 30 and 40 nm spacer samples, the width of all the diffuse maxima in q_z-direction is nearly the same as the FWHM of the coherent peaks. This is a clear indication that the vertical correlation along the growth direction is nearly 100 %. As discussed above, the diffuse maxima of the samples with thinner spacer layers (10 and 20 nm) are broader in q_z-direction. This decrease in the vertical correlation is not due to a variation in the dot positions but rather caused by changes in the dot shape. This conclusion was confirmed by TEM studies showing an increase in the dot radius and height towards the free surface. This change is most pronounced for the sample with the thinnest spacer layers [16].

CONCLUSIONS

In summary, we have investigated the structural properties of self-organized Ge-islands on silicon using high resolution x-ray diffraction and reflection. The vertical correlation (stacking) of the dots and their lateral ordering has been studied.

ACKNOWLEDGEMENTS

Work supported by FWF project 10083 and 11557, ESRF Grenoble, HASYLAB Hamburg, by the Grant Agency of the Czech Republic (project 202/1994/1871), by BMfWVK, and by the BMBF (No. 01M2953B2).

REFERENCES

1. D. J. Eaglesham, M. Cerullo, Phys. Rev. Lett. **64**, 1943 (1990).
2. Y.-W. Mo, D. E. Savage, B. S. Swartzentruber, and M. G. Lagally, Phys. Rev. Lett. **65**, 1020 (1990).
3. Q. Xie, A. Madhukar, P. Chen, and N. P. Kobayashi, Phys. Rev. Lett. **75**, 2542 (1995).
4. J. Tersoff, C. Teichert, and M. G. Lagally, Phys. Rev. Lett. **76**, 1675 (1996); and C. Teichert, M. G. Lagally, L. J. Peticolas, J. C. Bean, and J. Tersoff, Phys. Rev. B **53**, 16334 (1996).
5. H. Sunamura, N. Usami, Y. Shiraki, and S. Fukatsu, Appl. Phys. Lett. **66**, 3024 (1995).
6. P. Schittenhelm, M. Gail, J. Brunner, J. F. Nützel, and G. Abstreiter, Appl. Phys. Lett. **67**, 1292 (1995).
7. R. Apetz, L. Vescan, A. Hauptmann, C. Dieker, H. Lüth, Appl. Phys. Lett. **66**, 445 (1995).
8. S. K. Theiss, D. M. Chen, and J. A. Golovchenko, Appl.Phys. Lett. **66**, 448 (1995).
9. J.-M. Baribeau, J. Cryst. Growth **157**, 52 (1995)
10. V. Holy, A. A. Darhuber, G. Bauer, P. D. Wang, Y. P. Song, C. M. Sotomayor-Torres, M. C. Holland, Phys. Rev. B **52**, 8348 (1995).
11. Q. Shen, C. C. Umbach, B. Weselak, and J. M. Blakely, Phys. Rev. B **53**, R4237 (1996).
12. P. Schittenhelm, G. Abstreiter, A. A. Darhuber, G. Bauer, P. Werner, and A. Kosogov, Thin Solid Films, in print.
13. A. A. Williams J. M. C. Thornton, J. E. Macdonald, R. G. van Sifhout, J. F. van der Veen, M. S. Finney, A. D. Johnson, C. Norris, Phys. Rev. B **43**, 5001 (1991).
14. A. J. Steinfart, P. M. Scholte, A. Ettema, F. Tuinstra, M. Nielsen, E. Landemark,

D.-M. Smilgies, R. Feidenhans'l, G. Falkenberg, L. Seehofer, R. L. Johnson, Phys. Rev. Lett.**77**, 2009 (1996).
15. V. Holy. T. Baumbach, Phys. Rev. B **51**, 10668 (1994) and references therein.
16. P. Schittenhelm *et al.*, to be published

ANTIMONY CLUSTER MANIPULATION ON THE Si(001) SURFACE BY MEANS OF STM

I.I.Kravchenko*, C.T.Salling**, and M.G.Lagally
Materials Science Department, University of Wisconsin, Madison, WI 53706
*Current address: Physics Department, University of Florida, Gainesville, FL 32611,
kravchenko@phys.ufl.edu
**Current address: Beckman Institute, Urbana, IL 61801, salling@tm.uiuc.edu

ABSTRACT.

We present results of the manipulation of antimony clusters on Si(001) by means of a scanning tunneling microscope. By adjusting tip-sample separation and pulse voltage, an antimony cluster can be removed from the sample surface without damaging it. The success rate of the removed-cluster redeposition from the tip back onto the surface is 30%. In the remainder of the attempts a square shaped structure is created that had a hillock in the center. The hillock exhibits a metallic-like I-V curve. Such a structure cannot be created without an Sb cluster previously removed from the surface and located on the tip.

INTRODUCTION.

The ability to create atomic-scale structures with the scanning tunneling microscope (STM) plays an important role in the development of nanoscale technology. For the investigation of nanoscale test structures and the development of practical nanoscale devices, it is desirable to produce atomically ordered and clean structures of any geometry. By using atomic manipulation with an STM, one can modify an already well ordered surface under conditions of UHV cleanliness.

In the STM, strong chemical forces can be exerted on sample atoms and enormous current densities and electric fields can be achieved using relatively small voltages and currents. These interactions have been utilized to create isolated nanoscale structures by a number of methods: mechanically scribing a surface [1-3], creating an etch mask [4-8], decomposing precursor molecules [9-13], and manipulating atoms [14-38].

Basically there are three modes of atomic manipulation: lateral motion, deposition, and removal. Reproducible atomic manipulation and atomic-scale fabrication of general structures was demonstrated by Eigler at al. [15-17], who utilized an STM to move adsorbed atoms on a metal surface at cryogenic temperatures. Other nanoscale structures were created by moving islands of Cs atoms [23] on GaAs(110) to make larger island complexes and by accumulating Si atoms under the tip to create nanocolumns [24], crystalline nanoscale pyramids [25], and nanoneedles [26]. Additional examples of lateral motion include displacing adsorbed atoms of Si on Si(111) [18] and Cl on Si(001) [19], rotating dimers of Sb on Si(001) [20], and moving vacancies on GaP(110) [21,22].

Deposition of a single Ge atom onto Ge(111) surface from the STM tip was demonstrated in 1987 [14]. In 1990, clusters of atoms from a gold tip were reliably deposited onto various surfaces [27,28]. Individual atoms of hydrogen were deposited onto Si(111) surfaces from a PtIr tip supplied with hydrogen molecules from an ambient gas [29]. Recently, the assembly of atomic-scale

Mat. Res. Soc. Symp. Proc. Vol. 448 © 1997 Materials Research Society

structures was accomplished at room temperature by depositing silicon atoms from a W tip onto the Si(111) surface [30]. The atoms were placed onto the tip by first extracting them from the surface.

EXPERIMENTAL.

Our experiments were conducted using a custom made STM in ultrahigh vacuum (UHV) with STM tips prepared from polycrystalline tungsten wire. The UHV chamber has a base pressure of $\sim 1*10^{-10}$ torr. Sb was deposited by evaporating from a thoroughly outgassed tantalum basket. The Sb exposure was controlled by a shutter that was placed directly in front of the basket, and exposure time was from 10 sec to 2.5 min. During evaporation, the pressure in the chamber was higher than the base pressure by $0.6*10^{-10}$ torr. The deposition rate was estimated as 0.01ML/sec. Samples were directly heated by passing a current through them. The experiments were carried out on many samples cut from different p-type wafers having a resistivity of 10 Ω *cm and cleaned in situ using a standard procedure [39]. The dopant level was 10^{15} cm^{-3}.

RESULTS AND DISCUSSION.

Initially, we were trying to induce Sb cluster diffusion with the tip's electric field. The STM tip was kept over the Sb cluster covered surface for periods of time from 1 sec to 2 min at different tip-sample separations in the range from 1 to 5 Å (the method for measuring the tip-sample separation has been described previously [34]), and the tip bias in the range from -6V to +10V. No changes in cluster positions were observed.

Previous studies [20,40] have shown that upon deposition of a small fraction of a monolayer of Sb onto Si(001) kept at room temperature, five distinct types of Sb clusters were produced. Four types can be converted to the final state and displacement of the final-state dimers can be frequently induced by the STM tip. The STM caused conversion and displacement of the clusters at high tip biases (more than 3V) when the tip was right on top or around the cluster, i.e. the conversion and displacement were local effects.

In our work we conclude that the tip influence over Sb clusters on Si(001) is strictly local, and it is impossible to assemble larger island complexes by means of long range electric-field-induced diffusion similar to that reported for Cs:GaAs(110) system [23].

To remove Sb clusters from the sample surface, the STM feedback circuit is disengaged, the tip is displaced (with zero volts on the tip) toward the sample by a known amount to give a desired tip-sample separation, and the tip voltage is pulsed to a selected value for 200 ms. It was possible to pick up an antimony cluster by means of tip displacement towards the surface by 2.4 Å from an initial tip-sample separation 4.4 Å, while the tip located directly over a cluster, and pulsing the tip voltage to -5V. The redeposition procedure for the cluster on the tip is identical to the procedure of removal, but needs the tip voltage reversed to +5V. We have been able to pick up an antimony cluster that contains four atoms and to redeposit it at any chosen destination. Close to 30% attempts to return a cluster onto silicon surface are successful while it is possible to remove a cluster from the surface in 90% of attempts . In 70% of redeposition attempts the nanostructures described below were created. Figure 1 and fig.2 illustrate the process of cluster manipulation.

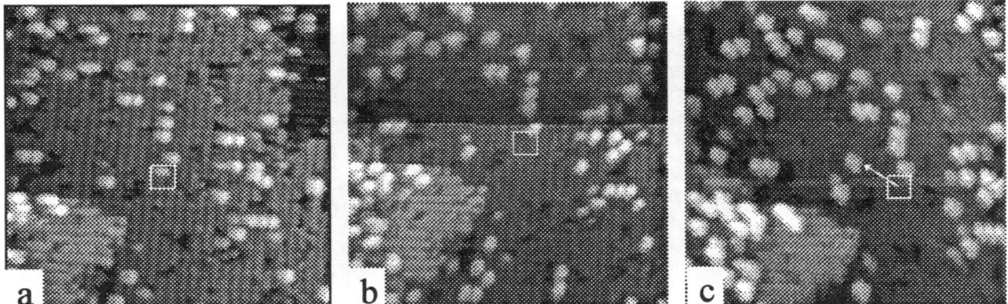

FIG. 1. Removal and subsequent deposition of a single Sb cluster on Si(100). (a) A STM image of the Si(100) surface covered by a small concentration of Sb clusters. An Sb cluster is chosen (framed) to be removed. (b) The cluster is removed and placed on the tip. While obtaining the image, the tip is unstable. The surface is not damaged at the spot where the cluster was located. (c) The cluster is redeposited at another site. The tip further deteriorates. The images size is 220x220 Å.

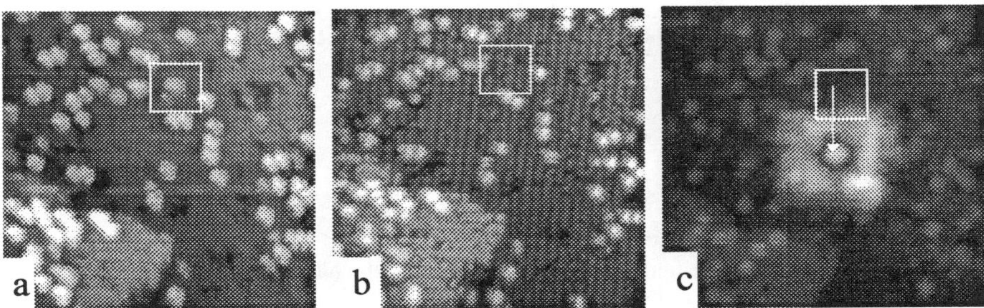

FIG. 2. The 70x70 Å structure creation. (a) A STM image identical to fig. 1c. Another Sb cluster is chosen. (b) The cluster is removed and located on the tip. After the procedure of removal the tip is improved and is capable of obtaining a better quality image. (c) While redepositing the cluster from the tip a nanostructure was created. The size of the structure is approximately 70x70Å and its sides are oriented perpendicular and parallel to substrate dimer rows. The hillock inside is approximately 10Å tall and about 17Å in diameter.

At the beginning, the tip is positioned over the surfaces (fig. 1a) at a tip-sample separation of 4.4 Å, with a tunneling current of 0.2nA. The surface is not damaged when a cluster is removed (fig. 1b). The cluster is redeposited at a deliberately chosen site (fig. 1c). In other removal and redeposition attempt (fig. 2a-2b), the return onto the surface leads to creation of a "sombrero-like" structure that has a rather square shape with a hillock surrounded by a moat inside (fig. 2c). It should be pointed out that the quality of the images is different after each of the pulses, implying tip atom rearrangement. It is impossible to predict which of the redeposition attempts can be successful or can create the structure, but we have not been able to obtain such a structure without an Sb cluster placed on the tip.

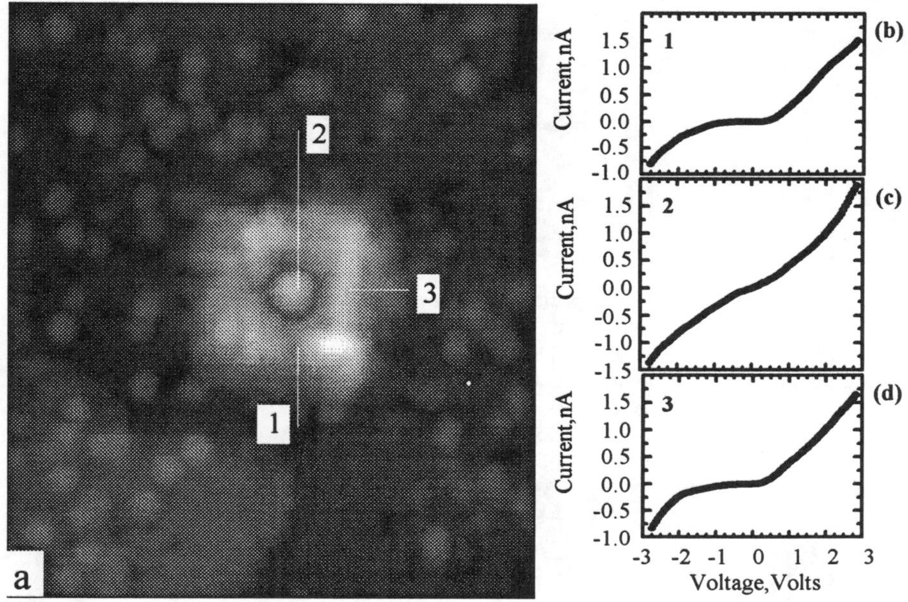

FIG. 3. Current-voltage characteristics of a nanostructure. (a) Each curve was acquired at a constant tip-sample separation while the tip was located over the hillock (1) in the center of the nanostructure, and over the outer ring at locations (1) and (3). (b), (d) I-V curves have a semiconductor type shape. (c) I-V curve of the hillock has a metallic behavior.

The size of the structure (fig. 3a) is approximately 70x70Å and its sides are always oriented perpendicular and parallel to substrate dimer rows. The hillock inside which is usually less than 10Å tall and about 17Å in diameter, can be higher or lower than the outer structure. It is impossible to estimate how deep the moat is, because the tip radius is apparently greater than the gap between the hillock and the outer structure.

Currently we are not able to propose a mechanism to explain the shape of the structure, but we can make some speculations based on results published previously. Nanoneedle structures have been created by slowly increasing the tip bias voltage to around -10V for about 3 to 30 s, and maintaining the tunneling current at 0.2nA [26]. According to the authors, the needle formation conditions do not depend on the tip material, and the sample surface becomes quite rough. That implies that silicon atoms move from their reconstructed position and diffuse under the tip apex area, resulting in formation of a nanoneedle. In work on Si(111) [31], three types of surface modifications were observed: a hill, a hill surrounded by a moat, and a pit. The pit corresponds to the complete removal of atoms from the surface and the other two types of features result from incomplete removal of atoms. Based on monitoring the tip-sample current during the formation of the hill-and-moat structure, it was suggested that during the voltage pulse a bridge of atoms between the tip and sample is being created [31]. The bridge stretches and breaks when the tip is retracted, during the return to the imaging mode at the end of a removal attempt. The hillock in the middle of the nanostructures described in this paper points to either the bridge mechanism or mechanism responsible for creation of the nanoneedles. It is possible that the outer structure is created by a silicon atom transfer from the tip back to the surface. Only the area of the surface under the tip is involved in the transfer, while areas outside of the structure are substantially unperturbed.

Fig.3b-3d presents current-voltage measurements on both the outer structure (fig.3b and fig.3d) and the inner hillock (fig.3c). Unexpectedly, we discovered that the I-V characteristics when the tip was positioned directly over the hillock had metallic behavior (fig.3c). The I-V curves of the outer structure are characteristic of semiconductors (fig.3a-c) and are similar to the I-V characteristics of the silicon surface, but we can not draw any conclusions about the density of states because the tip could have been contaminated previously.

In the case of degenerate doping of Si by Sb, it is possible to obtain metallic behavior of a silicon surface [41]. If we take into account the volume of the hillock and assume that one Sb cluster has four atoms, we can evaluate the doping level which in this case is close to degenerate, 10^{-20} cm^{-3}.

We can not rule out the possibility that some tungsten atoms from the tip have also been deposited, but we have not found any evidence for such a transfer either from published results or from our experience.

CONCLUSIONS.

We have demonstrated the use of the STM for the manipulation of antimony clusters on the Si(001) surface. Further studies are needed to improve redeposition process of Sb clusters already located on the STM tip back onto the surface. Without such an improvement, it is highly problematic to utilize this technology for assembling larger cluster complexes. Square shaped nanostructures, which had metallic behavior in the middle, were created. These structures might be of interest for the quantum dot physics. We can confirm previous studies that Sb clusters have negligible thermal diffusion at room temperature, which gives an experimenter long enough time to create nanostructures on the surface without investment in cryogenic equipment. It is possible to induce Sb cluster diffusion on the Si(001) surface at room temperature by STM tip voltage pulses, but the probability of the process is too low to have any practical significance.

ACKNOWLEDGMENT. This work was funded by AFOSR.

REFERENCES.

1. M.A.McCord and R.F.W.Pease, Appl.Phys.Lett.**50**,569 (1987).
2. E.J.van Loenen, D.Dijkkamp, A.J.Hoeven, J.M.Lenssinck, and J.Dieleman, Appl.Phys.Lett.**55**, 1312 (1989)
3. Y.Kim and Lieber, Science **257**, 375 (1992).
4. M.A.McCord and Pease, J.Vac.Sci.Technol.**B5**,430 (1987).
5. J.A.Dagata, J.Schneir, H.H.Harary, C.J.Evans, M.T.Postek, J.Bennet, Appl.Phys.Lett. **56**, 2001 (1990).

6. For a recent survey of STM -based lithography see *The Technology of Proximal Probe Lithography*; ed. C.R.K.Marrion (SPIE, Bellingham,WA 1993).

7. J.W.Lyding, T.-C.Shen, J.S.Hubacek, J.R.Tucker, and G.C.Abeln, Appl.Phys.Lett.**64**, 2010 (1994).

8. E.S.Snow and P.M.Campbell, Appl.Phys.Lett. **64**, 1932 (1994).

9. R.M.Silver, E.E.Ehrichs, A.L.de Lozanne, Appl.Phys.Lett. **51**, 247 (1987).

10. F.Thibaudau, J.R.Roche, and F.Salvan, Appl.Phys.Lett. **64**, 523 (1994).

11. J.S.Foster, J.E.Frommer, and P.C.Arnett, Nature **331**, 324 (1988).

12. W.Li, J.A.Virtanen, R.M.Penner, Appl.Phys.Lett. **60**, 1181 (1992).

13. G.Dujardin, R.E.Walkup, and Ph.Avouris, Science **255**, 1232 (1992).

14. R.S.Becker, J.A.Golovchenko, B.S.Swartzentruber, Nature **325**, 419 (1987).

15. D.M. Eigler and E.K. Schweizer, Nature **344**, 524 (1990).

16. J. A. Stroscio and D. M. Eigler, Science **254**, 1319 (1991).

17. M.F.Crommie, C.P.Lutz, and D.M.Eigler, Science **262**, 218 (1993).

18. H.Uchida, D.H.Huang, J.Yoshinobu, and M.Aono, Surf.Sci. **287/288**, 1056 (1993).

19. J.J.Boland, Science **262**, 1703 (1993).

20. Y.W.Mo, Science **261**, 886 (1993); J.Vac.Sci.Technol.B **12**, 2231 (1994).

21. Ph.Ebert and K.Urban, Ultramicroscopy **49**, 344 (1993);

22. Ph.Ebert, M.G.Lagally, and K.Urban, Phys.Rev.Lett.**70**, 1437 (1993).

23. L.J.Whitman, J.A.Stroscio, R.A.Dragoset, and R.J.Celotta, Science **251**, 1206 (1991).

24. R.M.Ostrum, D.M.Tanenbaum, and A.Gallagher, Appl.Phys.Lett.**61**, 925 (1992).

25. M.Iwatsuki, S.Kitamura, T.Sato, and T.Sueyoshi, Appl.Surf.Sci. **60/61**, 580 (1992).

26. S.Heike, T.Hashizume, Y.Wada, Jpn.J.Appl.Phys **34,** L1061-1063 (1995).

27. H. J. Mamin, P. H. Guethner and D. Rugar, Phys. Rev. Lett. **65**, 2418 (1990);

28. H.J.Mamin, S.Chiang, H.Birk, P. H. Guethner, D.Rugar, J.Vac.Sci.Technol.**B9**,1398 (1991).

29. H.Kuramochi, H.Ushida, and M.Aono, Phys.Rev.Lett.**72**, 932 (1994).

30. D. Huang, H. Uchida, and M. Aono, J. Vac. Sci. Technol. **B12**, 2429 (1994).

31. I.W.Lyo and Ph. Avouris, Science **253**, 173 (1991).

32. A. Kobayashi, F. Grey, R. S. Williams, M. Aono, Science **259,** 1724 (1993).

33. M. Aono, A. Kobayashi, F. Grey, H. Uchida, D. Huang, Jpn. J. Appl. Phys. **32**, 1470 (1993).

34. C.T. Salling and M. G. Lagally, Science **265**, 502 (1994).

35. S. Hosoki, S. Hosaka, T. Hasegawa, Appl. Surf. Sci. **60/61**, 643 (1992).

36. J.-L. Huang, Y.-E. Sung, and C.M. Lieber, Appl. Phys. Lett. **61**, 1528 (1992).

37. S.E. McBride and G.C.Wetsel, Jr., Appl. Phys. Lett. **57**, 2782 (1990); **59**, 3056 (1991).

38. J.P. Rabe and S. Buchholz, Appl. Phys. Lett. **58**, 702 (1991).

39.B.S.Swartzentruber, Y-W.Mo, M.B.Webb, M.G.Lagally J.Vac.Sci.Technol. **A7**, 2901 (1989).

40. Y.W.Mo Phys. Rev. Lett. **69**, 3643 (1992); Phys. Rev. B **48**, 17233 (1993).

41.D.H.Rich,G.E.Franklin, F.M.Leibsle,T.Miller, and T.-C. Chiang, Phys.Rev.**B40**,11804 (1989).

DIRECT FORMATION OF FINE STRUCTURE
BY LOW ENERGY FOCUSED ION BEAM

T.Chikyow, A.Shikanai*, and N.Koguchi

*National Research Institute for Metals,
1-2-1 Sengen, Tsukuba Ibaraki 305, JAPAN
Tel:+81-298-53-1055 fax:+81-298-53-1093
e-mail: tchikyo@momokusa.nrim.go.jp*

**School of Science and Engineering, Waseda University
2-3-4 Ohkubo Shinjyuku, Tokyo 169, JAPAN*

ABSTRACT

GaAs micro crystals in line were grown on a sulfur-terminated GaAs surface by low energy focused ion beam. Ga ions, picked out from a liquid Ga ion source, were accelerated up to 10 KV to obtain a focused ion beam. The ions were given a positive bias to reduce their kinetic energy by retarding lens.The Ga ions landed on the surface softly and formed a series of Ga droplets. By subsequent As molecule supply to the Ga droplet, GaAs micro crystals in line were grown. This method was found to be useful to make fine structures directly on the semiconductor materials.

INTRODUCTION

Recently carrier confinement in low dimension has been attempted for practical device application as well as fundamental interests[1]. Especially the quantum dot has been studied intensively because of its potential for laser application[2]. To fabricate the quantum dots, various type of methods had been proposed[3-5]. As a sophisticated fabrication method, a " Droplet Epitaxy", where GaAs micro crystals grow from Ga droplets on Se or S terminated GaAs surface, was proposed and successful results have been reported[6,7]. However, the position of GaAs micro crystals could not be controlled. In addition,to improve optical properties, a position control of GaAs micro crystal is required. For this purpose, a low energy focused ion beam system (LE-FIB) with new idea was proposed. Here, Ga droplet position is controlled by low energy focused ion beam at first. Subsequently Arsenic molecules are supplied to the droplets to grow GaAs micro crystals. This concept is shown in the figure 1.

In general, a serious problem in LE-FIB has been a difficulty in making fine structures due to its expanded beam diameter in retarding. However, a combination of LE-FIB and the " Droplet Epitaxy" brings a great advantage. Namely it is possible to form Ga droplets which have smaller diameter than ion beam size. This leads to an expectation of fine GaAs micro crystal growth from the Ga droplets. By this method, a direct GaAs micro crystal growth on the S-terminated GaAs surface was challenged.

LOW ENERGY FOCUSED ION BEAM SYSTEM (LE-FIB)

In the past, several types of focused ion beam with retarding system has been proposed for

Mat. Res. Soc. Symp. Proc. Vol. 448 © 1997 Materials Research Society

etching and deposition by kinetic energy control [8-11]. The newly proposed LE-FIB system with a liquid Ga ion source had four static lens groups. The first one was used for Ga ion to be picked up from the source to the acceleration column at 7.0-8.0 kV. The second lens was biased up to 10.0 KV for making a fine beam. The third lens was biased for focusing around 4.0 kV. The characteristic point of this system is that the last one had four isotropic electronodes for retarding.

The isotropic retarding in the cylindrical column suppressed the ion beam expanding to some extent.

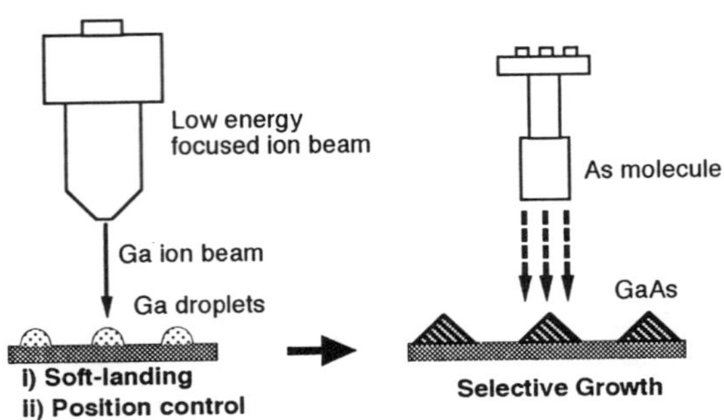

Fig.1   Position control of GaAs micro crystal by combination of low energy focused ion beam system and " Droplet Epitaxy".

The retarding bias was applied up to 9.97KV. Reflectors in middle of the collum scanned the beam and secondary electrons caused by ion bombardment were detected by multi-channel plate to obtain surface images in the same manner of scanning electron microscopy (SEM).The end of the retarding part was set 2.0 cm apart from the sample in the ultra high vacuum (UHV) chamber. To keep the sample at the grand level electrically, the LE-FIB system is insulated from the earth level. This LE-FIB system is illustrated schematically in the figure 2.

The sample holder was shared commonly between a scanning tunneling microscopy (STM), which was equipped in anther UHV chamber, and a conventional molecular beam epitaxy (MBE) system to supply Arsenic molecules. The three system (LE-FIB, STM, and MBE) were connected in UHV. The sample holder was transferred in three systems without breaking vacuum.The sample heating was carried out by direct current supply .

EXPERIMENTS

GaAs buffer layer was grown on GaAs(001) substrate by another MBE system. The surface showed (2×4) reconstructed structure after the growth. Subsequently sulfur molecules were supplied to the surface at 400 °C to form a sulfur-terminated (2×6) reconstructed surface[12]. Then the surface was covered by Arsenic at room temperature to form a protection layer for carrying the sample from the MBE chamber to the LE-FIB system. In the UHV chamber of the LE-FIB system, the As protection layer was removed. On the surface, Ga ions were supplied at 30 eV to form Ga droplets at room temperature in line scanning mode. The ion beam current was about 1 nA. The scanned width was 500 μm. The scanning speed was 25 cm/sec and total scanning time per line was 7.0 min. As molecules were supplied to the Ga droplets to grow GaAs.

Surface morphologies were observed by SEM and the surface states were characterized by

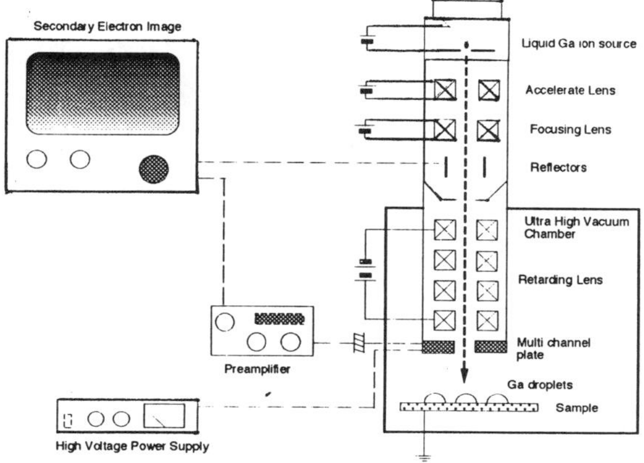

Fig.2 A schematic illustration of the proposed low energy focused ion beam system.

scanning tunneling spectroscopy (STS) of STM.

RESULTS

A series of Ga droplets 800 nm in size were observed on the sulfur-terminated GaAs surface when the kinetic energies of ions were around 30 eV. The Ga droplets are shown in Fig.3. The Ga droplets were observed in the center region of the scanned ion beam line. The spacing between droplets was about 6.3 μm. In this case, the beam diameter was 10 μm because of the ion repulsion in the column in retarding, though the original ion beam size is about 300 nm before retarding. Here, it was found that fine Ga droplets, which had smaller diameter size than that of the ion beam size, were formed on the Sulfur-terminated GaAs surface.

After the As molecule supply, three dimensional growth of micro crystal was observed.

The size of GaAs micro crystal was about 500 nm and the spacing was 14.5 μm. From the past results in our " Droplet Epitaxy", the micro crystals were thought to be GaAs. The size and spacing difference are thought to be due to ion beam current fractuation (ion flux fractuation) and different focusing condition in two samples.

To estimate a surface damage caused by ion bombardment and soft-landing of Ga ions, a STS measurement was carried out. The initial sulfur terminated GaAs surface was found to have little surface states because a significant band bending was not observed. However, when the Ga ions were supplied at 30 eV to cover 20% of the surface with Ga atoms, the Fermi level began to move to the center of the GaAs bandgap. This means that some of the Ga ions made surface states, resulting in the Fermi level pinning. Surface defect of a few percent for surface sites is enough for the Fermi level to be pinned. From the results, most of the Ga ions were thought to land on the surface softly.

DISCUSSION

The Ga ion beam has an energy distribution when it is taken from the liquid ion source. This distribution is kept after the retarding. The value is estimated to be +-10 eV for the liquid Ga ion source [13]. If the higher energy of Ga ions is focused to the beam center to hit the sulfur terminated GaAs surface, defect such as vacancies will be formed by ion bombardment. These defects will become nucleation sites for Ga to form droplets. Another Ga ions, which have lower kinetic energy, are under focused, landing on the surface softly , and migrate on the surface. Some of them may be trapped by the defects to form Ga droplets. As a result, a series of Ga droplets are thought to be formed in the center of the scanned region. This model is illustrated schematically in Fig.5

Defect density at the surface may affect the Ga diffusivity in the scanning direction. If the scanned region has more defect density, a diffusion length of Ga becomes shorter. In this case, spacing of Ga droplets becomes narrow.

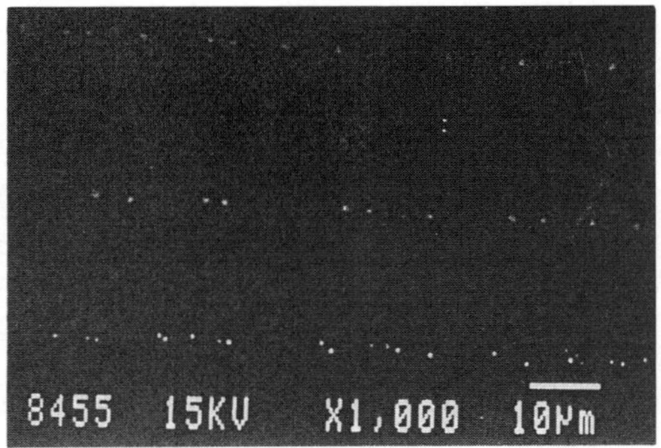

Fig.3   *Ga droplets in line formed by 30 eV Ga droplets 800 nm in size were observed.*
The spacing between droplets is 6.3 μm in average.

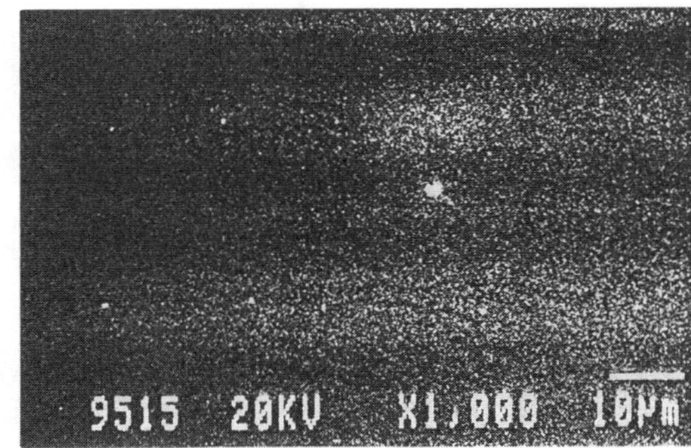

Fig.4   *GaAs micro crystals in line after Arsenic molecule suppy. Micro crystal 500 nm in size were observed.*

On the other hand, less density of defects leads to the longer diffusion length, resulting in the relatively separated Ga droplets. If the proposed model would be correct, the ion beam current density affects the defect density at the surface, leading to the different spacing as observed in Fig.3 and Fig.4. The focusing condition changes the Ga ion flux, leading to the droplet size difference as observed in the two figures.

From the STS analysis, it was found that the most of Ga ions landed on the surface softly without generating surface states but some of them made surface defects, leading to the Fermi level pinning. These results agree with the proposed model that higher energy ions make surface defects for nucleation and lower energy ions land softly on the surface to grow Ga droplets as speculated in Fig.5.

An expected advantage of this method was that fine structures might be formed even though the beam diameter was expanded during retarding. This expectation was confirmed correct because the smaller size micro crystals were observed, compared with the ion beam size. From the obtained results, the LE-FIB deposition combined with the "Droplet Epitaxy" seems to have a great potential for fabricating fine structures, such as the quantum dot.

CONCLUSION

As a summary, a series of GaAs micro crystal was grown on the sulfur terminated GaAs surface when the kinetic energy of the supplied Ga ions was 30 eV. This method could provide an idea that fine structures could be realized by low energy focused ion beam system.

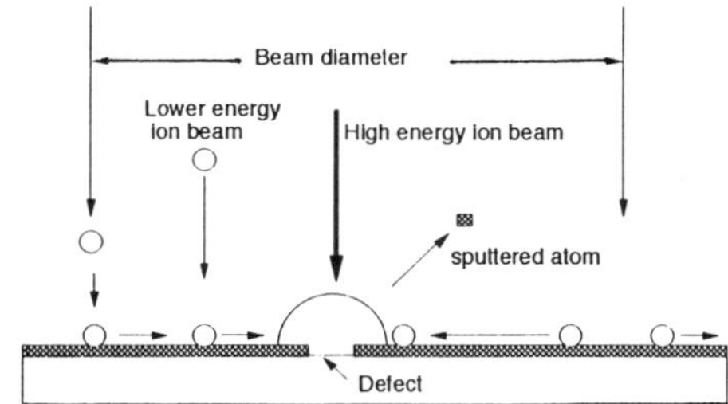

Fig.5   A schematic illustration of defect formation by ion bombardment and growth of Ga droplets by surface diffusion.

ACKNOWLEDGEMENTS

The authors are grateful to Dr.Tsukamoto for his fruitful discussion about the sulfur terminated GaAs surface.

REFERENCES

[1] H.Sakaki, Surface Science **267,** p.623 (1992).

[2] Y.Arakawa and H.Sakaki, Appl.Phys.Lett **40,** p939 (1982).

[3] H.Temkin, G.J.Dolan, M.B.Panish, and N.G.Chu, Appl.Phys.Lett.,**50,** p413 (1987).

[4] Y.Nagamune, M.Nishioka, S.Tsukamoto, and Y.Arakawa, Appl.Phys.Lett.,**64,** p2495 (1994).

[5] R.Notzel, J.Temmyo, T.Tamura, Jpn.J.Appl.Phys.,**33,** pL275 (1994).

[6] T.Chikyow and N.Koguchi, Appl.Phys.lett.,**61,** p2431 (1992).

[7] N.Koguchi and K.Ishige, Jpn.J.Appl.Phys.,**32,** p2052 (1993).

[8] R.Aihara, H.Kasahara, and H.Sawaraki, J.Vac.Sci.Technol.**B7,** p79 (1989).

[9] K.Gamo, in the Beam Solid Interactions:Fundamentals and Applications,
edited by M.Nastasi, L.R.Herbots, and R.S.Averback (Mater.Res.Soc.Proc. 30,
Pittsburgh, 1993) p.577.

[10] S.Nagamachi, Y.Yamakage,H.Maruno,M.Ueda, S.Sugimoto, and M.Asari,
Appl.Phys.Lett., **62,** p2143 (1993).

[11] K.Pak, I.Saitoh, N.Ohshima, and H.Yonezu, J of Crystal.Grwoth.,**140,** p244(1994).

[12] S.Tsukamoto and N.Koguchi, Jpn.J.Appl.Phys.,**33,** pL1185 (1994).

[13] L.W.Swanson, G.A.Schwind, and A.E.Bell, J.Appl.Phys.,**51,** p3453 (1980).

SURFACE ADSORPTION KINETICS OF GA WIRE ARRAYS ON SI(112)

S.M. Prokes and O.J. Glembocki
Naval Research Laboratory, Washington D.C. 20375.

ABSTRACT

The growth and energetics of Ga chains on a Si(112) facet surface have been investigated. Low energy electron diffraction (LEED) and Reflectance Difference Anisotropy (RDA) experiments have been performed in UHV to examine the kinetics of the Ga atom chain formation process along the Si steps. This is a self-limiting process for substrate temperatures in the 400°C to 500°C range, since the binding energy of the metal atoms at the step edges is greater than that on the (111)-type terraces. At saturation coverage, the Ga chains exhibit sixfold periodicity along the step edges, but below saturation, these chains exhibit fivefold periodicity. A model has been suggested from these results in which the periodicities of the chains are determined by the metal atom vacancies, which balance the tensile stress created by the larger metal atoms. We have used RDA to study the time evolution of the surface kinetics and changes in the surface reconstruction of Ga deposited on these facet (112) Si surfaces, as a function of substrate temperature and Ga deposition rate. The RDA measurement is shown to be sensitive to the formation of both the 5x1 as well as 6 x 1 reconstructions and to island formation at low substrate temperatures. The diffusion-controlled Ga line formation process at low temperatures has also been characterized as well as the evaporation controlled regime, which is evident at substrate temperatures over 400°C.

INTRODUCTION

The areas of nanostructure formation and their resultant properties have been receiving considerable attention recently. This is due to the fact that as device structures decrease, there is a need to understand novel ways of defining nanostructures and studying their materials, optical and electrical properties. One approach to form lower dimension structures involves the use of lithography, combined with various dry etching techniques. However, in many cases, the dry etching step can lead to significant damage, resulting in poor electronic and optical properties of such structures. Issues such as passivation of the etched surfaces to reduce damage are very important and under current investigation. Due to the surface damage problems, an alternate method of creating reduced-dimension structures is also under investigation, involving the direct formation of nanostructures by properly chosen growth conditions and surfaces. One example involves the growth of nanoscale structures on facet surfaces. This technique relies on the fact that certain thermodynamically stable surfaces exhibit ordered steps and ledges, which can be used as templates for metal atom deposition. This is the case for the Si(112) facet surface, which consists of (111) terraces and (100) steps [1] and which exhibits more stable bonding of certain metal ad atoms at the step edges. This can lead to a stabilized stepped surface structure with specific adatom periodicities along these steps. For example, in the case of Ga and Al, higher binding energies of these metal adatoms at the step edges have led to the formation of Ga or Al chains along these steps, exhibiting a 5x1 or 6x1 reconstruction [2], depending on the coverage. A model, in which the periodicities of the chains are determined by the density of metal atom vacancies along these steps, has been proposed [3] (model shown in Fig. 1) and confirmed experimentally using angle-resolved Auger electron spectroscopy [4]. Although steady state data

Mat. Res. Soc. Symp. Proc. Vol. 448 © 1997 Materials Research Society

is available in this system, there exists no real time information about the changes in surface structure and hence details of the kinetic processes that are responsible for the formation of these

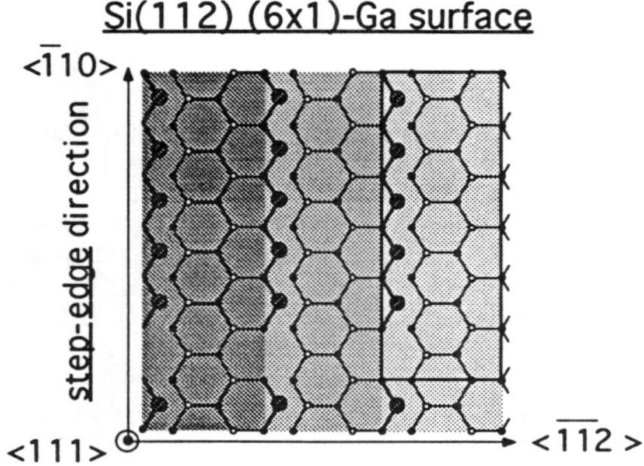

Fig. 1 Model for Ga adatom self assembly along step edges of a facet Si (112) surface. From ref. 3.

chains. Thus, we have used an optical technique, reflectance difference anisotropy (RDA), to monitor the behavior of this surface in real time, as a function of Ga atom deposition and desorption [5]. We show that we can easily observe submonolayer coverage of the Si surface by Ga and a change from a 5x1 to a 6x1 periodicity, as well as observe kinetic regimes for Ga line formation ranging from diffusion to evaporation limited regimes, depending on substrate temperatures and Ga atom deposition rates.

EXPERIMENTAL

All experiments were performed in situ in a UHV chamber having a base pressure of 3×10^{-10} torr and which included Auger and LEED spectrometers, and optical access to the RDA set up. The Ga deposition was performed using a Knudsen evaporation cell and the Ga depositions were performed at substrate temperatures between 300°C and 550°C, at a pressure of 3×10^{-9} torr. The samples were initially cleaned using a modified Shiroki etch, involving alternate dips in hydrofluoric acid (HF), boiling Nitric acid, followed by boiling in a solution of peroxide and ammonium hydroxide.

The RDA apparatus used in this work [5] included two HeNe lasers operating at either 632.8nm or 543.0nm, a polarizer and a Hinds International photoelastic modulator. The modulated light was reflected at near normal incidence from the sample, which was inside the vacuum chamber and located at the focal point of the Ga and Al K-cells. The reflected light was detected with a Si photodiode and the modulated component, ΔR, was detected at twice the modulation frequency with a PAR 5210 lock-in amplifier.

RESULTS

In Fig. 2, the change in surface reflectivity as measured by RDA is shown as a function of Ga atom deposition time at a substrate temperature of 402°C. The time at which the shutter was opened and closed is marked in the figure. The reflectivity consists of a very steep initial rise, followed by two distinct peaks in the signal. Since the RDA signal remains stable upon the closure of the deposition shutter, it was possible to examine the surface structure at the maxima of the two peaks using LEED. In this way, it was determined that the first signal maximum corresponds to a 5x1 surface reconstruction and the second maximum corresponds to a 6x1 reconstruction. The drop of the RDA intensity between these two maxima corresponds to a local rearrangement of the Ga chains from a four atom to a five atom average length [3,5].

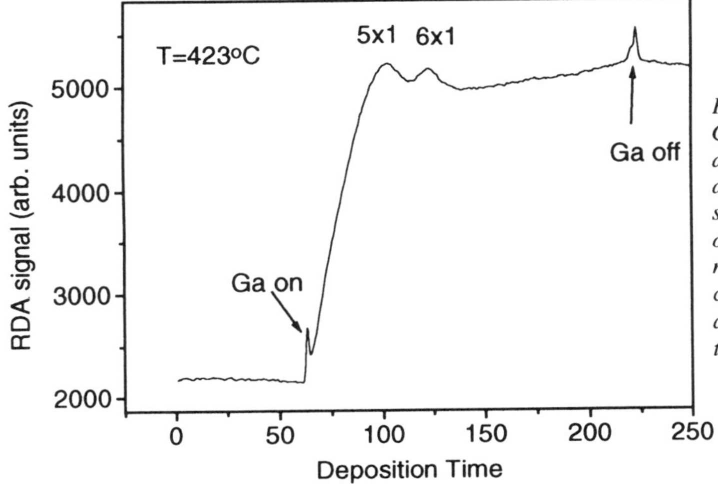

Fig. 2 RDA signal of Ga wire formation as a function of Ga deposition time at a substrate temperature of 423 °C. The two maxima shown correspond to the 5x1 and 6x1 reconstructions.

The behavior of the surface can also be monitored in real time as a function of substrate temperature as well as a function of different Ga atom fluxes, as shown in Fig. 3 a,b. In Fig. 3a, the changes in the surface RDA signal are shown for the same Ga atom deposition rate and various substrate temperatures. At the lowest substrate temperature studied (T = 350°C), the RDA signal rises and then continuously drops, indicating that Ga island formation on the terraces is occurring [5], which leads to a reduction of the surface anisotropy and thus the RDA signal. As the substrate temperature is increased to 500°C, the signal reaches a maximum, after which it remains relatively stable as a function of deposition time, indicating that the Ga line formation is a self-limiting process under these conditions. At this temperature, the Ga adatom surface diffusion and desorption rate from the (111) terraces are high enough to preclude Ga island formation on the terraces. This leads to Ga chain formation at the step edges only, since these sites have been shown to have a higher activation energy for desorption than the (111) terrace sites [2]. At higher temperatures, such as 550°C, the Ga line formation never achieves maximum coverage due to the high rate of evaporation. If the Ga atom flux is reduced, the time necessary to fill the maximum allowed step edge sites increases, resulting in a slower initial signal rise and a shift of all the RDA curves to longer times (Fig. 3b). The total step edge coverage obtained at 500°C, however, is the same after longer deposition time, since the RDA maximum signal is the same in Fig. 3a and 3b.

It is also possible to monitor the Ga line self-assembly from a diffusion-dominated regime at low substrate temperatures to an evaporation dominated regime at higher temperatures, which is evident from Fig. 4. At low substrate temperatures (300°C), the rate of the RDA signal is proporational to $Dt^{1/2}$, characteristic of a diffusion controlled regime. At higher temperatures, the atoms which land evaporate quickly, so that only atoms which land directly on the step edge sites can form the wire arrays. In this case, the wire formation should be proportional to (F-E)t, where F is the atom flux and E is the evaporation rate. Note that in this case, the RDA signal is linear with time t, as is evident from Fig. 4.

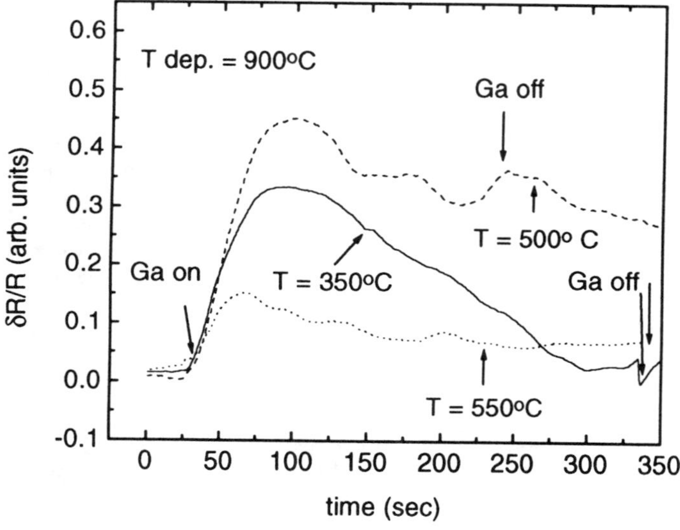

Fig. 3 RDA signal as a function of Ga deposition time for three substrate temperatures.

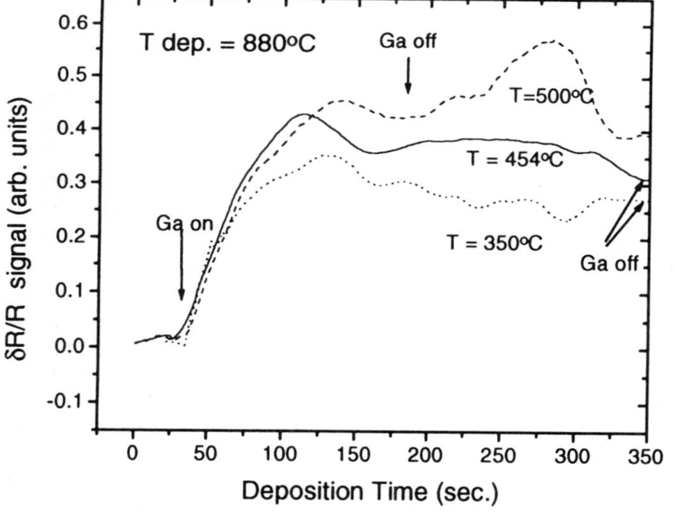

Fig. 4 *RDA signal as a funcion of Ga deposition time for three substrate temperatures. The Ga flux rate is lower than in Fig.3.*

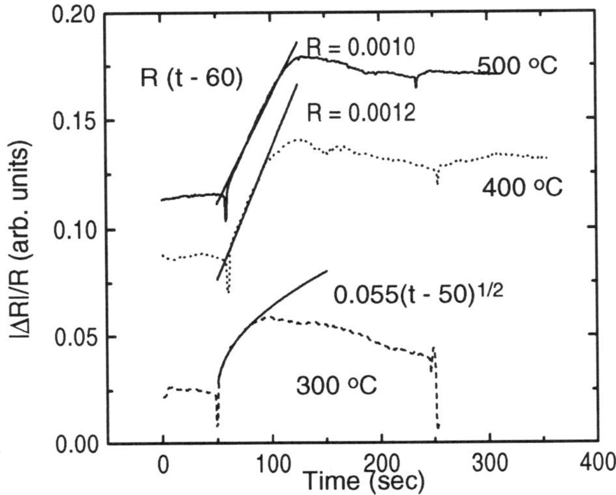

Fig. 5 The time evolution of the RDA signal as a function of sample temperature. The solid lines through the data are fits to $\Delta R/R = R(t-t_0)$ (400°C and 500°C) and to $\Delta R/R = [D(t-t_0)]^{1/2}$

CONCLUSION

The formation kinetics of Ga atom line self assembly on facet Si (112) surfaces have been studied using RDA and LEED. The RDA technique can examine the Ga line formation and the 5x1 and 6x1 reconstructions directly in real time. Furthermore, at very low substrate temperatures, island formation begins to overtake Ga line formation, and at very high temperatures, the desorption rate overwhelms the Ga line formation process, leading to small Ga line coverages and thus a much reduced RDA signal. The diffusion-controlled Ga line formation process at low temperatures has also been characterized as well as the evaporation controlled regime, which is evident at substrate temperatures over 400°C.

REFERENCES

1.	Th. Berghaus, A. Brodde, H. Neddermeyer and St. Tosch, Surf. Sci. **184**, 273 (1987).
2.	Timothy M. Jung, R. Kaplan and S.M. Prokes, Surf. Sci. Let. **289**, L577 (1993).
3.	Timothy M. Jung, S.M. Prokes and R. Kaplan, J. Vac. Sci. Technol. **A12**, 1838 (1994).
4.	J.E. Yater, A. Shih and Y.U. Idzerda, Phys. Rev. **B51**, 7365 (1995).
5.	O.J. Glembocki and S.M. Prokes, Phys. Rev. Lett., submitted.

SURFACE MORPHOLOGY OF NANOSCALE TiSi$_2$ EPITAXIAL ISLANDS ON Si(001)

Woochul Yang, F.J. Jedema, H. Ade and R.J. Nemanich
Department of Physics, North Carolina State University, Raleigh, NC 27695-8202

ABSTRACT

The morphologies of nanoscale epitaxial islands of TISi$_2$ are studied. The islands are prepared by deposition of ultrathin Ti (3-20Å) on both smooth and roughened Si(001) substrates. The island formation is initiated by annealing to 800-1000°C. The roughened substrates are prepared by etching with atomic H produced in a plasma. The morphologies of the substrate before and after island formation are examined by atomic force microscopy (AFM). In particular, the influence of surface-roughness on both the formation of islands and the size distribution of islands is investigated. On a rough substrate islands with a lateral dimension of ~350Å and a vertical dimension of ~25Å were observed with size uniformity of ~20%. Also it was observed that the roughness of the surface reduced the island size and affected the island distribution. The results are discussed in terms of surface energy and the strain field around the islands.

INTRODUCTION

Previous studies have shown the tendency of TiSi$_2$ on Si to form epitaxial island structures. These islands can be easily prepared from very thin Ti (8-50Å) deposition on clean Si substrates[1,2,3] followed by high temperature annealing. There is great potential in these nanoscale island structures for a wide range of possible electronic and optoelectronic device structures. Previously, CoSi$_2$ islands in Si have been shown to exhibit strong response in the IR and ultrafast response for above gap light[4]. Furthermore electrical probing of single nanoscale islands, could display quantum effects in the current-voltage characteristics[5,6]. In order to observe these unique transport and optical properties, the characteristic dimensions of the islands should be less than 10nm.

The formation of TiSi$_2$ by reaction of Ti with Si is an important process in Si IC technology. The process proceed through the formation of C49 TiSi$_2$ for annealing at less than 600°C followed by the transition to the low resistivity C54 phase at higher annealing temperatures. The C49 phase is metastable, meaning that it is not found in the bulk binary phase diagram. However, for very thin films produced with <5nm of Ti on Si, the transition to C54 TiSi$_2$ is not observed for annealing up to 1000°C and epitaxial island structures are often observed.

The growth of nanoscale Ti silicide islands involves an understanding of island formation mechanisms. Since the sample preparation only involves deposition and annealing, the processes employed could be described as self assembled or even as self organized. Our previous studies have shown that the Ti silicide island size and shape is very dependent on the growth conditions (Ti layer thickness and annealing temperature) [1]. The island structures reported previously were ~100nm.

In this study , the effect of the substrate surface morphology is considered. The formation of nanoscale Ti silicide islands is investigated with ultrathin Ti deposition on smooth and H-plasma induced rough Si(001) substrates. The rough substrates are still highly crystalline, but should exhibit preferred nucleation sites and also an increase in the overall substrate surface energy. After island formation the surfaces are analyzed with AFM. In particular, It is shown that the H-plasma exposed rough surface, which has the same high quality crystallinity as smooth surface, affects the island nucleation. Also the size and distribution of islands is investigated and related to the surface and interface energies and the strain field around the islands.

EXPERIMENT

Silicon (001) wafers (n-type, P-doped, resistivity 0.8-1.2 Ω-cm, 25 mm diameter) were used as substrates. The wafers were cleaned first by uv-ozone exposure and then by an HF based spin etch (HF:H$_2$O:ethanol=1:1:10). After ex-situ cleaning, the wafers were introduced into UHV

Mat. Res. Soc. Symp. Proc. Vol. 448 © 1997 Materials Research Society

through a load lock and then transferred to the UHV-MBE chamber which has a base pressure of $<1 \times 10^{-10}$ Torr. Before Ti deposition, the smooth cleaned Si substrates(RMS value: $<3\text{Å}$) were prepared by *in situ* heat cleaning at a temperature of 550°C for 10 minutes to desorb the residual contamination and hydrogen. Following heat cleaning, the LEED showed a 2x1 diffraction pattern typical of the Si(100) reconstructed surface.

To obtain roughened Si substrates, the wafers were transferred to the H-plasma chamber (base pressure 2×10^{-9} Torr). They were exposed to the H-plasma under the following conditions : process pressure=20mTorr, flow rate of H_2 gas=80sccm, rf-power=20W, substrate temperature=150°C and exposure time=1-60min. The variation of roughness can be controlled by H-plasma exposure time and substrate temperature. Titanium was deposited from an *in situ* hot-filament titanium source in the MBE chamber with the sample at room temperature. The titanium layer thickness was 3-20Å with a deposition rate of 0.2Å/S. Following the Ti deposition no LEED pattern was observed, indicating uniform metal coverage of the silicon substrate. To form silicide islands the substrates were annealed in UHV at temperatures ranging from 800°C to 1000°C for 10 minutes. After annealing, a 2x1 diffraction pattern reappeared. This was interpreted as evidence of island formation and attributed to exposure of the silicon substrate between the islands. After the substrates were unloaded, ex-situ AFM was performed to investigate the surface morphologies. The size, shape and density of the Ti silicide islands were obtained from an image processing program included with AFM. Also the RMS value of the surface roughness was obtained both for the roughened substrate and after island formation.

RESULTS

Figure 1 presents AFM images of the C49 $TiSi_2$ islands grown by deposition of 3-20Å Ti on smooth Si(100) substrates, and annealed in UHV at 900°C. For 3Å Ti deposition, the small cap-shaped islands were randomly distributed with a mean diameter of ~350Å and vertical height of ~20Å (Fig. 1-a). For 5Å deposition we observe the coexistence of small rounded islands and larger coalesced islands with evidence of faceting (Fig. 1-b). These faceted large islands are well aligned along the <110> directions of silicon substrate. When the Ti thickness is increased beyond 10Å most of the islands are coalesced and transformed into large faceted islands (Fig. 1 c-d). Furthermore, it is observed that the islands are well separated. As Ti deposition thickness increases, the average size of the islands increases laterally (from 350 to 860Å) and vertically (from ~20 to 120Å). These results are consistent with our previous studies [1,3,7].

To observe the influence of surface roughness on the formation of the islands, roughened substrates were prepared with H-plasma exposure from 1-60min. The roughness of the substrate depends on the H-plasma exposure time[8]. Fig. 2-a shows the morphology of islands of the rough surfaces (RMS value: 100Å before Ti deposition). Compared with the island distribution on the smooth surface (Fig. 1-c), the islands on the rough surface are distributed more uniformly and are smaller. Figure 2-c displays the diameter distribution of the islands and it is evident that the diameter distribution of the islands on the rough surface both narrows and shifts toward a smaller size than on the smooth surface. Also the shape and location of islands are affected by the surface roughness. For a very rough surface from a long H-plasma exposure (RMS value: 220Å), we find (Fig. 2-b) that large islands form near the peaks while the islands on the slopes are more flat and smaller.

To investigate the annealing temperature dependence of the size distribution and shape of the islands, a series of samples of 5Å Ti were annealed at temperatures from 800-1000°C. This was repeated for both smooth and rough surface were prepared. It was found that increasing the annealing temperature resulted in the areal density of islands to decrease by approximately one order of magnitude while the diameter increases by a factor of 2. This trend is observed in both the rough and smooth substrates (Fig. 4). At all temperatures the average diameter of islands on the rough surfaces are smaller than those on smooth surfaces. The smallest islands were observed for Ti (5Å) on rough surfaces annealed at 800°C. The islands exhibited a mean diameter of ~180Å, and they were isolated yet very dense. While at 1000°C annealing temperature, a very uniform size distribution of larger islands was found (standard deviation of island diameter: 16%) and the island separation increased (average spacing between the islands: 1200Å).

DISCUSSION

Consider first the difference in the island density for Ti deposited on smooth and roughened substrates. The results indicate a larger density of smaller islands for the roughened surfaces. Both surfaces are highly crystalline while the roughened surfaces will have an increased surface area, more surface steps, peaks and valleys that may act as nucleation sites and probably more surface defects. The increased surface area itself will result in a thinner Ti film coverage which could be expected to result in smaller islands. Furthermore, diffusion across peaks, valleys, and edges may limit the growth and consequently the island size. Lastly, the surface energy of the roughened surface will be increased just due to the increased area, but locally an opposite effect may occur due to the fact that low energy <111> surfaces are exposed in the H-plasma process[8]. Previous results have, however, indicated that the interface energy of $TiSi_2$ islands on <111> surfaces is also lower.[2,3] Therefore the surface energy terms will likely not play a strong role. All of the other effects (thinner Ti layer, limited diffusion range and increased nucleation sites at steps and defects) would tend to form smaller islands for the roughened surfaces. Unfortunately at this time it is impossible to determine which if any of the effects dominates the process.

For the very rough surfaces there appeared to be a strong preference to forming relatively large islands at the peaks. The surface energies may be expected to be highest locally at the peaks of the rough surface due to the high density of dangling bonds. Thus it would be expected that the $TiSi_2$ would nucleate preferentially at the peaks. The formation of these larger islands minimizes the total surface energy.

While many effects may contribute to the island formation process on the roughened substrates, it is apparent that control of the substrate roughness can be an important factor in obtaining the highest density of small islands.

It is interesting to note that most islands have similar size and relatively uniform separation for the high density structures. The similar size and shape indicates that the surface and interface energetics are relatively uniform. The uniform separation suggests another effect. It has recently been suggested that the strain field in the substrate around the epitaxial islands could lead to a repulsive interaction between the islands[9]. We have previously shown that the C49 islands are indeed epitaxial with a small lattice mismatch[1]. Thus it is likely that the relatively uniform spacing of the island structures represents a self organized effect with the island separation due to interactions through the strain field.

We observed that as the annealing temperature is increased the average size and spacing of islands increases while the areal density of the islands decreases (Fig. 3). Two effects may contribute to this observation. One possibility is that the islands diffuse on the surface and coalesce when they come into close proximity. The nearly sperical shape would suggest a high surface energy. The other possibility is that the islands remain relatively in place, but that the island growth is due to surface diffusion of Ti from one island to the next. We note that the surfaces between the islands exhibit a 2x1 LEED typical of clean Si which may argue against this model. It may be possible to model the two effects to determine is either is the dominant effect.

CONCLUSION

In this study it was observed that surface roughness of the Si substrate affects the formation of Ti silicide islands. The size of the islands is reduced and a relatively high density of uniformly spaced structures are observed. The smallest islands (lateral dimension of ~ 350Å and vertical dimension of ~25Å) were obtained with 5Å Ti deposition on the roughened surface followed by annealing at 800°C. As the annealing temperature increases, the size of islands increases and the density decreases. We suggest that several effects including increased surface area, limited diffusion and more nucleation sites contribute to the higher density of small islands. We also propose that the strain field associated with the small lattice mismatch plays a role in obtaining uniformaly separated islands.

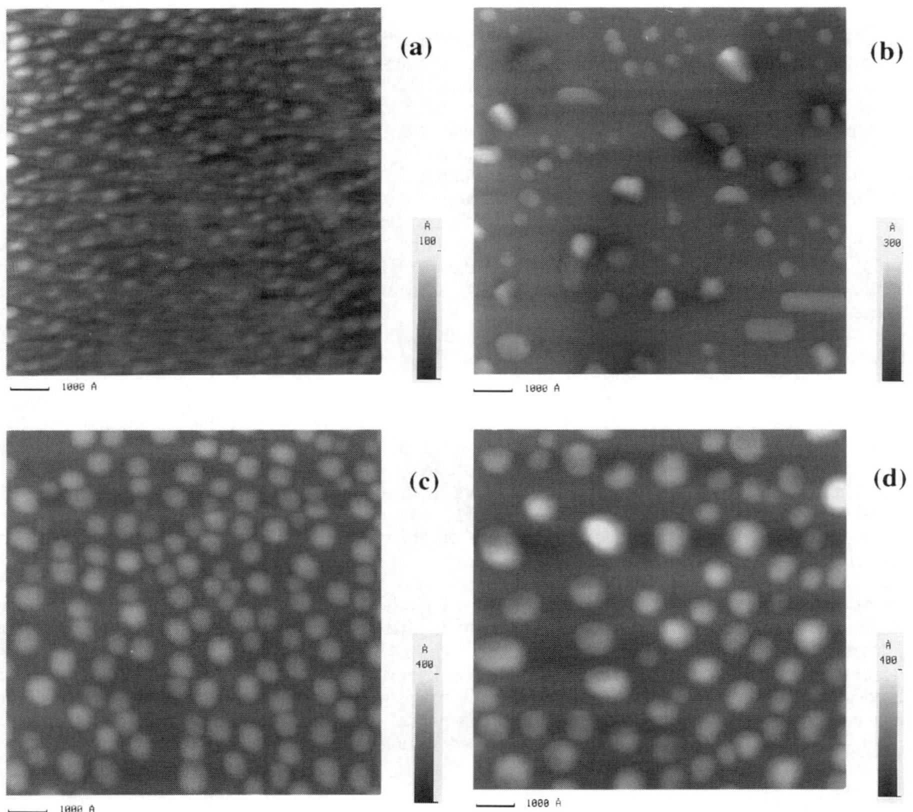

Fig. 1 A series of 1x1 μm^2 atomic force microscopy images of Ti silicide islands formed with varying deposition, annealed at 900 °C. Ti deposition thickness is 3, 5, 10, 20 Å for (a) through (d), respectively.

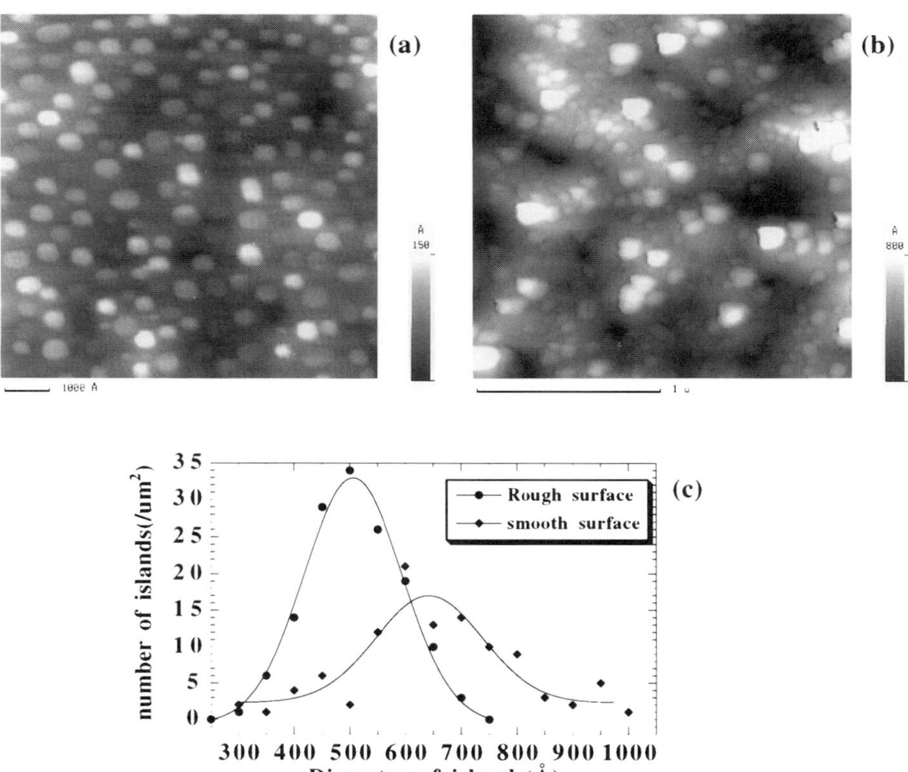

Fig. 2 (a) A 1x1 μm² AFM image of 10 Å Ti annealed at 900 °C on 10 min. H-plasma exposed rough Si substrate; (b) A 2x2 μm² AFM image of 10 Å Ti annealed at 900 °C on very rough surface (H-plasma exposed for 60 minutes); (c) The size distribution of islands for 10 Å annealed at 900 °C on smooth and rough substrate (H-plasma exposed for 10 minutes)

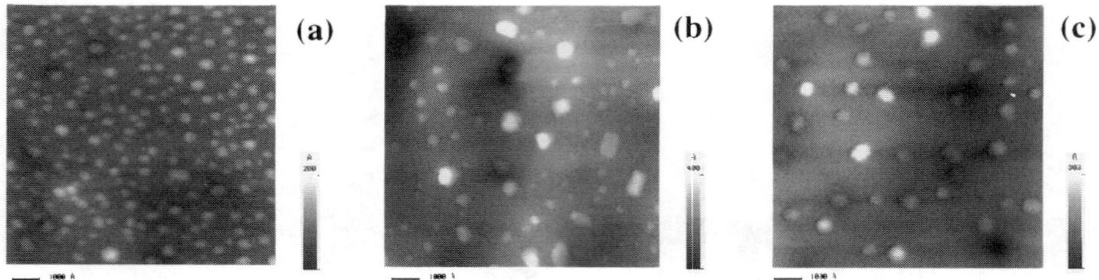

Fig. 3 A series of 1x1 μm^2 AFM images of 5 Å Ti deposition on rough substrate (H-plasma exposed for 10 minutes) varying annealed temperature (a) 800 °C (b) 900 °C (c) 1000 °C

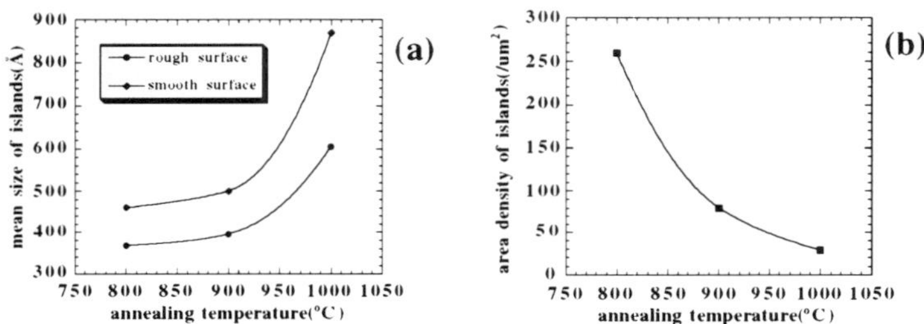

Fig. 4 (a) the meam diameter of islands on both rough and smooth surface as a function of annealing temperatures (b)number density of islands grown on the rough substrate as a function of annealing temperature

ACKNOWLEDGMENTS

The authors would like to thank B. Boyanov and P. Goeller for insightful scientific discussions. This work was supported in part by the NSF under grant DMR 9633547 and the ONR through grant N00014-95-1-1141.

REFERENCES

[1]   B.L. Kropman. C.A. Sukow and R.J. Nemanich. Mat. Res. Soc. Symp. Proc. vol 280, 589 (1993).
[2]   Hyeongtag Jeon. C.A. Sukow. J.W. Honeycutt, G.A. Rozgonyi, and R.J. Nemanich, J. Appl. Phys. 71, 4269(1992).
[3]   C.A. Sukow and R.J. Nemanich, J. Mater. Res., vol 9,1214 (1994).
[4]   R.W. Fathauer. J.M. Iannelli, C.W. Nieh, Shin Hashimoto, Appl. Phys. Lett. 57, 1419 (1990).
[5]   G. Medeiros-Ribeiro, D. Leonard, and P.M. Petroff, Appl. Phys. Lett. 66, 1767(1995).
[6]   K. Nakazato, R.J. Blaikie, and H. Ahmed, J. Appl. Phys., 75, 5123(1994).
[7]   A.W.Stephenson and M.E. Welland, J. Appl. Phys., 77, 563(1995).
[8]   T. Schneider Ph.D. thesis, North Carolina State University, (1994).
[9]   J.Tersoff, C. Teichert and M.G. Lagally, Phys. Rev. Lett. 76, 1675 (1996)

STRUCTURAL DEFECTS IN THICK InGaAs LAYERS GROWN BY LPEE ON PARTIALLY MASKED GaAs SUBSTRATES

T. BRYSKIEWICZ, Crystar, A Johnson Matthey Company, 721 Vanalman Ave. Victoria, BC V8Z 3B6, Canada

ABSTRACT

Defect formation in thick $In_xGa_{1-x}As$ layers, $0.06 \leq x \leq 0.17$, grown by liquid phase electroepitaxy (LPEE) on partially masked GaAs substrates has been studied. The $In_xGa_{1-x}As$ ingots were 6 mm in thickness and, in some cases, more than 53 mm long in one direction. Prior to the LPEE growth, a thin SiO_2 layer was deposited on the (100) oriented GaAs substrate by plasma enhanced chemical vapor deposition (PECVD), and patterned by photolithography with oxide free seeding windows of different geometry. The growth of a ternary crystal from a slightly supercooled melt originated in the form of islands within the window areas and extended laterally over the SiO_2 layer to result, as the growth proceeded by LPEE, in a continuous layer with no apparent boundaries between merging islands. The two main structural defects have been identified in these compositionally uniform ternary crystals: microtwins and microcracks. They were found to result from the crystal/substrate misfit stress, when insufficiently suppressed by the SiO_2 layer, and/or from improper seeding window geometry. The microcrack and microtwin formation in large diameter $In_xGa_{1-x}As$ crystals has been suppressed by the right choice of the seeding window geometry as well as by minimizing damage to the SiO_2 layer caused by the H_2 atmosphere and high growth temperature.

INTRODUCTION

Alloy semiconductors such as $In_xGa_{1-x}As$, $0 < x < 1$, in the form of bulk crystals or strain relaxed buffer layers grown on commercially available, lattice mismatched substrates are of interest as substrates for novel device structures [1,2]. The growth of ternary crystals by liquid encapsulated Czochralski (LEC) and float zone methods is actively pursued by Fujitsu [3] and Johnson Matthey [4]. Recently, liquid phase electroepitaxy (LPEE) has been identified as the most cost effective tool for growing and studying structural defects in these novel electronic materials [5]. It has been demonstrated that highly uniform, compositional inhomogeneity $| \Delta x | < 0.001$ over several millimeter wide area [5,6], and strain relaxed $In_xGa_{1-x}As$ ingots, $0 < x < 0.2$, can be grown by LPEE on partially masked GaAs substrates despite the substrate/crystal lattice mismatch [7-9]. Prior to the LPEE growth, the GaAs substrate is coated with a 0.1- 0.2 μm thick SiO_2 layer by plasma enhanced chemical vapor deposition (PECVD) and patterned by photolithography with oxide free seeding windows. The growth of the $In_xGa_{1-x}As$ islands from a slightly supercooled In-Ga-As melt originates within seeding

Mat. Res. Soc. Symp. Proc. Vol. 448 © 1997 Materials Research Society

windows and extends laterally over the SiO_2 layer. This nearly two dimensional growth occurs on terraces formed by both the dislocations and small misorientation of the GaAs substrate from the (100) plane. In order to assure an efficient overgrowth of the SiO_2 layer, the longer edge of seeding windows must be tilted away from the [110] and [100] directions. Subsequently, as the growth proceeds by LPEE, the islands merge to create a continuous, monocrystalline layer with no apparent boundaries between merging ternary islands [7-9].Thus, both the substrate preparation and initial stages of growth are similar to the previously developed liquid phase epitaxy (LPE) of Si [10-13], GaAs [14-17], GaP [18] and SiGe [19,20] layers on Si, GaAs and GaP substrates. A selective LPEE growth of isolated GaAs islands on partially masked Si substrates has also been reported [21,22].

In this work, the impact of different seeding window geometry on the structural perfection of the LPEE grown $In_xGa_{1-x}As$ ingots, as well as the impact of other factors such as strength of the SiO_2 layer, is studied.

LPEE GROWTH PROCEDURE

$In_xGa_{1-x}As$ crystals, $x = 0.06 - 0.17$, were grown in a vertical LPEE system described elsewhere [23]. At the beginning of each growth run, the In-Ga-As melt made of In, Ga and InAs components was kept apart from both the GaAs substrate and ternary source material, while saturating at a constant temperature in the flow of high purity hydrogen. Both the growth temperature, which was in the $710\text{-}780^{\circ}C$ range, and composition of the melt appropriate for the growth of a ternary crystal of expected composition were evaluated from the In-Ga-As phase diagram [24]. In order to initiate fast lateral growth of terraces within seeding windows but to avoid the two dimensional nucleation, the melt was slowly supercooled by $2\text{-}3^{\circ}C$ and subsequently put in contact with the GaAs substrate from the bottom and with a ternary source material from the top. Immediately, the growth of a ternary crystal originated in the form of islands within oxide free seeding windows and extended over the SiO_2 layer, as shown in Fig.1a. After an hour, a direct electric current was turned on and the growth proceeded by LPEE at a constant temperature. The islands gradually merged to create a continuous layer with no apparent boundaries between the islands [7-9], as shown in Fig.1b. The $In_xGa_{1-x}As$ crystals grown for this study were 6 mm in thickness and in some cases extended by more than 53 mm in one direction [25]. High compositional homogeneity, $|\Delta x| < 0.001$ [6], of the LPEE grown ternary crystals has been assured by choosing the right composition of the source material.

DEFECT FORMATION VERSUS WINDOW GEOMETRY

A cross section of the $In_xGa_{1-x}As$ /SiO_2/GaAs structure is shown in Fig.2. In this case, an impact of the crystal/substrate lattice mismatch on defect formation in the ternary crystal is expected to be largely suppressed by the SiO_2 layer and strictly limited to the seeding window area [9], in accordance with Saint-Venant's principle [26-28].

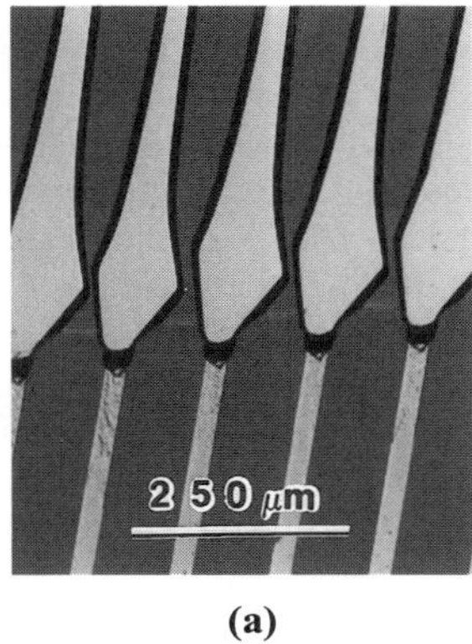

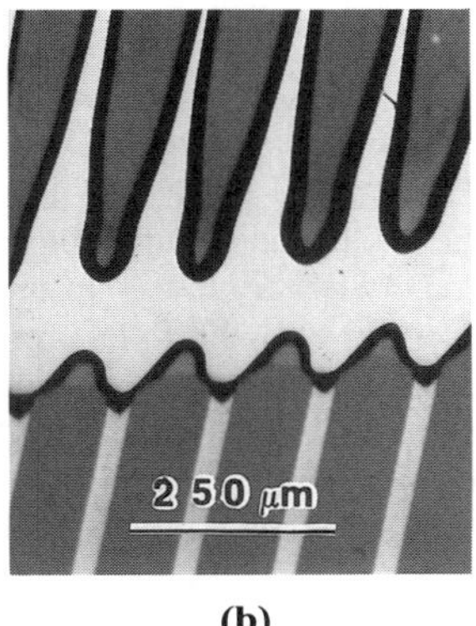

(a)   (b)

Fig. 1. Top view of an epitaxial layer at the initial stages of the selective LPE(E) growth: **(a)** an overgrowth of the SiO_2 layer by isolated islands prior to merging, **(b)** the onset of island merging. The growth originated in 20 μm wide seeding windows. Note lack of any apparent boundaries between merging islands.

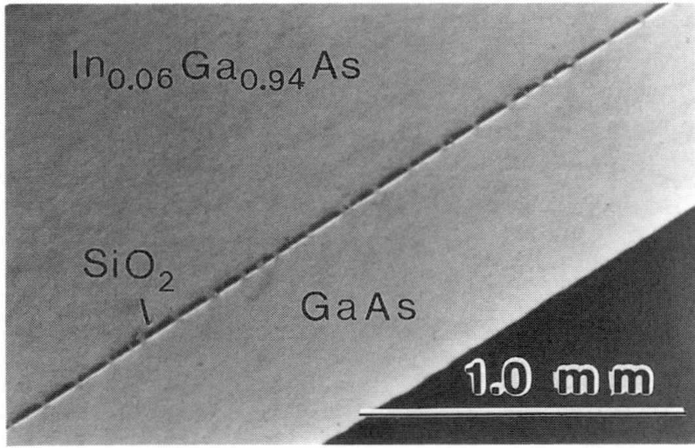

Fig.2. Microphotograph of a cross section of the $In_{0.06}Ga_{0.94}As/SiO_2/GaAs$ structure. The seeding windows are 20 μm wide and extend across the entire length of the GaAs substrate. Note lack of any apparent boundaries between merging islands.

Indeed, calculations based on the Suhir and Luryi model [26,28] for long oxide free windows and for the window separation much larger than the window width have shown that the crystal/substrate misfit stress exponentially decays with a distance from the window area. It turns out that an effective height of the strained area within the $In_x Ga_{1-x}As$ crystal does not exceed 15% of the window width [28]. For narrow enough seeding windows, in the range of a few microns or less, the strained zone height may become smaller than the critical layer thickness for a massive misfit dislocation formation. In this case, the Suhir and Luryi model predicts feasibility of very thick and low dislocation density ternary layers despite the layer/substrate lattice mismatch.

A successful growth of 18 mm diameter, $In_xGa_{1-x}As$ crystals by LPEE, $x < 0.2$, on the GaAs substrates patterned with 2-20 μm wide and long seeding windows has been demonstrated [7-9]. However, we have found this particular window geometry (see Fig.3a) insufficient for a successful growth of large diameter ternary crystals. The dominant structural defect revealed in such crystals was a microcrack, as shown in Fig.4a. Apparently, the crystal/substrate misfit stress was suppressed only in one direction, as recently predicted by A. Atkinson et al [27]. The best seeding window geometry for an effective suppression of the crystal/substrate stress in all possible directions seems to be a square or a circle. However, our crystal growth experiments with such windows (see Fig.3b) have been unsuccessful so far. Ternary crystals revealed very high concentration of microtwins, a clear indication of inefficient lateral overgrowth. This finding is consistent with the results of the selective LPE growth of Si layers in circular seeding windows [13]. Apparently, the oxide free windows have to be elongated in one direction to achieve enough lateral overgrowth of the SiO_2 layer and to prevent from the facet formation at the front of expanding ternary islands prior to merging [10-13,15-20].

The above conclusion is supported by our crystal growth experiments with the rectangular, 10 μm x 200 μm windows separated by 75 μm from each other (see Fig.3c). $In_{0.08}Ga_{0.92}As$ crystals grown by LPEE on such seeding windows were microcrack free and revealed only a few microtwins with a small stacking fault width. Apparently, the crystal/substrate mismatch stress has effectively been suppressed in these ternary crystals, while the window segments were still too short for an efficient lateral overgrowth. This finding is consistent with the results of Suzuki et al [13] with the selective LPE growth of Si layers. Finally, we have found that for the seeding window segments somewhat longer than 200 μm, the microtwins were eliminated and microcracks did not form even in ternary crystals which extended by more than 53 mm in the lateral direction.

DEFECT FORMATION DUE TO OTHER FACTORS

We have found the mechanical and chemical strength of the SiO_2 layer crucial to a successful growth of $In_xGa_{1-x}As$ crystals by the selective LPEE growth method. The SiO_2 layer turns out to be exposed to two damaging factors: decomposition by hydrogen,

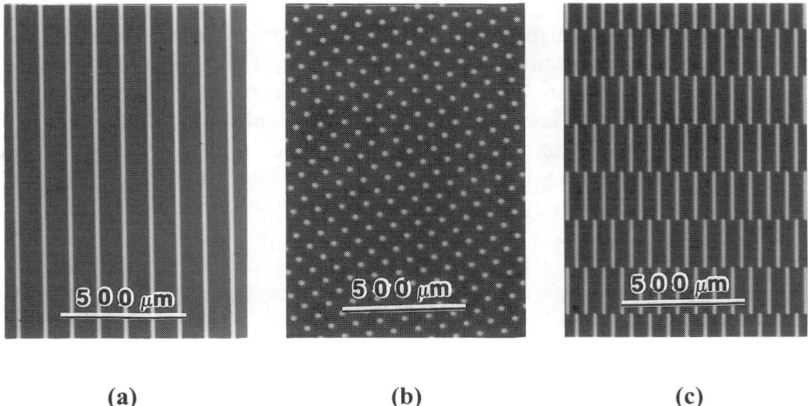

Fig.3. The GaAs substrate coated with the SiO_2 layer and patterned with oxide free seeding windows of different geometry: **(a)** 20 μm wide trenches, **(b)** 20 μm diameter circles, and **(c)** 10 μm x 200 μm windows.

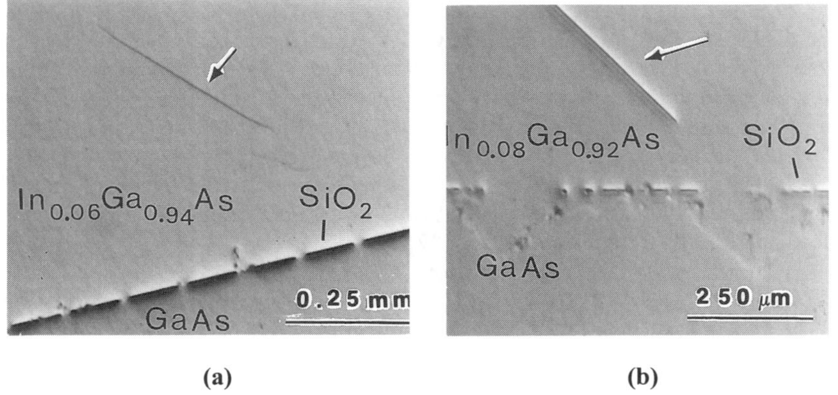

Fig. 4. The main structural defects revealed in large diameter $In_xGa_{1-x}As$ crystals: **(a)** a microcrack formed while employing long, oxide free seeding windows, **(b)** a microtwin formed due to the partial dissolution of the SiO_2 layer.

indispensable during electroepitaxy to assure high purity environment, and mechanical damage at a comparatively high LPEE growth temperature due to a considerable difference in the thermal expansion coefficients between SiO_2 and GaAs. Partial damage to the SiO_2 layer resulted in quite large uncovered areas on the GaAs substrate. In this case, the crystal/substrate misfit stress was not sufficiently suppressed and has led to the enhanced microtwin formation, as shown in Fig.4b. Note high stacking fault width of the stress induced microtwin. An effort to compromise purity with the strength of the SiO_2 layer, by growing ternary crystals in the He atmosphere for example, may lead to both the microtwin and microcrack formation as shown in Fig.5. We have learned that in terms of the defect formation suppression in the LPEE grown $In_x Ga_{1-x}As$ crystals, the best results were achieved when the SiO_2 coating sustained the H_2 atmosphere and no other gases were involved.

Both the theoretical calculations [26-28] and experimental data [7,8] reveal a clear impact of the seeding window width on the effectiveness of the crystal/substrate stress suppression in the LPEE grown ternary crystals. The crystal/substrate lattice mismatch is the prime cause of an enhanced dislocation formation in these compositionally uniform materials. Such dislocations may also climb or glide to a growing ternary crystal through the oxide free windows, as shown in Fig.6. It turns out that very narrow seeding windows, presumably well below 1 μm in width, are required to be able to grow low dislocation density $In_xGa_{1-x}As$ crystals in the $x \geq 0.1$ range [28]. Such narrow seeding windows constitute a challenge not only to the photolithography but most of all to the LPEE growth process itself. Simply, poor wetting of such windows by the In-Ga-As melt is anticipated. In addition to this, the LPE(E) method is known to be inferior to the gas phase epitaxy in growing low dislocation density, lattice mismatched layers [29]. Therefore, a ternary buffer layer grown by MBE or MOCVD on the GaAs substrate, prior to coating with the SiO_2 layer and patterning, is always expected to result in a great improvement. Indeed, the misfit dislocations were found to be well confined to the buffer layer and only slightly affected the $In_{0.09}Ga_{0.91}As$ crystals [8,30], while employing seeding windows as wide as 10 μm. The dislocation density in the $In_{0.09}Ga_{0.91}As$ crystals was in the high 10^4 cm^{-2} range [8], the lowest value achieved for the LPEE grown ternary crystals so far, and much lower than in the buffer layer. Without a buffer layer such a confinement would have not been possible for this type of windows and for this compositional range [7].

However, our attempts to grow $In_xGa_{1-x}As$ crystals, $x \approx 0.2$, on patterned GaAs substrates have revealed that the misfit dislocation confinement to the ternary buffer layer may become difficult to achieve, as shown in Fig.7. High indium content in the buffer layer inevitably leads to an enhanced dislocation formation within the layer as well as at layer/substrate interface. Both the residual stress within the layer and the electric current that drives the LPEE growth are expected to contribute to uncontrolled dislocation migration. In this case, not only narrow windows (2 μm in width or less) but presumably a new sort of a ternary buffer layer are expected to become indispensable in order to effectively suppress the glide or climb of threading dislocations into the ternary crystal.

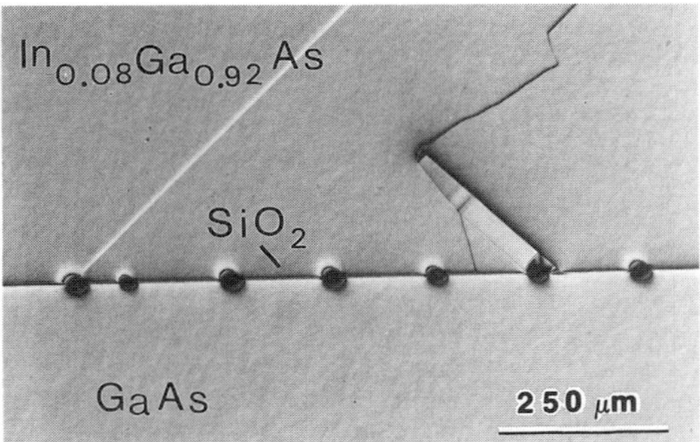

Fig.5. Microphotograph of a cross section of the In$_{0.08}$Ga$_{0.92}$As/SiO$_2$/GaAs structure. The ternary crystal has been grown in the He atmosphere. Note both the microtwin and microcrack formed in the contaminated seeding windows.

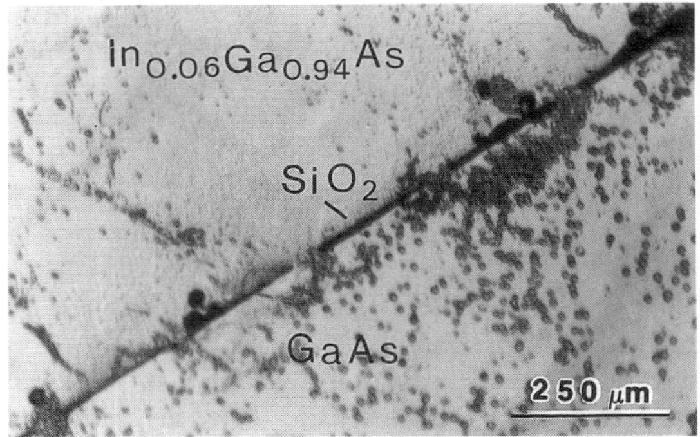

Fig.6. Microphotograph of a cross section of the In$_{0.06}$Ga$_{0.94}$As/SiO$_2$/GaAs structure and dislocation etch pits revealed in hot KOH. The seeding windows are long, 20 μm wide trenches. Note the dislocations gliding and/or climbing through the oxide free windows.

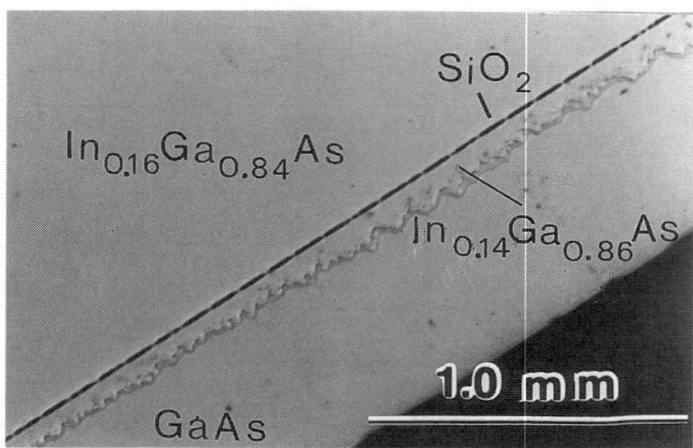

Fig.7. Microphotograph of a cross section of the In$_{0.16}$Ga$_{0.84}$As/SiO$_2$/ In$_{0.14}$Ga$_{0.86}$As/ GaAs structure. The seeding windows are 10 μm wide and longer than 200 μm. Note the misfit dislocations "diffusing" from the In$_{0.14}$Ga$_{0.86}$As buffer layer into the GaAs substrate.

CONCLUSIONS

In$_x$Ga$_{1-x}$As crystals, $0 < x < 0.17$, which extend by more than 53 mm in one direction can successfully be grown by LPEE on the GaAs substrate partially masked by a thin SiO$_2$ layer, despite the crystal/substrate lattice mismatch.

In order to achieve efficient enough overgrowth of the SiO$_2$ layer by ternary islands that originate within oxide free seeding windows and subsequently merge to create a continuous layer and to avoid the microtwin formation at boundaries of the merging islands, the seeding window segments have to be as long as possible. On the other hand, in order to avoid crack formation in large diameter ternary crystals, the oxide free windows have to be as short as possible. Thus, an optimum seeding window geometry exists for a given composition x of the In$_x$Ga$_{1-x}$As crystals that fulfills both conflicting requirements.

The chemical and mechanical strength of the SiO$_2$ layer, resistant to a detrimental influence of both the high growth temperature and H$_2$ atmosphere, is essential to the successful growth of In$_x$Ga$_{1-x}$As crystals by LPEE.

ACKNOWLEDGMENTS

The work has been supported by DARPA under the contract No. MDA972-91-C-C0046. The author is grateful to Steven K. Brierley and William E. Hoke of Raytheon for the encouragement to present this work at the 1996 MRS Fall Meeting. The author is also indebted to Alan Moore and his coworkers at EG&G Canada for SiO_2 coating and patterning, to Jinsheng Hu of Simon Fraser University for PL measurements, as well as to Barbara Bryskiewicz of Nortel for TEM studies.

REFERENCES

1. K. Otsubo, H. Shoji, T. Kusunoki, T. Suzuki, T. Uchida and H. Ishikawa, <u>Proc. Int. Conf. Solid State Devices and Materials</u>, Yokohama, 1996, pp. 649-651, and references therein.

2. W.E. Hoke, P.S. Lyman, J.J. Mosca, H.T. Hendriks, A. Torabi, W.A. Bonner, B. Lent, L.-J. Chou, and K.C. Hsieh, J. Appl. Phys. 81, 15 January, 1997.

3. T. Kusunoki, K. Nakajima, T. Kusunoki, and T. Katoh, J. Cryst. Growth, 25, 1996, pp. 357-361, and references therein.

4. W.A. Bonner, B. Lent, D.J. Freschi and W. Hoke, in: <u>Producibility of II-VI Materials and Devices</u>, SPIE 2228, 1994, pp. 33-43.

5. T. Bryskiewicz and P. Edelman, J. Appl. Phys. 68, 1990, pp. 3018-3020.

6. J.-S. Hu, unpublished.

7. T. Bryskiewicz, E. Jiran, B. Bryskiewicz and M. Buchanan, Mat. Res. Soc. Symp. Proc. 340, 1994, pp. 367-372.

8. B. Bryskiewicz, T. Bryskiewicz and E. Jiran, J. Electron. Materials 24, 1995, pp. 203-209.

9. J.P. McCaffrey, B. Bryskiewicz, T. Bryskiewicz and E. Jiran, Appl. Phys. Lett. 64, 1994, pp. 2344-2346.

10. R. Bergmann, J. Cryst. Growth 110, 1991, pp.823-834, and references therein.

11. Y. Suzuki and T. Nishinaga, Jap. J. Appl. Phys. 28, 1989, pp. 440-445.

12. Y. Suzuki and T. Nishinaga, Jap. J. Appl. Phys. 29, 1990, pp. 2685-2689.

13. Y. Suzuki, T. Nishinaga and T. Sanada, J. Cryst. Growth 99, 1990, pp. 229-234.

14. S. Sakai, R.J. Matyi and H. Shichijo, J. Appl. Phys. 63, 1988, 1075-1079.

15. T. Nishinaga, T. Nakano and Suian Zhang, Jap. J. Appl. Phys. 27, 1988, pp. L964-L967.

16. Y. Ujiie and T. Nishinaga, Jap. J. Appl. Phys. 28, 1989, pp. L337-339.

17. S. Zang and T. Nishinaga, J. Cryst. Growth 99, 1990, pp. 292-296.

18. S. Zhang and T. Nishinaga, Jap. J. Appl. Phys. 29, 1990, pp. 545-550.

19. P.O. Hansson, R. Bergmann and E. Bauser, J. Cryst. Growth 114, 1991, pp. 573-580.

20. P.O. Hansson, A. Gustasson, M. Albrecht, R. Bergmann, H.P. Strunk and E. Bauser, J. Cryst. Growth 121, 1992, 790-794.

21. S. Sakai, Y. Ohashi and Y. Shintani, J. Appl. Phys. 70, 1991, pp. 4899-4902.

22. S. Sakai and Y. Ohashi, Jap. J. Appl. Phys. 33, 1994, pp. 23-27.

23. T. Bryskiewicz, C.F. Boucher, Jr., J. Lagowski and H.C. Gatos, J. Cryst. Growth 82, 1987, pp. 279-288.

24. S.M. Bedair, C. Morrison, R. Fang and N.A. El-Masry, J. Appl. Phys. 51, 1980, pp. 5413-5418.

25. T. Bryskiewicz and W.A. Bonner, unpublished.

26. S. Luryi and E. Suhir, Appl. Phys. Lett. 49, 1986, pp. 140-142.

27. A. Atkinson, S.C. Jain and A.H. Harker, J. Appl. Phys. 77, 1995, pp. 1907-1913.

28. T. Bryskiewicz, Appl. Phys. Lett. 66, 1995, pp. 1237-1239.

29. M. Ettenberg, C.J. Nuese, J.R. Appert, J.J. Gannon, and R.E. Enstrom, J. Electron. Materials 4, 1975, pp. 37-66.

30. B. Bryskiewicz and R. Mallard, unpublished.

Facet Formation in Submicron Selective Growth of Si/SiGe

K. L. Wang and Dawen Wang, Device Research Laboratory, Department of Electrical Engineering, University of California, Los Angeles, Los Angeles, CA 90095

Abstract

The paper reviews the work in mostly Si and SiGe epitaxy and some III-V work on patterned substrates. Results of metalorganic chemical vapor deposition (MOCVD), low pressure chemical vapor deposition (LPCVD), gas source molecular beam epitaxy (GSMBE), and solid source molecular beam epitaxy (MBE) were discussed in the context of facet formation and mass accumulation. A model was shown to explain the facet formation and its evolution in the process of growth. Further work on surface diffusion and nucleation processes as functions of temperature and other growth parameters will provide needed information for accurate modeling of the facet growth process.

I. Introduction

As the scale of integration of electronic devices continues to increase, the feature size must be reduced. It will approach below 100 nm at the turn of the Century. Thus there is a great deal of interest in studying structures of reduced dimensions and in fabricating nanometer scale structures. In the past, advanced crystal growth techniques, such as molecular beam epitexy (MBE) and chemical vapor deposition (CVD), have made it possible to fabricate two-dimensional structures such as heterojunctions, quantum wells and superlattices with layer thickness on the atomic scale. However further reduction of lateral dimensions in forms of one-dimensional quantum wires, zero-dimensional quantum dots and nanometer scale structures in general has faced technological difficulties.

Most widely used methods for the fabrication of nanostructures such as quantum wires and quantum dots are based on ex-situ lateral patterning techniques which combine lithography with dry or wet chemical etching to achieve desired patterns. Despite of the fact that e-beam lithography can achieve structures with dimensions smaller than 10 nm [1], there exist a lot of problems such as surface damages and irregularities in the shape and size of nanostructures. Therefore, there is an impetus for the development of new alternative techniques to form nanometer scale structures. Among them, the direct growth of nanostructures has a potential to form damage-free structures, making this technique desirable for studying the properties of nanostructures.

To directly grow nanostructures, the substrate should be structured, i.e. there is a controlled starting surface pattern present on the substrate. There have been several kinds of structured substrates used and they are, for example, nonplanar patterned substrates[2], planar masked substrates [3], and planar high index surfaces [4]. When films are selectively grown on these

Mat. Res. Soc. Symp. Proc. Vol. 448 © 1997 Materials Research Society

structured substrates, mesas are often formed on the substrates. Independent of growth techniques, the sidewalls of these mesas inevitably form facets. More than one facet may result This is because the adatoms migrate on the top surface of the mesa and the sidewalls to minimize the total free energy during growth. The possible facets and their development depend on the substrate orientation, surface pattern alignment, and growth conditions.

In this paper, we review selective growth (SG) of semiconductors using CVD and MBE, especially gas source MBE (GSMBE), focusing on the facet growth and the facet evolution of Si-based films. SG is often used for pattern growth. Compared with III-V materials, there have only been a few pieces of published work on the SG of SiGe nanostructures. Some of III-V facet growth results were also discussed in the paper.

II. Selective growth by CVD

The most frequently used SG technique is CVD, such as metalorganic chemical vapor deposition (MOCVD) and low pressure chemical vapor deposition (LPCVD) for III-V materials. The growth mechanism is mainly determined by adsorption-desorption processes and surface reactions.

Kim et al., [5] have investigated the facet evolution of $Al_{0.5}Ga_{0.5}As/GaAs$ multilayers grown on mesa- or V-groove patterned GaAs substrates by MOCVD as a function of growth temperature and the flow rate of the p-type dopant (CCl_4). For parallel stripes patterned along the <01T> direction on (100) oriented n^+ GaAs substrate, a (511)A or a (411)A facet was observed, instead of the conventional (433)A sidewall facet for undoped $Al_{0.5}Ga_{0.5}As/GaAs$ multilayers on a mesa. For InAlGaAs selective growth, there are only a few papers reported in low pressure metalorganic vapor phase epitaxy, because the growth is difficult due to the high reactivity of the Al species with dielectric masks used on patterned substrates [6,7,8].

For Si-based materials, usually CVD techniques with chlorine containing gases are used for selective deposition, in which the presence of HCl prevents the Si deposition on SiO_2 and full selectivity can be obtained. HCl is either introduced intentionally along with SiH_4, or SiH_2Cl_2 or is formed from the decomposition of SiH_2Cl_2.

SEG on oxide windows patterned on Si (100) substrates has been performed by LPCVD [9]. Under the conditions which result in uniform and planar selective growth, facets usually develop around the corners of the grown mesas. For the sidewalls aligned with the <110> direction, the {111} and {311} facets can be observed at the edges of the windows with the size down to 400 nm width. The {111} facet began first and then the {311} facet developed afterward. For small oxide windows of 140 to 200 nm, addition facets, such as {411} and {433}, were observed. For the sidewalls aligned along with the <100> direction, the {110} facet developed. Because of the facet growth, the dimension of the top area of the grown mesa usually decreases as the growth continues. So far, sizes of the (100) top parts as small as 50 nm for {311} facets and 200 nm for {110} facets have been achieved, respectively.

III. Selective growth by GSMBE

GSMBE uses gas sources instead of solid sources to grow crystal. The growth mechanism is the dissociative adsorption of gas molecules, followed by incorporation of dissociated products which leave such as Si or Ge atoms to be incorporated into the epitaxial layer on substrates. The most important feature of GSMBE is selective growth of semiconductor layers, that is, the epitaxial growth takes place only on the semiconductor, i.e., on the Si and Ge surfaces where epitaxy occurs on patterned windows with a dielectric film such as SiO_2 [10] and Si_3N_4 [11] as masks. For Si-based GSMBE, the most common used gases are Si_2H_6 and GeH_4. This selectivity comes from the difference in the sticking probabilities of Si_2H_6 molecules between Si and SiO_2. The molecule dissociatively adsorbs on Si surfaces. The Si atoms are incorporated into the grown film while this process does not occur on SiO_2 surface. For CVD, the selectivity is found to improve with increasing GeH_4 flow both due to an increase in incubation time and in growth rate [12]. The faceting occurs on the sidewalls of the grown mesas to minimize the total free energy of the film, thus resulting in changes of the size and the shape of mesas.

1. SG on V-groove patterned substrates

SEG of GaAs, AlAs and AlGaAs on V-groove patterned GaAs substrates by MBE has been reported [13]. It is found that the growth of AlAs leads to form a very smooth surface, both on (111)A sidewalls of the V-groove and on (111)A flat substrates. For the V-groove case, a sharpen structure at the bottom of the V-groove is formed. On the other hand, no such effect occurs for GaAs and AlGaAs growth. By using the sharpening effect of AlAs at the bottom of the V-groove, GaAs/AlAs multiple quantum wires have been fabricated [13].

Similar to GaAs SEG [13], successful growth of Si quantum wires (QWR) on a V-groove patterned Si substrate has been demonstrated by GSMBE [14]. Because the sticking probability of Si_2H_6 molecules depends on the orientation of Si substrates, the growth rate varies for different crystal orientations, and this fact makes it possible to obtain a crescent-shaped Si and SiGe layer. When V-groove patterns are fabricated on a Si(100) surface by chemical etching, (111) sidewall facets are formed. When the growth is carried out on this type of patterned substrate, the pattern is considerably deteriorated with increasing growth temperatures, as shown in fig. 3-1. This is because at higher temperatures, the growth rate on the (100) surface is higher than that on the (111) side wall; therefore low temperature growth is necessary to maintain the sharpness of the V-groove shape and to achieve desired optical properties of QWR formed on the V-groove.

2. SG on Si substrates with oxide windows

Facet formation and inter-facet mass transport in SEG of Si on SiO_2-masked Si (100) and Si(110) substrate were studied using GSMBE [10, 15]. For Si(100) surface, the smallest lateral feature size of the SEG mesa obtained to date is about 80 nm with a height of about 85 nm, as shown in fig. 3-2.

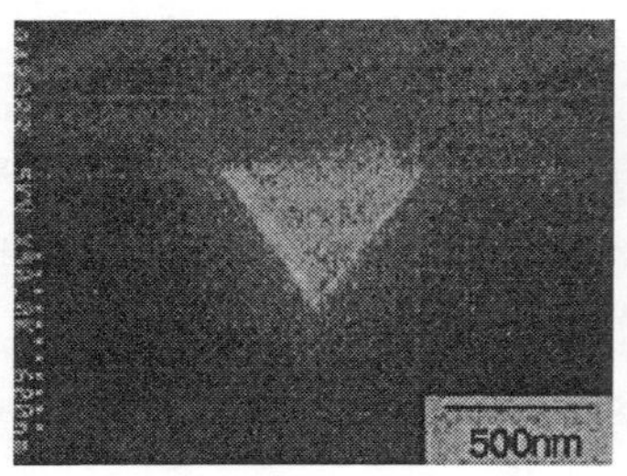

(a). Before Growth

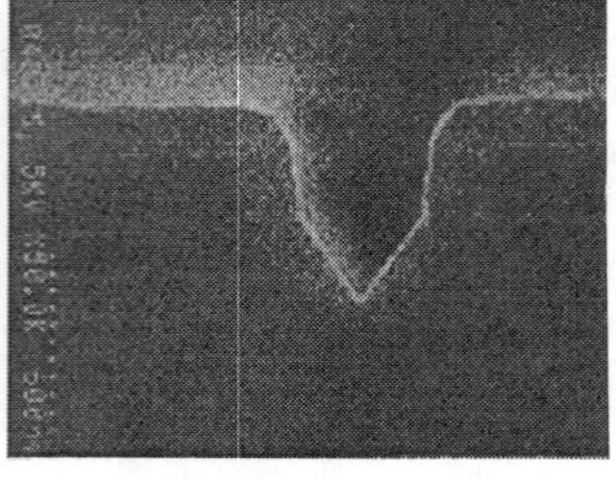

(b). After Growth,
200 nm thick, 600 C

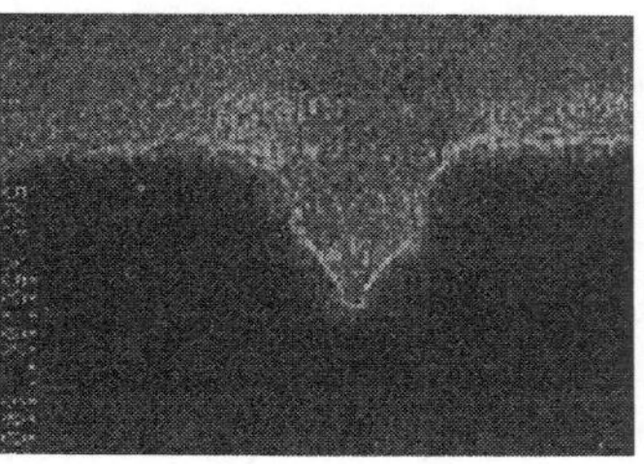

(c). 200 nm, 740 C

Fig. 3-1. Cross-sectional SEM image of Si layers: (a) before growth, (b) 200 nm film grown at 600 °C, and (c) 700 °C.

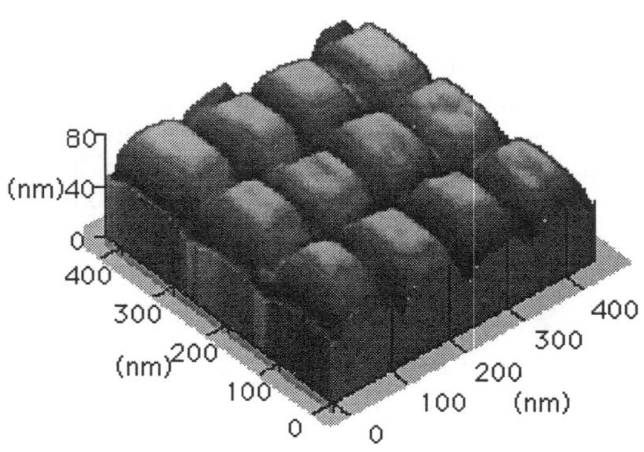

Fig. 3-2. Three dimensional AFM image of an Si array of the 80 nm dots grown by GSMBE. The picture was obtained after the SiO_2 layer used for patterned growth was removed [10].

For the (100) Si surface, when windows were patterned along the <110> direction, (311), (111) and (110) facets were observed. Because of different growth rates for the (100), (311), and (111) substrates, faceting was observed to evolute from the initial (311) to (111). For Si(110) surface [15], when the sidewalls were aligned along with the [110] baseline, both (311) and (111) facets were observed. For the sidewalls aligned along with the [100] baseline, only the (100) facet was observed. Fig. 3-3 (a) shows the line analyses for the beginning stage of growth (top) and for the final stage of growth (bottom) on Si(100) surface after the SiO_2 layer used for patterned growth was removed. Fig.3-3 (b) and (c) show the AFM line analyses of SEG mesa edges along the [100] and [110] directions on Si(110) substrates, respectively, after the SiO_2 layer was removed. For all the cases, mass accumulation around the edges of the top surface was observed. The mass accumulation is believed to due to mass transport from the sidewalls to the top surfaces, similar to that observed in III-V growth on patterned substrates [16]. The driving force for mass transport may be attributed to surface migration in minimizing the total free energy during the SEG growth. In other words, the mass transport will be directed to the energetically favored facet to minimize the total free energy.

IV. SEG of Solid Source MBE

Solid source MBE is a physical process and offers little selectivity, in contrast with the cases of CVD and GSMBE. However the faceting can still occur. Using a micro shadow mask to limit the deposition area [17], self-assembled mesa growth can be performed. Faceting was reported for growth on ridge-type patterned nonplanar substrates [18] and V-grooved substrates [13].

1. Self-assembled growth on micro shadow masks

The fabrication process of the micro shadow masks and the subsequent growth process of Si are illustrated in fig. 4-1 [17]. The silicon nitride shadow mask was formed by over etching the underlying SiO_2 as shown in fig. 4-1 (a), (b) and (c). After growth, free-standing mesa islands were obtained as shown in (d).

The shape and quality of the mesa strongly depend on the growth temperature [19]. Only at an intermediate temperature (400 °C ~ 700 °C) region, perfect faceted (111) sidewalls can occur. Fig. 4-2 shows the facet evolution as the growth continues for a 500 °C growth temperature. The deposition of MBE growth aligns with the mask opening initially (fig. 4-2 (a)). As the growth continues, (111) facets are formed at the sidewalls as the result of the process in minimizing the total free energy. Because of the high diffusion length of the atoms on the (111) sidewalls, the vertically impinging adatoms diffuse along the facet underneath the shadow mask (fig. 4-2 (b)). The maximum length of the (111) mesa sidewall underneath the shadow mask is thus limited by the diffusion length of the deposited species on the (111) sidewall. When the diffusion limit is reached, vertical (011) sidewalls are formed at the end of the (111) surfaces, while the size of the (111) facets remains constant (fig. 4-2 (c)). For higher growth temperatures, e.g. 730 °C, the (111) facet vanishes and the (311) facet appears. For a moderate temperature, e.g. 570 °C, a crossover from the (111) to the (311) facet was clearly observed.

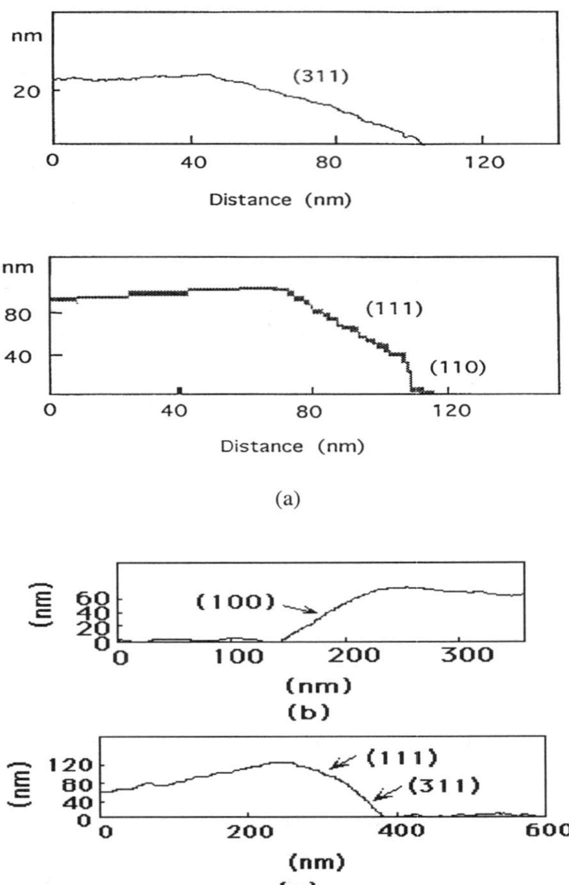

Fig. 3-3. AFM line analyses of SEG grown mesa edges: On (100) surface, (a) for baseline along <110> direction. The top fig. is for the beginning stage of growth, i.e., for thin film and the bottom one is for the final stage of a relatively thick film growth. On (110) surface, (b) for baseline along <100> direction and (c) for baseline along < 110> direction. All the lines were obtained after the SiO_2 layer used for patterned growth was removed.

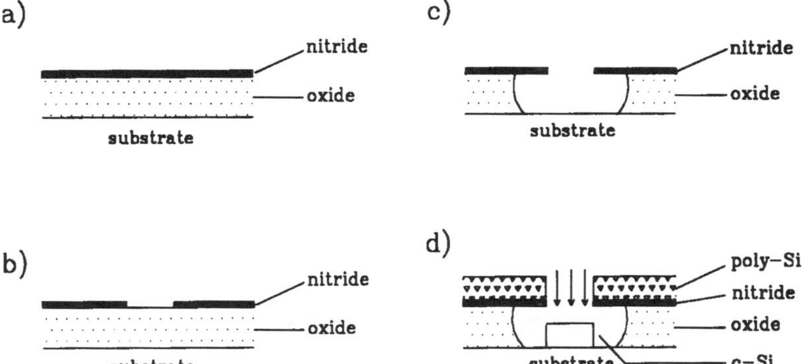

Fig. 4-1. Fabrication process of micro shadow masks for self-assembled Si MBE growth: (a) deposition of masking material, (b) patterning of the masking layer, (c) etching of oxide spacer, and (d) MBE growth [17].

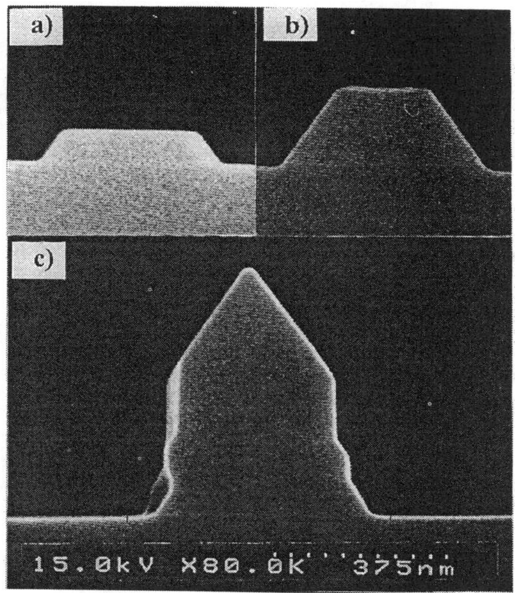

Fig. 4-2. SEM micrograph of mesa side walls grown at different stages with a 310 nm mask opening (T = 500 °C) [17].

2. Self-assembled growth on ridge-type patterned nonplaner substrates [18]

The concept of self-assembled growth has been used in Si/Ge MBE growth on nonplanar substrates. The growth process takes the advantage of the diffusion mechanism to reduce the lateral dimensions of a mesa. On the substrate consisted of a truncated pyramid template with {111} sidewalls and the (100) tops, the MBE growth showed that {113} facets began to form on the (100) surface at a growth temperature between 650 ~ 700 °C. Lateral dimensions of the (100) top as small as 20 nm have been reported [18]. At higher temperatures (~ 800 °C), the {113} facets remain stable but the {111} facets no longer exist. Thus by using small mesas on substrates, quantum dots were successfully fabricated by strained-layer epitaxy, as shown in fig. 4-3.

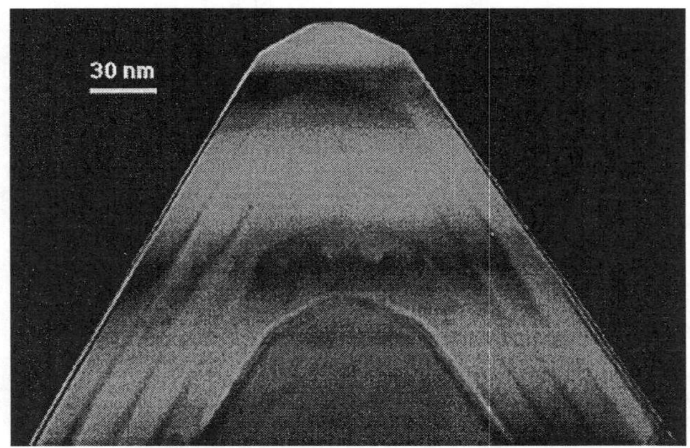

Fig. 4-3. TEM micrograph of Si/Si$_{1-x}$Ge$_x$ MQWs grown on a pyramid with 10 nm top mesa. Note that two pyramids are shown in the image.

V. Modeling of SEG

Despite of the infancy of the research in facet growth, theoretical modeling has been done. A model for SEG on small windows was developed by taking both surface migration and surface free energy into consideration [20]. In the model, each facet is divided into several small divisions with the dimension of the migration length. The adatoms migrate between adjacent divisions. Two kinds of migrations are considered: (1) interfacet migration between the divisions at the edge of two facets and (2) intrafacet migration between the divisions on the same facet. The intrafacet migration is diffusion-like, which is determined by the adatom spatial density gradient. The interfacet migration depends on the free energy, the adatom density of the division, and the growth temperature.

The results show that on (100) surface for a window size on the order of microns, the (311) is the only observable sidewall facet at the beginning of the growth (thin film). However, as the growth continues (and the film becomes thicker), the (111) facet dominates, as illustrated in fig.5-1 (a) and (b). The calculated mesa cross section in fig. 5-1 (a) and (b) also show ridges near the edge of the top (100) surfaces, indicating the mass transport from the sidewall facets to the top surface. These simulation results are in good agreement with our experimental observation shown in the insets. Similar results for the SEG growth on (110) surface are shown in fig. 5-1 (c).

Although this mode illustrates the special case of Si-base GSMBE, the basic idea in the mode is also suitable for the other growth methods controlled by the diffusion and migration of the adatoms. To minimize the total free energy, the adatoms migrate between different facets and thus the energy favorable facets are finally formed. For accurate modeling, we also need to know the surface diffusion and other growth parameters at different growth environments, especially the growth temperature and growth rate. In this mode, only (111), (110) and (311) facets were taken into consideration, thus no information about higher index facets is included for explaining the experimental results.

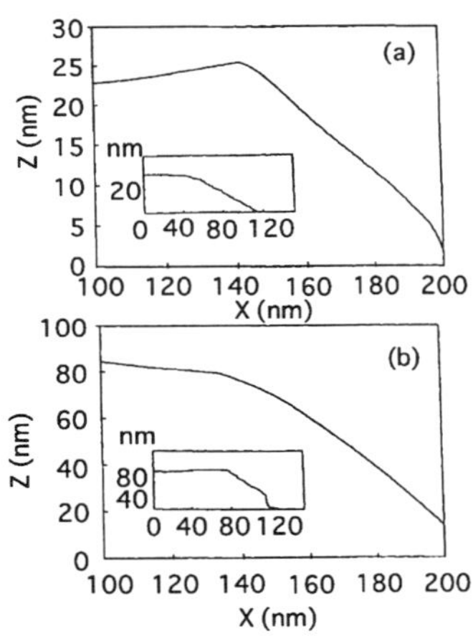
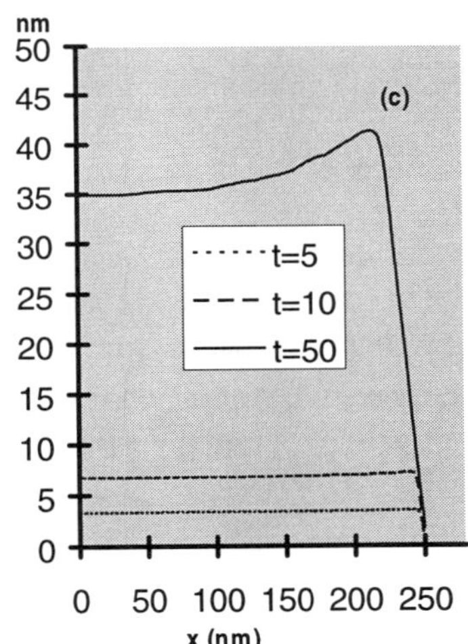

Fig. 5-1 Simulated cross sections of mesas on a large pattern for (a) $t_g = 25$, (100) surface, (b) $t_g = 100$, (100) surface, and (c), $t_g = 5$, 10, and 50, (110) surface. (The insets in fig. (a) and (b) are the AFM scanned mesa profiles, t_g is given in an arbitrary scale).

For III-V materials, simulation has also been done [21, 22]. W. Wegscheider et al., [22] used a model, which describes kinetically limited MBE growth over patterned GaAs surfaces. The simulation is structured in terms of differential equations that describe the time dependence of Ga or Al adatoms on the growth surface. By assigning each type of adatoms a life-time before epitaxial incorporation for each orientation, the results show the details of experimental observations, such as the spontaneous development of new facets and the sharpening of facet edges.

VI. Application

There have been a great deal of work on the application of patterned growth using the selective growth and faceting features. Self passivation is an important characteristics in selective growth. Using these techniques described above, passivated nanostructures such as quantum wire structures may be grown. Potential applications also include the formation of ordered "self-organized" quantum dots on high index substrates for advanced electronics devices and low threshold current lasers [4]. Future work on surface diffusion and nucleation process, both in theoretically and experimentally, will help understand the facet growth and its control and thus new applications.

VII. Acknowledgment

This work was in part supported by NSF and SRC. One of the authors (KLW) also acknowledges the help of all his colleagues in the performance of this research.

References

[1]. X. Pan, D. R. Allee, A. N. Broers, Y. S. Tang, and C. W. D. Wilkinson, Appl. Phys. Lett., **59**, 3157 (1991).

[2]. K. Fujita, H. Ohnishi, M. Hirai, K. Shimada, and T. Watanabe, Solid-State Electronics, **40**, 633 (1996).

[3]. T. Fukui, and S. Ando, Superlattices and Microstructures, **12**, 141(1992).

[4]. R. Notzel, Semicond. Sci. Technol., **11**, 1365 (1996).

[5]. Y. Kim, Y. K. Park, M. S. Kim, J. M. Kang, S. I. Kim, S. M. Hwang, and S. K.Min, J. Crystal. Growth, **156**, 169 (1995).

[6]. M. Kushibe and K. Takaoka, J. Crystal Growth, **145**, 263 (1994).

[7]. T. Takeuchi, M. Tsuji, K. Makita, and K. Taguchi, in: Proc. 7th Int. Conf. on InP and Related Materials, Sapporo, Hokkaido, 1995, p.341.

[8]. M. Tsuji, K. Makita, T. Takeuchi, K. Taguchi, J. Crystal. Crowth, **162**, 25 (1996).

[9]. L. Vescan, in Low Dimensional Structures Prepared by Epitaxial Growth or Regrowth on Patterned Substrates, edited by K. Eberl, P. M. Petroff and P. Demeester (Kluwer Academic Publishers, Dordrecht/Boston/London, 1995) pp.173.

[10]. Qi Xiang, Shaozhong Li, Dawen Wang, Kang L. Wang, J. Greg Couillard, and Harold G. Graighead, J. Vac. Sci. Technol., **14**, 1 (1996).

[11]. S. Yoshida, M. Sasaki, and H. Kawanishi, J. Crystal. Growth, 136 (1994) 37.

[12]. M. Racanelli and D. W. Greve, Appl. Phys. Lett., **58**, 2096 (1991).

[13]. X. Q. Shen, M. Tanaka, K. Wada, and T. Nishinaga, J. Crystal. Growth, **135,** 85 (1994).

[14]. Y. Shiraki, N. Usami, T. Mine, N. Akiyama, and S. Fukatsu, in Low Dimensional Structures Prepared by Epitaxial Growth or Regrowth on Patterned Substrates, edited by K. Eberl, P. M. Petroff and P. Demeester (Kluwer Academic Publishers, Dordrecht/Boston/London, 1995) pp.151.

[15]. Qi Xiang, Shaozhong Li, Dawen Wang, Kunihiro Sakamoto, K. L. Wang, Greg U'Ren, and Mark Goorsky, J. Crystal Growth, 175, (1997).

[16]. A. Madhukar, Thin Solid Films, **231**, 8 (1993).

[17]. I. Eisele, H. Baumgartner and W. Hansch, in Low Dimensional Structures Prepared by Epitaxial Growth or Regrowth on Patterned Substrates, edited by K. Eberl, P. M. Petroff and P. Demeester (Kluwer Academic Publishers, Dordrecht/Boston/London, 1995) pp.161.

[18]. K. D. Hobart, F. J. Kub, H. F. Gray, M. E. Twigg, D. Park, and P. E. Thimpson, Journal of Crystal Growth, **157**, 338 (1995).

[19]. H. Gossner, G. Fehlauer, W. Kiunke, I. Eisele, M. Stolz, M. Hintermaier, and E. Knapek, Mat. Res. Soc. Symp. Proc., **351**, 393 (1994).

[20]. S. Li, Q. Xiang, Dawen Wang, and K. L. Wang, J. Crystal. Growth, **157**, 185 (1995).

[21]. M. Ohtsuka and S. Miyazawa, J. Appl. Phys., **108**, 55 (1988).

[22]. W. Wegscheider, L. Pfeifer, and K. Wsat, W. Hansch, in Low Dimensional Structures Prepared by Epitaxial Growth or Regrowth on Patterned Substrates, edited by K. Eberl, P. M. Petroff and P. Demeester (Kluwer Academic Publishers, Dordrecht/Boston/London, 1995) pp.283.

GaInP Selective Area Epitaxy for Heterojunction Bipolar Transistor Applications

S. H. PARK, S.-L. FU, P. K. L. YU, P. M. ASBECK
Department of ECE, University of California, San Diego, La Jolla, CA 92096

ABSTRACT

A study of selective area epitaxy (SAE) of GaInP lattice matched to GaAs is presented. The selectively regrown GaInP is used as the emitter of a novel heterojunction bipolar transistor (HBT) device structure. Successful SAE of GaInP on both dark field (mostly covered) and light field (mostly open) SiO_2 masks is compared. To characterize the critical regrown heterojunction, diodes and HBTs were fabricated and measured. It is found that a pre-growth pause of either TEGa or PH_3 results in forward bias characteristics with low leakage and an ideality factor of ~1.25, indicating low interfacial defect density. Non-self aligned regrown emitter HBTs grown with a dark field mask scheme have been fabricated. Devices with an emitter area of 3x12 μm exhibit small signal current gain up to 80 with an f_T and f_{MAX} of 22 GHz and 18 GHz, respectively. To further improve the performance of these devices, a structure with a self-aligned refractory metal base contact and light field regrowth is proposed.

INTRODUCTION

A key issue for future generations of III-V HBTs is the reduction of device dimensions both vertically and horizontally.[1] The motivation to scale HBTs stems from the desire to reduce parasitic components such as base resistance, R_B, and junction capacitances, C_{BE} and C_{BC}, which in turn leads to lower power consumption, higher integration density, and higher device performance. However in recent years the minimum feature size for production III-V HBTs has hovered just above 1 μm while that for Si devices has continued to march well into the sub-micron range. In addition to the photolithographic limitations of the current mesa device technology, the effects of emitter edge recombination are exacerbated for small dimension GaAs HBTs.[2-5] State of the art technology requires a depleted emitter ledge[6-8] of finite dimension to counter the effects of surface recombination,[9] thus putting a lower limit on device dimensions as well as increasing overall base resistance and device size. Moreover, the vertical scaling of base thickness is a trade-off between reducing minority carrier transit time and increased base sheet resistance.

In order to move beyond these current limitations, a scheme utilizing selective area epitaxy to customize the lateral as well as vertical structure can be devised. While regrowth technology has been widely used for opto-electronic applications, it has only recently found its way towards use in electronic devices such as HBTs.[9-15] Here, a novel HBT structure that incorporates multiple epitaxy to selectively regrow the emitter is demonstrated. This HBT seen in

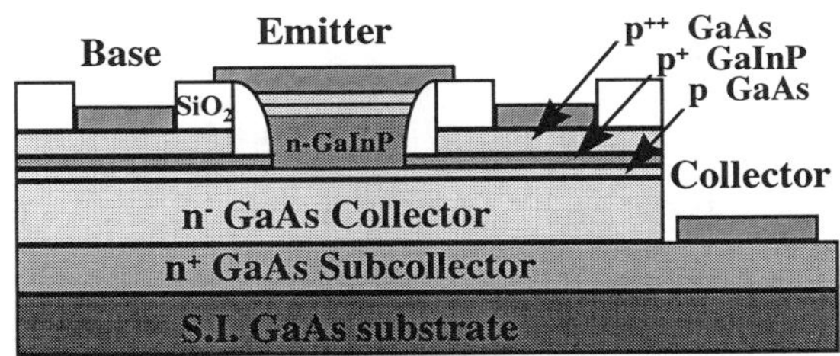

Figure 1. Schematic of proposed regrown emitter HBT.

figure 1 has a number of advantages over a conventionally fabricated device:

- The wide bandgap extrinsic base produces a larger turn-on than the intrinsic base, suppressing emitter edge current. This passivation scheme can be significantly smaller in dimension than the depleted emitter ledge.
- The use of GaInP instead of AlGaAs precludes the problems with DX center defects[16-18] as well as facilitating the process through the use of wet etchants selective between GaInP and GaAs. In addition, it has been shown that regrowth with phosphorus containing materials has superior compositional uniformity than arsenic compounds.[19-22]
- The three layer extrinsic base stack gives a lower overall base resistance, leading to high f_{MAX} with low base transit time.
- The emitter dimension is reduced from the photolithographic stripe size with SiO_2 sidewall spacers, similar to Si bipolar technology.[23,24]

To successfully fabricate this device, a significant amount of regrowth technology must be demonstrated. First, the selective growth of high quality $Ga_{.52}In_{.48}P$ lattice matched to GaAs in the presence an SiO_2 dielectric mask must be demonstrated. Then it is necessary to investigate the forward biased I-V characteristics of the critical regrown base-emitter junction. As a final measure of the SAE quality, complete devices should be fabricated and measured.

<u>EXPERIMENT</u>

All sample growths were done on [001] S.I. GaAs substrates. The base and collector layers of the diode and device samples prior to regrowth were grown by MBE. The collector consists of a 6000Å n+ GaAs buffer Si doped at ~$5x10^{18}$ cm^{-3}, a lower collector layer 3000Å thick doped to $3x10^{17}$ cm^{-3}, and an upper collector layer 3000Å thick doped to $3x10^{16}$ cm^{-3}. The base layers are composed of a three layer stack: 500Å p+ GaAs intrinsic base layer Be doped to $1x10^{19}$ cm^{-3}, 500Å p+ GaInP wide bandgap extrinsic base layer Be doped to $1x10^{19}$ cm^{-3}, and 1500Å p+ GaAs base contact layer Be doped to $1x10^{19}$ cm^{-3}. Approximately 3500Å of SiO_2 was deposited over these by PECVD at 300°C. The SiO_2 was patterned by photolithography with either a wet buffered HF etch solution or an anisotropic dry etch using CHF_3 plasma.

Test diodes for regrowth evaluation had the simplified structure shown in figure 2. The top p+GaAs layer was etched off prior to SiO_2 deposition. After etching the SiO_2 pattern, the exposed GaInP was selectively wet etched in an $HCl:H_3PO_4:H_2O$ solution. Prior to regrowth samples were solvent degreased then dipped in a dilute $NH_4OH:H_2O$ surface etch. A detailed description of the complete HBT device process has been discussed previously.[25]

Selective area regrowth was accomplished by low pressure (20 Torr) metal-organic vapor phase epitaxy (LP-MOVPE). The system consists of a horizontal two inch diameter quartz reactor with a tilted graphite susceptor. Details involving the growth of high quality III-V semiconductors with this system have been published elsewhere.[26] GaInP lattice matched to GaAs was done at 600°C at a rate of ~2.8Å/second. The wide bandgap n- emitter layer was nominally doped by SiH_4 at $5x10^{17}$ cm^{-3} with a thickness of 1000Å. N+ GaAs, grading, and InGaAs cap layers for contact were grown at maximum doping level

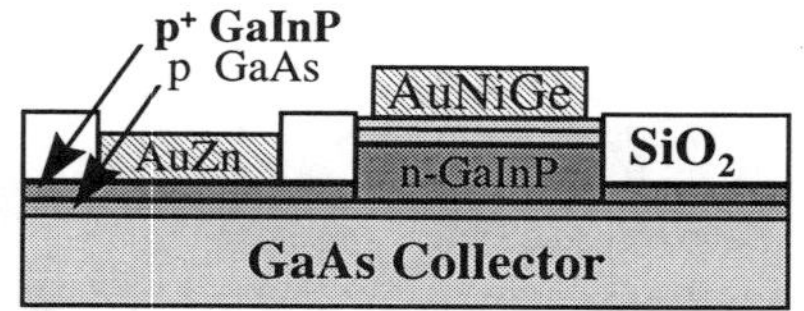

Figure 2. Schematic of simplified regrown p-n junction diode.

with thicknesses of 500Å, 300Å, and 1000Å, respectively.

After regrowth, metallization of the diodes and HBTs was accomplished by conventional liftoff techniques. The diode contacts for n- and p-type material were AuNiGe and AuZn, respectively. Photoluminescence (PL) and X-ray diffraction (XRD) rocking curves were used to characterize the regrown GaInP material. An HP 4145B semiconductor parameter analyzer was used to measure the I-V characteristics of the p-n diodes and HBTs. Finally the high speed performance of microwave HBTs were measured by an HP 8510B network analyzer.

RESULTS

Two basic growth schemes were studied in characterizing the selective regrowth of GaInP. These involve the use of either a dark field (mostly covered) or a light field (mostly open) SiO_2 mask. In the former, only the emitter areas (~10%) are open to growth. In the latter only the base mesa is covered by SiO_2 so that the emitter as well as all areas outside the base mesa are open for growth (~90%). As expected, dark field (DF) epitaxy exhibited a growth rate enhancement of five to six-fold compared to unpatterned deposition. In addition, the growth rate enhancement increased with proximity to the mask edges. These phenomena occur due to lateral gas phase diffusion of unused growth species above the mask and into the growth windows.[22,27,28] With the DF scheme, the increased partial pressure of unincorporated species caused nucleation on the mask surface and hence non-selective growth. To prevent this the group III flow rates were reduced such that the growth rate dropped back to the nominal value of ~2.8Å/second. Indeed this rate represents the upper limit for selective epitaxy at the given growth conditions.

On the other hand, selective epitaxy on a light field (LF) mask was more easily accomplished. Such growth compares closely to conditions of unpatterned deposition such that growth rate enhancement, both overall and near mask edges, is minimized. Moreover the possibility of compositional shifts and variations is also reduced. X-ray rocking curves of the LF patterned GaInP show virtually no change composition from lattice matched unpatterned GaInP. No x-ray signal could be obtained from the DF growth, most likely due to the small surface area. Low temperature PL spectra were taken to further investigate the compositional dependence of the patterned growths. As seen in figure 3, other than a slight decrease in intensity, the overall composition and linewidth of the LF growth is almost identical to the reference unpatterned GaInP. In contrast the DF spectrum is significantly shifted to lower energy as well as having a wider FWHM. The peak shift corresponds to approximately a 7% (atomic) increase in indium composition. This increase in indium content can be attributed to the higher gas phase diffusion mobility of TMIn compared to TEGa.[29] This leads to a increased relative rate of indium incorporation dependent on the amount of mask coverage and distance from the mask edge. Hence to grow

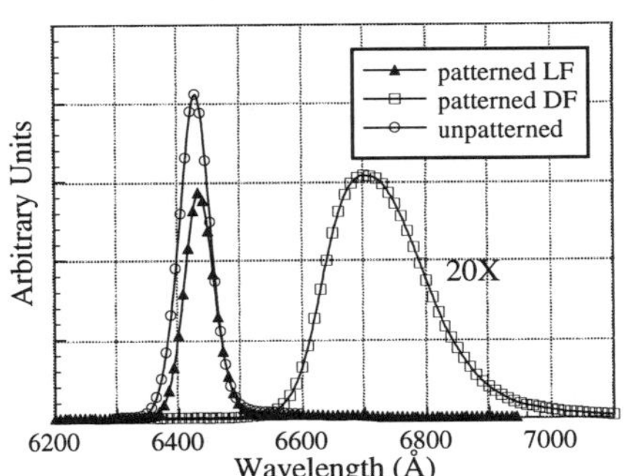

Figure 3. Low temperature PL at ~8K

lattice matched GaInP on a dark field mask, the flowrate ratio of TMIn to TEGa must be adjusted from the unpatterned calibration.

The forward bias I-V characteristics of regrown p-n junction diodes were measured to evaluate interface defects that can cause excess recombination. In figure 4, the forward biased base-emitter logarithmic I-V characteristics of the various samples are plotted. Most samples had a one second PH_3 pause before commencing GaInP growth. One remaining sample, P66, incorporated a one second TEGa pause before growth. This strategy was attempted to further improve the quality of the interface, as implied in the literature.[30] The unpatterned reference diode (U67) shows a well

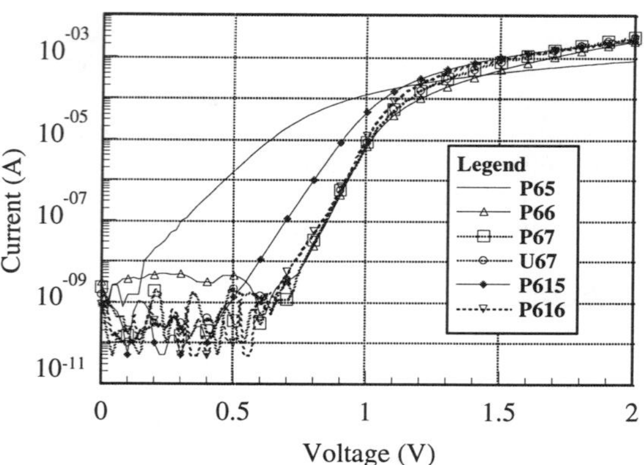

Figure 4. Forward bias I-V characteristics of regrown B-E junction.

behaved curve with an ideality factor n_f of ~1.25. The baseline patterned diode (P67) with the same PH_3 pause shows identical characteristics, indicating that the interface is not compromised by the presence of a dielectric mask. It is interesting to note that sample P66 with the TEGa pause did not noticeably improve the I-V characteristics. Though it may improve the interface from an optical or x-ray point of view[30], the I-V was minimally affected. The most noticeable excursion from the other samples is the extremely high leakage from the sample with RIE SiO_2 (P65). Dry etching to obtain a sidewall is crucial to the device fabrication. To remove the deleterious effects of the plasma etch, the sample underwent O_2 plasma treatment for five minutes and the GaInP was wet etched for 10X the nominal time (P616). After this, the I-V was recovered to nearly the same curve as the reference. A sample with the O_2 plasma but only a 2X GaInP wet etch recovered only partially (P615) while the sample without O_2 plasma but with a 10X GaInP wet etch did not improve at all (not shown).

As an ultimate test for the quality and feasibility of the regrown interface, HBT devices were fabricated and analyzed. The devices fabricated to date originate from the dark field regrowth scheme. Shown in figure 5 are the I_C-V_{CE} family of curves for a 3x12μm emitter device. A maximum current gain of β~80 is obtained at a collector current density of $J_C = 4x10^4$ A/cm^2. The cutoff frequency f_T for the 3x12μm device was 22 GHz and f_{MAX} was 18 GHz. From extracted scattering parameters, the overall base resistance R_B was estimated to be ~17 Ω and base-collector capacitance C_{BC} was ~160fF. The reason for these high parasitic values can be attributed to a non-optimized maskset and process. The high R_B is partially due to the non-self aligned nature of the base metal contact and also possibly from additional interface resistance between the three base layers. Calculations show that the contact

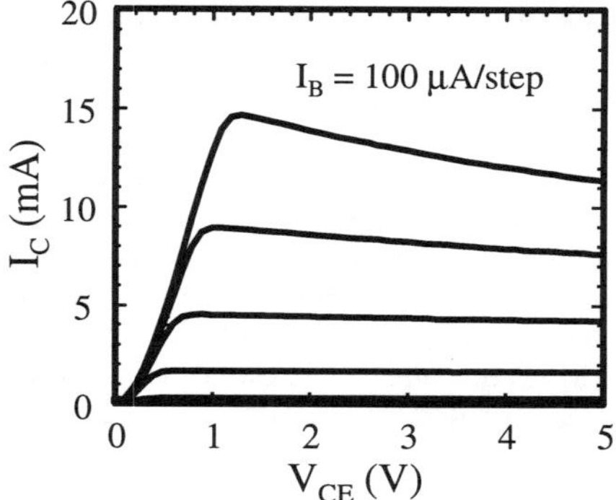

Figure 5. I_C-V_{CE} plot of 3x12μm regrown emitter HBT.

resistance through the two interfaces with the current doping scheme is a significant $\sim 1.5 \times 10^{-6}$ $\Omega{*}cm^2$. Higher doping of the wide bandgap p+GaInP up to 3×10^{19} cm^{-3} reduces this value to an negligible 8×10^{-8} $\Omega{*}cm^2$. The large value of C_{BC} comes from a combination of the non-self aligned base contact and a junction area that incorporates the collector contact area. Minimization of R_B and C_{BC} is paramount for improving f_T and f_{MAX} and can be accomplished with a self-aligned base contact process. This can be done by using a refractory metal such as tungsten as a self-aligned contact deposited prior to regrowth. The thermal stability of this contact will allow it to withstand the heat cycle of the regrowth without significant degradation. With these improvements, a state of the art overall base resistance of less than 10 Ω should be attainable. Finally, the self-aligned contact also shrinks the overall device size hence decreasing C_{BC} and increasing potential packing density.

CONCLUSIONS AND FUTURE WORK

In summary, the feasibility of selectively regrown GaInP for HBTs has been demonstrated. The material characteristics for SAE GaInP on light field and dark field masks were studied. While the LF growth was very similar to unpatterned bulk growth, the DF growth exhibited a high degree of growthrate enhancement resulting in an increase of indium incorporation. Regrown emitter p-n diodes were fabricated to study the I-V characteristics of the interface. Samples with either a PH$_3$ or a TEGa pause scheme both showed acceptable forward bias characteristics with an ideality factor of ~ 1.25. Finally, HBTs fabricated from a DF regrowth demonstrated a current gain of up to 80 with an f_T and f_{MAX} of 22 GHz and 18 GHz, respectively.

As mentioned earlier, there are several points of improvement for the devices. Growth on a LF mask should improve the device uniformity and base-emitter characteristics. Doping the wide bandgap p+GaInP up to 3×10^{19} cm^{-3} should help improve the base resistance. Self-aligned processing should significantly reduce R_B and C_{BC}. With these improvements, the successful application of SAE to HBTs can advance device technology to the next level of performance and scaling.

ACKNOWLEGEMENTS

We would like to gratefully acknowledge the valuable contributions from Y. M. Hsin, T. P. Chin, and T. Nakamura.

REFERENCES

1.   P. M. Asbeck, <u>High-Speed Semiconductor Devices</u>, edited by S. M. Sze (Wiley-Intersciences, New York, 1990), p. 389.

2.   N. Chand, R. Fischer, T. Henderson, H. Morkoc, and A. Neugroschel, *Appl. Phys. Lett.* **48** (5), 367-369, 1986.

3.   S. Tiwari, D. J. Frank, and S. L. Wright, *J. Appl. Phys.* **64** (10), 5009-5012, 1988.

4.   K. Mochizuki, H. Masuda, M. Kawata, K. Mitani, and C. Kusano, *Japanese J. Appl. Phys.* **30** (2B), L266-L268, 1991.

5.   W. Liu, D. Costa, and J. Harris, Jr., *Solid-State Electronics* **34** (10), 1119-1123, 1991.

6.   H. H. Lin and S.-C. Lee, *Appl. Phys. Lett.*, **47** (8), 839-841, 1985.

7.   R. J. Malik, L. M. Lunardi, R. W. Ryan, S. C. Shunk, and M. D. Feuer, *Elec. Lett.* **25** (17), 1175-1177, 1989.

8.   N. Hayama and K. Honjo, IEEE *El. Dev. Lett.* **11** (9), 388-390, 1990.

9.   Y.-F. Yang, C.-C. Hsu, and E. S. Yang, *54th Annual Device Research Conference Digest*, p.34, 1996.

10.   C. Dai, W. Liu, A. Massengale, A. Kameyama, J. Harris, Jr., *52nd Annual Device Research Conference Digest* **52**, paper IVB-1, 1994.

11.   Y. Miyamoto, A. G. Dentai, J. M. M. Rios, and S. Chandresekhar, *Elec. Lett.* **31** (17), 1510-1511, 1995.

12.   Y. Miyamoto, A. G. Dentai, J. M. M. Rios, and S. Chandresekhar, *Elec. Lett.* **31** (17), 1510-1511, 1995.

13.   Y. Amamiya, C.-W. Kim, N. Goto, S. Tanaka, N. Furuhata, H. Shimawaki, and K. Honjo, *International Electron Devices Meeting Technical Digest*, 199-202, 1994.

14.   R. A. Hamm, A. Feygenson, D. Ritter, Y. L. Wang, H. Temkin, R. D. Yadvish, and M. B. Panish, *Appl. Phys. Lett.* **61** (5), 592-594, 1992.

15.   T. Tanoue, H. Masuda, K. Washio, and T. Nakamura, *International Electron Devices Meeting Technical Digest*, 87-90, 1992.

16.   W. Liu, T. Kim, A. Khatibzadeh, *International Journal of High Speed Electronics and Systems* **5** (3), 411-471, 1994.

17.   K. Krynicki, M. A. Zaidi, M. Zazoui, and J. C. Bourgoin, *J. Appl. Phys.* **74** (1), 260-266, 1993.

18.   D. Costa and J. Harris, *IEEE Trans. El. Dev.* **39**, 2383-2394, 1992.

19.   J. S. C. Chang, K. W. Carey, J. E. Turner, and L. A. Hodge, *J. Elec. Mat.* **19** (4), 345-348, 1990.

20.   J. Thompson, A. K. Wood, N. Carr, P. M. Charles, A. J. Mosely, R. Pritchard, B. Hamilton, A. Chew, D. E. Sykes, and T.-Y. Seong, *J. of Cryst. Growth* **124**, 227-234, 1992.

21.   Y. D. Galeuchet, P. Roentgen, and V. Graf, *J. Appl. Phys.* **68** (2), 560-568, 1990.

22.   C. Caneau, R. Bhat, M. R. Frei, C. C. Chang, R. J. Deri, and M. A. Koza, *J. of Cryst. Growth* **124**, 243-248, 1992.

23.   H. Nakashiba, I. Ishida, K. Aomura, T. Nakamura, *IEEE Trans. El. Devices* **27**, 1390-1394, 1980.

24.   D. D. Tang, P. M. Solomon, T. H. Ning, R. D. Isaac, R. E. Burger, *IEEE J. Solid-State Circuits* **17**, 925-931, 1982.

25.   S. L. Fu, S. H. Park, Y. M. Hsin, M. C. Ho, T. P. Chin, P. K. L. Yu, C. W. Tu, P. M. Asbeck, T. Nakamura, *52nd Annual Device Research Conference Digest*, paper IVB-2, 1994.

26.   X. S. Jiang, A. R. Clawson, and P. K. L. Yu, *J. of Cryst. Growth* **124**, 547-552, 1990.

27.   R. Bhat, *J. of Cryst. Growth* **120**, 362-368, 1992.

28.   M. Gibbon, J. P. Stagg, C. G. Cureton, E. J. Thrush, C. J. Jones, R. E. Mallard, R. E. Pritchard, N. Collis, and A. Chew, *Semicond. Sci. Technol.* **8**, 998-1010, 1993.

29.   M. Maassen, O. Kayser, R. Westphalen, F. E. G. Guimares, J. Geurts, J. Finders, and P. Balk, *J. of Elec. Mat.* **21** (3), 257-264, 1992.

30.   T. Nittono, S. Sugitani, and F. Hyuga, *J. Appl. Phys.* **78** (9), 5387-5390, 1995.

Selective growth of MOVPE on AlGaAs/GaAs patterned substrates for quantum nano-structures

Makoto SAKUMA, Takashi FUKUI, Kazuhide KUMAKURA and Junichi MOTOHISA
Research Center for Interface Quantum Electronics, Hokkaido University,
North 13 West 8, Sapporo 060, Japan

Abstract

We propose and demonstrate a new mask material of AlGaAs native oxide for selective area metalorganic vapor phase epitaxy (MOVPE) which has several advantages over conventional SiNx or SiO_2 masks. GaAs selective area growth occurs on masked substrate of AlGaAs native oxide whose Al composition is 0.4, and the wire structures with trapezoidal cross section are formed along [100] direction on (001) GaAs substrates with line & space mask pattern. Furthermore, after annealing the selectively grown GaAs wire samples, GaAs layers can be regrown with atomically smooth surface, in which GaAs wires are perfectly buried. The results show that this novel selective area MOVPE technique using AlGaAs native oxide masks are promising for quantum nano-structure device fabrication.

Introduction

Recently, low dimensional quantum structures have been fabricated by various methods, in order to apply to future generation devices which are expected to exhibit novel electrical and optical properties. Selective area MOVPE on masked substrates is one of the most promising method because smaller size quantum structures compared to the substrate patterns can be fabricated, and it does not introduce any damage nor contamination during their fabrication processes. For example, quantum wires on V-grooves[1], facet quantum wires[2] and tetrahedral quantum dots[3] were formed on patterned or masked substrates by MOVPE. Furthermore, the highly uniform quantum structures were performed utilizing self-limited growth mechanism on SiN_x masked substrates[4-6].

However, when SiN_x or SiO_2 films deposited by thermal CVD or plasma CVD were used as mask materials, many defects exist at mask-substrate interface . Such defects may fatally influence on the quality of small quantum structures which are formed close to the substrate. In addition, although quantum structures should be buried planarly by the subsequent overgrowth process for the device application, it is very hard to bury them because no crystal growth occurs on the masked region.

In this paper, we propose and demonstrate selective area MOVPE using AlGaAs native oxide[7] as a new mask material. Stable thin oxide layer can be formed on AlGaAs surface simply by exposing to air. After AlGaAs and GaAs thin layer growth, GaAs top layer is partially removed by lithography and etching processes as defined by the mask pattern. The underlying AlGaAs exposed area works as a mask for next GaAs/AlGaAs selective area growth because stable oxide are formed on AlGaAs surface. We also confirm the possibility of the regrowth on the masked region of the air-exposed AlGaAs surface after annealing under AsH_3/H_2 or H_2[8] ambient.

Experimental procedure

A low-pressure horizontal MOVPE system was used. The working pressure during crystal growth was 76 Torr. Purified hydrogen (H_2) was used as a carrier gas, and the total gas flow rate was 3.0 ℓ /min. The source materials were trimethylgallium (TMGa), triethylaluminum (TEAl) and 20% arsine (AsH_3) in H_2. The partial pressure of TMGa and TEAl were 1.9×10^{-6}atm and 4.5×10^{-7}atm, respectively. AsH_3 partial pressure and growth temperature were 6.7×10^{-5}atm, 650℃ for selective area growth, and 5.0×10^{-4}atm and 650℃ for regrowth on the masked region. Growth rate of GaAs was 1.4 Å/sec.

The preparation procedure of masked substrates is schematically shown in **Fig.1**. First, GaAs buffer layer, 100nm $Al_xGa_{1-x}As$ and 10nm GaAs top layer were grown on (001) GaAs substrate by MOVPE. Al compositions of the AlGaAs layer were 0.1, 0.2, 0.3 and 0.4. Next, line and space

Mat. Res. Soc. Symp. Proc. Vol. 448 © 1997 Materials Research Society

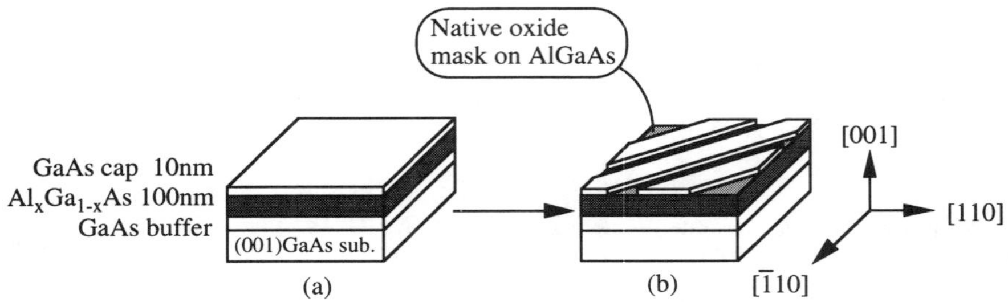

Fig.1. Schematic view of preparation procedure of substrates.
(a) First, GaAs/AlGaAs/GaAs layers were grown by
MOVPE. (b) Next, native oxide mask with line & space
patterns was formed on AlGaAs for selective area growth.

patterns were defined by electron beam lithography. Then, GaAs top layer was etched off using Citric acid/H$_2$O$_2$/H$_2$O solution. The solution has high selectivity for etching rate between GaAs and Al$_x$Ga$_{1-x}$As[9]. **Figure 2** shows an atomic force microscope(AFM) image of AlGaAs native oxide mask. The widths of GaAs line and AlGaAs space were 400nm and 70nm, respectively, and the pattern was aligned along the [100] direction. Air-exposed AlGaAs areas (space) are expected to work as a mask in selective area MOVPE growth.

In order to investigate the selectivity, GaAs layers were grown on these masked substrates. Before the growth, the substrates were annealed for 5 minutes within the furnace under AsH$_3$/H$_2$ atmosphere, which was the standard thermal cleaning process.

500nm

Fig.2. AFM image of AlGaAs native oxide masked substrate.
Gray and black areas show GaAs and AlGaAs layer, respectively.

Results and discussion

The dependence on Al composition of Al$_x$Ga$_{1-x}$As native oxide mask was investigated. **Figure 3** shows the AFM images of GaAs grown on the masked substrates with Al$_x$Ga$_{1-x}$As native oxide masks with Al compositions, x, of (a) 0.3 and (b)0.4. As shown in **Fig.3**, the pattern structures on masked substrate for x=0.3 were buried planarly and selective area growth did not occur. For Al$_{0.4}$Ga$_{0.6}$As native oxide mask, GaAs wires with a trapezoidal cross section consisting of {110} facet sidewalls were formed as shown in **Fig.3**(b), which is similar to that for SiN$_x$ mask. The nominal thickness of the epitaxial layer for planar substrate was 200nm. The SEM image of wire structures on Al$_{0.4}$Ga$_{0.6}$As native oxide mask is shown in **Fig.3**(c). These results show that native

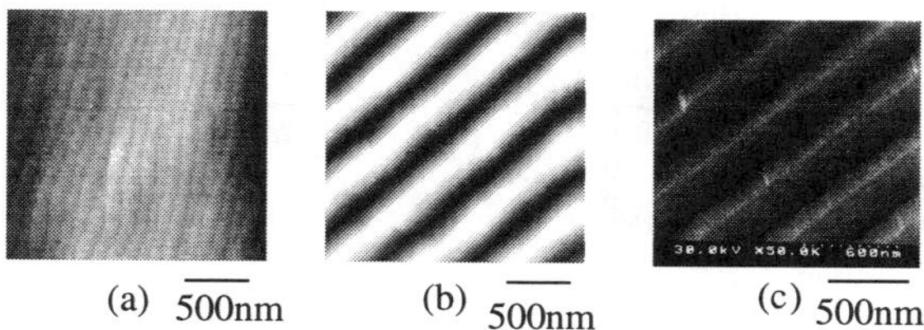

(a) <u>500nm</u>   (b) <u>500nm</u>   (c) <u>500nm</u>

Fig.3. (a)AFM images of selective area growth on $Al_xGa_{1-x}As$ native oxide masks for x=0.3 and (b)x=0.4, and (c)SEM image of GaAs selective area growth on the $Al_{0.4}Ga_{0.6}As$ native oxide masked substrate.

oxide on $Al_{0.4}Ga_{0.6}As$ surface can be used as a mask for selective area growth.

Next, in order to investigate the possibility of regrowth on the $Al_{0.4}Ga_{0.6}As$ masked region, the substrates were annealed to remove the oxide film prior to GaAs growth. **Figures 4**(a), (b) show AFM images of GaAs surfaces grown on the masked substrates after annealing under (a)AsH_3/H_2 (850℃, 30min.) and (b)H_2 (400℃, 15min.) ambient. The nominal thickness of GaAs layer for planar substrate was 60nm. Selective area growth did not occur and the mask patterns were perfectly buried for both cases. When the annealing was performed under AsH_3/H_2 ambient at 850 ℃, partly rough surfaces were observed, while under H_2 ambient at 400℃, atomically flat surfaces were obtained. The results suggest that the oxide on $Al_{0.4}Ga_{0.6}As$ was partly removed during thermal annealing processes, hence, the regrowth was performed. The experimental results are summarized in **Table 1**. For x=0.4 of $Al_xGa_{1-x}As$, both selective area growth and planar growth can be performed on the native oxide masks. This means that selective area growth and planar growth on the masked substrate can be controlled using the thermal annealing process. For the device application the quantum nano-structures should be buried planarly by subsequent overgrowth process. However, when the conventional SiNx mask was used, the buried surface was wavy and the poly-crystals were often deposited on the mask area. By using the AlGaAs native oxide mask, the quantum structures are expected to be buried planarly by subsequent overgrowth process.

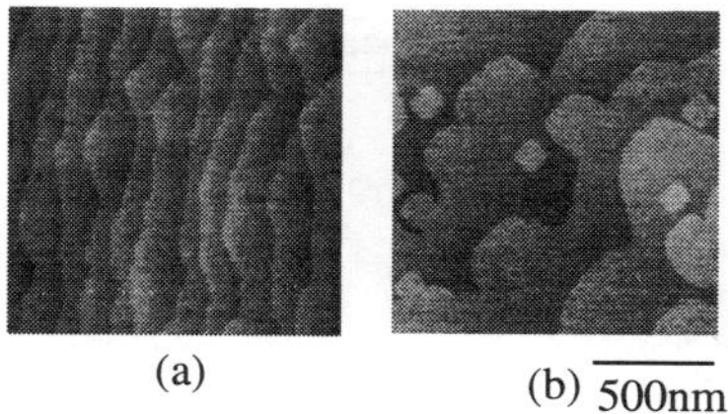

(a)   (b) <u>500nm</u>

Fig.4. AFM images of GaAs grown on $Al_{0.4}Ga_{0.6}As$ native oxide masked substrates after (a)AsH_3/H_2 annealing at 850℃ and (b) H_2 annealing at 400℃.

Table 1. GaAs growth modes on the masked substrates as mask materials and annealing conditions. Thick frame for x=0.4 shows the possibility of both selective area growth and planar growth.

mask materials / annealing	$Al_xGa_{1-x}As$ native oxide		SiN_x
	$0.1 \leqq x \leqq 0.3$	$x=0.4$	
none	●	○	○
850℃(AsH_3/H_2) 30 minutes	●*	●*	○
400℃(H_2) 15minutes	●	●	○

○ selective area growth

● planar growth

(* rough surface)

Next, we performed the H_2 annealing and GaAs regrowth processes for GaAs wire structures with a trapezoidal cross section grown on masked substrates. The result is shown in **Fig.5**. GaAs wires were perfectly buried, and atomically smooth planar surface was observed by AFM. In this case, average thickness of regrown GaAs layer was 100nm.

These results indicate that the native oxide on $Al_{0.4}Ga_{0.6}As$ layer is effective as a mask material as well as SiN_x or SiO_2. Furthermore, since crystal growth is also possible on masked region using pre-annealing technique under H_2 ambient, quantum nano-structures can be buried planarly by using these characteristics.

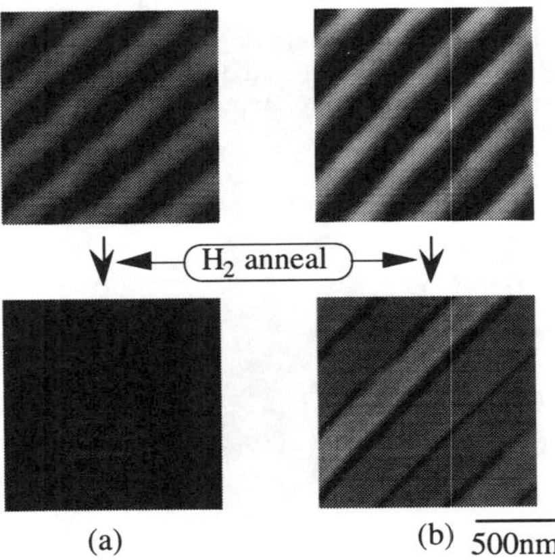

Fig.5. AFM images of selectively grown wire structures and GaAs regrown surfaces on (a)AlGaAs native oxide mask and (b)SiN_x mask. GaAs regrowths were performed after H_2 annealing at 400℃ for 15 minutes.

Conclusions

We proposed and demonstrated the new mask material of AlGaAs native oxide which has several advantages over conventional SiN_x or SiO_2 mask. When Al composition of AlGaAs native oxide was 0.4, the selective area growth of GaAs occurred for AlGaAs native oxide mask, which indicates that native oxide on AlGaAs is effective as mask materials. Furthermore, it is also possible to grow on the masked region of AlGaAs native oxide using pre-annealing technique under AsH_3/H_2 or H_2 atmosphere. Therefore, quantum wire or quantum dot structures formed by selective area growth using AlGaAs native oxide mask can be buried planarly by the regrown GaAs within the same growth run. The results show that selective area MOVPE using AlGaAs native oxide mask is promising method for quantum nano-structure device application.

Acknowledgement

The authors are grateful to Prof. H. Hasegawa and Mr. S. Hara and Mr. K. Nakakoshi for fruitful discussion for this work.

References

1) E.Kapon, S.Simhony, R.Bhat and D.M.Hwang: Appl. Phys. Lett. 55(1989)2715.
2) T. Fukui and S. Ando: Electronics Letters 25(1989)410.
3) T.Fukui, S.Ando, T.Honda, and T.Toriyama: Surf. Sci. 267(1992)236.
4) K.Yamaguchi. and K.Okamoto: Appl. Phys. Lett. 59(1991)3580.
5) S.Ando, T.Honda and N.Kobayashi: Jpn. J. Appl. Phys. 32(1993)104.
6) K.Kumakura, K.Nakakoshi, J.Motohisa and T.Fukui: Jpn. J. Appl. Phys. 34 (1995) 4387.
7) M.Notomi, Y. Kadota and T. Tamamura: Jpn. J. Appl. Phys. 34(1995)1451.
8) S.Gotoh and H.Horikawa:To be published in Appl. Phys. Lett.
9) M.Tong, D.G.Ballegeer, A.Ketterson, E.J.Raon, K.Y.Cheng and I.Adesida: Journal of Electronic Materials 21(1992)9

SELECTIVE EPITAXIAL GROWTH OF STRAINED SILICON-GERMANIUM FILMS IN TUBULAR HOT-WALL LOW PRESSURE CHEMICAL VAPOR DEPOSITION SYSTEMS

I. -M. Lee[†], W. -C. Wang[*], M. T. K. Koh[††], J. P. Denton[*], E. P. Kvam[††], G. W. Neudeck[*], and C. G. Takoudis[†,#]

Schools of Chemical[†], Electrical[*], and Materials[††] Engineering, Purdue University, Indiana 47907

ABSTRACT

Selective epitaxial growth (SEG) of silicon-germanium (SiGe) films on patterned-oxide silicon substrates, using a tubular hot-wall low pressure chemical vapor deposition (LPCVD) system, is demonstrated in this study. This conventional system is proposed as a low cost alternative for SiGe epitaxial growth. Three process improvements needed to achieve quality growth are discussed. First, the hydrogen bake process is modified to eliminate Ge-outgassing. Secondly, a Si SEG buffer layer is deposited prior to SiGe SEG. Finally, a small flow of dichlorosilane is introduced during the temperature ramp-down period prior to SiGe SEG. The growth results are discussed in terms of growth selectivity, thickness uniformity, growth rate, defect density, SiGe film composition, and electrical properties.

INTRODUCTION

Conventional tubular hot-wall LPCVD systems are a very good low cost alternative for SiGe epitaxial growth. While these systems may offer as good material quality as UHV/CVD systems do, LPCVD reactors appear to be far more amenable to industrial application. LPCVD systems are widely used in the extremely cost competitive silicon semiconductor industry, offering high wafer capacity, good temperature control, and excellent growth rate uniformity across the wafer. More importantly, selective epitaxial growth conditions can be achieved in an LPCVD system. Selective epitaxial growth on patterned-oxide substrates has been developed to extend the range of silicon device applications [1]. In the case of SiGe devices, SEG provides much more flexibility in the design of novel high-speed device structures. In addition, by not using blanket growth, SiGe SEG limits defect propagation and reduces misfit and threading dislocation densities [2].

The objective of this work is to explore the possibility of using a conventional high-capacity, tubular hot-wall LPCVD system to selectively grow electronic quality SiGe epitaxial layers. By utilizing LPCVD, SiGe selective epitaxial growth can be easily integrated with current silicon processes so that heterojunction structures, for example, can be more easily manufactured. The LPCVD system, the process improvements, and the results of SiGe SEG are reported in this paper.

EXPERIMENTAL

The LPCVD system for this study composed of a Tempress 280 three zone furnace which had a 15 cm diameter quartz tube, a system control panel, a Leybold Dryvac M100S dry-compression vacuum pump, and a Controlled Dissociation Oxidation (CDO) system connected to the exhaust line. The 2.1 m long quartz tube could accommodate 50 or more wafers per run. The wafers stood vertically, parallel to each other in the quartz boat and perpendicular to the gas flow. Dichlorosilane (DCS) and dilute germane (9% GeH_4 in H_2) were used as the silicon and germanium source gases, respectively, while H_2 was used as the carrier gas. The gases flowed into the reactor via two injection tubes through the water-cooled stainless steel front door. Rear injection of N_2 prior to the pump inlet port not only helped to control the system pressure, but also effectively diluted the remaining reactant gases flowing into the pump. A more detailed description is available in [3].

Selective epitaxial growth of SiGe was initiated on 3-inch (100) silicon substrates. Wafers were cleaned using a piranha cleaning procedure ($H_2SO_4:H_2O_2=$ 2:1), and then 2000 Å of SiO_2 was grown at 1050 °C in wet oxygen. Photolithography and BHF (buffered HF) wet etch were used to define oxide patterns, which were aligned to the <100> directions on the wafers. The

#Presently with the Department of Chemical Engineering, University of Illinois at Chicago

Mat. Res. Soc. Symp. Proc. Vol. 448 © 1997 Materials Research Society

wafers were cleaned using a piranha cleaning procedure again, followed by a 5 second BHF dip and a thorough deionized water rinse, prior to being loaded into the hot-wall reactor. The capability of growing quality Si SEG was first demonstrated in the system [3]. For Si SEG, the typical growth conditions studied were 900-950 °C, 0.75 Torr, and flow rates of 25 and 500 sccm for DCS and H_2, respectively. Then the SiGe SEG followed based on the process flow as determined from Si SEG. The typical SiGe SEG conditions were 650-800 °C, 0.8 Torr, with flow rates of DCS and H_2 of 27 and 750 sccm, respectively. GeH_4 (9% in H_2) flow rate was varied between 20 and 50 sccm. However, several process improvements were found to be needed and were implemented in chganging from Si SEG to high quality SiGe SEG. Figure 1 outlines the typical process temperature profiles of the experiments for Si and SiGe SEG. The process improvements are discussed in the next section.

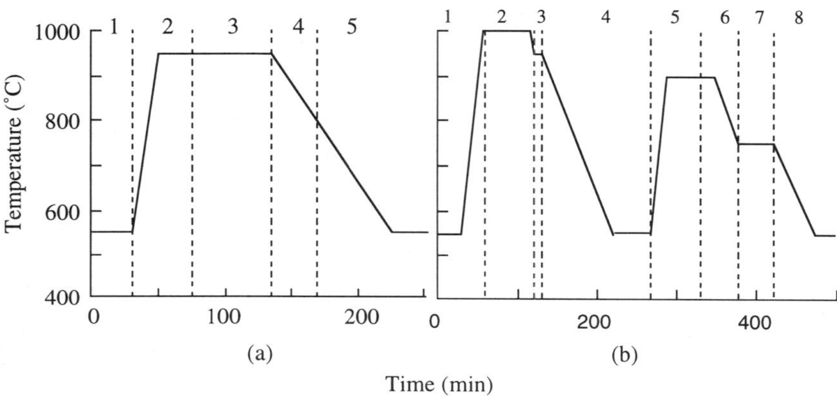

Figure 1. LPCVD reactor temperature profiles used for the selective epitaxial growth of Si (a) and SiGe (b). Steps in (a): 1. Load wafers, N_2 purge; 2. H_2 bake; 3. Si SEG; 4. N_2 purge; and 5. Unload wafers. Steps in (b): 1. N_2 purge; 2. H_2 bake; 3. Si deposition; 4. N_2 purge, load wafers; 5. H_2 bake; 6. SEG buffer layer of Si; 7. SiGe SEG; and 8. N_2 purge, unload wafers.

PROCESS IMPROVEMENTS

Two-Step Hydrogen Bake - Ge Outgassing

The hydrogen bake conditions used for Si SEG in the reactor of this study were T=950 °C and P=0.9 Torr for 30 minutes. They resulted in a good initial growth surface and good electronic quality was achieved [3]. However, these hydrogen bake conditions were found to be ineffective for SiGe SEG growth. Substrate surface was contaminated by Ge particles after the hydrogen bake as observed by scanning electron microscopy (SEM) and energy dispersive spectroscopy (EDS) analysis [4]. Thermodynamic analyses revealed that Ge outgasses at conditions of high temperature and low pressure [4]. Because the system had hot-walls, there was deposition of SiGe on the reactor wall from previous runs. The deposition on the wall thus provided a Ge source for Ge outgassing. Therefore, fundamental knowledge-driven modifications in the hydrogen bake process were devised and adopted. A hydrogen bake of the empty reactor at 1000 °C, in 0.75 Torr of flowing hydrogen for 1 hour prior to substrate wafer insertion was found to significantly improve the quality of the SiGe SEG films later on. These conditions favored high Ge/Si ratios in the gas phase, and Ge from the deposits on the reactor wall thus entered the gas phase was carried away by the H_2. After this bake, a DCS flow was introduced at the same conditions for 10 minutes. This resulted in a thin Si film over everything in the reactor, thus further reducing the possibility of Ge outgassing, if any was left. A N_2 purge and loading of the wafers were then done followed immediately by a second hydrogen bake. The typical conditions for this process were 900 °C, 0.95 Torr, for 40 min, with the objective to eliminate native silicon oxide from the wafer surfaces (steps 2-5 in Figure 1 (b)).

Si SEG Buffer Layer Prior to SiGe SEG

In order to eliminate surface imperfections after the high temperature hydrogen bake, and to reduce the possibility of impurity accumulation at the Si/SiGe interface, a Si buffer layer was grown immediately after the second hydrogen bake procedure. Cross-sectional transmission electron micrograph (XTEM) illustrated the purpose of Si SEG buffer layer (Figure 2). An interface between the Si substrate and the Si SEG buffer layer was visible. A plausible explanation is the following: to minimize the Ge outgassing problem and avoid high substrate temperatures, the second (lower temperature) hydrogen bake might not completely remove all native oxide. Therefore, the growth of SiGe films immediately after the bake, without the growth of a Si SEG buffer layer, might not result in a quality SiGe film because of these remaining oxide patches. Since the growth conditions were chosen so that selectivity was achieved, the Si buffer layer grew only from the oxide-free surface, and then proceeded to bury the residual oxide through an epitaxial-lateral-overgrowth (ELO) fashion [1].

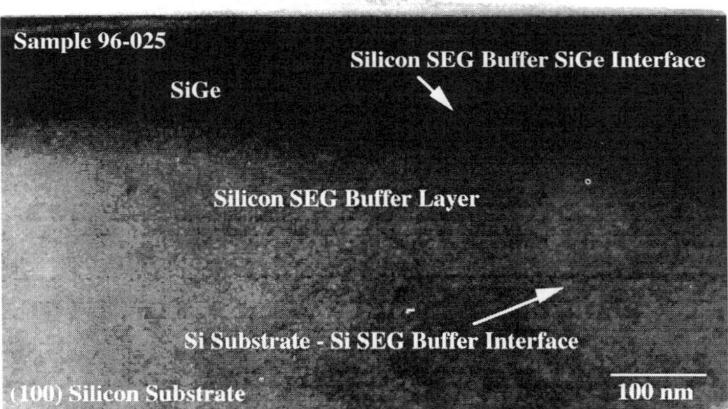

Figure 2. Cross-sectional TEM micrograph of the $Si_{0.93}Ge_{0.07}$ SEG deposited at 800 °C and 0.8 Torr using 27 sccm DCS, 25 sccm 9% GeH_4, and 750 sccm H_2 in the tubular hot-wall LPCVD system. Si selective buffer layer was grown between SiGe epitaxial layer and Si substrate.

Suppression of Surface Contamination During Temperature Ramp Down

In CVD processes, the reactor environment has to be maintained at a very low oxygen and water vapor contamination levels throughout the whole process in order to achieve excellent epitaxial growth. In the reactor used in this study, when process temperature was lowered from 900 °C for the Si SEG buffer layer to below 750 °C for SiGe SEG, the process environment could change from an oxide-free condition to an oxide-forming condition [4,5]. Therefore, during the temperature ramping down from the Si SEG conditions to the ones for SiGe SEG, accumulation of oxide on the growth surface could take place. To eliminate this problem, a small amount of DCS was introduced during this period (Step 6 in Figure 1(b)). The surface oxide could be removed by a reaction like $SiO_{2(s)} + Si \rightarrow 2SiO$, where Si could come from any Si source [5]; in this case, a small flow of DCS was, therefore, introduced into the reactor. This was also suggested by Gaynor *et al.*. [6,7], and utilized by Lou *et al.* [8].

With these three process improvements, high quality strained Si SEG films were produced in the conventional LPCVD reactor of this study. A discussion of the growth results and evaluation are presented next.

RESULTS

The growth of SiGe layers at temperatures from 700 to 800 °C in this work was determined to be epitaxial and selective with respect to SiO_2. Addition of HCl was not necessary to achieve selectivity. Experiments at temperatures lower than 700 °C resulted in poly-crystalline growth. A Nomarski micrograph of SiGe SEG grown at 750 °C is shown in Figure 3. Good selectivity of SiGe epitaxial growth was confirmed with good surface smoothness and no nucleation occurring on the SiO_2 areas.

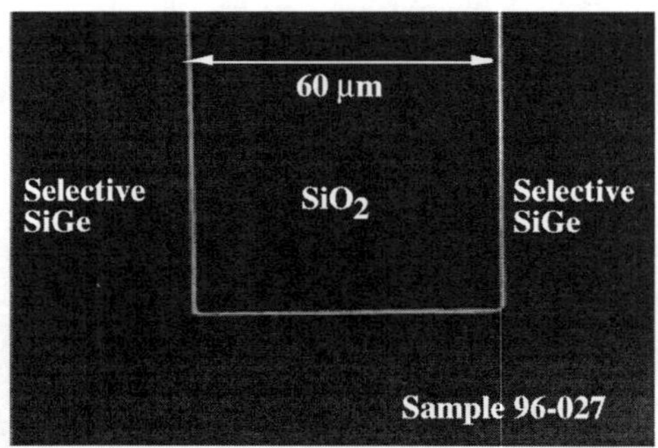

Figure 3. Nomarski micrograph (750x) of the SiGe SEG deposited at conditions shown in Table I.

The SiGe growth rates at 700, 750, and 800 °C were 15 Å/min, 19 Å/min, and 24 Å/min, respectively, when 27 sccm SiH_2Cl_2, 20 sccm dilute GeH_4 (9% GeH_4 in H_2), and 750 sccm H_2 were used. The growth rates at different temperatures yielded an activation energy of 0.42 eV at the aforementioned reactant flow rates and pressure. The Ge concentration decreased with increasing temperature, and the measured Ge fraction of these samples are presented in Table I. As for uniformity, the variations in total thickness across the wafer, including Si SEG buffer layer and SiGe SEG were less than 2.2%.

Table I. Measured Ge content and film thickness for SiGe SEG deposited at 0.8 Torr using 27 sccm DCS, 20 sccm 9% GeH_4, and 750 sccm H_2 in the tubular hot-wall LPCVD system. The layer thickness was measured by cross-sectional TEM.

Sample Number	Growth Temperature	Ge Content*			SiGe Layer Thickness
		By AES	By EDS	By XRC	
96-026	800 °C	5%	6%	6%	1095 Å
96-027	750 °C	8%	11%	12%	850 Å
96-028	700 °C	17%	20%	24%	663 Å

*AES: sputtering Auger electron spectroscopy, EDS: energy dispersive spectroscopy, XRC: x-ray rocking curve.

Plan-view TEM analysis of the grown samples revealed occasional long misfit dislocations. In the samples of higher Ge content (20% by EDS), the dislocations were traced as far as possible in the samples but no threading termination was observed. The total length of misfit dislocation lengths examined in this way was about 200 μm in each sample studied, suggesting a threading dislocation density of no more than 10^6 cm^{-2}. In the samples of lower Ge content, the misfit dislocations were spaced even wider, giving the maximum threading dislocation density at or below 10^4 cm^{-2} (Figure 4). This analysis followed from considering the observed length and spacing between misfit dislocations [9]. Use of SEG in small seed windows should further reduce or eliminate even the interfacial misfit dislocations [10]. In contrast, threading dislocations of about 10^9 cm^{-2} were visible in samples grown in the earlier runs before the process improvements were implemented.

With the addition of a Si buffer layer, it is necessary to investigate the material quality of this layer. Si SEG test runs were carried out at the same H_2 bake temperature (900 °C) and same growth conditions (900 °C, 0.75 Torr, 500 sccm H_2, 25 sccm DCS) as for the growth of the Si buffer layer. Then test diodes were fabricated in these Si SEG films. There were four types (SEG-0, SEG-3, SEG-5, and SEG-10) of diodes fabricated and each type consisted of 8 diodes with different area and perimeters. SEG-0 represented the diodes with implant areas, or junctions, that intersected the seed windows, or sidewall oxide, while SEG-3, SEG-5, and SEG-10 diodes were 3, 5, and 10 μm away from the sidewall oxide. For the diodes with the junctions intersecting the sidewall oxide (SEG-0), the perimeter effect was observed and the average minimum ideality factor was the worst among the four types. When the junctions were away from the sidewall oxide, not only the perimeter effect which caused by sidewall defects was eliminated but the average minimum ideality factors were much improved. By reducing the perimeter effect, the average minimum ideality factor of 1.01 and an average reverse bias leakage current density of 1.45×10^{-7} A/cm^2 (average of 80 diodes) were obtained. The reverse bias leakage current density was measured at -1 V. One thing to be noted, the sidewall defects in Si SEG could be drastically reduced or eliminated by using thermally nitrided silicon oxide (NOX) as the field insulator. From the electrical characteristics of these diodes, it is concluded that the quality of Si SEG buffer layer used in the SiGe/Si heterostructure should be excellent.

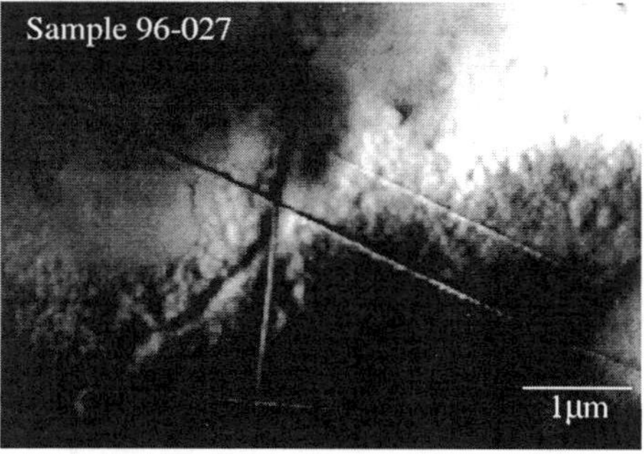

Figure 4. Plan-view TEM of a low Ge content SiGe thin film.

SUMMARY - CONCLUSIONS

Selective epitaxial growth of SiGe and Si in a conventional tubular hot-wall low pressure chemical vapor deposition system using DCS, GeH$_4$, and H$_2$ was demonstrated. Three process improvements were shown to be essential for the success of the overall process. First, the hydrogen bake conditions were modified so that Ge outgassing was minimized. A two-step hydrogen bake was, therefore, implemented and proved to be effective. Secondly, the growth of a Si buffer layer under selective growth conditions helped to improve the initial growth surface for later SiGe SEG. Lastly, the addition of a small amount of DCS during the temperature ramp down from the Si buffer layer growth temperature to the desired SiGe SEG temperature helped retain a clean surface prior to SiGe SEG.

Selective growth with respect to the SiO_2 mask was verified by Nomarski microscopy with good surface smoothness. Growing epitaxial SiGe selectively provides isolation structures for fabricating heterojunction devices. The growth rate, SiGe film composition, thickness uniformity as well as defect density were reported to be good for SiGe films deposited using this reactor. This research work on SiGe SEG paves the way for the possibility of high volume production of silicon-based heterostructures while maintaining the low cost nature of silicon processing.

ACKNOWLEDGMENT

The authors gratefully acknowledge the financial support from the National Science Foundation (Grant number 9306696-DMR).

REFERENCES

1. P. J. Schubert, J. L. Glenn Jr., J. A. Friedrich, W.A. Claasen, R. D. Zing, J. P. Denton, and G. W. Neudeck J. Electronic Materials, **19**, 1111 (1990).

2. D. B. Noble, J. L. Hoyt, C. A. King, J. F. Gibbons, T. I. Kamins, and M. P. Scott, Appl. Phys. Lett., **56**, 51 (1990).

3. W. -C. Wang, J. P. Denton, G. W. Neudeck, I. -M. Lee, C. G. Takoudis, M. T. K. Koh, and E. P. Kvam, accepted for publication in J. Vac. Sci. Tech. B (1996)

4. I. -M. Lee, C. G. Takoudis, W. -C. Wang, J. P. Denton, G. W. Neudeck, M. T. K. Koh, and E. P. Kvam, submitted to J. Electrochem. Soc. (1996)

5. I. -M. Lee, A. Jansons, and C. G. Takoudis, submitted to J. Vac. Sci. Tech. B (1996)

6. W.H. Gaynor, Research Thesis, School of Chemical Engineering, Purdue University (1989).

7. W. H. Gaynor, C. G. Takoudis, G. W. Neudeck, J. Vac. Sci. Tech. A, in press (1996).

8. J. -C. Lou, C. Galewski, and W. G. Oldham, Appl. Phys. Lett., **58**, 59 (1991).

9. E. P. Kvam and F. Namavar, Appl. Phys. Lett., **58**, 21 (1991)

10. E. A. Fitzgerald, G. P. Watson, R. E. Proano, D. G. Ast, P. D. Kirchner, G. D. Pettit, and J. M. Woodall, J. Appl. Phys., **65**, 2220 (1989)

ULTRA THIN SiO$_2$ MASK LAYER FOR NANO-SCALE SELECTIVE-AREA PECVD OF Si

J. W. PARK*, T. YASUDA**, K. IKUTA**, L.H. Kuo*,
S. YAMASAKI**, and K. TANAKA**
* Joint Research Center for Atom Technology (JRCAT)-Angstrom Technology
Partnership (ATP), 1-1-4 Higashi, Tsukuba, Ibaraki 305, Japan
** Joint Research Center for Atom Technology (JRCAT)-National Institute for Advanced
Interdisciplinary Research (NAIR), 1-1-4 Higashi, Tsukuba, Ibaraki 305, Japan

ABSTRACT

We discuss the applicability of ultrathin SiO$_2$ layers as a mask for low-temperature selective-area deposition of Si.   Thin oxide layers with estimated thickness ranging from 4 to 20 Å were formed by oxidizing H-terminated Si(100) surfaces by a remote plasma exposure at room temperature.   Low-temperature selective-area deposition was carried out using two different techniques: flow-modulated plasma-enhanced chemical vapor deposition (FM-PECVD) using SiH$_4$ and H$_2$, and very low pressure CVD (VLPCVD) using Si$_2$H$_4$.   We show that the ultra-thin plasma oxide layers exhibit good properties for a use as a passivating mask layer, and that the oxide layer can be patterned directly by E-beam irradiation.   These results open up a possibility to realize Si-nanostructures formation by selective-area processing.   Degradation of the oxide layer by plasma processing is also discussed.

INTRODUCTION

There is an increasing number of proposals and trials to realize novel devices incorporating three-dimensional and nano-scale structures.   Future semiconductor processing to fabricate such advanced structures with an ultimate precision will require the technology for selective-area deposition and etching at a nanometer scale.   Among the prerequisites to realize the selective deposition at reduced lateral dimensions are (i) formation and patterning of an ultrathin (< nm) mask and (ii) deposition chemistry compatible with the thin mask as well as with the reduced dimension.   One important consideration in developing these techniques is the thermal budget: if the processing can be performed at low temperatures, it may be applied at later stages in a device-processing flow when doping profiles and other constituents in the device have been already completed.

In this paper, we investigated the applicability of ultra-thin SiO$_2$ layers as a mask for selective-area deposition of Si.   The oxide layer was formed by remote-plasma oxidation of a H-terminated Si(100) surface at a room temperature.   We show   that the ultrathin oxide layers thus prepared are compatible with two different chemistries for selective deposition: flow-modulated plasma-enhanced chemical vapor deposition (FM-PECVD) at 473 K,   and   very low pressure CVD (VLPCVD) using Si$_2$H$_6$ at 823 K.   We will also report an initial result of E-beam patterning of the ultra-thin mask layers.

Mat. Res. Soc. Symp. Proc. Vol. 448 © 1997 Materials Research Society

EXPERIMENTAL

Figure 1 shows a schematic illustration of the multi-chamber CVD system constructed for this study.   It is equipped with Auger electron microscope(AES), scanning electron microscope(SEM), and scanning tunneling microscope(STM) for in-line characterization of the processed samples.

The substrates used in this study were p-type, boron doped, Si(100) wafers.   The samples were first RCA-cleaned and were thermally oxidized to form a sacrificial oxide layer.   Prior to each experiment,   a H-terminated surface was prepared by etching off the sacrificial oxide in a 30:1 HF solution.   A thin SiO_2 mask layer was formed by exposing the H-terminated surface to remote plasma of 0.2 % O_2 (He balance) in the loadlock chamber.   The O_2/He was fed through a remote-plasma tube (quartz, 20 mm I.D.), and plasma was generated at a r.f.(13.56MHz) power of 20 W.   The sample was located 13 cm downstream the plasma tube and maintained at a room temperature.   After oxidation the sample was annealed at 823 K for 15 minutes to reduce the H concentration in the oxide.

The selective depositions of Si were performed using two different methods.   One is flow-modulated PECVD (FM-PECVD) by using SiH_4 and H_2[1], and the other is thermal CVD operated at a very low pressure (VLPCVD) using Si_2H_6 as a source gas[2,3]. Figure 2 shows a typical flow sequence for FM-PECVD, where deposition/etching cycles were repeated by feeding SiH_4 intermittently into H_2 plasma.   FM-PECVD takes advantage of the fact that Si nuclei on the mask surface are readily etched by H_2 plasma.   Typical conditions for selective deposition were: H_2 flow = 90 sccm; SiH_4 flow =10sccm; substrate temperature = 473 K; r.f.power = 10W; and pressure = 150 mTorr.   For VLPCVD, typical conditions were: Si_2H_6 flow rate = 10 sccm; pressure = 0.4 mTorr; and substrate temperature = 823 K.   The base pressure of the CVD chamber was low 10^{-8} Torr.

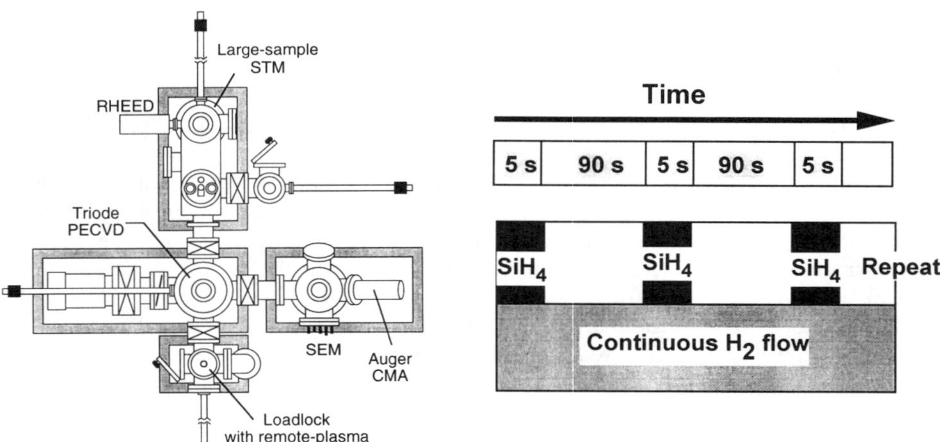

Fig. 1.   Schematics of the multi-chamber CVD system constructed for this study.

Fig. 2.   Flow sequence for FM-PECVD.

RESULTS and DISCUSSION

<u>Oxide Formation</u>

Figure 3 shows the variations in Si_{LVV} AES spectra with the plasma oxidation time.   It is well established that elemental Si exhibits a sharp peak at   92 eV while SiO_2 shows peaks at 76, 63, and 59 eV[4].   One can clearly see that the oxide thickness increases with the oxidation time. From the Si_{LVV} spectra, thickness of the oxide layer were approximately estimated to be 4, 6, 9, 20 Å for oxidation time of 8, 15, 30, and 300 s, respectively.   Thickness of the oxide can be controlled reproducibly at Å-scale precision by adjusting the oxidation time.   In addition, no C_{KLL} peak was detected at 272eV in the AES spectra[4], indicating the cleaning effect of the plasma-oxidation process.

<u>FM-PECVD on Thermal Oxide: A Reference Experiment</u>

We first discuss the selective deposition by FM-PECVD.   As an reference experiment, FM-PECVD was carried out on a thermal-oxide layer of approximately 20 Å in thickness.   This surface was prepared by etching back a 100 Å thick thermal-oxide layer in a diluted HF solution. Figure 4 shows Si_{LVV} spectra taken every 30 cycles of FM-PECVD.   The deposition rate under the conditions   identified above was 2.6 Å/cycle on the H-terminated Si(100).   We find that the surface of the thermal oxide remained almost unchanged up to the 60 cycles of FM-PECVD processing.   With repeating the cycles further,   the peak of elemental Si increased and the surface was finally covered by Si after 120 cycles.

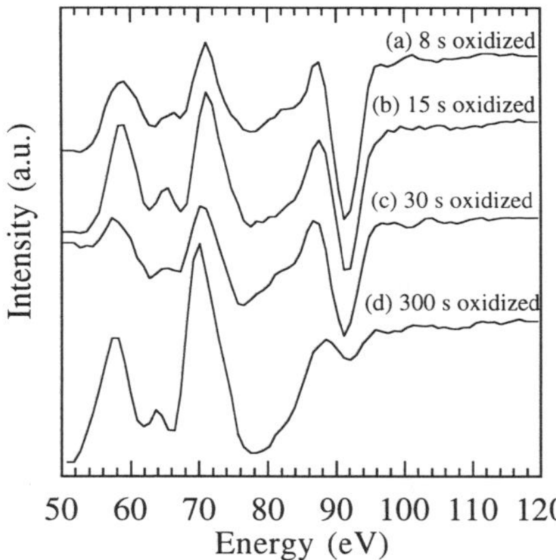

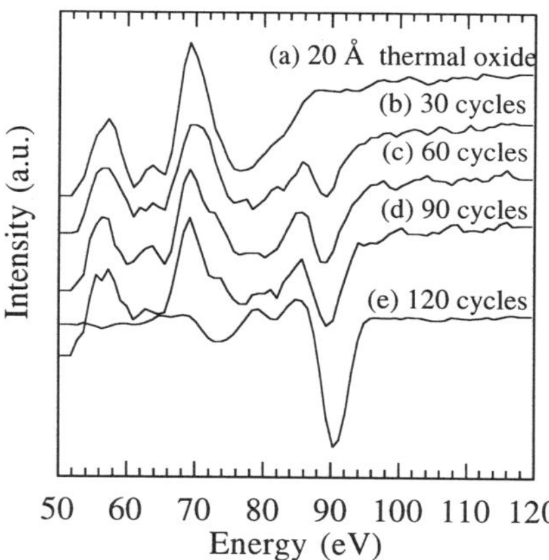

Fig. 3.   Variations of AES spectra with oxidation time.   H-teminated   Si(100) surfaces were oxidized by remote plasma at a room temperature.

Fig. 4.   Evolution of the AES spectra during FM-PECVD processing: (a) as-cleaned thermal oxide with a thickness of ~20 Å, (b), (c), (d), and (e) were measured at 30, 60, 90, and 120 cycles, respectively.

There are two causes for the observed loss of selectivity.   One has to do with the processing condition.   Since the selective deposition by FM-PECVD relies on the etching of the Si nuclei formed on the SiO_2 mask[1],   insufficient etching time would result in Si residues at the end of an etching period.   As the deposition/etching cycles are repeated, such residues are accumulated, leading to the loss of selectivity.   This mechanism can be suppressed, for example, by making the etching period longer.   However, there is the other cause for the selectivity degradation which imposes a limit on the etching period the length of the etching period.   In a separate experiment, we found an extensive exposure to pure H_2 plasma by itself degrades the thermal oxide.   For example, when the thermal oxide was exposed to pure H_2 plasma for 1 hr and then was processed by FM-PECVD for only 10 cycles,   the sample surface showed an AES spectra similar to the 120 cycle data in Fig. 4.

<u>FM-PECVD on Plasma Oxide</u>

Curve (a) in Fig. 5 shows the Si_{LVV} peaks of 20 Å thick plasma oxide exposed to FM-PECVD for 30 cycles.   Comparing to the starting surface shown in (d) of Fig. 3,   there is an increase in the elemental Si peak.   Selectivity degradation took place earlier for the plasma oxide than for the thermal oxide (see Fig. 4).   After additional 30 cycles on the surface (a) in Fig. 5, the surface was completely covered by Si (AES spectrum not shown).   One may suspect that the loss of selectivity in this case was caused by removal (or thinning) of the plasma oxide layer during the FM-PECVD processing.   However, this is not the case, as evidenced by the etch-back experiment shown in (b).   By etching the Si overlayer in pure H_2 plasma, the SiO_2 surface could be recovered.

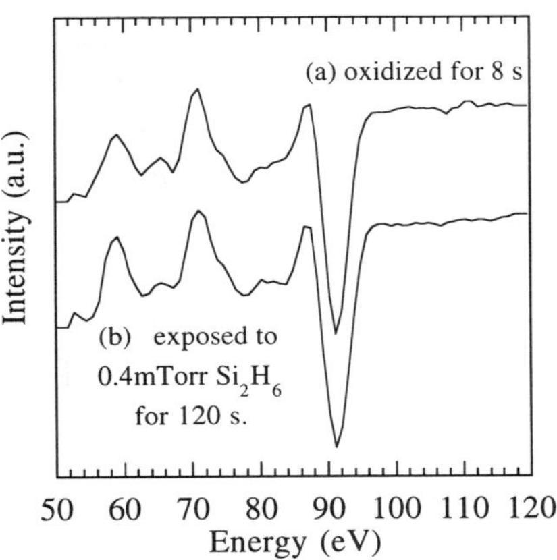

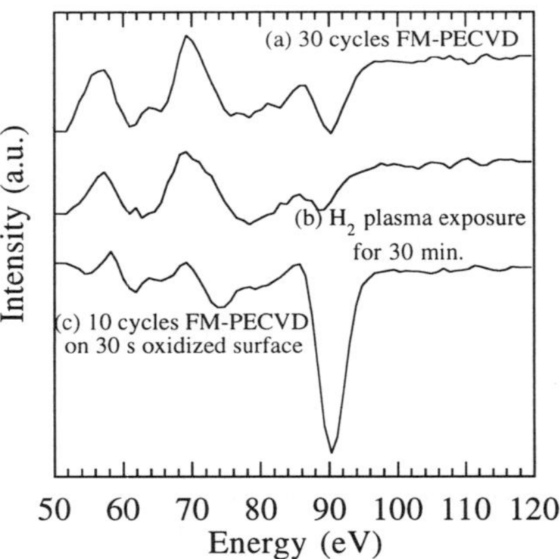

Fig. 5.   AES spectra of (a) after 30 cycle of FM-PECVD processing on (d) of Fig. 3.; (b) after exposing the surface in (a) to H_2 plasma at 473 K; and (c) after 10 cycles' FM-PECVD processing on (c) of Fig.3.

Fig. 6.   AES spectra of (a) the surface plasma-oxidized for 8 s., (b) after exposing the surface in (a) to 0.4 mTorr Si_2H_6 for 120 s..

In the prospect of nanoscale application, it is important to figure out how thin an oxide layer works as the mask for selective deposition.   Curve (c) in Fig. 5 is the surface of 9 Å thick plasma oxide exposed to FM-PECVD for 10 cycles.   The AES spectrum of the starting surface is shown in (c) of Fig. 3.   We can still recognize the SiO_2 features in Fig. 5(c),   indicating that the 9 Å SiO_2 layer can serve as the protecting mask if the total FM-PECVD cycles are limited.   Further efforts are necessary to realized thinner mask layers with improved stability in the plasma-processing environments.

VLPCVD

We also examined the performance of the plasma oxide mask for the selective VLPCVD.   In the experiment shown in Fig. 6, an oxide layer as thin as 4 Å was exposed to 0.4 mTorr Si_2H_6 at 823 K for 120 s.   Under these conditions, deposition rate of Si on a H-terminated Si(100) surface was 0.35 Å/s.   In Fig. 6,   we cannot find any sign of the Si deposition on the oxide surface after the 120 s exposure.   Thus the plasma oxide is an excellent mask particularly for VLPCVD, where ions and atomic H are absent.   At temperatures higher than 1000 K,   a thin SiO_2 layer is desorbed through formation of micro-voids [5].   Due to a relatively low substrate temperature in our experiment, we found no evidence of such void formation which would presumably result in the loss of selectivity.

E-Beam Patterning

It is well known that E-beam irradiation on SiO_2 (bulk glass, CVD and thermal oxides) induces electron-stimulated desorption of O which changes the stoichiometry of the irradiated area to Si-rich

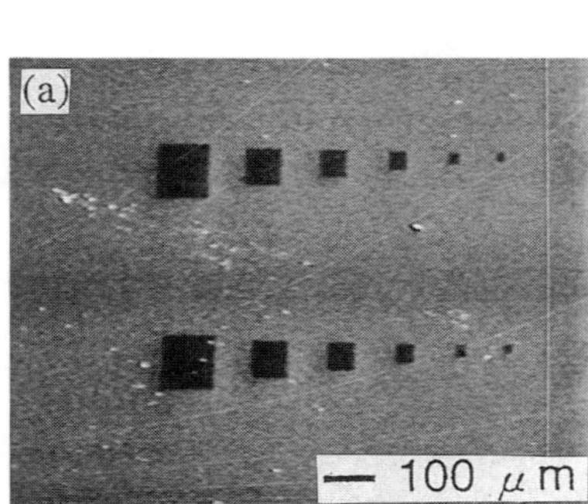

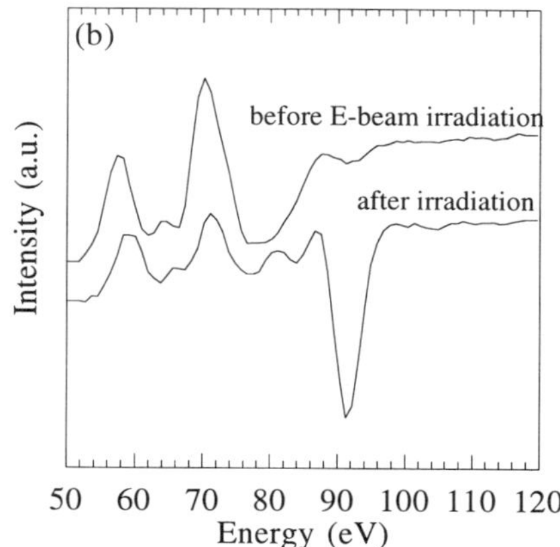

Fig. 7.   Direct patterning of a 20 Å SiO_2 layer by 15keV E-beam irradiation: (a) SEM image of irradiated are, (b) AES spectra before and after irradiation at $120C/cm^2$.

composition[6].   A recent study by a separate group at our research center(JRCAT) showed that desorption of thermal SiO_2 takes place selective at an E-beam irradiated area, exposing a clean Si surface[7].   If the same phenomenon takes place for the present ultrathin plasma oxide, it can be utilized to pattern the mask layer directly by an E-beam.

Figure 7 shows the results of our initial trials in this regard,   taking advantage of the E-gun of the SEM installed in the AES chamber.   In this experiment the acceleration voltage of the E-beam was 15keV and the dose was about   120 C/cm^2.   In the secondary-electron image shown in Fig. 7(a),   the irradiated regions show a dark contrast.   In Fig. 7(b) the AES spectra of the irradiated and non-irradiated regions are compared.   Clearly the composition of the irradiated area is Si-rich,   reproducing the effect reported in the literature.   Thus ultrathin plasma oxide can be directly patterned by E-beam irradiation.   We are continuing our study to achieve selective deposition on the E-beam patterned area.

CONCLUSIONS

We have demonstrated the formation of ultra thin oxides(less than 20Å) using a remote-plasma oxidation at room temperature.   The results presented here establish that the ultra thin oxide is a good mask layer for low-temperature selective-area deposition by FM-PECVD and VLPCVD. We also presented that the plasma oxide can be patterned directly by E-beam irradiation.

ACKNOWLEDGMENTS

This work, partly supported by NEDO, was performed in the Joint Research Center for Atom Technology (JRCAT) under the joint research agreement between the National Institute for Advanced Interdisciplinary Research (NAIR) and the Angstrom Technology Partnership (ATP).

REFERENCES

1. G.N. Parsons, Appl.Phys.Lett., **5**9(20), 2546 (1991).
2. T. Tasumi, K. Aketagawa, M. Hiroi, and J. Sakai, J. Cryst.Growth, **120**, 275 (1992).
3. K.E. Violette, M.K. Sanganeria, M.C. Ozturk, G. Harris, and D.M. Maher, J.Electrochem.Soc., **141**, 3269 (1994).
4. L.E. Davis, N.C. MacDonald, P.W. Palmberg, G.E. Riach, and R.E. Weber, "Handbook of Auger Electron Spectrosopy-2nd Edition."(Physical Electronics Industries Inc., Eden Prairie, Minnesota, 1995).
5. Y. Wei, R.M. Wallace, and A.C. Seabaugh, Appl.Phys.Lett., **69** (9), 1270 (1996).
6. J.W. Coburn and H.F. Winters, J.Appl.Phys., **5 0**, 3189 (1979).
7. S. Fujita, S. Maruno, H. Watanabe, and M. Ichikawa, Appl. Phys. Lett., **69** (5), 638 (1996).

SOLID PHASE CRYSTALLIZATION OF LPCVD AMORPHOUS SI FILMS BY NUCLEATION INTERFACE CONTROL

Eui-Hoon Hwang and Jae-Sang Ro

Department of Metallurgy and Materials Science, Hong-Ik University
Seoul, 121-791, Korea

ABSTRACT

A novel method for the fabrication of poly-Si films with a large grain size is reported using solid phase crystallization (SPC) of LPCVD amorphous Si films by nucleation interface control.   The reference films used in this study were 1000 Å-thick a-Si films deposited at 500°C at a total pressure of 0.35 Torr using Si_2H_6/He.   Since the deposition condition changes the incubation time, i.e. nucleation rate, and since nucleation occurs dominantly at a-Si/SiO_2 interface, we devised the following deposition techniques for the first time in order to obtain the larger gain size.   A very thin a-Si layer (~ 50 Å) with the deposition conditions having long incubation time is grown first and then the reference films (~ 950 Å) are grown successively.   Various composite films with different combinations were tested.   The crystallization kinetics of composite films was observed to be determined by the deposition conditions of a thin a-Si layer at the a-Si/SiO_2 interface.   Nucleation interface was also observed to be modified by interrupted gas supply resulting in the enhancement of the grain size.

INTRODUCTION

Polycrystalline silicon films have received a great deal of attention since they can be applied to various applications such as a gate electrode for MOSFET, the emitter in bipolar transistors and interconnects.   Recently polycrystalline silicon thin film transistors (poly-Si TFT's) have attracted considerable interest for the fabrication of active-matrix-liquid-crystal-display (AMLCD).   In these applications the large grain size of poly-Si films is desirable since the presence of grain boundaries lead to a degradation in device performance.   Directly deposited LPCVD (low pressure chemical vapor deposition) poly-Si films exhibit a very fine grain size with a columnar microstructure [1].   Enhancement of the grain size is generally achieved by solid phase crystallization (SPC) of LPCVD amorphous silicon (a-Si) films.

Solid phase crystallization of a-Si films proceeds by nucleation and growth.   Deposition conditions such as deposition temperature and deposition pressure can affect the initial state of as-deposited a-Si films[2]. The degree of structural disorder and the purity of as-deposited a-Si films can in turn influence nucleation and growth kinetics during solid state transformation [3].   To obtain poly-Si films with the larger gain size the nucleation rate should be retarded and the growth rate should be enhanced during phase transformation [4,5].   Since the temperature at which solid phase crystallization is usually conducted is at around 600°C the phase transformation occurs at a temperature far from the equilibrium point between amorphous and crystal phases [6].   Since crystallization is carried out at a relatively low temperature regime the phase transformation is limited by kinetic control.   In other words both the nucleation rate and the growth rate decrease as the crystallization temperature decreases.   The activation energy for nucleation in the case of a-Si, however, is greater than that for growth.   As a result the low annealing temperature is desirable for the enlargement of the grain size because the nucleation rate has a steep temperature-dependence while the growth rate has not.   Therefore the final grain size is determined by two main variables such as deposition conditions and SPC temperature.

The system in interest is a-Si films deposited on the top of SiO_2.   The possible nucleation sites are the a-Si/ SiO_2 interface, the top surface covered with the native oxide and the bulk of a-Si films.   In

Mat. Res. Soc. Symp. Proc. Vol. 448 © 1997 Materials Research Society

this paper we report the critical role of the initial state of as-deposited a-Si films near, or, at the a-Si/ SiO$_2$ interface.   We also report the deposition techniques for the nucleation interface control, resulting in the enhancement of the grain size.

EXPERIMENTAL

1000 Å -thick thermally grown SiO$_2$ was deposited on 4 " Si wafers. 1000 Å -thick amorphous silicon films were deposited in a cold-wall-type LPCVD reactor.   The precursor used was disilane diluted with helium.   The LPCVD chamber was pumped by a turbomolecular pump backed by a mechanical pump.   The deposition pressure was set for a given flow by adjusting the opening of a throttle valve. The deposition pressure was measured using a capacitance manometer.   The film thickness was measured using a surface profilometer.   After the film deposition a-Si films were isothermally annealed at 600°C in a nitrogen ambient using a quartz tube furnace. The fraction of crystallization as a function of annealing time was determined by X-ray diffraction using diffractometer. Transmission electron microscopy was employed for direct observation of microstructure in crystallized films.

RESULTS AND DISCUSSIONS

Deposition behavior was first checked using the LPCVD reactor employed in this study. Temperature dependence of deposition rate was observed to show Arrhenius-type behavior.   Deposition temperature was varied from 430°C to 550°C under a constant total pressure of 0.35 Torr.   The partial pressure of Si$_2$H$_6$ was 0.30 Torr.   The activation energy for Si deposition using Si$_2$H$_6$ was found to be ~ 42 kcal/mole.   This value is very similar to that using SiH$_4$.   The deposition rate for disilane deposition, however, increased by a factor of ~10 compared to that for silane deposition.   The deposition rate had a linear relationship with the deposition pressure.   Deposition was thus carried out in the regime of surface phase reaction control.

To investigate the effect of deposition pressure on the crystallization behavior, a-Si films were deposited at different pressures of 0.15 Torr, 0.35 Torr and 0.60 Torr respectively at a constant deposition temperature of 550°C.   These samples were annealed at 600 °C in a N$_2$ ambient.   The crystallized samples exhibit 3 main x-ray peaks such as (111), (110) and (311).   Since (111) is the dominant texture (111) x-ray intensity was used to monitor the crystallization kinetics.   Fig. 1 shows (111) x-ray intensity vs. annealing time for those samples.   As the deposition pressure increases the incubation time increases and the saturation intensity increases.   Since the deposition rate linearly increases with the increase of the deposition pressure time to form a monolayer decreases linearly with the increase of the deposition pressure.

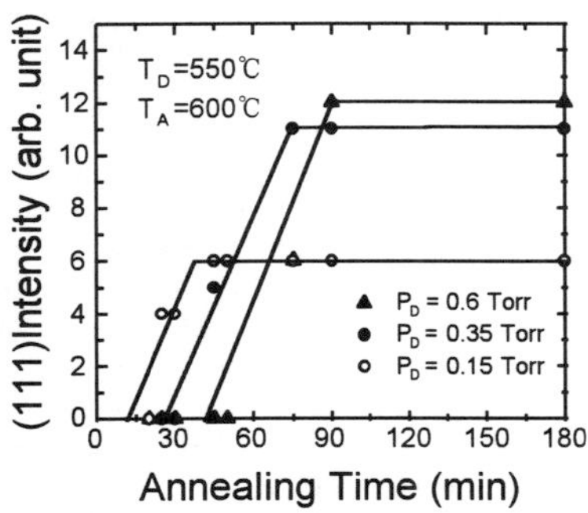

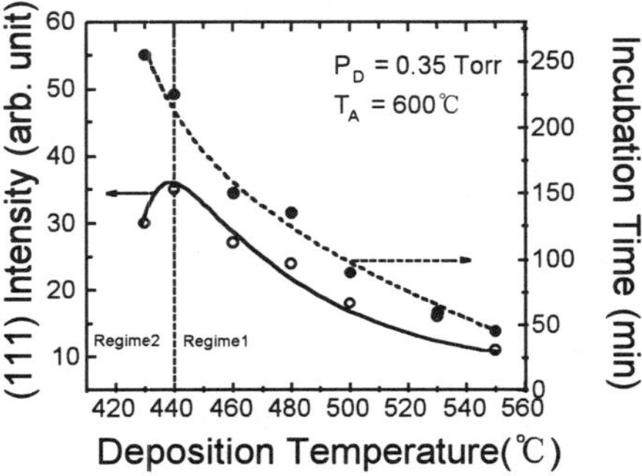

Fig. 1   (111) x-ray intensity vs. annealing time as a function of the deposition pressure

Fig. 2   (111) x-ray intensity and incubation time as a function of the deposition temperature

As a result the surface atoms on the substrate surface, or, on the growing film do not have enough time for surface diffusion as the deposition pressure increases.   During SPC, if the phase transformation occurs via preexisting supercritical crystal nuclei formed during deposition, the density of them will be decreased as the deposition pressure increases.   Fig. 2 shows (111) x-ray intensity and incubation time as a function of the deposition temperature.   A-Si films were grown at a constant total pressure of 0.35 Torr and then they were annealed at 600°C in a N_2 ambient.   As indicated in Fig. 2 the incubation time increases as the deposition temperature decreases.   (111) intensity, however, has two regimes.   The intensity increases as the deposition temperature decreases to 440°C, while it decreases as the temperature decreases below 440°C.   Generally as the deposition temperature decreases the density of prenuclei decreases since the surface diffusion rate decreases exponentially, thus resulting in the larger grain size. The deposition rate, however, also decreases exponentially as the deposition temperature decreases.   The amount of impurity incorporation such as oxygen into a growing film can be significant at the lower deposition temperature.   Although the incubation time increases at 430 °C the grain size decreases since impurities such as oxygen impede both the nucleation and the growth.

As indicated in Fig. 1 and Fig. 2, the deposition conditions greatly affect the crystallization behavior.   The effect of the initial state of as-deposited a-Si films on the crystallization kinetics, we believe, is mainly due to the density of prenuclei and the amount of impurity incorporation.   Wu et. al. reported the effect of ion implantation on the crystallization kinetics [7].   They found that the most effective grain size enhancement occurred by so called "deep silicon implantation" with the projected range beyond a-Si/SiO_2 interface to allow the maximum kinetic energy transfer at or near that interface. Their implantation condition was such that the damage peak by Si self-implantation is located at a-Si/SiO_2 interface.   We also observed the results similar to theirs.   Moreover, when we conducted 30 keV Si^+ self-implantation to 1000 Å-thick a-Si films with various doses we could not observe the grain size enhancement.   The incubation time and the grain size were not changed at all compared to those of unimplanted ones.   The implantation condition was such that the amount of damage induced in a-Si films is comparable to the case of deep implantation but the damage, or, the amount of recoiled oxygen at the a-Si/SiO_2 interface is negligible according to the TRIM-code simulations.   Therefore we may draw some significant conclusions about the dominant nucleation sites.   If nucleation occurs largely in the bulk of the films, 30 keV Si^+ self-implantation to 1000 Å-thick a-Si films should have affected the crystallization kinetics.   And the experimental results from deep implantation may exclude the top surface of a-Si films to be the dominant interface for nucleation.   The remaining possible site is only the a-Si/SiO_2 interface.

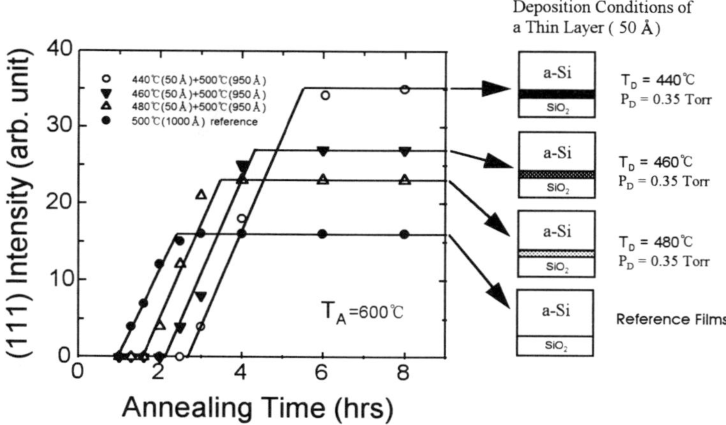

Fig. 3   (111) x-ray intensity vs. annealing time for various composite films

Based on the above argument we intentionally manipulated the nucleation interface. The reference films used in this study were 1000 Å-thick a-Si deposited at 500°C at a total pressure of 0.35 Torr. When these films were annealed at 600 °C in a N_2 ambient the grain size was ~ 1 μm. Since the deposition condition changes the incubation time, i.e. nucleation rate, and since nucleation is assumed to occur dominantly at the a-Si/SiO$_2$ interface, we devised the following deposition techniques for the first time in order to obtain the larger gain size. A very thin amorphous layer (~ 50 Å) with the deposition conditions having long incubation time is grown first and then the reference films (~ 950 Å) are grown successively. If phase transformation is governed by nucleation crystallization kinetics such as incubation time and time to complete crystallization would be limited by the deposition conditions of a thin layer.

Fig. 3 shows (111) x-ray intensity vs. annealing time for the reference films and the composite films respectively. Various 1000 Å-thick composite films were deposited and then were annealed at 600°C in a N_2 ambient. Each thin layer of 50 Å-thick a-Si in the three composite films was deposited at 440°C, 460°C and 480°C respectively at a constant pressure of 0.35 Torr. The three composite films have the same 950 Å-thick reference films successively deposited at 500°C at a constant pressure of 0.35 Torr on the top of each thin a-Si layer. As shown in Fig. 3 the incubation time of the composite films increases as the deposition temperature of the thin layer decreases. For example, the incubation time and the final grain size of the reference films are ~ 1.3 hrs. and ~ 1 μm, while those of the composite films having the thin layer deposited at 440°C are ~ 6 hrs. and ~ 2-3 μm respectively. Fig. 4 shows a series of TEM micrographs for those 4 samples. As the deposition temperature of the thin layer decreases to 440 °C the grain size increases by a factor of two or three compared to that of the reference films. These experimental results clearly show that the deposition conditions near the a-Si/SiO$_2$ interface determine the nucleation kinetics. If nucleation occurs dominantly in the bulk of a-Si films, or, on the top surface covered with the native oxide, 4 different samples would show the similar crystallization behavior. Moreover, interestingly enough, 1000 Å-thick monolayer films deposited at 440°C, 460°C and 480°C exhibit the almost exactly the same crystallization kinetics as the composite films having the 50 Å-thick thin layer deposited at 440°C, 460°C and 480°C respectively according to the XRD analyses and the TEM observations. In other words the growth kinetics is not sensitive to the deposition conditions of the rest of a-Si layer. The initial state of as-deposited a-Si films in the vicinity of the a-Si/SiO$_2$ interface is thus believed to determine the whole crystallization kinetics.

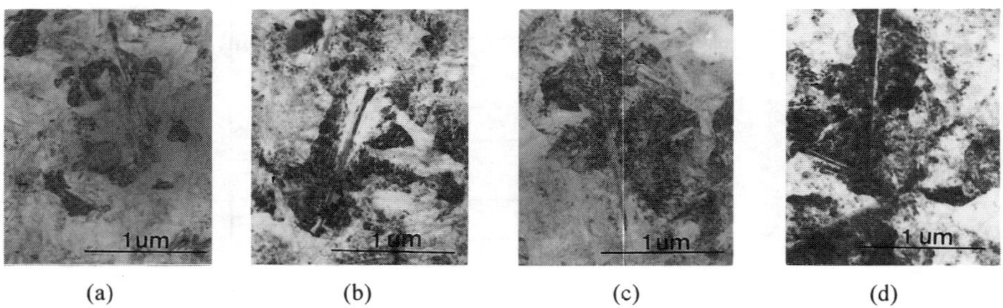

(a)　　　　(b)　　　　(c)　　　　(d)

Fig. 4　A series of TEM micrographs for various composite films　(a) Reference films (b) T_D (thin layer) = 480 ℃　(c) T_D (thin layer) = 460 ℃　(d) T_D (thin layer) = 440 ℃

We conducted another experiment for nucleation interface control. As shown in Fig. 2, the impurity incorporation such as oxygen impede both the nucleation and the growth. In the regime 2 in Fig. 2, although the incubation time increases the final grain size decreases since the impurities such as oxygen are distributed uniformly from the bottom to the top surface. If we localize the oxygen incorporation only into the thin a-Si layer on the a-Si/SiO$_2$ interface we may expect the grain size enhancement. During the deposition runs for producing the composite films shown in Fig. 3 the gas supply of Si$_2$H$_6$/He was continued when the deposition temperature was raised to 500°C from 440°C, 460°C and 480°C respectively. In another experiment for nucleation interface control, the thin layer of 50 Å-thick a-Si films was deposited at 500°C at a pressure of 0.35 Torr and then the gas supply was shut off. After around 3 min. the MFC switch is turned on and the deposition continued at the same deposition conditions as the thin a-Si layer. Fig. 5 shows (111) x-ray intensity vs. annealing time for the reference films and the films deposited by gas supply control. The incubation time and the saturation intensity of the films deposited by gas supply control increases compared to those of the reference films. The TEM micrograph in Fig. 6 shows that the final grain size greatly increases compared to that of the reference films as shown in Fig. 4-(a). Fig. 7 shows XTEM micrographs for two composite films having the 50 Å-thick a-Si layer deposited at 440°C at a pressure of 0.35 Torr. The rest of 950 Å-thick a-Si layer were the reference films deposited at 500°C at a pressure of 0.35 Torr. Fig. 7-(a) is the case for continuous gas supply and Fig. 7-(b) is the one for gas supply control by shutting off the gas flow for a few min. after the thin layer deposition. As shown in Fig. 7-(a), no trace between the thin layer and the reference films can be found after complete crystallization. However, Fig. 7-(b) indicates that 50 Å-thick a-Si films presumably containing oxygen impurities by discontinuous gas supply does not crystallize yet while the 950 Å-thick reference films are fully crystallized. In the case of the composite films by continuous gas supply, nucleation appears to occur at a-Si/SiO$_2$ interface and the nucleation rate is controlled by the deposition conditions of the thin layer. However, the thin layer containing the larger amount of oxygen by interrupted gas supply seems to have a very long incubation time while the reference films already complete the crystallization.

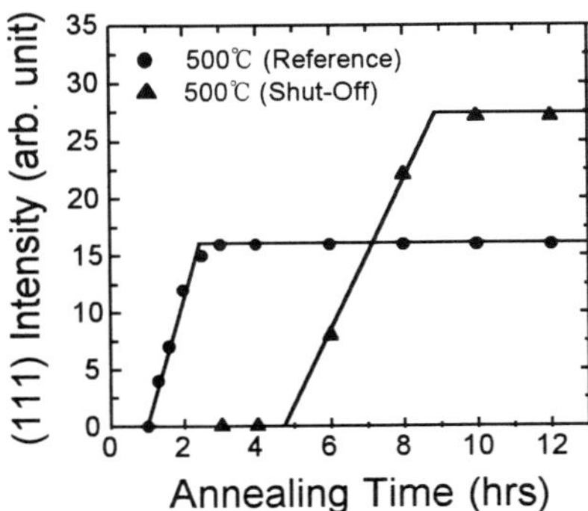

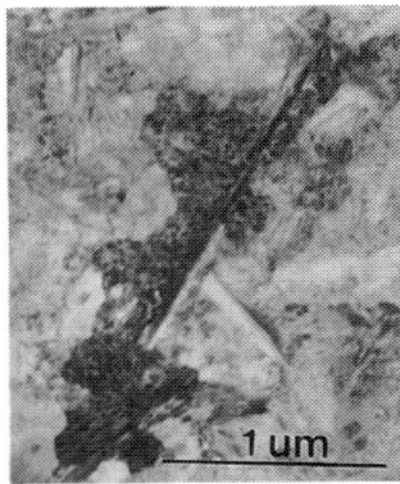

Fig. 5 (111) x-ray intensity vs. annealing time The circles represent the reference films and the triangles represent the sample by interrupted gas supply.

Fig. 6 TEM micrograph for the sample by interrupted gas supply

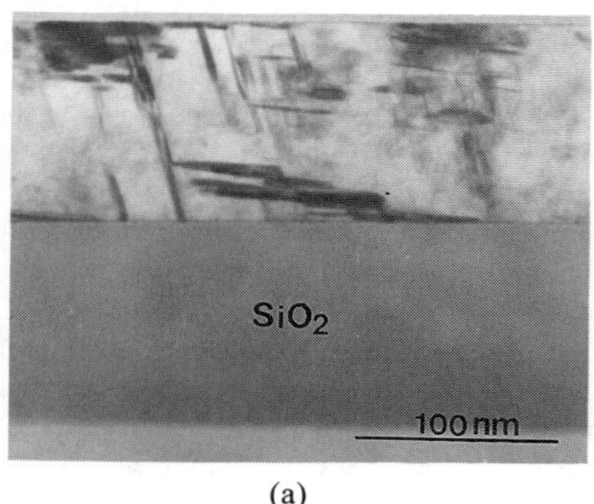

(a)

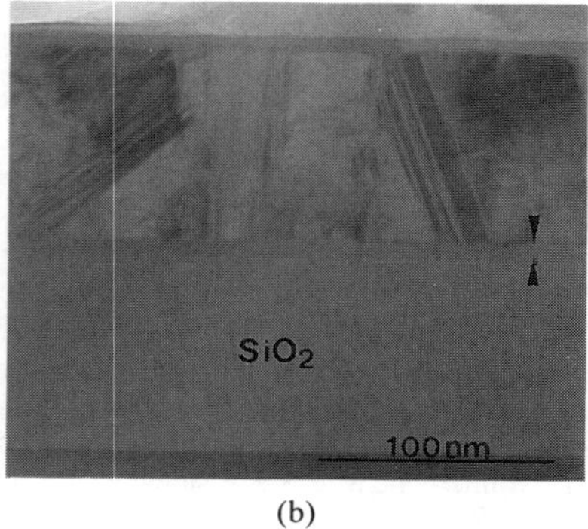

(b)

Fig. 7 XTEM micrographs for the composite films with the deposition sequence of 50 Å (T_D=440 °C) + 950 Å (T_D=500 °C) (a) Continuous Gas Supply (b) Interrupted Gas Supply

CONCLUSIONS

Through the model experiments using various nucleation interface controlled a-Si films, a-Si/SiO$_2$ interface was found to be the dominant nucleation site for crystallization. The initial state of a-Si near the a-Si/SiO$_2$ interface was observed to be critical to crystallization kinetics. The initial state of a-Si in the vicinity of the interface was artificially controlled either by introducing the thin a-Si layer with the different deposition conditions, or, by impurity incorporation into a film by interrupted gas supply. Nucleation interface controlled films fabricated by both the methods exhibited the enhancement of the grain size upon annealing.

ACKNOWLEDGMENTS

This work was supported by KOSEF (Korea Science & Engineering Foundation) under Grant 941-0800-006-2.

REFERENCES

1. T. Kamins, <u>Polysilicon for Integrated Circuit Application</u>, Kluwer, Boston, 1988

2. Apostolos T. Voutsas and Miltiadis K. Hatalis, J. Electrochem. Soc. **140**, p.871 (1993)

3. Hideya Kumomi and Takao Yonehara, J. Appl. Phys. **75**, p.2,884 (1994)

4. M.K. Hatalis and D.W. Greve, J. Appl. Phys. **63**, p.2,260 (1988)

5. Kenji Nakazawa, J. Appl. Phys. **69**, p.1,703 (1991)

6. E.P. Donovan, F. Spaepen, D. Turnbull, J.M. Poate and D.C. Jacobson, Appl. Phys. Lett. **42**, p.698 (1983)

7. I.-W. Wu, A. Chiang, M. Fuse, L. Ovecoglu and T.Y. Huang, J. Appl. Phys. **65**, p.4,036 (1989)

Part V

Dielectric and Semiconductor Interfaces

INTERFACIAL ARSENIC FROM WET OXIDATION OF $Al_xGa_{1-x}As$/GaAs: ITS EFFECTS ON ELECTRONIC PROPERTIES AND NEW APPROACHES TO MIS DEVICE FABRICATION

CAROL I.H. ASHBY, JOHN P. SULLIVAN, PAULA P. NEWCOMER, NANCY A. MISSERT, HONG Q. HOU, B.E. HAMMONS, MICHAEL J. HAFICH, ALBERT G. BACA
Sandia National Laboratories, P.O. Box 5800, Albuquerque, NM 87185-0603

ABSTRACT

Three important oxidation regimes have been identified in the temporal evolution of the wet thermal oxidation of $Al_xGa_{1-x}As$ ($1 \geq x \geq 0.90$) on GaAs: 1) oxidation of Al and Ga in the $Al_xGa_{1-x}As$ alloy to form an amorphous oxide layer, 2) oxidative formation and elimination of elemental As (both crystalline and amorphous) and of amorphous As_2O_3, and 3) crystallization of the oxide film. Residual As can result in up to a 100-fold increase in leakage current and a 30% increase in the dielectric constant and produce strong Fermi-level pinning and high leakage currents at the oxidized $Al_xGa_{1-x}As$/GaAs interface. The presence of thermodynamically-favored interfacial As may impose a fundamental limitation on the application of AlGaAs wet oxidation for achieving MIS devices in the GaAs material system.

INTRODUCTION

There is a continuing quest to find a good insulator for metal-insulator-semiconductor (MIS) devices to enable the GaAs-equivalent of Si CMOS technology. Earlier approaches involving dry thermal oxides and anodic oxides have not yielded interfaces of sufficient quality (interface-state densities in the mid-10^{10} /cm^2-eV) for MIS applications. A relatively new approach to forming a technologically useful oxide involving oxidation of high-Al-content AlGaAs layers at elevated temperature during exposure to wet nitrogen [1] has shown promise for forming readily manufacturable stable oxides that might prove suitable for MIS applications [2]. We have studied the temporal evolution of AlGaAs films undergoing wet oxidation using Raman spectroscopy to identify product species that are present during the intermediate and final stages of the oxidation and have determined electrical properties of the Al-oxide/GaAs materials system using DC transport and capacitance measurements. These studies reveal a fundamental limitation on the application of AlGaAs wet oxidation for achieving MIS devices in the GaAs material system.

EXPERIMENT

Samples with the following structure were grown by MOCVD: 300 Å GaAs cap/ variable thickness $Al_xGa_{1-x}As$/1000 Å GaAs/GaAs substrate with x= 1.0, 0.98, and 0.90. Raman studies employed 0.25- and 2-μm $Al_xGa_{1-x}As$ layers, while diffraction (X-ray and electron) and bulk electrical measurements were performed on 0.25-μm $Al_xGa_{1-x}As$ samples. MIS diode samples consisted of 300-500 Å of $Al_xGa_{1-x}As$ on 2-5x10^{16}/cm^3 n-GaAs.

Prior to wet oxidation , the GaAs cap was removed by a selective etch in a citric acid/peroxide mix (5:1 of [1g citric monohydrate/1g H_2O]:30% H_2O_2) and the sample was loaded into the furnace under a dry nitrogen purge. The sample was then heated under flowing nitrogen. When the desired reaction temperature was reached (400<T<455 °C), the nitrogen flow (typically 0.4 slm in a 2-in. diam. tube) was switched to bubble through water at a temperature of

Mat. Res. Soc. Symp. Proc. Vol. 448 © 1997 Materials Research Society

80 ± 1 °C for a timed reaction. Dry nitrogen flow was restored at the completion of the reaction time and continued during sample removal from the furnace.

Raman spectra were measured using an excitation wavelength of 514.5 nm. The 500-mW probe beam was line focused to apply less than 85 W/cm^2 on the sample. Scattering was measured in the $x(y',y'+z')x^-$ backscattering configuration with y' and z' parallel to (110) planes. A triple monochromator with an internal depolarizer and a CCD array detector were employed. Reference Raman spectra were obtained for α-Al_2O_3 (99.997%, 110 μm crystalline, 9 m^2/g surface area) and γ-Al_2O_3 (99.999%, 110 μm powder, 55 m^2/g surface area). Peak positions in cm^{-1} and relative intensities for α-Al_2O_3 were as follows: 379 (0.34), 417(1), 428(0.28), 458(0.06), 576(0.11), 644(0.24), 750(0.23). X-ray measurements on γ-Al_2O_3 showed a broad amorphous background with broad peaks; using the Scherrer formula gives a crystallite size of 70 Å. Weak Raman peaks from γ-Al_2O_3 were observed at 385, 468, 624, 644, 750 cm^{-1}. This small crystallite size might cause red-shifting and asymmetric broadening of the peaks [3]; low signal intensity prevented evaluation of this possible effect.

X-ray diffraction measurements were performed on the films using Cu K-alpha radiation from a fixed anode source and a curved graphite crystal detector. Conventional Θ-2Θ scans were taken between 10° and 70°, corresponding to lattice spacings of 8.8 Å-1.34 Å. Cross-section samples for high resolution transmission electron microscopy were prepared by mechanical thinning and ion milling with 5 kV Ar^+ and examined at 200 keV using a JEOL 2010 at the University of New Mexico. Samples for electrical characterization were prepared by shadow mask deposition of Ti/Au circular contacts (200-1000 μm) on the oxide surface followed by deposition of a non-alloyed ohmic backside contact. For bulk dielectric measurements, 0.25 μm thick oxidized $Al_xGa_{1-x}As$ layers on top of near-degenerately n-doped GaAs were used in order to minimize the interfacial depletion layer capacitance. For interface electrical characterization, thinner $Al_xGa_{1-x}As$ layers were measured on $2-5 \times 10^{16}$ cm^{-3} n-doped GaAs buffer layers. Capacitance-voltage measurements (Fig. 2) were performed over the frequency range 100 kHz to 1 MHz, typically 1 MHz, at a sweep rate of 150 mV/min. DC transport measurements were performed over the range -5 V to 5 V under dark conditions.

RESULTS AND DISCUSSION

Progressive oxidation of the AlGaAs layer reduces the phonon peak intensity at 400 cm^{-1}, as shown in Fig. 1 for 2-μm AlGaAs samples. Substrate GaAs is manifest at 292 cm^{-1} due to the low absorption coefficient of indirect-gap $Al_{0.98}Ga_{0.02}As$ at 514.5 nm. The Raman spectrum of a partially oxidized film is dominated by crystalline $As°$ peaks at 198 and 257 cm^{-1} and broad feature between 200 and 250 cm^{-1} peaking near 227 cm^{-1} ascribed to amorphous $As°$ [4]. The $As°$ peaks are very strong due to resonant enhancement of the Raman scattering at this excitation energy; the detection limit is on the order of 10-20-Å layers [5]. The broad feature centered at 475 cm^{-1} is due to amorphous As_2O_3 [5]. Specific peaks associated with amorphous Al-oxides are not observed. Gallium oxide films of 8000-Å thickness are not detectable by Raman because of low cross section [5], and the less polarizable Al-oxide films should be even weaker scatterers. Since the relative intensities of the various peaks cannot be construed as a quantitative measure of the different compositions due to large and quantitatively unknown differences in their scattering cross sections, the following discussion is qualitative.

As the oxidation front advances into films of either AlAs or high-Al-fraction AlGaAs, there is established a relatively constant Raman signal from As and a-As_2O_3 while the intensity of the

AlAs-like phonon decreases (Fig 1a-c.) Growth of the amorphous Al-oxide film is manifested by the gradually rising baseline. As oxidation of the $Al_xGa_{1-x}As$ layer nears completion, the AlAs-like phonon peak drops below our detection limit. The a-As_2O_3 peak intensity also drops below the detection limit. There remains barely detectable scattering from As near the oxide/GaAs interface and a steeply rising baseline (Fig. 1d). With continued exposure to the oxidizing atmosphere or with annealing in vacuum or dry nitrogen, crystallization of the films occurs (Fig. 1e), with most of the strong peaks attributable to Rayleigh scattering of laser plasma or background light or to GaAs-associated features typically observed on rough GaAs, where polarization selection rules that are operative with specular surfaces break down. Weak peaks at 750 and 644 cm^{-1} ascribed to γ-Al_2O_3 are observed in crystallized films. Delamination of the 2.0-μm films routinely occurs after complete oxidation, suggesting relatively weak adhesion between the oxide and semiconductor. This is consistent with the relative fragility of laterally oxidized optoelectronic structures when subjected to cleaving operations.

We propose the following reaction scenario for Al(Ga)As films undergoing wet oxidation. Al(Ga)As undergoes reaction with water, hydroxyl, or other oxygen-containing species derived from water to form a mixture of aluminum(gallium) oxides and/or hydroxides and amorphous As_2O_3, which forms a glassy matrix and prevents wide-scale crystallization. Elemental As (As^0) is present at the interface between the oxide and the semiconductor, and a relatively constant amount is observed throughout most of the oxidation. This amount is determined by the kinetic balance between its formation and subsequent loss through further oxidation or volatilization and escape through the permeable oxide film; this balance will vary with specific reaction conditions. TEM studies of laterally wet-oxidized AlGaAs layers have shown the oxide layer to be far from fully dense and should have adequate permeability for ready escape of volatile reaction products [6]. Both As (P(vap) = 760 Torr at 407 °C) and a-As_2O_3 (P(vap) = 760 Torr at 320 °C) may serve as volatile products for removal of As from the growing oxide film. It has been postulated that arsine is the major product of AlGaAs wet oxidation. We cannot rigorously exclude a role for arsine as a minor product, since our furnace was not equipped for identification of volatile products. However, our observation of two volatile As-species in more highly oxidized states than the original AlGaAs suggests that formation of major quantities of the lower oxidation state arsine as the volatile reaction product is unnecessary for As removal from the oxidized film.

Figure3 shows high-resolution transmission electron micrographs (TEM) of the interfacial region between GaAs and $Al_{0.90}Ga_{0.10}As$ as the reaction approaches completion. Moiré fringes due to a 2-4 nm crystallite at the interface and lattice fringes associated with unreacted $Al_{0.90}Ga_{0.10}As$ are observed near the oxide/GaAs interface. Selected area diffraction taken within the oxidized $Al_{0.90}Ga_{0.10}As$ layer show diffuse polycrystalline rings with d-spacings corresponding to the prominent reflections of γ-Al_2O_3. X-ray diffraction measurements on similar films reveal only a diffuse background with no apparent crystalline Al_2O_3 phases, thus putting an upper limit on crystallite size of ~ 10 nm. The lower magnification image of Fig. 3 reveals that the interfaces are not abrupt, but exhibit interfacial roughness on the order of 2-4 nm.

The temporal evolution of the structure of these surface-oxidized films correlates with that observed along the length of a laterally-oxidized buried layer[6]. At the oxide/semiconductor interface of a laterally oxidized layer, an amorphous meniscus is observed. Farther back from the oxidation front the oxidized layer consists of increasingly crystalline γ-Al_2O_3. Since increasing distance from the oxidation front in a laterally oxidized sample correlates with increasing oxidation time, the evolution from amorphous to crystalline structures in both laterally and surface-oxidized samples is probably due to progressive loss of the stabilizing glassy matrix provided by a-As_2O_3 or to conversion of amorphous ρ-Al_2O_3 to crystalline γ-Al_2O_3.

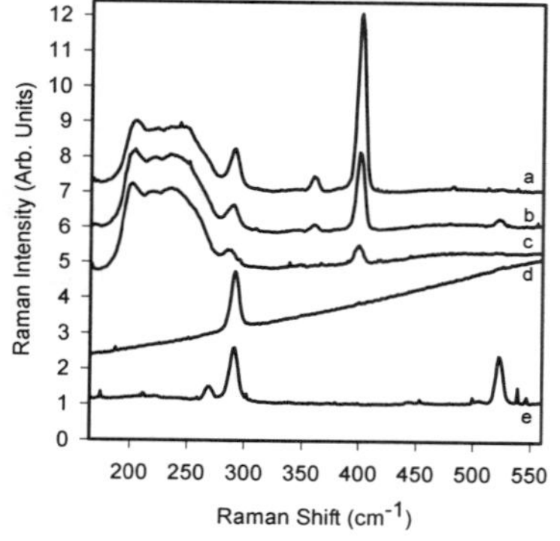

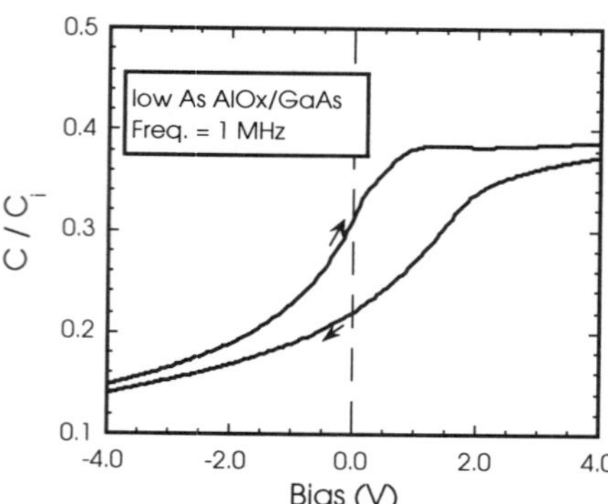

Fig. 1: Raman spectra of progressively oxidized 2.0-μm layer of $Al_{0.98}Ga_{0.02}As$ on GaAs: a-c) partially oxidized; d) oxidized "to completion"; e) layer crystallized and delaminating.

Fig. 2: Capacitance-voltage curve for MIS diode with 300-Å of oxidized AlAs on 2×10^{16} n-GaAs.

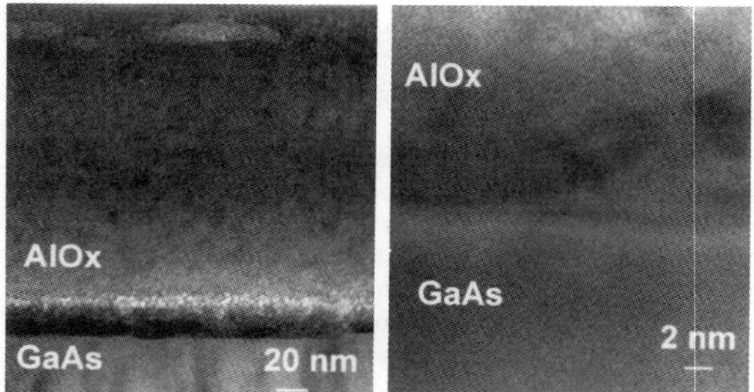

Fig. 3: High-resolution transmission electron micrograph of interfacial region of oxidized $Al_{0.90}Ga_{0.10}As$ as oxidation front approaches GaAs. Arsenic precipitates (2-4 nm) present at interface.

The results of capacitance and DC transport measurements on 0.25-μm wet-oxidized films as a function of residual As and Ga content are shown in Table 1. Samples labeled as "high" arsenic exhibit As^0 Raman signals on the order of those shown in Fig. 1 while "low" arsenic samples exhibit As^0 signals below the Raman detection limit. Residual As^0 can result in up to a two order of magnitude increase in leakage current and up to a 30% increase in the dielectric constant. The increase in dielectric constant with increasing As concentration may be partly due to dielectric heterogeneity associated with metallic As^0 precipitates in the insulating aluminum oxide matrix. With the assumption that this represents the entire increase and a Maxwell-Wagner form for the effective dielectric constant [7], the required volume fraction of elemental As would be approx. 6% for the high As content oxidized AlAs layer. The increase in bulk dielectric constant with increasing Ga content is likely due to increased bond polarizability associated with the Ga-O bonds. The dielectric constant of deposited Ga_2O_3 films has been reported to be up to 14.2 [8].

Table 1: Bulk Electrical Properties of Wet-Oxidized AlGaAs Films

Starting Composition	Arsenic	Leakage (A/cm^2 @ 5V)	Bulk Resistivity (Ohm-cm)	Bulk Dielectric Constant (1MHz)
AlAs	Low	7.5×10^{-8}	2.7×10^{12}	5.8
	High	7.5×10^{-7}	2.7×10^{11}	7.0
$Al_{0.98}Ga_{0.02}As$	Low	2.9×10^{-9}	6.9×10^{13}	6.2
	High	2.5×10^{-7}	8.0×10^{11}	8.3
$Al_{0.90}Ga_{0.10}As$	Low	3.7×10^{-9}	5.4×10^{13}	8.6
	High	1.0×10^{-7}	1.2×10^{11}	8.2

Capacitance-voltage (CV) measurements on thin oxidized $Al_xGa_{1-x}As$ layers (Fig.2) with residual As^0 at levels below the Raman detection limit ($\approx$ 10-20-Å layers) [5] reveal partial Fermi-level pinning while measurements on layers with higher As^0 content exhibit strong Fermi-level pinning and no CV modulation. The presence of weak CV modulation and failure to reach the full accumulation capacitance in the low-As^0 samples would be consistent with the existence of an interface-state band located approximately 0.3 to 0.6 eV below the GaAs conduction band (CB) with a state density exceeding 10^{12} cm^{-2} and deep depletion at inversion biases. Given the ubiquitous presence of As^0 as detected near the interface by Raman in a large number of the samples, it is likely that local pinning of the Fermi-level at As^0 precipitates at the oxide/GaAs interface may be the source of the apparent high interface state density. The Schottky barrier height observed for the As^0/n-GaAs interface is 0.67 - 0.72 eV [9], which would appear as an apparent interface-state band below the level found here. However, if the As precipitates do not fully cover the oxide/GaAs interface, a situation may arise in which the Fermi-level at the interface may only be locally pinned and unpinned at other areas. The observed pinning positions of 0.3 to 0.6 eV below the GaAs CB may, therefore, alternatively be explained by non-uniform Fermi-level pinning at As crystallites at the interface (e.g., the formation of local barriers) with an areal coverage of As at the interface from about 83 - 98%. This non-uniform pinning of the Fermi-level at the oxide/GaAs interface would prevent the formation of a spatially uniform accumulation layer near the oxide interface and could explain the poor observed transconductance in field effect transistor (FET) devices employing oxidized $Al_xGa_{1-x}As$ gates.[2].

The continuing presence of residual As^0 at the oxide/GaAs interface even after lengthy oxidation is to be expected on thermodynamic considerations [10] when As_2O_3 is an oxidation product, as we observe. Even if the As^0 formed from Al(Ga)As oxidation were totally removed, further generation of interfacial As^0 from oxidation of GaAs could proceed from the reaction

$$2\,GaAs + As_2O_3 = 4\,As + \beta\text{-}Ga_2O_3 \quad \Delta G^0_f = -64\ kcal,\ \Delta H^0_f = -62\ kcal \quad (1a).$$

Since interfacial As^0 even at the 10^{-4} monolayer level will degrade MIS performance, this poses a serious problem for achieving a viable GaAs MIS technology based on wet oxidation.

SUMMARY

Elemental As^0 and a-As_2O_3 form during AlGaAs wet oxidation. Their volatility is sufficient for them to serve as the primary species responsible for arsenic removal from the relatively porous oxidized film. Elemental As^0 is concentrated at the oxide /semiconductor interface in both crystalline and amorphous forms, with crystalline As^0 requiring longer reaction time for removal than amorphous As^0 when the oxidation front reaches the substrate GaAs. Thick (2-μm) Al-oxide films crystallize and delaminate after As^0 and a-As_2O_3 are removed by continued oxidation and/or heating, suggesting relatively weak adhesion that may be responsible for the fragility of some laterally oxidized optoelectronic structures. Both capacitance-voltage and dc transport behavior are degraded by the interfacial As^0. Non-uniform pinning of the Fermi-level at the oxide/GaAs interface by As precipitates may explain the strong Fermi-level pinning and have serious negative implications for FET devices. Good MIS devices will require As^0 removal without film crystallization. However, the fundamental limit imposed by the thermodynamically favored formation of elemental As at the interface makes this a formidable challenge.

ACKNOWLEDGEMENTS

This work was performed at Sandia National Laboratories and supported by the U.S. Department of Energy under Contract No. DE-AC04-94AL85000. Sandia is a multiprogram laboratory operated by Sandia Corporation , a Lockheed Martin Company, for the United States Department of Energy. The authors wish to thank Kent Geib and Denise Tibbetts-Russell for technical assistance.

REFERENCES

1. J.M. Dallesasse, N. Holonyak, Jr., A.R. Sugg, T.A. Richard, and N. El-Zein, Appl. Phys. Lett, **57**, 22844 (1990).
2. E.I.Chen, N.Holonyak, Jr., and S.A Maranowski, Appl. Phys. Lett. **66**, 2688, (1995).
3. H. Richter, Z.P. Wang, L. Ley, Solid State Comm. **39**, 625 (1981).
4. G.P. Schwartz, B. Eschwartz, D. DiStefano, G.J. Gualtieri, and J.E. Griffiths, Appl. Phys. Lett. **34**, 205 (1979).
5. G.P. Schwartz, G.J. Gualtieri, J.E. Griffiths, C.D. Thurmond, B. Schwartz, J. Electrochem. Soc. **127**, 2488 (1980).
6. R.D. Twesten D.M. Follstaedt, K.D. Choquette, and R.P. Schneider, Jr. Appl. Phys. Lett. **69**, 19 (1996).
7. B. K. P. Scaife, Principles of Dielectrics (Clarendon Press, Oxford, 1989.)
8. M. Passlack, M. Hong, J. P. Mannaerts, Appl. Phys. Lett. **68**, 1099 (1996).
9. J.M. Woodall, P.D. Kirchner, J.F. Freeouf, D.T. McInturff, M.R. Melloch, F.H. Pollak, Phil. Trans. Roy. Soc. Lond. A **344**, 521 (1993).
10. C.D. Thurmond, G.P. Schwartz, G.W. Kammlott, and B. Schwartz, J. Electrochem Soc. **127**, 1366 (1980).

MICROSTRUCTURE AND INTERFACIAL PROPERTIES OF LATERALLY OXIDIZED $Al_xGa_{1-x}As$

R.D. TWESTEN, D. M. FOLLSTAEDT, AND K. D. CHOQUETTE
Sandia National Laboratories, Albuquerque, NM. 87185-1056 rdtwest@sandia.gov

ABSTRACT

The oxidation of high Al content $Al_xGa_{1-x}As$ has received much attention due to its use in oxide-aperture, vertical-cavity surface emitting lasers (VCSELs) and for passivating AlAs against environmental degradation. We have recently identified the spinel, gamma phase of Al_2O_3 in layers laterally oxidized in steam at 450°C for x=0.98 & 0.92 and have seen evidence for an amorphous precursor to the gamma phase. At the interface with the unoxidized $Al_xGa_{1-x}As$, an ~17nm amorphous phase remains which could account for the excellent electrical properties of oxide-confined VCSELs and help reduce stress concentrations at the oxide terminus.

INTRODUCTION

It has long been known that AlAs is subject to environmental degradation due to its high reactivity. It was discovered that intentional oxidation of AlAs can passivate the surface against further oxidation[1] and that varying the Ga content of the film will drastically reduce the oxidation rate for this process[2]. This selective and passivating nature of oxidized AlGaAs has been incorporated into vertical-cavity surface emitting lasers (VCSELs) to form current-confining and optical-mode defining apertures, which have allowed the development of high-performance devices[3,4,5].

In this paper, we discuss the microstructure of the oxide phase formed from the wet oxidation of 2%- and 8%-Ga AlAs in actual working VCSEL device structures (Figure 1). It is found that the oxide phase formed is fine grained (~4nm) γ-Al_2O_3 surrounded by an amorphous matrix. This is consistent with early work of Sugg et al.[6] and the crystalline phase identified in wet oxidation of pure AlAs[7]. It is also found that the lateral oxidation front terminates with an ~17nm amorphous zone. Lattice images show that the oxide terminus has a smooth transition from crystalline to amorphous, and lower magnification, strain-contrast images reveal no defects associated with the oxidation front. This lack of defects occurs in spite of the fact that measured contraction of the oxidized layers is 6.3%, straining the unoxidized layers[8].

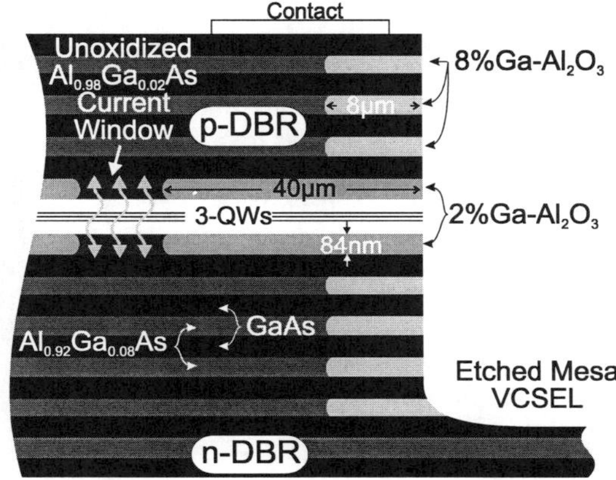

Figure 1: Schematic diagram of an oxide-confined, 980nm VCSEL device. Not to scale.

Mat. Res. Soc. Symp. Proc. Vol. 448 © 1997 Materials Research Society

EXPERIMENTAL

Layers containing 2%- and 8%-Ga AlAs were laterally oxidized in a steam furnace at 450°C by bubbling N_2 through 80°C H_2O. The layers were part of working 980nm VCSEL devices. TEM cross-sectional samples were made by dicing the devices out of the wafer and epoxying them in a Si support stack. The stack was mechanically polished and then Ar-ion thinned so that the unoxidized current aperture of the device was thinned to electron transparency (see Figure 1). The final sample contained the oxide terminus of the 2%-Ga AlAs layers that define the current aperture in addition to the 8%-Ga AlAs layers that formed the remainder of the distributed Bragg reflector mirrors. The 8%-Ga layers only oxidize 8μm inward as compared to the 40μm oxidation of the 2%-Ga layers, due to the strong dependence of oxidation rate on Ga content[2]. TEM images were obtained using 200keV electrons and recorded either directly on to film or using a cooled, slow-scan video camera.

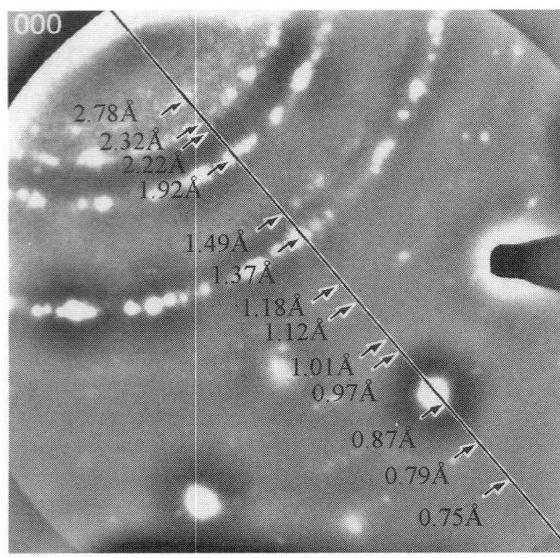

Figure 2: A portion of a selected area diffraction pattern of oxidized 2%Ga-AlAs. The image has been filtered to show detail. The sharp spots are from the unoxidized GaAs layers, while the rings are from the oxidized layer. The d-spacing for each ring is indicated.

RESULTS

Figure 2 shows a portion of an electron diffraction pattern which has been high-pass filtered to show the detail of the pattern. Clear polycrystalline rings can be seen in the pattern as well as an array of spots from the surrounding GaAs layers, which act as an internal calibration for the pattern. By measuring the ring spacing, the polycrystalline phase is identified as γ-Al_2O_3[9]. This is the cubic phase of Al_2O_3 and is based on the spinel ($MgAl_2O_4$) prototype with 32 O-atoms forming an FCC sublattice, 8 Al-atoms on the tetrahedral sites, and the remaining $13\frac{1}{3}$ Al-atoms on the octahedral sites leaving $6\frac{2}{3}$ octahedral vacancies[10]. The degree of ordering of

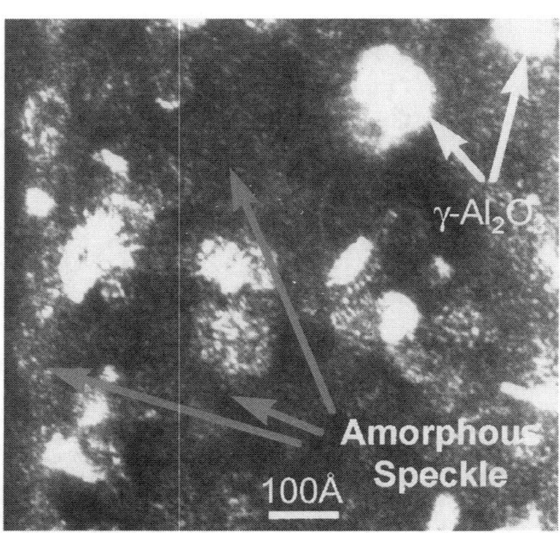

Figure 3: Dark-field g=(311) γ-Al_2O_3 image of oxidized 8%-Ga AlAs. The bright grains are γ-Al_2O_3 while the fine speckle between the grains is an amorphous phase.

the octahedral Al-atoms distinguishes the various forms of the cubic, transitional aluminas that we group together as the gamma phase. The gamma phase is formed by dehydration of $Al(OH)_3$ and $AlO(OH)$. Upon high temperature processing, the gamma phase transforms into the hexagonal, alpha phase (sapphire).

At low concentrations, Ga_2O_3 forms solid solutions with Al_2O_3[11] and Ga_2O_3 has corresponding cubic and hexagonal phases. It is expected, therefore, that the Ga in the layers simply forms a solid solution of Ga_2O_3/Al_2O_3. This is supported by energy-dispersive x-ray spectroscopy (EDXS) that shows the Ga retained in the layer (see below) and the lack of any additional Ga-related phases in the diffraction pattern or images.

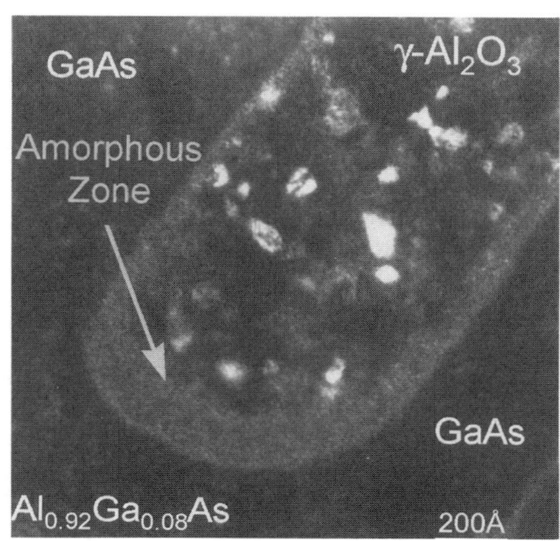

Figure 4: Dark-field image of the oxide terminus. Between the polycrystalline γ-Al_2O_3 and the unoxidized $Al_xGa_{1-x}As$, there is an ~17nm thick amorphous zone.

While the diffraction patterns show that the gamma phase is the only oxide present with long-range order, dark-field images suggest an amorphous matrix surrounding the crystalline phase. Figure 3 shows a dark-field image obtained using a portion of the (311) ring of the γ-Al_2O_3. The image shows the fine grain nature of the oxide with an average diameter of ~4nm. The orientation of the grains appears random, indicating precipitation from a random precursor rather directly from the AlGaAs lattice. In addition to the crystalline γ-Al_2O_3 grains, a fine speckle can be seen between the grains, which is typical of dark-field images of amorphous material. This amorphous phase could be an uncrystallized hydroxide, $Al(OH)_3$, or the amorphous ρ-alumina phase which forms in the vacuum dehydration of $Al(OH)_3$. It is likely that this amorphous phase forms as a precursor to the γ-Al_2O_3.

The suggestion that an amorphous oxide phase acts as a precursor to the gamma phase is also supported by the

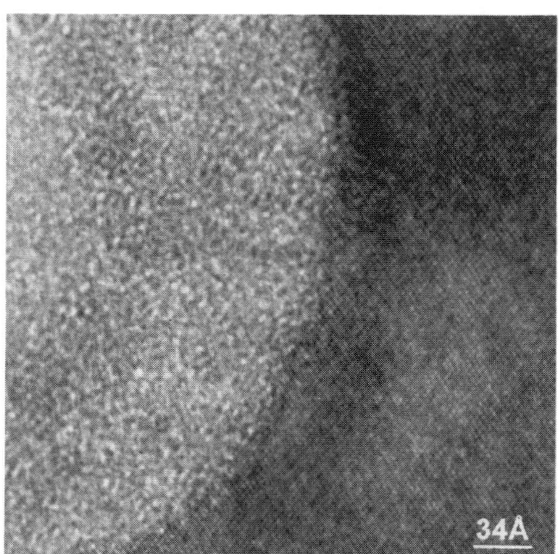

Figure 5: [110] Lattice image of the of the oxide terminus. The smooth transition from single crystal $Al_xGa_{1-x}As$ to amorphous oxide is seen. The transition from amorphous to polycrystalline oxide is seen on the left.

existence of an amorphous interface region at the oxide terminus. Figure 4 shows a dark-field image of the oxide terminus of an 8%-Ga AlAs layer. The polycrystalline γ-Al_2O_3 can be seen in addition to the amorphous matrix. Also, at the interface with the unoxidized crystal, a 17nm-thick amorphous band can be seen. Bright-fields images also show granular amorphous contrast in this zone (see Figure 5). The tapering of this amorphous layer near the GaAs interfaces is a result of the intentional grading of the group III content at the $Al_xGa_{1-x}As$ /GaAs interfaces in the DBRs, which lowers the series resistance of the mirror stacks. Since oxidation rate is strongly dependent on the Al content, the grading of the interfaces causes the oxidation rate to be reduced near the interfaces.

Figure 6: Bright-field strain contrast image of the oxidized portion of the DBR mirror stack. Some strain associated with the oxidation front is seen, but it does not propagate into the rest of the mirror. Also, no extended defects are seen emanating from the oxidation front.

The lattice image in Figure 5 shows a smooth transition from the unoxidized crystal to the oxidized amorphous zone. The exact position of the interface cannot be determined because the oxidation front is not exactly aligned with the crystal axes. However, it can be seen that the interface is smoothly varying without any sharp spikes into the crystal. In lower magnification strain contrast images (Figure 6), strain can be seen around the oxide fronts. However, the strain is confined to approximately one DBR period within the mirrors, and no extended defects have been seen emanating from the oxide front.

To form a fully dense γ-Al_2O_3 layer from AlAs would require a 20% linear contraction of the layer[12]. However, internal strains and a possible amorphous matrix would reduce this requirement. We have directly measured this contraction by comparing oxidized and unoxidized portions of a DBR mirror[8]. By measuring the change in position of the 18th mirror pair, the linear contraction is determined as 6.3% for the 8%-Ga Al_2O_3. The contraction occurs over an approximately 2μm-wide transition region behind the oxidation front. Thus, the region nearest the oxidation front is under a higher tensile strain than the rest of the DBR. While some relaxation is expected to occur during TEM sample preparation, the contraction of the oxidized layers is seen prior to sample preparation on the surface of the VCSEL devices using differential-interference contrast optical microscopy. The amount of contraction in the single, 2%-Ga Al_2O_3 layer was not measured due to the difficulty in assigning the exact position of the $Al_{0.98}Ga_{0.02}As$/GaAs interfaces, which were intentionally graded to lower the resistivity of the DBR mirrors. Also, the amount of contraction is expected to be less due to the rigidity of the surrounding, unoxidized material.

As stated earlier, the Al-Ga-O system is known to form solid solutions of Al_2O_3 /Ga_2O_3. Hence, it is expected that the Ga will remain in the in the oxidized layer $(Al_xGa_{1-x})_2O_3$. The

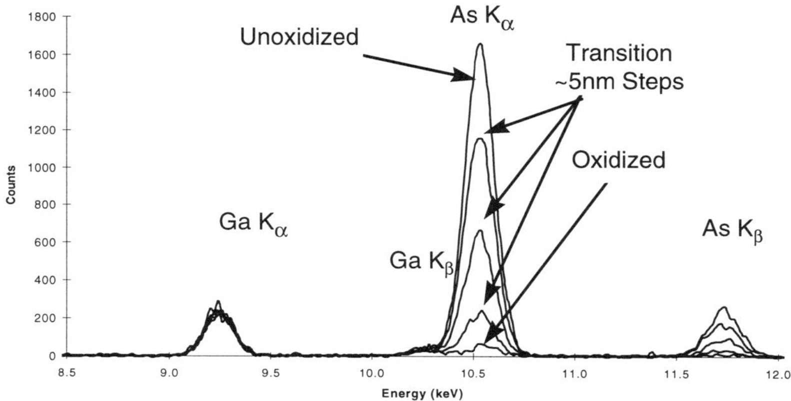

Figure 7: Portion of EDXS scan from the region around the oxide terminus. The Ga signal remains constant (as does the Al signal, not shown) while the As signal smoothly decreases to almost zero in the oxide.

disposition of the As is not as clear. It has been reported that AsH$_3$ is detected in the waste stream during bulk oxidation of AlAs[13], but during lateral oxidation, it is not clear that the As can escape. Using EDXS, we have measured the constituents in the region of the oxide terminus. The Al and Ga signal remained constant; however, the As signal was reduced to less than 2% in the oxidized portion of the layer as seen in Figure 7. In the transition from oxidized to unoxidized, we see a smooth decrease in the As signal, indicating the As is not enriched at the oxidation front. However, the probe size used (~10nm) is not small enough to resolve any fine structure at the reaction front.

We had recently reported that there were amorphous inclusions associated with the rapid oxidation of the 2%-Ga AlAs layers[8]. We have determined that these inclusions are, in fact, electron-beam-induced artifacts. They probably represent local, e-beam-induced crystallization of the amorphous matrix, causing a local contraction of the layer. This local contraction then leaves behind the low density pockets observed. This is consistent with the amorphous matrix being a lower density hydroxide phase.

CONCLUSIONS

While oxide confined VCSELs have recently demonstrated record device performances[3,4,5], the microstructure of the oxide layer defining the aperture has not been fully studied. We have performed TEM studies of the microstructure of the oxidized layers in VCSEL device structures and have begun to obtain a more complete understanding.

The crystalline phase observed in the oxidized layers is γ-Al$_2$O$_3$, which is stable against further oxidation. In addition to the crystalline γ-Al$_2$O$_3$, an amorphous matrix is present in the layers. This matrix is presumably a hydroxide or the amorphous rho-phase of alumina, and probably acts as a precursor to the γ-Al$_2$O$_3$ crystallites. This is supported by images of the oxide terminus, which show a purely amorphous zone between the oxidized and unoxidized portion of the layers. This zone may act to passivate dangling bonds at the interface and help account for the high efficiencies of these devices.

The $Al_{0.92}Ga_{0.08}As$ layers contracted linearly by 6.3% after oxidation, but no extended defects were observed eminating from the oxidation front. This contraction of the oxide may have the biggest impact in high index-contrast DBR mirror applactions utilizing GaAs/oxidized-AlAs layer pairs. Since the reflectivity band of these mirrors depends strongly on the layer thickness, a contraction of the oxidized layers would affect these mirrors. The oxide contraction does not seem to be a large problem in the present VCSEL application since the oxidized portion of the device is not dimensionally critical to its performance and the strain associated with the contraction does not generate defects in the device. This may not be the case when using pure AlAs rather than 2%Ga-AlAs as the current confining layer, since these devices are exceptionally vulnerable to mechanical failure[14].

ACKNOWLEDGEMANTS

The authors wish to gratefully acknowledge the support of the United States Department of Energy under contract DE-AC02-94AL85000.

REFERENCES

[1] J. M. Dallessasse, N. Holonyak, Jr., A. R. Sugg, T.A. Richard and N. El-Zein, Appl. Phys. Lett. **57**, 2844 (1990).

[2] K. D. Choquette, K. L. Lear, R. P. Schneider, Jr., K. M. Geib, J. J. Figiel, and R. Hull, IEEE Photonics Technol. Lett. **7**, 1237 (1995)..

[3] D.L. Huffaker, D.G. Deppe, K. Kumar, and T.J. Rogers, Appl. Phys. Lett. **65**, 97 (1994).

[4] K. D. Choquette, R. P. Schneider, Jr., K. L. Lear, and K. M. Geib, Electron Lett. **30**, 2043 (1994).

[5] G.M. Yang, M.H. MacDougal, P.D. Dapkus, Electron Lett. **31**, 886 (1995).

[6] A. R. Sugg, E.I. Chen, N. Holonyak, K.C. Hsieh, J.E. Baker and N. Finnegan, J. Appl. Phys. **74** 3880 (1993).

[7] S. Guha, F. Agahi, B. Pezeshki, J.A. Kash, D.W. Kisker and N.A. Bojarczuk, Appl. Phys. Lett. **68**, 906 (1996).

[8] R.D. Twesten, D.M. Follstaedt, K. D. Choquette, and R. P. Schneider, Jr., Appl. Phys. Lett. **69**, 19 (1996).

[9] JCPDS file 10-425.

[10] B.C. Lippens and J.J. Steggerda in *Physical and Chemical Aspects of Adsorbents and Catalysts,* edited by B.G. Linsen (Academic Press, London, 1970) p.171-213.

[11] E.M. Levin, C.R. Robbins, and H.F. McMurdie, *Phase Diagrams for Ceramists,* (American Ceramic Society, Columbus, OH, 1970) Figure 310.

[12] The volume per Al atom in AlAs is $(3.57Å)^3$, while in γ-Al_2O_3 it is $(2.85Å)^3$ (for gibbsite, $Al(OH)_3$, it is $(3.49Å)^3$).

[13] H. Holonyak, private communication.

[14] K. D. Choquette, K. M. Geib, H.C. Chui. H.Q. Hou, and R. Hull, Mat. Res. Soc. Symp. Proc. **412** (1996) 53.

NITRIDATION OF Si(111)-7x7 SURFACE BY LOW ENERGY NITROGEN IONS : STM INVESTIGATION

JEONG SOOK HA*, KANG-HO PARK*, WAN SOO YUN*, EL-HANG LEE *, SEONG-JU PARK**
*Research Department, Electronics and Telecommunications Research Institute, Taejon, 305-600, Korea
**Department of Materials Science and Engineering, Kwangju Institute of Science and Technology, Kwangju, 506-303, Korea

ABSTRACT

The surface structure of Si(111) post-annealed at 980 °C after nitrogen ion induced nitridation has been investigated by using a scanning tunneling microscope (STM) and low energy electron diffraction (LEED). The LEED and STM results indicated the formation of ordered domain of quadruplet structure in the silicon nitride layer. The LEED pattern taken from the nitrated Si(111) surface showed a coexistence of 7x7 domain with quadruplet one. In the STM image taken from the same surface, a three directional periodicity with a periodic arrangement of white protrusions was observed in the local area of silicon nitride island and its symmetry directions were rotated about $10°$ with respect to those of Si(111) surface. In addition to the quadruplet structure of the silicon nitride island, meta-stable structures such as 9x9, c(4x2), and 2x2 as well as 7x7 phase boundaries were observed to have been formed on the Si(111) surface during the rapid cooling of nitrated surface from the post-annealing temperature of 980 °C. The investigation of the surface structure of nitrated Si(111) showed that the surface nitrated at high temperature had better epitaxial silicon nitride layer than that post-annealed after nitridation at room temperature.

INTRODUCTION

The continuing requirement of the miniaturization of silicon devices has motivated the research on the atomic-scale investigation of the surface reactions, which are closely related to the thin film growth on silicon surface. In particular, there have been extensive efforts to make ordered silicon nitride layers since the silicon nitride is a versatile material which is used for microelectronic device as well as for high temperature structural ceramics applications. Among the extensive works done on the nitridation of silicon surfaces using various source materials [1-14], it was recently reported that the nitrogen ion beams with hyperthermal energy, N^+ and N_2^+, have a high reactivity with Si surfaces and the nitride layer thus formed has stoichiometry and local bonding configuration close to those of Si_3N_4 [14].

The high potential of nitrogen ions as a source material for the growth of high quality silicon nitride layer stimulated the atomic-scale investigation of the low energy nitrogen ion induced nitridation of Si(111) surface in this study. STM and LEED results indicated the formation of quadruplet structure in the silicon nitride layer. Meta-stable structures such as 9x9, c(4x2), and 2x2 as well as 7x7 phase boundaries were also observed to have been formed on the Si(111) surface during the rapid cooling of nitrated surface after post-annealing at 980 °C. The dependence of the surface morphology on the nitridation temperature showed that the surface nitrated at high temperature turned out to have higher quality epitaxial silicon nitride layer than that post-annealed after nitridation at room temperature. Based on these experimental results, we will discuss the initial growth mechanism of silicon nitride layer.

EXPERIMENTAL

The experiments were performed in a ultrahigh vacuum chamber equipped with a STM, a LEED/Auger electron spectroscopy, a quadrupole mass spectrometer, a gas inlet system, an ion sputtering gun, and a sample manipulator with a resistive heating stage. The base pressure was below 3.0×10^{-10} torr. A clean and well ordered Si(111)-7x7 surface was obtained by repeated cycles of resistive heating up to 1230 °C followed by slow cooling to room temperature. Nitrogen ions were generated by a cold cathode ion sputtering gun after back-filling the chamber with N_2 gas at a partial pressure between $5x10^{-7}$ and $1x10^{-5}$ torr. Nitrogen ions with an energy of 100 eV were used and almost no physical sputtering was expected with this energy. The angle between the surface normal and the ion gun was about 30°. During the exposure of nitrogen ions, the ion current at the sample surface was monitored to estimate the dose even though there always existed a nonuniform spatial distribution of ions on the surface. After nitridation both at room temperature and at 950 °C, the samples were post-annealed at 980 °C. The sample temperatures were measured with an optical pyrometer. All the STM measurements were done at room temperature in a constant current mode.

RESULTS and DISCUSSION

Figure 1(a) shows the LEED pattern taken from the Si(111) surface post-annealed at 980 °C after nitridation by 100 eV nitrogen ions at room temperature. It clearly shows the coexistence of 7x7 with the quadruplet patterns. The quadruplet pattern had been observed in the previous works using various source materials and interpreted as resulting from domains having a triangular lattice whose lattice vectors are about 25 % shorter than those of Si(111) [9, 14]. Those domains are rotated ±5° or ±10° with respect to the silicon surface [9]. As shown in Fig. 1(a), the diffraction peaks from the ordered domains rotated ±10° with respect to the symmetry direction of the silicon substrate were more intense than those from domains rotated ±5°. The origin of the quadruplet structure had been related to the carbon contamination on the nitrated surface even though the role of carbon in forming such structure had not been understood [9, 15]. In contrast to the previous arguments on the strong relationship between the quadruplet structure and the carbon contamination, a recent work on the ion beam induced nitridation of Si(111) surface suggested that the quadruplet structure they observed should not be related to the chemical composition but to the intrinsic characteristics of the ion reaction on the surface [14]. Unfortunately, we can not make any clear suggestion on the chemical nature of the quadruplet structure since it was not possible to get an information on the chemical composition of the sample by STM in this work. Fig. 1(b) is the STM image taken from the same surface under the tunneling conditions of I_t = 0.5 nA and V_{sample} = -3.0 eV. It showed two different regions ; the area with 7x7 periodicity and the island surrounded by protruding wall. It was very hard to obtain a stable STM image of the island due to unstable tunneling current under imaging conditions for clean Si(111) surface, implying the poorer conductivity of the island than that of the clean Si(111)-7x7 surface. The density of the island increased with increase in the nitrogen ion exposure indicating that the islands consist of silicon nitride layer. Inside the silicon nitride island, some periodic arrangements of white protrusions are observed along three symmetry axes. The symmetry axes are rotated about $10°$ with respect to the major symmetry directions of the 7x7 unit. The distance between white protrusions on the symmetry axis was estimated to be about 10-11 Å, which is very close to the value obtained in the previous STM work on the 8x8 structure of silicon nitride [7]. These LEED and STM results suggest that the silicon nitride islands formed by low energy nitrogen ions have ordered domain of quadruplet structure even though the domain size seems to be fairly small judging from the diffuse quadruplet LEED pattern and the relatively poor STM image of the silicon nitride island. The surrounding protruding wall is considered to be unreacted Si atoms, which is evidenced by the existence of the Si adatoms forming the reconstruction structure of Si(111) surface as indicated as A in Fig. 2.

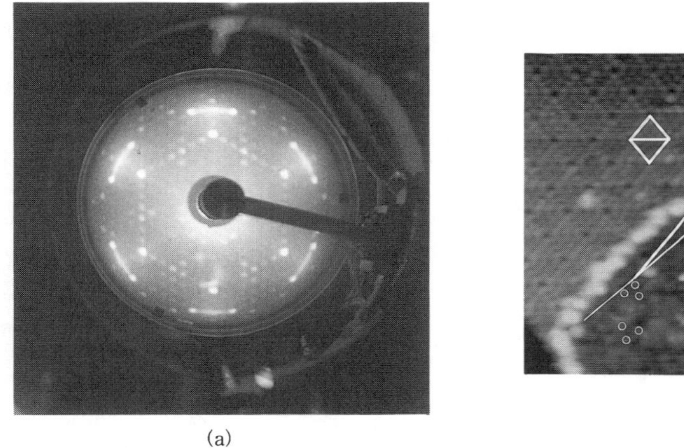

(a)   (b)

Figure 1. LEED pattern and STM image taken from the Si(111) surface post-annealed at 980 °C after nitridation by 100 eV nitrogen ions at room temperature. (a) LEED pattern (b) STM image under scan area = 280 x 230 Å^2, I_t = 0.5 nA, V_{sample} = −3.0 eV.

In addition to the quadruplet structure of silicon nitride islands, several kinds of interesting atomic structures were observed on the Si(111) surface between the adjacent silicon nitride islands. Fig. 2 is the STM image taken from the Si(111) surface post-annealed at 980 °C after nitridation at room temperature. As mentioned in the last paragraph, the Si adatoms forming the ordered arrangement on the protruding wall is shown in the figure. On the Si(111) surface between the neighboring silicon nitride islands, small areas of 9x9, 2x2, and c(4x2) reconstruction structures, which are indicated as B, C, and D, respectively, are observed. Those meta-stable structures were previously observed on the Si(111) surface after laser quenching and the thermal annealing at the temperature lower than that to give perfectly ordered 7x7 structure [16, 17]. Observation of those structures were explained in terms of the lower thermal barrier for forming intermediate 2x2 structure than that for 7x7 structure and the extension of the Takayanagi's DAS model to have 5x5 , 7x7, and 9x9 by shrinking or expanding the unit cell along the two primitive translation directions [16]. Those meta-stable structures observed in this work seem to have been formed during the fast cooling of the high temperature Si(111)-1x1 phase and stabilized by the strain due to nearby lattice-mismatched silicon nitride islands. In addition to these meta-stable structures, there appeared a 7x7 phase boundary between the adjacent silicon nitride islands in Fig. 2(b). In the previous work by Hadley et al. [18], the phase boundary was considered to increase the areal density of dangling bond in the interface over that of the 7x7 reconstructed surface. The resultant increase in the surface energy was expected to make the phase boundary unfavorable be formed unless it is either ; (a) stabilized by contamination, (b) initiated on cooling by some defect structure and propagated as the 7x7 reconstruction grows, or (3) formed on the area between two independently growing 7x7 regions. The phase boundaries observed on the Si(111)-7x7 surface in between the silicon nitride islands are considered to be formed during the cooling process of the sample after formation of lattice-mismatched silicon nitride islands at 980 °C ; the Si atoms are still very mobile and the ordering process during surface cooling can be disturbed by the existence of immobile nitride islands resulting in the formation of 7x7 phase boundary.

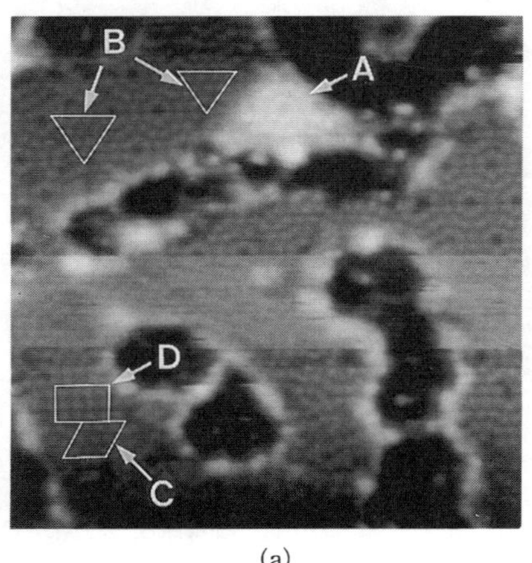

(a)

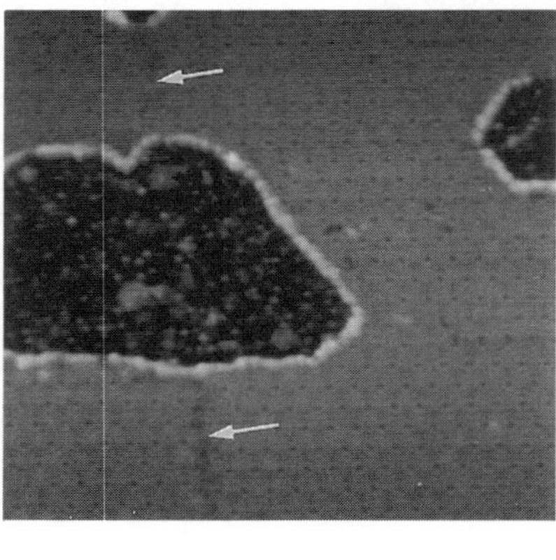

(b)

Figure 2. STM images taken from the Si(111) surface post-annealed at 980 °C after nitridation at room temperature. (a) The Si adatoms consisting of the protruding wall of the silicon nitride island, 9x9, 2x2, and c(4x2) reconstruction structures are indicated as (A), (B), (C), (D), and (E), respectively. (b) Si(111)-7x7 phase boundary is indicated as an arrow.

Figure 3 shows the change of the surface morphology with nitridation temperature. Fig. 3(a) and (b) are the STM images taken from the Si(111) surface post-annealed at 980 °C after nitridation by 100 eV nitrogen ions at room temperature and at 950 °C, respectively. As clearly distinguished in two images, the sizes of the flat silicon nitride islands were increased in Fig. 3(b) compared to those in Fig. 3(a). When the sample was exposed to nitrogen ions at room temperature, it was observed that the amorphous silicon nitride clusters were formed. With post-annealing at high temperature, the silicon nitride nucleus would be formed and grow into larger islands on prolonged annealing. Since the thermal diffusivity of N is quite low and the silicon atoms are strongly bonded to nitrogen atoms, the size of thus formed silicon nitride island would be restricted to be small unless sufficient annealing is performed. The thermal diffusion constant of nitrogen in Si_3N_4 was estimated to be less than 10^{-23} cm^2/s even at 900 °C [19] and that in Si to be 6×10^{-17} cm^2/s at 750 °C [20]. Therefore, the increase of the mobilities of Si and N due to elevation of surface temperature could enlarge the epitaxial silicon nitride islands. These results suggest that the surface nitrated at high temperature have better silicon nitride layer than that post-annealed after nitridation at room temperature.

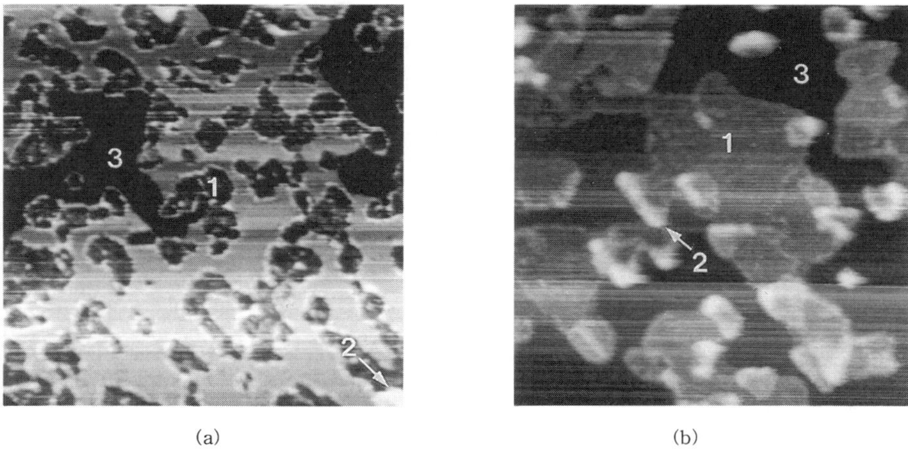

(a)   (b)

Figure 3. STM images taken from the Si(111) surface post-annealed at 980 °C after nitridation by 100 eV nitrogen ions (a) at room temperature and (b) at 950 °C, respectively. Scan area = 1120 x 1200 Å^2, I_t = 0.5 nA, V_{sample} = 3.0 eV.

Based on these experimental results, we suggest an initial growth mechanism of silicon nitride islands on Si(111)-7x7 surface when low energy nitrogen ions were used as a source material. The rough surface morphology observed in the STM image taken from the Si(111) exposed to nitrogen ions at room temperature is considered to be resulted from the clustering of silicon nitride whose stoichiometry does not satisfy the bulk Si_3N_4 due to low thermal diffusivity of Si and N at room temperature. If the surface is post-annealed at high temperature after exposure to nitrogen ions, the nitrogen atoms would be released from over- or under-coordinated Si, and then diffuse to find proper sites and form properly coordinated Si_3N_4. During this process, the excess Si atoms will remain as protruding walls which surround the silicon nitride islands as observed in the STM image. Here, the size and ordering of the silicon nitride islands would strongly depend on the annealing temperature and time since the thermal diffusivity of the reacting species would influence the epitaxial growth. During the cooling period of the surface from post-annealing temperature of 980 °C down to room temperature, the silicon nitride islands are quite immobile whereas Si atoms are quite mobile resulting in the formation of 7x7 phase boundaries and meta-stable surface structures as shown in Fig. 2. The immobile silicon nitride islands would also act as step pinning sites so that many of them would reside on the Si step edges.

CONCLUSIONS

The surface structure of Si(111) post-annealed at 980 °C after nitrogen ion induced nitridation has been investigated by using a STM and LEED. The LEED and STM results indicated the formation of ordered domain of quadruplet structure in thus formed silicon nitride layer. Meta-stable surface structures such as 9x9, c(4x2), and 2x2 as well as 7x7 phase boundaries were observed to have been formed on the Si(111) surface during the rapid cooling of nitrated surface after post-annealing at 980 °C. The dependence of the surface morphology on the nitridation temperature showed that the surface nitrated at high temperature turned out to have higher quality epitaxial silicon nitride layer than that post-annealed after nitridation at room temperature. With these LEED and STM results, we suggested an initial growth mechanism of silicon nitride layer on Si(111) surface using low energy nitrogen ions.

ACKNOWLEDGMENTS

This work has been supported by the Ministry of Information and Communications, Korea.

REFERENCES

1. R. Heckingbottom and R. Wood, Surf. Sci. **36**, 594 (1973).
2. A. Glachant, D. Saidai, and J. F. Delord, Surf. Sci. **168**, 672 (1986).
3. Ph. Avouris, J. Phys. Chem. **94**, 2246 (1990).
4. M. Yoshimura, E. Takahashi, and T. Yao, J. Vac. Sci. Technol. B **14**, 1048 (1996).
5. M. D. Wiggins, R. J. Baird, and P. Wynblatt, J. Vac. Sci. Technol. **18**, 965 (1981).
6. M. Nishijima, H. Kobayashi, K. Edamoto, and M. Onchi, Surf. Sci. **137**, 473 (1984).
7. B. Rottger, R. Kliese, and H. Neddermeyer, J. Vac. Sci. Technol. B **14**, 1051 (1996).
8. T. Ito, S. Hijiya, T. Nozaki, H. Arakawa, M. Shinida, and Y. Fukukawa, J. Electrochem. Soc. **125**, 448 (1978).
9. A. G. Schrott and S. C. Fain, Jr., Surf. Sci. **111**, 39 (1981) ; *ibid* **123**, 204 (1982).
10. K. Edamoto, S. Tanaka, M. Onchi, and M. Nishijima, Surf. Sci. **167**, 285 (1986).
11. H.-C. Wang, R.-F. Lin, and X. Wang, Surf. Sci. **188**, 199 (1987).
12. R. Karcher, L. Ley, and R. L. Johnson, Phys. Rev. B **30**, 1896 (1984).
13. J. A. Taylor, Appl. Surf. Sci. **7**, 168 (1981).
14. B. C. Kim, H. Kang, C. Y. Kim, and J. W. Chung, Surf. Sci. **301**, 295 (1994).
15. E. A. Khramtsova, A. A. Saranin, and V. G. Lifshits, Surf. Sci. Lett. **280**, L259 (1993).
16. R. S. Becker, J. A. Golovchenko, G. S. Higashi, and B. S. Swartzentruber, Phys. Rev. Lett. **57**, 1020 (1986).
17. M. D. Pashley, K. W. Haberern, and W. Friday, J. Vac. Sci. Technol. A **6**, 488 (1988).
18. M. J. Hadley and S. P. Tear, Surf. Sci. Lett. **247**, L221 (1991).
19. K. Kijima and S. Shirasaki, J. Chem. Phys. **65**, 2668 (1976).
20. O. C. Hellman, O. Vancauwenberghe, and N. Herbot, Nucl. Instrum. Methods B **67**, 301 (1992).

The effect of processing conditions on the structure of buried interfaces between silicon and silicon dioxide

XIDONG CHEN and J. MURRAY GIBSON
Department of Physics
University of Illinois at Urbana-Champaign
1110 W. Green St., Urbana, IL 61801

ABSTRACT

A transmission electron microscope technique is used to image atomic steps at buried interfaces between silicon and silicon dioxide. We have studied the effect of processing conditions on the interfacial structure of Si/SiO_2 with this technique. We observed a dramatic effect of post-oxidation annealing on silicon (100). Roughening of Si(111) interfaces due to chemical preparation is also reported.

INTRODUCTION

The structure of Si/SiO_2 interfaces is considered an important parameter for electrical properties of Ultra-Large-Scale-Integrated Metal-Oxide-Semiconductor (ULSI MOS). The imaging technique of transmission electron microscopy (TEM) has the capability of directly imaging buried interfaces in this system without removal of silicon dioxide. By developing image processing techniques and theoretical calculation, we are able to trace down individual atomic steps and determine the root-mean-square roughness and the step spacing without any pre-assumption. We studied the effect of post-oxidation annealing on the interfacial structure of Si/SiO_2. Our result shows that post-oxidation annealing at 900 °C can smoothen the interfaces between silicon and silicon dioxide for silicon (100). However, there is no obvious difference for silicon(111), with or without post-oxidation annealing. We also observed that chemical processing of silicon wafers can also change the interfacial structure.

EXPERIMENT

A silicon disc of diameter 3mm is first cut from a wafer and then chemically etched from both sides, to produce a hole with surrounding electron transparent thin areas near the center of the disc. A high-temperature sacrificial oxidation (typically at 1100°C with 1 atm O_2) is first performed to get microscopic flat interfaces. Then the sacrificial oxide is removed by chemical etching in HF solution and a thin oxide is grown at 900 °C for 30 minutes. In some cases a post-oxidation anneal is performed, by changing from O_2 to N_2 at the same temperature, for 15 minutes. Samples were examined with a CM-12 transmission electron microscope at 120kV. Dark field images were recorded by a cooled CCD camera and then analyzed by Gatan Digital Micrograph ^{TM} and NIH Image^{TM} software.

One pronounced advantage of TEM study is that buried Si/SiO_2 interfaces can be imaged with dark-field imaging techniques, without removal of the silicon dioxide layers, thanks to the amorphous structure of SiO_2. Dark-field images are obtained when only a diffracted beam is chosen by an objective aperture to form the image. The image intensity I(t) as a function of the local specimen thickness t(x,y) can well described, using the two-beam dynamical column approximation [1], as

Mat. Res. Soc. Symp. Proc. Vol. 448 © 1997 Materials Research Society

$$I = \frac{1}{s_{eff}^2 \xi_g^2} \sin^2(\pi t(x,y) s_{eff}) \tag{1}$$

Here $s_{eff} = \sqrt{s^2 + \frac{1}{\xi_g^2}}$ s is the deviation parameter that measures the distance between the Ewald sphere and the Bragg condition in reciprocal space when the image is taken, and ξ_g is the extinction distance. It can be proved that even for a large deviation parameter (s > 0.1nm^{-1}) and thickness variations at a scale smaller than a unit cell, equation (1) still holds with a little modification[2]. From equation (1), we can see that the same thickness change will introduce a larger relative intensity change with a larger deviation parameter. The intensity is therefore more sensitive to interface or surface configurations with increased deviation parameter. The penalty paid here is a decrease in the intensity as shown in figure 1. Practically, it is possible to image atomic steps if an appropriate deviation parameter is chosen and the background noise is reduced by controlling the size of the objective aperture [2,3].

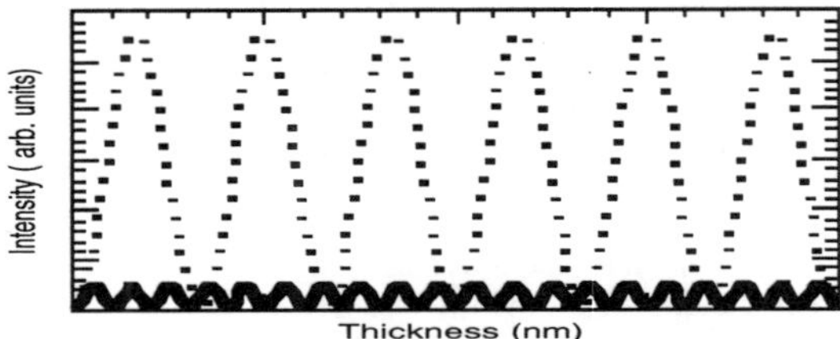

Figure 1. A theoretical calculation of intensity as a function of the local specimen thickness with different deviation parameter s is shown here. The dashed line has a deviation parameter of 0.2nm^{-1}. The solid line has a deviation parameter of 0.6 nm^{-1}. The graph shows that although the intensity is decreased with the increase of the deviation parameter s, the sensitivity to thickness changes is increased as shown in the solid line. More thickness fringes show up for a larger deviation parameter.

Equation (1) shows that although the intensity is sensitive to step positions, it is not directly related to the step distribution on the interfaces. Step distributions have to be retrieved from the original transmission electron microscope images. First we filter the images in Fourier space to remove the noise. We combine high-frequency filtering and amplitude filtering[4]. Values with a frequency higher that a certain value and an amplitude smaller than a certain threshold are removed in the Fourier transformed image in reciprocal space. Then the image is transformed back to real space to obtain the new filtered image. From this filtered image, we trace each fringe manually with NIH Image^{TM}. The maximum number of steps that each fringe corresponds to and the step height can be found from the deviation parameter. Images with different deviation parameters are compared to determine if the steps are up-going or down-going. With all this information available, a step distribution and a height map of the interfaces can be retrieved. Figure 2 gives an example of a step profile obtained from an original TEM image. From the step distribution, the roughness frequency spectrum can be found and detailed analysis, such as calculating the auto-correlation function of step positions can be performed. From equation (1), the root-mean-square roughness can also be estimated from the deviation parameter of a dark-field image in which thickness fringes start to disappear due to roughness [2].

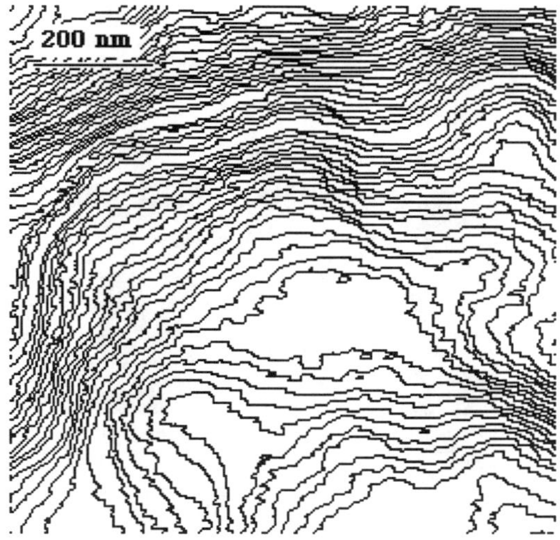

Figure 2. A step distribution map retrieved from a TEM image of an annealed silicon(100) sample with our method. The average height is about 15 Å

RESULTS

Effects of post-oxidation annealing were studied for both silicon(100) and silicon(111). For silicon (100), a dramatic smoothening by post-oxidation annealing was observed. The results are shown in figure 3. The left images are from an unannealed sample. It is shown that the unannealed sample has macroscopically rougher interfaces than the annealed one in image a) and b), with a small deviation parameter (s) of 0.2 nm^{-1}. There are many hillocks and dents for the unannealed sample, as shown in image a) but they are mostly removed after annealing as shown in image b). When the deviation parameter   s is increased to 0.5 nm^{-1}, no thickness fringes can be detected for the unannealed sample because of roughness, as shown in image c). However, there are still fringes for the annealed sample, as shown in image d).

The effect of post-annealing on silicon (111) is not so obvious, although interfaces of silicon (111)/SiO$_2$ were observed to be much flatter than those of silicon(100)/SiO$_2$. The results are shown in figure 4.

HF acid etching is an important step in producing silicon surfaces. It is used to remove the native silicon oxide. Different chemical solutions have been tested in an effort to generate an ideal silicon surface [5, 6]. We tried a different chemical processing approach with silicon (111) in our experiments. After sacrificial oxidation, sacrificial oxide layers were removed by HF acid etching, but in some cases, samples were further processed by a basic solution of NH$_4$F:NH$_4$OH = 1:0.2. Then both groups of samples were oxidized to grow a layer of silicon dioxide of thickness ~ 5nm. The Si(111)/SiO$_2$ interfaces from the samples treated with by this basic solution before oxidation were much rougher than those from the samples which were only etched by HF acid. The results are shown in figure 5. Image b) is from the interfaces of a sample that was further processed by a basic solution of $NH_4F:NH_4OH = 1:0.2$ after HF acid etching. Image a) shows the interfaces from a sample that was only etched by HF and then oxidized. The interfaces in image b) are so rough that some speckle patterns show up, which is a well-known phenomenon for scattering from very rough surfaces[7].

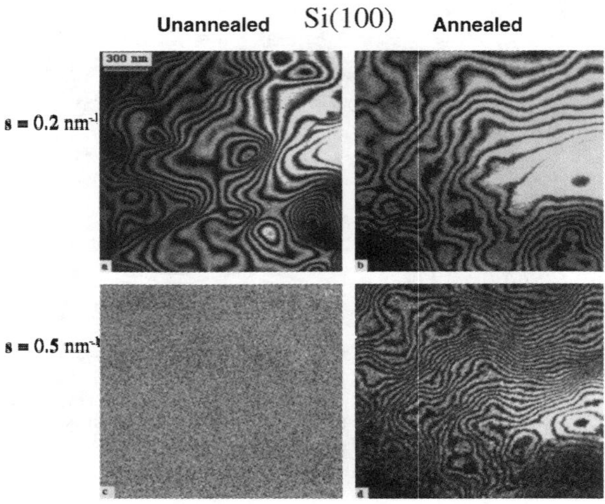

Figure 3 The effects of post-oxidation annealing on the interfacial structure of $Si(100)/SiO_2$ a) unannealed, s = 0.2 nm^{-1}. b) annealed, s=0.2 nm^{-1}. c) unannealed, s = 0.5 nm^{-1} d) annealed, s = 0.5 nm^{-1}

A limitation of our method is that there are two interfaces actually observed by the transmission electron microscope. Hence, it is hard to determine if steps are from the top interface or the bottom one. However, on the reasonable assumption that the interfaces are independent, the detailed statistical properties of steps can be easily retrieved and our conclusions are not affected.

One of our future plans is to develop a kinetic model of silicon oxidation to further understand our data. This model will be compared with experiments under different controlled conditions such as oxidation and annealing temperature, time, etc. The oxidation process is believed to be an intrinsically roughening process. However, there is a thermodynamic smoothening process to compete with this roughening process[8]. By modeling these two processes and the balance between them more quantitatively, we hope that we can further understand this problem.

CONCLUSIONS

An image technique of transmission electron microscopy is developed to study buried interfaces of Si/SiO_2. We found that $Si(100)/SiO_2$ interfaces are very rough but this roughness can be dramatically removed by a post-oxidation annealing at 900 °C. Silicon (111) interfaces are not dramatically influenced by this lower temperature post-oxidation annealing, yet they are quite flat. We also observed that starting surfaces processed further by a basic solution of $NH_4F:NH_4OH$ after HF acid etching can result in much rougher interfaces after oxidation.

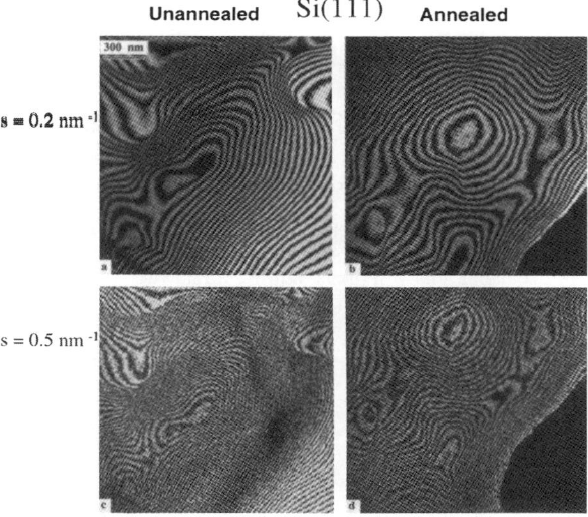

Figure 4 The effects of post-oxidation annealing on the interfacial structure of $Si(111)/SiO_2$ a) unannealed, s = 0.2 nm⁻¹. b) annealed, s=0.2 nm⁻¹. c) unannealed, s = 0.5 nm⁻¹ d) annealed, s = 0.5 nm⁻¹. No obvious difference in roughness can be observed here and the interfaces are flatter than those of silicon (100).

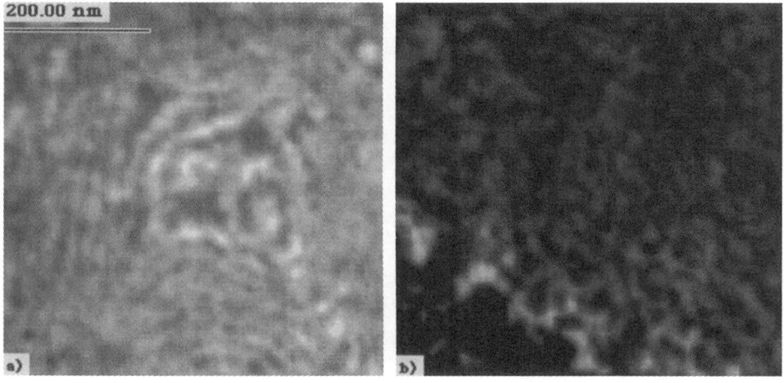

Figure 5. Two dark field images of different $Si(111)/SiO_2$ interfaces oxidized with starting surfaces from different chemical procedures a) HF acid etching b) HF acid etching and $NH_4F:NH_4OH$ = 1.0:0.2 treatment. Interfaces in image b) are much rougher and speckle patterns appear.

ACKNOWLEDGMENTS

This work is funded by the Semiconductor Research Cooperation under Grant No. SRC 95-BJ-361. The experiments were carried out using facilities in the Frederick Seitz Materials Research Laboratory, University of Illinois at Urbana-Champaign, which is supported by the U.S. Department of Energy under Grant No. DEFG02-91-ER45439

REFERENCES

[1] P. Hirsch, A. Howie, R.B. Nicholson, D.W. Pashley and M.J. Whelan, Electron Microscopy of Thin Crystals, Robert E. Krieger Publishing Co., INC, Florida, 1977, pp.158.
[2] X. Chen, J. M. Gibson, in preparation for publication.
[3] X. Chen, J. M. Gibson, Phys. Rev. B **54,** 2846 (1996).
[4] J. C. Ross, The Image Processing Handbook, CRC Press, Inc. 1995, pp. 312 - 346.
[5] K. Utani, T. Suzuki, and S. Adachi, J. Appl. Phys. **73** 3467 (1993)
[6] G.S. Higashi, Y. J. Chabal, G. W. Trucks, and K. Raghavachari, Appl. Phys. Lett. **56,** 656 (1990).
[7] J. W. Goodman, in Laser Speckle and Related Phenomena, edited by J. C. Dainty (Spring-Verlag, Berlin, 1984), pp. 9 - 75
[8] X. Chen, J. M. Gibson, submitted for publication in Appl. Phys. Lett.

SPECTROSCOPIC INVESTIGATION OF LITHIUM INTERCALATION IN THIN FILMS OF ANATASE TITANIUM DIOXIDE

R. VAN DE KROL, A. GOOSSENS, AND J. SCHOONMAN
Laboratory for Applied Inorganic Chemistry, Delft University of Technology, P.O. Box 5045
2600 GA Delft, The Netherlands, R.vandeKrol@stm.tudelft.nl

ABSTRACT

Thin films of anatase TiO_2 have been deposited on tin-doped indium oxide (ITO) and Sb-doped tin oxide using electron beam evaporation. Subsequently these samples have been mounted into an electrochemical cell. Assuming a composition of Li_xTiO_2 with x being 0.5 or 1, voltammetric measurements show that all lithium is present within the first 15 or 7.5 nm of the TiO_2 film, respectively. Stepped potential experiments in combination with optical transmission measurements show that de-intercalation is much faster than intercalation. Differential absorption spectra as a function of intercalation potential suggest that the observed dark coloring of intercalated anatase can be attributed to electron traps at Li^+ sites. This contradicts assumptions made in literature that the coloring mechanism of anatase is based on intervalence charge transfer from Ti^{4+} to Ti^{3+}.

INTRODUCTION

Anatase TiO_2 is a promising candidate for devices based on lithium intercalation, such as electrochromic windows[1-4] and rocking-chair lithium-ion batteries.[5,6] Porous nanostructured films can be used to increase the specific surface area of the electrode, thereby decreasing the diffusion path length for the lithium ions. This results in a large enhancement of the efficiency and speed of the device.[1]

Though a considerable amount of research has already been dedicated to lithium intercalation in titanium dioxide,[1-9] some fundamental questions remain unanswered. The mechanism of lithium intercalation has not yet been fully resolved. Another point of discussion is the rate-determining step of the intercalation process, for which both diffusion limited[2,9] as well as surface-adsorption limited[7] processes have been suggested. There is also the question what will happen to the electrons once lithium is intercalated into the titanium dioxide. These electrons can be donated to the conduction band, or be trapped at Li^+, Ti^{4+}, or positively charged oxygen vacancies.

In this work lithium intercalation in electron-beam evaporated thin films of anatase TiO_2 is studied. Films deposited with electron beam evaporation are extremely smooth, which enables a more accurate interpretation of the results than is possible with sputtered films.[2] Evaporated films can be made with approximately the same donor density as porous films, i.e. in the order of $\approx 10^{17}$ cm^{-3}. Several aspects of the intercalation process are described using voltammetric, stepped-potential and spectroscopic techniques. Results are related to electronic properties of the intercalated material.

EXPERIMENTAL

Electron beam evaporation of reduced TiO_2 was used to deposit thin films of anatase TiO_2 on both ITO coated glass (80 nm ITO on glass, Glastron, $30\Omega/\square$) and $Sb:SnO_2$ coated quartz substrates. The $Sb:SnO_2$ films, having a thickness of 100 nm, were made by electron beam

Mat. Res. Soc. Symp. Proc. Vol. 448 © 1997 Materials Research Society

evaporation of a mixture of Sb_2O_3 and SnO_2 powders. The target material for TiO_2 deposition was anatase powder (Acros Chimica, 99.95+%) reduced in a hydrogen atmosphere at 1000°C for 6 hours. Deposition took place under a controlled O_2 atmosphere of $1x10^{-4}$ mbar, with an argon background pressure smaller than $2x10^{-6}$ mbar. During deposition the substrates were heated to 375°C. Typical growth rates, monitored with a quartz microbalance, were 2 nm/min. After deposition a heat treatment of 12 hours at 450°C in air was carried out to restore stoichiometry. The TiO_2 film thickness is 80 nm in all experiments.

Electrochemical and spectroscopic measurements were carried out in a conventional three-electrode electrochemical cell. Glassy carbon was used as a counter electrode, while a saturated calomel electrode (SCE) served as a reference electrode. All potentials mentioned will be referred to as versus SCE. As an electrolyte a 1 M $LiClO_4$ solution in propylenecarbonate was used. Since water absorption by the electrolyte could in our case not completely be avoided, 1% water was added deliberately in order to get a constant water concentration.

RESULTS AND DISCUSSION

A convenient method to follow the Li-intercalation in titanium dioxide is by standard voltammetric and optical transmission measurements, shown in Figure 1. The peaks at -1.35 V and -0.78 V can be attributed to lithium intercalation and de-intercalation respectively, since these peaks are not observed if the same measurement is performed in an electrolyte not containing $LiClO_4$. The intercalation process shows a good reversibility after several scans.

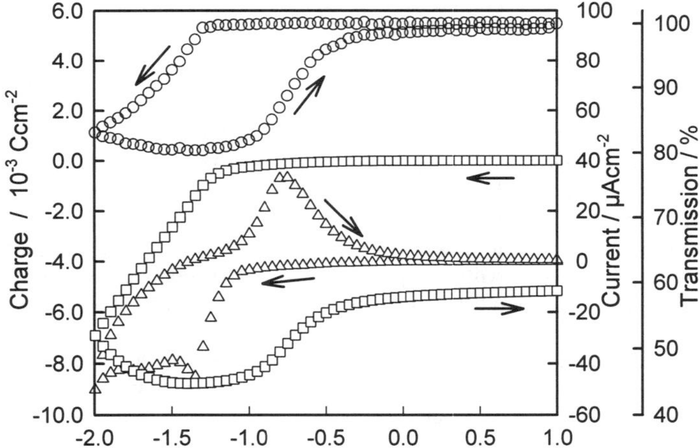

Figure 1. Current (Δ), charge ($\square$), and optical transmission (O) curves recorded as function of potential during lithium intercalation and de-intercalation of thin film anatase TiO_2. The surface area under the de-intercalation peak at -0.78 V corresponds to a charge of $3.4x10^{-3}$ Ccm^{-2}.

When monitoring the transmission with 650 nm light, coloring and bleaching of the electrode occurs at the same potentials as where the peaks are located. At cathodic potentials side-reactions are present, which are probably caused by the reduction of water. To determine the amount of lithium that has been intercalated, the current has been integrated to yield the total charge passed through the system (Figure 1). Due to the occurrence of side-reactions at very negative potentials, only the de-intercalation peak should be integrated, yielding a total charge of 3.4×10^{-3} Ccm^{-2}, corresponding with 3.5×10^{-8} molcm^{-3}. If one assumes a constant stoichiometry of $Li_{0.5}TiO_2$,[7] this corresponds to a diffusion length of 15 nm. A stoichiometry of $Li_{1.0}TiO_2$ has also been suggested,[9] which in our case would yield a diffusion length of 7.5 nm. These results are in good agreement with values found in literature.[7,9] It should be noted, however, that no conclusive data concerning the actual composition of Li_xTiO_2 has yet been presented in literature.

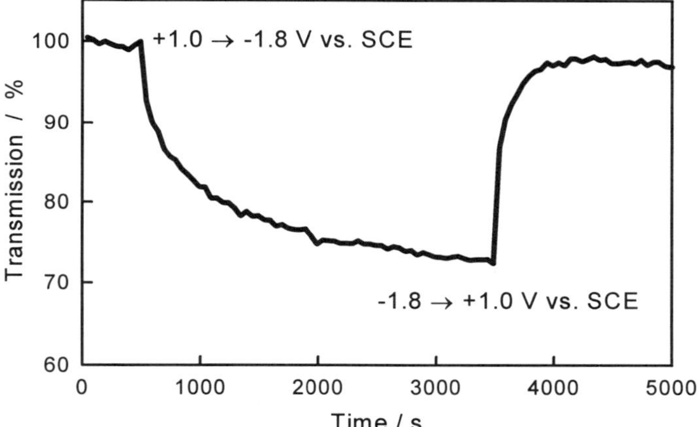

Figure 2. Optical transmission at 650 nm after potential steps to intercalation and de-intercalation conditions. Intercalation at -1.8 V vs. SCE results in a transmission decrease of approximately 30% after 1 hour. De-intercalation is a much faster process, a return to the bleached state is accomplished in less than 15 minutes.

To determine the rate of intercalation, stepped-potential measurements have been performed while measuring the optical transmission at 650 nm. Potential steps from +1.0 V to -1.8 V vs. SCE and vice versa have been made to start intercalation and de-intercalation, respectively. The results are presented in Figure 2, showing that intercalation is a much slower process than de-intercalation. This phenomenon has also been observed by other authors,[3,7] and can be explained in terms of lattice relaxation and expansion effects, and/or energy barriers due to ionic solvatation energies of the lithium ions in the electrolyte.[7] It is interesting to note that for porous nanostructured films both processes are reported to occur with the same rate.[1] The nature of this discrepancy is unclear and calls for further investigation.

Figure 3 shows the differential absorption spectra of the intercalated TiO_2 at various intercalation potentials. All spectra shown are calculated with respect to the absorption spectrum recorded at 0 V vs. SCE, i.e. under de-intercalation conditions. Under intercalation conditions, a bleaching at 320 nm is observed, while at 750 nm a dark coloring occurs. The bleaching at 320 nm can be attributed to a Moss-Burstein shift, i.e. an apparent increase of the bandgap caused by free electrons filling up the energy levels at the bottom of the conduction band. These free electrons compensate the positive charge carried by the intercalated lithium ions. At potentials between -1.2 V and -1.3 V vs. SCE, a sudden decrease in absorption at 320 nm is observed, accompanied by a increase in absorption at 750 nm. This is shown in Figure 4, where

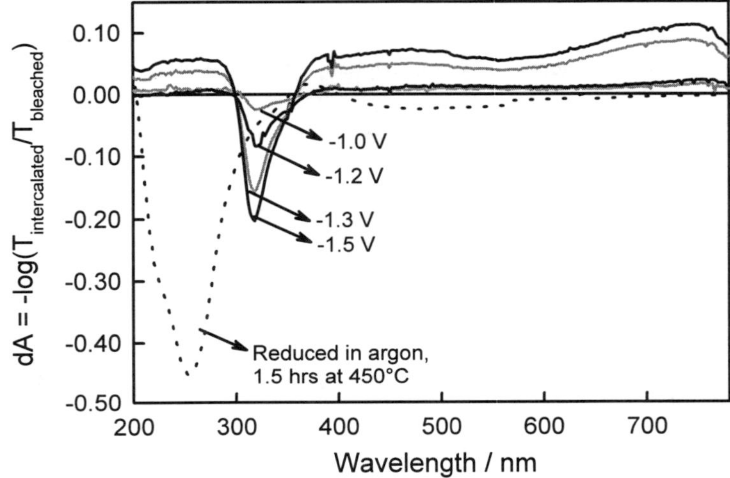

Figure 3. Differential absorption spectra at various intercalation potentials. All spectra are calculated with respect to the de-intercalated state at 0 V vs. SCE. The spectrum of the sample reduced in argon was recorded using a Cary 2400 spectrophotometer. Measurements on the lithium intercalated samples were performed in a wet electrochemical setup, using a photomultiplier to measure the transmission.

the peak maximum of the differential absorption at 320 nm is plotted as a function of intercalation potential. A possible explanation may be found in a decrease of the density of states around a specific energy in the conduction band, giving rise to a local sharpening of the absorption band edge. Figure 4 also shows the position of the peak maxima, and the peak widths at half height as a function of intercalation potential. The peak maximum shifts to higher energies at stronger intercalation potentials, as would be expected. The peak width, however, does not change significantly. This contrasts with the results of Enright et al.[10], who calculated the effective hole mass for nanocrystalline anatase from the broadening of the peak. So far no satisfactory explanation has been found for this discrepancy, though it points to a difference in electronic structure between porous nanostructured anatase and thin film anatase. Specifically, a higher density of states from some point in the conduction band of thin film anatase is expected to sharpen the absorption band edge, which can lead to a increase in peak height without additional broadening.

The darkening of the intercalated TiO_2 observed at 750 nm indicates the presence of trapped electrons. Possible trap sites are Ti^{4+}, Li^+ and oxygen vacancies. To determine the nature of the sites at which the electrons are trapped, absorption spectra of anatase TiO_2 before and after reduction at 450°C in argon were recorded. The resulting differential absorption spectrum is shown in Figure 3. An almost identical spectrum is obtained when the reduction is carried out in a hydrogen atmosphere. Comparing reduced TiO_2 to lithium intercalated TiO_2, a much larger Moss-Burstein shift is present, while no trapped electrons are observed at 750 nm. This suggests that in the case of lithium insertion a significant fraction of the free conduction band electrons is trapped at Li^+ sites. This contradicts previous assumptions made in literature that the absorption band in the near-infrared region can be attributed to intervalence charge transfer between Ti^{4+} and Ti^{3+} species.[8,9]

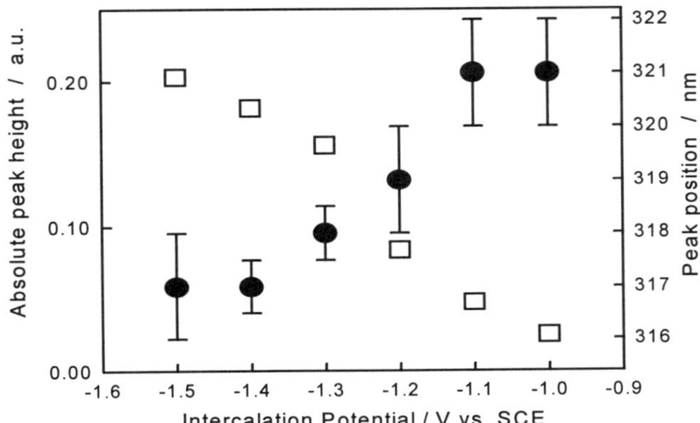

Figure 4. Data for the peaks centered around 320 nm in the differential absorption spectrum from Figure 3. Under stronger intercalation conditions the peak position shifts to higher energies, and the peak height increases. Both the peak position (filled circles) as well as the peak height (open squares) show a large change at intercalation potentials around -1.25 V vs. SCE. No significant peak broadening was found as a function of potential.

CONCLUSIONS

Lithium intercalation in thin films of anatase TiO_2 is studied. Assuming a composition of Li_xTiO_2 with x between 0.5 and 1.0, the lithium intercalates over a distance of 7.5 to 15 nm into the TiO_2. However, since the actual composition across the film cannot be measured directly, assumptions on the composition should be treated with caution. Stepped-potential experiments show the intercalation process to be much slower than the de-intercalation of lithium, which contrasts with the results reported for porous nanostructured anatase. Differential absorption spectra suggest that the coloring of anatase observed on intercalation should be attributed to electron trapping at Li^+ instead of intervalence charge transfer from Ti^{4+} to Ti^{3+}. Furthermore, the spectra for thin film anatase are different from those reported for porous nanostructured anatase. Further investigations are necessary to clarify these interesting discrepancies between thin film and porous anatase TiO_2.

ACKNOWLEDGMENT

A.G. wishes to thank the Royal Netherlands Academy of Arts and Sciences (KNAW) for his fellowship.

REFERENCES

1. A. Hagfeldt, N. Vlachopoulos, M. Grätzel, *J. Electrochem. Soc.*, 141, L82 (1994).
2. M. Ottaviani, S. Panero, S. Morzilli, B. Scrosati, *Solid State Ionics*, 20, 197 (1986).
3. T. Ohzuku, T. Hirai, *Electrochim. Acta*, 27, 1263 (1982).
4. C.G. Granqvist, *Solid State Ionics*, 53-56, 479 (1992).
5. W.J. Macklin, R.J. Neat, *Solid State Ionics*, 53-56, 694 (1992).
6. T. Ohzuku, Z. Takehara, S. Yoshizawa, *Electrochim. Acta*, 24, 219 (1979).
7. L. Kavan, M. Grätzel, S.E. Gilbert, C. Klemenz, H.J. Scheel, *J. Am. Chem. Soc.*, 118, 6716 (1996).
8. M.P. Cantão, J.I. Cisneros, R.M. Torresi, *Thin Solid Films*, 259, 70 (1995).
9. M.P. Cantão, J.I. Cisneros, R.M. Torresi, *J. Phys. Chem.*, 98, 4865 (1994).
10. B. Enright, D. Fitzmaurice, *J. Phys. Chem.*, 100, 1027 (1996).

IMPROVEMENT OF ULTRATHIN OXIDES
BY POST-OXIDATION ANNEALING

Tomoyuki SAKODA, Mieko MATSUMURA, and Yasushiro NISHIOKA

Advanced Materials & Processes

Texas Instruments   Tsukuba Research & Development Center Ltd.

17 Miyukigaoka, Tsukuba, Ibaraki 305, Japan

Abstract

The low temperature oxidation is effective for the atomically-controlled gate oxide growth. We focused the effects of post-oxidation annealing (POA) and attempted to improve the properties of the low-temperature-grown ultrathin oxides with a thickness of 3nm by POA. POA abruptly reduced the leakage current at a low gate voltage below 1.5V and the interface trap density. The correlation between the interface trap and the leakage current at a low applied voltage region were confirmed. We found that the stressing immunity of the ultrathin oxides grown at a low temperature, 650°C, is drastically improved by POA at 850°C.

Introduction

Ultrathin gate oxides are required to be thinner than 3nm and to have a higher reliability than those of conventional oxides for sub 0.1μm devices. The exact control of oxide thickness is essential because even a thickness deviation of one atomic layer results in over 10% of device parameter fluctuations. The low temperature oxidation is effective for the atomically-controlled gate oxide growth. We have been interested in SiO_2 films grown at a low temperature, and have been studying if they meet the requirement for sub 0.1μm devices. Ultrathin SiO_2 films grown at low temperatures, however, have poor electrical properties possibly due to a large stress in the SiO_2 films at the Si/SiO_2 interface. It has been previously reported the post-oxidation annealing (POA) improves effectively the electrical properties of thin SiO_2 [1,2]. S.S.Cohen showed that oxides grown in an oxygen ambient, which are post-annealed for a long period of time in an inert atmosphere and then reoxidized for a short time, have considerably improved values for both the magnitude and distribution of the electric field at the dielectric breakdown [1]. In addition, Z.A.Weinberg et al. reported that short anneal in Ar or in N_2 can substantially reduce water-related electron traps and short anneal in O_2 have an optimal time for hole trapping reduction [2].

Mat. Res. Soc. Symp. Proc. Vol. 448 © 1997 Materials Research Society

However, there is little information on the effects of POA on low-temperature-grown ultrathin SiO_2 less than 3nm.

In this paper, we focused the effects of POA and attempted to improve the properties of the low-temperature-grown ultrathin oxides with a thickness of 3nm by POA. The electrical and interfacial properties are investigated for 3nm SiO_2 grown at a low temperature with or without a high temperature POA in N_2.

Experimental

The MOS capacitors with 3nm oxides were fabricated on p-type (100) silicon substrates with a resistivity of 2Ωcm. The silicon substrates were cleaned in a boiling solution of $HCl/H_2O_2/H_2O$ and subsequently dipped in $NH_4OH/H_2O_2/H_2O$ solution at a room temperature, followed by 5% HF dipping and DIW rinse. After the cleaning, the thermal oxidation was performed in a vertical furnace at 650°C and 750°C in pure O_2, and the oxidation at 800°C and 850°C was done in 33% and 5% O_2 diluted with N_2, respectively. The oxidation and POA conditions of each sample were listed in Table I. The oxide thickness was measured by the ellipsometer with the fixed refractive index of 1.46. Following oxidation with or without POA in N_2, the aluminum electrodes with the area of 2×10^{-3}cm^2 were evaporated through a mask. We didn't perform the post-metallization annealing using a forming gas because of investigating the influence of the interface trap.

Table I.   Conditions of oxidation and post-oxidation annealing for each sample

Sample No.	Oxidation	N_2 Annealing
#650-1	Temp=650°C,pure O_2	No
#650-2	Temp=650°C,pure O_2	850°C,20min
#750-1	Temp=750°C,pure O_2	No
#750-2	Temp=750°C,pure O_2	750°C,20min
#750-3	Temp=750°C,pure O_2	850°C,20min
#800-1	Temp=800°C,33% O_2	No
#800-2	Temp=800°C,33% O_2	800°C,20min
#800-3	Temp=800°C,33% O_2	850°C,20min
#850-1	Temp=850°C,5% O_2	No
#850-2	Temp=850°C,5% O_2	850°C,20min

Current-voltage (I-V) and time dependent dielectric breakdown (TDDB) characteristics were measured using HP4145B Semiconductor Parameter Analyzer. The constant current density bias of -0.1A/cm^2 was used for the charge-to-breakdown (Q_{bd}) measurements. The constant voltage stressing for TDDB was performed at a room temperature. The interface properties were investigated from AC conductance method [3,4] using HP4284A LCR Meter because the influence of the DC-tunneling component can be separated.

Results and Discussion

I-V characteristics under a negative gate bias for each sample are shown in Figure 1. All I-V curves show that the leakage current due to the direct tunneling from the gate electrode to the substrate is the predominant conduction mechanism in 3nm oxides above 1.5V. As seen in Fig.1, the oxides grown at 850°C (#850-1, #850-2) showed a lower leakage current whether annealed or not. In contrast to these oxides, a larger leakage current at a lower applied voltage than 1.5V is observed in the oxides grown at 650°C without POA (#650-1). The oxides grown at a lower temperature than 800°C without POA have similarly a large leakage current. However, the leakage current of the oxides grown at 650°C with POA at 850°C (#650-2) are lower compared to that of the oxides without the annealing. The reduction of the leakage current at a lower applied voltage (<1.5V) is larger than an order of magnitude. This drastic change of the leakage current seems that the relaxation of the oxide's structure at the SiO$_2$/Si interface occurs and the effective thickness of the oxides increases during POA in N$_2$.

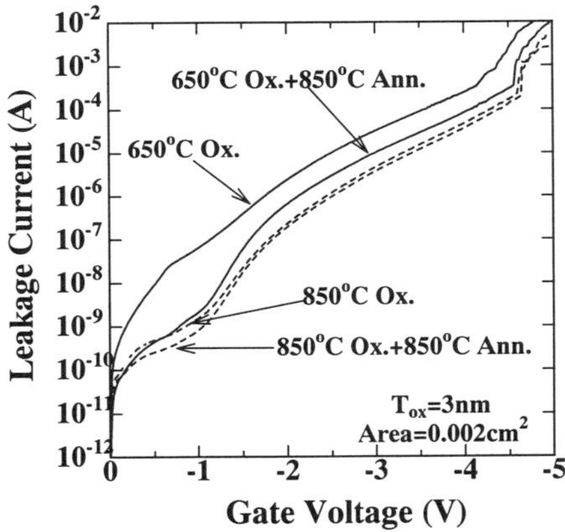

Figure 1   Current-voltage characteristics of 3nm oxides
with various oxidation and annealing conditions

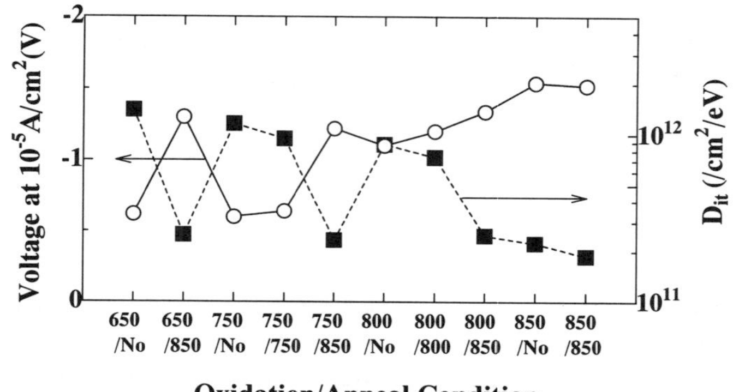

Figure 2   Voltage at 10^{-5}A/cm^2 and interface trap density
of various samples

Figure 2 shows the sustaining voltage, V_{limit}, at the leakage current level of 10^{-5}A/cm^2 and the interface trap density, D_{it}, for each sample. The larger the values of V_{limit}, the smaller the leakage current is caused. The values of V_{limit} for the oxides grown at 850°C (#850-1, #850-2) are about -1.5V whether annealed or not. This indicates that the oxides grown at 850°C have the excellent electrical and interfacial properties without POA. However, the oxides grown at a lower temperature than 850°C have the different trends. For example, V_{limit} for the oxides grown at 650°C without the annealing (#650-1) was -0.62V. By contrast, V_{limit} for the oxides with 850°C annealing (#650-2) was -1.30V, about two times larger than that for the oxides without the annealing. The values of Dit for the oxides grown at 650°C reduced from 1.40×10^{12} to 2.52×10^{11}/cm^2/eV by 850°C annealing. The magnitudes of V_{limit} and D_{it} for the oxides grown at 650°C with the annealing get close to those of the oxides grown at 850°C.

As seen in Fig.2, there is the clear correlation between V_{limit} and D_{it}. That is to say that the leakage current for 3nm-thick oxides is closely related to the interface trap density. As shown in the previous reports [5,6,7], it seems that the existence of the interface traps introduces "a trap-assisted leakage mechanism" and leads to a significant increase in the leakage current at a low applied voltage. This leakage mechanism is conceptually shown in Figure 3. In this figure, the electrons injected from the gate electrode tunnel directly through the gate oxide and are trapped by the interface traps. Then the trapped electrons are emitted into the conduction band of the silicon or recombined with the holes emitted from the valence band of the silicon. For the 5nm oxides, the correlation as seen in Fig.2 disappeared. Since the electrons cannot directly tunnel to the interface traps for the oxides thicker than 5nm, it seems that these oxides don't have the correlation between the leakage current and the interface trap density. That is to say that from the results shown in Figs.1 and 2, the oxides with the large interface trap density have the large leakage current at a low applied voltage when the oxide thickness is thinner than 3nm.

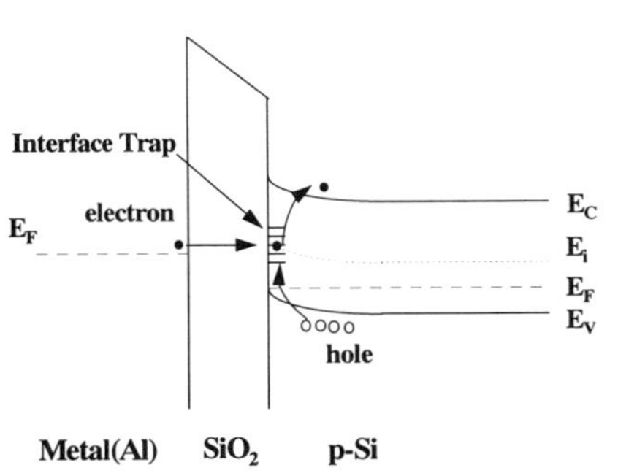

Figure 3   Interface trap-assisted leakage mechanism

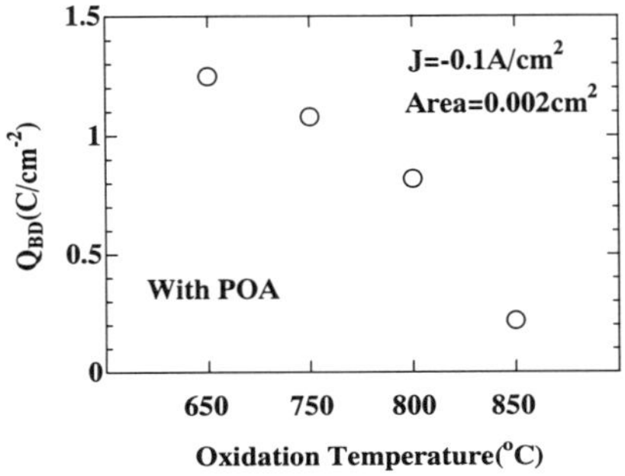

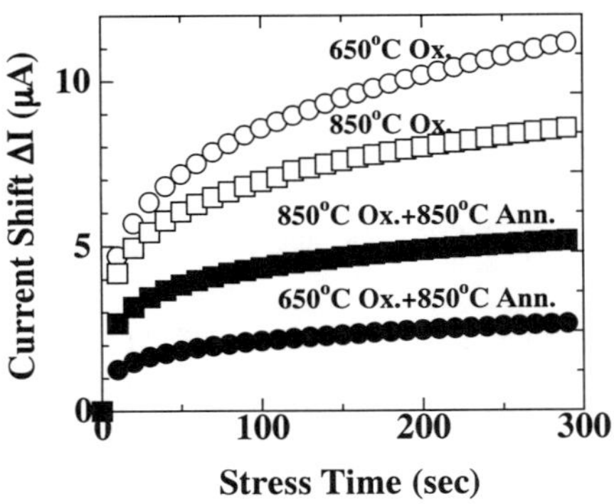

Figure 4   Charge-to-breakdown for the oxides grown at various temperatures

Figure 4 shows the values of charge-to-breakdown (Q_{bd}) for the oxides grown at various temperatures with POA. As seen in Fig.4, the oxides grown at a lower temperature have a larger value of Q_{bd} compared to the oxides grown at a higher temperature. For example, the increase of Q_{bd} for the oxides grown at 650°C is about an order of magnitude in comparison with the oxides grown at 850°C.

Figure 5   Current shift during a constant voltage stressing of -4.1V

The current shift, ΔI, during a constant voltage stressing is shown in Figure 5. In this figure, the gate bias is -4.1V, and the results of 4 samples, #650-1, #650-2, #850-1 and #850-2 are shown. The increase in a current may be related to a hole trapping observed all samples. However, ΔI was reduced by POA and the effect was prominent for the oxides grown at 650°C. Furthermore, ΔI of the oxides grown at 650°C with POA is smaller than that of the oxides grown at 850°C. That is to say that 3nm-thick oxides grown at a low temperature (650°C) with a high temperature annealing (850°C) have superior stressing immunity to the oxides grown and annealed at a high temperature (850°C). The similar trend was obtained for the other stressing voltage.

Conclusion

The effects of the post-oxidation annealing were investigated for 3nm-thick oxides grown at a very low temperature (650°C). As a results, we confirmed that post-oxidation annealing at 850°C improves the characteristics of ultrathin oxides grown at a very low temperature of 650°C. Since the oxides with a large interface trap density have a large leakage current at a low voltage when the oxides is thinner than 3nm, it is important to reduce the interface trap density. We found that 3nm-thick oxides grown at a low temperature with a high temperature anneal have superior stressing immunity to the oxides grown at a high temperature.

References

1. Simon S. Cohen, J. Electrochem. Soc. **130**, 929 (1983)
2. Z.A.Weinberg, D.R.Young, J.A.Calise, S.A.Cohen, J.C.Deluca and V.R.Deline, Appl. Phys. Lett. **45**, 1204 (1984)
3. E.H.Nicollian and J.R.Brews, <u>MOS Physics and Technology</u> (John Wiely & Sons, New York, 1982)
4. T.P.Ma and R.C.Barker, Solid-State Electronics **17**, 913 (1974)
5. I.C.Chen, C.W.Teng, D.J.Coleman and A.Nishimura, IEEE Electron Device Lett. **10**, 216 (1989)
6. T.Wang, C.Huang, T.E.Chang, J.W.Chou and C.Y.Chang, IEEE Trans. Electron Devices **41**, 2475 (1994)
7. M.H.White and J.R.Cricchi, IEEE Trans. Electron Devices **19**, 1280 (1972)

Charge Trapping and Degradation of High Permittivity TiO$_2$ Dielectric Metal-Oxide-Semiconductor Field Effect Transistors

Hyeon-Saeg Kim*, S.A. Campbell*, D.C. Gilmer**, D.L. Polla*
* Dept. of Electrical Eng. , **Dept. of Chemistry University of Minnesota, Minneapolis, Minnesota 55455, hskim@ee.umn.edu, campbell@ee.umn.edu

ABSTRACT

Suitable replacement materials for ultrathin SiO$_2$ in deeply scaled MOSFETs such as lattice polarizable films, which have much higher permittivities than SiO$_2$, have bandgaps of only 3.0 to 4.0 eV. Due to these small bandgaps, the reliability of these films as a gate insulator is a serious concern. Ramped voltage, time dependent dielectric breakdown, and capacitance-voltage measurements were done on 190Å layers of TiO$_2$ which were deposited through the metal-organic chemical vapor deposition of titanium tetrakis-isopropoxide. Measurements of the high and low frequency capacitance indicate that virtually no interface states are created during constant current injection stress. The increase in leakage upon electrical stress suggests that uncharged, near-interface states may be created in the TiO$_2$ film near the SiO$_2$ interfacial layer that allow a tunneling current component at low bias.

INTRODUCTION

The preparation of high permittivity thin films have received considerable interest for the fabrication of charge storage insulators for new generations of dynamic random access memories (DRAMs) [1], [2]. In another application, the use of high permittivity materials is considered as one of the best alternative to conventional ultra-thin silicon dioxide [3], [4]. For MOSFETs scaling into the deep submicron, there is a strong need to reduce the gate oxide thickness. Tunneling currents however, limit the scaling of SiO$_2$ to approximately 25 Å. Therefore, scaling of the gate oxide must end or an alternate material must be used. Suitable replacement materials such as lattice polarizable films which have much higher permittivities than SiO$_2$ have bandgaps of only 3.0 to 4.0 eV [5], [6]. Due to these small bandgaps, the reliability of these films as a gate electrode is a serious concern. Although wearout information is now being obtained on some storage capacitor materials, very little is known regarding charge trapping and breakdown for high permittivity films when used as a gate dielectric and therefore when deposited directly on silicon. The interface of the gate electrode/gate insulator is also important for the reliability of the MOS system. A poor electrode/gate oxide interface with high interface state density results in high current through the gate film, and a high probability of charge trapping in the gate film, reducing breakdown strength and the charge to breakdown in time dependent dielectric breakdown (TDDB) tests. These states will also lead to low mobilities and poorly controlled threshold voltages when used for FETs. In this paper we report on leakage currents and the effect of electrical stress on TiO$_2$ and the performance of TiO$_2$ MOSFETs using one such high permittivity material.

EXPERIMENT

The TiO$_2$ MOS capacitors and TiO$_2$ MOSFETs were fabricated on 9-15 Ω-cm boron doped substrates. The backside of the wafers received a 10^{15} cm^{-2} boron implant at an energy of 50 KeV to ensure a good ohmic contact of aluminum to the substrate body. 190Å layers of TiO$_2$ were deposited through the thermal decomposition of titanium tetrakis-isopropoxide on 2" p-type (111) silicon wafers which had undergone a standard LOCOS isolation [4]. After deposition of TiO$_2$ the wafer was annealed for 30 minutes in dry oxygen at 750 °C. This anneal was found to reduce leakage currents through the film. After TiO$_2$ deposition, 2500 Å of platinum was sputter deposited for the gate electrode. The wafers then received a 30 minute nitrogen anneal at 850 °C after the transistor source/drain implant to activate the implanted impurities and to densify a

Mat. Res. Soc. Symp. Proc. Vol. 448 © 1997 Materials Research Society

deposited oxide layer. Transmission electron microscope (TEM) results indicate that this deposition anneal process produces an interfacial silicon dioxide layer approximately 25 Å thick. After the fabrication was complete a 450 °C hydrogen anneal was performed. Current - Voltage and time dependent dielectric breakdown (TDDB) with capacitance-voltage (C-V) measurements were done with an HP4156 semiconductor parameter analyzer and an HP4194 impedance analyzer.

RESULTS AND DISCUSSION

The layers were analyzed for chemical and structural composition by X-ray diffraction (XRD) and Rutherford backscattering spectrometry (RBS). XRD results indicated that the films were polycrystalline anatase with no evidence of any rutile phase, and RBS showed a [Ti]/[O] stoichiometric ratio of 1:2, within the sensitivity of the analysis [4].

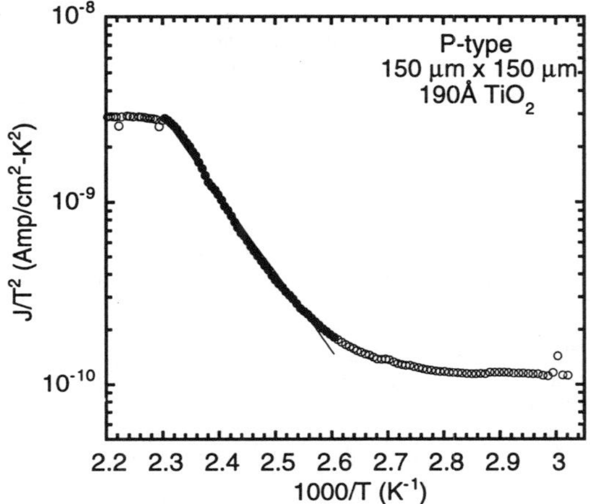

Figure 1 J/T^2 plot for an dry O$_2$ annealed film

The leakage current through the device has been measured as a function of temperature for both accumulation and inversion biases. The anneal was found to significantly reduce the leakage current and J/T^2 is found to follow standard thermionic emission behavior. Forward Recoil Spectroscopy showed that the concentration of hydrogen in these films is correlated with increased interface state densities and leakage current, and that the hydrogen concentration of films decreases sharply with post annealing at 750 °C. The effective barrier height was approximately 0.5 eV for the as-deposited film, while the O$_2$ annealed film had a barrier height of 1.0 eV. X-ray diffraction measurements of the TiO$_2$ films on (111) substrates show no significant change in crystallization with the 750 °C anneal, although atomic force microscopy shows a reduction in surface roughness. We therefore believe that this increase in the barrier height with O$_2$ annealing is due to a change in the band alignment, perhaps due to a change in the charge state or strain at the TiO$_2$/Si interface. When biased into accumulation as shown in figure 1, the current initially is temperature independent, suggesting that the current in this regime is dominated by tunneling. Above 100 °C however it shows standard thermionic emission behavior up to 160 °C where it again becomes temperature independent. The leakage under inversion bias was independent of temperature up to 70 °C while it shows standard thermionic emission behavior. The absence of temperature dependence at 25 °C to 70 °C suggests that the post-stress current is dominated by tunneling over the temperature range studied as shown in figure 2. This would be consistent with the formation of tunnel states in the TiO$_2$. The barrier for emission was between 0.94 and 1.0 eV for all samples measured [7].

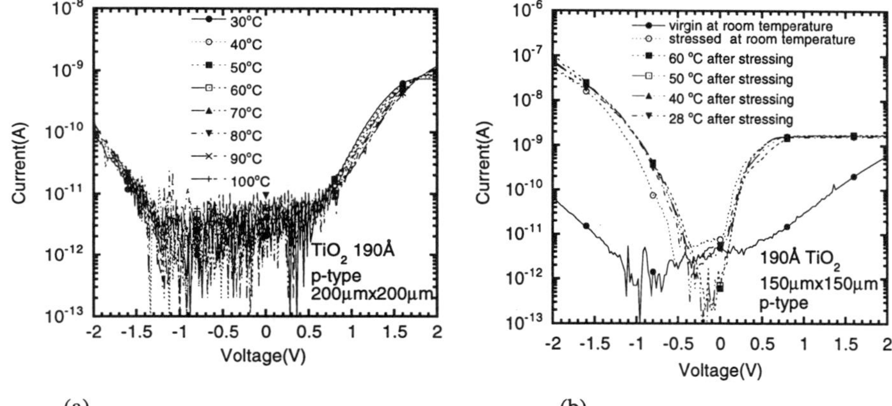

(a)   (b)

Figure 2. thermal leakage current before stress (a) and after stress (b)

The ramped voltage measurements with a 1 V/sec ramp rate showed Fowler-Nordheim current-voltage characteristics of 150×150 μm^2 capacitors. To minimize the voltage drop in the substrate, negative biases were used in order to keep the capacitors in accumulation. A clear irreversible breakdown occurred at an applied field of approximately 3.2 MV/cm, and the charge to breakdown Q_{BD} is ~6 C/cm^2 as shown in figire 3. A distribution of breakdown fields is reasonably tight.

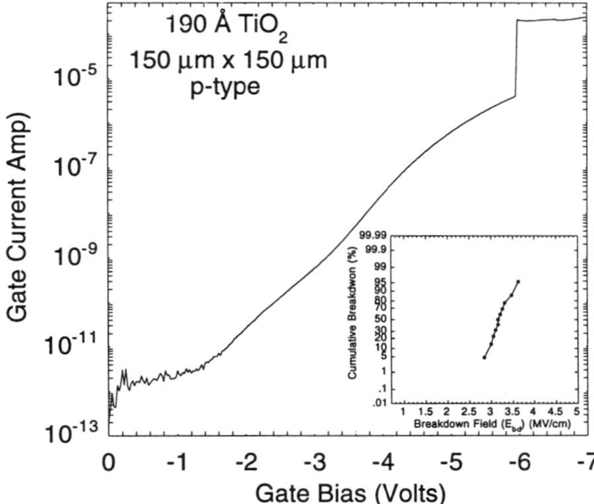

Figure 3. Ramped voltage measurements

323

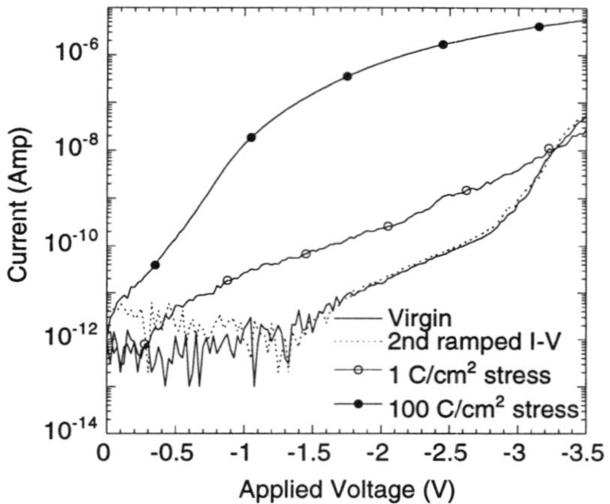

Figure 4. I-V characteristics after TDDB stress

No clear breakdown was seen in time dependent dielectric breakdown (TDDB) tests. The leakage current through the TiO_2 increased after electrical stressing. This effect was most pronounced at low bias conditions. At a 2 volt accumulation bias the leakage current increased by a factor of 10 after 1.0 Coul/cm² and another factor of 50 after 100 Coul/cm² as shown in figure 4, but no clear discontinuity in the voltage-time (constant current) or current-time (constant voltage) curves are seen. Measurements of the high frequency capacitance indicate that little charge trapping occurs during electrical stress as shown in figure 5. By assuming an N_{eff} of order 10^{14} cm^{-2}, the capture cross section σ was estimated to roughly 10^{-23} cm². This is much smaller than conventional SiO_2 trap values. TiO_2 insulator may be much less vulnerable to TDDB stress.

From figure 6, both 1 MHz and 100 Hz capacitance voltage curves measured using capacitors made with 190 Å layers of TiO_2 displayed a pronounced voltage dependent accumulation capacitance, which may be related to hole trapping in the TiO_2 very near the SiO_2 interfacial layer. Taking this interfacial layer into account, the permittivity of the TiO_2 films is approximately 30, in good agreement with other published papers [5]. The interface state density extracted from the C-V curves with Castagne and Vapaile's method fell below 10^{11} cm^{-2}-eV^{-1} at midgap, but rose sharply on either side, unlike the "U" shaped behavior in thermal oxide capacitors. Measurements of the high and low frequency capacitance with constant current TDDB stress with 0.4, 4.2, 20, and 38 C/cm² indicate that virtually no interface states were created during stress. Very little bulk charge trapping exists after electrical stress suggesting that these films have few intrinsic traps and are highly resistant to current induced trap creation. It is also possible that an presenting interfacial SiO_2 layer may reduce the strain gradient across the dielectric significantly making the MOS system less susceptible to interface defect generation. The increase in leakage upon electrical stress suggests that uncharged, near-interface states may be created in the TiO_2 film near the SiO_2 interfacial layer that allow a tunneling current component at low bias. Alternatively, the increase in current may be due to the creation of neutral raps near the platinum gate electrode.

Finally, TiO_2 FETs were fabricated from these layers over a standard LOCOS structure. Figure 7 shows a family curves for 10 (gate width) x 1.25 (gate length) μm^2 transistor. Nearly ideal device performance is seen.

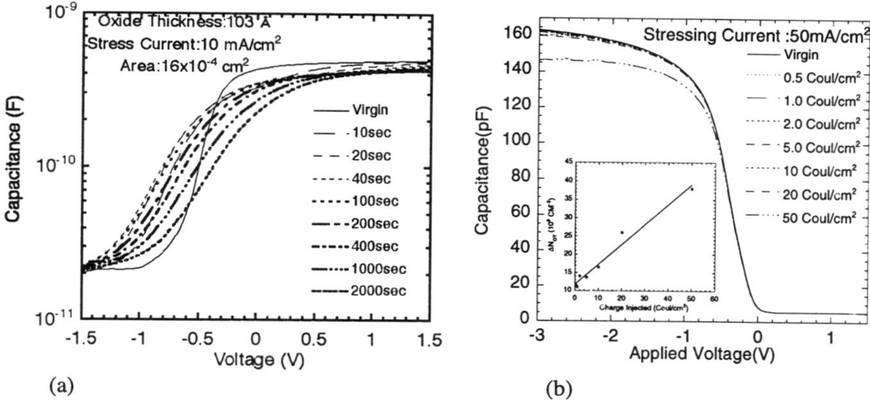

Figure 5. High frequency C-V characteristics of 103 Å SiO_2 (a) and 190 Å TiO_2 films (b) after TDDB stress.

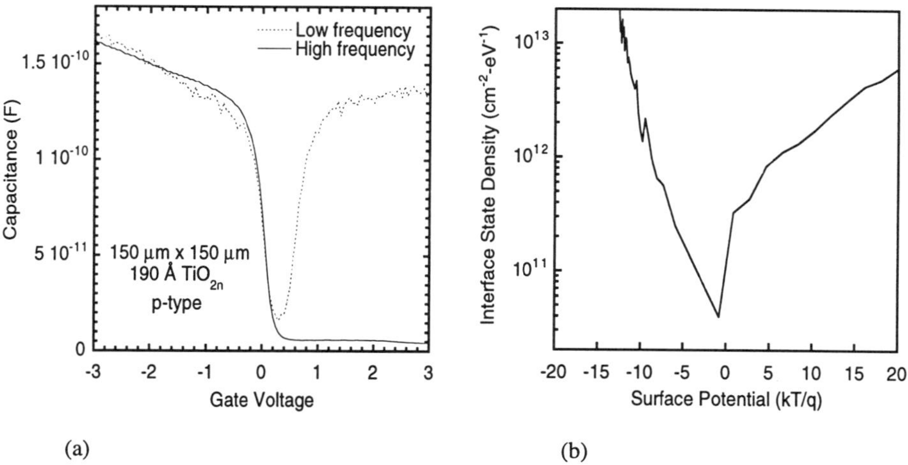

Figure 6. (a) low and high frequency C-V curve of a MOS capacitor with a 190 Å TiO_2 gate and (b) the mid gap interface state density was in the low of 10^{11} cm^{-2}-V^{-1} range.

CONCLUSION

A MOSFET processed with TiO_2 gate oxide was fabricated and tested. The barrier for emission was between 0.94 and 1.0 eV for all high temperature anneal samples in the oxygen ambient. The absence of temperature dependence suggests that the post-stress current is dominating by tunneling over the temperature range studied. Ramped current and voltage studies showed a clear breakdown at an applied field of approximately 3.2 MV/cm, but TDDB measurements showed very little effect after stressing at more moderate fields. Leakage current through the TiO_2 increased after electrical stressing. At a 2 V accumulation bias the leakage current increased by a factor of 10 after 1.0 Coul/cm^2 and another factor of 50 after 100 Coul/cm^2, but no clear discontinuity in the constant current or constant voltage curves are seen. Measurements of the high and low frequency capacitance indicate that virtually no interface states are created during stress. The increase in current may be due to creation in leakage upon electrical stress suggests that uncharged, near-interface states may be created in the TiO$_2$ film near the SiO$_2$ interfacial layer that give rise to increased tunneling leakage. The lack of a breakdown event however, suggests that these films may be extremely robust as gate dielectrics.

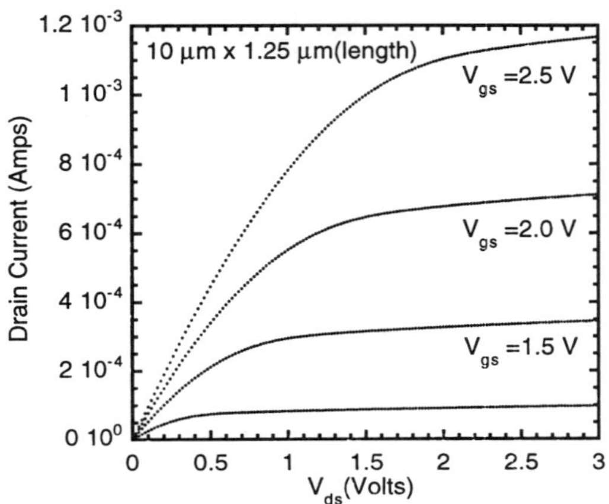

Figure 7. A family curves for 10 (gate width) x 1.25 (gate length)µm^2 transistor.

ACKNOWLEDGMENT

One of the authors would like to thank Sei-Kee Kim, Yun-Kyu Yang, and Hyeong-Ran Choi for encouragement and financial support to put this research to paper.

REFERENCES

1. S. Kamiyama, T. Saeki, H. Mori, and Y. Numasawa, IEEE Inter. Electron Device Meeting *Tech. Dig.*, pp. 827 (1991)

2. P. Y. Lesaicherre, S. Yamamichi, H. Yamaguchi, K. Takemura, H. Watanabe, K. Tokashiki, K. Satoh, T. Sakuma, M. Yoshida, S. Ohnishi, K. Nakajima, K. Shibahara, Y. Miyasaka, and H. Ono, IEEE Inter. Electron Device Meeting *Tech. Dig.*, pp. 831 (1994)

3. J. Yan, D.C. Gilmer, S.A. Campbell, W.L. Gladfelter, and P. G. Schmid, J. Vac. Sci. Techn. B 14, pp. 1704 (1996)

4. S.A. Campbell, D.C. Gilmer, X. Wang, M.T. Hsieh, H.-S. Kim, W.L. Gladfelter, and J. Yan, vol. **44**, pp. 104 IEEE Trans. Electron Devices (1997)

5. T. Fuyuki and H. Matsunami, Jpn. J. Appl. Phys. **25**, pp. 1288 (1986)

6. J. A. Duffy, in <u>Bonding Engergy Levels and Bands in inorganic Solids,</u> John Wiley, New York (1990)

7. H.-S. Kim, D.C. Gilmer, S.A. Campbell, and D.L. Polla, Appl. Phys. lett. **69** (25), pp. 3860 (1996)

Interface properties of Si$_3$N$_4$/Si/n-GaAs metal-insulator-semiconductor structures grown on GaAs(111)B substrate

D.G. Park, D. M. Diatezua, Z. Chen, S. N. Mohammad, and H. Morkoç*

University of Illinois at Urbana-Champaign, Materials Research Laboratory and Coordinated Science Laboratory, 104 South Goodwin Avenue, Urbana, Illinois 61801
*Also with Wright Laboratory, Wright Patterson Air Force Base

ABSTRACT

We present characteristics of Al/Si$_3$N$_4$/Si/n-GaAs metal-insulator-semiconductor (MIS) interfaces grown on GaAs(111)B prepared with a combination of *in situ* molecular beam epitaxy (MBE) and chemical vapor deposition (CVD) techniques . The density of the surface states in the high 10^{10} eV^{-1} cm^{-2} near the GaAs midgap for the GaAs grown at 575°C and 625°C was obtained. The MIS structure with GaAs homoepitaxial layer grown at 625°C, showing smoother surface morphology than the surface grown at 575 °C, exhibited small hysteresis which was as small as 30 mV under a field excursion of 1.5 MV/cm. The presence of a 1 MHz frequency response at 77 K suggests that the traps be within 60 meV of the conduction band edge of GaAs .

INTRODUCTION

Unpinning the GaAs surface Fermi level is key to the realization of viable GaAs metal-insulator-semiconductor field-effect-transistors (MISFETs). As observed recently, this is achieved fairly well by utilizing an epitaxial Si or Ge interlayer grown on GaAs.[1-3] The pseudomorphic Si interlayer is thought to prohibit the formation of an intervening native oxide prior to the insulator deposition and the outdiffusion of the volatile group V element.[2,3] The optimum thickness of the Si interlayer for the best C-V characteristics in Si$_3$N$_4$/Si/GaAs capacitor grown on GaAs(001) is about 10 Å with minimum interface trap density D$_{it}$ ~ mid 10^{10} eV^{-1} cm^{-2}.[3] This trap density may be considered low enough to yield unpinned Fermi level in the GaAs MIS structure. Almost all of the GaAs based MIS structures were prepared on GaAs(001) substrate presumably because the surface morphology is smoother with GaAs(001) than with GaAs (110) and GaAs(111). However, the growth on GaAs(111) surfaces with features is not uncommon. These growths at various temperatures and related parameters (e.g., surface morphology, doping, etc.) have been studied extensively.[4-7] Owing to the importance of low threshold current lasers[8] and electro-optical devices[9] with strong internal fields generated piezoelectrically, significant attention has been paid to epitaxial growth on GaAs(111)B substrates.[4-7] No attempts have, however, been made to investigate the interface properties of MIS structure grown on GaAs(111)B substrate. Rougher surface morphology of the GaAs(111)B surface as compared to the GaAs(100) surface, and an incomplete growth mechanism of epitaxial growth of GaAs on GaAs(111)B surfaces are probably the causes behind it. Since the growth of prepatterned GaAs(100) results in the evolution of (h11) type facets including (111)B,[6] and the energy levels in the quantum well of Si$_3$N$_4$/Si/GaAs strutures grown on GaAs (111) are almost unconfined by the theoretical calculation,[10] a study of MIS interface on GaAs (111) substrate is crucial to gauge the availability of the inverted charges from GaAs channels.

In this article, we report, for the first time, the application of a pseudomorphic Si interlayer to ideal GaAs MIS diodes on n-GaAs(111)B substrate. The results from *in situ* deposited Si$_3$N$_4$/Si/n-GaAs MIS devices clearly demonstrate the accumulation, depletion

Mat. Res. Soc. Symp. Proc. Vol. 448 © 1997 Materials Research Society

and inversion regions with a minimum trap density of 8×10^{10} cm^{-2} eV^{-1} as determined by the conductance method.

EXPERIMENT

The GaAs(111)B substrate with 5° miscut toward [100] direction was degreased and etched in $NH_4OH:H_2O_2:H_2O = 3:1:150$ for about 3 min. prior to loading into the system. The formation of GaAs MIS capacitors was initiated by the growth of a 0.5 μm, 5×10^{16} cm^{-3} Si-doped GaAs buffer layer on a n$^+$ GaAs(111)B substrate in Perkin Elmer molecular beam epitaxy (MBE) system. The surface reconstruction pattern and the surface morphology were recorded with reflection high electron energy diffraction (RHEED) and Nomarski microscope, respectively, as a function of growth temperature. The ratio of As to Ga fluxes during the growth of GaAs buffer layer was kept about 5 during the entire growth. The sample was transferred via an ultrahigh vacuum (UHV) transfer tube ($< 2 \times 10^{-9}$ Torr) to a UHV chamical vapor deposition (CVD) system. Then a Si interlayer about 10 Å was employed, followed by a 200 Å-thick Si_3N_4 layer deposited at 250 W and 350°C. Employing an *in situ* X-ray photoelectron spectroscopy (XPS), measurement of very thin (~ 10-20 Å) Si interface layer thickness could be made with a reasonable accuracy.[3] Experimental details concerning the CVD deposition, the MBE growth, and MIS capacitor fabrication have been presented elsewhere.[2,11,3] Following the thermal desorption of the surface oxides at 600 ± 10°C, the GaAs(111)B surface yielded a faint ($\sqrt{19} \times \sqrt{19}$) surface reconstruction. GaAs homoepitaxial layers of 0.5 μm thickness on GaAs(111)B showed mirrorlike features in the surface morphology when grown at the substrate temperatures of 575 and 625 °C, respectively. However, the sample grown at 530 °C depicted rough surface morphology, displaying three fold pyramids and some other types of defects. During the growth, the surface reconstruction of the film grown at 625°C changed to $(1 \times 1)_{HT}$,[6] while that of the film grown at 575°C maintained ($\sqrt{19} \times \sqrt{19}$). Nomarski observation demonstrated better surface morphology on the as-grown layer when grown at 625°C rather than at 575 °C. Since the higher substrate temperature may reduce the relative amount of As and/or increase the cation (Ga atoms) migration length on GaAs surface, the smoother surface morphology was thought to be obtained at elevated growth temperatures, which is also reported elsewhere.[5,6]

RSULTS AND DISCUSSION

Fig. 1 depicts the typical C-V characteristics of the Si_3N_4/Si / n-GaAs capacitors grown on GaAs(111)B. The high frequency C-V (HFCV) and quasi-static C-V (QSCV) curves of MIS diodes with a GaAs growth temperature of 575 °C are shown in Fig.1 (a). These curves exhibit ideal C-V characteristics. Experimental C-V curves as a function of gate voltage and GaAs growth temperatures are presented in Fig. 1(b). For the sake of comparison, the theoretical C-V profile is also shown in this figure. Typical hysteresis values of about 60 mV, as shown in Fig. 1(a), were obtained at 1 kHz and 1 MHz under a field swing of about ± 1 MV/cm (bias 2 V). A reduction in hysteresis by about 30 mV (Fig. 1 (b)), corresponding to a charge of 5.4×10^{10} cm^{-2}, for the GaAs grown at 625 °C with an extended field swing of ± 1.5 MV/cm, is attributed to the improved surface morphology. The hysteresis for a field swing of +3 to - 3 MV/cm is about 150 mV. A frequency dispersion of 20 mV between the 1 kHz and the 1 MHz C-V curves implies that the density of fast interface traps with response times shorter than 1 ms corresponds to a charge density of ~ 3.6×10^{10} cm^{-2} (Fig. 1 (b)). The density of slow interface traps with response times longer than 1 ms is also quite small as determined by the small frequency dispersion (~20 mV) between 20 Hz and 1 KHz (data not shown).

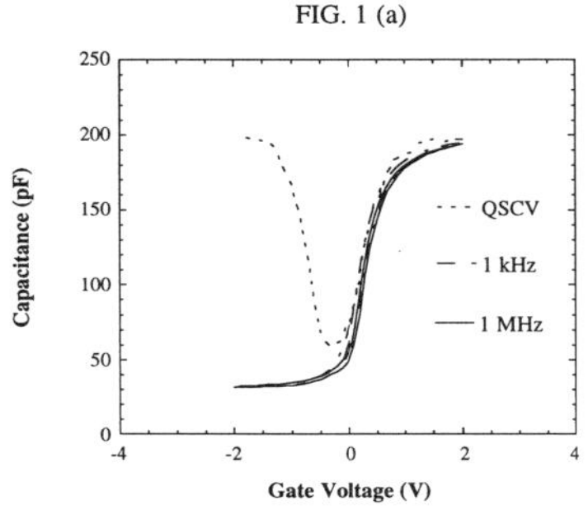 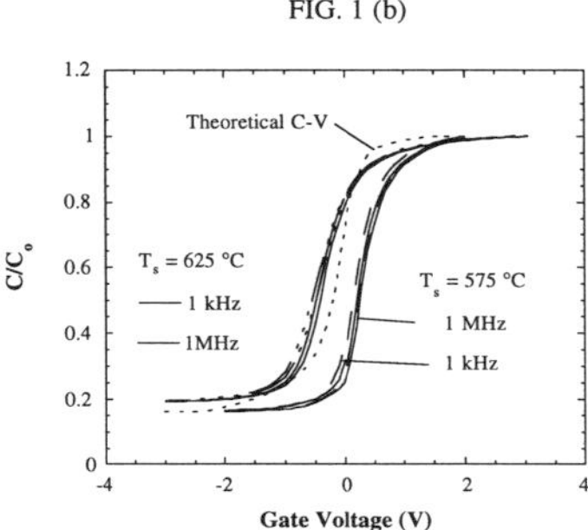

FIG. 1 (a) Measured 1 MHz, 1 kHz C-V and quasi-static (QS) C-V curves of the Si3N4/Si/n-GaAs capacitors grown on GaAs(111)B substrate. (b) Hysteresis and frequency dispersion of the GaAs MIS diode are plotted as a function of substrate growth temperature (T_s). The theoretical C-V curve is superimposed.

Superimposed on the HFCV curves in Fig. 1 (a) is the QSCV curve. The nature of this deep QSCV curve confirms the existence of high-quality gate insulator as well as small amount of interface defects which will be discussed later. The flat-band (FB) voltage shift (ΔV_{FB}) as determined by FB capacitance method,[12] assuming that the channel comprised of only GaAs, is about 0.42 V for the MIS diode prepared at T_s of 575 °C. This amount of ΔV_{FB} corresponds to a fixed or trapped charge density of ~ 7.5×10^{11} cm^{-2} at the interfacial region. On the other hand, the ΔV_{FB} of the MIS capacitor grown at a temperature $T_s = 625$ °C is shifted by -0.2 V, corresponding to a positive charge density of 4×10^{11} cm^{-2}. This is a topic that currently deserves our attention.

Fig. 2 shows the conductance loss (G_m/ω) versus angular frequency (ω) curves which allow us to extract the density of surface states. The interface trap density was determined using the parallel conductance loss (G_p/ω):

$$G_p/\omega = C_i^2 \, (G_m/\omega) \, / \, [(G_m/\omega)^2 + (C_i - C_m)^2], \quad (1)$$

which may be obtained by removing the series resistance and the capacitance effects from the measured conductance peaks (G_m/ω)[12] since the measured conductance G_m and the measured capacitance C_m are not equal to the equivalent conductance G_p and the equivalent capacitance C_p. The C_i is the insulator capacitance in accumulation. The loss peaks in depletion were observed to follow the statistical fluctuation model.[12] The interface trap density was calculated from[12]:

$$D_{it} = (G_p/\omega)_{fp} \, [\, f_D \, (\sigma_s) \, q \, A]^{-1}, \quad (2)$$

where f_p is the frequency corresponding to the peak value of (G_p/ω), f_D is the universal function as a function of the standard deviation of band bending (σ_s), and A is the area (7×10^{-4} cm^2) of the capacitor. The $f_D(\sigma_s)$ values determined from our samples were about 0.35 to 0.40. The minimum conductance peak of 3.2 pF (Fig. 2) is obtained at a gate bias of -0.35 V, corresponding to $D_{it} \sim 8 \times 10^{10}$ eV^{-1}cm^{-2} at 0.10 eV above the GaAs midgap. Beyond this bias, the conductance peak occurs at frequencies lower than 100 Hz, which is an apparatus limit. As shown in Fig. 2, with increasing surface potential, there occurs also an increase in trap density in the upper-half of the bandgap. The D_{it}, as determined by *low* methods using QSCV and 1 MHz C-V was, however, slightly higher, $D_{it} \sim 2 \times 10^{11}$ eV^{-1}cm^{-2}.

FIG. 2

FIG. 3

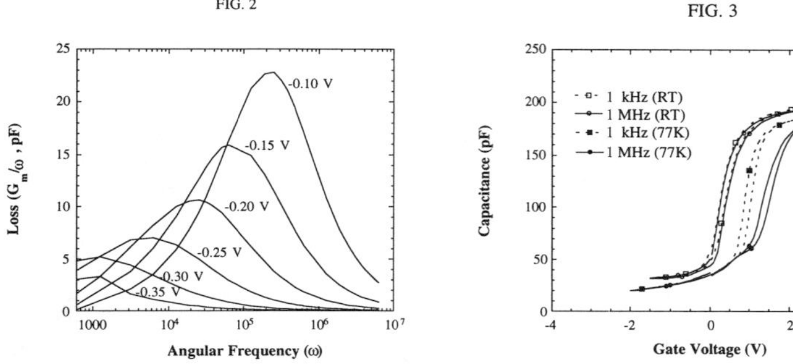

FIG. 2 The conductance loss (G$_m$/ω) - *log* angular frequency (ω) of the Si3N4/Si/n-GaAs capacitors to extract the density of surface states. The applied gate voltages are shown.

FIG. 3 Temperature dependent C-V characteristics of Si3N4/Si/ n-GaAs capacitors on GaAs(111)B substrate. The GaAs epilayers are grown at 575 °C. The samples are measured at room temperature (RT) and liquid nitrogen temperature (77 K).

Temperature dependent C-V measurement for Si$_3$N$_4$ /Si/ n-GaAs MIS diode at various frequencies is performed to confirm the unpinned GaAs Fermi energy, representing full accumulation even at 77 K (Fig. 3). It is noted that the capacitance frequency response relates to the separation of the interface pinning levels from the majority carrier band edge.[1,13] The time constant (τ) of the interface states is

$$\tau_{trap} = (\sigma_n \, \upsilon_{th} \, N_c)^{-1} \exp [\, (E_c - E_t) \, / \, k \, T \,], \qquad (3)$$

where N_C is the density of states in the conduction band ($\sim 4.7 \times 10^{17}$ for GaAs). Taking the conservative value of the capture cross section (σ_n) as 10^{-15} cm^2, the average thermal velocity (υ_{th}) as $\sim 10^7$ cm/s, and the Boltzmann's constant (k), we claim that 1 MHz frequency response at 77 K requires that the traps be within 60 meV of the conduction band of GaAs. This is possible only when the GaAs Fermi energy within GaAs band gap is unpinned. It was noted that the peak values of the measured loss for Si3N4/Si/n-GaAs structure increase rapidly with decrease in temperature, although the trap level is quite low, on the order of low 10^{11} eV^{-1} cm^{-2} (not shown). There is a noticeable frequency dispersion of C-V curves between 1 MHz and 1 kHz measured at 77 K. This dispersion suggests the existence of fast traps which are unable to follow the applied small signal at high frequencies, and thus cause conducatance loss.

CONCLUSION

Based on an *in situ* deposition approach promising interface properties of MIS diode gated on GaAs(111)B surface have been achieved for the first time with minimum interface state densities in the high 10^{10} eV^{-1} cm^{-2}. The low density of surface states as well as the presence of 1 MHz C-V at liquid nitrogen temperature confirm the realization of unpinned GaAs surface Fermi level. The MIS structures with these characteristics may be suitable for successful fabrication of GaAs-based MISFET devices. It would likely be possible even on the prepatterned GaAs(100) substrate.

ACKNOWLEDGMENTS

This work is supported by AFOSR Contract No. F49620-95-1-0298, NSF Contract No. DMR 93 -12422, and DOE Contract No. DEFG02-96-ER45439. The authors wish to thank A. Botchkarev, G. Martin and D. Jeffers for technical assistance and illuminating discussions.

REFERENCES

1. H. Hasegawa, M. Akazawa, H. Ishii, and K. Matsuzaki, J. Vac. Sci. Technol. **B7**, 870 (1989).
2. D. S. L. Mui, S. F. Fang and H. Morkoç, Appl. Phys. Lett. **59**, 1887 (1991); J. Reed, M. Tao, D. G. Park, K. Suzue, A. E. Botchkarev, Z. Fan, D. Li, S.N. Mohammad, S. J. Chey, J. V. Nostrand, D. G. Cahill, and H. Morkoç, Solid State Electron. **38**, 1351 (1995).
3. D. G. Park, Z. Chen, S. N. Mohammad, A. E. Botchkarev, and H. Morkoç, Phil. Mag. B **73**, 219(1996); Z. Chen, D.G. Park, S. Franck, S.N. Mohammad and H.Morkoç, Appl. Phys. Lett. **69**, 230 (1996).
4. A.Y. Cho, J. Appl. Phys. **41**, 2780 (1970).
5. P. Chen, K.C. Rajkumar, and A. Madhukar, Appl. Phys. Lett. **58**, 1771 (1991), J. Vac. Sci. Technol. **B 9**, 2312 (1991); K. Yang and L.J. Schowalter, Appl. Phys. Lett. **60**, 1851 (1992).
6. D.A. Woolf, D.I. Westwood, and R.H. Williams, Semicond. Sci. Technol. **8**, 1075 (1993).
7. L. Cong, F. Williamson, and M.I. Nathan, J. Electron. Mater. **25**, 305 (1996).
8. T. Hayakawa, M. Kondo, T. Morita, K. Takahashi, T. Suyama, S. Yamamoto, and T. Hijikara, Appl. Phys. Lett. **51**, 1705 (1987); Jpn. J. Appl. Phys. **26**, L302 (1987).
9. C. Mailoit and D.L. Smith, Phys. Rev.B **35**, 1242 (1987); E.A. Caridi, T.Y. Chang, K.W. Goossen, and L.F. Eastman, Appl. Phys. Lett. **56**, 659 (1990).
10. Z. Chen, S.N. Mohammad, D.G. Park, and H.Morkoç, Appl. Phys. Lett., (in press).
11. K. Adomi, S, Strite, H. Morkoç, Y. Nakamura and N. Otsuka, J. Appl. Phys. **69**, 220 (1991).
12. E.H. Nicollian and J. R. Brews, <u>MOS Physics and Technology</u>, John Wiley & Sons (1982).
13. J. l. Freeouf, D. A. Buchanan, S. L. Wright, T. N. Jackson, and B. Robinson, Appl. Phys. Lett. **57**, 1919 (1990).

INTERFACIAL LAYER FORMATION OF A HEAT TREATED TEOS BASED OXIDE PREPARED BY A PECVD TECHNIQUE

T.J. Lee, D.S. Jeong, C.S. Song, S.Y. Lee, and C.H. Park,
Samsung Electronics Co., LTD, Micro Pilot Operation Group, 82-3 Dodang-Dong, Wonmi-Ku, Buchun, Kyunggi-Do, 421-130, Korea

ABSTRACT

An interfacial layer is formed between the TEOS(tetraethylorthosilicate) based oxide and silicon substrate when the oxide is thermally treated and it shows unique properties by leaving a unremovable layer in an HF(hydrofluoric acid) dipping. While the interface formed in the temperature range of 950 - 1100 °C demonstrates hydrophilic state, it renders the surface hydrophobic in the higher range of 1150 - 1200 °C. Its characteristics are analyzed by ESCA and the measurement of water contact angle on the silicon surface after stripping the oxide.

INTRODUCTION

Oxide has been widely used as a capping material for preventing out-diffusion of dopant atoms. In the diffusion process, therefore, the oxide is typically undergone a thermal treatment at a relatively high temperature. After driving-in of dopants, the oxide layer has to be completely stripped in an HF dipping and the removal of oxide is monitored by hydrophobic nature of bare Si surface. Typically oxide layers for this purpose have been prepared by several techniques such as LP(low pressure), AP(atmospheric pressure), and PE (plasma enhanced) chemical vapor deposition. In this study, silane$\{SiH_4\}$is used as a source gas for LP, AP and PECVD, and TEOS$\{Si(C_2H_5O)_4\}$ is for PECVD. Among the oxide films produced by these methods, however, TEOS based oxide prepared by a PECVD technique shows distinct characteristics in a thermal treatment.

When the TEOS oxide films are annealed at high temperatures in the range of 850 - 1200 °C and stripped in an HF dipping, a certain unremovable layer is appeared to be remained on the top of Si surface. The occurrence of interfacial layer has been noticed by its unusual appearance of the water contact angle and a high sheet resistance on the doped silicon substrate. It is generally known that the surfaces of bare silicon wafer stripped in an HF solution and organic compounds tend to be hydrophobic while silica and silicate surfaces exhibit hydrophilic [1]. In this sense, the quantitative comparison analyses are necessary to understand the origin of thin interfacial layer of the heat treated TEOS based oxide and its properties.

EXPERIMENTAL PROCEDURE

A process sequence for the sample preparation and the measurement is shown in Fig.1. All oxide layers prepared by several different methods were annealed at the temperature of 1075 °C in a nitrogen ambient while TEOS based oxides were undergone various thermal treatments additionally. Among the samples the TEOS based oxide shows distinct behavior in the measurement of water contact angle as demonstrated in Fig.2. After removing the

Mat. Res. Soc. Symp. Proc. Vol. 448 © 1997 Materials Research Society

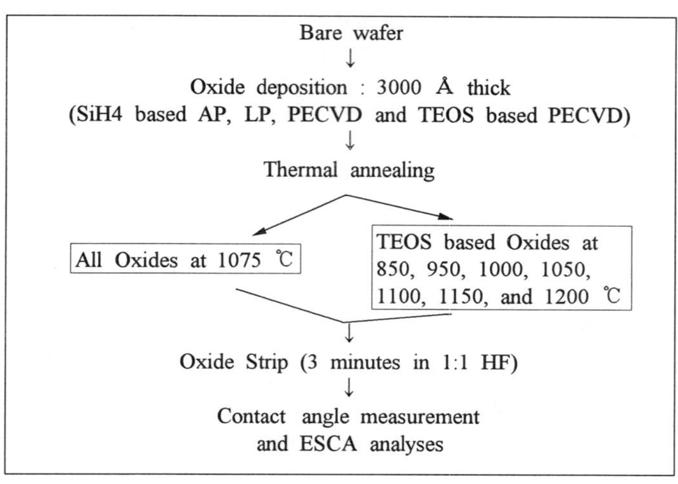

Figure 1. Process flow for sample preparation

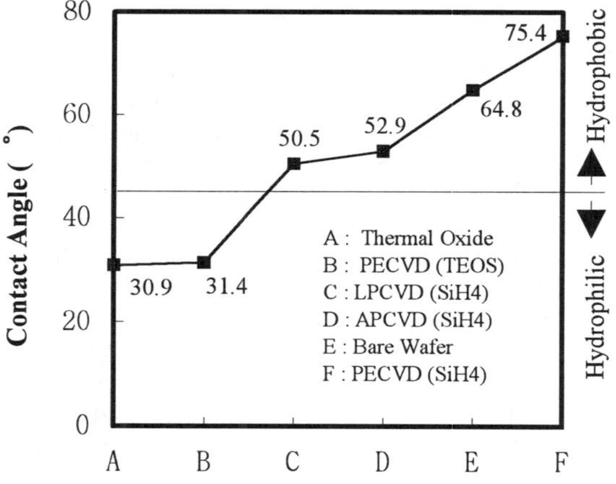

Figure 2. Contact angle measurements on the stripped silicon surface after annealing of oxides at 1075℃ and stripping in an HF solution
(* Where, the contact angle of thermal oxide (A) is measured on the oxide surface for reference.)

TEOS based oxide in an HF solution, the bare silicon surface has been appeared to be hydrophilic state with the contact angle of 31.4° which is close to the contact angle of 30.9° for the surface of thermal oxide obtained in this test.

RESULTS AND DISCUSSION

Water contact angle measurements are also performed on the TEOS oxide stripped Si surfaces which have been annealed at several different temperatures. It demonstrates that the surface state changes from hydrophilic in the temperature range of 950 - 1100 ℃ with a minimum contact angle of 31.4° at 1075 ℃ to hydrophobic in the range of 1150 - 1200 ℃ as shown in Fig.3. The water contact angles for a bare Si wafer and a thermal oxide measured in this test are represented as dotted lines for reference. Difference in surface states of oxide stripped silicons is easily discernible and it is obviously indicative of variations in elemental compositions of the interface at the given annealing temperature.

Since carbon atoms are contained in a TEOS source gas, the movement of carbon atoms in high temperature environment is expected to be important to understand the interfacial layer [2,3]. Therefore, surface analyses using ESCA have been performed to investigate the details of composition of the thin interfacial layer formed by annealing of TEOS based oxide on the silicon substrate.

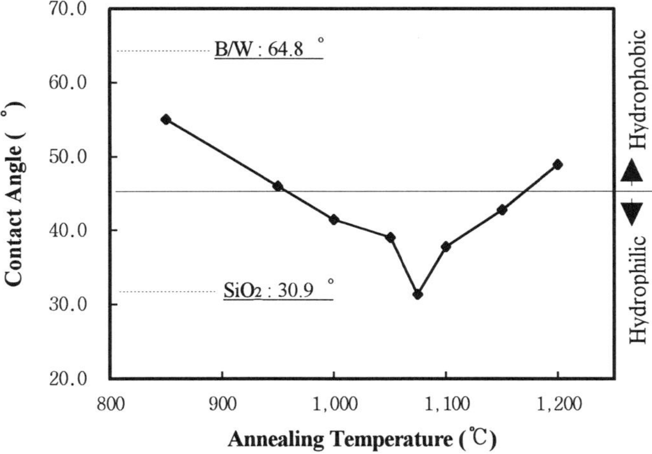

Figure 3. Contact angle measurement on the TEOS oxide stripped silicon surface after thermal treatment at various temperatures
(* The minimum value of water contact angle is obtained at the temperature of 1075 ℃.)

To enhance the resolution of depth profile of each atomic constituent, a shallow detection angles of 15° and a nomal angle of 30° to the photoelectron analyzer were used separately. With a nomal angle of detection, relative atomic compositions shown in Fig.4 of the interfacial layers are obtained for each different thermal condition. It is represented as the ratio of SiO_x + SiO_2 to SiC and the data show 1.8:1, 2.2:1, and 0.9:1 for 850 ℃, 1075 ℃, and 1200 ℃, respectively. From this result, it is believed that an oxygen-rich phase is found at the interface as the annealing temperature is increased and its formation reaches a maximum amount at 1050 ℃. This also indicates that oxygen rich phase at the surface results in a hydrophilic state.

In addition, variations in the amount of SiC composition in the interlayer is also noticed by the depth profiling of atomic constituents with a shallow detection angle of 15° as shown in Fig.5. Each raster scan indicates depth profiling of silicon and carbon atoms from the top surface of the interfacial layer to the bulk silicon with a total information depth of about 50 Å. The depth profiles on the stripped Si surface show the prominent increase of the carbon peaks at the position of 283.4 eV for the higher temperature annealing of 1200 ℃. The peaks at the position of 283.4 eV are attributed to the carbon atoms originated from the TEOS oxide and caused by the SiC composition in the layer. It represents that carbon atom diffusion from the TEOS oxide to the silicon surface and carbidization at the silicon surface is enhanced by high temperature annealing. It also shows a gradual change of the interlayer to a carbon rich phase and it renders the surface hydrophobic by the contribution of organic compounds [4].

CONCLUSIONS

We have shown that the relative quantity of each element in the interfacial layer depends upon the annealing temperature of TEOS based oxide while the interfacial layer is a mixed

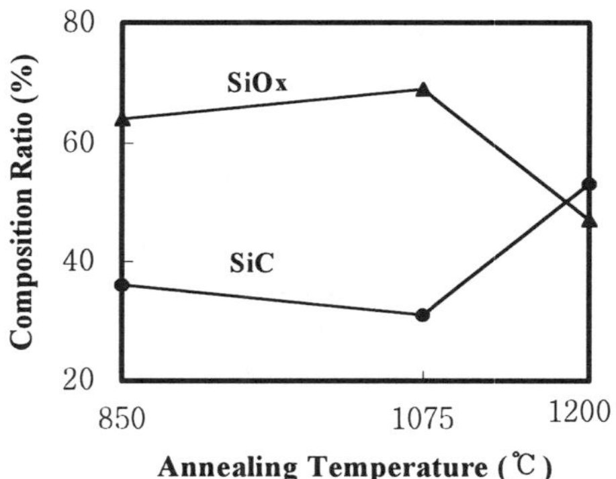

Figure 4.   The relative compositions of the interfacial layer at different annealing temperatures of 850 ℃, 1075 ℃, and 1200 ℃.

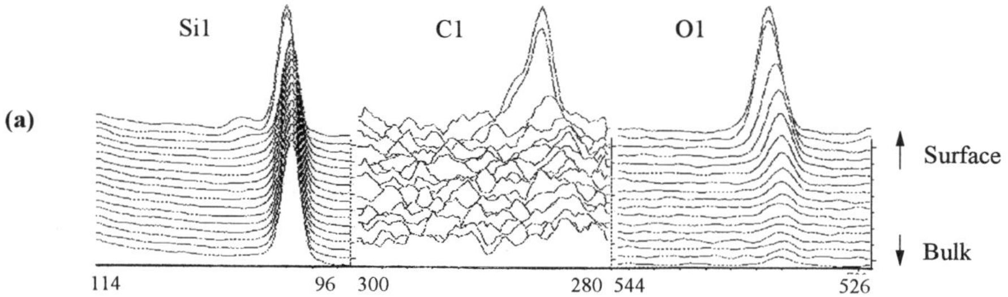

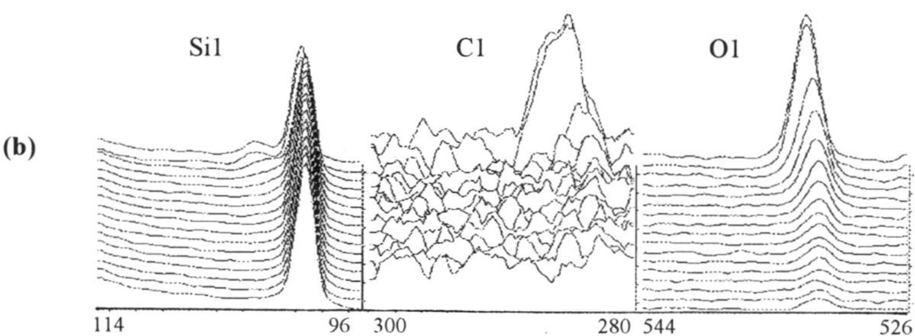

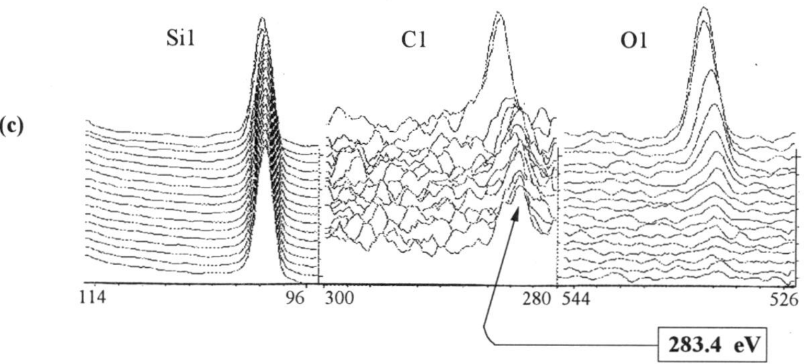

Figure 5. Depth profiles obtained by ESCA on the stripped silicon surface indicate prominent increase of carbon peaks at 283.4 eV. These peaks are originated from the SiC composition in the thin layer. In this figure (a), (b), and (c) represent samples annealed at different temperatures of 850℃, 1075℃, and 1200℃, respectively.

337

state of silicon, oxide, and carbon. At the annealing temperature range of above 950 ℃, the formed interlayer is unremovable in an HF dipping, which is due to incorporation of carbon atoms diffused from TEOS oxide to the surface of silicon substrate. In addition, the relative ratio of oxygen and carbon elements in the layer determines the surface state to be either hydrophilic or hydrophobic.

ACKNOWLEDGMENTS

The authors wish to thank H.S. Yang and H.K. Song for ESCA analyses and J. W. Kim of Dong-Jin Chemicals for the water contact angle measurements.

REFERENCES

1. J.H. Shen and K. Klier, J. Coll. Interface Sci. 75, 5667 (1980).

2. S. Nguyen, D. Dobuzinsky, D. Harmon, R. Gleason and S. Fridmann, J. Electrochem. Soc. 137, 2209 (1990).

3. Ismail T. Emesh, Giulio D'Asti, Jacques S. Mercier and Pak Leung, J. Electrochem. Soc. 136, 3404 (1989).

4. Gregg S. Higashi and Yves J. Chabal in <u>Handbook of Semiconductor Wafer Cleaning Technology</u>, edited by W. Kern, Noyes Publication, New Jersey, 1993, pp. 433-496.

Field emission through diamond/Mo interfaces

W.B. Choi, A.F. Myers, J.J. Cuomo, J.J. Hren

Department of Materials Science and Engineering
North Carolina State University, Raleigh, NC 27695, USA
E-mail: wchoi@eos.ncsu.edu

Abstract

The influence of interfacial nanostructure on electron emission from diamond coated Mo emitters is presented. Diamond coating is known to increase electron emissivity, but interfacial oxides and an amorphous layer change the magnitude. After annealing, emissivity was enhanced further and Mo_2C was formed at the interface, as well as a decrease in the oxides and amorphous layers. The shape of the FEM image changed but no individual emitting sites were observed. Possible mechanisms causing these changes are discussed.

Introduction

Recently there has been increased interest in vacuum microelectronic devices, especially for flat panel displays. To improve field emission properties, diamond field emitters are being developed and show promising results [1-4]. Diamond possesses many desirable properties, such as a wide band gap (5.4 eV), current stability, and negative electron affinity (NEA) on several faces after certain surface treatments [5]. Previous results indicate that high current emission at a low electric field is possible from diamond coated field emitters, but the reasons are not clear. Several studies explain the high emissivity as due to defects state in the gap[6], others due to conducting channels in the diamond [7] and high n-type doping [2].The high emissivity from diamond coatings can be explained by the low electron effective mass in diamond and by substitutional nitrogen doping, which produces a state located 1.7eV below the conduction band minimum [8].

Even if the diamond/vacuum surface exhibits NEA, the diamond/metal interface remains a barrier to the electron supply. A lower interfacial barrier could provide higher electron emissivity and more stable electron emission. However, it is well known that the electrical properties are sensitive to the electronic structure of interfaces. Research has been reported on metal/diamond interfaces. Carbide formation tends to show ohmic contact at the interface such as Moazed et al. [9] reported an ohmic contact on Mo/diamond interfaces after annealing. Tachibana et al.[10] and Gildebalt et al. [11] suggested a titanium carbide formed during annealing may alter the Schottky barrier to form an ohmic contact. S. Mikhailov et al.[12] reported the formation of Mo_2C carbide at a temperature as low as 400 °C on a Mo deposited CVD diamond.

In this study, diamond coating and subsequent annealing effects on the interface between diamond and Mo are correlated with the electron emission properties. To clarify the mechanisms, I-V characteristics, FEM images, and electron energy distribution data were taken before and after annealing on the same emitters and the same interface was examined later by TEM.

Mat. Res. Soc. Symp. Proc. Vol. 448 © 1997 Materials Research Society

Experimental Procedure

Mo emitters were fabricated by electrochemical etching. High pressure high temperature (HPHT) type Ib diamond powder was deposited on the emitters by dielectrophoresis [13]. The current-voltage (I-V) characterization was carried out in a combined field emission microscope (FEM) and anode Faraday cup system under an ultra high vacuum (10^{-9} Torr) with a tip-to-anode distance of 10 mm. The emitting current was measured by an electrometer and the data was recorded by computer through an IEEE-488 interface, with each data point sampled 60 times at time intervals of 0.25sec. Before measuring I-V data, the tip was conditioned at a current of 100nA to remove the absorbed gas on the tip surface. Diamond coated emitters were annealed in a hydrogen atmosphere at 400 - 600 °C for times varying from 30 min. to 2hrs. After I-V characterization, field emission images from the same emitter were observed on a Chevron microchannel plates-fiber optic imaging assembly under a vacuum of 10^{-9} Torr. The diamond coating morphology and the diamond/Mo interface microstructure were analyzed by both scanning electron microscopy (SEM) and high resolution transmission electron microscopy (HRTEM).

Results and Discussion

The field emission I-V characteristics of each Mo field emitter was measured before and after diamond deposition (Fig.1). After diamond coating (coating thickness 0.2μm) the electron emissivity of each emitter was found to increase several orders of magnitude, which is comparable to the data presented earlier [3,4]. In conjunction with I-V data, changes in the spatial distribution of the field emission distribution were viewed on imaging assembly as shown in Fig.2. Field emission was confined to one area, which appeared more uniform and considerably brighter after coating. After coating the image was also somewhat compressed compared with the uncoated emitters. However , several emitting spots, which would relate to certain crystal faces or small protrusions, were not observed.

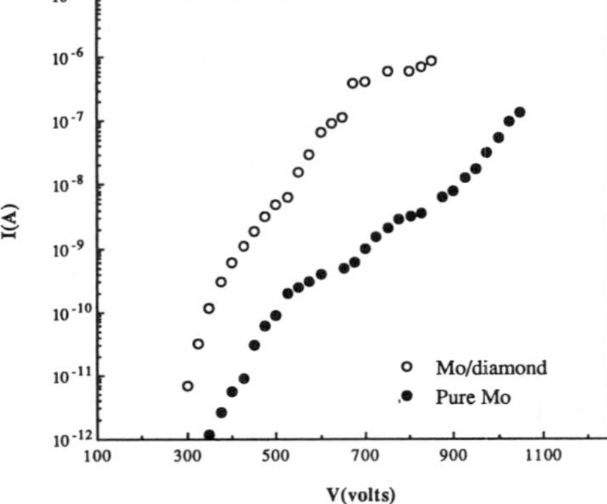

Fig. 1 I-V characteristics of a diamond coated Mo emitter before and after diamond coating. The tip-to anode distance was 10mm. The tip height was 1.5mm and the Mo tip radius was 80nm.

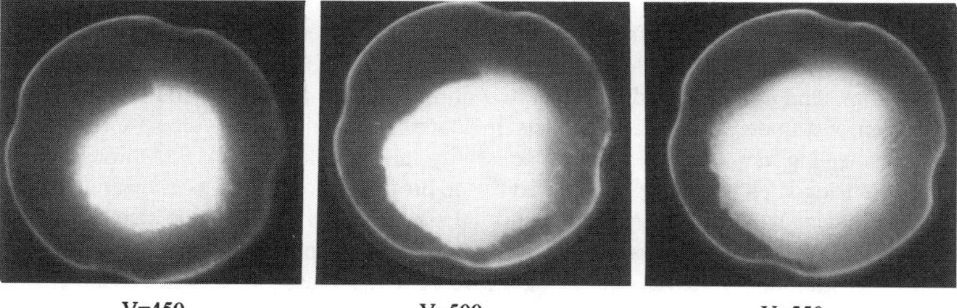

Fig.2 Field emission images from a diamond coated Mo emitter corresponding to the I-V data in Fig.1. After diamond coating, the electron beam was confined to on one area, whereas before coating, several area showed emission. The size and brightness of the beam increased when the applied field was increased.

It was found from HRTEM analysis that a native oxide MoO_3 (~100Å thick) and an amorphous layer exist at the diamond/Mo interface. The oxides were naturally present after specimen preparation and amorphous layers are probably due to impurities surrounding the diamond powder [15]. However, electrons can tunnel through such a thin amorphous interfacial layer, even though it may produce a potential drop. In fact, the result of our previous electron energy analyses showed that the position of the energy distribution peak dropped linearly according to the bias applied to the diamond coated emitters [16]. This peak shift was explained in terms of a potential drop which could be related to the Mo/diamond interfaces and diamond bulk (defects, grain boundary, etc.). If electrons are emitted at the minimum of the conduction band of the diamond, then the peak position is related to the potential drop across the interfaces and across the diamond bulk. Field penetration seems the most plausible means for electrons from the metal tip to be transferred through the interface and the diamond and then emitted at the diamond surface (Fig.3).

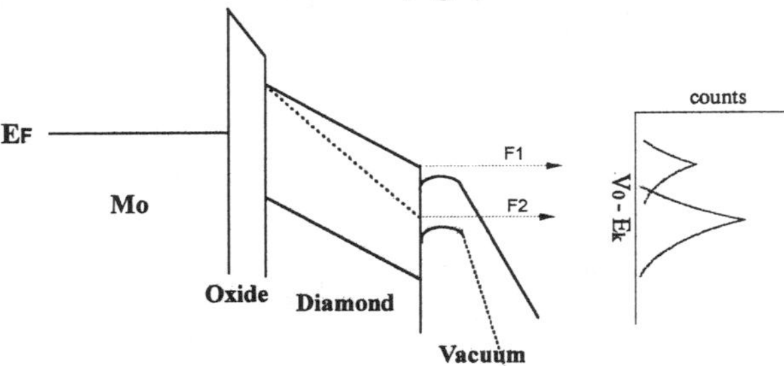

Fig. 3 The energy band diagram of a diamond coated Mo emitter. Electrons tunnel through a thin oxide layer (100Å) to the conduction band of the diamond, travel through the diamond, and emitted into the vacuum. The electron energy distribution is shown on the right. The peak shift and height increased when the applied field increased.

Our studies also found from HRTEM results that regions of small Mo_2C carbide particles were formed at the Mo/diamond interface during annealing at ~435 °C (Fig.4). The lattice constant of diamond is a=3.561Å and that of Mo_2C is a=2.61Å. Thus the lattice mismatch between diamond and Mo_2C leads to defect formation at the interface during carbide formation. Tachibana et al.[10] suggested that such defects could be generated during carbide formation and thereby reduce the depletion layer of diamond at the diamond/metal interface. The result of on I-V characterization show clearly that annealing increased the electron emissivity substantially (Fig.5). FEM image shows that no specific emitting sites which may be related to the protrusion or second phase carbides were not observed on the annealed emitter. The beam brightness increased and emitting shape changed after annealing (Fig.6). It seems likely that carbide formation plays a prominent role. Carbide can enhance the electron emission in several reasons: First, a carbide may increase local field enhancement at the diamond / Mo interface. Second, defect generation at the diamond / Mo interface during carbide formation may produce defect subband in the diamond band gap and may reduce the depletion layer. Finally, the carbide creates a good chemical contact between diamond particles and the Mo substrate.. Our previous electron energy distribution measurements on an annealed diamond/Mo emitter, showed that the potential drop decreased and showed non-linear behavior with increasing applied bias. The potential drop reduced by 2eV, compared to an unannealed diamond coated emitter [14]. This reduction seems to be related to the interfacial structure change as follows. When the electron barrier is reduced after annealing, electrons are injected into the diamond conduction band with higher energies than the minimum of the conduction band. In the gradient of the penetrating field, the hottest electrons can ionize impurities, leaving holes, even leading to avalanche breakdown [17]. The reduced potential drop and enhanced emissivity can be explained by the interfacial structure change and change in the electron transport mechanism (Fig.7).

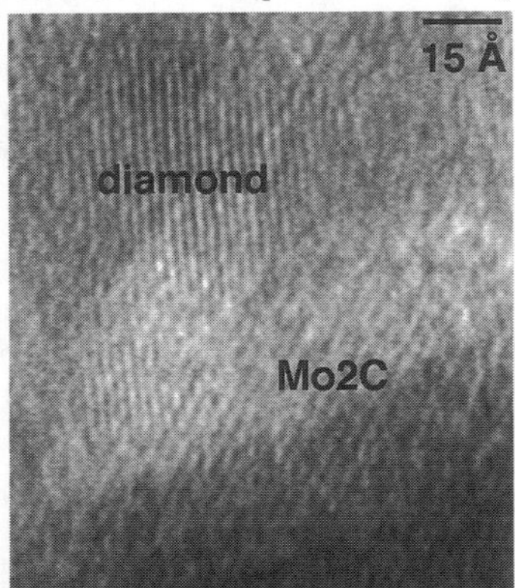

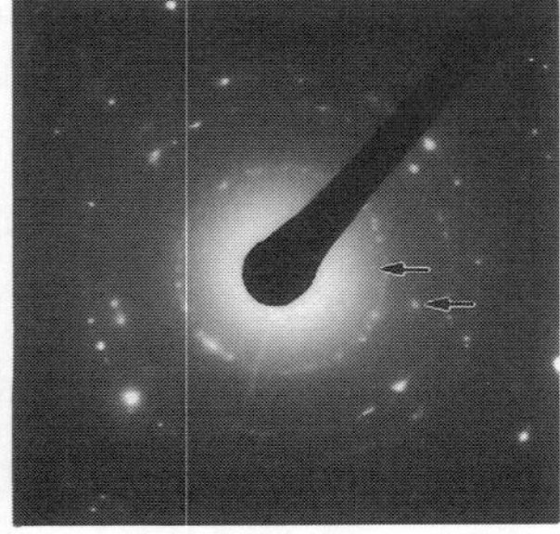

Fig. 4 HRTEM image and selected area electron diffraction of a diamond powder/Mo interface after annealing. Lattice fringes with a spacing of 2.61 Å are visible in this image; d_{021} (α-Mo_2C)=2.613 Å and d_{100} (β-Mo_2C)=2.608Å, so the 2.61 A lattice fringes could correspond to either one of these carbides. Faint Mo_2C rings are visible on SAED, as indicated by the arrow.

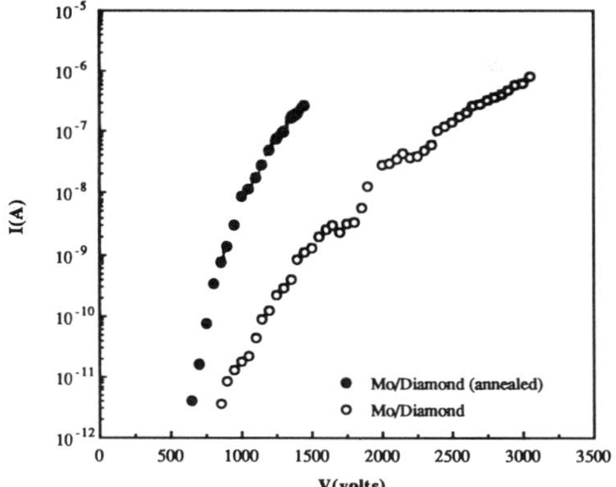

Fig. 5 I-V characteristics of a diamond coated Mo emitter before and after annealing. The Mo tip radius was 100nm. The thickness of the diamond coating was 1 μm.

Fig.6 Field emission image from a diamond coated Mo field emitter after annealing. The thickness of the diamond was 1μm. After annealing, the beam brightness was increased but, no other emitting sites which may relate to the protrusion or second phase were not observed.

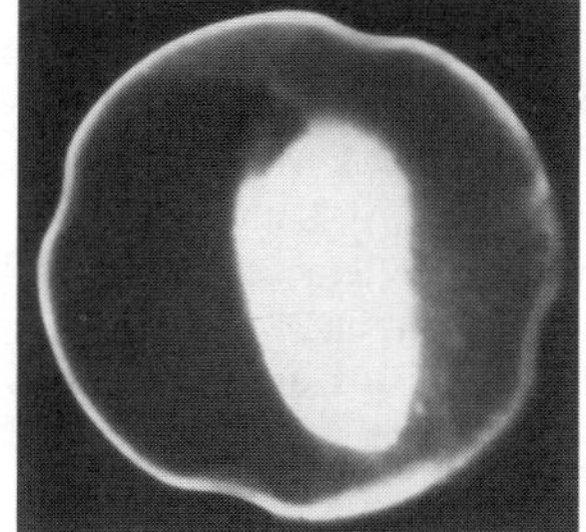

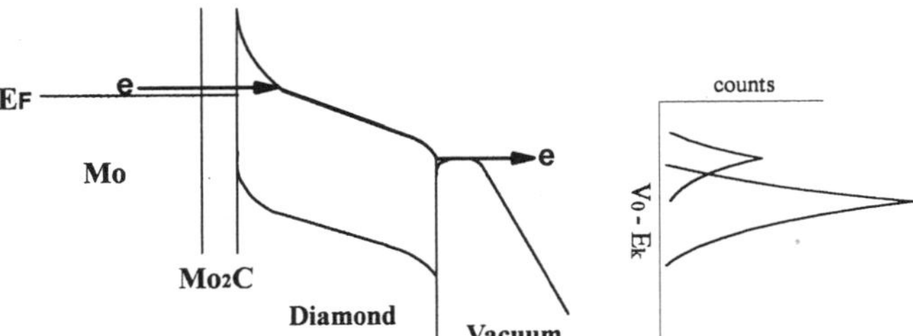

Fig.7 The energy band diagram of a diamond coated Mo emitter after annealing. Mo$_2$C phase was formed at the Mo / diamond interfaces. The energy barrier for electron tunneling was lowered by carbide formation and field induced hot electrons are emitted. The electron energy distribution shows that the peak shift is reduced and peak height is increased after annealing.

In summary, FEM images of diamond coated emitters confirm the greatly increased brightness seen in I-V measurements and showed stable, relatively uniform emission, somewhat compressed compared with uncoated emitters. Mo_2C precipitates along with Mo oxides were found at the Mo/diamond interface of the annealed emitter, but amorphous layers were removed. The change in emission mechanism of the diamond coating and following annealing was discussed: a field penetrating model for the diamond coated emitter and hot electron emission model for the diamond coated annealed emitter seems the most plausible.

Acknowledgments

The authors wish to thank Dr. D. Ming, G. Wojak and Dr. V. Zhirnov for useful discussion.

References

1. K. Okano, S. Koizumi, S. Ravi, P. Silva and G.A. Amaratunga, Nature 381, 140 (1996)
2. M.W. Geis, J.C. Twichell, and T.M. Lyszczarz, J. Vac. Sci. Technol. B 14 2060 (1996)
3. W.B. Choi, J.Liu, M.T. McClure, A.F. Myers, J.J. Cuomo and J.J. Hren, J. Vac. Sci. Technol. B14, 2050 (1996)
4. E.I. Givargizov, V.V. Zhirnov, A.N. Stepanova, E.V. Rakova, A.N. Kiselev, and P.S. Plekhanov, Appl. Surf. Sci. 87/88 24 (1995)
5. J. Van der Weide, Z. Zhang, P.K. Baumann, M.G. Wensell, J. Bernhloc, and R.J. Nemanich, Phys. Rev. B 50, 5803 (1994)
6. Z.H. Huang, P.H. Cutler, N.M. Miskovsky, and T.E. Sullivan, Appl. Phys. Lett. 65, 2562 (1994)
7. N.S. Xu, Y. Tzeng, and R.V. Latham, J. Phys. D 26. 1776 (1993)
8. Victor, J. de Physique, 1996 (accepted for publication)
9. K.L. Moaxed, J.R. Zeidler, and M.J. Taylor, J. Appl. Phys. 68, 2246 (1990)
10. T. Tachibana, B.E. Williams, and J.T. Glass, Phys. Rev. B 45, 11 975 (1992)
11. G.Sh. Gildenblat, S.A. Grot, C.W. Hatfield, A.R. Badzian and T. Badzian, IEEE Electron Device Lett. 11, 371 (1990)
12. S.Mikhailov, D. Ariosa, J. Weber, Y. Baer, W. Hanni, X.M. Tang and P. Alers, Diamond Relat. Mater., 3, 61 (1993)
13. W.B. Choi, J.J. Cuomo, V.V. Zhirnov, A.F. Myers, and J.J. Hren, Appl. Phys. Lett. 68, 720 (1996)
14. F. Charbonnier, Appl. Surf. Sci. 94/95 26 (1996)
15. A.F. Myers, Ph. D thesis, North Carolina State University, (1996)
16. W.B. Choi, M.T. McClure, R. Schlesser, Z. Sitar and J.J. Hren, J. de Physique, 1996 (acccepted for publication): The electron energy distribution was analyzed on Mo emitter, after diamond coating , and subsequent annealing. The peak position of the pure Mo was not changed at 4.2eV but diamond coated Mo changed linearly from 4.2eV to 8.5eV. After annealing, the degree of shift reduced and changed non-linearly.
17. T.H. Distefano and M. Shatzkes, J. Vac. Sci. Technol., 13, 50 (1976)

SUPPRESSION OF SURFACE SiO₂ LAYER AND SOLID PHASE EPITAXY OF AMORPHOUSLY DEPOSITED SI FILMS USING HEATING-UP UNDER Si₂H₆ ENVIRONMENT

TAE-HEE CHOE, SE-JUNE KIM, WOON CHOI, and HYOUNG-JUNE KIM
Department of Metallurgy & Materials Science, Hong-Ik Univ.,
72-1 Sangsu-Dong, Mapo-Gu, Seoul, Korea 121-791

ABSTRACT

A novel technique to realize a selectively grown Si epitaxy layer has been developed. This technique consists of deposition of amorphous Si with oxide-free Si surface, selective solid phase epitaxial (SPE) growth on Si windows, and etching of uncrystallized amorphous Si on SiO_2 layer. Formation of surface oxide can be effectively suppressed by flowing Si_2H_6 gas during heating-up stage to the deposition temperature. This method enables us to grow epitaxial layer without any high temperature cleaning procedures. Substantially higher growth rate of vertical SPE on Si windows over lateral SPE on SiO_2 regions allows the growth of thick SPE layer with a minimized lateral overgrowth. With a proper etching solution, the remaining amorphous Si on SiO_2 layer can be readily etched to form a selectively defined epitaxial layer on Si windows.

I . INTRODUCTION

Fabrication technique to realize a selectively grown Si epitaxial layer at low temperatures has great potentials for novel device structures for ultra-large-scale integrated (ULSI) circuits. The most common technique to achieve this has been the selective epitaxial growth (SEG) which facilitates an in-situ selective growth during the vapor-phase-epitaxial process. Usually, SEG could be obtained at high temperatures above 900 ℃. Recently, there have been extensive efforts to lower the SEG temperatures[1-4]. Main scheme to achieve this is the rigorous reduction of H_2O or O_2 contamination in epitaxial reactor as well as process gases. The low temperature SEG was reported by atmospheric pressure CVD (APCVD) using high purity $Si_2H_2Cl_2$ gas, ultrahigh vacuum CVD (UHVCVD) and gas source molecular beam epitaxy (gas-source MBE) using SiH_4 or Si_2H_6 gas[4-7]. Even though SEG could be obtained at low temperatures by these methods, process regimes were extremely limited.

We report an alternative way of achieving SEG layer at low temperatures. This process uses the solid-phase-epitaxy (SPE) of amorphously deposited Si layer. After the deposition of amorphous Si at low temperatures, annealing is performed in the regime where SPE growth of amorphous Si on Si surfaces selectively occurs and amorphous Si on SiO_2 remains as amorphous state. Finally, the amorphous Si on SiO_2 is selectively etched using a proper etching solution.

Normally amorphously-deposited films on Si substrates become polycrystalline Si referred as solid phase crystallization(SPC), due to the presence of oxides at amorphous Si/Si substrate interfaces[8]. Therefore, for SPE of amorphous Si, oxide-free interfaces are prerequisite, which also requires a high temperature cleaning process. It is known that air-stable, oxide-free surface can be obtained by passivating Si surface with hydrogen using proper surface treatments such as HF dipping[9-12]. However, the passivating H atoms are known to be desorbed from Si surfaces at low temperatures of 400 ~ 500 ℃, resulting reactive bare Si surfaces[13]. These reactive Si surfaces are rapidly reoxidized by absorption of oxygen atoms from water vapor or oxygen in the reaction chamber. Meyerson et al.[14] suggested the oxide-free surfaces and thus

345

Mat. Res. Soc. Symp. Proc. Vol. 448 © 1997 Materials Research Society

successful epitaxial growth can be obtained either in low temperature regime (<600 ℃) where H passivation persists, or in high temperature regime (>800 ℃) where oxide-free surface is thermodynamically stable. While epitaxial growth of high temperature regimes strongly depends on vacuum conditions, that of low temperature for H passivation will be somewhat independent of environment.

An effective method utilizing the low temperature epitaxy regime has been reported by Kobayashi et al.[13]. They could realize the oxide-free surface and epitaxial growth using the conventional LPCVD by flowing SiH_4 gas during the heating-up stage for epitaxial growth. This method introduces SiH_x environment in the temperature ranges where extensive H desorption occurs, allowing Si surfaces to be adsorbed by SiHx species instead of oxygen. This method seems effective in realizing amorphous Si films without interface oxides at low temperatures. We applied this method using Si_2H_6 gases, taking account of high decomposition rate of Si_2H_6 at these temperatures (400 ~ 500 ℃).

The purposes of this work are twofold. First is to explore the possibility of realizing selective epitaxial layer using SPE and selective etching. Second is to investigate the effect of in-flowing Si_2H_6 gases during heating-up stage on the suppression of interface SiO_2.

II. EXPERIMENTS

P-type (100) Si wafers with SiO_2 patterns using conventional LOCOS processes were used as substrates. These substrates were RCA-cleaned and UV ozone-cleaned. The UV ozone cleaning is known to be effective in removing the hydrocarbon contamination. Subsequently Si surfaces were passivated by HF dipping (7:1 diluted). Amorphous Si films were deposited using the high vacuum chemical vapor deposition system (HVCVD) with a load lock. The initial vacuum pressure of CVD reactor was 10^{-8} Torr. The deposition was carried out at 480 ℃ at 0.5 Torr using a Si_2H_6 gas. In order to examine the sensitivity of SPE growth on O_2 or H_2O contamination, deposition was carried out with various deposition set-up such as base pressure, uses of load lock, and heating-up under Si_2H_6 gas, as shown in table 1. For SPE growth samples were heat treated at 600 ℃ under N_2 ambient. Film microstructures were investigated using transmission electron microscope (TEM), x-ray diffractometer (XRD), and Raman spectrometer.

III. RESULTS

Occurrence of SPE was investigated by XRD studies for various deposition processes, as referred in table 1. When deposited amorphous Si is crystallized to polycrystalline Si, films show strong (111) textures. Fig. 1 shows the (111) peaks in the samples after anneals at 600 ℃ for 20 hrs for complete crystallization. Samples without flowing Si_2H_6 gases during heating-up (processes D and E), show the (111) peaks, indicating that amorphous Si becomes polycrystalline Si. In contrast, samples with Si_2H_6 gas (processes A~C) show absence of (111) peaks indicating the occurrence of SPE. The observation of SPE for A~C processes reveals that SPE can be obtained without high vacuums or use of load lock.

Detail microstructures were observed by TEM, as shown in Fig. 2a and b, for the process E. The films have polycrystalline structure in agreement with the XRD result. A high resolution TEM micrograph reveals that about 20 Å-thick oxide layer exists at the interfaces. Fig. 3a and b are for process C. Amorphous Si is SPE-grown to form an epitaxial layer. Any oxide layers are not observed, even though some oxide clusters still exist at interfaces. These results lead to conclusions that the suface oxide layer forms in the course of heating-up stage toward the temperature of Si deposition (480 ℃). Formation of these oxides seems unavoidable even in high vacuum (10^{-8} Torr) and high heating rate (100 ℃/min.) applied in this work.

Table Ⅰ. Various process conditions of amorphous Si depositions.

Process	Heating-up under Si$_2$H$_6$ gas environment	Base Pressure	Load-Lock
A	◯	10^{-3} Torr	◯
B	◯	10^{-8} Torr	×
C	◯	10^{-8} Torr	◯
D	×	10^{-3} Torr	×
E	×	10^{-8} Torr	◯

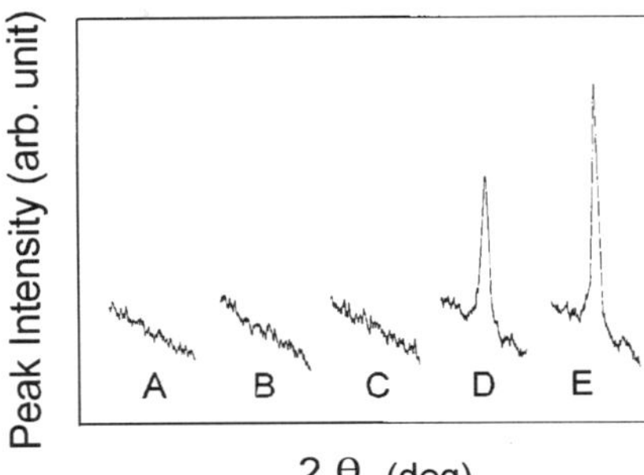

Fig. 1. X-ray (111) diffraction peaks of crystallized Si films anneals at 600 ℃.

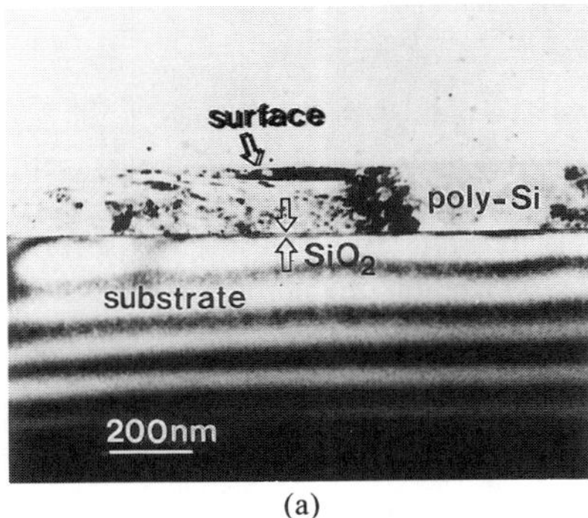

(a)

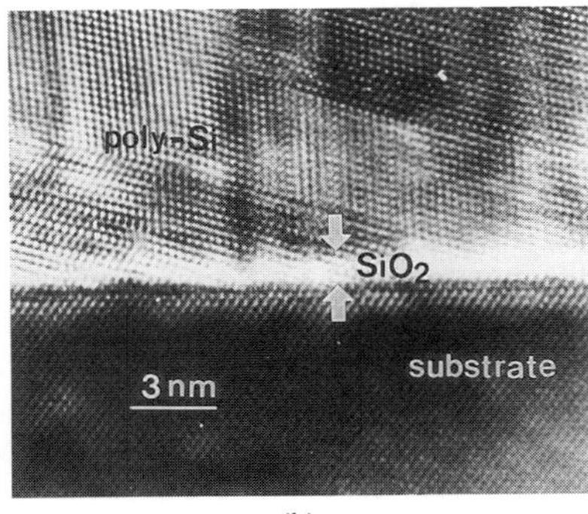

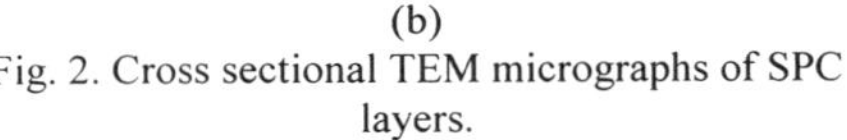

(b)

Fig. 2. Cross sectional TEM micrographs of SPC layers.

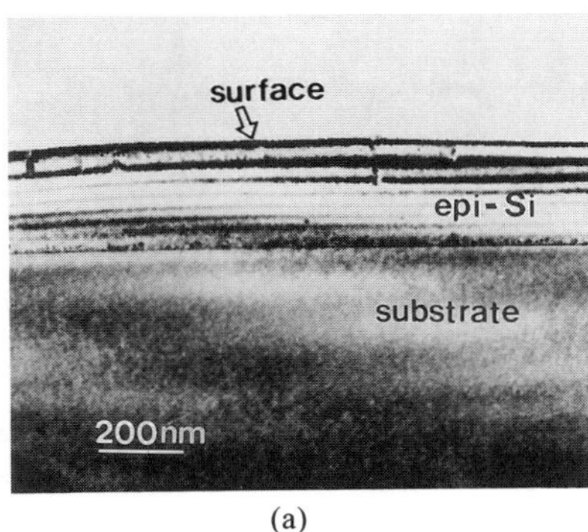

(a)

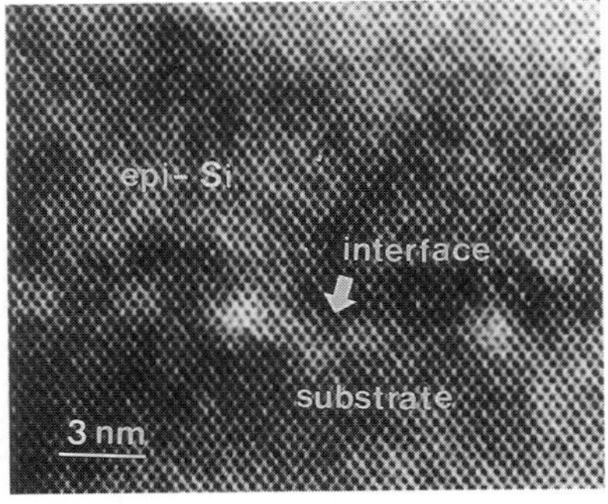

(b)

Fig. 3. Cross sectional TEM micrographs of SPE layers.

Crystallinities of SPE layer were evaluated by Raman spectrocopy. For this study 5000 Å-thick amorphous layers were deposited with process C, and SPE-annealed. Fig. 4 shows the Raman spectra. The SPE layer shows a sharp peaks at 521 cm^{-1} with a FWHM (full width at half maximum) of 5.4 cm^{-1}. These values are close to those of single crystalline Si wafer, 521 cm^{-1} and 5.7 cm^{-1}, indicating a high crystalline quality of the SPE layer.

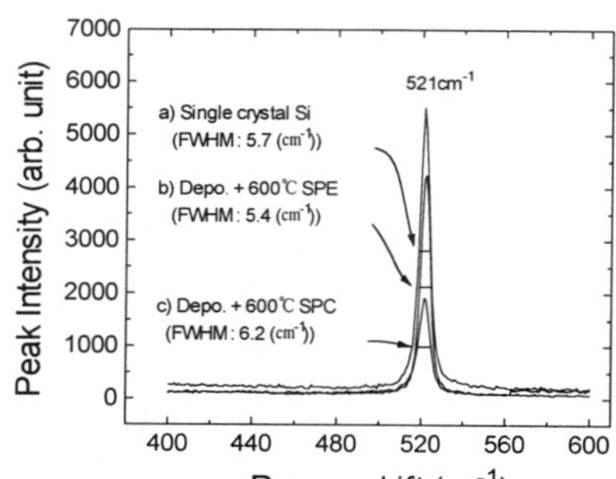

Fig. 4. Raman spectra for a) Si wafer, b) SPE films, and c) SPC films

Anneals for Selective SPE Growth

In order to obtain SPE layers grown exclusively on the exposed Si windows, vertical SPE growth should be completed with a minimized lateral growth over SiO$_2$ regions. Also, amorphous Si on SiO$_2$ layer should remain as amorphous state after SPE anneals. We performed SPE annealing at 600 °C. Incubation time for random crystallization of amorphous Si on SiO$_2$ is estimated to be approximately 3 hrs at 600 °C. Fig. 5 shows the TEM micrographs showing the SPE growth on LOCOS regions. While high quality SPE layer is observed for vertical growth, extensive twin defects are observed in the regions of lateral growth. Growth rates of vertical and lateral growth were measured by TEM works and shown in Fig. 6. For SPE growth at 600 °C, vertical growth rate is about 360 Å/min., which is 9 times higher than that of lateral growth (40 Å/min.). Taking account of these SPE growth rates and incubation times, we can grow the selective SPE layer to the maximum thickness of 4 μm with lateral overgrowth of 0.5 μm before random crystallization of amorphous Si on SiO$_2$.

Selective Etching for Amorphous Si on SiO$_2$

Realization of self-aligned formation of SPE layer required a selective etching of uncrystallized amorphous Si layer on SiO$_2$. We used a etching solution composed of 40% HNO$_3$, 49% CH$_3$COOH, and 1% HF. Using this solution we could etch the amorphous Si layer without noticeable etching of epi-Si layer, as shown in Fig. 7. This solution does not attack the underneath SiO$_2$ layer either. Fig. 8 shows a SEM micrograph showing the selectively defined SPE-epi layer after etching. This SPE layer were grown at 600 °C for 1hr. Well-defined SPE regions can be observed exclusively on Si windows. TEM micrograph after selective etching reveals that the irregularities of edge regions of SPE layer adjacent to SiO$_2$ layer can be minimized within 500 Å (Fig. 9.).

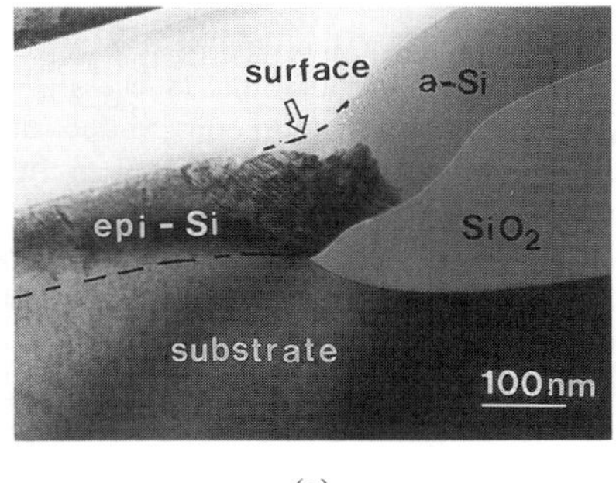

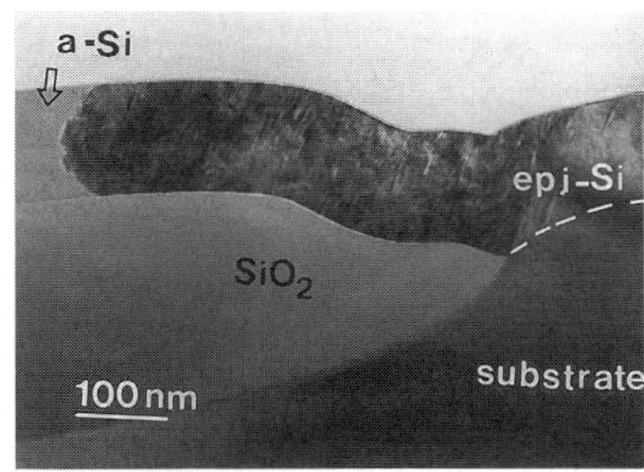

(a)   (b)

Fig. 5. Cross sectional TEM micrographs of SPE growth after anneals at 600 ℃ for a) 1hr and b) 2hr

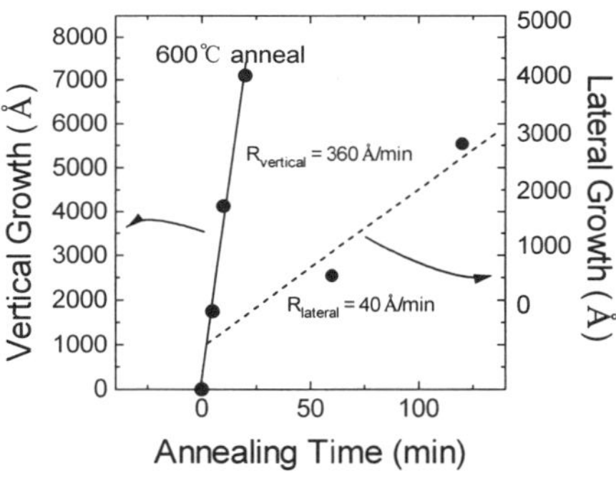

Fig. 6. Rates of vertical and lateral growth of amorphous Si films plotted vs. annealing time.

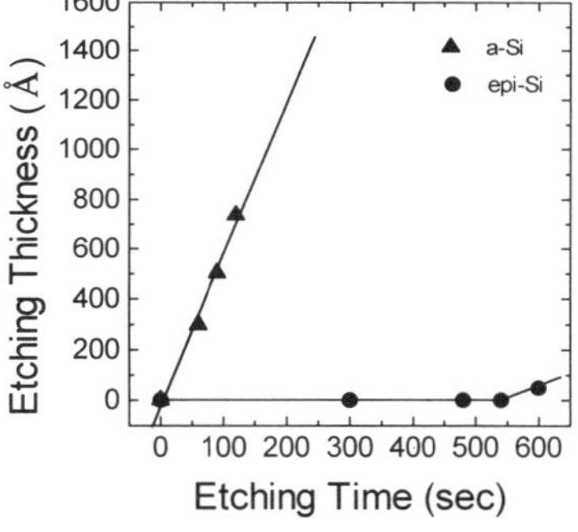

Fig. 7. Etching selectivity of selective etching solution (HNO_3 : CH_3COOH : HF = 40 : 49 : 1).

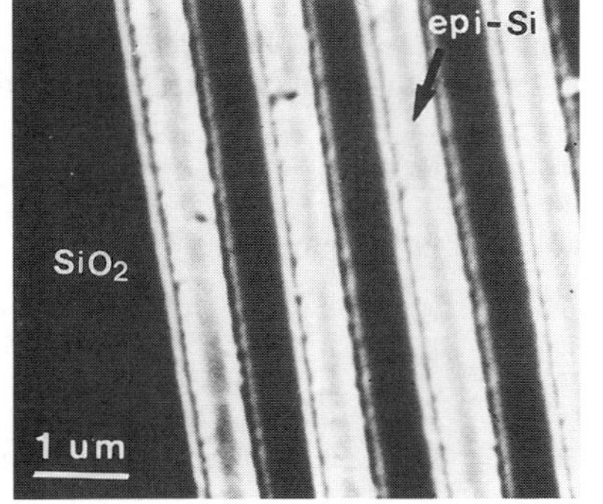

Fig. 8. SEM micrographs showing selective SPE layers after selective etching.

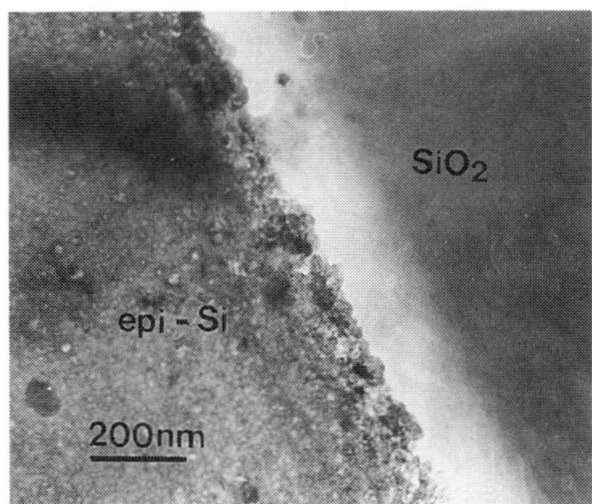

Fig. 9. TEM micrographs showing selective SPE layers after selective etching.

IV. CONCLUSIONS

A novel technique for fabricating selective epitaxial layer through SPE growth and selective etching has been investigated. Suppression of surface oxides and, thus, successful SPE growth were obtained by heating-up with Si_2H_6 environments during amorphous Si deposition. Selective SPE growth on Si windows avoiding random crystallization of amorphous Si on SiO_2 is possible in wide SPE-anneal process window. A substantially higher growth rate of vertical SPE on Si windows than that of lateral SPE over SiO_2 layer allows growth of thick-SPE layer without significant lateral growth. The amorphous Si layer on SiO_2 can be readily removed by a proper etching solution. Through these processes a selective epitaxial layer can be successfully fabricated at low temperatures (<600 ℃) and without ultra-clean environment.

ACKNOWLEDGMENTS

This work is supported by Hyundai Electronics Industries Co., LTD.

REFERENCES

1. T.O. Segwick, M. Berkenblit, and T.S. Kuan, Appl. Phys. Lett. 54, 2689 (1989).
2. T.O. Segwick, P.D. Agnello, D. Nguyen Ngoc, T.S. Kuan, and G. Scilla, Appl. Phys. Lett. 58, 1896 (1991).
3. G.R. Srinivasan and B.S. Meyerson, J. Electrochem. Soc. 134, 1518 (1987).
4. B.S. Meyerson, Appl. Phys. Lett. 48, 797 (1986).
5. H. Hirayama, T. Tatsumi, A. Ogura, and N. Aizaki, Appl. Phys. Lett. 51, 2213 (1988).
6. H. Hirayama, T. Tatsumi, and N. Aizaki, Appl. Phys. Lett. 52, 2242 (1988).
7. Ken-ichi Aketagawa, Toru Tatsumi, and Junro Sakai, J. Cryst. Growth 111, 860 (1991).
8. C.H. Hong, C.Y. Park, and H.J. Kim, J. Appl. Phys. 71, 5427 (1992).
9. T. Takahagi, I. Nagai, A. Ishitani, H. Kuroda, and Y. Nagasawa, J. Appl. Phys. 64, 3516 (1988).
10. L.A. Zazzera and J.F. Moulder, J. Electrochem. Soc. 136, 484 (1989).
11. Karin Ljungberg, Ylva Backlund, Anders Soderbarg, Mats Bergh, Mats O. Andersson, and Stefan Bengtsson, J. Electrochem. Soc. 142, 1297 (1995).
12. Hiroki Ogawa, Kenji Ishikawa, Masaru Aoki, Shuzo Fujimura, Nobuo Ueno, Yasuhiro Horiike, and Yoshiya Harada, Extended Abstracts of the 1995 International Conference on SSDM, Osaka, 13 (1995).
13. K. Kobayashi, K. Kukumoto, T. Katayama, T. Higaki, and H. Abe, Extended Abstracts of the 1992 International Conference on SSDM, Tsukuba, 17 (1992).
14. B.S. Meyerson, Franz J. Himpsel, and Kevin J. Uram, Appl. Phys. Lett. 57, 1034 (1990).

## Part VI

Metal/Semiconductor Interfaces: Structural and Electrical Properties

HRLEED AND STM STUDY OF MISORIENTED Si (100) WITH AND WITHOUT A Te OVERLAYER

SALIMA YALA AND PEDRO A. MONTANO
Materials Science Division, Argonne National Laboratory, Argonne IL 60439,
pedro_montano@qmgate.anl.gov
Department of Physics, University of Illinois-Chicago, Chicago, IL

ABSTRACT

The growth of high quality Te on misoriented Si(100) is important as an intermediate phase for epitaxial growth of CdTe. The misorientation angle plays a key role in the growth quality of CdTe/Si(100); this incited our curiosity to investigate the effect of the misorientation angle on the topography of the surface structure of Si(100). Our main goal is to show the relation between the misorientation angle, the terrace width and the step height distributions. HRLEED (High Resolution Low Energy Electron Diffraction) provides information in reciprocal space while STM gives real space topographic images of the surface structure. STM and HRLEED measurements were performed on Si(100) with misorientation angle $\vartheta = 0.5°$, $1.5°$ and $8°$ towards the [110] direction and $\vartheta = 4^0$ towards the [130] direction. Except for the $8°$ misorientation in which case a regular step array with diatomic step height was observed, for the other misorientations the terrace width was variable. The average terrace width decreased with increasing misorientation angle. A mixture of diatomic and monatomic step heights was observed on the $0.5°$ and $1.5°$ misoriented Si(100) samples. It proves that one can not assume purely monatomic step height for low misorientation angles. Our results do not agree with the belief that at low miscut angle A and B terraces are equal and that as the misorientation angle increases the B terrace tends to be wider than the A terrace. In fact, pairing of terraces was not observed at all. Te was deposited at a substrate temperature of 200 C. We observed a significant reduction in the terrace widths for all miscut angles.

INTRODUCTION

In recent years, the growth of high quality thin films became first priority because of the need to miniaturize microelectronics. Among all the substrates, Silicon is the most attractive one not only for its high crystalline properties but also for its availability and low cost in the market. The growth of high quality CdTe on misoriented Si(100) represents a challenging and yet promising project for technological applications (substrate fror growth of MCT, for x-ray detectors) [1-3]. The misorientation angle plays a key role in the growth quality of CdTe/Si(100); this incited our curiosity to investigate the effect of the misorientation angle on the topography of the surface structure of Si(100). Our main goal is to show the relation between the misorientation angle, the terrace width and the step height distributions. Si(100) with $\vartheta = 0.5°, 1.5°, 4°$ and $8°$ misorientation angles towards [110] and [130] were systematically investigated using HRLEED (High Resolution Low Energy Electron Diffraction) and STM (Scanning Tunneling Microscopy).HRLEED provides information in reciprocal space while STM gives real space topographic images of the surface structure. Te was evaporated from a Knudsen cell at a substrate temperature of 200 C. We investigated the effect of coverage and temperature on the Te growth.

EXPERIMENT

The experiments were performed in a standard UHV-chamber at a base pressure of approximately 10^{-10} torr. The chamber is equipped with a HRLEED or SPA-LEED [4,5]

Mat. Res. Soc. Symp. Proc. Vol. 448 © 1997 Materials Research Society

and an STM [6] system. Prior to introduction to the vacuum chamber, samples receive an *ex-situ* treatment that is they are first boiled in a NH_3OH solution to remove organic contaminants then submerged in a dilute HF solution to strip off the native oxides; finally, they are boiled in an HCl solution to grow a thin protective oxide layer which is removed upon annealing . *In-situ*, samples are annealed at 1100 C° or until bright diffraction spots are observed. Both STM and HRLEED measurements were performed on Si(100) with misorientation angle $\vartheta = 0.5°$, 1.5° and 8° towards the [110] direction and $\vartheta = 4$ towards the [130] direction. For HRLEED measurements, profiles of the specular beam (along the [011] and [01$\bar{1}$]) at room temperature for different electron energies were recorded by means of an electrostatic deflecting system that scans the electron beam across the aperture of a channeltron detector [5]. By varying the energy of the incident electrons one changes the interference condition which is described by, S, the scattering phase. S is defined as the phase difference between electrons scattered by first and second layer atoms in multiples of the wavelength: integer values of S correspond to an *in-phase* scattering condition while half-integer values correspond to an out-phase scattering condition. STM measurements were also performed on the different Si(100) misorientations at room temperature.
Te was evaporated from an effusion cell, and the deposition rate was monitored by measuring the ion current.

RESULTS

One can gather qualitative and quantitative information from HRLEED. The first information is just a basic visual inspection; the existence of spots implies periodicities of the substrate. The shape of a spot profile provides information on atomic steps and island growth; a splitting indicates a regular step array, a broadening means a random step arrangement etc..... . The second information is extracted (in the case of a broadening) from the equation:

$$\frac{a}{D} = \frac{\Delta K_{//}}{K_{10}} \tag{1}$$

where a is the lattice spacing, $\Delta K_{//}$ is the full width at half maximum of the broadened beam, K_{10} is the distance in reciprocal space between the specular beam and the (10) beam and D is the average terrace width. In the case of a regular step array the terrace width is simply inversely proportional to the splitting distance [7]. From the energy (or $K_{\perp}$) dependence of the spot profile, the step height can be deduced. For $\vartheta = 0.5°$ and 1.5°, the out-phase profile of the specular beam showed a broadening which was fitted with a Lorentzian then the FWHM was deduced. A model based on the kinematic approximation and random array distribution was developed [8] with γ being the total probability of meeting a step such as:

$$\gamma = \sum_{q=1}^{Q} \gamma_q \tag{2}$$

where γ_q is the probability of meeting an atom displaced vertically (either up or down) by qd (q is an integer) and Q is the maximum step height. If the occurrence of a step a one site is independent of its occurrence at another one, then one can assume a geometric distribution for the terrace width given by:

$$P(M) = (1-\gamma)^{(M-1)} \gamma \tag{3}$$

where the average terrace width is then given by:

$$\langle L \rangle = \frac{a}{\gamma} \qquad\qquad\qquad (4)$$

In the case of Si(100), the (2×1) surface reconstruction had to be taken into account. For this reason a two dimensional model was considered [9] with γ_1 and γ_2 being respectively the step probabilities in the $\vec{a}_1$ and $\vec{a}_2$ directions ($\vec{a}_1$ and $\vec{a}_2$ are the horizontal unit vectors in the [011] and [01$\overline{1}$] directions). On neighboring terraces with monatomic step height the two probabilities have to be exchanged because of the (2×1) surface reconstruction. At the out-phase condition, the total intensity is the incoherent addition of the intensities scattered from different terraces. Close to the specular reflection (for $\gamma_1 = \gamma_2$) the resulting intensity profile at out-phase reduces to a Lorentzian along the [011] direction:

$$I(u) = C \left[\frac{(1 - \rho^2)}{(1 - \rho)^2 + \rho u^2} \right] \qquad\qquad (5)$$

where $\rho = f$, f is called the "boundary structure factor" or average phase factor for pairs of neighboring sites which is a function of the step probability [8].

For $\vartheta = 0.5°$ and $1.5°$, the profile of the specular beam changed from intense and narrow at in-phase to weak and broad at out-phase. Figure 1 gives a typical three dimensional measurement (only a selected number of spectra are shown).

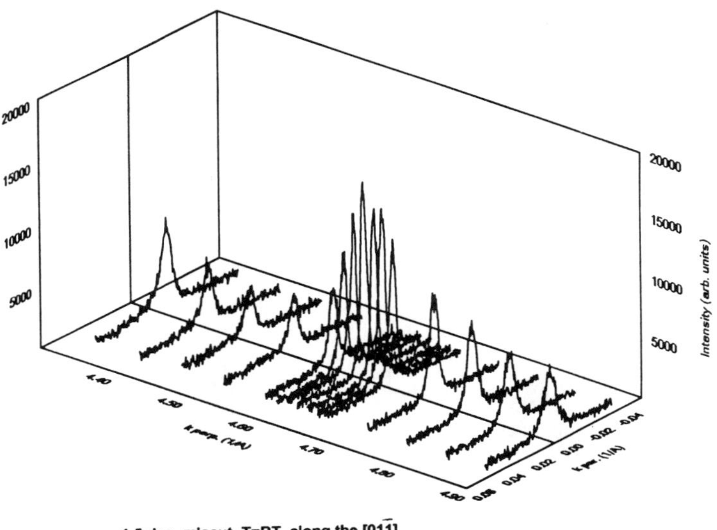

Figure 1. In/out of phase intensity profiles for the (00) beam at room temperature (1.5 ^{0})

The broadening was fitted with a Lorentzian and from the FWHM the step probability was deduced. The average terrace width decreases with increasing angle (from 133 A to about

70 A). STM measurements on Si(100) with $\vartheta = 0.5°$ and $1.5°$ misorientation angle showed monatomic and diatomic steps; the terrace width has a distribution with an average value close to that deduced from HRLEED measurements. For $\vartheta = 8°$, the LEED pattern showed only one domain with a splitting of the spots at out-phase; the terrace width was deduced from the splitting distance [7] and amounted to 18 A, (Figure 2) and the step height was calculated and was found to be diatomic (which is expected since only one single domain was observed in the LEED pattern).STM measurements confirmed the HRLEED result. For $\vartheta = 4°$ misorientation towards [130], the LEED pattern showed a

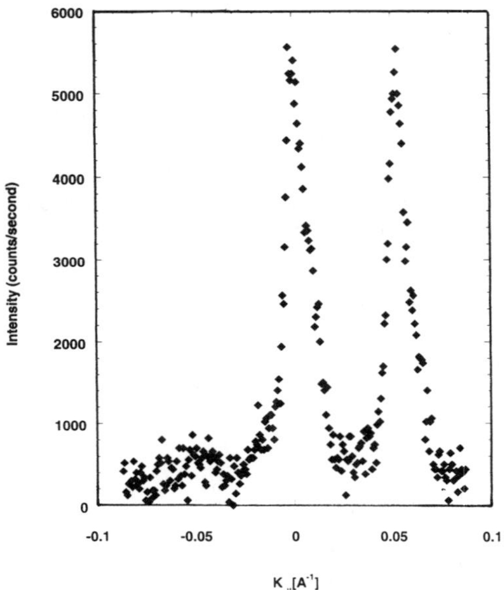

Figure 2. Splitting of the (00) beam at out of phase conditions (8 ^{0}).

splitting (typical to a regular step array) and not only a (2×1) reconstruction in the $[011]$ direction but also an additional weak (2×1) reconstruction in the $[01\overline{1}]$ direction. The splitting was not along one of the principal axes ($[011]$ or $[01\overline{1}]$); this was no surprise since the misorientation is towards [130] (the step edge is neither parallel nor perpendicular to the dimerizational direction). The average step height deduced from the energy dependence of the specular beam profile amounted to d = 2.7 A; the terrace width amouted to 36 A. This suggests a diatomic step height. However, the presence of a weak (2×1) surface reconstruction in the $[01\overline{1}]$ direction leads one to think that the surface is not purely diatomic but the diatomic character overwhelms the monatomic one. An important point that needs to be mentioned is the (2×8) reconstruction on the $1.5°$ misoriented Si(100); from a previous publication [10], this was attributed to a nickel contamination which would make the surface Si atoms rearrange in such a fashion where single dimers would be missing periodically leading to a (2×8) surface reconstruction .The question that we are addressing is why in the series of experiments we ran this reconstruction appeared only on the 1.5° misoriented Si(100) even though all the samples were treated the same way. So, if there was a nickel contamination because of the way samples were handled then the (2×8) reconstruction should have been seen in all the other misoriented Si(100) samples but that was not the case. That leads us to one conclusion: the misorientation angle and stress trigger the (2×8) reconstruction. We also performed HRLEED measurements on the 1.5° misoriented Si(100) sample at T=523 K°; no major change has been noted concerning the terrace width and the step height.

The effect of Te is quite significant. The measurements were performed at 200 C after annealing to 275 C. After this process there is only a monolayer of Te on the surface and the structure changes from 2x1 to 1x1. We observed a decrease in the terrace width for the 0.5^0 misorientation from about 133 A to 17 A. A similar phenomenon is observed for the 1.5^0 misoriented sample, the terrace width is reduced to 18 A. For the 4^0 sample we measured a terrace width of 36 A that is reduced to 16 A when Te is present on the surface. The splitting disappears, indicating no regular array of the terraces. The sample with an 8^0 misorientation retains the structure after 1ML of Te, with diatomic step heigths. The splitting of the (00) spot is present. For low coverages where the surface structure is 2x1 shows a great amount of disorder in terrace widths is shown. More measurements are necessary before we can reach any conclusions about the low coverage samples. We did not obtain very good STM images with the Te samples; this work is still in progress.

CONCLUSION

HRLEED and STM combined represent a powerful tool for defect characterization. While HRLEED gives information on the cleanness and periodicity of the surface atomic arrangement, STM allows direct imaging of the surface. Except for the 8° misorientation in which case a regular step array with diatomic step height was observed, for the other misorientations the terrace width was variable (it justifies our use of the geometric distribution of the terrace width) but its average value was agreed-upon by both STM and HRLEED. The average terrace width decreased with increasing misorientation angle and with Te coverage. Another point is that a mixture of diatomic and monatomic step heights was observed by STM on 0.5° and 1.5° misoriented Si(100) which proves that one can not assume purely monatomic step height for low misorientation angles. Our results do not agree with the belief that at low miscut angle A and B terraces are equal and that as the misorientation angle increases the B terrace tends to be wider than the A terrace. In fact, pairing of terraces was not observed at all. As a matter of fact this just proves that terrace and step distributions are not only affected by the amount of misorientation angle but also by the preparation procedure (in-situ and ex-situ).

ACKNOWLEDGMENT

The authors acknowledge the support of the U.S. DOE BES #W-31-109-ENG-38

REFERENCES

[1] Y. Lo, R.N. Bicknell, T.H. Myers, and J.F. Schetzina, J. Appl. Phys. 54 (1983) 4238
[2] R. Sporken, S. Sivananthan, K.K. Mahavadi, G. Monfroy, M. Boukerche, and J.P. Faurie, Appl. Phys. Lett. 55 (1989) 1879

[3] Y.P. Chen,) Ph.D. Thesis Chicago (1995)
[4] K.D. Gronwald and M. Henzler, Surface Sci. 117 (1982) 180
[5] U. Sheithauer, G. Meyer, and M. Henzler, Surface Sci. 178 (1986) 441
[6] C.J. Chen, Introduction to Scanning Tunneling Microscopy, Oxford Series in Optical and Imaging Sciences
[7] K. Besocke and H. Wagner, Surface Sci. 52 (1975) 653
[8] T.M. Lu and M.G. Lagally, Surface Sci. 120 (1982) 47
[9] S. Heun, J. Falta and M. Henzler, Surface Sci. 243 (1991) 132
[10] H. Niehus, U.K. Kohler, M. Copel,and J.E. Demuth, J. Microscopy, 152 (1988) 735

MICROSTRUCTURAL STUDIES OF Co SILICIDE LAYERS FORMED ON SiGe AND SiGeC

S. JIN[*], H. BENDER[*], R.A. DONATON[*], K. MAEX[*], A. VANTOMME[**], G. LANGOUCHE[**], A.ST. AMOUR[***], J.C. STURM[***]

[*]IMEC, Kapeldreef 75, B-3001, Leuven, Belgium
[**]Instituut Kern- en Stralingsfysica, University of Leuven, B-3001 Leuven, Belgium
[***]Department of Electrical Engineering, Princeton University, Princeton, NJ08544, USA

ABSTRACT

Transmission electron microscopy is used to investigate the structural development as a function of the annealing temperature of Co-silicides prepared on SiGe and SiGeC. The transition temperature from Co(SiGe) into Co(SiGe)$_2$ is higher for SiGeC than for SiGe.

INTRODUCTION

Metal-silicides have been widely used in Si based devices because of their low resistivity, low contact resistance and high thermal stability [1]. Compared to the previously used C54-TiSi$_2$, CoSi$_2$ has attracted much interest due to its better properties for ultra large scale integrated circuits (ULSI) fabrication [2].

SiGe/Si alloys have been used for high performance electron device applications, such as heterojunction bipolar transistors, heterojunction field effect transistors, and infrared detectors [3]. However, owing to the lattice-mismatch between Si and Ge (~ 4%), misfit dislocations are formed if the layer thickness exceeds a critical value, which leads to the limitation of device design flexibility. Recently, it was shown that the strain introduced by the Ge atoms can be compensated by the addition of substitutional carbon [4].

Metal reaction with SiGe alloys has been reported for Pt, Pd, Ti and Co [5-12]. Among the various silicides, CoSi$_2$ is a very attractive material because of its low resistivity and possibility of self-aligned formation at relatively low temperature.

This paper presents the investigation of the structural development as a function of the annealing temperature of Co-silicides prepared on SiGe and SiGeC.

EXPERIMENTS

The SiGe and SiGeC layers used in this work are grown by rapid thermal chemical vapour deposition on n-type Si(100) substrates. The thickness of SiGe and SiGeC is 35nm and 32nm, respectively. The Ge contents as measured by Rutherford backscattering spectrometry (RBS), are 20% for the SiGe layer and 18% for the SiGeC layer. The nominal C content is 0.9%. The wafers are dipped in HF solution before loading in a DC magnetron sputtering system for deposition of a 18nm thick Co layer. Silicidation of Co with SiGe and SiGeC is performed by

Mat. Res. Soc. Symp. Proc. Vol. 448 © 1997 Materials Research Society

rapid thermal processing (RTP) in a N_2 ambient at temperatures from 500 to 1000°C. The annealing time is always 30s.

The microstructure of these implanted samples is investigated with a JEM 200CX transmission electron microscope operating at 200 kV. Energy-dispersive x-ray spectroscopy (EDX) analysis is performed with a CM200-FEG TEM. Cross-sectional specimens for transmission electron microscopy (TEM) studies are prepared by mechanical thinning and subsequent ion milling. The normal of the TEM thin foil is parallel to the [011] direction of the silicon substrate.

RESULTS

The x-ray diffraction ($\theta/2\theta$) patterns taken from the as-deposited layers on the SiGe and SiGeC samples [12] show strong (002) texture of the Co films. The cross-sectional high resolution electron microscopy (XHREM) images with the beam direction parallel to the [011] direction of the silicon substrate, such as figure 1, indicate that a pronounced (002) texture with columnar Co grains is formed: $(002)_{Co}$ // $(100)_{SiGeC}$ // $(100)_{Si}$. The presence of an intermediate layer with a thickness of 2 ~ 3nm is believed to consist of a mixture of cobalt, silicon, germanium and carbon formed by the sputtering during the deposition of Co films. In the case of the SiGe, the observations are similar as for the SiGeC.

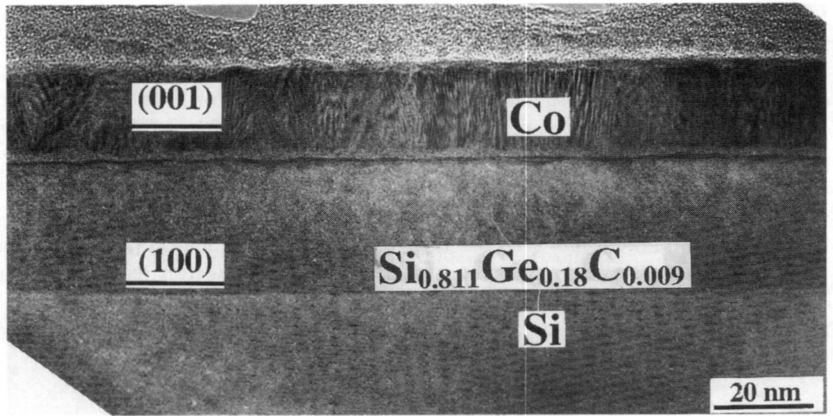

Figure 1: XTEM micrograph of 18nm Co on top of 25nm Si$_{0.811}$Ge$_{0.18}$C$_{0.009}$ on Si substrate.

Preliminary results indicated that the dominant phase in both cases is Co_2Si after annealing at 500°C [12]. Si-rich $Co(Si_{1-y}Ge_y)$ is observed in RBS and AES spectra and x-ray diffraction patterns after annealing at 650°C. The formation of Co-silicides on SiGe and SiGeC is retarded compared to pure silicon. Moreover, the delay is found to be more pronounced for SiGeC [12]. Figure 2 is a XHREM image of the Co silicide on the SiGe after annealing at 650°C. It can be seen that a $Co(Si_{1-y}Ge_y)$ film with 30 - 40nm in thickness has been formed. Furthermore, below the $Co(Si_{1-y}Ge_y)$ film lies a region with small epitaxial domains on the silicon. Stacking faults are generally found in these grains.   The composition is identified by EDX measurements as $Si_{1-x}Ge_x$. For these samples the interface roughness and the surface undulations are on the order of several nanometers. The $Co(Si_{1-y}Ge_y)$ phase has no texture or orientation relationship with the silicon matrix. In the case of SiGeC the morphology shows the same features as for the SiGe samples.

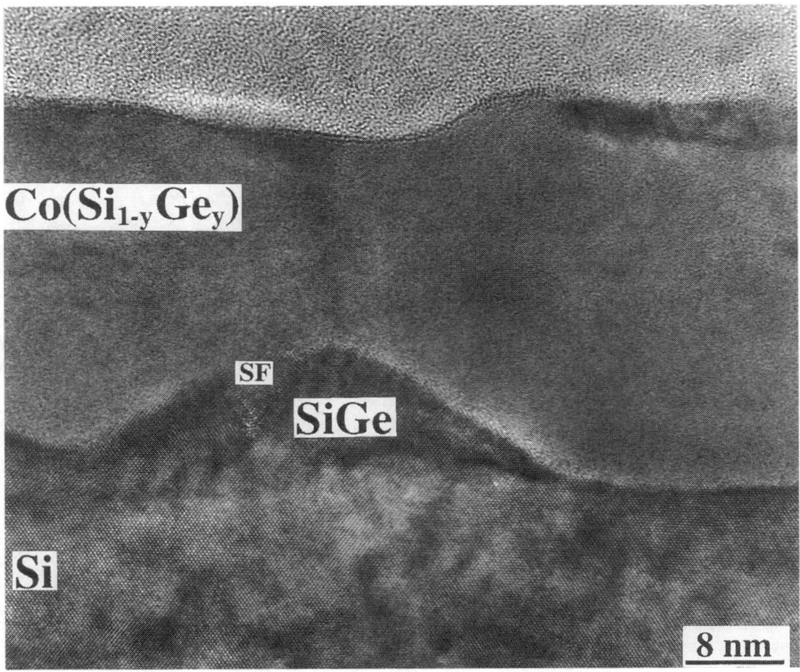

Figure 2: The XHREM image obtained from a typical region of 18 nm Co on top of SiGe after annealing at 650°C for 30s shows the presence of a small SiGe domain. The $Co(Si_{1-y}Ge_y)$ layer is continuous with a nonuniform thickness of 30 - 40nm. (SF: stacking fault)

XRD analysis of the layers formed at 800°C show that in the case of SiGe the dominant phase is the disilicide, while for SiGeC still the monosilicide dominates [12]. XTEM micrographs (Fig. 3) of the samples annealed at 800°C show that continuous silicide layers with rough surfaces and interfaces are formed both on SiGe (a) and on SiGeC (b). XHREM images (Fig. 4(a)) show that in the case of SiGe, the $Co(Si_{1-x}Ge_x)_2$ possesses grains which have twin (B-type) orientation relationship with respect to the silicon, i.e., the $Co(Si_{1-x}Ge_x)_2$ grains are rotated 180° along [111]. For the samples on SiGeC, small SiGe domains are distributed on the silicon substrate (see e.g. arrow on fib 3b and fig. 4). Fig. 5 shows EDX line scans through a typical silicide / SiGe domain / Si region and through a silicide / Si region.

If one further increases the annealing temperature to 950°C, a full conversion into $Co(Si_{1-x}Ge_x)_2$ occurs also for SiGeC. Except of randomly oriented grains, there are grains with

A-type epitaxy as the preferred orientation. The surface and interface are still rough for both the SiGe and SiGeC cases. Figure 6 is a XHREM image taken from the interface between the Co-disilicide film and the silicon substrate for the SiGeC sample. The interface step is bound by $(11\bar{1})$ and $(1\bar{1}1)$ planes.

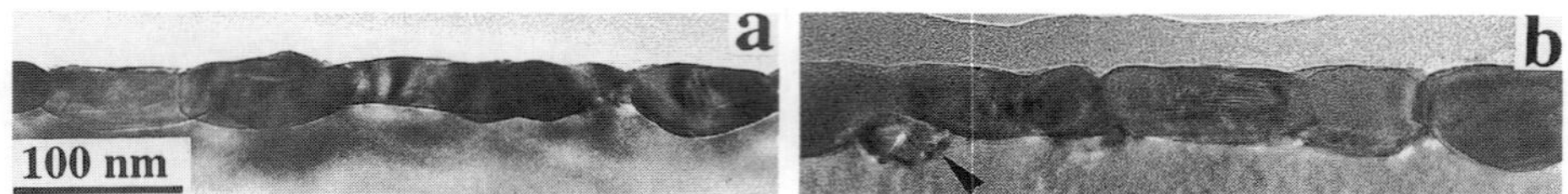

Figure 3: XTEM micrographs of the samples annealed at 800˚C. a: SiGe, b: SiGeC

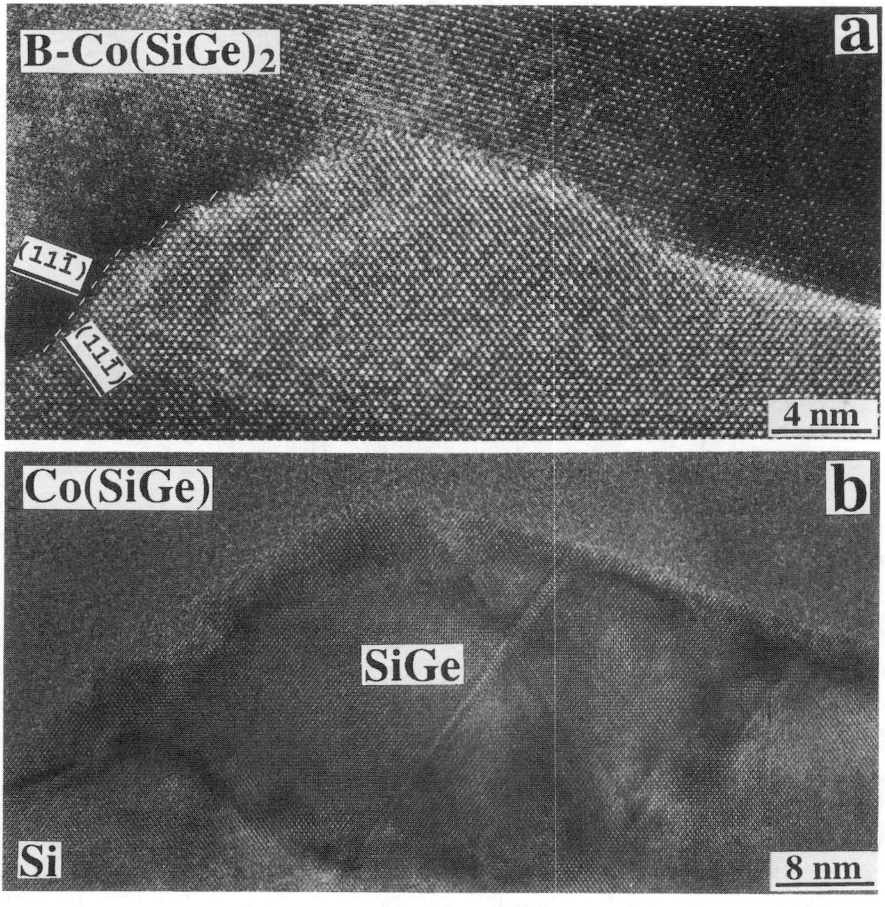

Figure 4: XHREM images of the samples from figure 3: a SiGe, b SiGeC

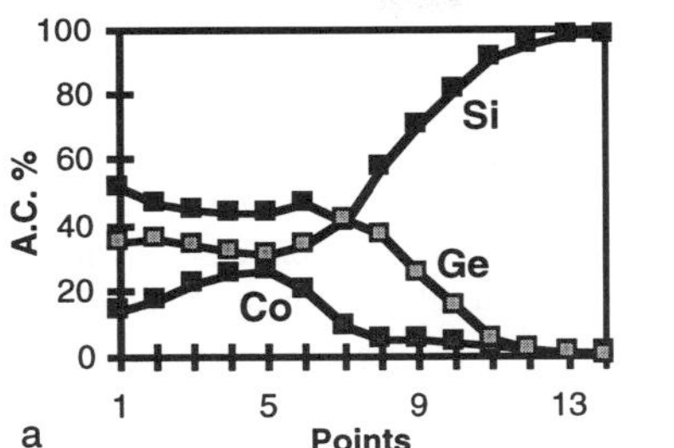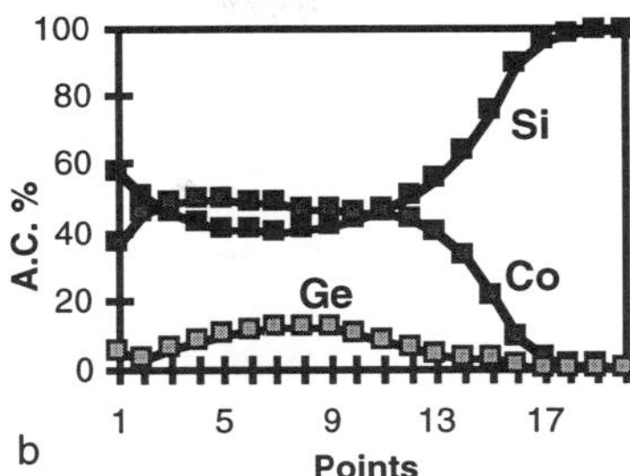

Figure 5: EDX linescan from top to substrate through a region of silicide/epitaxial domain/substrate (a) and through a region with silicide directly on the substrate (b) for the Co on SiGeC annealed at 800˚C.

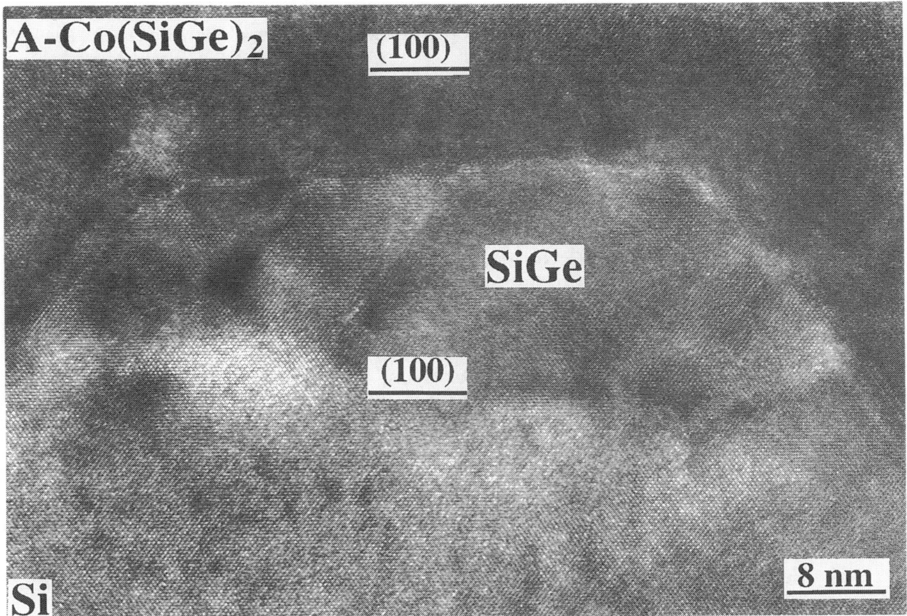

Figure 6: XHREM image taken from the interface between the Co(SiGe)₂ film and the silicon substrate of the SiGeC samples annealed at 950˚C for 30s.

The silicide formation sequence in SiGe and SiGeC is the same as the Co reaction on monocrystalline Si, however, the transition temperature from Co(SiGe) into Co(SiGe)₂ is higher for SiGeC than for SiGe, and is in both cases higher than for Co on pure silicon. The role of Ge and C atoms in the phase transition is currently under investigation.

CONCLUSION

In the as-deposited samples, an intermixed layer with a thickness of 2-3 nm is present between the (002) textured Co film and the SiGe or SiGeC layer. Annealing at 650°C mainly results in the presence of a $Co(Si_{1-y}Ge_y)$ layer with some epitaxial SiGe domains at the interface. After annealing at 800°C for 30 s, $Co(SiGe)_2$ is the dominant phase in the case of $Si_{0.8}Ge_{0.2}$ whereas mainly CoSi is formed on the $Si_{0.811}Ge_{0.18}C_{0.009}$ alloy. For the $Si_{0.8}Ge_{0.2}$, the epitaxial growth of B-type oriented $Co(SiGe)_2$ is observed. For the $Si_{0.811}Ge_{0.18}C_{0.009}$ lies also at 800°C still a discontinuous layer of SiGe domains below the Co-monosilicide film. A full conversion to $Co(SiGe)_2$ occurs at 950°C in both SiGe and SiGeC.

ACKNOWLEDGEMENT

The authors greatly acknowledge Dr A.E.M. De Veirman (Philips Semiconductors, Nijmegen) and Dr J.J.T.M. Donkers (Philips CFT, Eindhoven) for the EDX analysis. XTEM and HREM investigations are performed with the microscopes at EMAT, Universiteit Antwerpen (RUCA).

REFERENCES

1   J.W. Mayer, S.S. Lau, in <u>Electronic Materials Science: For integrated circuits in Si and GaAs</u>, (Macmillan, New York, 1990), p. 284.

2   K. Maex, Mater. Sci. Eng. Rep. **11**, 2 (1993).

3   M. Arienzo, J.H. Comfort, E.F. Crabbe, D.L. Harame, S.S. Iyer, V.P. Kesan, B.S. Meyerson, G.L. Patton, J.M.C. Stock and Y.C. Sun, Microelectronic Engineering **19**, 519 (1992).

4   K. Eberl, S.S. Iyer, S. Zollner, J.C. Tsang and F.K. LeGoues, Appl. Phys. Lett. **60**, 3033 (1992).

5   X. Xiao, J.C. Sturm, S.R. Parihar, S.A. Lyon, D. Meyerhofer, S. Palfrey and F.V. Shallcross, IEEE Electron Device Letters **14**, 199 (1993).

6   A. Buxbaum, M. Eizenberg, A. Raizman and F. Schaffler, Appl. Phys. Lett. **59**, 665 (1991)

7   H.K. Liou, X. Wu, U. Gennser, V.P. Kesan, S.S. Iyer, K.N. Tu and E.S. Yang, Appl. Phys. Lett. **60**, 577 (1992).

8   O. Nur, M. Willander, H.H. Radamson, M.R. Sardela, Jr., G.V. Hansson, C.S. Petersson and K. Maex, Appl. Phys. Lett. **64**, 440 (1994).

9   D.B. Aldrich, Y.L. Chen, D.E. Sayers, R.J. Nemanich, S.P. Ashbun and M.C. Öztürk, J. Mater. Res. **10**, 2489 (1995).

10. M.C. Ridgway, R.G. Elliman, N. Hauser, J.-M. Baribeau and T.E. Jackman, Mat. Res. Soc. Symp. Proc. **260**, 857 (1992).

11. R.A. Donaton, S. Kolodinski, M. Caymax, P. Roussel, H. Bender, B. Brijs and K. Maex, Appl. Surf. Sci. **91**, 77 (1995).

12. R.A. Donaton, K. Maex, A. Vantomme, G. Langouche, Y. Morciaux, A. St. Amour and J.C. Sturm, Appl. Phys. Lett. (submitted).

THIN FILMS OF CoSi$_2$ CO-DEPOSITED ONTO Si$_{1-x}$Ge$_x$ ALLOYS

Peter T. Goeller[2], Boyan I. Boyanov[1], Dale E. Sayers[1], Robert J. Nemanich[1,2]
[1]Department of Physics and [2]Department of Materials Science
North Carolina State University, Raleigh, NC 27695

ABSTRACT

Cobalt disilicide films have been formed on strained epitaxial Si$_{0.80}$Ge$_{0.20}$/Si(100) alloys via co-deposition of silicon and cobalt. Co-deposition is shown to improve the epitaxy and prevent the phase segregation commonly observed with the formation of Co/SiGe contacts using other methods such as the direct deposition of cobalt onto SiGe or the sequential deposition of a silicon sacrificial layer and cobalt onto SiGe. EXAFS measurements at the cobalt K edge indicate that co-deposited films annealed at 500–700° C are indeed crystalline CoSi$_2$ throughout this temperature range. The XRD patterns of the co-deposited films do not exhibit any of the CoSi$_2$ (111), (220) or (311) peaks normally associated with other preparation methods. The sheet resistance and r.m.s. roughness of the CoSi$_2$ films increase monotonically with annealing temperature. These results indicate that co-deposited films are epitaxial to the (100)-oriented SiGe substrate and suggest that low thermal budget, low resistivity contacts to strained SiGe can be grown with this method. Issues related to the presence of Ge at the CoSi$_2$/substrate interface will be discussed.

INTRODUCTION

The recent advent of commercial Si-Ge based devices enhances the need to develop optimized methods for growth of low thermal budget, low resistivity metal-semiconductor contacts. The thermally induced metal/SiGe thin film reaction has been previously studied for Ti, Co, Zr, Ni, Pd, and Pt [1–7]. Due to the low resistivity and epitaxial nature of CoSi$_2$, Co is a promising contact material. Previous work has shown that when deposited directly on SiGe, Co reacts preferentially with Si and requires temperatures in excess of 700° C to form the low resistivity CoSi$_2$ phase [3,6,7]. The high processing temperature has been shown to result in Ge segregation, roughening of the film, and strain relaxation in the SiGe substrate [6,7].

This work reports recent progress towards the stabilization of CoSi$_2$/Si$_{0.20}$Ge$_{0.80}$ contacts. The bulk ternary phase diagram of the Co-Si-Ge system indicates that CoSi$_2$ exists in bulk equilibrium with Si$_{1-x}$Ge$_x$ for x≤0.71 [9]. This suggests that it may be possible achieve a stable CoSi$_2$/SiGe contact by co-depositing Co and Si on the SiGe substrate. It will be shown that it is indeed possible to form low resistivity CoSi$_2$ films on Si$_{0.20}$Ge$_{0.80}$ at ~500°C without Ge segregation. Issues related to the presence of Ge at the CoSi$_2$/substrate interface will be discussed.

EXPERIMENTAL

Samples used in this work were prepared on 25 mm wafers of p-type Si (100) with a maximum misorientation of ±0.5° and resistivity of 0.8–1.2 Ωcm (Virginia Semiconductor). The wafers were pre-cleaned by the manufacturer and were further cleaned by spin etching [11] with a solution of HF:H$_2$O:ethanol (1:1:10) followed by thermal desorption at >900 °C in UHV for 10 minutes to remove any remaining contaminants [12]. After thermal desorption the substrate

Mat. Res. Soc. Symp. Proc. Vol. 448 © 1997 Materials Research Society

temperature was reduced at 40 °C/min and was held at 550 °C for the deposition of a homoepitaxial 200 Å Si buffer layer and a heteroepitaxial 800 Å $Si_{0.80}Ge_{0.20}$ layer. Sharp 2×1 RHEED patterns were observed along the Si [110] azimuth indicating a well ordered surface. Cobalt metallization was achieved with one of three methods: Co and Si co-deposited at 450° C in a system described elsewhere [10] (co-deposited samples); Co deposited on a sacrificial Si layer at room temperature ("sequential" samples); Co deposited directly onto the SiGe film at room temperature ("direct" samples). In all cases Co deposition was followed by a thermal anneal for 10 minutes at 400–700° C. Results for the "direct" samples have been reported previously [6,7]. Co-deposited and sequential samples were prepared by depositing ~47 Å of Co and 153 Å of Si to produce a stoichiometric $CoSi_2$ layer ~150 Å thick. An additional co-deposited sample was prepared at 400 °C without annealing. In order to study the bonding at the Co/SiGe interface a sample was also prepared by depositing 2 Å of Co onto $Si_{0.80}Ge_{0.20}$ at room temperature followed by annealing at 450 °C. Similar structures were employed by Tung *et al.* as $CoSi_2$ templates for the growth of $CoSi_2$ on Si(100) [13,20].

Structural properties of the co-deposited films were determined with EXAFS and XRD. EXAFS data were collected at room temperature in total electron yield (TEY) mode at the Co K-edge at beamline X-11 at the NSLS. The thick (~150 Å) $CoSi_2$ layers were measured *ex situ*, while the Co template was transferred under UHV to the analysis chamber of our growth system and was measured *in situ*. Fourier filtering was performed with the MacXAFS package [14] and structural parameters were extracted by non-linear fitting to the standard EXAFS equation [15] with a program written in-house. The edge shift E_0, coordination number N, bond-length R, and EXAFS Debye-Waller factor σ^2 were used as adjustable parameters in the fits. XRD data were collected in the θ–2θ mode with Cu K_α radiation on a Rigaku Geigerflex diffractometer equipped with a (0001) graphite monochromator.

Surface and interface composition and morphology of the films were examined with AES, SEM, and AFM. AES data were collected for the Co and Ge LMM lines in pulse-count mode with a Physical Electronics CMA spectrometer. The primary beam voltage was 3 keV. AFM data were collected with a Park Scientific BD2-210 instrument and 400 Å cantilevers. SEM images were acquired with a JEOL 6400 field-emission scanning electron microscope. Sheet resistance was measured with a Magne-Tron M-700 four-point probe on rectangular samples. The sample size was between 15×9 and 18×12 mm, and the probe spacing was 1.588 mm. Appropriate geometrical correction factors were applied [16].

RESULTS

The Fourier transforms of the EXAFS data for the co-deposited are shown in Fig. 1. Corresponding structural parameters are given in Table 1. Unlike films obtained by annealing Co deposited directly on $Si_{1-x}Ge_x$ [7], co-deposited films exhibit a $CoSi_2$-type crystal structure at all annealing temperatures above 500° C. Higher coordination shells identical to those of bulk crystalline $CoSi_2$ are visible [7], which indicates that long-range order is present in the films. No measurable CoSi signal was detected and, within the resolution of our measurement, the first coordination shell around Co consists entirely of Si atoms. The only temperature-dependent structural evolution in the first Co–Si shell appears to be due to ordering, as evidenced by the monotonic decrease in the EXAFS Debye-Waller factor (Table 1). These results are consistent with previous findings which demonstrated the preference for Co–Si bonding (over Co–Ge) in the $Co/Si_{0.80}Ge_{0.20}$ system [6,7]. EXAFS data for the 2 Å template layers (not shown) are identical

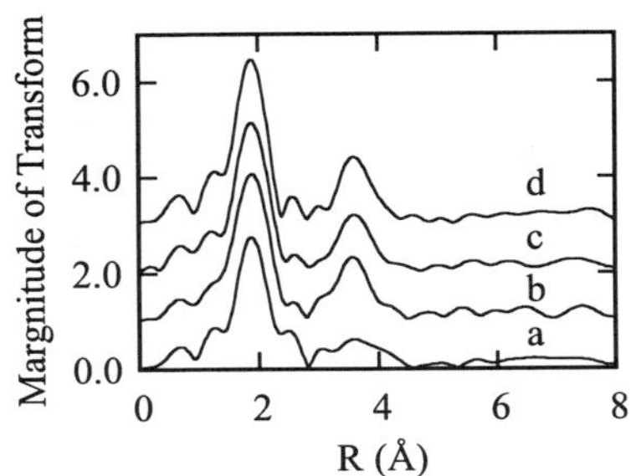

Figure 1. Fourier transforms of k^2-weighted EXAFS data for co-deposited samples annealed at: (a) 400 °C; (b) 500 °C; (c) 600 °C; (d) 700 °C

Table 1. EXAFS fitting results for co-deposited samples. The 700 °C data were used as a CoSi$_2$ reference: N=8.0, R=2.32 Å. Typical error bars are: E_0=±1.5 eV, N=±1.0, R=±0.01 Å, $\Delta\sigma^2$=±0.0015 Å^2.

T (°C)	E_0 (eV)	N	R (Å)	$\Delta\sigma^2\times10^4$ (Å^2)
400	0.7	8.1	2.33	32
500	1.9	8.6	2.32	21
600	2.2	7.6	2.32	7
700	–	8.0	2.32	–

to those for the thicker co-deposited samples. Fits that incorporated first-shell Co–Ge interaction in addition to the dominant Co–Si interaction failed to converge for both the template and bulk-like systems. We conclude that in co-deposited CoSi$_2$/Si$_{0.80}$Ge$_{0.20}$ films annealed between 500 and 700 °C Co atoms bond exclusively to Si in a CoSi$_2$-type structure, including at the Co/SiGe interface. As suggested by Wang *et al.*, the driving force for this effect is probably the large difference in the enthalpy of formation of CoSi$_2$ and CoGe$_2$ [6].

X-Ray diffraction measurements indicate that co-deposited samples exhibit better epitaxial alignment to the (100)-oriented SiGe substrate than either "direct" or "sequential" samples. Figure 2 shows a comparison of θ-2θ scans for the three types of samples. The crystalline nature of the co-deposited films has already been established with EXAFS measurements. Since the "sequential" sample is of identical thickness and composition as the co-deposited ones we interpret the lack of (111), (220) and (311) CoSi$_2$ lines in the co-deposited films as an indication of better epitaxial alignment of the CoSi$_2$ overlayer to the SiGe substrate.

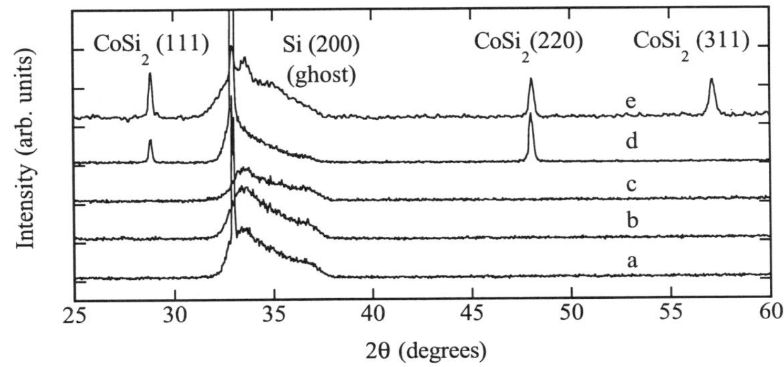

Figure 2. XRD scans of CoSi$_2$ films: (a,b,c) co-deposited film annealed at 500, 600, and 700 °C, respectively; (d) sequential film annealed at 700 °C; (e) "direct" film annealed at 700° C.

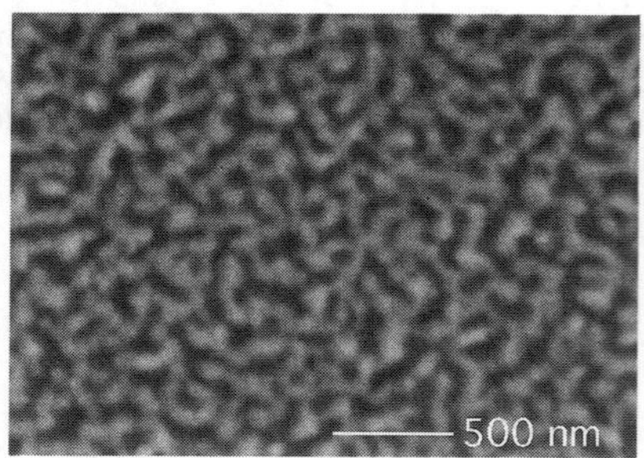 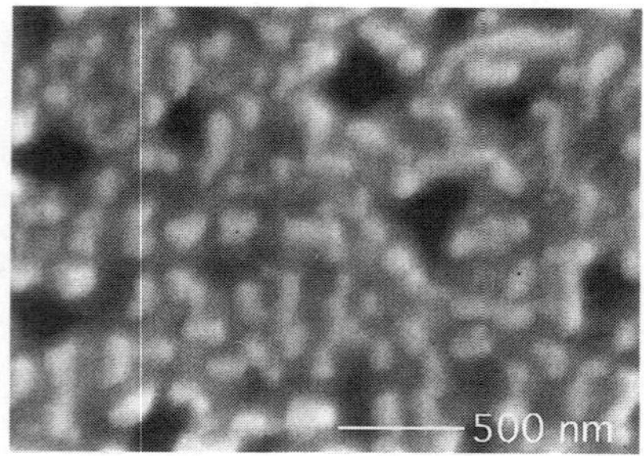

Figure 3. SEM images of co-deposited samples annealed at 500° C (left) and 700° C (right)

The lack of any peaks attributable to a CoSi-type structure in the co-deposited films is noteworthy. Similarly prepared "direct" samples exhibit (210) and (211) $Co(Si_{1-y}Ge_y)$ peaks when annealed at temperatures up to 600° C [6]. Within the sensitivity limits imposed by the instrument we have failed to detect such peaks, as well as any peaks attributable to Co-Ge compounds. We conclude that the co-deposited films form directly $CoSi_2$ and do not undergo a $CoSi \rightarrow CoSi_2$ transition. The $CoSi \rightarrow CoSi_2$ transition is typically observed in the bilayer reaction of Co with both Si and SiGe [18]. The fact that $CoSi_2$ is the desirable low-resistivity phase makes this finding significant, as it implies that co-deposited films can be subjected to milder annealing, which lowers the thermal budget of the process and reduces islanding in the film and the increase in resistivity that accompanies such islanding (see below).

Representative SEM images of co-deposited samples annealed at 500 and 700 °C are shown in Fig. 3. The image for the sample annealed at 600° C is virtually identical to that of the 500 °C sample and is not shown. The surface of the $CoSi_2$ film is rough at all temperatures and the SiGe substrate appears to be exposed for the sample annealed at 700° C. The roughness of co-deposited and "direct" Co/SiGe films, as measured with AFM, is plotted as a function of annealing temperature in Fig. 4. The transition from the high-resistivity CoSi phase to the low-resistivity $CoSi_2$ phase in the "direct" samples is accompanied by a seven-fold increase in the surface roughness. In contrast, surface roughness in the co-deposited samples evolves similarly to that of the Co/Si(100) bilayer system. Analysis of the AFM images suggests that the SiGe

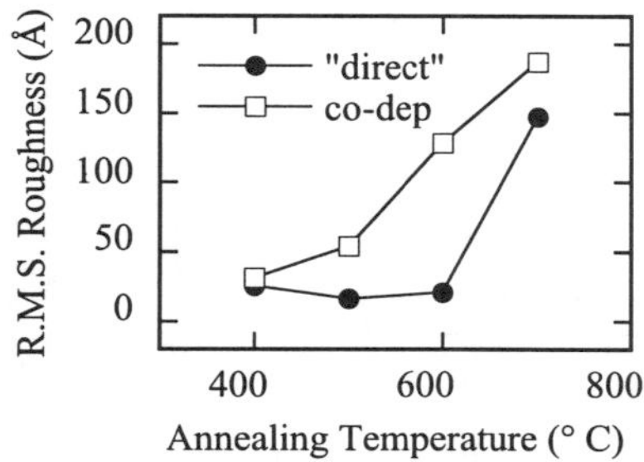

Figure 4. Surface roughness as a function of annealing temperature

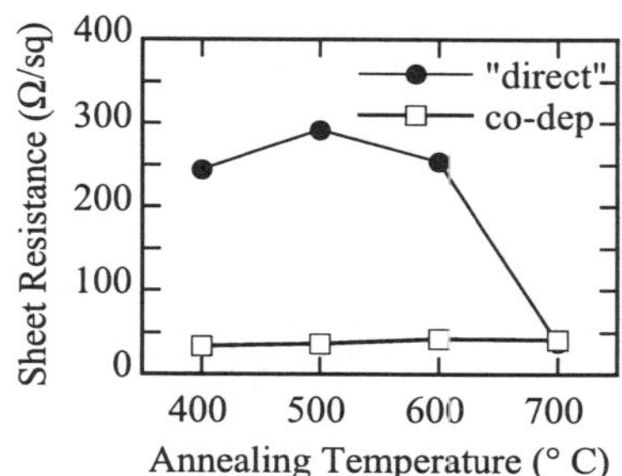

Figure 5. Sheet resistance as a function of annealing temperature [19]

substrate is exposed even for the co-deposited films annealed at 500° C. Pinhole formation at similar temperatures has also been detected in $CoSi_2/Si(100)$ films [20]. AES measurements indicate that the Co/Ge ratio in the surface layer is approximately the same in the co-deposited films annealed at 500–700° C. Since no Co-Ge or Co-Si-Ge compounds were detected with EXAFS and XRD, the Ge signal originates from either the SiGe substrate, or Ge segregation at grain boundaries.

Pinhole formation has been successfully controlled in $CoSi_2/Si(100)$ films through the use of $CoSi_2$ templates [20,23]. Two types of reconstruction have been observed for $CoSi_2$ surfaces on Si(100)—$(\sqrt{2}\times\sqrt{2})R45$ and $(\sqrt{2}\times3\sqrt{2})R45$ [20–22]. We were unable to obtain either reconstruction in the Co/SiGe system with any preparation method, including templates [20,23]. This may be due to the strong preferential Co–Si bonding at the SiGe surface, as discussed earlier in this section, which precludes the formation of a continuous epitaxial $CoSi_2$ template.

The sheet resistance of the co-deposited and "direct" films is plotted as a function of the annealing temperature in Fig. 5. Not surprisingly, the monotonic increase in the roughness of the co-deposited films is paralleled by a corresponding increase in the sheet resistance. At 500° C the resistivity of the co-deposited films is at least three times lower [19] than that of comparably annealed "direct" films measured by Wang [8] and Ridgway *et al.* [3].

SUMMARY

Cobalt disilicide films have been formed on strained epitaxial $Si_{0.80}Ge_{0.20}/Si(100)$ alloys via co-deposition of silicon and cobalt. Co-deposition was shown to improve the epitaxy and prevent the phase segregation commonly observed with the formation of Co/SiGe contacts using other methods such as the direct deposition of cobalt onto SiGe or the sequential deposition of a silicon sacrificial layer and cobalt onto SiGe. EXAFS measurements at the cobalt K edge indicated that co-deposited films annealed at 500–700° C were indeed crystalline $CoSi_2$ throughout this temperature range. The XRD patterns of the co-deposited films did not exhibit any of the $CoSi_2$ (111), (220) or (311) peaks normally associated with other preparation methods. The sheet resistance and r.m.s. roughness of the $CoSi_2$ films increased monotonically with annealing temperature. These results indicate that co-deposited films are epitaxial to the (100)-oriented SiGe substrate and suggest that low thermal budget, low resistivity contacts to strained SiGe can be grown with this method. However, pinhole formation similar to that observed in the $CoSi_2/Si(100)$ system was detected in the co-deposited films at temperatures as low as 500 °C. Pinhole formation and islanding in thin films is usually attributed to the interplay of surface and interface energies, so as to minimize the total free energy of the system. An additional driving force for pinhole formation in the $CoSi_2/SiGe$ system may be the strong preference for Co–Si bonding over Co-Ge bonding, which in effect increases the energy of the $CoSi_2/SiGe$ interface.

ACKNOWLEDGMENTS

Funding for this work was provided by the Department of Energy under contracts DE-FG05-93ER79236 (D.E.S., instrumentation), DE-FG05-89ER45384 (D.E.S., X-11 Operations at NSLS), and by the National Science Foundation under contract DMR-9633547 (R.J.N.). The authors gratefully acknowledge use of beamline X-11 at the National Synchrotron Light Source. The NSLS is funded by the Department of Energy under contract DE-AC02-76CH00016.

REFERENCES

[1]   O. Thomas, F. M. d'Heurle, and S. Delage, *J. Mat. Res.* **5**, 1453–1461 (1990).

[2]   R. D. Thompson, K. N. Tu, J. Angillelo, S. Delage and S. S. Iyer, *J. Electrochem. Soc.* **135**, 3161 (1988).

[3]   M. C. Ridgway, R. G. Elliman, N. Hauser, J.-M. Baribeau and T. E. Jackman, *Mat. Res. Soc. Symp. Proc.* **260**, 857–861 (1992).

[4]   D. B. Aldrich, Y. L. Chen, D. E. Sayers, R. J. Nemanich, S. P. Ashburn, and M. C. Öztürk, *J. Mater. Res.* **10**, 2849–2863 (1995).

[5]   D. B. Aldrich, F. M. d'Heurle, D. E. Sayers and R. J. Nemanich, *Phys. Rev.* **B53**, 16297–16282 (1996).

[6]   Z. Wang, D. B. Aldrich, Y. L. Chen, D. E. Sayers and R. J. Nemanich, *Thin Solid Films*, **270**, 555–560 (1995).

[7]   Z. Wang, D. E. Sayers and R. J. Nemanich, *Physica* **B208&209**, 567–568 (1995).

[8]   Z. Wang, unpublished.

[9]   F. Wald and J. Michalik, *J. Less Comm. Met.* **24**, 277–289 (1971).

[10] Z. Wang, P. T. Goeller, B. I. Boyanov, D. E. Sayers and R. J. Nemanich, in *Proc. XAFS IX*, to be published in *J. Physique*.

[11] D. B. Fenner, D. K. Biegelsen and R. D. Bringans, *J. Appl. Phys.* **66**, 419, 1989.

[12] H. Jeon, C. A. Sukow, J. W. Honeycutt, T. P. Humphreys, R. J. Nemanich and G. A. Rozgonyi, *Mater. Res. Soc. Symp. Proc* **181**, 559 (1991).

[13] R. T. Tung, J. L. Batstone and S. M. Yalisove, *Mater. Res. Soc. Symp. Proc.* **102**, 265 (1987)

[14] C. E. Bouldin, W. T. Elam and L. Furenlid, *Physica* **B208&209**, 190–195 (1995).

[15] D. E. Sayers and B. A. Bunker, in *X-Ray Absorption: Principles, Applications, Techniques of EXAFS, SEXAFS, and XANES*, D. C. Koningsberger and R. Prins, Eds. (Wiley, New York, 1988) ch. 4.

[16] S. P. Weeks, in *Quick Reference Manual For Silicon Integrated Technology*, W.E. Beadle, J.C.C. Tsai and R.D. Plummer, Eds. (Wiley, New Tork, 1985) ch. 4.

[18] H. Ying, Z. Wang, D. B. Aldrich, D. E. Sayers and R. J. Nemanich, *Mat. Res. Soc. Symp. Proc.* **320**, 335–340 (1994).

[19] Resisitivities have been compared on the basis of the nominal film thickness reported here and in Refs. [3] and [8].

[20] S. M. Yalisove, R. T. Tung, and J. Balstone, *Mater. Res. Soc. Symp. Proc.* **116**, 439 (1988).

[22] R. Stalder, C. Schwartz, H. Sirringhaus, and H. von Känel, *Surf. Sci.* **271**, 355–375 (1992).

[21] P. T. Goeller, Z. Wang, D. E. Sayers, J. T. Glass and R. J. Nemanich, *Mat. Res. Soc. Symp. Proc.* **402**, 511–515 (1996).

[23] L. Haderbache, P. Wetzel, C. Pirri, J. C. Peruchetti, D. Bolmont, and G. Gewinner, *Thin Solid Films* **184**, 317–323 (1990).

**SEGREGATION OF COPPER TO (100) AND (111) SILICON SURFACES
IN EQUILIBRIUM WITH INTERNAL Cu₃Si PRECIPITATES**

W. R. WAMPLER, Sandia National Laboratories, Albuquerque, NM 87185-1056
wrwampl@sandia.gov

ABSTRACT

The energetics of copper segregation to silicon surfaces were examined by measuring the Cu coverage after equilibration between Cu on the surface and internal Cu_3Si, for which the Cu chemical potential is known. For oxide-free surfaces the Cu coverage was close to one monolayer on (111) surfaces but was much smaller on (100) surfaces. The Cu coverage was greatly reduced by oxide passivation of the surface. LEED showed the 7x7 structure of the clean (111) silicon surface converted to a quasiperiodic 5x5 structure after equilibrating with Cu_3Si. The 2x1 LEED patterns for (100) surfaces indicated no change in surface structure due to the Cu_3Si. These results show that the free energy of copper in Cu_3Si is higher than that of copper on (111) surfaces but lower than that of copper on (100) surfaces.

INTRODUCTION

The high electrical conductivity of copper makes it an attractive material for interconnections in advanced microelectronic devices [1]. An obstacle to this application is that the high solubility and fast interstitial diffusion of copper in silicon make it a particularly troublesome impurity in silicon devices [2]. Gettering is used to remove impurities from critical regions of devices. Internal gettering is a routinely used process in which oxide precipitates and associated lattice defects provide sites for internal precipitation of metal-silicide phases [3]. Segregation of impurities onto surfaces of internal microcavities has also been proposed as a mechanism for gettering impurities [4,5]. It has been observed that gettering to cavities can dissolve pre-existing internal metal silicide precipitates [5]. This transfer of metal from silicide to cavities is driven by a lower chemical potential for the metal at cavities than in the silicide phase. A lower chemical potential implies that gettering by cavities should be capable of reducing concentrations of mobile impurities to lower levels than can be achieved by conventional internal gettering. The efficiency of impurity gettering by cavities depends on the energetics of impurity segregation onto silicon surfaces, about which little is known. Measurements and modeling of impurity redistribution in silicon with cavities and silicide have yielded values for the free energy of binding to cavities, and for coverages after equilibration with silicide, for copper and other metals [5]. However, these results are averages over the differently oriented facets of the cavities [4,6]. Here we examine the energetics of Cu segregation onto external surfaces of Si where the orientation is known.

The structure and growth modes of copper evaporated onto silicon surfaces have been the subject of many previous studies. However, the chemical potential of copper on the surface may not be well defined under conditions of vapor deposition since metastable phases might form. Here we examine the energetics of copper segregation to silicon surfaces by measuring the coverage of copper in thermodynamic equilibrium with internal Cu_3Si which has a known chemical potential. With internal silicide as the source of copper, there is no excess free energy to drive nucleation of silicide or other less stable phases at the surface. Measurements were made on both (100) and (111) surfaces at the same chemical potential, namely that of Cu_3Si.

Mat. Res. Soc. Symp. Proc. Vol. 448 © 1997 Materials Research Society

EXPERIMENTAL METHOD

The basic method in these experiments was to prepare clean ordered surfaces by thermal desorption of a thin oxide from (100) and (111) surfaces [7], or hydrogen from (111) surfaces [8], then anneal the samples to allow copper on the clean surface to come into thermodynamic equilibrium with internal silicide. After equilibrating, the samples were rapidly cooled to preserve the equilibrium copper coverage on the surface, and then analyzed.

The surface copper coverage was measured by Rutherford backscattering (RBS). Auger electron spectroscopy (AES) was used to indicate the presence of oxygen, carbon and copper on the surface. The atomic structure of the surface was monitored using low energy electron diffraction (LEED). The samples were radiatively heated by a tungsten filament behind the sample, and the sample temperature was measured by an infrared thermometer viewing the front of the sample.

Samples were (100) and (111) oriented float zone silicon 250μm thick with a resistivity > 1 kΩ cm. Samples were chemically cleaned leaving a thin chemical oxide using the procedure described by Ishizaka and Shiraki [7]. In addition, some (111) samples were prepared with hydrogen terminated surfaces by rinsing in NH_4F [8]. After chemical cleaning the samples were implanted with copper on one side at 150 keV to a fluence of 1000 Cu/nm^2. This implant produced a peak Cu concentration of 10 atomic % about 100 nm beneath the surface. Transmission electron microscopy has shown that the η-Cu_3Si equilibrium silicide phase is formed by annealing at 600°C [4,9]. Directly after the Cu implant the samples were transferred into a UHV chamber for analysis of the side not implanted with copper. The base pressure of the UHV system was 1×10^{-10} Torr, low enough that samples remained free from contamination by residual gasses as shown by AES analysis of surface composition.

Samples were first heated to 750°C for ten minutes, then cooled to room temperature and analyzed. They were then heated to 850°C for 10 minutes, cooled and analyzed again. The heating has three important effects. First, it insures that the implanted Cu is in the equilibrium η-Cu_3Si phase [4,9]. Second, it establishes thermodynamic equilibrium between copper in the silicide, copper in solution and copper bound to the front surface of the sample. Finally, heating to 750°C desorbs hydrogen from the hydrogen terminated (111) surfaces [8]. The thin chemical oxide is desorbed from the oxide terminated surfaces by heating to 850°C, but not by heating to 750°C [7]. These procedures for preparing clean, well-ordered surfaces were tested using samples which were not implanted with copper. H terminated Si(111) surfaces had sharp 1x1 LEED patterns before heating and 7x7 LEED patterns after heating to 750°C or higher. AES on the oxide terminated surfaces showed a large peak from oxygen which was unchanged after heating to 750°C but was entirely absent after heating to 850°C. LEED showed sharp 2x1 and 7x7 patterns for the (100) and (111) surfaces, respectively, after heating the oxide terminated surfaces to 850°C. These tests show that without copper, (100) and (111) samples prepared with oxide terminated surfaces have clean ordered (2x1) and (7x7) surfaces, respectively, after heating to 850°C, but still have oxide terminated surfaces after heating to 750°C, and samples prepared with H terminated (111) surfaces had clean ordered (7x7) surfaces after heating to 750°C or higher. These results agree with published procedures for producing clean well-ordered silicon surfaces [7,8].

The annealing time of ten minutes was chosen to be much longer than the time to reach equilibrium between Cu on the front surface and the silicide near the back surface. Equilibrium is reached by diffusion of Cu through the silicon lattice from the silicide to the front surface. The equilibration time was determined by solving the diffusion equation with saturable traps at the front surface and with the concentration at the back fixed at the solid solubility. Using published values for the diffusivity and solubility [2,4], equilibration times are estimated to be about 10 seconds at 850°C and 90 seconds at 750°C. The cooling rate at the end of the anneal was sufficiently fast to retain the copper coverage on the front surface present during the anneal. The initial cooling rate was measured by the IR thermometer to be ~ 100°C/sec. This is close to the rate expected for cooling by radiative heat loss.

EXPERIMENTAL RESULTS

The coverage of copper on the front surface was measured by RBS using an analysis beam of 1 MeV ^{4}He. The samples were aligned with the analysis beam for axial channeling to reduce scattering from the silicon. Table 1 gives the main results from the RBS measurements on (100) and (111) samples prepared with oxide terminated surfaces, denoted by (100)-O and (111)-O respectively, and on (111) samples prepared with H terminated surfaces, denoted by (111)-H. The experimental uncertainty indicated in table 1 is the relative uncertainty due to the counting statistics. In addition, there is a systematic uncertainty of about ± 10% in the absolute values of copper coverages from RBS.

Samples prepared with oxide terminated (100) and (111) surfaces have much more Cu after annealing at 850°C than after annealing at 750°C. This shows that the coverage of copper is greatly reduced by the surface oxide, which is still present after heating to 750°C but absent after heating to 850°C. This conclusion is reinforced by comparing the Cu coverage on the (111) samples after heating to 750°C, which desorbs the H but not the oxide. Again the oxide terminated surface has much less Cu than the oxide free surface.

The Cu coverage also depends strongly on the surface orientation. After annealing to 850°C the areal density of Cu is about five times higher on (111) surfaces than on (100) surfaces. Dividing the measured Cu areal density by the Si atom areal density (6.8/nm^2 for (100) and 7.8/nm^2 for (111) surfaces) gives Cu coverages of 1.6 monolayer (ML) on (111) surfaces and 0.35 ML on (100) surfaces after annealing at 850°C.

The copper can also be seen in the AES spectra. Figure 1 shows AES spectra for (111)

TABLE I. The coverage of copper on the surface measured by RBS after annealing.

Surface	750°C Anneal (Atoms/nm^2)	850°C Anneal (Atoms/nm^2)
(111)-H	9.9 ± 0.3	12.2 ± 0.4
(111)-O	1.0 ± 0.2	12.3 ± 0.4
(100)-O	0.4 ± 0.2	2.4 ± 0.2

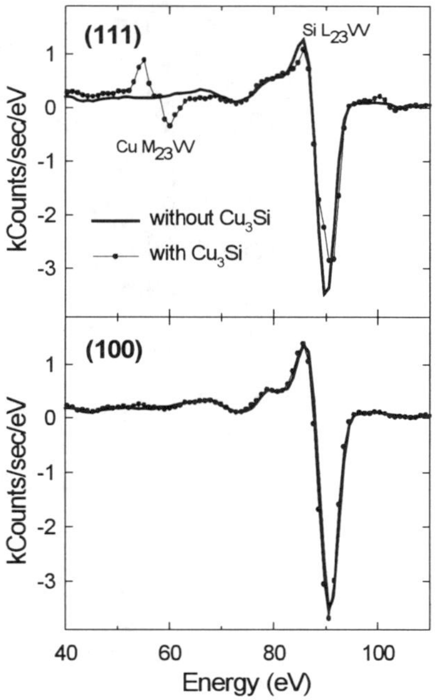

Fig. 1. Auger electron spectra on oxide free surfaces after annealing at 850°C, for (111) and (100) surfaces in samples with and without implanted copper.

and (100) surfaces after annealing at 850°C. Spectra are shown for samples both with and without internal silicide. (111) samples with Cu have a Cu $M_{23}VV$ Auger peak at 60 eV which is absent in samples not implanted with Cu, whereas the Cu Auger peak is much smaller for Cu implanted (100) samples. The Cu Auger peak was not present on (100) or (111) samples prepared with oxide terminated surfaces after heating to 750°C, i.e. which still had oxide terminated surfaces after the anneal. The Cu Auger peak for Cu implanted H terminated (111) samples, after annealing at 750°C or 850°C, was the same as for Cu implanted (111)-O samples after the 850°C anneal. The conclusions from the Auger analysis are qualitatively the same as those from the RBS analysis. In particular, for oxide-free surfaces there is much more Cu on (111) surfaces than on (100) surfaces, and there is much less Cu on oxide terminated surfaces than on oxide-free surfaces.

The surface structure was determined by LEED. Figure 2 shows LEED images from (100) and (111) samples prepared with oxide terminated surfaces both with and without internal Cu_3Si. The LEED images in figure 2 were recorded on samples near room temperature after annealing at 850°C which desorbs the oxide. (100) samples show the same (2x1) LEED pattern from the

(100) 2x1   (111) 7x7

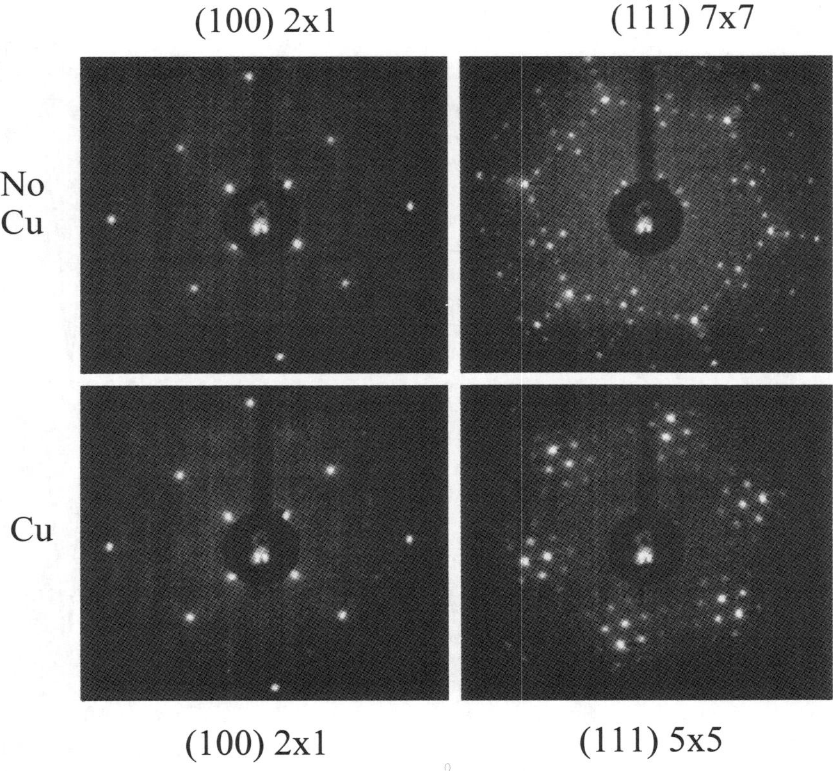

(100) 2x1   (111) 5x5

Fig. 2. LEED images on oxide free surfaces after annealing at 850°C, for (100) (left) and (111) (right) surfaces in samples with (below) and without (above) implanted copper.

dimer reconstruction both with and without silicide. (111) samples without silicide had (7x7) LEED patterns whereas (111) samples implanted with Cu showed a very different LEED pattern with an apparent 5x5 periodicity. (111) samples prepared with H terminated surfaces instead of oxide terminated surfaces gave the same LEED patterns after annealing at 750°C or 850°C as those shown in figure 2, i.e. 7x7 patterns without silicide and 5x5 patterns with silicide. The LEED results show that for surfaces free of oxide, (100) surfaces are unchanged by the presence of internal silicide, whereas (111) surfaces are changed from the (7x7) to a (5x5) structure after annealing to equilibrate copper on the surface with the internal silicide.

DISCUSSION AND CONCLUSIONS

The structure of copper evaporated onto silicon surfaces has been the subject of many studies [10-18]. The conclusions can be summarized as follows. Copper deposited onto Si(111) surfaces at temperatures between 200°C and 650°C forms an ordered two dimensional (2D) phase which saturates at a copper coverage of 1.3 monolayer when the layer completely covers the surface [10-13]. At higher coverages 3D islands of η''-Cu_3Si grow on the surface while the surface between the islands remains covered by the 2D phase [10,11]. This 2D layer has a complex quasiperiodic structure which has been elucidated by X-ray standing wave [12], helium diffraction [14] and scanning tunneling microscopy [14,16] studies. Copper deposited onto Si(100) surfaces at temperatures between 130°C and 500°C condenses into 3D islands of Cu_3Si. The (100) silicon surface between the islands has very little copper and retains its 2x1 dimer structure [13,18].

The ordered 2D phase on (111) surfaces gives the 5x5 LEED pattern [13] as we observed, and Auger electron spectra very similar to those we observed [11]. The copper coverage on (111) surfaces equilibrated with internal Cu_3Si which we measured (table I) is close to the value previously reported for full coverage by the quasiperiodic 2D phase. Our measurements of the copper coverage on surfaces after equilibration with internal Cu_3Si, and the previously reported studies of copper deposited onto silicon surfaces, both lead to the conclusion that the free energy of copper in Cu_3Si is higher than that of copper on (111) surfaces but lower than that of copper on (100) surfaces.

Segregation of copper and other impurities onto surfaces of internal microcavities has been investigated as a potential gettering mechanism [4,5]. It was observed that gettering to cavities dissolved pre-existing internal Cu_3Si precipitates at 450°C and 600°C [5]. This transfer of copper from silicide to cavities occurs via the solution state and is driven by a lower chemical potential for the copper at cavities than in the silicide phase. Measurements and modeling of copper redistribution between cavity layers at different depths, and between cavities and internal silicide give a value for the free energy of copper bound to cavities which is about 0.8 eV/atom below that of copper in Cu_3Si at 630°C [5]. With this strong binding, the copper at the cavities after equilibration with silicide should be saturated at the equilibration temperatures used in our study. In the cavity experiments the saturation coverage of copper was estimated to be 6.5 Cu/nm^2 from the areal density of copper measured by RBS, and measurements of total cavity surface area from transmission electron microscopy (TEM) [4]. The results from our investigation suggest that gettering of Cu at cavities may be mainly on the {111} facets. The saturation coverage found in the study of copper gettering at cavities is consistent with the coverage we observe on external (111) surfaces over approximately half of the cavity surface area. This is consistent with TEM which showed that the cavities were faceted with predominately {111} faces [4,6].

ACKNOWLEDGEMENTS
This work was funded by the Office of Basic Energy Sciences, Division of Materials Science of the U.S. Department of Energy under contract DE-AC04-94AL85000. Sandia is a multiprogram laboratory operated by Sandia Corporation, a Lockheed Martin Company, for the U.S. Department of Energy.

REFERENCES

1. Mater. Res. Soc. Bull., Vol.19, No. 8, Aug. 1994.

2. E. R. Weber, Appl. Phys. **A30**, 1 (1983).

3. D. Gilles and E. R. Weber, Phys. Rev. Lett. **64**, 196 (1990).

4. S. M. Myers and D. M. Follstaedt, J. Appl. Phys. **79**, 1337 (1996).

5. S. M. Myers G. A. Petersen and C H. Seager, J. Appl. Phys. **80**, 3717 (1996).

6. D. M. Follstaedt, Appl. Phys Lett. **62**, 1116 (1993).

7. A. Ishizaka and Y. Shiraki, J. Electrochem. Soc. **119**, 666 (1986).

8. Le Thanh Vinh, M. Eddrief, C. A. Sebenne, P. Dumas, A. Taleb-Ibrahimi, R. Gunther, Y. J. Chabal and J. Derrien, Appl. Phys Lett. **64**, 3308 (1994).

9. D. M. Follstaedt and S. M. Myers, Mat. Res. Soc. Symp. Proc. **316**, 27 (1994).

10. M. Mundschau, E. Bauer, W. Telieps and W. Swiech, J. Appl. Phys. **65**, 4747 (1989).

11. E. Daugy, P. Mathiez, F. Salvan and J. M. Layet, Surface Sci. **154**, 267 (1985).

12. J. Zegenhagen, E. Fontes, F. Grey and J. R. Patel, Phys. Rev. B **46**, 1860 (1992).

13. H. Kemmann, F. Müller and H, Neddermeyer, Surface Sci. **192**, 11 (1987).

14. R. B. Doak and D. B. Nguyen, Phys. Rev. **40**, 1495 (1989).

15. J. E. Demuth, U. K. Koehler, R. J. Hamers and P. Kaplan, Phys. Rev. Lett. **62**, 641 (1989).

16. K. Mortensen, Phys. Rev. Lett. **66**, 461 (1991).

17. T. Ikeda, Y. Kawashima, H. Itoh, and T. Ichinokawa, Surface Sci. **336**, 76 (1995).

18. P. Mathiez, E. Daugy, F. Salvan, J. J. Metois and M. Hanbücken, Surface Sci. **168**, 158 (1986).

SURFACE AND INTERFACE ANALYSIS OF THIN-FILM/Si(SUBSTRATE) CONTACTS BY SXES

C.HECK*, M.KUSAKA*, M.HIRAI*, H.NAKAMURA+, M.IWAMI* and H.WATABE**
* Research Laboratory for Surface Science, Faculty of Science, Okayama University, Okayama 700, JAPAN
+ Osaka Electro-Communication University, Hatsu, Neyagawa 572, JAPAN
** Matsushita Research Institute Inc., 3-10-1 Higashi-mita, Tama-ku, Kawasaki 214, JAPAN

ABSTRACT

A soft X-ray emission spectroscopy(SXES) study under an energetic electron irradiation has been applied to a nondestructive buried interface analysis of a thin-film(e.g., Cr)/Si(substrate) contact system, where the energy of primary electrons, E_p , is less than 20keV. An interesting point of this method is that we can have a specific signal for an element to be used as a finger print, otherwise it is difficult. By using this e-beam excited SXES, we can study an interface buried deep in a rather thick overlayer, e.g., more than a hundred of nm, which is due to the fact that a mean free path of a soft X-ray or an X-ray production depth is much larger than the mean free path of an energetic electron in solids. Electronic structural study of silicides by SXES is also shown.

INTRODUCTION

Surprising development in the field of electronic devices is largely dependent on the achievement in the fabrication of well defined interfaces, because most of devices function at their interfaces. Among them, semiconductor devices have been increasing the packing density due to the increment of the integration, which asks very thin layers to be fabricated in high quality and reliability. In order to answer to this kind of requirement, it is necessary to have methods which can characterize such a buried very thin layer effectively and, hopefully, nondestructively.

A soft X-ray emission spectroscopy(SXES) is an interesting method for studying valence band density of states(VB-DOS), because a core level involved in the soft X-ray emission has little wave number dependence[1]. Also it can provide information corresponding to different electronic states, e.g., s-, p- and/or d-electronic DOS, separately for different elements constructing a material to be studied due to the dipole selection rule in electron transition to give rise to a photon emission[2]. The SXES signals give information on VB-DOS complemental to a photoelectron spectroscopy(PES), e.g., X-ray photoelectron spectroscopy(XPS), which reflects the total VB-DOS of a material under study.

EXPERIMENT

SXES experiments were carried out in a vacuum chamber with an electron gun, an X-ray spectrometer and an X-ray detector in an ambient pressure of $\sim 10^{-4}$Pa or better during analysis at room temperature. A monochromated X-ray was directed to either a detector through a fine slit with a width of 60 μ m or a multichannel plate. Specimens to be analyzed were put on a metal specimen holder. The energy resolution of the spectrometer is $\sim$ 1eV or better.

Mat. Res. Soc. Symp. Proc. Vol. 448 © 1997 Materials Research Society

A thin-film/Si(substrate) contact specimen was prepared by either a solid phase reaction(SPR) or a chemical vapor deposition(CVD). A silicon compound film was prepared by the former method, where a thin film of a component material was first deposited on a Si substrate to be followed by a heat treatment below melting points of either constituent materials or the compound to be grown.

XPS study was done with a commercially available apparatus.

RESULTS AND DISCUSSION

Si-K β emission band spectrum is shown in Fig.1(a) for $CrSi_2$, where the signal is normalized at the highest peak and plotted as a function of photon energy(h ν). The features to be noted in the spectrum for $CrSi_2$[Fig.1(a)] is that it has a peak at h ν $\sim$1836.6eV with a shoulder at $\sim$1834.2eV. This is a clear difference from the fact that the Si-K β emission band spectrum of a Si crystal has a peak at $\sim$1836.9eV. Cr_3Si, another compound phase in the Cr-Si binary system, showed a Si-K β emission band spectrum with a peak at $\sim$1834.2eV, which is different from either that of a Si crystal or the one of $CrSi_2$. This kind of difference in the spectral shape among a Si crystal and Cr-silicides can be used as finger prints to identify what kind of material exists on top or at the interface of a thin-film/substrate contact system.

Cr(thin-film, 50$\sim$100nm)/Si(100) specimens heat treated at different temperatures have been studied. In Fig.1 Si-K β emission band spectra are shown for $CrSi_2$(a) and the specimens heat treated at 450(b) and 425°C(c), where the incident electron energy is 5keV. The spectra of Fig.1(a)$\sim$(c) are quite similar. Therefore, one can claim that $CrSi_2$ like compounds are formed for the specimen heat treated at 450 and 425 °C.

Si- $L_{2,3}$ emission band spectra are shown in Fig.2 for $CrSi_2$(a) and Cr(thin-film)/Si(substrate) specimens heat treated at 450(b) and 425°C(c), where detected photons are plotted as a function of the photon energy(h ν). Here, it should be noted that the Si $L_{2,3}$ emission band spectrum reflects the VB-DOS with s- and/or d-symmetry according to the dipole selection rule, because the signal of the photon emission cor-

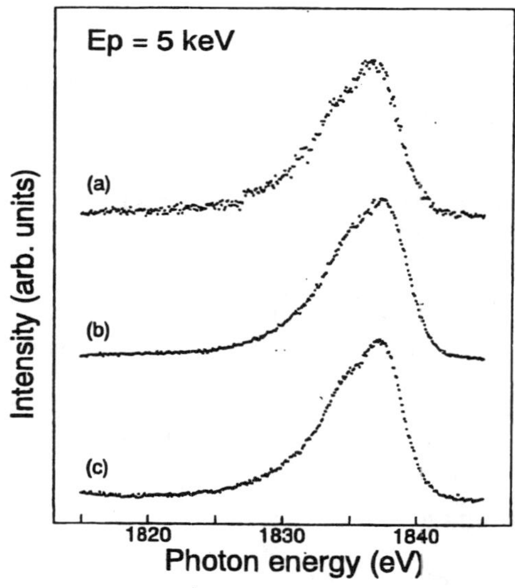

Fig.1 Si-K β emission band spectra: $CrSi_2$(a), and specimens of Cr(film)/Si(100) heat treated at 450°C(b) and 425°C(c).

responds to the electron transition from the valence band to the Si(2p) core holes. On the other hand, the Si-K β emission band spectrum gives information on the partial VB-DOS of Si-p electronic states. The spectrum of Fig.2(a) for CrSi$_2$ has the following characteristics, i.e., it has two broad peaks at h ν ~91.4 and 96.7eV, which is a clear difference from the fact that Si-L$_{2,3}$ spectrum for a Si crystal have a double peak at h ν ~89 and 92eV and a terrace that extends to the top of the valence band[3]. There are clear differences between the spectrum for CrSi$_2$ and the one for a Si crystal. Si-L$_{2,3}$ emission band spectrum for Cr$_3$Si has a peak at ~90eV[3]. These differences in the spectral shape in the Si-L$_{2,3}$ emission band spectra can also be used as finger prints to identify an unknown material.

The fact that the spectrum of Fig.2(b) is quite similar to that of Fig.2(a) again suggests the fact that the material formed on the substrate by the 450°C heating of Cr(thin-film)/Si(substrate) specimen is CrSi$_2$. This is consistent with what we have claimed from the experimental results shown in Fig.1. By the way, the spectrum of Fig.2(c) does not show any special feature at all. On the other hand, the Si-K β emission band spectrum can clearly be observed for the same heat treatment as is shown in Fig.1(c). A question is what the origin of this kind of difference is, i.e., no features in the Si-L$_{2,3}$ emission band[Fig.2(c)] and a clear structures in the Si-K β emission spectrum[Fig.1(c)]. This can be explained as follows. Namely, the above observation is due to the fact either that the mean free path of the Si-K β X-ray in a solid is almost 10 times larger than that of Si-L$_{2,3}$[4] or that the X-ray production depth is much less for 5keV electrons to create K-holes than that for 10keV electrons to create L-holes. The specimen heat treated at 425°C must have an unreacted Cr layer on top, which gives a barrier for the Si-L$_{2,3}$ X-rays leaving the buried layer beneath to be detected. In other word, it can be said that we have successfully carried out a nondestructive analysis of a buried interface structure without adopting any destructive tool like ion sputtering.

Clear modifications in SXE spectra of an element in different networks, e.g., those in Si L$_{2,3}$ spectra in different compounds , or alloys, shown above, can be used for a nondestructive depth profiling of a thin-film/substrate contact system, where energetic electrons are used to excite core electrons. In the above case, we have utilized the difference in the mean free path for Si-K β and Si-L$_{2,3}$. We can also carry out such a nondestructive study by varying the incident electron energy or the incident angle to the specimen surface.

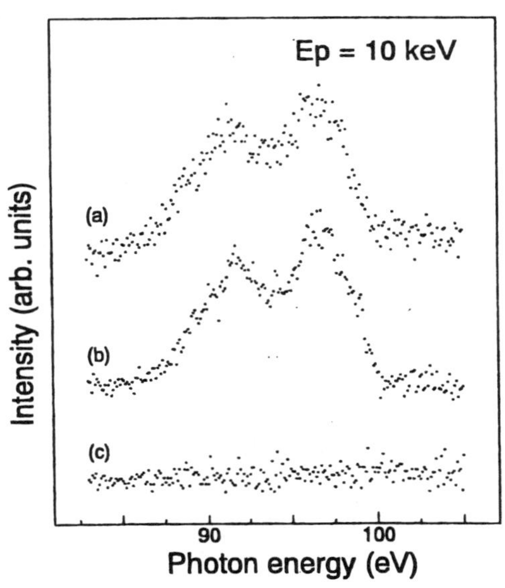

Fig.2 Si-L$_{2,3}$ emission band spectra corresponding to specimens shown in Fig.1.

An electron with energy higher than ~0.1keV can reach deep into a solid with increasing its energy. In other words, the X-ray production depth, d_x, of an energetic electron becomes larger with its energy. This physics leads to a possible application of the electron excited SXES to a nondestructive depth profiling method of a thin film contact system with the help of a certain modification of the X-ray emission spectrum of an element in different chemical bonding states, i.e. an incident energy variation(IEV) method[5]. An important point for the depth profiling using an energetic electron as primaries is that the X-ray production depth(d_x) is much larger than the mean free path (λ) of an energetic electron, i.e., $d_x \gg \lambda$. Therefore, we can carry out nondestructive depth profiling using the electron excited SXES for a specimen with thicker overlayer than in the case to utilize the mean free path of an energetic electron like electron spectroscopies.

An example of the SXES-IEV method is as follows: a Cr(thin-film)/Si(substrate) specimen heat treated at 450°C has been studied by changing incident electron energy between 3 and 20keV, which gives no essential difference in the Si-L$_{2,3}$ emission band spectrum as that of Fig.2(b). This fact indicates that the specimen heat treated at 450°C has a uniform CrSi$_2$ layer on top.

Nondestructive depth profiling can also be carried out using electron excited SXES method by varying the angle of incident electron to the specimen surface, i.e., the incident angle variation (IAV) method[6]. That is, one can tune the X-ray production depth, d_x, of an incident energetic electron by varying the incident angle, where the energy of an incident electron is kept constant.

Until now, we have been concentrated our attention to utilize SXE spectra as finger prints for the applications of the nondestructive structural analysis of a buried interface layer. As mentioned at the beginning, a SXE spectrum due to the transition from the valence band to a shallow core holes carries a wave function selected information on the VB-DOS for each element constructing a material. Therefore, we can clarify an electronic state of a buried interface layer nondestructively.

Si-K β and Si- L$_{2,3}$ emission band spectra for a CrSi$_2$ crystal are shown in Fig.3 together with theoretical Si-s,p VB-DOS[7]. Here, it should be noted that the Si L$_{2,3}$ emission band spectrum reflects the valence band(VB) density of states(DOS) with s- and/or d-symmetry according to the dipole selection rule, because the signal of the photon emission corresponds to the electron transition from the valence band to Si(2p) core holes. On the contrary, the Si-K β emission band reflects Si-p VB-DOS. The Fermi

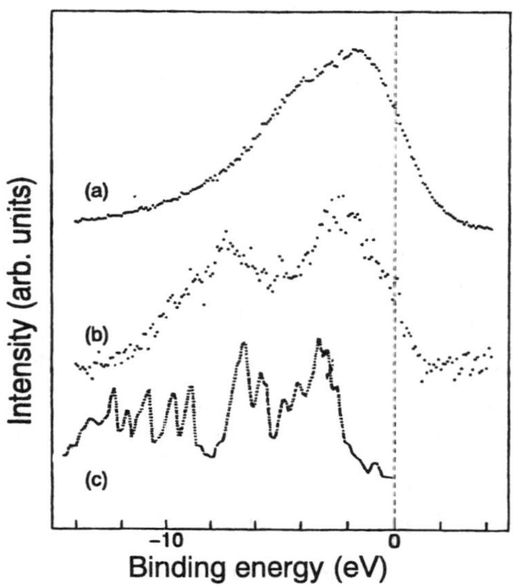

Fig.3 Experimental[(a) Si-K β and (b) Si-L$_{2,3}$] and theoretical[Si-s, p(c)] spectra.

energy(E=0) is determined by using information on the binding energy for Si(1s) and (2p) deduced from the XPS study. The Si-K β signal is considered to correspond to the theoretical VB-DOS peaked at $E_b \sim 3eV$. Si-$L_{2,3}$ signal peaked at $E_b \sim 7eV$ may correspond to the lower part of the theoretical VB-DOS. The upper part of the Si-$L_{2,3}$ emission is due to either Si-s or -d states considering above selection rule. Therefore, we can again safely conclude that the Cr(film)/Si(substrate) specimen heat treated at 425 °C has a buried interface layer with electronic structure of $CrSi_2$ from the experimental result shown in Fig.1(c) together with that of Fig.2(c).

So far, experimental results were obtained using a detector to collect a spectrum point by point. In order to perform experiments more quickly with enough energy resolution, we have adopted a position sensitive detector(PSD) for obtaining soft X-ray emission spectrum. An example is shown in Fig.4 for the Si-$L_{2,3}$ emission band spectrum of $NiSi_2$ which has a sharp line at the Fermi edge. The spectral shape is almost the same as the one reported previously[8]. The data collection time is less than one tenth, or even one hundredth, which enables us to do experiments in shorter time, or to examine even an atomic surface layer of a solid.

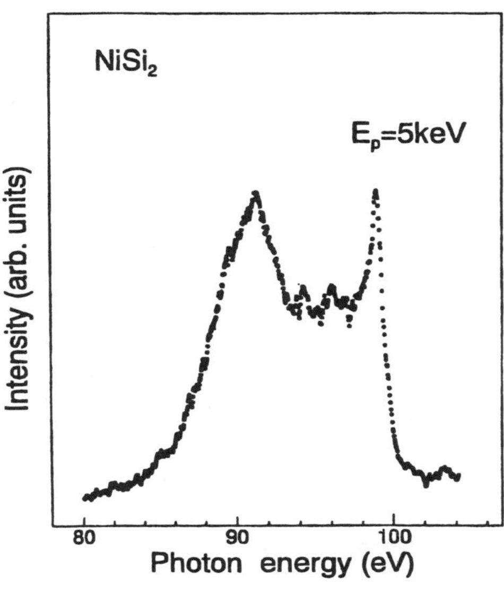

Fig.4 Si-$L_{2,3}$ emission band spectrum of $NiSi_2$ obtained using PSD.

SUMMARY

Structural studies using soft X-ray emission spectroscopy(SXES) have been carried out on heat treated thin-film(e.g., Cr)/Si(substrate) system. They are summarized as follows:

(1) Si-K β and $L_{2,3}$ SXE spectra certainly give characteristic information for different Si-compounds, or alloys. Namely, a SXE spectrum can be used as a finger print for a material to be identified.

(2) The SXES is useful to clarify structures of a buried interface of a thin-film/Si(substrate) contact system nondestructively, where an energetic electron is used for the core hole excitation to give rise to a photon emission. The nondestructive depth profiling can be carried out both by utilizing the difference in the mean free path of soft X-rays with different energy and by varying either the energy or the angle with respect to specimen surface normal of incident energetic electrons. The sampling depth is much larger than conventional electron spectroscopies, which is due to the fact that the former utilizes the energy dependence of an X-ray production depth in a solid in contrast to the latter depends on the inelastic mean free path of energetic electrons.

(3) The usefulness of the SXES for the study of the VB-DOS of a compound, or an alloy, is also

shown, which is due to the selection rule for the photon emission caused by electron transition from valence band to core holes. This fact enables us to identify the electronic structure of a buried layer formed after thin film formation on a solid surface. Also important is the fact that a SXE spectrum is specific to an element and an electronic state of an element constructing a material.

REFERENCES

[1] L.V.Azaroff, X-ray Spectroscopy (McGraw-Hill, Inc., New York, 1974).
[2] L.I.Schiff, Quantum Mechanics (McGraw-Hill, New York, 1968).
[3] A.Simunek, M.Polcik and G.Wiech, Phys. Rev. B52, 11865 (1995).
[4] E.B.Saloman, J.H.Hubbell and J.H.Scofield, Atomic Data and Nuclear Data Tables 38, 65 (1988).
[5] M.Iwami, M.Hirai, M.Kusaka, M.Kubota, S.Yamamoto, H.Nakamura, H.Watabe, M.Kawai and H.Soezima, Jpn. J. Appl. Phys. 29,1353 (1990).
[6] S.Yamauchi, M.Hirai, M.Kusaka, M.Iwami, H.Nakamura, S.Minomura, H.Watabe, M.Kawai and H.Soezima, Jpn. J. Appl. Phys. 31, 395 (1992).
[7] A.Fransiosi, J.H.Weaver, D.G.O'Neill, F.A.Schmidt, O.Bisi and C.Calandra, Phys. Rev. B28, 7000 (1983).
[8] H.Nakamura, M.Iwami, M.Hirai, M.Kusaka, F.Akao and H.Watabe, Phys. Rev. B41, 12092 (1990).

APPENDIX

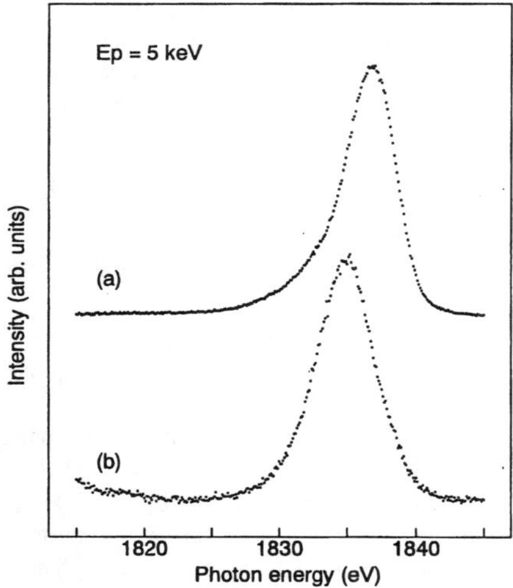

Fig.A1 Si K β SXE spectra for Si crystal(a) and Cr_3Si(b).

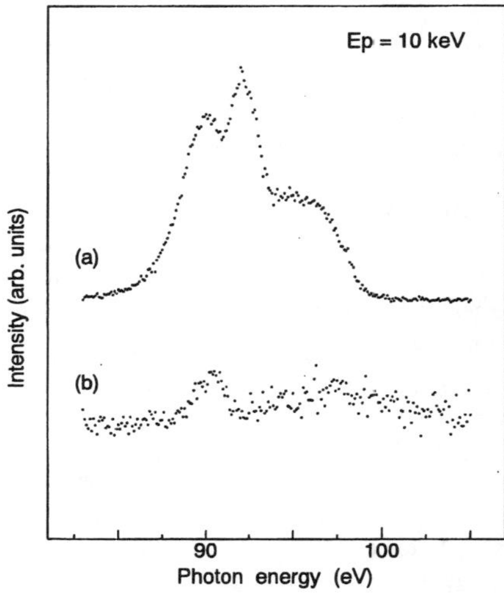

Fig.A2 Si $L_{2,3}$ SXE spectra for Si crystal(a) and Cr_3Si(b).

INVESTIGATION OF Cu-Ge/GaAs METAL-SEMICONDUCTOR INTERFACES FOR LOW RESISTANCE OHMIC CONTACTS

SERGE OKTYABRSKY, M.A. BOREK, M.O. ABOELFOTOH, AND J. NARAYAN,
Department of Materials Science and Engineering, North Carolina State University, Raleigh, NC 27695-7916.

ABSTRACT

Chemistry and interfacial reactions of the Cu-Ge alloyed ohmic contacts to n-GaAs with extremely low specific contact resistivity (6.5×10^{-7} $\Omega \cdot cm^2$ for n~10^{17} cm^{-3}) have been investigated by transmission electron microscopy, EDX and SIMS. Unique properties of the contact layers are related to the formation (at Ge concentration above 15 at.%) of a polycrystalline layer of ordered orthorhombic ε_1-Cu_3Ge phase. Formation of the ε_1-phase is believed to be responsible for high thermal stability, interface sharpness and uniform chemical composition. The results suggest that the formation of the ζ- and ε_1-Cu_3Ge phases creates a highly Ge-doped n^+-GaAs interfacial layer which provides the low contact resistivity. Layers with Ge deficiency to form ζ-phase show nonuniform intermediate layer of hexagonal β-Cu_3As phase which grows epitaxially on Ga{111} planes of GaAs. In this case, released Ga diffuses out and dissolves in the alloyed layer stabilizing the ζ-phase which is formed in the structures with average Ge concentration of as low as 5 at.%. These layers also exhibit ohmic behavior.

INTRODUCTION

Development of ohmic contacts to GaAs and related compound semiconductors that exhibit a low contact resistivity, a high thermal stability, and at the same time, possess a wide process window has been a long-standing challenge in micro- and optoelectronics. Recently, we have found that Cu-Ge compound forms an ohmic contact to n-type GaAs with a specific contact resistivity of 6.5×10^{-7} $\Omega \cdot cm^2$ [1], lower than that reported for Au-Ge-Ni [2], and also Ge/Pd and Ge/Pt/Au contacts on n-type GaAs with comparable doping concentrations (~10^{17} cm^{-3}). The Cu_3Ge alloy layers exhibit remarkably low bulk metallic resistivity of ~6 $\mu\Omega \cdot cm$ at room temperature [3], only a factor of 3 larger than pure Cu, high electrical stability during annealing at temperatures up to $450^{\circ}C$ after contact formation, and also, interface sharpness with GaAs substrate [4]. In addition, it has been recently shown that Cu_3Ge phase has excellent oxidation resistance [5]. All these results suggest a great potential of Cu-Ge alloys for III-V contact technology.

In the temperature range below 570 ^{0}C the equilibrium phase diagram of Cu-Ge binary system consists of the following four phases [6]: (i) α-phase (isomorphous with Cu) exists in the range of composition upto ~10 at.% of Ge; (ii) hexagonal closed-packed ζ-phase with average composition of Cu_5Ge and a wide range of homogeneity from 10 to 18 at.% of Ge. This structure has closed-packed layers similar to those in the f.c.c. structure of pure Cu; (iii) orthorhombic ε_1-Cu_3Ge phase with a quite narrow range of homogeneity, from 24.3 to 25.5 at.%. This phase can be treated as slightly deformed ζ-phase; (iv) Ge with solubility of Cu upto ~10^{-5} at.%.

We have shown recently [1], that the compound with the lowest specific resistivity which also provides the lowest contact resistivity to n-type GaAs is ε_1-phase. This paper is primarily focused on the chemistry of formation of CuGe/GaAs interface as a function of Ge concentration.

EXPERIMENT

The Cu_3Ge contacts with average Ge concentration from 5 to 30 at.% were formed by sequential deposition of Cu and Ge layers on 1 μm-thick n-type GaAs homoepitaxial layers (doped with Si to a concentration of 1×10^{17} cm^{-3}) at room temperature, followed by an *in situ* anneal at $400^{\circ}C$ for 30 min to produce Cu-Ge layers of about 0.2 μm in thickness. The Cu and Ge layers were deposited using electron-beam evaporation in a vacuum 10^{-7} Torr at a rate of 1 nm/s. The average

Mat. Res. Soc. Symp. Proc. Vol. 448 © 1997 Materials Research Society

composition of the layers was determined from the thickness of the elemental Cu and Ge layers assuming the bulk density of these layers. The crystal structure and microstructure of the layers were examined by transmission electron microscopy (TEM) with 200 kV Topcon 002B electron microscope equipped with Noran energy - dispersive x-ray (EDX) detector with superthin Norvar window which allowed us to analyze low-energy x-ray lines.

RESULTS

The results on electrical properties of the Cu-Ge alloy contacts including specific contact resistivity and specific bulk resistivity measurements were described elsewhere [1,4,7]. Here we just want to emphasize that pure Cu forms Schottky contact, and the layers with Ge concentration in the range 5 to 40 at.% form ohmic contact on n-type GaAs. The minimum value of a specific contact resistivity, 6.5×10^{-7} $\Omega \cdot cm^2$, on n-GaAs (1×10^{17} cm^{-3}) was obtained for Cu-Ge alloy contact with 30 at.% of Ge.[5] The n-channel GaAs metal-semiconductor field-effect transistors using the ohmic contacts with 30 at.% of Ge have exhibited a considerably higher extrinsic transconductance (145 mS/mm) compared to the devices with AuGeNi and Ge/Pd contacts [1].

<u>Layers with Ge deficiency to form ζ-phase (<15 at.%)</u>

When pure Cu is deposited on GaAs and alloyed at 400^0C for 30 min, a distinct region of chemical reaction between Cu and GaAs is formed (Fig. 1). This intermediate layer has been identified as hexagonal Cu_3As phase with lattice parameters, a = 7.14 Å and c = 7.32 Å, known in the literature as "artificial" or β-domeykite [8]. The Cu_3As layer is very nonuniform and appears as faceted protrusions grown inside the GaAs substrate. An almost perfect 3:1 composition of the phase is also confirmed by EDX spectroscopic analysis. It shows also rather low solubility of Ga in the Cu_3As phase which does not exceed 1 at.%.

The alloyed layers with low Ge concentration in the range 5-10 at.% (with Ge deficiency to form ζ-phase) exhibit the microstructure illustrated by Fig. 2. Here the cross-sectional images are shown in the (01 $\bar{1}$) projection of the GaAs substrate. The top layer has a hexagonal close-packed structure of ζ-Cu_5Ge phase. Similar to the case of pure Cu, the intermediate layer of β-Cu_3As appears between of Cu-Ge layer and GaAs substrate. Electron diffraction and HRTEM analysis shows that the formation of Cu_3As layer occurs through a solid state epitaxy. As in the case of pure Cu (Fig. 1), the (0001) base plane of hexagonal Cu_3As phase is aligned parallel to two of the four possible {111} planes of GaAs. Specifically, that Cu_3As grains grow epitaxially on GaAs with the following orientation relationship between Cu_3As and GaAs: (0001) // Ga(111), [10 $\bar{1}$ 0] // [01 $\bar{1}$] [9]. For (100) GaAs substrate the basal plane of Cu_3As can be parallel to either of the two equivalent Ga(111) and Ga ($\bar{1}$ 1 $\bar{1}$) planes resulting in two possible orientations of Cu_3As grains. It should be also noted that both β-Cu_3As and ζ-CuGe phases are in contact with the GaAs substrate.

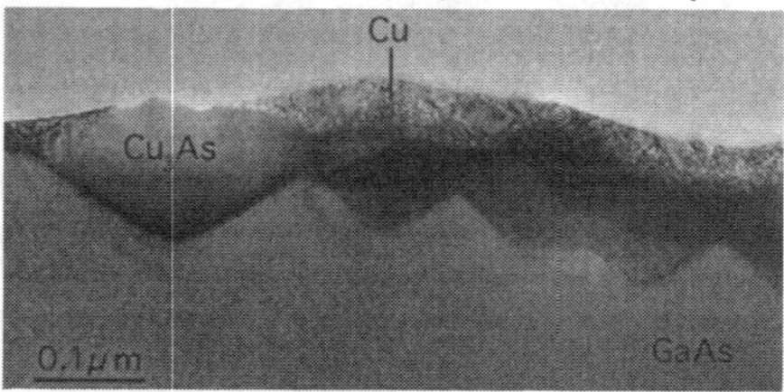

Fig. 1. Cross-sectional TEM micrograph showing the pure Cu layer deposited on GaAs and alloyed at 400^0C for 30 min. The image is taken in (01 $\bar{1}$) projection of the substrate. The top Cu layer is partially removed during ion milling

Fig. 2. Cross-sectional TEM micrographs of the alloyed layer with 5 at.% of Ge in (01 $\bar{1}$) projection of GaAs.

Layers with deficiency of Ge to form ε_1-phase (15 - 25 at.%)

Usually CuGe films appear as polycrystalline with the average grain size of the order of 200 - 300 nm which is almost independent of the composition in the range of Ge concentration from 5 to 40 at.%. Cross-sectional TEM image of the contact formed with a Cu-20 at.% Ge layer (Fig. 3) shows that the film is laterally uniform in thickness and consists of grains of both the orthorhombic ε_1-phase and the hexagonal close-packed ζ-phase. Some faceted protrusions are formed at the interface and consist of the same ε_1- and ζ-phases. No traces of the Cu_3As phase has been observed. We have shown that protrusions at the Cu-Ge/GaAs interface are formed in all the samples with Ge deficiency for the formation of ε_1-Cu_3Ge phase (<25 at.% Ge) [9].

Layers with excess of Ge to form ε_1-phase (> 25 at.%)

The contact layer with average concentration of 30 at.% Ge is laterally very uniform in thickness and contains ε_1-Cu_3Ge (large) and Ge (small) grains (Fig. 4). The Ge grains were found to grow epitaxially on the GaAs substrate [10] and were always present when the average Ge concentration in the alloy layer exceeded 25 at.%. The interface between the ε_1-Cu_3Ge grain and the GaAs substrate is planar and structurally abrupt to within atomic scale, and neither transition layer nor protrusions are observed at the interface. With the increase of Ge concentration to 40 at.%, the microstructure remains similar to that of the contact with a 30 at.% Ge alloy layer, except that the proportion of Ge epitaxial grains is found to increase.

DISCUSSION

Chemistry of CuGe-Cu$_3$As-GaAs system

The results of studying of the chemical composition of the alloyed Cu/Ge/GaAs structures are summarized in Table 1. Four crystal phases were observed: Cu, β-Cu_3As, ζ-CuGe and GaAs. In the case of pure Cu, the metal is not consumed completely by the alloying reaction:

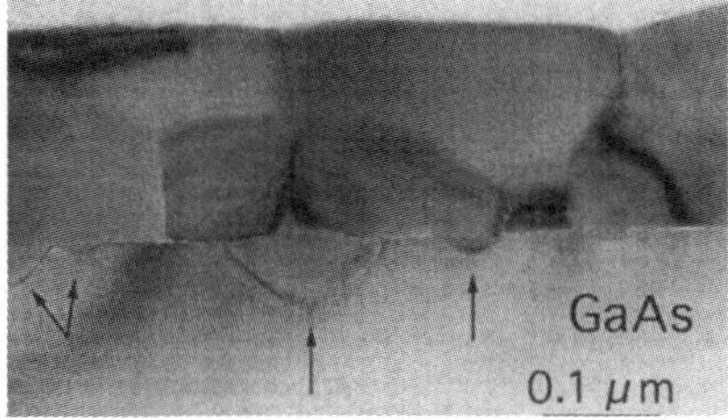

Fig. 3. a) Cross-sectional TEM micrograph showing the microstructure of the Cu-20 at.% Ge layer after annealing at 400˚C for 30 min. Arrows indicate Cu-Ge protrusions.

Fig. 4. Cross-sectional TEM micrograph showing the microstructure of the Cu-30 at. % Ge layer

Table I. Thickness and composition of Cu-Ge contact layers with different average Ge concentration

Average Ge concentration, at.%	0%	5%	10%	15%
Average thickness of Cu$_3$As layer, nm	230*	65	38	< 5
Thickness of ζ-CuGe(Ga) layer, nm	0	210	210	220
Measured (Ge+Ga)/Cu atomic ratio in ζ-CuGe(Ga) layer, at.%	-	15%	19%	23%
Measured Ga/Ge atomic ratio in ζ-CuGe(Ga) layer, at.%	-	50%	25%	<20%
Ga/Ge atomic ratio estimated from Ge-balance, at.%	-	65%	42%	25%
Ga/Ge atomic ratio estimated from Ga-As balance, at.%	-	42%	22%	-

*incomplete reaction with Cu

$$GaAs + 3Cu \longrightarrow Cu_3As + Ga. \qquad (1)$$

and the structure consists of a thick intermediate Cu_3As and top Cu layers (Fig.1). When the average Ge content is increased to 5 at.% the top layer of ζ-CuGe phase appears in the structure. The transitional Cu_3As layer, a product of the reaction (1), remains in the structure with Ge concentration up to 15 at.% though its average thickness keeps decreasing. In the sample with 15 at.% of Ge, only small grains of the Cu_3As phase have been found at the interface. An EDX spectrum of β-Cu_3As layer is shown in Fig. 5a. Although Ga-L line appears in the spectrum, we believe that Ga concentration in Cu_3As is still very low as in the case of pure Cu alloying, and Ga-L peak is tentatively attributed to a secondary x-ray emission from GaAs substrate. This assumption is rather

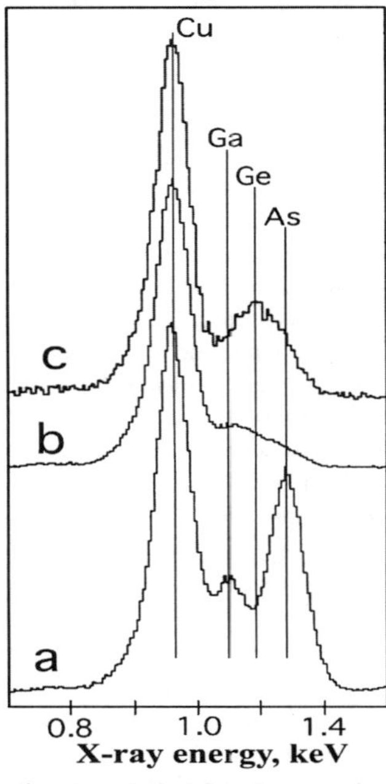

Fig. 5. TEM EDX L - region spectra of (a) β-Cu_3As layer in an alloyed layer with 5 at.% of Ge, (b) ζ-CuGe(Ga) phase in the same layer, and (c) ζ-phase formed in the structure with 15 at.% of Ge.

obvious since the thickness of Cu_3As layer decreases with the increase of Ge concentration in the contact layer and the secondary emission from the adjacent material contributes largely to the spectrum. On the other hand, the spectrum in Fig. 5a gives a much higher As-to-Cu ratio than for stoichiometric Cu_3As compound. We can, therefore, explain this fact consistently by contribution of As-L secondary emission from the substrate. In the layers with 5 at.% of Ge, the Ge concentration is not high enough to form ζ-phase which can be obtained in the composition range 10-18 at % of Ge [6]. The x-ray EDX analysis of the ζ-phase (Fig. 5b) indicates the presence of both Ge and Ga at a total concentration of 15 at. %. The presence of Ga can stabilize the hexagonal ζ-phase because both Cu-Ge and Cu-Ga compounds belong to the same type of Hume-Rothery electron phases and Cu-Ge-Ga ternary system has a wide homogeneity range [11]. For comparison, Fig. 5c shows the EDX spectrum of ζ-phase formed in the structure with 15 at.% of Ge which contains relatively low concentration of Ga.

We have estimated the Ga/Ge composition ratio in the ζ-phase using the measured average thicknesses of the layers and assuming mass conservation of the elements involved in the reaction. The two bottom lines of Table 1 summarize these estimates performed for Cu and Ge conservation as well as Cu, Ga and As conservation, respectively. Both results look rather consistent with the EDX composition measurements within the experimental accuracy which does not exceed 30 % due to uncertainties in thickness measurements of the nonuniform layers and composition. The two estimates give somewhat different results mainly due to the loss of As during the alloying reaction.

Relatively high Ga concentration in the ζ-phase is obviously due to outdiffusion of the excess Ga released in the reaction (1). With the increase of Ge concentration two tendencies are obvious. Firstly, the thickness (or volume) of the Cu_3As layer decreases, and this layer disappears almost completely at 15 at.% of Ge. Secondly, Ga concentration in the ζ-layer falls indicating that Ga content is determined primarily by the reaction (1). Clearly, copper consumed in the ζ-phase can not be released any more to form β-Cu_3As phase indicating a much lower chemical potential of Cu in the ζ-phase. These results strongly suggest that Cu_3As phase is formed exclusively when the free copper is present in the sample. The direct reaction in the film:

$$Cu + Ge \longrightarrow \zeta\text{-}Cu_xGe \qquad (2)$$

consumes existing Ge, and also, if the Ge content is high enough, As is pulled out from the initially formed Cu_3As phase :

$$\beta\text{-}Cu_3As + Ge \text{ --> } \zeta\text{-}Cu_xGe + As \qquad\qquad (3)$$

In this case, germanium can be either in the elemental form or in the form of Ge-rich ζ-phase consuming more and more Cu from the Cu_3As layer. In addition, excess Ga can help to stabilize the ζ-phase, as has been shown above. Arsenic released in the reaction (3) is probably removed from the sample in a molecular form since no signs of the As-containing phases have been observed in the alloyed structures with high (>15%) Ge concentration. The suggested scheme for the formation of the alloyed structure involving the reactions (1), (2) and (3) also explains the roughness of the ζ-CuGe/GaAs interface as a result of initial growth of the Cu_3As compound.

<u>Chemistry of Cu_3Ge -GaAs system</u>

We should emphasize a very simple chemistry of a resultant alloyed contact layer with Ge concentration higher than 15 at.%, i.e. containing enough Ge to form ζ,-phase. The layers consist exclusively of ζ, ε_1-phases and some Ge grains when the average Ge:Cu composition is varied from 15:85 to 40:60. It is a remarkable feature of Cu-Ge layer that no chemical reaction products with GaAs have been found. It should be also noted that interfacial protrusions typical for the layers with Ge concentration <25 at.%, are not observed in the contact layers containing excess of Ge to form ε_1 -phase. This provides an evidence for a strong kinetic suppression of the reaction (1) in the presence of high Ge concentration. In fact, though Cu was deposited first followed by Ge layer, thermal annealing did not provide a rough interface typical for Cu_3As-phase formation. The rate of the reaction (2) (or similar reaction forming ε_1-Cu_3Ge phase) is much higher than that of reaction (1). In addition, the ordered ε_1-Cu_3Ge phase has very low reactivity with the GaAs substrate, thus providing excellent homogeneity and uniformity of the contact.

<u>Ohmic contact formation</u>

As was emphasized above, the contact resistivity is effected only slightly by Ge concentration in the layers in the range of 15-30 at.%. No significant differences in electrical properties were found in the samples with epitaxial Ge grains (when Cu is deposited first) or discontinuous epitaxial layer (when Ge is deposited first). Consequently, the presence of the Ge/GaAs heterostructure at the interface has minor effect on the contact properties. The most probable mechanism for ohmic contact creation is a formation of highly Ge-doped n^+-layer at the interface.

Since Ge is an amphoteric impurity, for n^+-layer to appear, Ge should occupy mainly the Ga sites. As was mentioned above, Cu-Ge phases certainly provide a huge drain for Ga atoms because of almost infinite solubility of Ga in the Cu-Ge alloys. On the other hand, As solubility in Cu-Ge is relatively low. Therefore, thermodynamic stability of the contact layer implies that a layer enriched with Ga vacancies should be formed in the GaAs substrate close to the Cu_xGe/GaAs interface. The Ga outdiffusion from the GaAs substrate can indeed be observed in the SIMS profile (Fig. 6). The SIMS profiles of a contact structu-res with 15 at.% Ge (Fig. 6) indicate strong diffusion of Ge into GaAs. The value of the diffusion coefficient is almost independent of Ge concentration in the 15-30

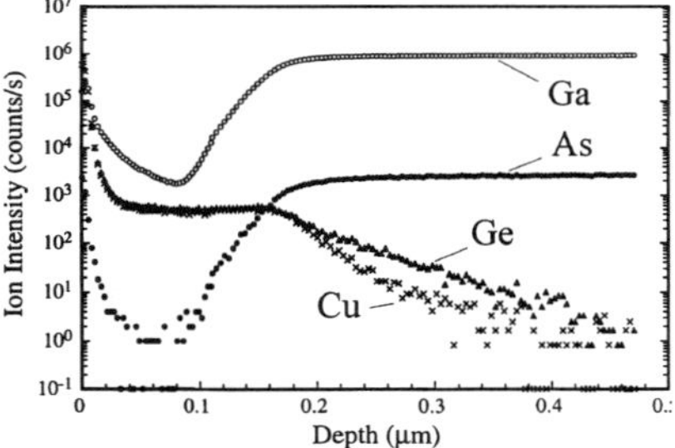

Fig. 6. SIMS profiles of contacts formed with 30 at.% Ge alloy layers and with Cu deposited first on the GaAs substrate after annealing at 400°C for 30 min.

at.% range and decreases only by a factor of 2-3 when the average concentration of Ge drops to 5 at.%. These results indicate that the diffusion of Ge into GaAs is enhanced at lower temperatures by

387

the presence of Cu as has been reported earlier for Ni [12]. Usually Cu is suggested to be a fast interstitial diffuser in GaAs with a diffusion coefficient of as high as ~10^{-6} cm^2/s at 400 ^{0}C. Occupying Ga sites copper becomes a double acceptor. The diffusion of Cu, thus, provides a competitive mechanism for doping the Ga sublattice with p-type impurities and will compensate Ge$_{Ga}$ donors. However, we did not find any effect of compensation. On the contrary, specific contact resistivity was stable in the range of Ge concentrations 15-30 at.% and thermal annealing at <400^{0}C. Therefore, concentration of compensating Cu$_{Ga}$ acceptors is much lower than the concentration of donors in the interfacial region of GaAs. The SIMS profiles (Fig. 6) also show that Cu diffuses into the GaAs not farther than Ge does. Consequently, we should emphasize that Cu diffusion is suppressed by the presence of Ge. Additional work, however, is needed to determine the mechanism responsible for significant change of diffusivity of both Ge and Cu and their influence on each other.

Finally, one more experimental fact should be explained. It was shown above that β-Cu$_3$As phase is formed at the CuGe/GaAs interface when the average Ge concentration is below 15 at.%. Formation of this phase is supposed to provide an excess of Ga-interstitials rather than Ga-vacancies (due to the reaction (1)) in the interfacial GaAs region. As a result, this should suppress the formation of Ge$_{Ga}$-doped n$^+$-layer and prevent the formation of Ohmic contact. However, ohmic contact is still formed with Ge concentration of as low as 5 at.%. This controversy could be explained by the fact that in the samples with low Ge concentration both β-Cu$_3$As and ζ-CuGe phases are in contact with the substrate. In this case Ge-doped n$^+$-layer still should be formed at the CuGe/GaAs interface due to Ga-outdiffusion from the substrate into CuGe phase. Thus the creation of the ohmic behavior of the contact is due to the regions of the substrate which occasionally appear in the direct contact with CuGe phase. The contact resistivity will be increased in this case owing to decreasing of the contact area but will still exhibit ohmic behavior.

CONCLUSIONS

Chemistry, interfacial reactions and diffusion characteristics of the Cu-Ge alloyed ohmic contacts to n-GaAs have been investigated by TEM and SIMS techniques. Unique properties of the contact layers are related to the formation (at Ge concentration above 15 at.%) of a polycrystalline layer of ordered orthorhombic ε_1-Cu$_3$Ge phase. The results suggest that the formation of the ζ- and ε_1-Cu$_3$Ge phases creates a highly Ge-doped n$^+$-GaAs interfacial layer which provides the low contact resistivity. Layers with Ge deficiency to form ζ-phase show nonuniform intermediate layer of hexagonal β-Cu$_3$As phase which grows epitaxially on Ga{111} planes of GaAs. In this case, released Ga diffuses out and dissolves in the alloyed layer stabilizing ζ-phase which is formed in the structures with average Ge concentration of as low as 5 at.%. These layers also exhibit ohmic behavior.

ACKNOWLEDGMENT

This work was supported by the National Science Foundation (ECS - 9525993).

REFERENCES

1. M.O. Aboelfotoh, C.L. Lin, and J.M. Woodall, Appl. Phys. Lett. **65**, 3245 (1994).
2. M. Murakami, Mater. Sci. Rep. **5**, 273 (1990).
3. M.O. Aboelfotoh, H.M. Tawancy, and L. Krusin-Elbaum, Appl. Phys. Lett. **63**, 1622 (1993).
4. S Oktyabrsky, M.O.Aboelfotoh, J.Narayan, and J.M.Woodall, J.Electron.Mat.**25**,1662 (1996).
5. H.K.Liou, J.S.Huang, and K.N.Tu, J. Appl. Phys., **77**, 5443 (1995).
6. M. Hansen, *Constitution of Binary Alloys* (McGraw-Hill, New York, 1958), pp.160, 585.
7. L. Krusin-Elbaum and M.O. Aboelfotoh, Appl. Phys. Lett. **58**, 1341 (1991).
8. B. Steenberg, Arkiv foer Kemi, 12A No.26, 1 (1938).
9. S. Oktyabrsky, M.O. Aboelfotoh, and J. Narayan, J. Electron. Mat. **25**, 1673 (1996).
10. M.O.Aboelfotoh, S.Oktyabrsky, J.Narayan, and J.M.Woodall, J.Appl.Phys. **75**, 5760 (1994).
11. G.V.Raynor and T.B.Massalski, J. Inst. Metals, **84**, 66 (1955))
12. M. Heiblum, M.I. Nathan, and C.A. Chang, Solid State Electron. 25, 185 (1985).

ALLOYING BEHAVIOR AND RELIABILTY OF Pt EMBEDDED METAL/n^+-GaAs THIN OHMIC CONTACT SYSTEM

C.Y. KIM *, W.S. LEE *, H.J. KWON *, Y.W. JEONG *, J.S. LEE *, and C.N. WHANG **
*LG Electronics Research Center, 16 Woomyeon-dong, Seocho-gu, Seoul 137-140, Korea
**Department of Physics, Yonsei University, 134 Sinchon, Sudaemoon-gu, Seoul 120-749, Korea

ABSTRACT

Pt embedded ohmic contacts to n^+-GaAs (AuGe-800 Å/ Ni-150 Å/Pt-200 Å/Au-500 Å and AuGe-800 Å/Pt-200 Å/Ni-150 Å/Au-500 Å/n^+-GaAs) have been developed for the advanced discrete devices and MMIC (monolithic microwave integrated circuit) applications. The specific contact resistance investigated by Transmission Line Method is 1×10^{-6} Ω cm^2. Ohmic contact reliability investigated by thermal storage test at 300 °C under N_2 ambient demonstrated nearly the same contact characteristics after 3000 hours. In both systems, X-ray diffraction results and Auger depth profiles show that the good ohmic contact is related to the formation of Au_7Ga_2, $PtAs_2$, and $Ni_{19}Ge_{12}$ phases. AuGa compound enhances the creation of Ga vacancies, allowing incorporation of Ge into Ga sites, and PtAs compound is piled up in the middle of AuGa layer to suppress As outdiffusion from GaAs substrate. TEM cross-sectional view indicates that metal/n^+-GaAs reaction layer is ~1200 Å beneath GaAs. Surface and interface are very smooth and abrupt in comparison to conventional AuGe/Ni/Au contact.

INTRODUCTION

Metal/n^+-GaAs ohmic contacts have been investigated for various GaAs integrated circuits. For the fabrication of MMICs (monolithic microwave integrated circuits) with submicron gate length, ohmic contact system needs to meet several requirements [1, 2, 11]. First of all, contact resistance should be nearly negligible and the contacts should be thermally stable. Moreover, the metallized layer must have smooth faced. Shallow vertical and lateral diffusion lengths are also required for large scale integration. Meanwhile, it is of paramount important to control all the process parameter because the microstructure of contact is sensitive to the doping level of GaAs substrate, the wafer cleaning method, the contact metallurgy, the deposition process, and the alloy temperature. Correlation between ohmic mechanism and microstructures at a metallurgical viewpoint has been studied for about a decade [3~5, 11].

The conventional AuGeNi contact has been the most widely used system for ohmic devices with low contact resistance. However, it has well known that its surface and interface exhibit a non-planar morphology. In order to avoid this drawback, non-gold contact systems and their microstructural changes were intensively studied [7, 8]. Additionally, low-temperature carrier transportation mechanism has been investigated and reported by some research groups [9].

In this paper, we embedded thin Pt metal layer into the conventional AuGeNi system to prepare the modified AuGeNiPt ohmic system for applications in advanced MMICs. The total thickness of the contact metal was kept to be 1650 Å. Electrical and physical properties were investigated by the various analytical techniques. Reliability of the contact was studied using the thermal storage test up to 3000 hr at 300 °C. We emphasized on the role of the embedded Pt layer as well as on the correlation between microstructural change and good ohmic behavior and reliability.

Mat. Res. Soc. Symp. Proc. Vol. 448 © 1997 Materials Research Society

EXPERIMENTAL

The 0.1 μm n^+-GaAs layer was grown by metalorganic chemical vapor deposition on the semi-insulating (100) oriented GaAs substrates. Undoped buffer layer of 1 μm thickness was deposited between these layers. Carrier concentration of the as-grown n^+-GaAs layer was 1×10^{18} cm^{-3}. Prior to metal deposition by electron-beam evaporator, the wafers were etched in a solution of pure HF for 90 sec and rinsed in D.I. water to remove any surface oxide. Transmission Line Method (TLM) pattern was made by photolithography. Prior to metal deposition the wafers were chemically cleaned again using a solution of HCl:H$_2$O=1:3 for 1 min.

Fig. 1 illustrates schematic cross-section of the samples. Sample A has the configuration of AuGe(12at%-Ge)-800 Å/Ni-150 Å/ Pt-200 Å/Au-500 Å/ n^+-GaAs. Sample B is identical to sample A except that the deposition order of Ni and Pt is reversed: AuGe-800 Å/ Pt-200 Å/Ni-150 Å/Au-500 Å/n^+-GaAs. For contact resistance and sheet resistance measurements, samples with TLM and van der Pauw patterns were prepared on the mesa etched n^+-GaAs layer. Ohmic metal pad and dot line were also fabricated on mesa etched n^+-GaAs by liftoff. Ohmic contact was obtained by rapid thermal annealing (RTA) for 10 sec at 430 ℃ in N$_2$ ambient.

Reliability test was carried out in N$_2$ ambient for 486 hr, 2500 hr, and 3000 hr at isothermal temperature of 300 ℃. The metals adhered well to GaAs without peeling.

Auger electron spectroscopy(AES) and X-ray diffraction(XRD) were utilized to measure the interfacial reactions and to study the phases formed. Atomic force microscopy (AFM), Scanning electron microscopy(SEM), and transmission electron microscopy(TEM) were employed to examine the surface and interface morphology and microstructural changes.

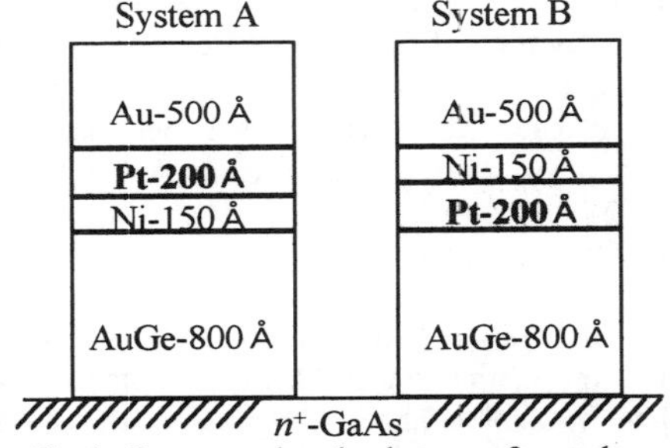

Fig.1. Cross-sectional schemes of samples

RESULTS

Specific Contact Resistance

Fig. 2 shows the variation of specific contact resistance as a function of aging time for both systems. Minimum specific contact resistance is 9×10^{-7} Ω cm^2. This specific contact resistance is higher than that for AuGeNi scheme: 10^{-7}-10^{-6} Ω cm^2 for n=10^{17} cm^{-3}. But the characteristics of specific contact resistance showed nearly same even after thermal storage test for 3000 hr.

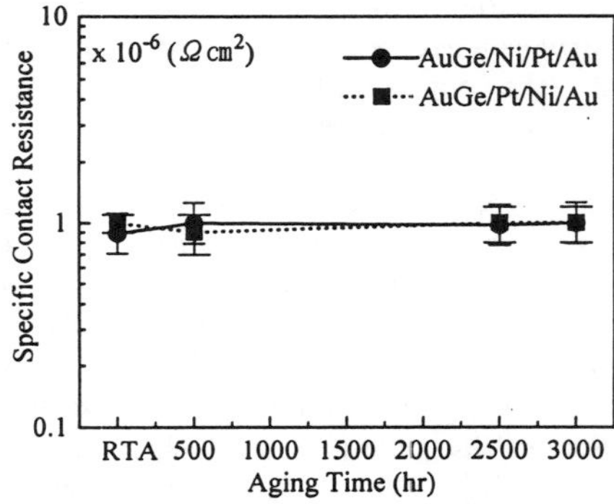

Fig.2. Specific contact resistance variations
as function of aging time.
(The solid line is system A and the dotted system B).

<u>Phase Identification and Atomic Redistribution</u>

Fig. 3 presents XRD profiles for system A. The result of system B was identical except for the difference of intensity ratio of double peak(Au_7Ga_2) at $2\theta \fallingdotseq 40°$. In the figure, XRD peaks corresponding to β-Au_7Ga_2(β-AuGa, hexagonal), $PtAs_2$(cubic), and $Ni_{19}Ge_{12}$(hexagonal) were clearly observed. Furthermore, these phases remained unchanged even after a 3000-hr thermal storage test. Small amount of γ-AuGe phase was detected after aging.

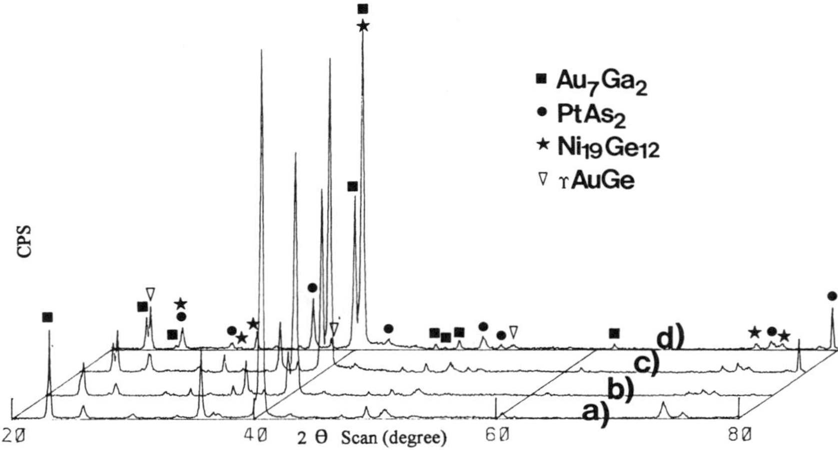

Fig.3. XRD profiles for system A with the AuGe/Ni/Pt/Au configuration. (a: RTA, b: 486 hr, c: 2500 hr, and d: 3000 hr aging).

Fig.4. shows the atomic redistribution investigated by Auger depth profiling for both systems. After RTA for 10sec at 430 ℃, the contacts undergo a reaction to form complex compounds. This reacted layer has been separated in a qualitative sense into the three layers: an AuGa layer on the top, a PtAs layer isolated in the middle (an arrow indicates in Fig. 4.d)), and a Au-enriched layer on the bottom adjacent to GaAs (a dotted arrow indicates). From XRD measurement, the Au-enriched bottom layer was considered to be AuGa compound. Additionally, Ni and Ge elements were found to be redistributed broadly from PtAs region toward surface region. The resulted redistribution behaviors coincided well with XRD phase formation. With increasing aging time, Ga at near surface was oxidized while in system B, Ni was oxidized because of its close vicinity to the surface. These atomic redistribution was not changed after aging test for 3000 hr.

Au element, the dominant moving species, diffuses fast into GaAs and mixes with the dissociated Ga. The dissociated Ga and As elements also out-diffuse, and then Ga spreads through the entire region to form AuGa alloy. The observation of piled-up As element is consequence of the immobility of As [1] and the formation of PtAs compound. From binary phase diagram [12] and from the factor that Ni element plays a role of wetting agent which can adhere to substrate materials and make surface smooth [6], it was considered that NiGe and PtAs compounds were very stable, and they respectively played important roles on the forming of ohmic contact by blocking the out-diffusion of excess Ge and As.

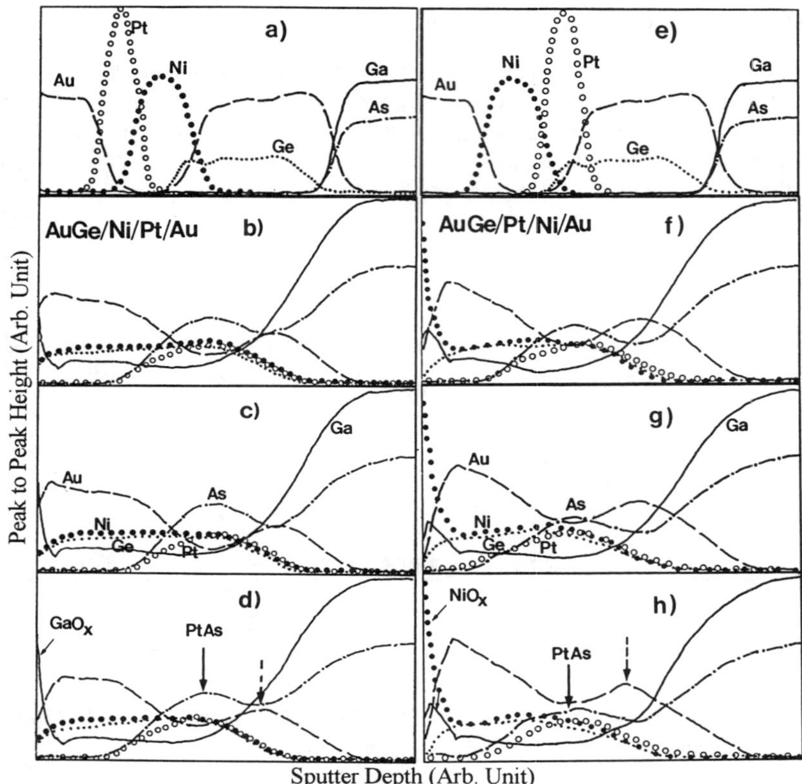

Fig.4. AES depth profile results for AuGe/Ni/Pt/Au system at left side and AuGe/Pt/Ni/Au at light side (a and e: as-deposited, b and f: RTA, c and g: 486 hr, and d and h: 3000 hr aging) : The solid arrow indicates the position of piled-up PtAs layer, the dotted AuGa layer.

Microstructural Changes

Fig.5 illustrates the cross-sectional views for both systems. Upper three images are for AuGe/Ni/Pt/Au system and lower three for AuGe/Pt/Ni/Au. Drastic microstructural changes were observed, but the layered structure did not coincide with that expected from AES and XRD results. In the interface of System A, long-periodic nodal morphology were observed. Such morphology typically appears in Au-based contact system because of fast diffusion of Au element. Surface morphology of system A was more flatter than that of system B. Both surfaces of system A and B, however, were relatively smooth compared to those of other systems. Fig. 6 shows the comparison of SEM surface images between systems: AuGe/Ni/Au system have a balling-up morphology and AuGe/Pt/Au have a dendrite. From dark field images, β-AuGa phase was found to spread through the entire reaction layer, and small grains were existed in this β-AuGa phase. From AES results, we conjectured that small PtAs grains was located preferentially at the middle and NiGe was located broadly from the surface to the middle region.

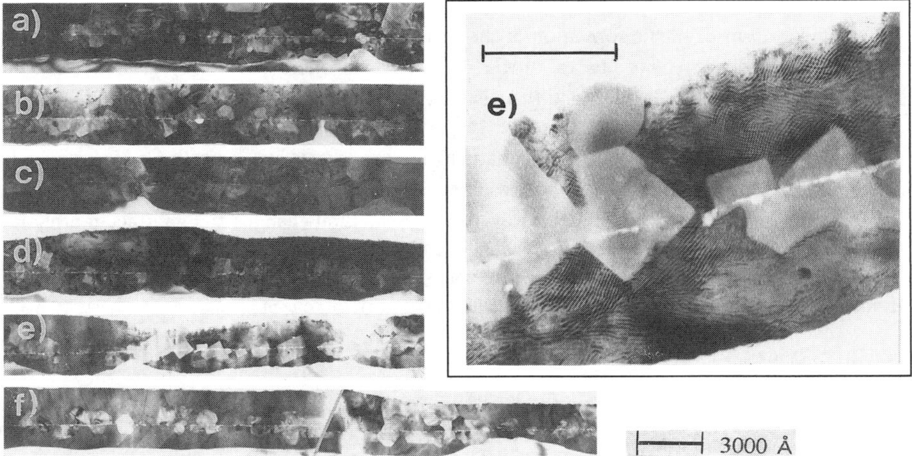

Fig.5. Cross-sectional TEM views for both systems annealed in N_2 ambient at 300 ℃ (a and d: RTA, b and e: 486 hr, and c and f: 2500 hr): Upper three image is AuGe/Ni/Pt/Au, lower three is AuGe/Pt/Ni/Au. Scale marker in low right corner applies to all images except the rectangle box.

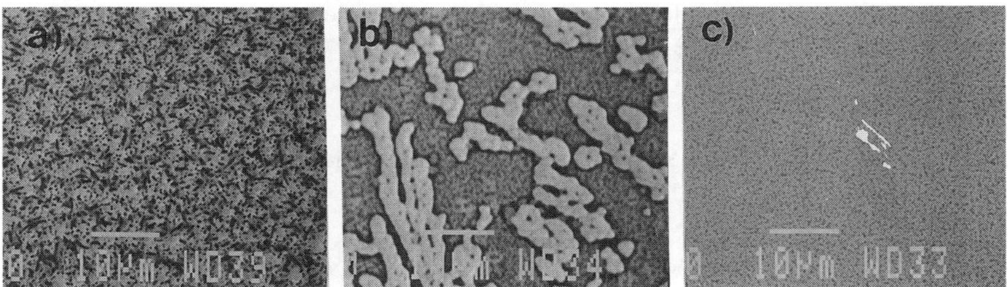

Fig.6. Comparison of surface morphologies of a) AuGe/Ni/Au, b) AuGe/Pt/Au and c) AuGe/Ni/Pt/Au system after RTA at 430 ℃ for 10sec.

In the case of system B, surface morphology and film thickness were irregular. Two images of Fig.5.f) were investigated respectively at other region of same 2500 hr-annealed sample. The left one was quite flat surface and interface, while the light one was shown also irregular. A dislocation networking was investigated in sample e), which likely formed on releasing the faceted $Ni_{19}Ge_{12}$ grain defects (the rectangle box shows). In the 2500-hr annealed sample f), it was observed that Au_7Ga_2 grain was divided to several smaller grains inclined to surface by 30° and to interface by ~55°. From the microstructures and image contrast, $PtAs_2$ grains were not found in this specimen because of the localized image information, but it can be found at other areas under a low magnification. In the middle of the metallized layer, a vacancy line with thickness of about 50 Å was observed, which was corresponded to initial native oxide layer(GaOx). The thickness of the metallized layer was extended up to average 1200 Å beneath initial GaAs surface, and these were reflected a shallow vertical and horizontal diffusion length. According to AFM result, surface rms roughness scanned on 2x2 μm^2 area using contact mode was below 50 Å.

DISCUSSION

The mechanism of ohmic formation of the Pt embedded system was related to the formation of Au_7Ga_2, $PtAs_2$, $Ni_{19}Ge_{12}$ phases. AuGa compound enhances more Ga vacancies, allowing incorporation of Ge into Ga sites, and PtAs is piled up in the middle of Au_7Ga_2 grain to suppress effectively As out-diffusion from substrate, respectively. Finally, amphoteric Ge element then is substituted to Ga vacancies(V_{Ga}), forming a heavily doped layer just beneath GaAs surface. The above reactions were clarified by the following equations: $7Au+2GaAs = Au_7Ga_2+2As$, $1Pt+2As= 1PtAs_2$, and $19Ni+12Ge = Ni_{19}Ge_{12}$. However, microstructural changes and interface morphology were less sensitive to ohmic characteristics in Pt embedded system. The Entire reaction layer was consisted of β-AuGa single phase, which was directly contacted to n^+-GaAs surface. Small grains were broadly distributed from the middle to the surface region in this single phase.

CONCLUSIONS

We have made the modified AuGeNiPt thin ohmic contact systems: AuGe/Ni/Pt/Au/, AuGe/Pt/Ni//Au/n^+-GaAs. The specific contact resistance remained 1×10^{-6} $\Omega\,cm^2$ after annealing for 3000 hr at 300 ℃. Both systems, with a total thickness of 1650 Å, were analyzed to have undergone a reaction up to about 1200 Å beneath GaAs surface.

Ohmic contact and reliability characteristics of the Pt embedded system were related to the formations of Au_7Ga_2, $PtAs_2$, and $Ni_{19}Ge_{12}$ phases.

Both systems were proven to be thermally stable, low contact resistive, and shallow ohmic contact system with smooth surface and interface.

ACKNOWLEDGMENTS

Authors gracefully thank Mr. J.S. Kwak of Yonsei University for his advice and discussion.

REFERENCES

1. T.C.Shen, G.B.Gao, and H.Morkoc, J. Vac. Sci. Technol. B, **10(5)**, 2113(1992).
2. M.Murakami, Mat. Sci. Rep., **5**, 273 (1990).
3. T.Sands, Materials Science & Engineering, **B1**, 289 (1989).
4. T.Sands, V.G.Keramidas, K.M.Yu, J.Washburn, and K.Krishnan, J. Appl. Phys., **62(5)**, (1987).
5. S.A.Marshall, S.S.Lau, C.J.Palmstrom, T.Sands, C.L.Schwarz, S.A.Schwarz, J.P.Harbison, and L.T.Florez, Mat. Res. Symp. Proc., **148**, 163, (1989).
6. M.N.Yoder, Solid-State Electronics, **23**, 117, (1980).
7. D.Y.Chen, Y.A.Chang, and D.Swenson, Appl. Phys. Lett., **68(1)**, 96, (1996).
8. J.S.Kwak, H.N.Kim, H.K.Baik, J.-L.Lee, H. Kim, H.M.Park, and S.K.Noh, Appl. Phys. Lett., **67(17)**, 2465, (1995).
9. C.C.Han, X.Z.Wang, S.S.Lau, R.M.Potemski, M.A.Tischler, and T.F.Kuech, J Appl. Phys., **69(5)**, 3124, (1991).
10. E.C.Marshall and M.Murakami, p. 21. and Z. Liliental-Weber and E. Weber, p. 114., in Conatct to Semiconductors, edited by L.J.Brillson (Noyes Pub. Park Ridge, NJ, 1993).
11. M. Pecht and P. Lall, p. 41. and F.Magistrali, C.Tedesco, E.Zanoni, and C.Canali, p. 114., Reliability of Gallium Arsenic MMICs edited by A.Christou(John Wiley & Son 1992).
12. Binary Alloy Phase Diagram, Vol. 1-2 (American Society for Metals, Metals Park, Ohio, 1986), p. 129, 258, and 1231.

IMPROVEMENT OF THE REFRACTORY METAL/n-GaAs INTERFACE BY LOW TEMPERATURE ANNEAL

A. Singh and L. Velásquez
Laboratorio de Semiconductores, Departamento de Física, Universidad de Oriente,
Apartado 188, Cumaná 6101 Sucre, Venezuela

ABSTRACT

The W/n-GaAs Schottky junctions A and B of area 1.75×10^{-2} cm^2 were fabricated by deposition of W on the chemically etched polished surfaces of n-GaAs samples by rf sputtering using a rf powers of 300 Watt for 30 min. The W contact B was subjected to a 90 min. thermal anneal at 390 °C. The room temperature I-V and C-V/f (with 200 Hz < f < 1 MHz) measurements were carried out for both the as-deposited and thermally annealed W/n-GaAs Schottky junctions A and B, respectively. From the direct I-V data, the values of 1.09 and 8.1×10^{-8} A for the ideality factor (n) and the reverse saturation current (I_o), respectively, were estimated for the diode B, compared to the values of n=1.70 and $I_o = 6.3 \times 10^{-6}$ A for the diode A. The observed frequency dispersion in the zero bias capacitance in the diode B was attributed to fast interface states with a time constant, $\tau_2 = 6$ μs and density, $Nss_2 = 5.8 \times 10^{10}$ eV^{-1}cm^{-2}, whereas, both the slow interface states (with $\tau_1 = 4$ ms and density, $Nss_1 = 7.8 \times 10^{12}$ eV^{-1}cm^{-2}) and fast states (with $\tau_2 = 1$ μs and density $Nss_2 = 8.6 \times 10^{10}$ eV^{-1}cm^{-2}) were responsible for the observed frequency variation of the zero bias capacitance in the diode A. For the forward bias values in the range 20-100 mV, the frequency dispersion in the measured capacitance suggested the presence of both the fast and slow interface states (with time constants differing by three orders of magnitude) in the as-deposited and the heat treated W/n-GaAs interfaces. Thermal anneal at 390 °C for 90 min. lowered the density of states at the W/n-GaAs interface by two orders of magnitude and resulted in the formation of a high quality rectifying W contact to n-GaAs with a rectification ratio of 1.4×10^4, a low I_o and an ideality factor close to unity.

INTRODUCTION

Thermally stable refractory metal/n-GaAs interfaces find important applications in high temperature electronic devices. Among the various refractory metals, W is an attractive choice as a Schottky contact metal because of its high temperature resistance, high work function and low electrical resistivity. The rf sputtering is one of the most commonly used technique for the fabrication of refractory metal/n-GaAs Schottky barriers. Our previous work[1-3] on the as-deposited W/n-GaAs junctions has revealed that the characteristic parameters (ideality factor (n), reverse saturation current (I_o), rectification ratio (r), barrier height (ϕ_{bo}) and energy density of the surface defects (N_{ss})) of the W/n-GaAs Schottky diodes were highly dependent on the fabrication conditions such as rf sputter power and GaAs surface treatment. The W/n-GaAs junctions fabricated by sputter deposition of W at a high rf power on the chemically etched n-GaAs surface as well as those fabricated on the sputter etched n-GaAs surface using a low rf power, possessed poor rectification properties. The quality of W contacts deposited by rf sputtering at a low rf power on chemically etched n-GaAs surface, was relatively better, however, it was not satisfactory for practical applications. In this work we present a comparative study of the electrical characteristics of an as-deposited and a thermally annealed

Mat. Res. Soc. Symp. Proc. Vol. 448 © 1997 Materials Research Society

W/n-GaAs junctions. The results deduced from the forward I-V and C-V/f data indicated that the thermal annealing enormously improved the quality of the W/n-GaAs interface.

EXPERIMENTAL TECHNIQUE

Prior to metal deposition, the (100) n-GaAs:Si single crystals samples were degreased with acetone and methanol at 40° C for 10 min., rinsed with deionized (DI) water and dried with N_2 gas. To remove the native oxide, the samples were then chemically etched with HCl : H_2O (1:1) solution at room temperature for 10 min., rinsed with DI water and dried with N_2. To obtain ohmic contact, In metal was thermally deposited on the unpolished surface of n-GaAs followed by a 90 min. anneal in Ar atmosphere at 390° C. Two W/n-GaAs Schottky junctions A and B of area 1.75 mm^2 were fabricated by deposition of W on the chemically etched polished surfaces of two n-GaAs samples by rf sputtering using a rf powers of 300 Watt for 30 min. The W contact B was subjected to a 90 min. thermal anneal at 390 °C.

The I-V characteristics of the W/n-GaAs/In Schottky diodes A and B were measured using a Keithley 617 electrometer, a Keithley 480 picoammeter and a Hewlett-Packard 6160A power supply. The 1 MHz C-V measurements were made using a Boonton 72B capacitance meter. The small signal ac capacitance (C_M) and conductance (G_M) of the W/n-GaAs/In Schottky diodes as a function of applied dc bias (V_a), in the frequency (f) range 0.2-100 KHz, were simultaneously measured with a two phase PAR-5204 lock-in amplifier.

EXPERIMENTAL DATA AND ANALYSIS

The I-V characteristics of the In ohmic contact to n-GaAs:Si described elsewhere[3] provided a value of 4 Ω for the ohmic resistance. The room temperature forward and reverse currents measured as a function of applied voltage (V_a) for the as-deposited and the thermally annealed W/n-GaAs Schottky diodes A and B, respectively, are plotted in Fig. 1. The values of the rectification ratio at V_a=0.3 V, obtained from this data are listed in Table I. To account for the ohmic resistance, the voltage drop across the rectifying barrier (V) was calculated from the relation: V=V_a - 4I. Under forward bias, the experimental values of I/(1-exp(-qV/kT)) vs V for the diodes A and B shown in Fig. 2 (discrete points), were very well described by Eq. (3.13) in Ref. 4 (Fig. 2, solid lines) and the values of the ideality n and I_o, obtained the from this data are also listed in Table I.

TABLE I

Summary of the electrical parameters obtained from the I-V data for the diodes A and B

Device	n	I_o (A)	r
Diode A	1.70	6.3×10^{-6}	70
Diode B	1.09	8.10^{-8}	1.4×10^4

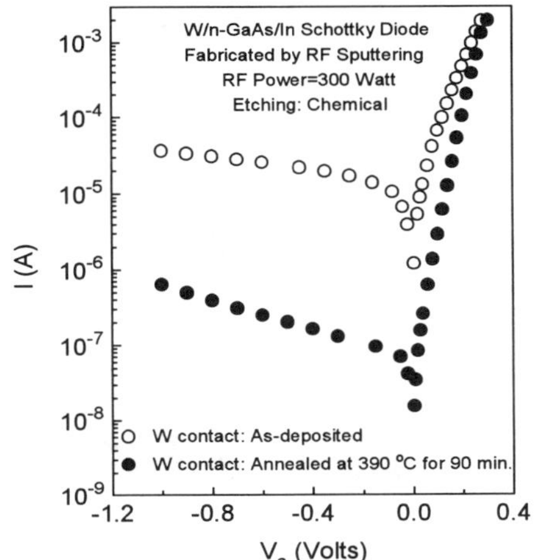
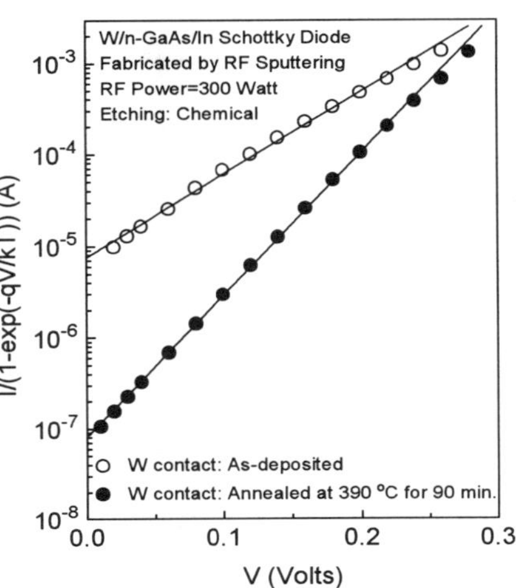

FIG. 1 Current-voltage characteristics of the W/n-GaAs/In Schottky Diodes A and B.

FIG. 2 Best fits of the measured values of I/(1-exp(-qV/kT)) to Eq. (3.13) in Ref. 4 for the W/n-GaAs/In Schottky diodes A and B.

The 1 MHz C^{-2} vs V data under reverse bias shown in Fig. 3, fitted well to Mott-Schottky relation[5] with a value of 1.2 V for the barrier heights (ϕ_{bo}) for both the diodes A and B with the values of 5.8×10^{16} cm^{-3} and 4.7×10^{16} cm^{-3} for their carrier concentrations (N_D-N_A), respectively.

The C_M-V data in the frequency range 0.2-100 KHz, was corrected for series resistance[6-7] to obtain the junction capacitance C. Some typical C-V/f plots for the thermally annealed W/n-GaAs Schottky diode B are shown in Fig. 4, and the C-V/f plots for the diodes A and B are compared in Fig. 5. The frequency variation of C with the forward bias as a parameter, for the diode B is shown in Fig. 6.

The surface defects capacitance, $C_p(\omega,V)$ in the W/n-GaAs junctions A and B was extracted from the measured values of $C(\omega,V)$ by using the procedure explained elsewhere[6-8]. Some typical C_p vs ω plots for the diodes A and B are shown in Fig. 7 (discrete pts.). The frequency dependence of the experimental values of C_p (Fig. 7, discrete pts.) was described reasonably well by the Lehovec's surface state model[9] with two time constants

$$C_p = qA[N_{ss1}\tan^{-1}(\omega\tau_1)/\omega\tau_1 + N_{ss2}\tan^{-1}(\omega\tau_2)/\omega\tau_2] \qquad (1)$$

where N_{ss1} and τ_1 are the energy density and relaxation time of slow surface defects, N_{ss2} and τ_2 the energy density and relaxation time of the fast surface defects, A the area of the rectifying contact, ω the angular frequency of the ac signal and q the magnitude of the electron charge. The measured values of C_p (Fig. 7, discrete pts.) fitted well to Eq. (1) (Fig. 7, solid curves) with $\tau_1 = 4$ ms for both the diodes and with the values 1 μs and 6 μs for τ_2 for the diodes A and B, respectively. The values N_{ss1} and N_{ss2} obtained from the fitting procedure for both the diodes are listed in Table II.

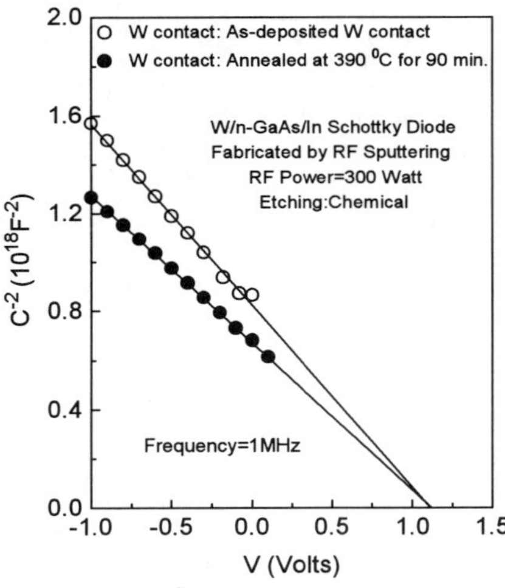

FIG. 3 C^{-2} vs V plots for the W/n-GaAs/In Schottky diodes A and B.

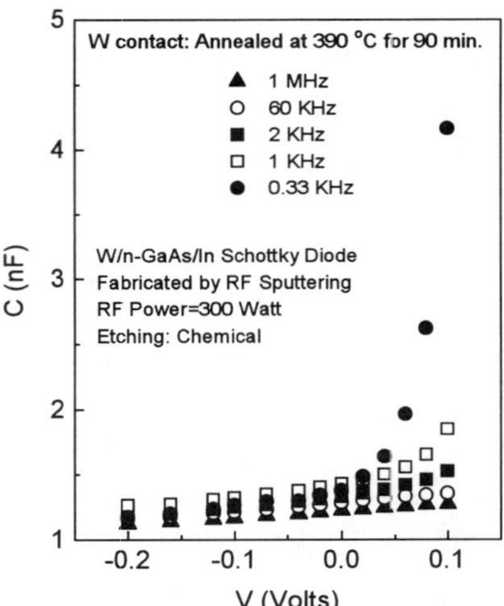

FIG.4 Variation of the junction capacitance with voltage at different frequencies for the W/n-GaAs/In Schottky diode B.

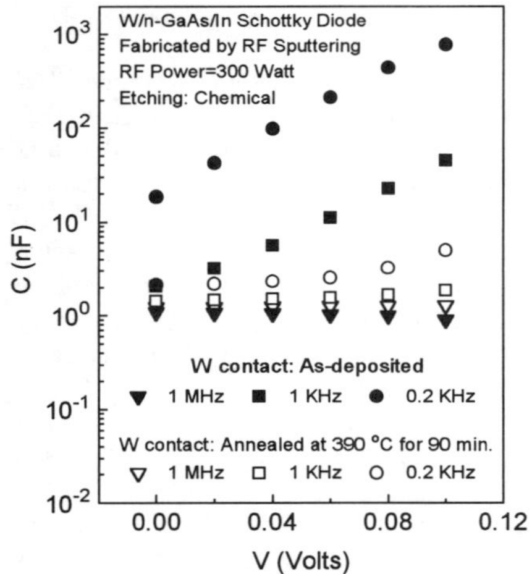

FIG. 5 Comparison of C-V/f plots for the W/n-GaAs/In Schottky diodes A and B.

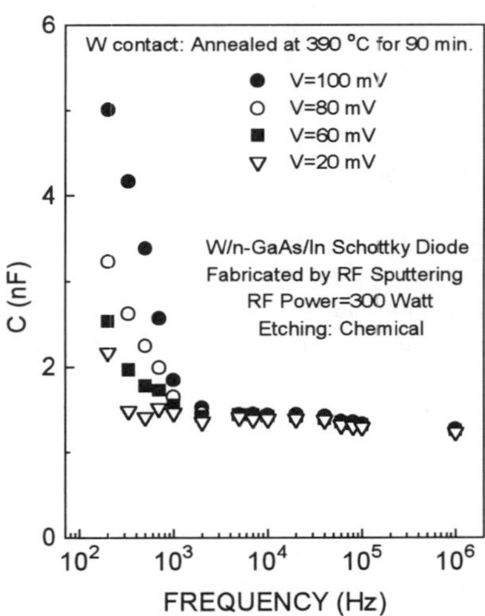

FIG. 6 Frequency dispersion of the junction capacitance for the W/n-GaAs/In Schottky diode B.

TABLE II

The energy density of the slow and fast surface defects at the W/n-GaAs interfaces A and B, obtained by fitting the experimental values of C_p to Eq. (1).

V (mV)	DIODE A		DIODE B	
	N_{ss1} (eV^{-1}cm^{-2})	N_{ss2} (eV^{-1}cm^{-2})	N_{ss1} (eV^{-1}cm^{-2})	N_{ss2} (eV^{-1}cm^{-2})
100	1.9×10^{14}	2.4×10^{11}	3.8×10^{12}	5.3×10^{10}
80	-	-	2.0×10^{12}	6.3×10^{10}
60	7.1×10^{13}	1.7×10^{11}	9.4×10^{11}	6.3×10^{10}
40	4.7×10^{13}	1.5×10^{11}	5.6×10^{11}	6.0×10^{10}
20	-	-	4.7×10^{11}	6.3×10^{10}
0	7.8×10^{12}	8.6×10^{10}	-	5.8×10^{10}

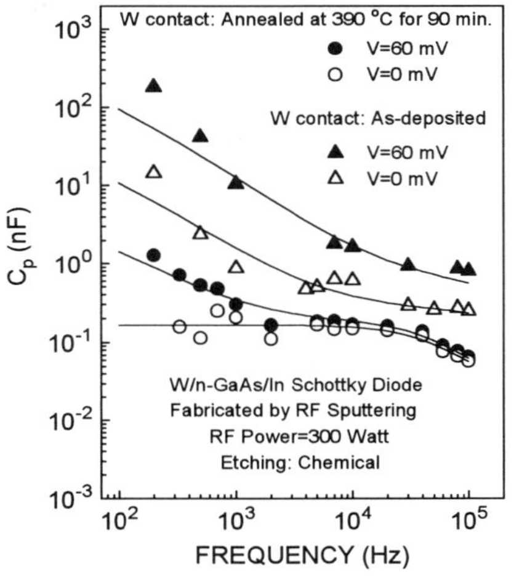

FIG. 7 Variation of C_p with frequency for the Schottky diodes A and B. The solid curves represent Eq. (1).

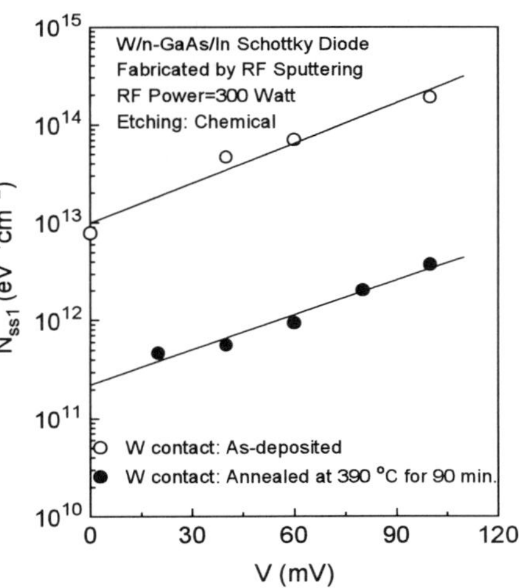

FIG. 8 Variation of the surface defects energy density with forward bias in the W/n-GaAs junctions A and B.

DISCUSSION

The value of the rectification ratio for the W/n-GaAs Schottky diode B obtained form the I-V data shown in Fig. 1 was 200 times greater than the one for the diode A (Table I). The plot of the I/(1-exp(-qV/kT)) vs V data for the diode B was linear over four decades of current (Fig. 2, solid circles) and the value of the reverse saturation current obtained from this data was 80 times smaller than the corresponding value for the diode A (Table I). The forward I-V characteristics for the thermally annealed W/n-GaAs Schottky diode (diode B) shown in Fig. 2 were nearly ideal with n=1.09 (Table I).

The 1 MHz C^{-2}-V data for the thermally annealed W/n-GaAs Schottky junction B, fitted very well to the Mott-Schottky theory over the voltage range -1.0V <V<0.1 V (Fig. 3), however, the value of 1.20 V for ϕ_{bo}, obtained from the C^{-2}-V data, was much higher than the value of 0.71 V, obtained from the forward I-V data[3]. Such disagreement between the values of ϕ_{bo} obtained from the two methods has been reported by other workers[10,11], and may be attributed to complicating effects of interfacial layer with fixed charges and unannealed surface defects.

An appreciable frequency dispersion in the junction capacitance was observed in both the diodes B (Fig. 4) and A (not shown) only at forward bias voltages, indicating the presence sputter induced defects near the surface of n-GaAs. At low frequencies, the measured values C (Fig. 5) and C_p (Fig. 6) under forward bias in the W/n-GaAs diode B were two orders of magnitude lower that those in the diode A. The observed frequency dispersion of C_p (Fig. 7) suggested the presence of both slow and fast surface defects near the W/n-GaAs interface and was described by the Leovec's surface states model with two time constants (Eq. (1)) which differed by three orders of magnitude. In both the diodes, the energy density N_{ss1} of the slow surface defects was about two orders of magnitude higher than that of the fast surface defects (N_{ss2}) (Table II). However, the values of N_{ss1} were two order of magnitude lower in the diode B that in the diode A (Fig. 8 and Table II). At a forward bias of 60 mV, the energy density N_{ss2} of the fast states in the diode B was three times lower than in the diode A.

In conclusion, thermal annealing at 390 °C for 90 min. reduced the density of the sputter induced defects at the W/n-GaAs interface by two orders of magnitude which enormously improved the quality of the W/n-GaAs Schottky diode as indicated by a very high rectification ratio, a low reverse saturation current and an ideality factor close to unity.

ACKNOWLEDGMENTS

This work was supported by the Consejo de Investigación de la Universidad de Oriente under contract No. CI-5-1002-776/96

REFERENCES

1.   A. Singh, G. Aroca and L. Valásquez, Mat. Res. Soc. Symp. Proc. **378**, 823-828 (1995).

2.   A. Singh, L. Velásquez and G. Aroca, Proc. 4th International Conference on Solid State and Integrated-Circuit Tech. October 24-28, Beijing, China, p 387-389 (1995).

3.   A. Singh and L. Velásquez, (submitted for publication to J. Appl. Phys.)

4.   E . H. Rhoderick and R. H. Williams, *Metal-Semiconductor Contacts*, Clarendon Press,Oxford (1988), p.100

5.   A. Singh, K. C. Reinhardt and W. A. Anderson, J. Appl. Phys. **68**, 3475 (1990)

6.   A. Singh, Solid St. Electron. **28**, 223, (1985)

7.   A. Singh, P. Cova and R. Masut, J. Appl. Phys. **74**, 6714 (1993)

8.   A. Singh, P. Cova and R. Masut, Mater. Res. Soc. Symp. Proc. **318**, 515 (1994).

9.   K. Lehovec, Appl. Phys.Lett. **8**, 48 (1966).

10.   C. Fontaine, T. Okumura and K. N. Tu, J. Appl. Phys. **54**, 1404 (1983).

11.   Y. Kuriyama, S. Ohfuji and J. Nagano, J. Appl. Phys. **62**, 1318 (1987).

RELATIONSHIP BETWEEN STRUCTURAL AND ELECTRICAL PROPERTIES OF Zn-BASED CONTACTS TO p-GaAs : TOWARDS THE MECHANISM OF THE OHMIC CONTACT FORMATION

E.KAMIŃSKA*, A.PIOTROWSKA*, S.KASJANIUK*, AND S.GIERLOTKA**
*Institute of Electron Technology, Al.Lotników 46, Warszawa, Poland, eliana@ite.waw.pl
** Unipress, PAS, Warszawa, Poland

ABSTRACT

The relationship between electrical properties and microstructure of pure Zn and Au(Zn) contacts to p-GaAs has been studied. Thermally activated changes in Φ_B correlate with structural processes at MS interfaces. For Zn/GaAs contacts, lowering of Φ_B from 0.63 eV to 0.35 eV corresponds to the penetration of Zn into the native oxide layer. In AuZn/p-GaAs contacts, β-AuZn phase is responsible for the formation of Φ_B=0.4 eV in as-deposited contacts. The onset of the ohmic behaviour of Au(Zn)/p-GaAs contacts (Φ_B=0.3 eV) coincides with the appearance of α_3-AuZn phase for Zn content less than 20 at.% or α_1-AuZn, for higher Zn concentrations.
The obtained results prove that the mechanism responsible for the formation of low-resistance Zn -based contacts to p-type GaAs is associated with the lowering of the Schottky barrier at the metal/semiconductor interface. We suggest that the ultimate properties of these contacts are determined by the presence of a single, specific phase in a direct contact with the semiconductor.

INTRODUCTION

Almost all theoretical attempts to explain the nature of a potential barrier at the metal/semiconductor interface tacitly assume that the microstructure of such an interface is homogeneous. However, the properties of real metal/semiconductor contacts may be affected by the crystallographic characteristics of the metal film, especially its grain size and orientation, as well as by its reactivity giving rise to the formation of multiple intermetallic phases. Furthermore, the recent studies of electrical behaviour of high-quality epitaxial transition metal disilicide/Si [1] and epitaxial metal/GaAs [2] interfaces have established the critical dependence of the Schottky barrier height on the atomic structure of the interface.

We have previously reported on low-resistance ohmic contacts to GaAs with improved microstructure. It was shown that the interface of Au(Zn)/p-GaAs contact may remain virtually intact after the thermal processing necessary to create an ohmic contact. The interaction between metallization and GaAs, as observed by HREM, was limited to the dispersion of the native oxide layer, which had remained at the surface of GaAs after chemical treatment [3]. Au and Zn were found to form a single Au-Zn phase in a direct contact with the semiconductor substrate [3,4]. Another important finding was the ohmic behaviour of pure, unreacted Zn contacts to p-type GaAs [4]. High crystalline quality and uniformity of these metal/semiconductor interfaces makes it possible to analyze the mechanism of the formation of ohmic contacts.

In the present work emphasis is placed on electronic properties of Zn-based contacts. We have determined the Schottky barrier heights (Φ_B) at different stages of the formation of pure Zn and Au(Zn) ohmic contacts, and the dominant transport mechanisms in these contacts, deposited on lightly and on highly doped p-type GaAs. In order to evaluate the effect of particular Au-Zn phases at MS interface on the contact properties, the Au(Zn) metallization with Zn concentration varying from 10 to 40 at.% was investigated in this study.

Mat. Res. Soc. Symp. Proc. Vol. 448 © 1997 Materials Research Society

EXPERIMENTAL PROCEDURE

A variety of Zn doped (100)-oriented GaAs substrates, ranging in doping from 5×10^{16} to 8×10^{18} cm^{-3}, were used in these studies. To follow the thermally activated changes in Φ_B, contacts were formed on lightly doped GaAs so that the thermionic emission controls the transfer of carriers over the barrier. Substrates with higher doping concentrations were used to verify the importance of Φ_B modifications for the formation of low-resistance metal/p-GaAs contacts. Zn and Au were deposited by thermal evaporation in oil-free vacuum, on unheated substrates. The AuZn metallization was formed by sequential deposition of the Au(40nm)/Zn(40nm)/("d"nm)Au structure. The thickness "d" of the second Au layer was adjusted to provide the desired concentration of Zn. For electrical characterization, the metallization was patterned by conventional photolithography, using AZ 1350 JSF resist and the soft- and postbake steps at 90^0C for 30 min. and at 120^0C for 20 min. respectively. Before annealing contacts were encapsulated with 200 nm thick, rf-sputtered SiO$_2$ film. The substrate temperature during sputter deposition was estimated not to exceed 100^0C. Heat treatments were carried out in flowing H$_2$ at temperatures ranging from 200 to 420^0C, for 3 min.

X-ray diffraction (XRD) was employed to identify phase transformations in the metallization. The contact resistivity r_c was measured by modified Terry-Wilson method and transmission line method (TLM). Φ_B at different stages of the formation of pure Zn and Au(Zn) ohmic contacts was determined from I-V characteristics under forward bias for rectifying contacts, and from r_c-T measurements between 80 and 423 K in the case of ohmic contacts.

RESULTS AND DISCUSSION

Phase transformations in the Au(Zn) metallization during the formation of ohmic contacts

The results of XRD analysis of Au(Zn)/p-GaAs contacts after particular steps of the fabrication of ohmic contacts i.e.: deposition of metallization, patterning, and annealing at 200, 320 and , 420^0C are presented in Table 1. The as-deposited Au(Zn) metallization, in spite of sequential deposition of Au and Zn, in the whole range of Zn concentration, consisted of two cubic phases: β-AuZn (approximately 50 at.%Zn) and α-AuZn (solid solution of Zn in Au), with lattice parameter varying from 0.3150 to 0.3165 nm and from 0.4052 to 0.4071 nm respectively. According to Elliot [5], at temperatures below 500^0C, for Zn concentration in Au up to 40 at.%, the equilibrium Au-Zn phase diagram consists of the following phases: α-AuZn, α$_3$-AuZn (Au$_4$Zn), α$_2$-AuZn or α$_1$-AuZn (low or high temperature Au$_3$Zn phase), Au$_5$Zn$_3$, and β-AuZn. Surprisingly, in AuZn thin film structures, in which the content of Zn ranges from 10 to 35 at.%, phase formation paths are similar. Thermally induced transformations start at around 100-200^{0}C and proceed through the gradual decomposition of β-AuZn phase and formation of α$_2$-AuZn phase. Rising the temperature to above 300^{0}C causes further transformations into α$_3$-AuZn for Zn content less than 20at.% or α$_1$-AuZn, for higher Zn concentrations. Both final phases are very similar: α$_3$-AuZn is orthorhombic with a=0.4026 nm, b=0.4034 nm, and c=0.4062 nm, while α$_1$-AuZn is tetragonal, with a=0.4019 nm, and c=0.4095 nm. When the concentration of Zn in Au(Zn) metallization exceeds 35 at.%, at around 100^{0}C β-AuZn phase forms and remains stable up to 450^{0}C.

Pure Zn shows no evidence of reaction with GaAs during annealing up to 320^{0}C. Our recent TEM study of Zn/GaAs interfaces revealed however, that at low temperatures Zn penetrates the residual oxide at the surface of GaAs [4]. Above 320^{0}C a tetragonal Zn$_3$As$_2$ phase with a=1.1778 nm and c=2.3643 nm forms.

Table I. Thermally induced phase transformations in the Au(Zn) metallization.

at.%Zn	unprocessed contact	photolithography (soft- & postbake)	Heat treatment		
			200^{0}C, 3 min.	320^{0}C, 3 min.	420^{0}C, 3 min.
10	β-AuZn(25%) α-AuZn(75%)	β-AuZn(25%) α-AuZn(75%)	α$_2$-AuZn α-AuZn	α$_3$-AuZn α-AuZn	α$_3$-AuZn α-AuZn
20	β-AuZn(50%) α-AuZn(50%)	β-AuZn(31%) α$_2$-AuZn (38%) α-Au Zn(31%)	α$_2$-AuZn α-AuZn	α$_1$-AuZn α-AuZn	α$_1$-AuZn α-AuZn
25	β-AuZn(50%) α-AuZn(50%)	β-AuZn(37%) α$_2$-AuZn(13%) α-Au Zn (50%)	α$_2$-AuZn	α$_1$-AuZn	α$_1$-AuZn
40	β-AuZn(91%) α-AuZn(9%)	β-AuZn	β-AuZn	β-AuZn	β-AuZn
100	Zn	Zn	Zn	Zn	Zn$_3$As$_2$

Electrical properties

Figure 1 shows the contact resistivity r_c of pure Zn and Au(Zn) (12 at.%Zn) contacts, deposited on lightly and moderately doped p-GaAs, as a function of annealing temperature. Pure Zn contacts become ohmic at temperature as low as 220^{0}C, and they loose their ohmicity at above 340^{0}C. For Au(Zn) contacts, the values of r_c do not depend on Zn concentration up to 35 at.%. Contacts with higher content of Zn show higher resistance.

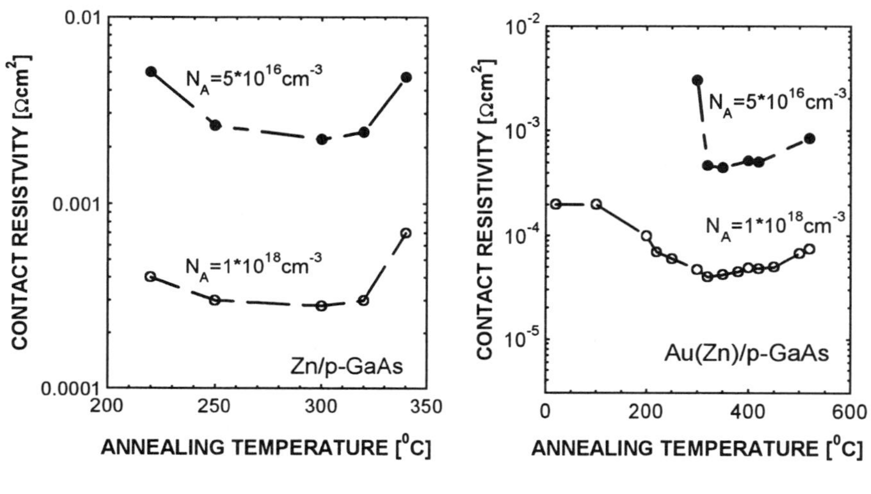

a)   b)

Fig. 1. Contact resistivity as a function of annealing temperature:
a) Zn/p-GaAs contact, b) Au(Zn)/p-GaAs contact.

The onset of the ohmic behavior for Au(Zn) contacts on lightly doped p-GaAs is observed at 300^0C, with a broad minimum at $320\div420^0$C. For substrates with dopant concentration N_A of $\sim 1 \times 10^{18} cm^{-3}$ unprocessed contacts exhibit linear I-V characteristics, and r_c has a similar temperature dependence.

The heights of potential barriers, as inferred from electrical measurements, for Au(Zn) (12 at.%Zn) and pure Zn metallizations at different stages of the formation of low-resistance contacts are listed in Table II. For comparison, pure Au contacts were also characterized.

Table II. Schottky barrier heights for Au(Zn), Zn and Au contacts to p-GaAs ($N_A=5 \times 10^{16} cm^{-3}$).

metallization	heat treatment	Φ_B[eV]
Au(Zn)	90^0C 30 min. + 120^0C 20 min.	0.40
(12 at.% Zn)	200^0C, 3 min.	0.38
	$320\text{-}450^0$C, 3 min.	0.30
Zn	90^0C 30 min. + 120^0C 20 min.	0.63
	$220\text{-}320^0$C, 3 min.	0.35
	360^0C, 3 min.	0.37
Au	90^0C 30 min. + 120^0C 20 min.	0.50
	320^0C, 3 min.	0.50

Out of three analyzed contact materials, Au(Zn) metallization forms on p-type GaAs the lowest potential barrier. Moreover, upon annealing the height of this barrier further decreases to 0.3 eV at the temperature of the onset of the ohmic behaviour. Relatively high value of Φ_B at the Zn/p-GaAs interface, rapidly diminishes under heat treatment, reaching the minimum value of 0.35 eV at the temperature when the contact becomes ohmic. In contrast, Φ_B of pure Au contact remains stable.

Thermally activated changes in Φ_B correlate with structural processes at MS interfaces. β-AuZn phase is responsible for the formation of the lowest Φ_B in as-deposited contacts, since it forms at the interface, as a result of interaction between the first Au layer and Zn during deposition of Au/Zn/Au metallization [3]. The onset of the ohmic behaviour of AuZn/p-GaAs contacts at $\sim300^0$C coincides with the appearance of α_3-AuZn or α_1-AuZn phase at the MS interface. For Zn/GaAs contacts, lowering of the Φ_B corresponds to the penetration of Zn into the native oxide layer, leading to formation of an intimate MS contact. Subsequent increase of the Φ_B is caused by the formation of a new phase in the interface.

Since the transport mechanism through the potential barrier at the metal/semiconductor interface depends essentially upon the ambient temperature, and upon the doping level of the semiconductor, r_c as function of the ambient temperature has been measured. The measurements were performed in two different temperature ranges: from RT to 423 K (Fig. 2) or from 80K to RT (Fig. 3). For moderately doped p-GaAs ($N_A=1 \times 10^{18} cm^{-3}$), the resistivity of Zn/GaAs and Au(Zn)/GaAs contacts, at different stages of the formation of ohmic contacts, is almost independent of the ambient temperature, which indicates that the thermionic-field emission through the potential barrier controls the current transport in these contacts.

In Figure 4 are presented the data points concerning contact resistivity for AuZn/p-GaAs ohmic contacts as a function of dopant concentration in p-type GaAs. The solid line is calculated for $\Phi_B=0.3$ eV, with $m_{lh}=0.074m_0$ and $m_{hh}=0.62m_0$, taking the functional dependencies of r_c on semiconductor doping level for thermionic, thermionic-field, and field emission given by Yu [6]. Experimental data agree very well with the theoretical model of carrier transport across the potential barrier at the metal/semiconductor interface, with thermionic emission over the barrier as a dominant transport mechanism for lightly doped semiconductor, and tunneling for higher dopant concentrations.

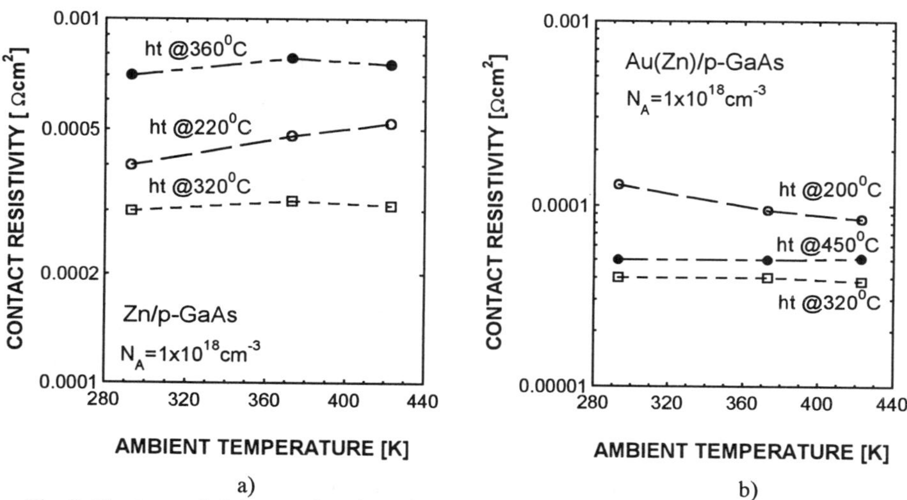

a)   b)

Fig. 2. Contact resistivity as a function of ambient temperature in the range 293÷423 K: a) Zn/p-GaAs contact, b) Au(Zn)/p-GaAs contact.

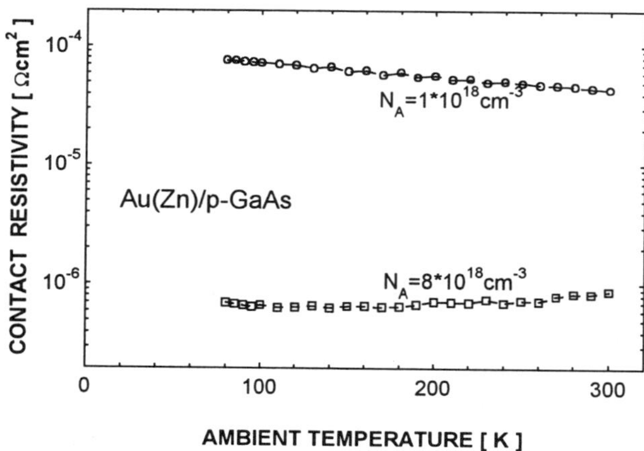

Fig. 3. Au(Zn)/p-GaAs contact resistivity as a function of ambient temperature in the range 80÷293 K.

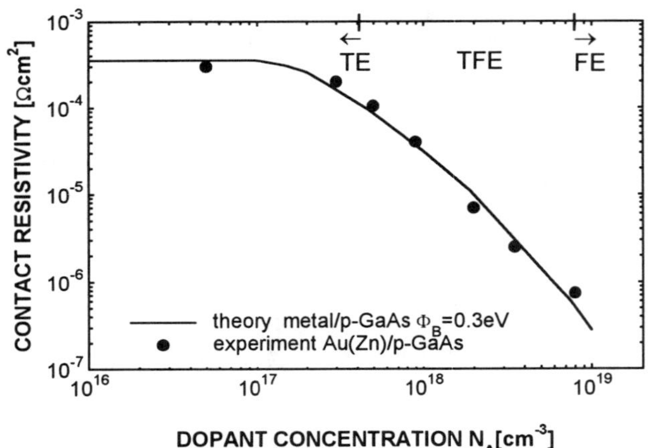

Fig. 4. The dependence of contact resistivity of AuZn/p-typeGaAs contacts on doping level of p-GaAs; circles - experimental data points, solid line - theoretical prediction assuming Φ_B =0.3 eV.

CONCLUSIONS

In conclusion, the obtained results prove that the mechanism responsible for the formation of low-resistance Zn -based contacts to p-type GaAs is associated with the lowering of the Schottky barrier at the metal/semiconductor interface. We suggest that the ultimate properties of these contacts are determined by the presence of a single, specific phase in a direct contact with the semiconductor.

ACKNOWLEDGMENTS

The authors would like to thank Victor Weizer (NASA Lewis Research Center, Cleveland, OH) and Navid Fatemi (Essential Research, Inc., Cleveland, OH) for valuable discussions and suggestions. This research was supported by the U.S. - Polish Maria Skłodowska-Curie Joint Found II MP/NASA-95-230.

REFERENCES

1. R.T.Tung, in Contacts to Semiconductors, (ed.L.J.Brillson), Noyes Publications, p.176 (1993).
2. C.J.Palmstrom, T.L.Cheeks, H.L.Gilchrist, J.G.Zhu, C.B.Carter, B.J.Wilkens, R.Martin, J.Vac.Sci.Technol., **A 10**, 1946 (1992).
3. E.Kaminska, A.Piotrowska, E.Mizera, A.Dynowska, Thin Solid Films **246**,143 (1994).
4. E.Kamińska, A.Piotrowska, E.Mizera, R.Żarecka, J.Adamczewska and E.Dynowska, Mat.Res.Soc.Symp.Proc. **300**, 237 (1993).
5. R.P.Elliot, Constitution of Binary Alloys, First Supplement, Mc Grow-Hill, p.107, 1965.
6. A.Y.C. Yu, Solid State Electron., **13**, 239 (1970).

NOVEL METAL-SEMICONDUCTOR-METAL PHOTODETECTORS ON SEMI-INSULATING INDIUM PHOSPHIDE

J.W. PALMER * and W.A. ANDERSON **
* Semiconductor Test Products, Keithley Instruments, Inc., 28775 Aurora Road, Cleveland, Ohio, 44139-1891
** Department of Electrical and Computer Engineering, State University of New York at Buffalo, 217C Bonner Hall, Amherst, NY 14260

ABSTRACT

Depositing Pd or Au on InP at substrate temperatures near 77K has previously been found to significantly reduce the interaction between the metal and semiconductor upon formation of the interface. In this work, this technique was used to fabricate metal-semiconductor-metal photodetectors (MSMPD's) on semi-insulating (SI) InP substrates with superior characteristics compared to detectors formed using standard room temperature (RT) metal deposition. The low-temperature (LT) metallizations were patterned using a polyimide/SiO_2 lift-off mask, and a SiO antireflection coating was used to attain near-zero reflection at λ=840nm. The detectors had an active area of 200μm x 200μm, and line widths and line spacings of 3μm. Detectors having a LT-Pd/SI-InP structure had a dark current of 80nA at 5V, which was a factor of 4 lower than the dark current of conventional MSMPD's. Additionally, LT-Pd/SI-InP MSMPD's exhibited excellent saturation characteristics and a responsivity of 0.75 A/W. Detectors with an indium-tin-oxide (ITO)/LT-Au(200Å)/SI-InP structure had a higher responsivity of 1.0 A/W, due to the relative transparency of this metallization. In contrast, MSMPD's with RT metallizations had poor saturation characteristics, consistent with the results of others. The difference in the illuminated characteristics of MSMPD's with RT and LT metallizations was due to a change in the internal photoconductive gain mechanism. In RT detectors, hole trapping at interface states near the cathode dominated the gain mechanism. In LT detectors, the difference in carrier transit-times dominated.

INTRODUCTION

Metal-semiconductor-metal photodetectors (MSMPD's) are prime candidates for integration into opto-electronic circuits. They are easy to fabricate, compatible with planar processing techniques, and have very high switching speeds. MSMPD's usually consist of two rectifying contacts with interdigitated fingers and have a transit-time limited photocurrent.[1] Ideally, when the applied voltage is high enough that the carrier transit-time is less than the carrier lifetime, the photocurrent saturates and the detector acts like a current source. However, most MSMPD's reported have a quasi-saturated regime in which the photocurrent increases slowly with the applied voltage. Such behavior indicates the presence of an internal gain,[2] and the extent to which this occurs varies depending on the details of the device fabrication.

Relatively little has been written about MSMPD's on InP. Standard metallizations to InP result in a low electron barrier height of about 0.5eV on undoped material ($N_D = 10^{15}cm^{-3}$) due to interactions between the metal and InP. One paper [3] reported a MSMPD using a standard Au metallization on semi-insulating (SI) InP to attain a dark current of 10nA at 10V for a 50x50μm

Mat. Res. Soc. Symp. Proc. Vol. 448 © 1997 Materials Research Society

device. However, this detector had a quasi-saturated photocurrent regime with a very strong bias dependence. Such a bias dependence is usually attributed to a photoconductive gain mechanism caused by hole accumulation near the cathode.[2, 4] The resulting build-up of positive charge lowers the barrier height to electrons at the cathode, causing electron injection. The accumulation of holes is due either to long-lifetime traps at the metal/InP interface or the difference in transit times between the electrons and holes, causing photogenerated electrons to be swept out of the semiconductor much faster than the holes. Standard metallizations to InP result in the creation of interface states because of chemical reactions between the metal and the semiconductor.[5] Therefore, hole trapping probably occurs at the cathode of InP MSMPD's fabricated in this manner. Recently, it was found that depositing Pd or Au on InP using substrate temperatures near 77K significantly reduces the interaction between the metal and InP,[6] increasing the electron barrier height.[7] The goal of this project was to fabricate InP MSMPD's using a low-temperature (LT) deposited metallization to reduce carrier trapping at the metal/semiconductor interfaces, and to measure how this affected the dark and illuminated device characteristics.

EXPERIMENT

MSMPD's were fabricated on SI-InP having a resistivity of about 10^7 Ω-cm. The active area of the devices was 200 x 200 μm with line widths and line spacings of 3μm. A plan-view sketch of the detectors is shown in Fig 1. The Pd or Au metallization was deposited at substrate temperatures near 77K. Standard lift-off could not be used for the LT metallization because photoresist cannot withstand the thermal stresses present during the deposition. Instead, a bi-layer polyimide/SiO_2 lift-off mask was used. First, the SI-InP was cleaned in acetone, methanol, and de-ionized H_2O, etched in H_2SO_4:H_2O_2:H_2O (2:1:1) for 10 min, etched in HF:H_2O (1:1) for 1 min, rinsed, and dried in N_2 gas. A polyimide layer, 0.6 μm thick, was then spun on and cured at 170°C, followed by deposition of a SiO_2 layer 0.1μm thick by plasma-enhanced-chemical-vapor-deposition (PECVD). Photoresist was then applied and standard lithography was used to pattern the grid lines. Reactive-ion-etching (RIE) in CHF_3 was used to transfer the metallization pattern to the SiO_2, followed by RIE in O_2, which simultaneously patterned the polyimide and removed the photoresist. Immediately prior to the LT metal deposition, the samples were etched in HCl:H_2O (1:10) for 1 min to remove any damage to the InP left by RIE. On some of the samples, LT-Pd was deposited by thermal evaporation using a base pressure in the 10^{-8} Torr range and a substrate temperature near 77K. Lift-off was done in Shipley 1165 remover, which dissolved the polyimide over a period of several hours. Detectors were also made using Pd metal and a substrate temperature of 300K (RT) for the purpose of comparison.

Additionally, a third type of detector was made having an ITO(800Å)/LT-Au(200Å)/SI-InP structure, shown in Fig 2. One of the major problems with MSMPD's is that the grid metallization usually prevents half of the incident light from entering the semiconductor. The structure shown in Fig 2 was designed to reduce the severity of this problem. First, a thin LT-Au deposition established a quality MS interface to the InP. The sample was then transferred from the LT deposition system to a sputtering system, where ITO was deposited using a substrate temperature of 150°C. The elevated substrate temperature was necessary to attain a conductive, transparent, ITO film. The result was a metallization which was at least 30% transparent to incoming photons, while maintaining the integrity of the metal/InP interface.

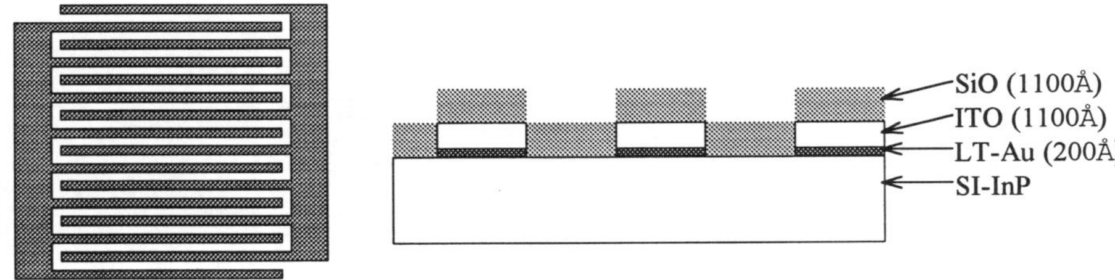

Fig 1 - Plan-view sketch of a MSMPD. Line width x line spacing = 3x3μm, device area = 200x200μm.

Fig 2 - Cross-section of a MSMPD with an ITO/LT-Au/SI-InP structure.

Finally, a SiO anti-reflection coating, 1100Å thick, was deposited on all three types of detectors to attain zero-reflection at λ=840nm. The photo-response of the detectors was measured using a continuous wave (CW) Spectra Physics Tsunami Ti:Sapphire laser operating at λ=840nm. The beam was passed through a half-wave-plate, a polarizer, and then a lens which focused it to a 100 μm full-width-half-max spot size on the center of a detector. By adjusting the half-wave-plate, the incident power of the polarized light was controlled.

RESULTS AND DISCUSSION

Fig 3 is the dark-current characteristics of MSMPD's having LT- and RT-Pd grids. The dark current of the LT-Pd detector is a factor of 4 lower than that of the RT-Pd device at all voltages, with a value of 400nA at 10V. The dark current in MSMPD's consists of a combination of electrons injected at the cathode and holes injected at the anode.[8] Since a standard metallization on InP has a low electron barrier height, the dark current in this case will be due primarily to electron injection at the cathode. However, use of a LT-Pd metallization increases the barrier height to electrons, or decreases the barrier height to holes. Therefore, the dark current of the LT-Pd devices should have a larger component due to hole injection at the anode. The overall effect was that the total dark current in the LT-Pd devices was slightly lower than in RT-Pd devices.

Fig. 4 is the illuminated characteristics of a detector having a LT-Pd metallization. Excellent saturation characteristics were obtained, with R=0.75A/W. This corresponds to an external quantum efficiency of 1.1. Since half of the incident photons were lost to grid shading, it is clear that the detector had a bias-independent gain of about 2. In comparison, devices having a RT-Pd metallization exhibited a strong bias dependence of the photocurrent in the quasi-saturated regime similar to SI-InP MSMPD's having RT-Au contacts.[3]

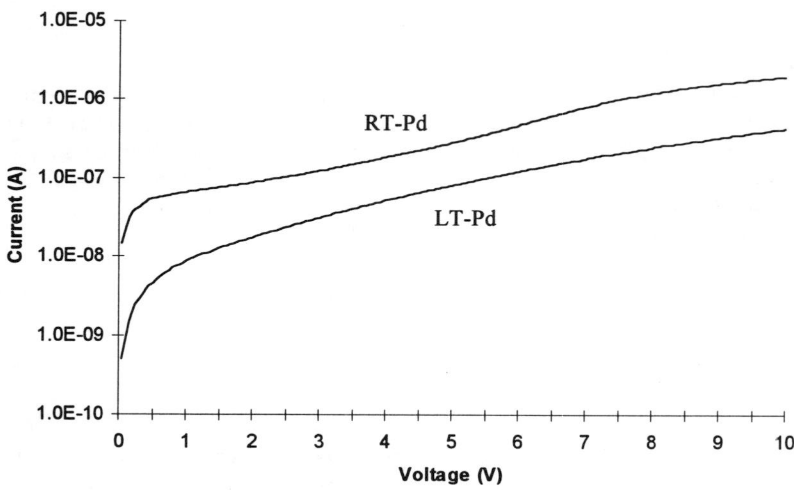

Fig 3 - Dark current vs. Voltage for SI-InP MSMPD's having LT-Pd and RT-Pd metallizations.

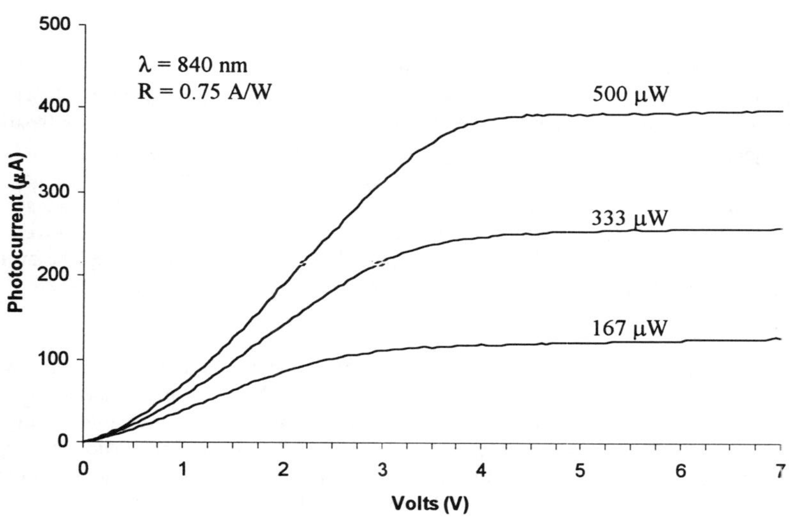

Fig 4 - Illuminated characteristic of a LT-Pd/SI-InP MSMPD.

Fig 5 shows the illuminated characteristics of a detector having an ITO/LT-Au/SI-InP structure. A high responsivity of 1.0 A/W was obtained, which is a 33% improvement over the responsivity of a detector with a LT-Pd metallization. This corresponds to a quantum efficiency of 1.5, indicating that there is an internal gain in this detector comparable to that of the LT-Pd MSMPD. In addition, the dark characteristics of the ITO/LT-Au/SI-InP detector were similar to those shown in Fig 4. However, the ITO/LT-Au devices did not saturate as well as the LT-Pd devices, probably because exposure to plasma and higher temperature during the ITO deposition slightly degraded the LT-Au/SI-InP interface.

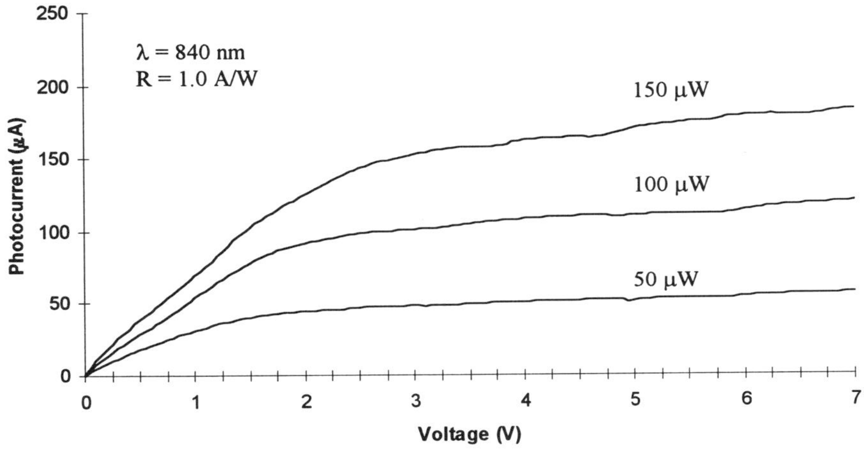

Fig 5 - Illuminated characteristic of an ITO/LT-Au/SI-InP MSMPD.

The difference between the illuminated characteristics of detectors having LT and RT metallizations is related to a change in the gain mechanism. The only difference in the fabrication of the RT and LT detectors was the substrate temperature used during metal deposition. The RT metallization results in a strong chemical interaction between the metal and InP, creating interface states which may trap holes in the vicinity of the cathode. This carrier trapping dominates the photoconductive gain mechanism, and the illuminated characteristic of MSMPD's having RT metal contacts. The LT metallization nearly eliminates the chemical reaction between the metal and InP, resulting in a very abrupt metal/InP interface.[6] In this case, the photoconductive gain mechanism is probably dominated by carrier transit-time effects.

CONCLUSIONS

MSMPD's were fabricated on SI-InP using a LT-Pd and an ITO/LT-Au metallization. The detectors had good saturation characteristics and high responsivities of 0.75 A/W and 1.0 A/W, respectively. The higher responsivity in the ITO case was due to a more transparent grid metallization, allowing more photons to enter the semiconductor. These improved characteristics, relative to RT fabricated devices, were attributed to the abrupt metal/InP interface that results when a LT deposition is used. It is believed that the photoconductive gain mechanism present in these detectors is dominated by carrier transit-time effects.

ACKNOWLEDGEMENTS

This project was funded by the National Science Foundation and monitored by D. Crawford of the ECS Division. Portions of the device fabrication were done at the Cornell Nanofabrication Facility with the assistance of M. Skvarla and L. Keim. A. Cartwright helped test the illuminated characteristics of the photodetectors.

REFERENCES

[1] R.B. Darling, J. Appl. Phys. 67, 3152 (1990).
[2] M. Klingenstein, J. Kuhl, J. Rosenzweig, C. Moglestue, A. Hülsmann, J. Schneider, and K. Köhler, Solid State Elec. 37, 333 (1994).
[3] D. Kuhl, F. Hieronymi, E.H. Böttcher, and D. Bimberg, IEEE Phot. Tech. Lett. 2, 574 (1990).
[4] S.F. Soares, Jpn, J. Appl. Phys. 31, 210 (1992).
[5] L.J. Brillson, Surf. Sci. 299/300, 909 (1994).
[6] J.W. Palmer, W.A. Anderson, D.T. Hoelzer, and M. Thomas, J. Elec. Mat. 25, 1645 (1996).
[7] Z.Q. Shi, R.L. Wallace, and W.A. Anderson, Appl, Phys, Lett. 59, 446 (1991).
[8] M. Ito and O. Wada, IEEE J. Quantum Elec. QE-22, 1073 (1986).

ENHANCED PHOTOYIELD WITH DECREASING FILM THICKNESS ON METAL-SEMICONDUCTOR STRUCTURES

V.Hoffmann, M.Brauer, M.Schmidt
Hahn-Meitner-Institut, Rudower Chaussee 5, 12489 Berlin, Germany, hoffmann-v@hmi.de

ABSTRACT

Experimental results of the internal quantum yield Y_i associated with the internal photoemission on Au/n-Si structures are presented. The samples were prepared on Si(100) and Si(111) substrates with photoemitter layer thicknesses ranging from 5 nm to 50 nm. The Y_i was measured at temperatures between 165 K and 300 K with the photoexciting energy varying from 0.72 eV to 1.07 eV. It was found that the Y_i increases with decreasing Au layer thickness with a strong enhancement (40 times) in regard to the conventional Fowler theory. This experimental result is in good agreement with model calculations taking account of hot carrier scattering in the photoemitter layer . Barrier energies are larger than deduced from the Fowler plot.

INTRODUCTION

The metal-semiconductor junction has an exemplary importance in the family of heterostructures. This refers to barrier energy, interface states and charge transfer processes. The internal photoemission (IP) or photoinjection is the charge transfer of optically excited carriers across the heterojunction. The photoemission current is frequently used as an analytical tool [1],[2] for barrier energy and band offset determination as well as a working principle for infrared photodetectors [3].

The essential characteristic in describing the IP is the internal quantum yield Y_i (hv) of the emitted carriers, electrons or holes, respectively. Y_i gives the ratio of the number of hot carriers emitted over the interfacial barrier to the number of absorbed photons. The exact absolute measurement of Y_i (hv) provides the basis for checking theoretical approaches and for an improvement of photoelectrical devices.

The IP is treated as a multi-step process: the optical charge carrier generation in the emitter layer (1), scattering processes of the excited carriers in the photoemitter including boundary reflections (2), scattering processes inside the image force region (3) and quantum-mechanical transmission of the hot carriers across the Schottky barrier maximum (4).

An approximation for Y_i (hv) is based on the Fowler relation [4]

$$Y \approx \frac{1}{8E_F hv} \cdot (hv - E_B)^2 \qquad (1)$$

(hv is the photon energy and E_B is the barrier energy) if $(hv-E_B)/k_B T > 3$ which is obtained from the photoemission of electrons from a metal into the vacuum. Some attempts have been made during the last decades to overcome the assumed simplifications, e.g. the neglect of carrier scattering processes inside the photoemitter [4],[5] and the quantum-mechanical reflection at the Schottky barrier [6]. But the lack of reliable absolute values of Y_i(hv) prevented refined theoretical models from being accepted. Therefore, the interpretation of IP measurements is still based on the Fowler relation [1], [2].

The present paper is focused on the experimental determination of the internal quantum yield on thin film metal-semiconductor heterojunctions thereby discussing the effects of carrier scattering in the emitter layer (process 2) . This is based on the measurement of the optical absorptance in the photoemitting layer (1) and an estimation of the processes (3,4).

Mat. Res. Soc. Symp. Proc. Vol. 448 © 1997 Materials Research Society

THEORETICAL MODEL

The theoretical model of IP in metal- semiconductor structures, including carrier scattering in the emitter and in the image force region, is only outlined in the present work. The detailed description is given in a previous publication [7].

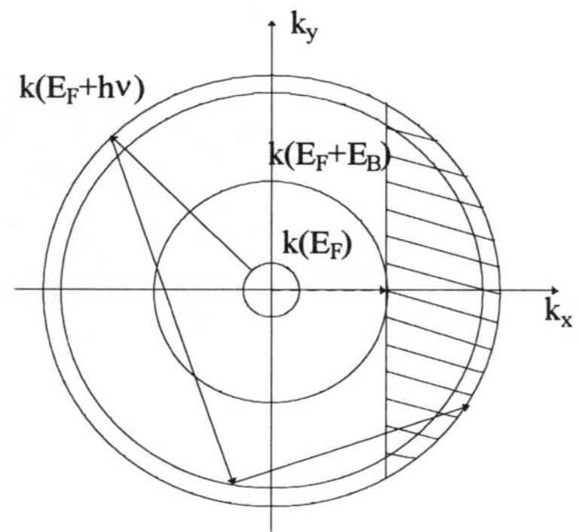

Fig. 1a: Momentum (k) space of photoexcited charge carriers with the escape cone for being injected (hatched region)

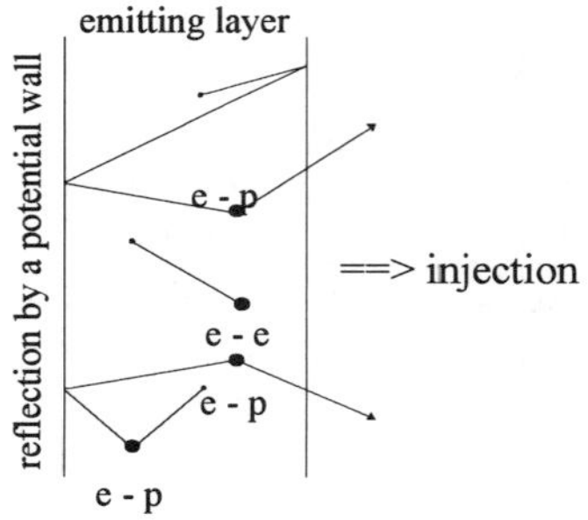

Fig. 1b: Elastic isotropic hot carrier scattering e - p at phonons and impurities, reflection at the metal layer boundaries and inelastic collisions e - e with cold carriers

The optical excitation of charge carriers in thin metal films is assumed to be homogeneous with isotropic carrier momentum distribution (free electron gas). The escape cone for IP determines that part of the excited carriers which have sufficient momentum normal to the barrier to cross E_B and it is given by the Fowler equation [8]

$$Y_F(E_F, E_B, h\nu, T) = \frac{k_B \cdot T}{4} \cdot \frac{\displaystyle\int_{E_F+E_B-h\nu}^{\infty} E^{-1/2} \cdot \ln(1 + e^{(E_F-E)/k_B \cdot T}) dE}{\displaystyle\int_{E_F-h\nu}^{\infty} E^{1/2} \cdot (1 + e^{(E-E_F)/k_B \cdot T})^{-1} dE} \tag{2}$$

T is the temperature, E_F the Fermi energy of the metal and k_B the Boltzmann's constant. Extending the theory of Fowler the scattering theory includes analytically the elastic scattering of photoexcited (hot) carriers at the layer boundaries and phonons, neglecting the energy transfer to the latter (Fig. 1) and inelastic collisions with thermalized (cold) carriers, as it was calculated by Kane[4] and Dalal[5]. The reflection at the layer boundaries is assumed to be diffuse and the scattering due to phonons to be isotropic. The attenuation lengths for inelastic carrier- carrier scattering l_e (E) depend on the energy of the hot carriers [9] and for elastic carrier-phonon scattering l_p (T) on temperature. The elastic scattering at grain boundaries in polycrystalline films is not included in this analytical theory.

Scattering in the image force region of the semiconductor (between interface and Schottky barrier maximum x_0) is considered by the lucky- electron concept which holds true for elastic isotropic as well as inelastic isotropic scattering events and results in an exponential damping characterized by the attenuation length L_S [10].

These assumptions lead to Y_i (hv)   (Eq. 3)   which is used in describing the experimental results.

$$Y_i(E_F, E_B, T, d, l_p, l_e, hv) = \frac{2 \cdot c \cdot Y_F \cdot e^{-f \cdot d} \cdot e^{-x_0/L_S}}{c \cdot Y_F \cdot (1 + e^{-2 \cdot f \cdot d}) + f \cdot (1 - Y_F) \cdot (1 - e^{-2 \cdot f \cdot d})} \tag{3}$$

with $c = a+b$, $a = l_p^{-1}$, $b = l_e^{-1}$, $f^2 = (c^2-ac)$, $L_S = 6.5$ nm, $x_0 = 7.5$ nm.

EXPERIMENTAL DETAILS

<u>Sample Preparation</u>

The substrates used in this work are n-type Si wafers which are mechanically polished on both sides. Au-films of 1 x 5 mm^2 area are formed on Si(100) and Si(111) substrates ($N_d = 3 \cdot 10^{14}$ cm^{-3}) by evaporation under HV conditions ($p = 10^{-6}$ mbar) at room temperature. Prior to the film deposition all the substrates are cleaned in a diluted HF solution to remove the native oxide and to passivate the surface with hydrogen. After film deposition ohmic contacts are formed on the substrates' backside by Al dc-magnetron sputtering. This deposition is made through a mask which leaves the center of the samples' backside blank for the optical measurements. A series of samples are prepared with layer thicknesses between 5 nm and 50 nm. The film thickness was in situ controlled by a vibrating quartz reference. AFM measurements reveal a polycrystalline film growth with a grain diameter of 50 - 100 nm.

<u>Experimental Setup</u>

The internal quantum yield Y_i is measured by detecting the optical reflectance and transmittance of the samples as well as the photoelectrical current in short circuit and the total incident photon flux. The experimental setup allows to measure these quantities simultaneously for each wavelength. The sample is mounted in a cryostat with CaF$_2$ windows for the incident light as well as for the reflected and transmitted beams. The light source is a 250W halogen bulb. The light is decomposed by a grating monochromator (H250 Ivon - Yobin) and focused by a CaF$_2$ optics on a 1 x 1 mm^2 area of the sample. A shutter cuts off the illumination after each wavelength position, allowing the sample to relax to the value of the dark current. The absolute incident photon flux is measured on a split reference beam which is focused on a calibrated Si/PbS tandem photodetector (Hamamatsu). The reflected and transmitted beams from the sample are focused on Si and PbS photodetectors as well. The absorptance is obtained as the missing rest from reflectance and transmittance to unity. Measurements are carried out under vacuum conditions (p=10^{-4}mbar) with light in a wavelength range from 250 nm to 2500 nm at sample temperatures between 165K and 300K.

RESULTS AND DISCUSSION

The Y_i of a series of Au/n-Si heterostructures with layer thicknesses d in the range of 5 nm to 50 nm is evaluated for photon energies hv from 0.72 eV to 1.07 eV at sample temperatures between 165K and 300K. A Fowler plot, i.e. $(Y_i * hv)^{1/2}$ vs hv is shown in Fig. 2. The experimental values exibit a linear dependence in this energy range in agreement with Fowlers theory, but they are significantly larger. An increase with respect to Fowler can be explained by the scattering processes involved in our theoretical approach as can be seen in the upper curve. The corresponding material specific scattering parameters are known for Au and Si from different experimental methods. Therefore, the calculated curve is completely determined with the supposed barrier E_B being the only free fitting parameter. The extrapolations of the theoretical and experimental curves intersect the energy axis at the same value. It has to be pointed out that for a given intersection E_{Bexp} with the energy axis usually interpreted as barrier energy both theories give larger barrier heights E_B. The difference to the fitted barrier energy increases with increasing temperature (Table I). E_{Bsc} is for both substrate orientations at room temperature approximately 45meV larger than the value E_{Bexp} extracted from the experimental Fowler plot; at 165K the difference amounts to 25meV.

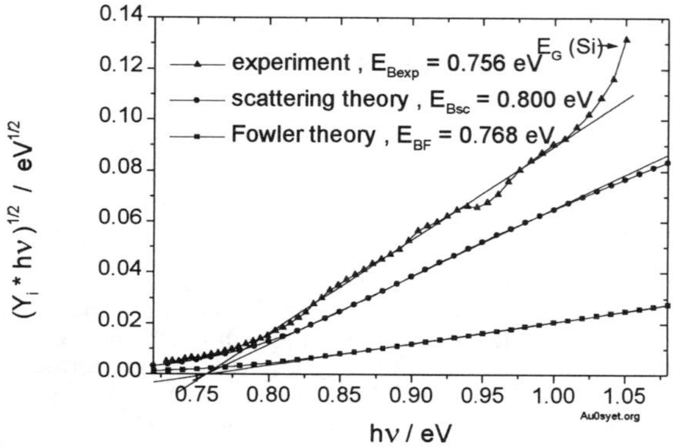

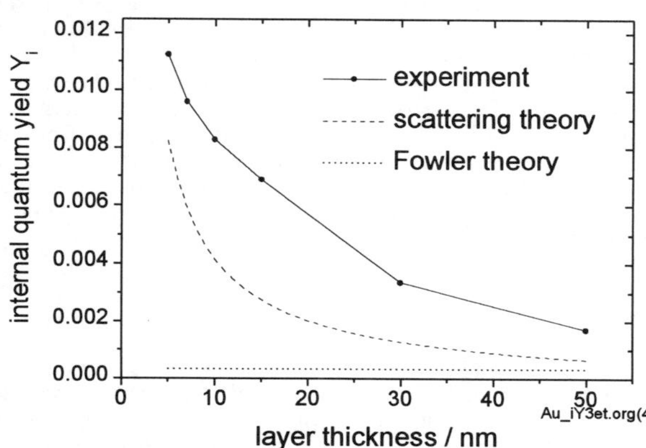

Fig. 2: Fowler plot of a 10 nm Au/n-Si(100) sample at 300 K in comparison with calculations. For the same intersection E_{Bexp} with the energy axis, both theories give larger barrier energies, E_{BF} =0.768eV and E_{Bsc}=0.800eV ($h\nu$ is the photon energy and Y_i the internal quantum yield).

T [K]	E_{Bexp} [meV]	E_{BF} [meV]	E_{Bsc} [meV]	E_G (Si) [meV]
165	810	815	835	1154
300	760	775	804	1125
$\Delta E(T)$[meV]	50	40	31	29

Tab. I
Apparent barrier energies from Fowler plot (E_{Bexp}), Fowler's theory (E_{BF}), scattering theory (E_{Bsc}) and gap energy of Si (E_G) vs temperature (T). $\Delta E(T)$ is the energy shift with temperature.

Fig. 3: Internal quantum yield Y_i at 300K and 1eV photon energy for a series of Au/n-Si(100) samples vs layer thickness and calculated curves, respectively

The Y_i increases with decreasing layer thickness (Fig. 3). On the average a five times increase of Y_i for the thinnest layers was measured with respect to the 50 nm films. Carrier scattering in the metal layer can explain qualitatively the growing Y_i with decreasing layer thickness (upper curve in Fig.3). Responsible for the larger Y_i, with respect to Fowlers approach is the redirecting of charge carriers into the depopulated escape cone by scattering events. The most effective one is the diffuse carrier reflection at the layer boundaries. This is because, in contrast to isotropic scattering processes, after the diffuse reflection at the metal-air interface the carrier momentum is directed always towards the metal-semiconductor interface,

416

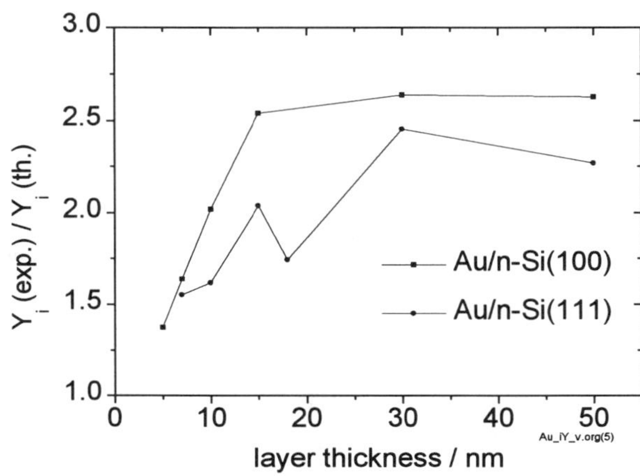

Fig. 4: Ratio of experimental Y_i (exp.) to theoretical Y_i (th.) vs d at T=300K and hν=1eV

where only those carriers are reflected which are not capable to escape. This process works the better, the thinner the film, explaining the increasing Y_i with decreasing layer thickness. The damping factor in this model is the inelastic carrier-carrier scattering, which leads to a vanishing Y_i for attenuation length $l_e \ll d$. Isotropic elastic carrier scattering with phonons and neutral impurities is found to have little influence on Y_i for attenuation length $l_p > d$, i.e. for thin layers and sufficiently pure material. Although isotropic scattering helps in filling up the depopulated escape cone, it extends the path of the excited carriers, too, which increases the possibility of an inelastic collision. The scattering theory agrees with the experimental values of the Y_i for the thinnest samples within the order of unity, whereas for the thicker samples the experimental values prove to be roughly 3 times larger than the theoretical ones (Fig. 4).

This discrepancy may be explained by additional carrier scattering at grain boundaries in polycrystalline layers, which are not included in the analytical model. These scattering processes can provide an enhancement of Y_i by a factor 2, if the film thickness is not much smaller than the grain diameter [2].

The quantum-mecanical transmission probability across a Schottky barrier turns out to be significantly smaller than unity for electron energies a few 10 meV above the barrier height maximum [6] and is not included in Eq (3).

The obtained Schottky barrier energies E_B vs Au layer thickness is plotted in Fig. 5 for different sample temperatures. E_B is independent of layer thickness as expected since the Schottky barrier formation is found to be completed within the first monolayers [11].

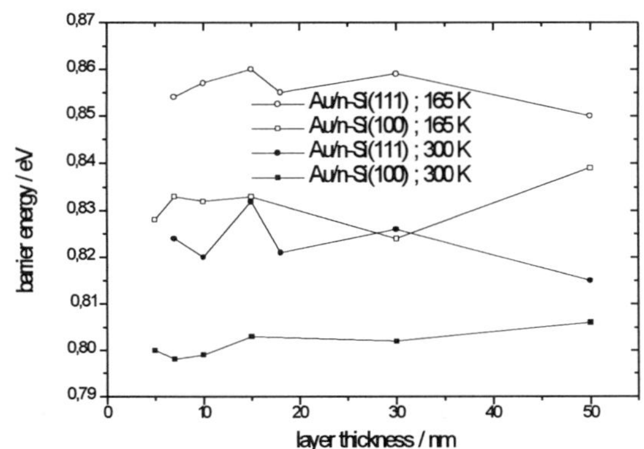

Fig. 5: Barrier energies E_B vs Au layer thickness at 165 K and 300 K for samples grown on Si(100) and Si(111) substrates, respectively

Variations in E_B of roughly 10 meV for different film thicknesses are consistent with microscopic barrier fluctuations detected in BEEM experiments on Au/n-Si (100) structures [12]. An average increase in E_B by 32meV with decreasing temperature from 300K to 165K is clearly shown. The band gap E_G (T) of Si increases in this temperature range by 29 meV [13]. The comparable dependence of E_B (T) and E_G (T) on temperature can be interpreted as a Fermi level pinning in relation to the valence band of Si [14]. Samples grown on Si(100) substrates reveal the same temperature dependence with a barrier energy approximately 23 meV lower than on structures with Si(111) substrates.

CONCLUSIONS

Basing upon simultaneous measurements of the internal photoemission current and the optical reflectance and transmittance on Au/n-Si structures reliable absolute values of the Y_i were obtained. The Y_i increases with decreasing emitter layer thickness, being for the thinnest layers 40 times larger than expected from Fowler's theory. This result is explained quantitatively by scattering processes of the excited charge carriers in the photoemitter. The scattering presents an improvement for the description of the IP with respect to Fowler's approach having consequences for the interpretation of Schottky barrier heights. The barrier energy is larger and the variation of the barrier energy with temperature is smaller than deduced from the Fowler plot.
The barrier energy of Au/n-Si heterostructures is found to be independent of layer thickness. Samples grown on Si(111) have a barrier energy 25 meV larger than those prepared on Si(100) substrates.The Fermi level of the Au/n-Si heterojunction is fixed relatively to the valence band of Si.
Further improvements of the theoretical model as to the grain boundary scattering, the energy loss due to the phonon scattering and the quantum-mechanical transmission are in progress by utilizing the Monte Carlo Method. Nevertheless, the present analytical theory including the description of charge carrier scattering in thin films already provides the right basis for analyzing IP measurements and for optimizing IP based applications, e.g. infrared detectors and photoinjection solar cells.

ACKNOWLEDGMENTS

The authors whish to thank B. Rabe for the accurate sample preparation and O. Nast for the helpful AFM measurements.

REFERENCES

[1] V.W. Chin, J.W.V.Storey and M.A.Green, Solid-St. Electron. **39**,277(1996)
[2] E.Roca, K. Kyllesbech Larsen, S. Kolodinski and R. Mertens, J. Appl. Phys. **79**,4426(1996)
[3] T.L.Lin, J.S.Park, S.D.Gunapala, E.W.Jones, and H.M.Del Castillo, IEEE Electron Device Letters **15**,103(1994)
[4] E.O.Kane, Phys. Rev. **147**, 335(1966)
[5] V.L.Dalal, J.Appl.Phys. **42**, 2274(1971)
[6] E.Y.Lee, L.J.Schowalter, J. Appl. Phys. **65**, 4903(1989)
[7] M.Schmidt, M.Brauer, V.Hoffmann, Appl.Surf.Science 102, 303(1996)
[8] R.H. Fowler, Phys. Rev. **38**,45(1931)
[9] J.J.Quinn, Phys.Rev.**126**, 1453(1962)
[10] Chung-Whei Kao, C.L.Anderson, and C.R.Crowell, Surf.Sci. **95**, 321(1980)
[11]A.D.Katnani, N.G.Stoffel, H.S.Edelmann, G.Margaritondo,
     J.Vac.Sci.Technol. **19**,290(1981)
[12] H. Palm, M. Arbes, M. Schulz, Phys. Rev. Lett. **71**,2224(1993)
[13] S.M.Sze, Physics of Semiconductor Devices, 2 nd Edn.
     Wiley, New York (1981), p. 15
[14] C.R.Crowell, S.M.Sze, W.G.Spitzer, Appl.Phys.Lett. **4**, 91(1964)

EFFECTS OF THE SUBSTRATE PRETREATMENTS ON THE LEAKAGE CURRENT IN THE LOW-TEMPERATURE POLY-Si TFTs

TAE-KYUNG KIM, BYUNG-IL LEE, TAE-HYUNG IHN, SEUNG-KI JOO
Division of Materials Science and Engineering, Seoul National University, Seoul, Korea

ABSTRACT

Poly-Si TFT's were fabricated on the glass substrate by the Metal Induced Lateral Crystallization(MILC). Before deposition of the active a-Si thin films(1000 Å), the glass substrate was pretreated in three different ways such as oxidation of a-Si(100 Å), oxide buffer layer deposition(1000 Å), and ion mass doping of the glass substrates. The leakage current at reverse bias could be reduced by one order of magnitude by the substrate pretreatment. Field effect mobility of n-channel TFT's was $108 cm^2/V \cdot sec$, subthreshold slope and on/off current ratio were 0.7 V/dec. and about 10^6, respectively.

I. INTRODUCTION

Poly-Si TFT's are attracting considerable attention for the replacement of a-Si TFT's for the AMLCD (Active Matrix Liquid Crystal Devices) requiring high resolution and high response speed. Conventional a-Si TFT's can switch the LCD pixels, but cannot drive the LCD peripheral circuits, so LCD should be connected to discrete IC's fabricated on a single crystal Si wafer. Since poly-Si TFT's show high speed, they enable the integration of LCD pixel and driving circuit on the same substrate, which result in high performance and high productivity. However, one of the major problems of the poly-Si TFT's is the large leakage current[1][2]. In this work, high performance poly-Si thin film transistors were fabricated on the glass substrate using a newly developed crystallization method, Metal-Induced Lateral Crystallization (MILC)[3] at 500 ℃ and the effects of interface states between the glass substrate and the poly-Si thin films on the leakage current of TFT's have been studied.

II. EXPERIMENTS

Substrate Treatment of the Glass Substrate
1) Buffer SiO_2 Layer Formation

Before the deposition of active a-Si thin films, 1000 Å thickness of SiO_2 Layer was deposited directly on the Corning 7059 glass substrate in ECR CVD chamber, using SiH_4 and O_2 as source gases. Alternatively, 100 Å thickness of amorphous-Si film was deposited and oxidized by ECR plasma.

2) Ion Mass Doping on the glass substrate

Phosphorus or Boron was doped into bare glass by Ion Mass Doping System. Source gases were PH_3 and B_2H_6 diluted in hydrogen gas, respectively. The plasma power was 100W and the acceleration DC voltage was about 18KeV. PH_3 and B_2H_6 diluted in hydrogen were doped for 10 minutes.

Mat. Res. Soc. Symp. Proc. Vol. 448 © 1997 Materials Research Society

Fabrication Method of poly-Si TFT by MILC

The fabrication steps are illustrated in Fig. 1. A 1000Å-thick a-Si film was deposited by LPCVD at 480℃ using pure Si_2H_6, and defined for active layer. For the gate dielectric, 300Å thick SiO_2 was formed by ECR plasma oxidation of active a-Si layer and then 700Å thick SiO_2 was deposited by ECR PECVD using [SiH_4(25%)+Ar] gas in ECR chamber[4]. The gate and gate dielectric was patterned and defined by conventional SF_6 plasma etching and wet chemical etching (diluted BHF solution), respectively. A 20Å thick Ni film was deposited without an additional mask as in Fig. 1(a). Ion mass doping was carried out to form n-type gate and source/drain for n-channel TFT's. Source/drain activation and crystallization of a-Si at the channel area were done by annealing at 500°C for 15hrs. After the crystallization, unreacted Ni films on oxide area were removed by wet etching. The process conditions are summarized in table 1.

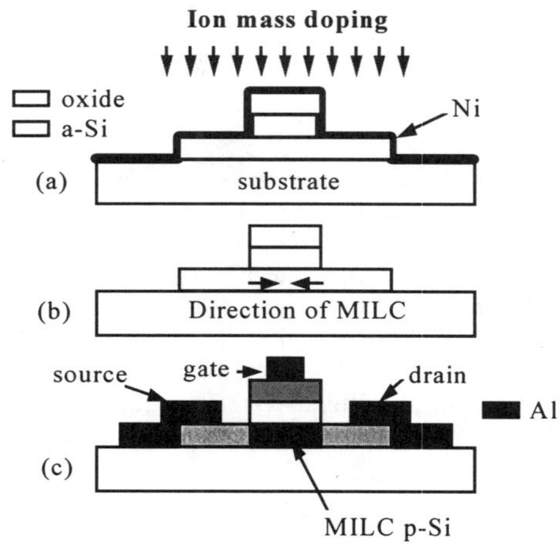

Fig.1. Fabrication steps of MILC poly-Si TFT's.
(a) Deposition of a 20Å Ni layer followed by dopant Ion Mass Doping
(b) Crystallization of the source, drain and channel area
(c) Metallization

Process	Method
a-Si deposition	LPCVD(1000Å, 480℃)
gate-oxide formation	ECR Oxidation (300Å) + ECR CVD (700Å)
gate(a-Si)	LPCVD (2000Å, 480℃)
source/drain doping	Ion Mass doping(RF=200W, DC=18keV)
Ni deposition	Sputtering (20Å)
Al deposition	Sputtering (2000Å)

Table 1. Key processes for the fabrication of poly-Si TFT's by MILC.

III. RESULT AND DISCUSSION

Metal Induced Lateral Crystallization

The channel area could be crystallization by Ni deposited on source and drain area. The characteristic in Fig.2(a) indicates a-Si, which yet to be crystallized. By annealing at 500°C for 15hrs, The channel area of 40μm in length was completely crystallized as shown in Fig. 2(b). Detailed discussion of MILC phenomenon can be found in the literature[3].

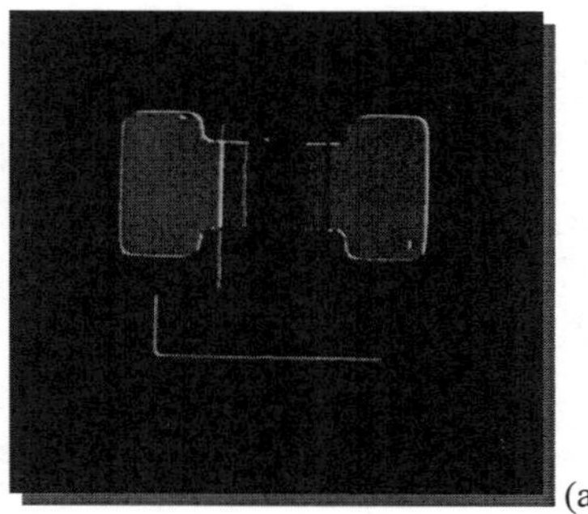 (a)
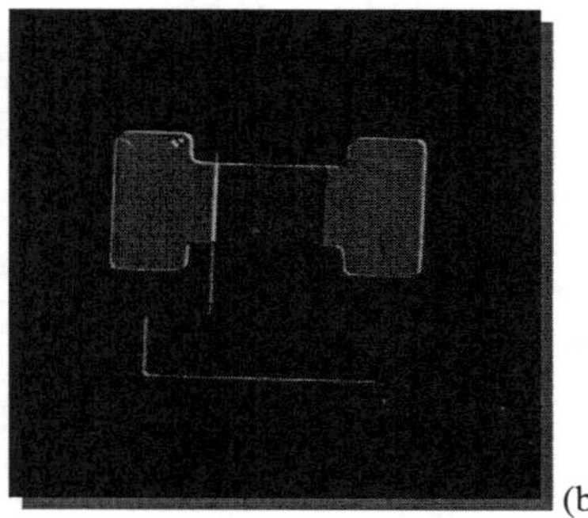 (b)

Fig. 2 Nomarski optical micrographs showing the progress of MILC in the channel area.
A 20Å-thick nickel layer was deposited after the source/drain definition and the width and length of the gate were 40μm/40μm. a) annealing at 500°C for 10hrs (b) for 15hrs

Poly-Si TFT's on the SiO$_2$ buffer layer

I-V characteristics of the buffer layered TFT's are shown in Fig. 3 and effects of the substrate pretreatment on device parameter are summarized in Table 2.
It can be noticed that the mobility was increased about two times by the formation of buffered SiO$_2$ between a-Si and the glass substrate. As far as the leakage current is concerned, the pretreatment of substrate turned out to reduced the leakage current to a certain extent. It has not been cleared yet but since the substrate was the common glass which may contain impurities very mobile at the crystallization temperature(500°C), it is very likely that the buffer layer blocks the diffusion of these impurities into the bulk silicon during annealing.

Ion Mass Doping Effects

Fig. 4 shows the effect of IMD of B or P into the glass substrates on I-V curves. N-channel poly-Si TFT's showed the mobility as high as 108cm^2/V·sec. Not much difference in the leakage current can be observed in comparison with SiO$_2$ buffer layered TFT's. The AFM that is shown in Fig.5, however, showed much charge in topology through the pretreatments. It is interesting to notice that the rougher interface caused by these pretreatments ends up with reduction of the leakage current. Also it is surprising that phosphorus or boron doping does not show appreciable difference in device performances. Investigation to understand the basic reason for the leakage current in poly-Si TFT's fabricated on the glass substrate is under way.

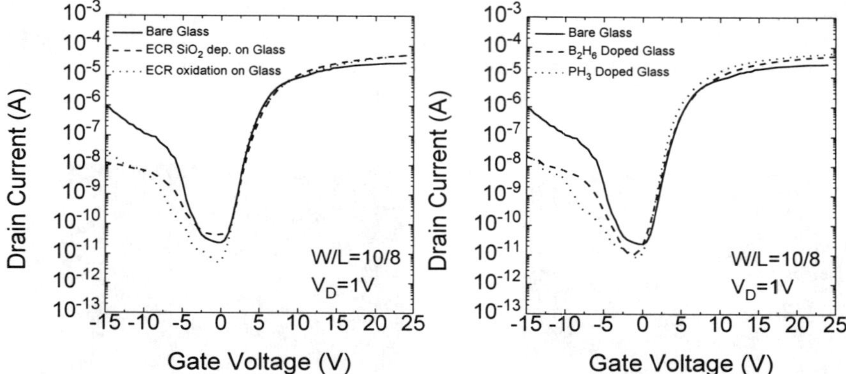

Fig 3. I-V characteristics of TFTs
on Buffer Layer treated glass substrate

Fig 4. I-V characteristics of TFTs
on Ion Mass Doped glass substrate

		SiO₂ Buffer Layer		Ion Mass Doping	
	Bare Glass	1000 Å	*100 Å	Phosphorus	Boron
Field Effect mobility	48	77	104	108	83
V_T (V)	4	4.2	3.8	3.3	3.7
Subthreshold slope (V/dec)	0.8	0.7	0.74	0.74	0.92
Min.leakage current (pA/μm)	2.5	4	0.6	0.9	1
On/Off ratio	8 E5	5E5	3E6	5E6	3E6

Table 2. Summary of the substrate pretreatment effects

For an N-channel TFT, the threshold voltage was defined at a normalized drain current $(I_D \times W/L)$ of 0.1μA at V_D = 1V. The field-effect mobility was calculated in the linear region at V_D = 0.1V. The maximum on/off current ratio was determined at V_D = 1V and V_G = -10 ~ 25V .

*100 Å was the thickness of a-Si before ECR oxidation.

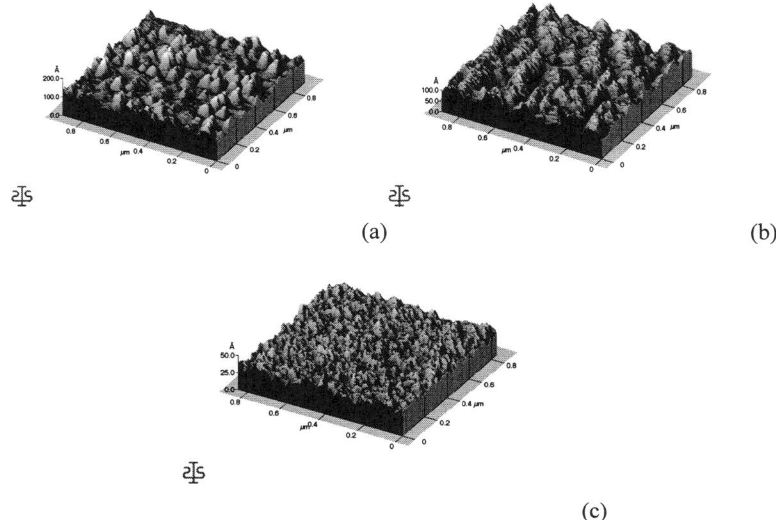

Fig. 5   AFM Images of the glass substrates after the various pretreatment
         (a) after the IMD of PH₃
         (b) SiO₂ buffer layered glass
         (c) the bare glass substrate

IV. CONCLUSION

High performance poly-Si TFT's were fabricated on Corning 7059 at 500 ℃ by a newly developed crystallization method ; Metal Induced Lateral Crystallization (MILC). Field effect mobility of the n-channel poly-Si TFT's by MILC at 500 ℃ was as high as 108cm²/Vs, which is a much larger value than that of conventional low temperature TFT's on glass substrate. SiO₂ buffer layer or IMD of B or P into the substrate reduced the leakage current. According to AFM analysis, it was found that caused the substrate surface rougher, but the leakage current could be reduced by these pretreatment.

V. ACKNOWLEDGEMENT

This work has been supported by LG electronics research center through RETCAM at Seoul National University in 1996.

VI. REFERENCE

1. J. G. Fossum, A. O. Conde, H. Schichijo, and S. K. Banerjee, IEEE Trans. Electron Devices, vol. ED-32, no.9, pp. 1878-1884
2. M. Yazaki, S. Takenaka and H. Ohshima, Jpn. J. Appl. Phys. vol. 31 (1992) pp.206-209
3. S.W. Lee, and S.K. Joo, IEEE Electron Device Lett., vol. 17, no.4, pp. 160-162, 1996
4. T. H. Ihn and S.K. Joo, 2nd Pacific Rim Inter. Conf. on Adv. Mat. and Proc., 1333, 1995.

MONITORING OF DOPANT ACTIVATION IN SUB-SURFACE P-TYPE Si USING THE SURFACE CHARGE PROFILING (SCP) METHOD

P. ROMAN, J. STAFFA, S. FAKHOURI, J. RUZYLLO and E. KAMIENIECKI *
The Electronic Materials and Processing Research Laboratory, Department of Electrical Engineering, The Pennsylvania State University, University Park, PA 16802, USA.
*QC Solutions, Inc., 150-U New Boston St., Woburn, MA, 01801, USA.

ABSTRACT

In this study the SCP (Surface Charge Profiling) method, based on non-contact, small-signal ac-SPV measurement is used to study thermal activation of boron in the near surface region of p-type Si wafers. Boron tends to form pairs with impurities such as hydrogen, iron and copper in the near surface region of Si substrates which render it inactive. During device processing, activation of boron may take place resulting in uncontrolled variations in active boron concentration in the near surface region.

In this work, both boron doped, polished CZ wafers and wafers with boron doped epitaxial layers are studied. In the former case, the concentration of active boron in the near surface region was initially up to an order of magnitude less than the bulk concentration determined from four-probe measurements, but increased with the temperature of an anneal in ambient air and approached the bulk value. In contrast, the wafers with epitaxial layers showed no consistent variations of surface dopant concentration with temperature. These results confirmed previous findings that the near surface region of the polished wafers is contaminated with metals introduced during polishing operations. The SCP method was found to be very effective in monitoring variations in active boron concentration in the near-surface region.

INTRODUCTION

It is known that unlike n-type dopants in silicon, the p-type dopant, boron, can be rendered inactive by pairing with selected contaminants. This effect is particularly pronounced in the shallow region immediately adjacent to the Si surface, which can be penetrated more readily than the bulk of the material. Two types of interaction can be identified in this regard. The first involves penetration of the surface by hydrogen, which is likely to occur under a variety of process conditions, for instance, gas-phase etching. The second involves interaction of boron with metals such as copper originating from chemomechanical polishing [1]. A brief anneal is typically sufficient to break the boron-contaminant pair and render the boron electrically active. In the case of hydrogen, featuring a high diffusivity in silicon, this also results in hydrogen leaving the silicon. In the case of metals, however, the metals remain in the silicon, where they can subsequently re-pair with boron.

These effects are well understood and can be used to determine the concentration of metallic impurities in the bulk of the Si wafer [2]. Until now, however, no convenient method that would follow boron deactivation/reactivation processes in the sub-surface region was available. The Surface Charge Profiling (SCP) method, a contactless ac-SPV method capable of measuring active doping concentration in the surface space-charge region, offers new possibilities in this regard. This work studies boron activation by hot-plate annealing in both p-type polished CZ wafers and wafers with p-type epitaxial layers using an SCP tool.

Mat. Res. Soc. Symp. Proc. Vol. 448 © 1997 Materials Research Society

EXPERIMENTAL

The operation of the Surface Charge Profiler (SCP) is based on the SPV effect, which in this case is induced by illumination of the surface with a beam of chopped, low-intensity, short-wavelength (450 nm) light having photon energy larger than the silicon bandgap. Due to the shallow penetration of the light, any information obtained is dominated by characteristics of the surface and shallow sub-surface regions of the silicon wafer. In contrast to other electrical methods, the SCP does not utilize any external bias voltage, corona charging or high-intensity illumination, and no contact is made to the wafer surface. The electrical characteristics are determined in this case from analysis of the real and imaginary components of the ac-SPV signal. The minority carrier recombination lifetime at the surface is determined from the ration of these components, or the phase of the ac-SPV signal. The width of the depletion layer, Wd, is simultaneously determined from the imaginary component corrected for the lifetime.

A schematic diagram of the SCP system is shown in Fig. 1. The wafer is positioned on the movable wafer support tray with insulating coating which then moves the wafer underneath the probe. The measurement is performed on the wafer in motion and the distance between the probe and the wafer surface, typically about 100 um, is controlled and maintained automatically. The results of the SCP measurement are presented in linear profile, map, trend chart and numerical form.

Under inversion conditions, the maximum depletion width can be used to determine the active doping concentration in the depletion region, Nsc. It is known from past work that immersion in dilute $HF:H_2O$ solutions for short times followed by rinsing and drying adds a large amount of positive charge to silicon wafer surfaces, which for p-type wafers is sufficient to drive the surface into inversion [3]. In this work, both boron doped polished wafers (<100>, CZ, 100 mm diam., 1-4 ohmcm.) and wafers with boron doped epitaxial layers (150 mm diam., 4.5-5.5 ohmcm.) were used. For the polished wafers, boron concentration in the bulk was determined using four-probe measurements. For the wafers with epitaxial layers, boron concentration in the bulk was determined from C-V measurements. Wafers were annealed on a hot plate in air at different temperatures, then immersed in $HF(1):H_2O(100)$ solution, rinsed and dried to invert the surface. Subsequently, the measurement of dopant concentration was performed using the SCP tool.

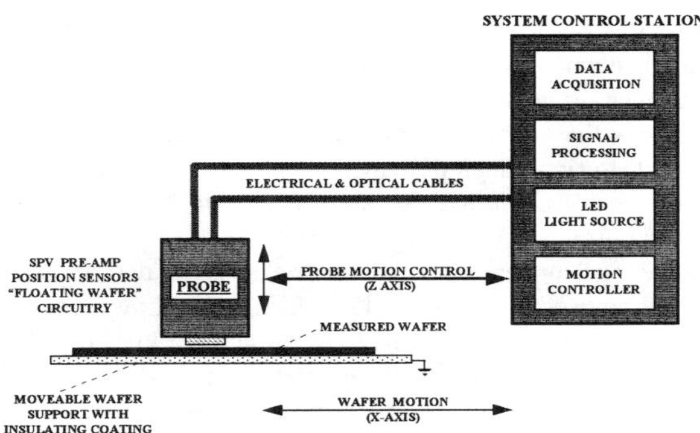

Figure 1. Schematic diagram of the SCP system.

RESULTS AND DISCUSSION

It was previously observed that the polished, boron doped p-type wafers coming from various manufacturers were typically displaying values of surface doping concentration measured using SCP up to an order of magnitude less than the bulk value determined from four-probe measurements. A similar effect was not observed in the case of n-type silicon. This observation led us to the conclusion that the effect of reduced near-surface concentration of active boron may be related to its interactions with elements such as hydrogen as well s copper and/or iron which may penetrate the near-surface region of the silicon and are known to render boron inactive. It was also shown in the past that low-temperature anneals are sufficient to break B-contaminant pairs and activate boron. Following this an attempt was made to determine how the surface doping concentration in the wafers used in this study would respond to a brief (15 min.) hot-plate anneal in ambient air. As Fig. 2 shows, the concentration of active boron at the surface of the polished wafers measured using SCP was initially over five times lower than the concentration of boron in the bulk determined by a four-probe measurement. With increasing temperature above 50°C an increasing concentration of active boron at the surface approaching the bulk concentration (200°C anneal) was observed. To various extents, the same effect was observed for all polished p-type wafers used in this study.

In contrast, the wafers with epi layers did not show consistent variations of surface dopant concentration with the anneal temperature (Fig. 3). Also, for each temperature the surface dopant concentration was slightly larger than the bulk value determined from C-V measurements. This is a clear indication that the near-surface region of the epi layers is not contaminated in the way the near-surface region of the polished wafers is. This result was expected since polishing processes are considered to be a main source of contamination leading to deactivation of boron. In addition, the plot in Fig. 3 suggests that the epitaxial process used to grow the layers studied in this experiment resulted in an excess of boron in the near-surface region.

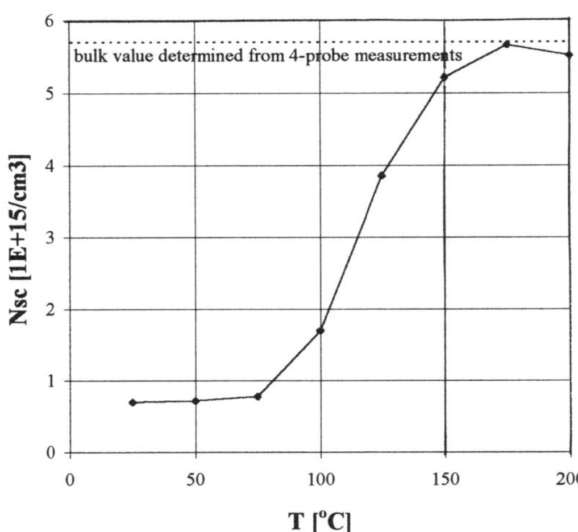

Figure 2. Surface dopant concentration vs. anneal temperature for polished wafers.

The data shown in Fig. 2 can be used to determine the activation energy for the effect observed. From 50°C up to 125°C, the rate of boron activation is observed to increase exponentially with temperature, and can be represented by the Arrhenius equation [4],

$$Nsc - Nsco = Ae^{-qEa/kT} \qquad (1)$$

where Nsco is the initial surface concentration of active boron, E_a is the boron activation energy, k is the Boltzmann constant, T is the Kelvin temperature, and the time of the anneal has been lumped into the constant, A. Consequently, for this temperature range an activation energy can be determined from the slope of a plot of ln(Nsc - Nsco) vs. 1/T (Fig. 4). Using the slope of this line results in the value, E_a = 0.8 eV for the boron activation energy. Based on previously reported binding energies for boron-metal pairs in the range of 1 eV [5], one may conclude the presence of metal contamination, most likely copper, in the near surface region of the polished wafers.

At temperatures larger than 125°C, the rate of boron activation becomes limited by the supply of available boron pairs, and the concentration of active boron in the near surface region saturates at a value very close to the bulk value.

SUMMARY

In this work boron activation by low-temperature annealing at different temperatures was studied using an SCP tool. Both boron doped polished wafers and wafers with boron doped epitaxial layers were studied. In the former case the concentration of active boron in the near surface region was initially much lower than the bulk concentration determined from four-probe measurements, but increased with temperature and approached the bulk value. This dependence on temperature was used to calculate a value for the boron activation energy which identifies the

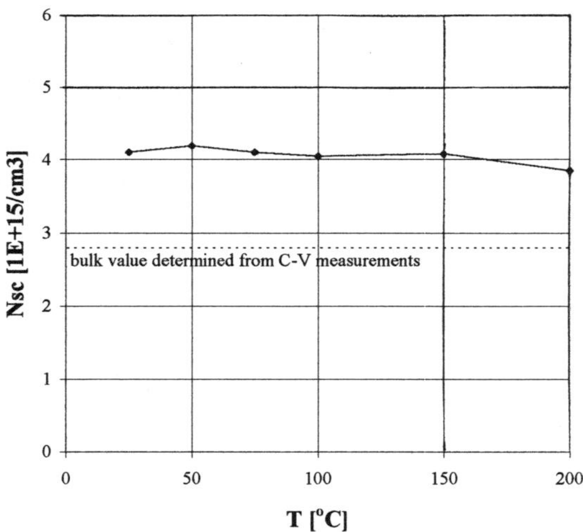

Figure 3. Surface dopant concentration vs. anneal temperature for epi layers.

428

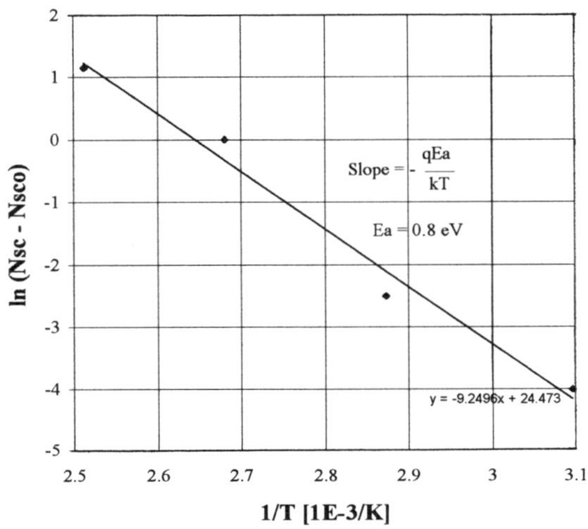

Figure 4. ln(Nsc - Nsco) vs. 1/T for polished wafers.

presence of metal contamination in the near surface region of these wafers. In the latter case the wafers did not show consistent variations of the surface dopant concentration with temperature. These experiments show that the contactless, non-invasive SCP method is very effective in controlling active boron concentration in the near surface region of both polished wafers and wafers with epitaxial layers.

REFERENCES

1 H. Prigge, P. Gerlach, P.O. Hahn, A. Schnegg and H. Jacob, J. Electrochem. Soc., **138**, 1385 (1991).

2 L. Jastrzebski, O. Milic, M. Dexter, J. Lagowski, D. DeBusk, K. Nauka, R. Witowski, M. Gordon and E. Persson, J. Electrochem. Soc., **140**, 1152 (1993).

3 P. Roman, D. Hwang, K. Torek, J. Ruzyllo and E. Kamieniecki in Ultraclean Semiconductor Processing Technology and Surface Chemical Cleaning and Passivation, edited by M. Liehr, M. Heyns, M. Hirose and H. Parks (Mater. Res. Soc. Proc. **386**, Pittsburgh, PA, 1995) pp. 401-406.

4 S. S. Zumdahl, Chemistry, 1st ed. (D.C. Heath and Co., Lexington, 1986), pp. 500-501.

5 K. Graff in Proceedings of the Fifth International Symposium on Si Materials Science and Technology, edited by Huff and Aloe (Electrochem. Soc. Proc., Pennington, NJ, 1986) pp. 751-765.

A STUDY OF THE INTERACTION BETWEEN Cu₃Ge AND (100) Si, AND ITS EFFECT ON ELECTRICAL PROPERTIES

M.A. Borek, S. Oktyabrsky, M.O. Aboelfotoh and J. Narayan
Department of Materials Science and Engineering, North Carolina State University, Raleigh, NC 27695

ABSTRACT

The reduction in the dimensions of advanced semiconductor devices has brought about the need for new metallization materials with low resistivity and high electromigration resistance. The Cu_3Ge, has been suggested as a contact and metallization material due to its low resistivity (6 $\mu\Omega$-cm), high electromigration resistance, and high chemical stability. We have grown thin films of Cu_3Ge on (100) Si by the sequential e-beam deposition of an amorphous Ge layer then a Cu layer, followed by a thermal anneal to crystallize the Cu_3Ge film. The Cu-Ge films maintain their low resistivity (10-15 $\mu\Omega$ cm) over a range of anneal temperatures, up to an anneal temperature of 600°C, where a significant increase in resistivity is observed. We have shown that this increase in resistivity is directly related to the structure of the Cu_3Ge film and interface. We have observed by cross-sectional transmission electron microscopy (TEM), that films of Cu_3Ge form a smooth, atomically sharp interface with (100) Si, up to an anneal temperature of 600°C, where the film agglomerates, and additional compounds are observed. In this paper, we discuss the correlation between microstructure, interface structure and electrical properties of these novel thin film structures.

INTRODUCTION

As the dimensions of semiconductor devices continue to be reduced, overall device performance and reliability will become a function of the materials which are used in metallizations. For next generation electronic devices to perform reliably, the metallization materials which are selected must have low resistivity and high electromigration resistance. Due to its low resistivity (ρ<2 $\mu\Omega$ cm) and excellent electromigration resistance, Cu has been proposed as a new metallization material in Si-based semiconductor devices.[1] However, many challenges must be overcome in order to integrate Cu with current Si -based devices. For example, Cu is considered to be one of the fastest diffusing elements in Si, diffusing interstitially with an activation energy of 0.43 eV.[2] Copper is also a deep level impurity, providing a trap and recombination center , and has also been shown to passivate boron acceptors in p-type Si by forming neutral B-Cu complexes, which results in an overall reduction in the electrically active acceptor concentration.[3] Copper also oxidizes in air at relatively low temperatures ~ 250°C, which leads to a significant increase in resistivity, and Cu adheres poorly to SiO_2, disqualifying it as a candidate for gate metallizations.[4]

Metal silicides have also received much attention as metallization materials due to their high temperature stability and relatively low resistivity, e.g. $TiSi_2$ (C54 phase) has a resistivity of 13-16 $\mu\Omega$ cm, and $CoSi_2$ with ρ=18-20 $\mu\Omega$ cm.[5] However, a high temperature anneal~800°C is required to form the low resistivity C54 phase of $TiSi_2$ resulting in an increase in the

Mat. Res. Soc. Symp. Proc. Vol. 448 © 1997 Materials Research Society

temperatures that are required during device processing.[6] Metal silicides also tend to form rough interfaces with the Si substrate, thus impeding the fabrication of shallow junction devices.

Metal germanides have recently received much attention as potential metallization materials, in particular the ε_1-Cu_3Ge phase of the Cu-Ge system has been shown to have a low resistivity of 5.5 $\mu\Omega$ cm, which is only approximately three times the resistivity of pure Cu films.[7-9] This low value of resistivity exists over a composition range of 25-40 at. % Ge, thus providing a processing window during device fabrication. The Cu_3Ge is chemically stable up to a temperature of 450°C, i.e. there is no out diffusion of Cu from the film into the substrate. The Cu_3Ge is stable against oxidation in air up to ~520°C, and adheres well to SiO_2.[4] Films of Cu_3Ge also form a sharp interface with the substrate, making it an attractive choice as a contact to shallow junction devices.

In order to succesfully integrate Cu_3Ge with current Si-based electronic devices, it is necessary to understand what effect elevated process temperatures will have on the film structure and properties. In this paper, we report the growth of Cu-Ge films on (100) Si substrates. We have previously shown that thin films of Cu_3Ge on (100) Si retain their low resistivity despite a large amount of Si incorporation that is induced by the post-deposition anneal treatments.[10] In this paper, we show that the change in resistivity of the films which arises at a specific anneal temperature can be directly related to the structure of the films and the substrate interface.

EXPERIMENT

Thin films of Cu_3Ge were formed on hydrogen-terminated (100) Si by the sequentail e-beam evaporation of an amorphous Ge film, followed by a Cu film. Films were deposited in a 1 x 10^{-7} Torr vacuum at room temperature at a rate of 10-20 Å/sec. Film composition was controlled by varying the thicknesses of the Cu and Ge layers. The average film composition could be determined by assuming a bulk atomic density of the Cu and Ge layers. In general, films of Cu_3Ge were grown to a total thickness of 2550 Å by depositing a 1000 Å layer of Ge, followed by a 1550 Å layer of Cu. The deposited films were annealed in flowing N_2 for 30 minutes over a temperature range of 150-600°C. These anneal treatments resulted in the complete recrystallization of the Cu_3Ge films. The room temperature resistivities of the films were determined by four-point probe and Van der Pauw methods. Cross sectional transmission electron microscopy (TEM) and selected area diffraction (SAD) were conducted with a 200 kV Topcon 002B electron microscope to study the film and interface microstructure, and identify the phases present in the films.

RESULTS

The room temperature resistivities of the Cu_3Ge films that were annealed in N_2 at 250°C, 400°C, and 600°C are displayed in Table I. It can be seen from these results, that the resistivity of the Cu_3Ge remains relatively constant until an anneal temperature of 600°C, where the resistivity increases significantly. We have previously shown that a large amount of Si diffuses into the Cu_3Ge film in varying amounts which are dependent on the anneal temperature.[10] Despite the greater concentration of Si in the Cu_3Ge film that was annealed at 400°C, the resistivity remains unchanged relative to the sample annealed at 250°C. It is therefore believed that the addition of Si in the Cu_3Ge film is not responsible for the resistivity increase that is

Sample	Anneal Temperature (°C)	Resistivity ($\mu\Omega$ cm)
Cu$_3$Ge/Si	250	10-15
Cu$_3$Ge/Si	400	10-15
Cu$_3$Ge/Si	600	7-8x10^3

observed at higher temperatures, rather it may be attributed to the structure and composition of the film.

Figure 1(a) shows a bright field cross section TEM image of a Cu-Ge film with an overall Ge concentration of 40 at.%, annealed in N$_2$ at 400°C. At this composition, the ε_1-Cu$_3$Ge phase is in equilibrium with a Ge-phase. The film consists of polycrystalline Cu$_3$Ge, with an average grain size of 200 nm, and smaller grains of Ge, having an average grain size < 100 nm. The micrograph displays the extremely sharp interface between the film and substrate. The Cu-Ge/Si interface is atomically sharp, with no faceting along the Si (111) planes, which is frequently observed in metal silicide contacts, e.g. TiSi$_2$ or CoSi$_2$ on (100) Si. This indicates that there are no chemical reactions between the Cu$_3$Ge and Si, making Cu$_3$Ge a strong candidate for contacts to shallow junction devices. The corresponding SAD patterns of the film are displayed in Fig; 1(b) & 1(c). Figure 1(b) was recorded from the grain marked "b" in Fig. 1(a), and is recorded along

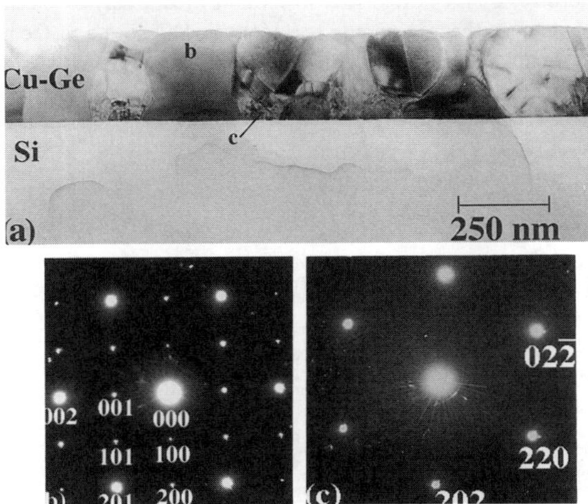

Figure 1. (a) cross sectional TEM bright field image of Cu-Ge film on (100) Si, annealed at 400°C. (b) SADP of Cu3Ge, b=[010]. (c) SADP of Ge, b=[111]

the [010] zone axis, and is consistent with the orthorhombic structure of Cu_3Ge, with lattice spacings of a=0.528 nm, b=0.422 nm, and c=0.454 nm. Figure 1(c) is from the region marked "c" on the micrograph, and was recorded along the [111] zone axis, and is consistent with the diamond cubic lattice of Ge with a=0.566 nm.

Figure 2 shows a bright field cross section TEM image of a Cu-Ge film with an overall Ge concentration of 40%, annealed in flowing N_2 at 600°C. As can be seen in Fig. 2, the Cu_3Ge film no longer forms a sharp interface with the Si substrate, rather there are many protrusions of the

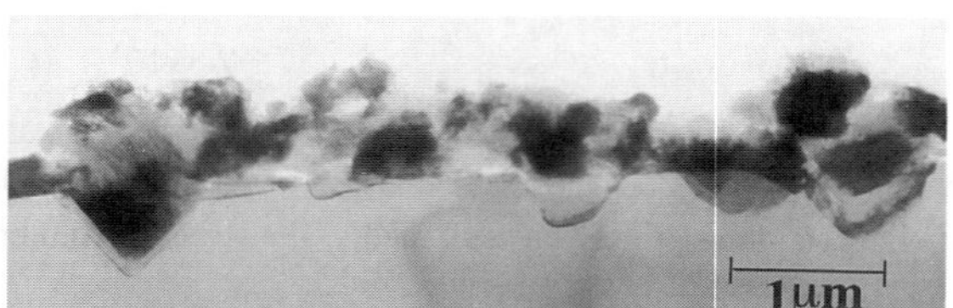

Figure 2. Cross section TEM bright field image of Cu-Ge on (100) Si, annealed at 600°C.

film deep into the substrate. Figure 3 focuses on a smaller area of the same Cu-Ge sample that was previously shown in Fig. 2. We have identified by electron diffraction the phases that are present in this region. The region labeled "a" in the micrograph has been identified as Cu_3Ge giving a SAD pattern identical to the one shown in Figure 1(b). The region labeled "b" in the micrograph gave an amorphous ring pattern when observed by electron diffraction, making this area impossible to determine by the lattice spacings, but is most likely an oxide of Si. The remaining region labeled "c" in the micrograph was identified as Si or a Si-Ge alloy. It is difficult to distinguish between Si and a low Ge concentration Si-Ge alloy due to the slight increase in lattice spacing that arises due to Ge incorporation in Si.

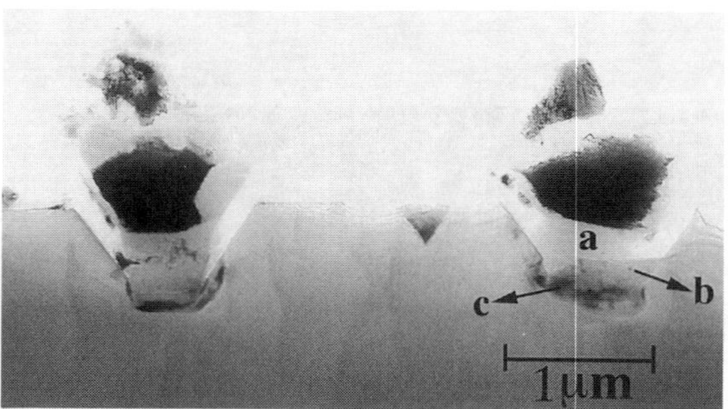

Figure 3. TEM bright field image of selected area of Cu-Ge film on Si, annealed at 600°C.

Based on the Cu-Ge phase diagram, ε_1-Cu_3Ge should remain in equilibrium with the Ge until a temperature of ~615°C, above this temperature, the Ge solid solution will be in equilibrium with the high-temperature ε_2 phase of Cu-Ge, thus at 600°C no changes in the structure of the Cu-Ge film should be expected.[11] It is of great interest to understand the reason for the restructuring of the Cu-Ge film that is annealed at 600°C. One possible explanation considers the effect of the Si that has diffused into the Cu-Ge film, thus creating a Cu-Ge-Si ternary alloy. This may cause agglomeration of the Cu-Ge film which allows oxidation of the underlying Si substrate, resulting in an overall increase in the bulk resistivity of the Cu-Ge film. Another possible explanation for the high-temperature restructuring of the Cu-Ge is that 600°C is greater than 90% of the melting point of Cu_3Ge. At such a high temperature, enhanced oxygen diffusion through the Cu-Ge film or grain boundaries may occur. This would allow oxidation of the underlying Si, which could also cause a restructuring of the Cu-Ge film.

CONCLUSIONS

We have deposited low resistivity films of Cu-Ge on (100) Si. The Cu-Ge maintains its low resistivity of approximately 10-15 $\mu\Omega$ cm over a range of anneal temperatures until an anneal temperature of 600°C, at which point the bulk resistivity increases substantially. Thin films of Cu-Ge form a sharp interface with the (100) Si until an anneal temperature of 600°C, at which point the film loses its structural integrity, i.e. the film becomes faceted, protruding into the Si substrate. These results establish a maximum anneal temperature that can be used to effectively form thin films of the low resistivity Cu_3Ge phase on (100) Si.

REFERENCES

1. H. Takasago, K. Adachi, and M. Takada, J. Electron. Mater. **18**, 319 (1989).
2. E.R. Weber, Appl. Phys. A **30**, 1-22 (1983).
3. M.O. Aboelfotoh and B.G. Svensson, Phys. Rev. B **44**, 12742 (1991).
4. H.K. Liou, J.S. Huang, and K.N. Tu, J. Appl. Phys. **77**, 5443 (1995).
5. S.P. Murarka, Silicides for VLSI Applications, (Academic, New York, 1983).
6. L.A. Clevenger and R.W. Mann, Mater. Res. Soc. Proc. **320**, 15 (1994).
7. L. Krusin-Elbaum and M.O. Aboelfotoh, Appl. Phys. Lett. **58**, 1341 (1991).
8. M.O. Aboelfotoh and H.M. Tawancy, J. Appl. Phys. **75**, 2441 (1994).
9. M.O. Aboelfotoh, K.N. Tu, F. Nava, and M. Michelini, J. Appl. Phys. **75**, 1616 (1994).
10. M.A. Borek, S. Oktyabrsky, M.O. Aboelfotoh, and J. Narayan, Appl. Phys.Lett. **69**, (1996).
11. M. Hansen, Constitution of Binary Alloys, (McGraw-Hill Book Company, New York, 1958).

ELECTRICAL CHARACTERIZATION OF In SCHOTTKY CONTACTS TO EPITAXIAL n-In$_{0.46}$Ga$_{0.54}$P GROWN ON n$^+$-GaAs BY MOCVD

N. Marcano and A. Singh,

Universidad de Oriente, Departamento de Física, Laboratorio de Semiconductores, Apartado 188, Cumaná 6101, Sucre, Venezuela.

ABSTRACT

In/n-In$_{0.46}$Ga$_{0.54}$P Schottky diode was fabricated by thermal evaporation of In on chemically etched surface of In$_{0.46}$Ga$_{0.54}$P:Si epitaxial layer grown on highly doped n type GaAs. The In metal formed a high quality rectifying contact to In$_{0.46}$Ga$_{0.54}$P:Si with a rectification ratio of 500. The direct current-voltage/temperature (I-V/T) characteristics were non-ideal with the values of the ideality factor (n) between 1.26-1.78 for 400>T>260 K. The forward I-V data strongly indicated that the current was controlled by the generation-recombination (GR) and thermionic emission (TE) mechanisms for temperature in the range 260-400 K. From the temperature variation of the TE reverse saturation current, the values of (0.75±0.05)V and the (4.5±0.5)x10^{-5} Acm^{-2}K^{-2} for the zero bias zero temperature barrier height (ϕ_{oo}) and modified effective Richardson constant were obtained. The 1 MHz capacitance-voltage (C-V) data for 260 K < T < 400 K was analyzed in terms of the C^{-2}-V relation including the effect of interface layer to obtain more realistic values of the barrier height (ϕ_{bo}). The temperature dependence of ϕ_{bo} was described the relation ϕ_{bo}=(0.86±0.03) - (8.4±0.7)x10^{-4}T. The values of ϕ_{oo}, obtained by the I-V and C-V techniques agreed well.

INTRODUCTION

The ternary compound semiconductor In$_{0.46}$Ga$_{0.54}$P lattice matched to GaAs[1], is of current interest for applications in optoelectronic, high speed and heterostucture electron devices. Its Schottky contacts have important applications in solar cells, metal-semiconductor field effect transistors (MSFETs)[2], high electron mobility transistors (HEMTs) and microwave mixer diodes. Because of its wide direct bad gap, this material offers visible wavelength operation for light-emitting diodes (LEDs) and laser diodes[3]. Due to the large valence-band offset of the InGaP/GaAs heterojunction, this material has potential applications in heterojunction bipolar transistors (HBTs)[4]. The knowledge of the characteristics parameters of the metal/InGaP Schottky diodes prior to optimum device fabrication is of vital importance.

From a technological point of view, the Schottky barrier height (ϕ_{bo}) is the most important device parameter because it controls both, the current-voltage (I-V) and capacitance-voltage (C-V) in metal-semiconductor (MS) and metal-thin interface layer-semiconductor (MIS) diodes. There are some reports in literature on barrier heights of Schottky contacts to InGaP epitaxial layers[5,6] determined from the room-temperature I-V and C-V data using standard techniques[7] valid for ideal Schottky diodes. However, ϕ_{bo} can not be determined accurately from simple room-temperature I-V characteristics when the value of the effective Richardson constant is not known exactly as in the case of InGaP[5].

In this paper, we report the results of a systematic investigation on the current-voltage/temperature (I-V/T) and high frequency 1 MHz capacitance-voltage/temperature (C-V/T) characteristics of an In/In$_{0.46}$Ga$_{0.54}$P MIS diode over the temperature range 260-400 K. The I-V/T data is used to examine the current transport mechanisms in the In/In$_{0.46}$Ga$_{0.54}$P and estimate its characteristic parameters[8,9]. The 1 MHz C^{-2}-V/T data is analyzed in terms of the MIS diode model of Hattori *et al.*[10] and the temperature dependence of ϕ_{bo} is established.

EXPERIMENTAL TECHNIQUE

Before deposition of metal contacts, the n-$In_{0.46}Ga_{0.54}P$/n^{+}-GaAs samples were degreased with acetone and methanol at 35° C for 10 min., rinsed with deionized (DI) water and dried with Argon. To remove the native oxide, the samples were then chemically etched with HCl : H_2O (1:1) solution at room temperature for 10 min., rinsed with DI water and dried with Argon. To obtain ohmic contact, In metal was thermally evaporated on the unpolished surface of n^{+}-GaAs, followed by a 10 min. anneal in vacuum at 360° C. An In/n-$In_{0.46}Ga_{0.54}P$ Schottky contacts of area 2.0×10^{-2} cm^{-2} was fabricated by thermal deposition of In onto the chemically etched surface of epitaxial n-$In_{0.46}Ga_{0.54}P$ layer.

The In/n-$In_{0.46}Ga_{0.54}P$ Schottky diode was mounted in a liquid-nitrogen cryostat and its temperature (T) was controlled and measured by a Stanton Red Croft 706 temperature programmer. The I-V characteristics were measured using a Keithley 480 picoammeter and a Keithley 617 electrometer. The 1 MHz C-V measurements were made using a Boonton 72B capacitance meter. The bias voltage was supplied by a EG&G PARC 175 voltage programmer.

EXPERIMENTAL DATA AND ANALYSIS

I-V/T Measurements

The forward I-V characteristics of the In/n-$In_{0.46}Ga_{0.54}P$ Schottky diode were measured over the temperature range 260-400 K. The $\{I/(1-\exp(-qV/kT))\}$ vs V plots at different temperatures shown in Fig. 1 are linear over the entire voltage and temperature range of our experiments showing that the series resistance effects were not important. The I-V data shown in Fig. 1 (discrete points) fitted very well to Eq. (3.13) in Ref. 11 (Fig.1, solid lines) with the values of the ideality factor (n) listed in Table I. The ideality factor varied between 1.26 and 1.78 over the temperature range 400-260 K (Table I).The non-ideal character in the forward I-V characteristics was introduced by the presence of an interface layer between the metal and semiconductor as well as by the participation of various current transport mechanisms.

Following the non-linear curve fitting procedure reported elsewhere[8,9], the experimental direct I-V values of the In/n-$In_{0.46}Ga_{0.54}P$ Schottky diode were fitted to the relation

$$I = I_1 + I_3 \tag{1}$$

where I_3 is the current due to the generation-recombination (GR) mechanism described by the Eq. (5) in Ref. 8, I_1 the current due to the thermionic emission (TE) mechanism given by

$$I_1 = I_o[\exp(qV/kT)-1] \tag{2}$$

the TE reverse saturation current, I_o is defined by

$$I_o = AA_{eff}T^2\exp[-q(\phi_{oo} - \Delta\phi_{bo})/kT] \tag{3}$$

with $\qquad\qquad A_{eff} = A^{**}\exp(\beta/k) \tag{4}$

and
$$\phi_{bo} = \phi_{oo} - \beta T \tag{5}$$

where A is the area of the Schottky contact, q the magnitude of the electron charge, A^{**} the modified effective Richardson constant which includes the effect of interface layer[12], ϕ_{bo} the zero bias barrier height, ϕ_{oo} the zero bias zero temperature barrier height, β the temperature coefficient of the zero bias barrier height and $\Delta\phi_{bo}$ the barrier lowering due to image force is defined by the Eq. (1.26a) in Ref. 11 and has a value of 54 mV for the In/n-In$_{0.46}$Ga$_{0.54}$P Schottky diode reported in this work.

The experimental values of direct current, I (Fig. 2, discrete points) fitted very well to the Eq. (1) (Fig. 2, solid curves) for temperatures in the range 260-400 K, which permitted the separation of the current contributions I_1 and I_3 of the TE and GR mechanisms, respectively, from the total current I. The currents I_1 and I_3 are also shown in Fig. 2 and the temperature dependence of the ratio I_3/I_1 is shown in Fig. 3.

The values of the TE reverse saturation current (I_o) obtained from the above fitting procedure are listed in Table I. The experimental values of $\{I_o/T^2\}$ vs 1000/T (Fig. 4, discrete points) gave a good fit to the Eq. (3) (Fig. 4, solid line) with

$$\phi_{oo} = (0.75\pm0.05) \text{ V}, \quad \text{and} \quad A_{eff} = (0.76\pm0.2) \text{ Acm}^{-2}K^{-2} \tag{6}$$

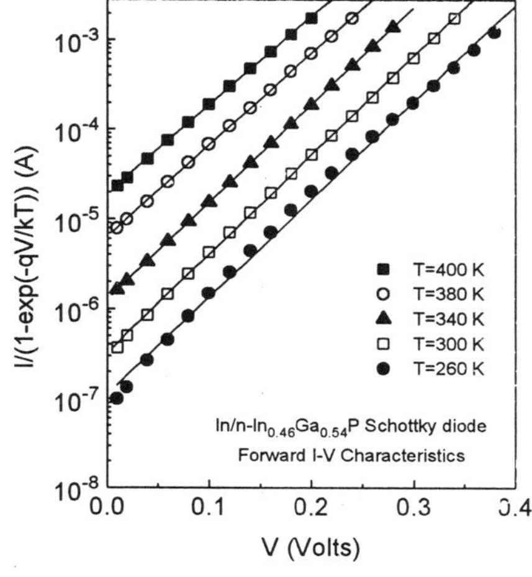

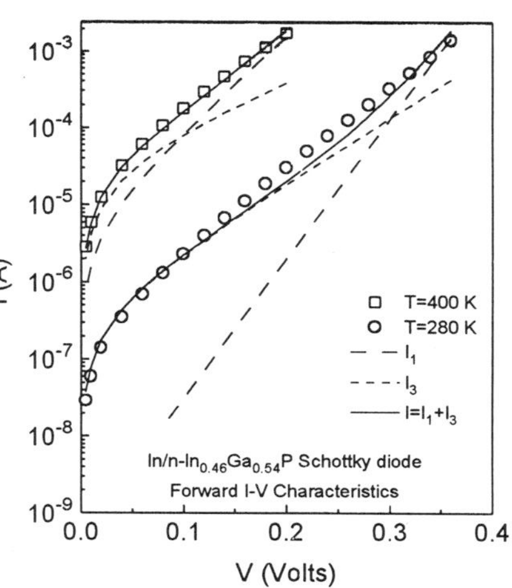

FIG. 1 Temperature variation of the forward I-V Characteristics for the In Schottky Contact to the epitaxial In/n-In$_{0.46}$Ga$_{0.54}$P layer.

FIG. 2 Typical fit of the experimental forward I-V data (discrete points) to Eq. (1) (Solid curve).

1 MHz C-V/T Measurements

The experimental reverse C^{-2} vs V characteristics at 1 MHz at different temperatures over the range 260-400 K shown in Fig.5. (discrete points) fitted very well to the MIS diode model[10]

$$C^{-2} = 2(V_o - V)/q\epsilon_s N_D A^2 \tag{7}$$

where V is the magnitude of the reverse voltage drop across the Schottky barrier, ϵ_s the permittivity of the semiconductor, N_D the density of the ionized donors and V_o the intercept of

TABLE I. Summary of the In/n-In$_{0.46}$Ga$_{0.54}$P Schottky diode parameter obtained from the I-V/T and C-V/T data.

T (K)	n	I_o (A)	V_o (V)	N_D (10^{17} cm^{-3})
400	1.26	4.71×10^{-6}	0.555	1.41
380	1.29	1.13×10^{-6}	0.549	1.37
360	1.33	3.14×10^{-7}	0.564	1.38
340	1.36	1.06×10^{-7}	0.586	1.39
320	1.43	2.07×10^{-8}	0.622	1.41
300	1.52	3.57×10^{-9}	0.667	1.45
280	1.63	4.94×10^{-10}	0.635	1.41
260	1.78	2.62×10^{-11}	0.728	1.52

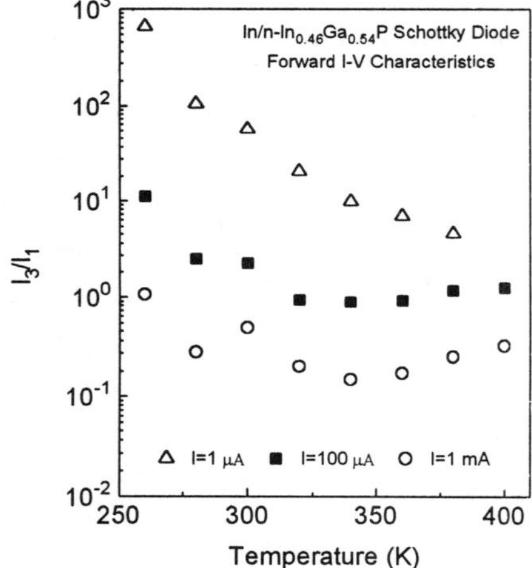

FIG. 3 I_3/I_1 vs T at different total current levels.

FIG. 4 Activation plot based on TE saturation reverse current (I_{te}).

the C^{-2} with voltage axis is related to the zero bias barrier height (ϕ_{bo}) through the relation[10]

$$\phi_{bo} = \{ V_o^{1/2} - (\delta/\epsilon_i)\, [q\epsilon_s\, N_D/2]^{1/2} \}^2 + (kT/q)\, \ln(N_c/N_D) + kT/q \tag{8}$$

where N_c is the effective density of states in the conduction band, δ/ϵ_i the thickness to permittivity ratio of the interface layer is obtained from the reverse I-V data has a value of 1.09×10^6 F^{-1}cm^2 in this work[13]. The values of V_o and N_D, obtained by fitting the C^{-2} vs V data (Fig. 5, discrete points) to the Eq. (7) are listed in Table I. The temperature dependence of the zero bias barrier height (ϕ_{bo}) values calculated from the Eq. (8) (Fig. 6, discrete points) was very well described by the Eq. (5) with

$$\phi_{oo} = (0.86 \pm 0.05)\ V, \quad \text{and} \quad \beta = (8.4 \pm 0.7) \times 10^{-4}\ V/K \tag{9}$$

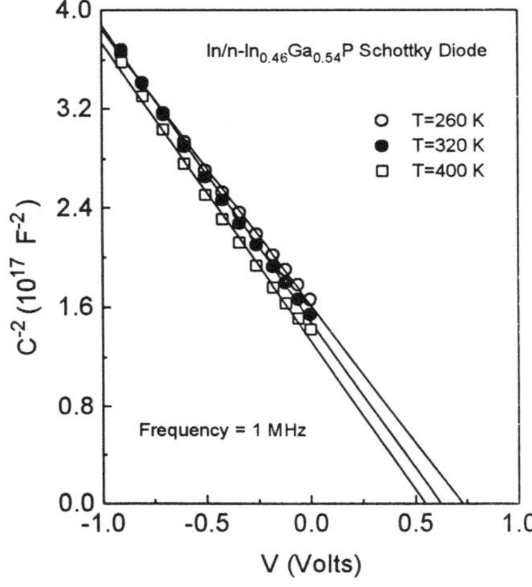

FIG. 5 High frequency C^{-2} vs V Characteristics of different temperaturs for the In/n-In$_{0.46}$Ga$_{0.54}$P epitaxial Schottky diode.

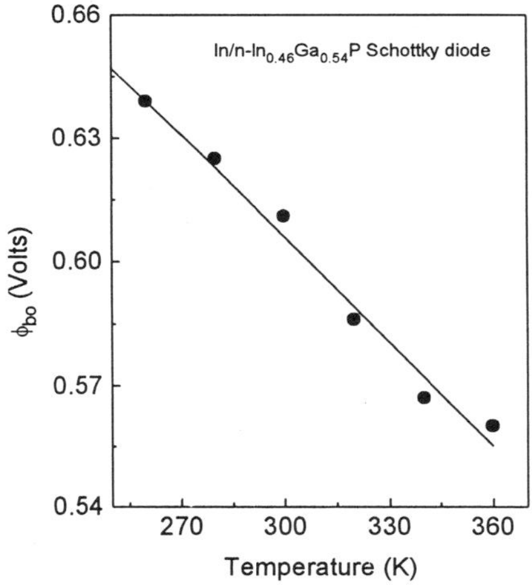

FIG. 6 Variation of ϕ_{bo} with temperature.

DISCUSSION

The forward I-V/T characteristics shown in Fig. 1 were non-ideal with the ideality factor in the range $1.78 > n > 1.26$ over the temperature range 260-400 K. This fact suggested that the pure TE mechanism was not responsible for the current transport across the In/n-In$_{0.46}$Ga$_{0.54}$P Schottky junction. The temperature dependence of n was not described by the thermionic field emission (TFE) model, therefore the pure TFE mechanism is also not responsible for the non-ideal behavior of the I-V/T characteristics. The experimental I-V/T data (Fig. 2, discrete points) was very well described by Eq. (2) (Fig. 2, solid curves) indicating that both the TE and GR mechanisms simultaneously contributed to the forward current transport in this device. From Fig. 2, it can be seen that at T=280 K, the GR contribution (I_3) to current transport dominates over the TE contribution (I_1) for $V \leq 300$ mV. However, at T=400 K, both the TE and GR mechanism were important. Two distinct regions are identified: (i) V<100 mV where $I_3 > I_1$ and (ii) $V \geq 100$ mV where the TE contribution to current transport dominates over the GE contribution.

The plot of I_3/I_1 vs T with total current as a parameter (Fig. 3) shows that in the temperature range 260-400 K, the relative importance of the GR and TE mechanisms is very nicely described by three distinct current levels: (1) $I \approx 1\ \mu A$, where the ratio I_3/I_1 increased exponentially from 5 to 700 with the temperature decrease from 400 K to 280 K; (2) $I \approx 100\ \mu A$, where the contributions of both the TE and the GR mechanisms remained nearly equal over the temperature range 280-400 K and (3)) $I \approx 1$ mA, where the ratio $I_3/I_1 \approx 0.25$, indicating that at this current level, the current transport across the In/n-In$_{0.46}$Ga$_{0.54}$P Schottky diode was dominated by the TE mechanism over the entire temperatures range of our experiments.

The values (0.86±0.05) V and (0.75±0.05) V for ϕ_{oo}, obtained from our C-V and I-V data, respectively, agreed within experimental errors. Using these values of ϕ_{oo} and the value of $\beta=(8.4\pm0.7)\times10^{-4}$ V/K in Eq. (5), the values of ϕ_{bo}(C-V)=(0.61±0.05) V and ϕ_{bo}(I-V)=(0.50±0.05) V for the room temperature barrier heights were calculated. Substituting the values A$_{eff}$ and β from Eqs. (6) and (9), respectively, in Eq.(4) a value of A**=(4.5±0.5)$\times10^{-5}$ Acm^{-2}K^{-2} was estimated. The small value of A** obtained from the I-V data suggested the presence of an interface layer between the metal and semiconductor. Our value of ϕ_{bo}=(0.50±0.05) V at 300 K obtained from the I-V/T data is much smaller than the value of ϕ_{bo}=0.79 V, reported by Chang et al.[5] for the In/n-Ga$_{0.51}$In$_{0.49}$P Schottky diode. Chang et al.[5] have not specified the value of A** used in the calculation of the barrier height, but suspect that they used the theoretical value of A** which is usually high and yields a higher value for the barrier height.

In summary, In/n-In$_{0.46}$Ga$_{0.54}$P Schottky junction formed by thermal deposition of In on chemically etched surface of n-In$_{0.46}$Ga$_{0.54}$P epitaxial layer, possessed a high rectification ratio ($\approx$500) and small reverse saturation current at room temperature. The non-ideality in the forward I-V/T characteristics was introduced by the simultaneous participation of the TE and GR mechanisms of current transport. The agreement between the values of zero bias zero temperature barrier height obtained from the analysis of the I-V/T and C-V/T data lends support to our method of our extracting the diode parameters from the non-ideal I-V/T characteristics.

ACKNOWLEDGMENT

This work was supported by the Consejo de Investigación de la Universidad de Oriente with Contract CI-05-1002-0776/96. We also want to thank Mr. K. C. Reinhardt for providing the epitaxial n-In$_{0.46}$Ga$_{0.54}$P samples.

REFERENCES

1. S. Loualiche, A. Ginudi, Le Corre, D. Lecroisnier, C. Vaudry, L. Henry amd C. Guillemot, IEEE Electron Device Letters **EDL-11**, 153 (1990).

2. Y. J. Chan, D. Pavlidis, M. Razeghi and F. Omnes, IEEE Trans. Electron Devices **ED-37**, 2141 (1990).

3. M. Ishikawa, Y. Ohba, H. Sugawara, M. Yamamoto and T. Nakanisi, Appl. Phys. Lett. **48**, 207 (1986).

4. T. Kobayashi, K. Taira, F. Nakamura and Kawai, J. Appl. Phys. **65**, 4898 (1989).

5. E. Y. Chang, Y.-L. Lai, K.-C. Lin, anad C.-Y. Chang, J. Appl. Phys. **74**, 5622 (1993).

6. S. D. Kwon, Ho Ki Kwon, B.-D Choe, H. Lim, and J. Y. Lee, J. Appl. Phys. **78**, 2482 (1995).

7. S. M. Sze, Physics of Semiconductor Devices, 2nd Edn. (Wiley, New York, 1981) Chap. 5

8. A. Singh and N. Marcano, Mater. Res. Soc. Symp. Proc. **378**, 829 (1995).

9. A. Singh and L. Velázques, in Surf. Sci., Vac. and their Appl. , Edited by I. Hernández-Calderón and R. Asomoza (AIP Conf. Proc. **378**, AIP Press, New York, 1996) p. 387.

10. K. Hattori, M. Yuito and T. Amakusa, Phys. Status Solidi A**73**, 157 (1982).

11 E. H Rhoderick and R. H. Williams, Metal-Semiconductor Contacts , 2nd Edn. Clarendon Press, Oxford (1988).

12 A. Singh, K. C. Reinhardt and W. A. Anderson, J. Appl. Phys.**68**, 3475 (1990).

13. N. Marcano and A. Singh (To be Published).

CONVERSION TUNNELING IN NON-IDEAL SCHOTTKY BARRIERS: VIRTUAL RESONANCE MANIFESTATION AND INTERFACE STATES INFLUENCE

D.A.ROMANOV, A.V.KALAMEITSEV, A.P.KOVCHAVTSEV, I.M.SUBBOTIN
Institute of Semiconductor Physics, Siberian Branch of the Russian Academy of Sciences,
630090 Novosibirsk, Russia, daroman@isp.nsc.ru

ABSTRACT

We have investigated experimentally and theoretically the elastic conversion tunneling of charge carriers in MOS structures Au on p^+-InAs with superthin (10-20 Å) oxide film, the structures used in infrared photodetectors. In these structures the Schottky barrier provides near-surface inversion layer. The tunnel current-voltage (I-V) curves obtained at helium temperatures demonstrate the negative differential resistance region (NDR). We develop semiclassical two-band transfer matrix approach to the conversion tunneling analysis in a multilayer structure and calculate on its base the I-V curves dependence on the structure parameters. The NDR occurs to be caused by the motion of the remote quantum level in the inversion layer. The calculated I-V characteristics agree with the experimental ones quite well. The very existence of NDR and the shape of I-V curves depends strongly on the nature of localized electron states at the semiconductor interface. The characteristics of these electron states are used in the calculations as fitting parameters. Therefore, we suggest a new method for the interface states diagnosis.

INTRODUCTION

The superthin-oxide-layer MOS structures with the Schottky barrier based on InAs are of wide use in various electro-optical applications. These structures demonstrate quite interesting electrophysical properties, which seem to be caused by the Fermi level pinning at the surface of InAs near the conduction band edge. This pinning creates an n-type layer near the surface, the inversion layer for p-type InAs. As early as in [1] (and, later, in [2]) the region of negative differential resistance (NDR) had been found in the current-voltage (I-V) curve of such structures. This NDR behavior cannot be connected with the resonant tunneling in effective two-barrier structure [3], because the quantized electron level in the inversion layer occurs to lie *higher* than the valence band top in the bulk. Neither competitive dependencies of the insulator and effective interband barriers on the applied voltage [2] nor thermostimulated tunneling via conduction band or phonon-assisted processes taking into consideration [4] occurred sufficient to pronounce this NDR.

In this paper we suggest new physical mechanism (we called it *virtual resonance*) leading to NDR behaviour. To describe the effect quantitatively, we develop a semiclassical transfer matrix approach and find analytical expressions for the structure transmission coefficients. Making use of these coefficients, we calculate the I-V curves for the structure parameters of our experiment. The calculated curves agree quite well with the experimental ones.

This new mechanism is highly sensitive to the localized electron states at the interface and may serve as a test for these states existence and nature.

Mat. Res. Soc. Symp. Proc. Vol. 448 © 1997 Materials Research Society

EXPERIMENT

As substrates for the MOS structure fabrication we used p-InAs plates of <111> crystallographic orientation and $1\text{-}5\cdot10^{17}$ cm^{-3} dopant concentration. Three groups of samples have been fabricated. The substrates of the first group were etched in KOH and then oxidized in a flow of dry oxygen under the temperature 200 °C during 2 hours. The substrates of the second group were etched in lactic acid ($C_3H_6O_3$, 80% H_2O, 20%) and then washed in deionized water. The substrates of the third group were annealed after etching up the 2×2 superstructure appearing on the surface.

After this chemical preposition the dots (gold for the first and the second group and platinum for the third one) of 100 μ diameter and 0,2 μ thickness were evaporated on the crystal surface through a mask in vacuum 10^{-6} Torr. For ohmic contact the solid In cladding of 0,5 μ width was deposited on the back side of the substrates.

The ellipsometry and tunneling spectroscopy data show the thickness of superthin insulator layer on the substrate surface in the first and the second group to be about 10-20 Å[5]. For given orientation both of the used etchers are polishing ones and provide quite good quality of the layer, which consists of the mixture of In and As oxides (with addition of the lactic acid remains in the second case).

Thus, the samples of the first two groups provided the MOS structures with superthin oxide layer, the structures which have so-called non-ideal Schottky barrier with sublayer. The main difference between the structures of the first and the second groups allegedly lies in the state of the semiconductor–insulator interface, because the lactic acid mediated etching procedure is supposed to create the interface electron states much more effectively.

The third group of samples had the Schottky-barrier structures without sublayer. We use this group as a control one.

The current-voltage curves have being measured at temperature 4.2 K (to avoid thermostimulated effects) and the voltage scanning rate 4 mV/s. The hysteresis phenomena for forward and backward I-V curves were absent, which indicates the absence of the charge trapping in the insulator. The obtained I-V characteristics for the first, the second, and the third groups of samples are shown in Figs. 2 to 4. They clearly demonstrate the NDR region for the first and the second groups and the absence of this for the third group.

THEORY

The key point for the I-V curve calculation is the transmission coefficient for the electron passing from metal to the valence band of semiconductor. Our approach is based on semiclassical analog of the transfer matrix method.

The band diagram is shown in Fig.1. The dopant concentration $\sim 2\cdot10^{17}$ cm^{-3} provides the depletion layer width $\sim 6.2\cdot10^{-6}$ cm. The smoothness of the depletion potential and large width of the effective interband barrier allow us to use the two-band Kane model, i.e., to neglect the heavy hole contribution [6]. The same smoothness of the potential $V(z)$ justifies the semiclassical approximation in finding the effective two-component wave function Ψ. Then we find the (2×2) matrices connecting the coefficients of semiclassical fundamental solutions on the sides of each of the four turning points between the neighboring classically allowed and forbidden regions, at which points the solutions diverge. We obtain these matrices by tracing the wave function transformation while rounding the turning point in the plane of complex variable. As at the point $z = 0$ the electron energy is close to the conduction band bottom, so $|\Psi_2(0)| \ll |\Psi_1(0)|$. Thus we find the connection matrices for both points of sharp boundaries ($z = 0$ and $z = -d$) from the same Bastard boundary conditions. Then we obtain the total transfer matrix $\hat{T}$ connecting the coefficients of the electron plane waves in the metal and in the valence band of semiconductor as the product of partial (diagonal) matrices and the above mentioned connecting matrices taken in sequence of positive direction in Fig.1.

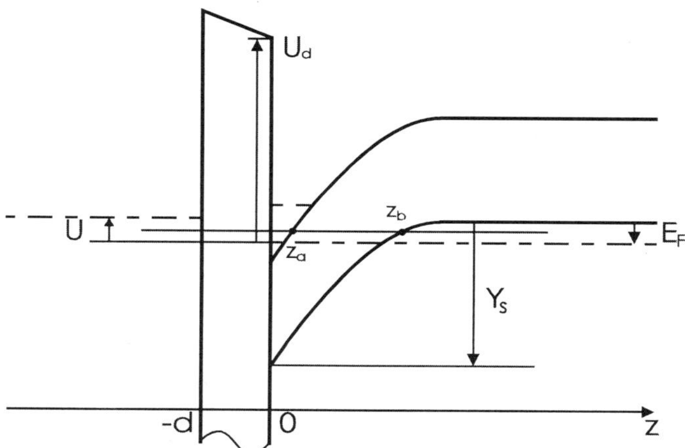

Fig. 1 The band diagram of the MOS structure under applied voltage; d U_d – the width of insulator layer and the height of its energy barrier; Y_s – the band bend.

The transmission coefficient is found as $D = (m_m/m_s) \cdot |t_{11}|^{-2}$:

$$
\begin{aligned}
D^{-1} &= 2|\alpha\beta|^2(\cosh\left(I(a,b)\right)\cosh\left(I(-d,0)\right) + \sinh\left(I(a,b)\right)\sin\left(I(0,a)\right)\cos\varphi_\alpha \\
&+ \sinh\left(I(a,b)\right)(\cosh\left(I(-d,0)\right)\sin\left(I(0,a)\right)\cos\varphi_\beta + \sinh\left(I(-d,0)\right)\cos I(0,a)\sin\varphi_\beta) \\
&+ \cosh\left(I(a,b)\right)\cos\varphi_\alpha\cos\varphi_\beta - \sin\varphi_\alpha\sin\varphi_\beta)
\end{aligned}
$$

where

$$
I(z_l, z_r) = \frac{1}{\hbar}\int_{z_l}^{z_r} p_z(z)dz,
$$

$$
\alpha = (1/2)\left(\sqrt{p_m/p_d} - i\sqrt{p_d m_m^2/p_m m_d^2}\right), \qquad \beta = (1/2)\left(\sqrt{p_d/p_s} + i\sqrt{p_s m_d^2/p_d m_m^2}\right)
$$

Here $p_z(z)$ is the classical momentum of the carrier (in the metal and the insulator it is determined by parabolic dispersion law, in the semiconductor $p_z^2 = (2m_s/E_g)((E-V(z))^2 - E_g^2) - p_\parallel^2$); p_m, p_d, p_s are the electron wave vectors in metal, insulator and semiconductor at the interfaces, m_m, m_d, m_s are the respective effective masses.

The elastic tunnel current is given [7] as

$$
j(U) = \frac{2\pi|e|m}{h^3}\int D(E_\perp, E_\parallel)[f(E+U) - f(E)]dE_\perp dE_\parallel
$$

The convenient integration variables are two components of the electron energy in metal $E_\perp$ and $E_\parallel$ ($E = E_\perp + E_\parallel$). The limits of the integration are determined by interplay of the Fermi level positions in metal and semiconductor and the forbidden gap edge.

NUMERICAL CALCULATIONS

We calculate the transmission coefficient and the tunnel current versus applied voltage for the structure parameters of all the tree groups of samples mentioned above.

As the inversion layer is empty of carriers, the depletion potential is only caused by the uniform distribution of charged acceptors and occurs to be parabolic. The voltage falls on the insulator layer and on the depletion layer have been found from the Poisson equation and the total electro-neutrality condition taking into account the charge on the surface states. At actual doping of $\sim 10^{17-18}$ cm^{-3} the impurity level broads to the impurity band which merges with the valence band [8]. We have taken this effect into account by shifting the valence band edge on the impurity band width, i.e. on ~ 50 meV. The transmission coefficient dependence on $E_\parallel$, which causes an effective increasing of the forbidden gap width, occurred essential for the I-V curve shape.

The results of numerical calculations and their comparison with the experimental data are shown in Figs. 2–4.

In the cases of the first two groups at the initial region of the voltage axis the current increases by two reasons. The first, trivial: the number of available final states for tunneling carriers increases along with the energy distance between the Fermi levels in the metal and the semiconductor. The second, specific: the voltage increasing causes the surface band bend flattening and makes the inversion layer wider. As a result, the quantum level gets closer to the Fermi energy (in metal), that causes sufficient enhancement of the tunneling via the level tail. When the metal Fermi energy approaches the valence band edge, the number of tunneling carriers ceases to grow, while the quantum level continues to withdraw into the depth of forbidden gap. This withdrawal from the energy region, where the tunneling takes place, leads to the tunneling suppression. Hence, the I-V curve falls down. The next growth of the tunnel current begins at the voltage corresponding to alignment of the conduction band bottom at the insulator–semiconductor interface and the valence band top in the bulk of semiconductor. Now the current increases simply due to decreasing of the effective barrier of the forbidden gap.

So, despite the quantum level in the inversion layer lies out of the actual energy region, its motion affects strongly on the tunneling process and provides the broad peak on I-V curve.

DISCUSSION

The fitting parameters, which give the best agreement between the theoretical and experimental curves, are shown in Figures. Note that these parameters are different for different Figures. From the physical standpoint they correspond to sufficiently different assumptions about the character of the surface states filling at the semiconductor–insulator interface. In Fig. 2 all the surface states are supposed to be empty, which means the surface charge remains constant. Fig. 3 corresponds to very high density of the surface states, by virtue of which their filling is strictly determined by the metal Fermi level ($D_{ins} \ll D_{inv}$). In

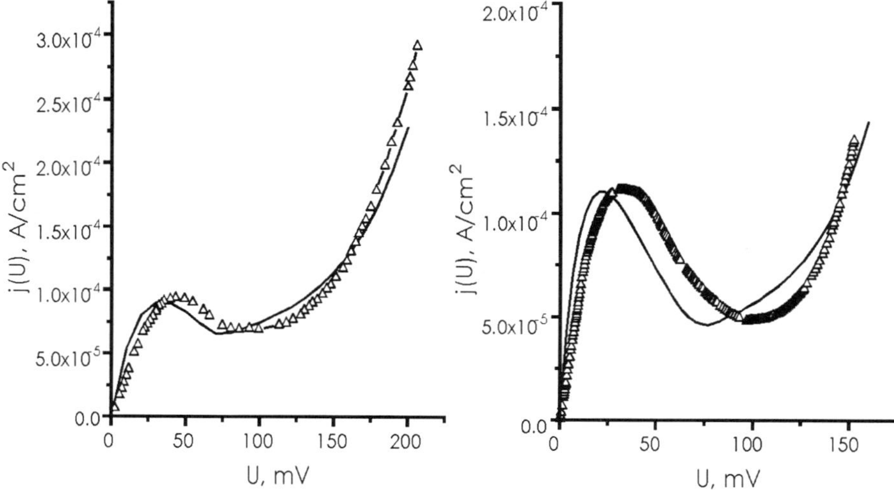

Fig.2 The I-V curves: triangles - the experiment (the first group of samples)
solid line - the theory (the constant embedded charge model d=11.2 Å, Y_s=0.49 eV, dielectric constant ε=10)

Fig.3 The I-V curves: triangles - the experiment (the second group of samples)
solid line - the theory (the pinned Fermy level model d=11.6 Å, Y_s=0.51 eV)

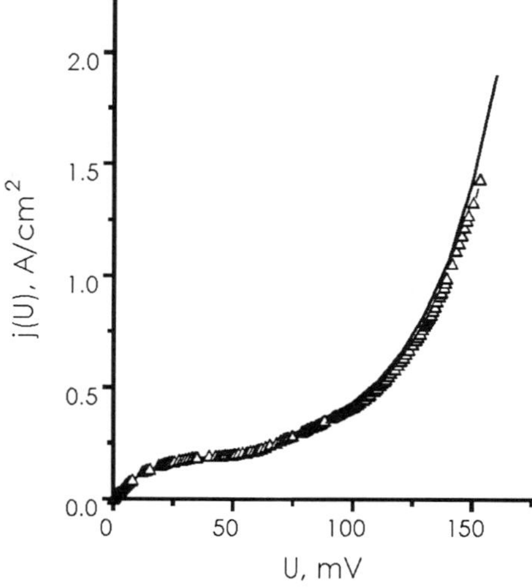

Fig.4 The I-V curves: triangles - the experiment (the third group of samples)
solid line - the theory (the absence of the insulator layer model Y_s=0.49 eV)

this case all the voltage applied to the structure drops at the space charge region in the semiconductor. Fig. 4 corresponds to the absence of the insulator layer. In this case the Fermi level is pinned at the surface of InAs and all the voltage drops on the depletion layer. Though there is no NDR in this case, the I-V curves follow the experimental ones, which justify our method of calculation.

CONCLUSIONS

We have made thorough theoretical and experimental investigation of the negative differential resistance effects in MOS structures based on p^+-InAs. Current-voltage characteristics were obtained at helium temperatures for different procedures of the insulator sublayer fabrication. Making use of the semiclassical analog of the transfer matrix method, we have calculated theoretical dependencies of the tunnel current on the structure parameters. The falling region of I-V curve is caused by the peculiarities of the tunneling via the tail of the quantum level in the inversion layer, which level lies apart the actual energy region. The theoretical I-V curves agree with the experimental ones quite well, provided different character of the electron surface states in dependence of the structure fabrication. Therefore, the I-V curve shape in the NDR region provides a test of the surface state nature.

ACKNOWLEDGMENTS

This work was supported in part by the Russian Universities Foundation (Grant No 95-0-7.2-151) and by the Russian Foundation for Basic Research (Grant No 96-02-19028)

[1] Mead C.A., Spitzer W.G., Phis. Rev. Letters **10**, *11*, 471 (1963)

[2] Esina N.P., Zotova N.V., Karandashev S.A., Philaretova G.M., Fiz. Techn. Polupr. **17**, 991 (1983) [Sov. Phys. Solid State]

[3] Solner T.C.L.G., Goodhue W.D., Tannenwald P.E. et al., Appl.Phys.Lett., **43**, p.588 (1983)

[4] Roy D.K., *Tunneling and Negative Resistance Phenomena in Semiconductors*, Int. Series in the Science of the Solid State, Pamplin B.R., ed., v.11 (1977)

[5] Kovchavtsev A.P., Kurishev G.L., Postnikov K.O., phys. stat. sol. **97**, 421 (1986)

[6] Kane E.O., J. Phys. Chem. Solids **12**, 181 (1959)

[7] *Tunneling phenomena in solids*, E.Burstein and S.Lundqvist, eds., Plenum Press, New York (1969).

[8] Baltensperger W., Phil.Mag. **44**, 1355 (1953)

Part VII

Optical Characterization: Real-Time and *Ex Situ*

MULTILEVEL APPROACHES TOWARD MONITORING AND CONTROL OF SEMICONDUCTOR EPITAXY

D. E. ASPNES*#†, N. DIETZ,*# U. ROSSOW,*§ and K. J. BACHMANN#
*Department of Physics, North Carolina State University, Raleigh, NC 27695-8202
#Department of Materials Science and Engineering, North Carolina State University, Raleigh, NC 27695-7907
†Corresponding author: aspnes@unity.ncsu.edu
§Now at Institute of Physics, Ilmenau, Germany

ABSTRACT

Various optical techniques have been developed over the last few years to allow real-time analysis of regions of importance for semiconductor epitaxy, in particular the unreacted and reacted parts of the surface reaction layer (SRL) and the near-surface region of the sample. When coupled with emerging microscopic methods of calculating optical properties, these approaches will allow several levels of control beyond that which has been currently demonstrated.

INTRODUCTION

Electronics and optoelectronics technologies are based on artificial materials and structures, which are becoming increasingly complex as new levels of performance are achieved [1,2]. This trend will clearly continue, with devices becoming smaller and more complex, tolerances becoming more stringent, and materials being tailored according to function rather than compatibility with a particular growth process. Present requirements also include lateral patterning, selective-area deposition, and scale-up to industrial production levels. This has generated a second trend away from physical deposition techniques such as molecular beam epitaxy (MBE) toward chemical-beam methods such as organometallic chemical vapor deposition (OMCVD) and chemical beam epitaxy (CBE), where the extra dimension of chemistry can be used for greater capabilities and flexibility [3,4]. At the same time, yields must be maintained at acceptably high levels.

These trends have created a strong incentive to develop a better understanding of growth processes and better methods of monitoring growth, ideally leading to sample-driven closed-loop feedback control of the growth process itself. Much present effort is being directed toward accurately measuring process parameters and accurately modeling growth, mainly because numerous probes are available to provide a relatively detailed assessment of the ambient. However, this strategy is clearly limited in its capability to deal with complex nonlinear processes, as for example OMCVD where the surface plays an integral role in decomposing precursor species and small changes of ambient compositions can affect growth substantially [3,5].

The device that best represents current challenges is the vertical-cavity surface-emitting laser (VCSEL) [6]. VCSELs may contain hundreds of layers of varying compositions and thicknesses, which are grouped into the distributed Bragg reflectors (DBRs) at the top and bottom and an active region in the middle. At deposition rates of 1 μm/h about 6 h are required to grow a VCSEL device structure. The probability that deposition rates determined at the beginning of growth will be the same at those at the end is relatively small, especially for MBE. For this reason considerable effort has been invested in adapting and extending reflectance monitoring techniques [7] that were developed over the last 20 years by the optical thin films industry to

Mat. Res. Soc. Symp. Proc. Vol. 448 © 1997 Materials Research Society

the growth of VCSEL substructures, specifically DBRs [8-10]. However, while reflectometry can deal adequately with layers that are thick enough to generate interference patterns, it cannot cope with the 20 Å thick quantum wells that comprise the active region of the VCSEL, or for that matter the thin layers characteristic of communications lasers in general. Here, thicknesses must be maintained to within a monolayer (ML) and compositions of $In_xGa_{1-x}As_{1-y}P_y$ maintained to within several percent, or the wavelength will be outside allowable tolerances [2]. These capabilities are completely beyond reflectometric techniques, and phase-sensitive approaches such as spectroellipsometry (SE) must be used.

Growth is a surface process, so it might appear that the numerous surface-diagnostic techniques that have been developed over the years could be used to advantage in semiconductor epitaxy. However, $In_xGa_{1-x}As_{1-y}P_y$ growth necessarily involves As and P, volatile constituents that are incompatible with surface-analytic equipment. In addition growth is often done in near-atmospheric pressure environments where standard surface diagnostics cannot be used. The only generally usable conventional technique is quadrupole mass spectrometry (QMS), which with suitable differential pumping can be used in high pressure environments. However, QMS measures only reaction byproducts and hence does not address growth surfaces directly. Reflection high energy electron diffraction (RHEED) can be used in ultrahigh vacuum (UHV), but RHEED provides information only about long-range order, which is of limited utility in establishing details of the growth chemistry. Until the recent development of the surface-optical approaches discussed below [11-13] and the recent application of grazing-incidence X-ray scattering (GIXS) [14,15] to OMCVD growth, OMCVD could only be investigated by gas-phase analysis of downstream reaction products.

This lack of suitable diagnostic tools has stimulated much of the activity behind the development of optical probes over the last 10 years as discussed in recent reviews [16-19]. Optical probes are nondestructive, noninvasive, and can be used in any transparent ambient. In addition, their diagnostic capabilities have been well developed in applications to thin films [7,20] However, optical measurements have not contributed significantly to surface analysis because photons interact only weakly with materials, and the interpretation of optical data is not straightforward. The sensitivity issue can be appreciated by noting that penetration depths of light are rarely less than 100 Å, whereas the surface is about 1 Å thick. Thus the information content in a reflected beam will be divided roughly as 99% bulk and 1% surface. However, various optical techniques have recently been invented or refined to improve our capabilities for isolating the 1% surface contribution. Relatively recently developed techniques include reflectance-difference (-anisotropy) spectroscopy (RDS/RAS) [11-13,21,22], surface photoabsorption (SPA) [23,24], and p-polarized reflectance spectroscopy (PRS) [25-27]. Techniques currently undergoing rapid refinement are second-harmonic generation (SHG) [18,28] and light/laser light scattering (LS/LLS) [29-31]. As will be discussed below, all rely on symmetry in one form or another to suppress the bulk response in favor of the surface component.

The interpretational issue consists of two parts. First, the spectral range of quartz-optics systems, about 1.5 to 6.0 eV, is severely limited relative to that attainable with electron spectroscopies. This translates into a lack of specificity regarding both materials and processes. Second, without a suitable atlas of spectra of surface species it has been necessary to rely on theoretical calculations to interpret structure in surface-optical spectra. This is an extremely difficult task. Recent experiments have demonstrated that many-body effects such as localization and propagation of the excited electron are involved as well as similar effects such as screening [32]. However, new approaches where the system is treated as a coherent superposition of scattering centers offer some prospect that this problem will be solved [33,34].

The emphasis on the use of surface-analysis techniques with semiconductor epitaxy stems not so much from practical considerations as from historical reasons, specifically from the fact that RHEED has been and still remains the primary diagnostic tool for MBE. Unfortunately, this emphasis misses an essential point. Growth chemistry is established in large measure by the choice of growth technique, so the main practical considerations deal with layer thicknesses and compositions. These *bulk* properties are more efficiently and accurately determined by bulk probes such as reflectometry and ellipsometry. Although reflectometry and ellipsometry have been used for over 100 years to determine average thicknesses and compositions of deposited films [35], in compositional control the interest lies in determining *fluctuations* from a target value, ideally over a region whose thickness is vanishingly small. This *near-surface* region has recently become accessible through the development of the virtual-interface (V-I) approach, where the near-surface composition is extracted by a suitable analysis of kinetic ellipsometric data. This has led to the first (and so far only) demonstration of sample-driven closed-loop feedback control of epitaxy [36].

GROWTH: WHAT WE KNOW, WHAT WE WOULD LIKE TO KNOW

For discussion purposes we can consider the main regions relevant for epitaxial growth to be the ambient, the surface reaction layer (SRL), and the sample, as shown in Fig. 1. The SRL can be subdivided into two parts: the (mainly) physisorbed species that are weakly bound to the substrate and not in registry with it, and the (mainly) chemisorbed species that are strongly bound to the substrate and acquire the symmetry of its outermost atomic layer. This distinction is important because different optical techniques have different sensitivities to these two regions. The sample can also be subdivided into two regions: a near-surface part that contains the most recently deposited material, and the underlying bulk material. This distinction is made because we must distinguish compositional fluctuations, which must be detected and corrected with as little delay as possible, from the average composition of the film. In practice the thickness of the near-surface region is determined by the signal-to-noise capabilities of the instrumentation. For thickness control the important quantity is the bulk.

The various regions are related as follows. The ambient is established by the process parameters and consists of the carrier gas (if any) and the nutrient species providing the constituent elements to the growth surface. Some growth chemistry, such as thermal cracking of precursor species, may occur in the ambient. The nature and composition of the weakly and strongly bound parts of the SRL depend on competition between that region and the two adjacent regions and on the reactions taking place within the regions. Thus the nature and composition of the weakly bound part of the SRL depend on its reactions with the ambient and the strongly bound part, while that of the strongly bound part depends on its reactions with the weakly bound part and the near-surface region of the sample.

It is also useful to summarize the information needed to describe growth. One possible list of parameters is provided in Table I. We identify 3 categories: primary, secondary, and tertiary. The primary category contains only bulk parameters: layer thicknesses, compositions, and uniformities, because if it is not possible to meet specifications at this level the remaining parameters do not matter. The secondary category contains parameters that are surface-determined: efficiency of dopant incorporation, possibility of atomic ordering of nominally random alloys, interface widths, etc. Tertiary parameters are those related to the process itself, including for example type, pressure, and flow rate of the carrier gas, the partial pressures of active species, sample temperature, etc.

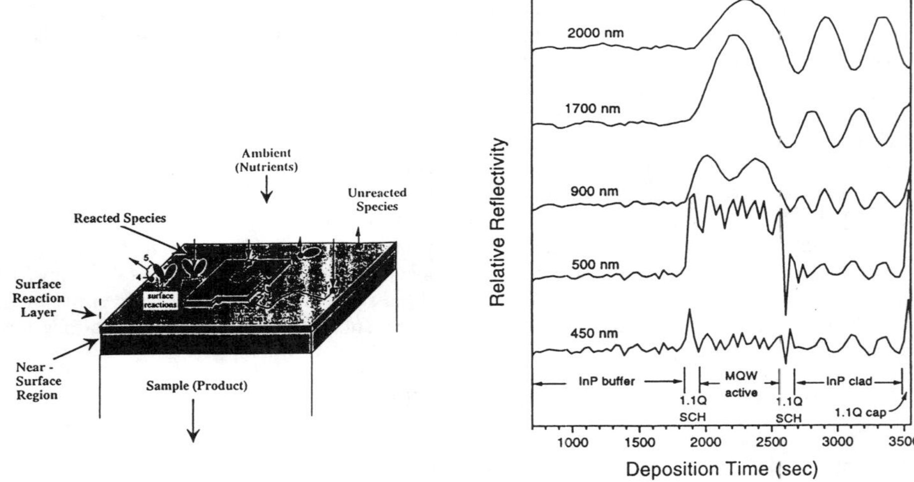

Fig. 1. Schematic diagram of the various regions of importance to epitaxial growth.

Fig. 2. Reflectance spectra obtained during the growth of a 1.3 μm InGaAsP multiquantum well laser device structure (after ref. 10).

Table I. Parameters involved in crystal growth.

Parameter:	Category:	Technique:
Primary:		
Thickness	Bulk	SR, SE
Composition	Near-surface	KE
Uniformity	Proximal	Imaging techniques
Secondary:		
Dopant incorp.	Surface-	RHEED,
Atomic ordering	determined	LS/LLS, RDS,
Interface widths		SPA, PRS, SHG
Tertiary:		
Sample temp.	Process-	TC, pyrometry
Ambient pressure,	related	ultrasonics, LIF,
temperature, species,		IRAS, CARS,
fluence...		UVAS, QMS, etc.

The tertiary parameters have received the most attention because of the expectation that sufficiently accurate measurements of process parameters together with sufficiently accurate modeling of the growth process will allow samples to be grown to the necessary degree of accuracy. However, there are several problems with this approach. First, the weakest link to the product is that formed by the ambient. Second, growth is inherently a nonlinear process and therefore difficult to model to needed levels of accuracy, especially for chemical beam methods where the surface plays an important part in the decomposition of precursor species [3,4]. Third, small changes in ambient conditions, for instance the addition of a dopant species, can modify active surface sites and thus change growth chemistry significantly [4]. Fourth, process parameters must be measured to unrealistically high accuracies, as evidenced by the fact that most growth stations are brought on line by producing test samples and modifying conditions empirically. Finally, every growth station is distinct with its own peculiar characteristics, which makes model portability problematic.

Given the above, it would appear that a far better strategy would be to determine as much as possible about the intervening stages and to use all available information, ambient through near-surface region, as input to the control process. This is the objective of our current program. By accessing the different regions directly it is not only possible to gain a better understanding of the growth process but also to detect and correct difficulties as they occur. Also, much of the complexity of modeling is bypassed, and the burden on process parameters is reduced to that of ensuring a reasonably stable growth environment. The near-surface region needs to be considered specifically because it provides the first opportunity to determine what is actually being produced.

The ability to do this is predicated on our ability to obtain information about these regions. Most of the listed optical techniques are discussed in connection with Table II and some examples of their use are given in a following section. The process-directed probes include thermocouple measurements (TC), laser induced fluorescence (LIF), infrared absorption spectroscopy (IRAS), coherent anti-Stokes Raman scattering (CARS), and ultraviolet absorption spectroscopy (UVAS).

OPTICAL APPROACHES

At present a number of optical probes are available for addressing the bulk, near-surface, and surface regions of Fig. 1, as summarized in Table II. Bulk probes include spectroreflectometry (SR), spectroellipsometry (SE), and photoreflectance (PR). At present photoreflectance is used mainly in off-line analysis of compositions, dopant levels, damage, etc. [37,38]. Near-surface analysis can be performed by kinetic ellipsometry (KE) in connection with virtual-interface (V-I) theory [39].

Regarding surface analysis, we distinguish between *surface-oriented* and *surface-specific* approaches. The former are sensitive to surface effects but cannot return unambiguous information with samples under steady-state conditions. Although SR and SE are bulk probes, surface conditions affect these data at the nuisance (few-percent) level, i.e., enough to affect accuracy adversely but not enough to be useful for surface analysis. The surface-specific techniques can return information about surfaces under steady-state conditions. Of these, IRAS is useful mainly at wavelengths where the sample is transparent. Grazing-incidence X-ray spectroscopy returns information about long-range order but at present it can only be performed with synchrotron sources.

Of the surface-specific approaches, LS/LLS is the simplest to implement since it can be done by monitoring nonspecularly scattered light from a high intensity source such as a laser. LS/LLS provides useful information ranging from growth kinetics through relaxation of defects. SHG isolate the contribution of the lower-symmetry surface by capitalizing on polarization selection rules that are more detailed than those used in linear optics [18,28]. However, SHG equipment is too bulky to be compatible with the space limitations near growth chambers. As compact, high-power, short-pulse solid-state lasers become available SHG will become an important

Table II. Optical probes for surfaces and interfaces.

Bulk-oriented:
SR	Spectroreflectometry
SE	Spectroellipsometry
PR	Photoreflectance

Near-surface oriented:
| KE | Kinetic ellipsometry with V-I theory |

Surface-oriented:
SDR	Surface differential reflectance
SPA	Surface photoabsorption
PRS	p-polarized reflectance spectroscopy
SR	(in kinetic mode)
SE	(in kinetic mode)

Surface-specific:
RDS/RAS	Reflectance-difference/anisotropy spectroscopy
SHG	Second-harmonic generation
LS/LLS	Light/laser light scattering
IRAS	Infrared absorption spectroscopy
GIXS	Grazing-incidence X-ray scattering

diagnostic tool for surfaces and interfaces. The present situation has been well summarized by McGilp [18].

SPA and PRS are both variations of the original surface differential reflection (SDR) spectroscopy work of the Rome group in the late 1970s [40]. SDR is done at normal incidence, while SPA and PRS are performed with p-polarized light at the pseudo-Brewster or Brewster angles, respectively. Suppression of the bulk reflectance enhances the relative surface contribution, and also the signal-to-noise ratio relative to SDR. For PRS, which is applied in the transparent or weak absorbent wavelength range, the surface related contribution is superimposed on interference fringes which allow one to obtain additional information about the deposition rate. However, since SDR and SPA data are obtained by modifying the state of the surface, it is not always possible to establish with certainty which (if either) surface termination gave rise to the observed surface component of the signal.

RDS/RAS avoids this "which-surface" ambiguity by taking advantage of the lower symmetry of the surface relative to the bulk of cubic semiconductors. Here, the optical anisotropy of the sample is determined at normal or near-normal incidence. Since cubic materials are optically isotropic, the main contribution to RD spectra is due to the surface. Since SPA, PRS, and RDS/RAS all involve surfaces it should not be surprising that these data are all connected, as has been shown both analytically and experimentally [41,42].

EXAMPLES

We illustrate the performance capabilities of some of the above approaches by several examples. Lum et al. recently extended a multiwavelength SR configuration introduced by Killeen and co-workers [8,9] to monitor the growth of a 1.3 μm InGaAsP laser device structure [10]. The reflectometer features optical fibers to transmit and collect the reflected light and a Si/PbS dual detector with a combined wavelength range of 400 to 2500 nm. Data of at several wavelengths are given in Fig. 2. The device structure consisted of an InP substrate, a 1.1 μm composition InGaAsP confinement layer, the multiquantum well stack with a repeat period of 17 nm, a top 1.1 μm confinement layer, an InP cladding layer, and a 1.1 μm cap. The data obtained at 450 and 500 nm wavelengths permit the individual MQW structures to be resolved, while those at longer wavelengths allow the thicknesses of the various regions to be determined. Interference oscillations in the envelope also provide information about layer thicknesses.

In Fig. 3 we show KE data obtained during sample-driven closed-loop feedback control of the growth of a 200 Å wide $Al_xGa_{1-x}As$ parabolic quantum well [36]. The structure was grown by beam epitaxy with triisobutyl (TIBAl), trimethylgallium (TMG), and cracked arsine (AsH_3) sources. The solid line in the top part of the figure shows the target composition as a function of position in the structure. The points represent the composition determined from the ellipsometrically measured pseudodielectric function $\langle\varepsilon\rangle$, where the time dependence of $\langle\varepsilon\rangle$ was analyzed by V-I theory for the dielectric function of the outermost region, which in turn was related to the composition x by a previously determined empirical expression. The middle is the difference between target and experimental values, and shows that composition was held to target values to within about 3%. The bottom shows the control voltage fed to the TIBAl mass flow controller while the experiment was in progress, the flow rate of which was used to establish the composition. The control voltage exhibits a noticeable asymmetry, probably due to the gettering of TIBAl as the flow increased. These results were obtained with a deposition rate was 0.95 Å/s, a sampling (including analysis) interval of 650 ms, and 5 points of averaging, meaning that the observed precision was obtained by the analysis of about 3.1 Å of material, approximately 1 ML.

The third example deals with the use of RDS to determine ML thicknesses during the OMCVD growth of a 30-period InGaAs/GaAs superlattice consisting of 5 and 10 ML thick InGaAS and GaAs layers, respectively [43]. Figure 4 shows the complete RD data set obtained at 2.6 eV, which corresponds to the peak of the As dimer structure of the (001) GaAs surface. The envelope exhibits interference oscillations, which allows the overall thickness to be determined. The 26th period is expanded at the bottom, and shows the RD oscillations with a thickness period of 1 ML.

The fourth example deals with temperature effects. Figure 5 shows SPA data obtained during exposure of an AsH_3-saturated (001) GaAs surface to trimethylgallium (TMG) at various substrate temperatures [24]. The objective is to determine conditions of atomic layer epitaxy (ALE), where growth becomes self-limiting at 1 ML such that extremely uniform layers can be deposited. Figure 5 shows that at 450 °C the growth rate decreases at 1 ML but does not stop. However, at 470 °C an ALE region is clearly defined. At 490 °C growth once again proceeds without interruption at the ML level. These data give direct information about *surface* temperatures that would be very difficult to obtain in any other way.

Finally, Fig. 6 illustrates diagnostic possibilities with the simultaneous use of various probes. Here, data are obtained for pulsed exposure to reactants during interruption of otherwise continuous heteroepitaxial growth of GaP at 350 °C [27]. The pulse sequence of tertiarybutylphosphine (TBP) and triethylgallium (TEG) is shown at the bottom. The PR and LLS data were obtained at 632.8 nm using the same HeNe laser source. The RD data were obtained at 2.6 eV. The relatively constant LLS signal indicates no substantial changes of surface morphology, such as the accumulation of Ga droplets, with the various exposures. Both the PR and RD signals show considerable changes that are not time-correlated with each other, and which continue to evolve even after exposure to the reactant has been terminated.

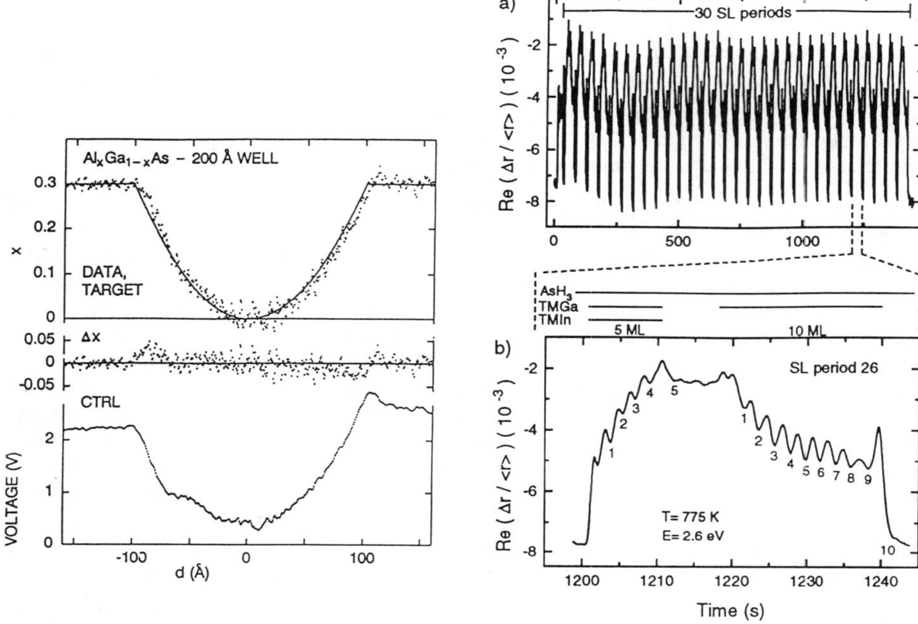

Fig. 3. Compositional data for a 200 Å wide parabolic quantum well grown by sample-driven closed-loop feedback control of epitaxy. Top: data and target values; middle: difference; bottom: control voltage. Growth and measurement parameters are given in the text (after ref. 36).

Fig. 4. Top: RD response measured during the growth of a 30-period InGaAs/GaAs superlattice consisting of 5 and 10 ML of InGaAs and GaAs, respectively. Bottom: expanded scale showing the RD oscillations allowing growth to be followed on a ML scale (after ref. 43).

The RD response, which follows primarily surface reconstruction, is much faster than the PR response, which is sensitive to composition as well, especially for TEG exposure. This shows that the reconstruction is established first, followed by a conversion of the reactant species to an intermediate which then decomposes further to donate a Ga atom to the growing crystal. When the surface is exposed to TBP the RD signal shows that the original reconstruction is obtained almost immediately. The slower PR response does not return to its original level, indicating deposition of GaP. For the double TEG sequence the PR signal is essentially repeated in both halves, while the RD signal shows an additional change indicating that further TEG exposure generates another reconstruction. The behavior of the PR signal with pulsed exposure shows directly that it contains information about both the dielectric response of the SRL and about the amount of material deposited.

A complete analysis of these data would require (1) a spectral capability and either (2) an atlas of spectra of the different species or (3) an accurate theoretical calculation of such spectra. As mentioned above, the latter is presently a difficult challenge, but one that needs to be solved if the full diagnostic power of surface-optical spectroscopy is to be realized.

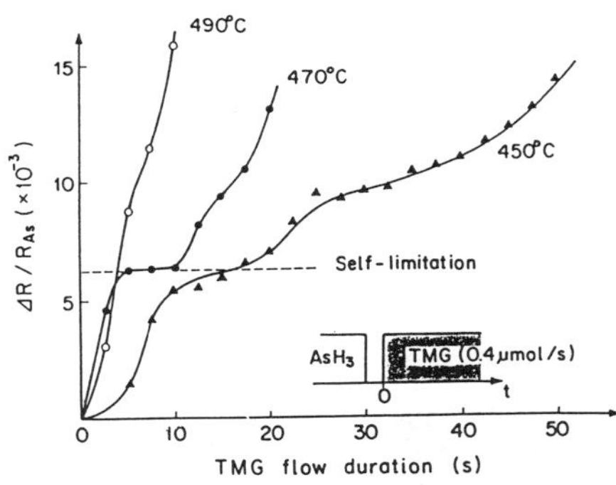

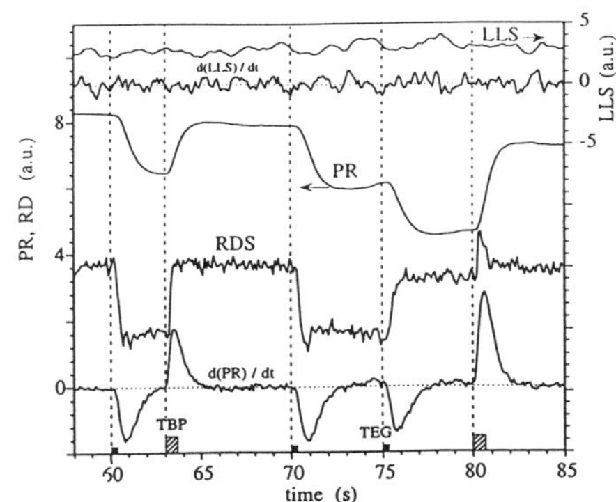

Fig. 5. Determination of a temperature range where atomic layer epitaxy occurs in the AsH₃-stabilized (001) GaAs — trimethylgallium system (after ref. 24).

Fig. 6. PR, RD, and LLS responses for single TEG and TBP pulses of 0.5 s duration during interruption of otherwise steady-state growth of GaP at 350 °C (after ref. 27).

A GENERAL CONTROL APPROACH

In general, control has meant post-mortem analysis of sample properties followed by correction of process parameters such that succeeding runs better meet specifications. While this is a reasonable first-line approach, it cannot be applied to compositionally graded structures nor can it achieve results comparable to real-time measurements and analysis, where errors can be corrected during growth. Although sample-driven closed-loop feedback control has been realized in simple situations involving ternary materials [36], improvements are needed especially for more complex systems. Ideally control decisions should be based on information obtained as close as possible to the elementary growth step where the constituent atoms are actually incorporated into the crystal lattice, typically at kinks in surface steps. This requires information about the chemical kinetics of the SRL, which is the intermediary between the process and the product, yet which to date has not been considered explicitly in the control process.

Based on the above it is clear that control of semiconductor epitaxy must evolve into a multifaceted process where information about each region is provided by specific probes and reduced by model calculations to required parameters. A general schematic is given in Fig. 7, with time increasing along the *region* axis from left to right and the *information* axis from top to bottom. The schematic diagram includes two analytic modules: scattering theory and surface chemistry; a control module; and a process module. The scattering theory and surface chemistry modules deal with the interpretation of the optical data and modeling the growth process, respectively. The surface-chemistry module includes and extends the conventional approach where growth is modeled on the basis of process information alone. The scattering-theory module deals with the analysis of the optical data. The scattering-theory model uses a database of the wavelength-dependent polarizabilities of the different species that in principle make up the unreacted and reacted parts of the SRL, allowing these concentrations to be determined. This model in principle involves calculations at the level of individual polarizable species, and can therefore accommodate local-field effects, chemical complexes, and nonspecular scattering

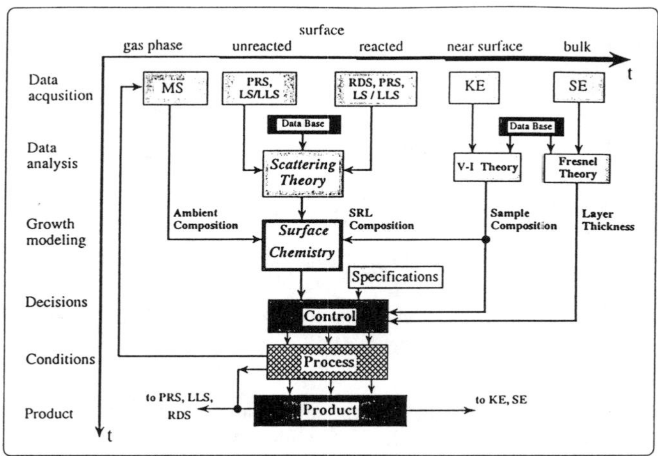

Fig. 7. General control system schematic, as discussed in the text.

by rough surfaces.

In Fig. 7 the information delivered by the probes are fed to the appropriate modules, analyzed, and sent to the control module, which compares this information to target values and adjusts process parameters accordingly. Here, we assume that information about the ambient is obtained by QMS, that concerning the unreacted part of the SRL by PRS, and that of the reacted part by a combination of RDS, PRS, and LLS. Near-surface information is obtained by KE in connection with V-I theory. Although this system remains to be implemented, the component capabilities have already been demonstrated to sufficient accuracy with the possible exception of the scattering theory and surface chemistry modules. The main challenge now is to develop these modules to the point where their capabilities match those of the probes.

Other challenges include the enhancement of signal-to-noise ratios, especially during operation in actual reactors where samples are rotated for uniformity and may experience mechanical vibration as well. Another issue is the runout of the optical beam that occurs if the surface of the sample is not perpendicular to the rotation axis. Some progress has already been made. Maracas et al. [44] have developed a piezoelectric-based manipulator for MBE where runout can be reduced to acceptable levels by applying appropriate voltages to the transducers. The NCSU group has developed a mechanical equivalent for OMCVD. Woollam and co-workers have demonstrated the capability of maintaining compositions x of $In_{0.53}Ga_{0.47}As$ constant to within 0.1% in an operating multiwafer OMCVD reactor, where the challenges to be met included not only having to average measurements over a number of wafers but also to deal with the dead time that occurs as a result of the gaps between wafers [45]. Even though the composition is not strictly controlled in real time, this accomplishment is still nontrivial.

Finally, techniques for probing the surface reaction layer, which is the closest region to the product and feeds bulk growth, are not yet fully developed and/or tested. Accessing and understanding the properties of this region will be one of the outstanding challenges of the next few years.

ACKNOWLEDGMENTS

It is a pleasure to acknowledge the support of the Defense Advanced Research Projects Agency and the Office of Naval Research under Contracts N-00014-95-1-0962 and N-00014-93-1-0255, respectively.

REFERENCES

1. E. H. A. Granneman, Thin Solid Films **228**, 1 (1993).

2. C. E. Zah, R. Bhat, F. J. Favire, Jr., S. G. Menocal, N. C. Andreadakis, K.-W. Cheung, D.-M. Hwang, M. A. Koza, and T.-P. Lee, IEEE J. Quantum Electronics **27**, 1440 (1991).

3. C. R. Abernathy, J. Vac. Sci. Technol. **A11**, 869 (1993).

4. S. M. Bedair, J. Vac. Sci. Technol. **B12**, 179 (1994).

5. A. C. Jones, J. Cryst. Growth **129**, 728 (1993).

6. M. G. Peters, B. J. Thibeault, D. B. Young, J. W. Scott, F. H. Peters, A. C. Gossard, and L. A. Coldren, Appl. Phys. Lett. **63**, 3411 (1993).

7. B. T. Sullivan and J. A. Dobrowolski, Appl. Opt. **32**, 2351 (1993).

8. S. A. Chalmers and K. P. Killeen, Appl. Phys. Lett. **62**, 1182 (1993).

9. K. P. Killeen and W. G. Breiland, J. Electron. Mater. **23**, 179 (1994).

10. R. M. Lum, M. L. McDonald, J. C. Bean, J. Vandenberg, T. L. Pernell, A. Robertson, and A. Karp, Appl. Phys. Lett. **69**, 928 (1996).

11. D. E. Aspnes, E. Colas, A. A. Studna, R. Bhat, M. A. Koza, and V. G. Keramidas, Phys. Rev. Lett. **61**, 2782 (1988).

12 I. Kamiya, D. E. Aspnes, H. Tanaka, L. T. Florez, J. P. Harbison, and R. Bhat, Phys. Rev. Lett. **68**, 627 (1992).

13. I. Kamiya, D. E. Aspnes, L. T. Florez, and J. P. Harbison, Phys. Rev. **46**, 15894 (1992).

14. D. W. Kisker, P. H. Fuoss, K. L. Tokuda, G. Renaud, S. Brennan, and J. L. Kahn, "X-ray analysis of GaAs surface reconstructions in H_2 and N_2 atmospheres," Appl. Phys. Lett. **56**, 2025 (1990).

15. F. J. Lamelas, P. H. Fuoss, P. Imperatori, D. W. Kisker, G. B. Stephenson, and S. Brennen, Appl. Phys. Lett. **60**, 2610 (1992).

16. W. Richter, Philos. Trans. R. Soc. London **A344**, 453 (1993).

17. C. Pickering, in **Handbook of Crystal Growth**, vol. 3, ed. D. T. J. Hurle (Elsevier, Amsterdam, 1994), 817.

18. J. F. McGilp, Progress in Surface Science **49**, 1, (1995).

19. D. E. Aspnes and I. Kamiya, SPIE Proc. **2730**, 306 (1996).

20. D. E. Aspnes, Thin Solid Films **89**, 249 (1982).

21. D. E. Aspnes, J. P. Harbison, A. A. Studna, and L. T. Florez, J. Vac. Sci. Technol. **A6**, 1327 (1988).

22. S. E. Acosta-Ortiz and A. Lastras-Martinez, Phys. Rev. **B40**, 1426 (1989).

23. N. Kobayashi and Y. Horikoshi, J. Appl. Phys. Jpn. **29**, L702 (1990).

24. N. Kobayashi and Y. Kobayashi, Thin Solid Films **225**, 32 (1993).

25. N. Dietz and H. J. Lewerenz, Appl. Surf. Sci. **69**, 350 (1993).

26. K. J. Bachmann, U. Rossow and N. Dietz, Mater. Sci. & Eng. **B35**, 472 (1995).

27. N. Dietz and K. J. Bachmann, Vacuum **47**, 133 (1996).

28. J. I. Dadap, B. Doris, Q. Deng, M. C. Downer, J. K. Lowell, and A. C. Diebold, Applied Physics Letters 64, 2139-2141 (1994).

29. A. J. Pidduck, D. J. Robbins, A. G. Cullis, D. B. Gasson, and J. L. Glasper, J. Electrochem. Soc. **136**, 3083 (1989).

30. F. G. Celii, E. A. Beam III, L. A. Files-Sesler, H.-Y. Liu, and Y. C. Kao, Appl. Phys. Lett. **62**, 2705 (1993).

31. J. E. Epler and H. P. Schweizer, Appl. Phys. Lett. **63**, 1228 (1993).

32. U. Rossow, L. Mantese, and D. E. Aspnes, Proc. 23rd Internat. Conf. Phys. Semicond. Berlin, M. Scheffler and R. Zimmerman, eds. (World Press, Singapore, 1996), p. 831.

33. C. M. J. Wijers and G. P. M. Poppe, Phys. Rev. **B46**, 7605 (1992).

34. B. S. Mendoza and W. L. Mochán, Phys. Rev. **B53**, R10473 (1996).

35. See, for example, R. M. A. Azzam and N. M. Bashara, **Ellipsometry and Polarized Light** (North-Holland, Amsterdam, 1977).

36. D. E. Aspnes, W. E. Quinn, M. C. Tamargo, M. A. A. Pudensi, S. A. Schwarz, M. J. S. P. Brasil, R. E. Nahory, and S. Gregory, Appl. Phys. Lett. **60**, 1244 (1992).

37. F. H. Pollak and H. Shen, J. Electron. Mater. **19**, 399 (1990).

38. O. J. Glembocki, J. A. Tuchman, K. K. Ko, S. W. Pang, A. Giordana, and C. E. Stutz, Mat. Res. Soc. Symp. Proc. **324**, 153 (1994).

39. D. E. Aspnes, J. Opt. Soc. Am. **10**, 974-83 (1993).

40. P. Chiaradia, G. Chiarotti, S. Nannarone, and P. Sassaroli, Solid State Commun. **26**, 813 (1978).

41. K. Hingerl, D. E. Aspnes, and I. Kamiya, Surface Sci. **287/288**, 686 (1993).

42. K. Uwai and N. Kobayashi, Appl. Phys. Lett. **65**, 150 (1994).

43. M. Zorn, J. Jönsson, A. Krost, W. Richter, J.-T. Zettler, K. Ploska, and F. Reinhardt, J. Cryst. Growth **145**, 53 (1994).

44. G. N. Maracas, C. H. Kuo, S. Anand, R. Droopad, G. R. L. Sohie, and T. Levola, J. Vac. Sci. Technol. **A13**, 727 (1995).

45. C. Herzinger, B. Johs, P. Chow, D. Reich, G. Carpenter, D. Croswell, and J. Van Hove, Mat. Res. Soc. Symp. Proc. **406**, 347 (1996).

STRAIN-RELATED EXCITONIC IN-PLANE OPTICAL ANISOTROPY IN   (100) InGaAs/InAlAs/InP MQW

A. DIMOULAS*, R. TOBER**, R. LEAVITT**, T. FENG*, A. CHRISTOU*
*University of Maryland, Dptm. of Materials & Nuclear Engineering, College Park, MD 20742.
**U.S. Army Research Laboratory, 2800 Powder Mill Road, Adelphi, MD 20783.

ABSTRACT

Strained (x=0.48) and lattice-matched (x=0.53) $In_xGa_{1-x}As$/InAlAs/InP (100) MQWs have been investigated by photoreflectance. In the strained sample the relative intensities of the light-hole and heavy-hole excitonic transitions is different for the two different polarizations of the incident light parallel and perpendicular to the [0-1-1] direction. This polarization anisotropy is explained in terms of the spontaneous formation of "quantum wires" and the presence of anisotropic strain due to spinodal-like phase decomposition of the InGaAs alloy in In-rich and Ga-rich regions.

INTRODUCTION

Multiple quantum wells (MQW) grown on (100) substrates possess tetragonal symmetry and are expected to exhibit isotropic optical properties with respect to the in-plane polarization of light. However, in the presence of ordering [1] or under the influence of an externally applied in-plane uniaxial strain [2] the symmetry is lowered and in such cases polarization anisotropy has been observed [1,2]. An understanding of the optical anisotropy may help to control the polarization instabilities [3] commonly observed in Vertical Cavity Surface Emitting Lasers (VCSELs) and may lead to new applications based on polarization-sensitive, very high contrast [2], very high speed spatial light modulators. In most of the cases [1] evidence of optical anisotropy has been obtained by using luminescence at low temperatures where only the lowest energy lying transition (either heavy-hole (*hh*) or light-hole (*lh*)) was observed. In the present study, by using photoreflectance, an absorption-based technique, we report on the optical anisotropy of both the *hh* and *lh* transitions as well as the higher excited transitions in strained QWs. This may help to better understand the reason that causes the anisotropy and may lead to useful applications.

MQW STRUCTURE AND EXPERIMENTAL SET-UP

The samples were grown by MBE at T_g = 540 ^{0}C and had the following structure : 10 nm $In_{0.53}Ga_{0.47}As$ cap layer/1 μm MQW multilayer structure/100 nm $In_{0.52}Al_{0.48}As$ buffer layer/(100) S.I. InP substrate. The MQW region for the lattice-matched (x=053) sample HDL597 consisted of 33 periods of 15 nm $In_xGa_{1-x}As$ wells followed by 15 nm InAlAs barriers. In this sample the wells were tensile strained while the barriers were under compression with a barrier composition such that the net strain was zero. Sample HDL377 (x=0.48), had 50 periods of 10 nm wells followed by 10 nm barriers. All layers were undoped, while the substrates were of exact (100) orientation.

Modulation Spectroscopy (photoreflectance) was used to probe the polarization anisotropy. This technique utilizes a monochromatic beam to probe small reflectivity changes $\Delta R/R$ caused by a chopped He-Ne laser light [4]. These changes occur only at the vicinity of interband optical

Mat. Res. Soc. Symp. Proc. Vol. 448 © 1997 Materials Research Society

transitions so that the method provides a means for a room temperature, high spectral resolution characterization of ground and excited states in QWs. A Glan Thomson polarizer was placed between the exit slit of the monochromator and the sample to determine the polarization state of the incident light.

EXPERIMENTAL RESULTS

<u>Description of the spectra</u>

The Photoreflectance spectra for two different in-plane polarizations along [0-1-1] and [0-11] crystallographic directions are given in Figures 1 (a) and (b) for the lattice-matched and strained MQW structures, respectively. In the lattice-matched structure, the ground state 11lh transition appears to be about 13 meV higher in energy compared to the 11h transition, due to quantum confinement. In the strained sample HDL 377, the light hole transition is positioned 10 meV below the heavy-hole one due to the presence of tensile strain equal to 3.6 X10^{-3}. In the latter sample, a strong polarization anisotropy is observed in all optical transitions, in contrast to the lattice-matched MQW where no anisotropy is present. As seen from Fig. 1 (b), the lh appears as a small, low-energy shoulder to the dominant hh transition when e // [0-11], but increases in intensity and becomes comparable to the heavy-hole transition when the polarization switches to the [0-1-1] direction.

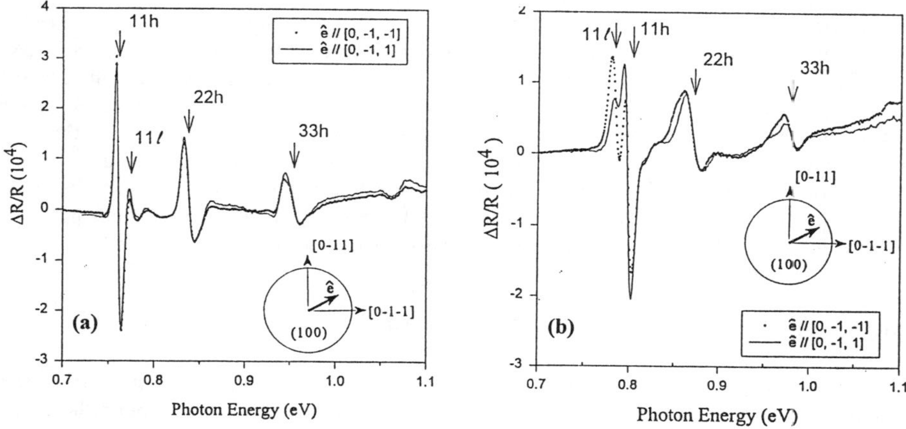

Fig. 1 *Photoreflectance spectra for two different polarizations for (a) lattice-matched, (b) strained InGaAs/InAlAs/InP(100) MQWs*

The photoreflectance spectra were analysed by employing a model [4] which assumes that the absorption coefficient α is described by a Gaussian and that DR/R is a function of the first derivatives of α with respect to the transition energy E, broadening parameter Γ and intensity I. The latter quantities were determined by fitting the spectra. The ratio I_{lh}/I_{hh} of the intensities of the *lh* and *hh* transitions as a function of the polarization angle ϕ is shown in Fig. 2. The ratio has a minimum value of ~ 1/3 when the polarization vector e is along [0-11] and a maximum value of ~1 when e //[0-1-1]. The data show a periodicity with a period of π.

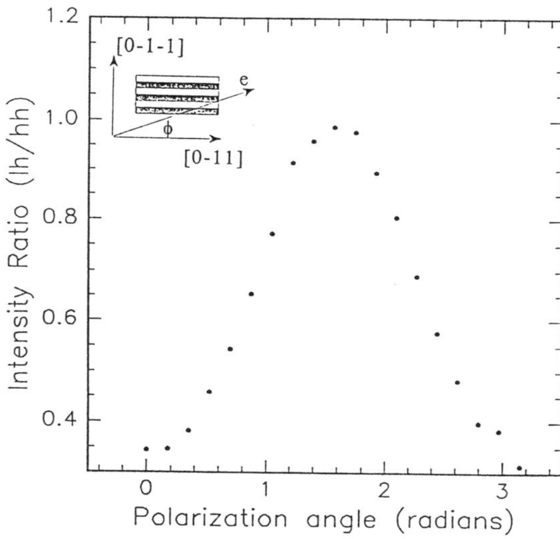

Fig. 2
Intensity ratio of the light-hole and heavy-hole optical transitions as a function of the polarization angle. The inset shows the orientation of the wire-like structure developed as a result of phase decomposition

DISCUSSION AND CONCLUSIONS

There can be several reasons for symmetry breaking which lead to optical anisotropy. One reason may be the linear electrooptic (Pockels) effect due to the surface built-in electric field. By shining a variable power DC laser light on the sample surface, we have been able [5] to screen the built-in electric field and measure the anisotropy ratio, which remains essentially the same for any value of the laser power. This result indicates that the bulk [6] linear electrooptic effect cannot account for the observed polarization anisotropies. Quantum confinement must be considered to correctly describe Pockels effects in quantum wells [7]. In any case, the absence of anisotropy in the lattice-matched sample HDL597 (Fig. 1 (a)) is not in favor of the explanation of our results in terms of the Pockels effect.

Spontaneous formation of "quantum wires" or the development of an in-plane anisotropic strain distribution due to phase decomposition is another possibility. Spinodal-like decomposition in ternary and quaternary III-V alloys [8] results in the formation of In-rich and Ga-rich regions in the form of a quasi-periodic modulation of composition (as shown

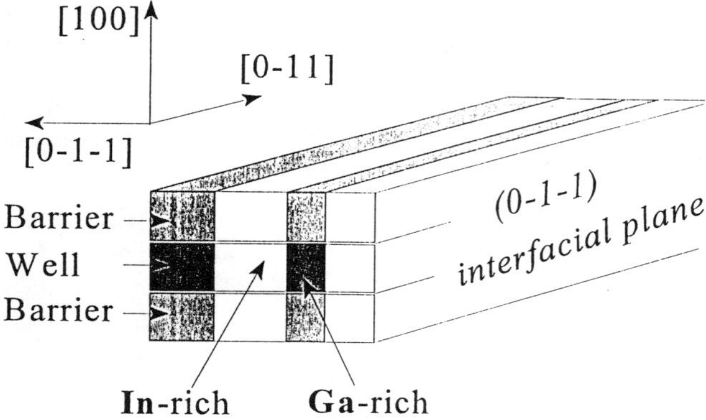

Fig. 3 *The succession of black and white contrast represents the modulation of composition and strain along [0-1-1]. The "quantum wires" are oriented along [0-11]. The growth direction is along [100].*

schematically in Fig. 3) which, remarkably, occurs along only one of the two <011> directions [1,5]. Our preliminary TEM results, in comparison with previous works [1,5], indicate that in our samples the composition modulation occurs along [0-1-1] suggesting that the quantum wires are formed along [0-11] direction as shown in Fig. 3. The resulting 2D quantum confinement (in both vertical and lateral directions) may explain qualitatively the polarization dependence of Fig. 2 in terms of the modification of quantum selection rules as has been previously discussed for the case of artificially made quantum wires [9].

In addition, the composition modulation is associated with a strain modulation since the adjacent In-rich and Ga-rich regions have different lattice constants. The nature of this strain for the In-rich sections is biaxial compressive in the (011) interfacial plane (Fig. 3), so that a state of anisotropic strain in the plane of the growth is formed. This in turn lowers the symmetry resulting in a valence band mixing of the states |3/2, 3/2> and |3/2, 1/2> having different quantum numbers of the z-component of the total angular momentum, which may also explain the behavior of Fig 2. In fact, 2D quantum confinement and anisotropic strain-induced valence band mixing, may both contribute to the observed anisotropy, however, identification of the dominant mechanism requires further investigation. Finally, it is important to note that the presence of a biaxial tensile strain in the plane of the growth due to lattice mismatch with the substrate, results in modifications of the critical temperature [10] for which the decomposition occurs. This may explain the presence of the anisotropy in the strained samples in contrast to the absence of it in the lattice-matched one.

REFERENCES

[1] Chou S T, Cheng K Y, Chou L J, Hsieh K C   J. Appl. Phys. **78,** 6270 (1995).
[2] Shen H, Pamulapati J, Wraback M, Taysing-Lara M, Dutta M, Kuo H C, Lu Y   IEEE Photon.Technol. Lett. **6,** 700 (1994).
[3] Jewell L, McCall S L, Lee Y H, Scherer A, Gossard A C, English J H   Appl. Phys. Lett. **54,** 1400 (1989).
[4] Dimoulas A, Leng J, Giapis K P, Georgakilas A, Michelakis C, Christou A   Phys. Rev **B 47,** 7198 (1993).
[5] Dimoulas A, Tober R, Leavitt R, Christou A, unpublished data.
[6] Acosta-Ortiz S E, Lastras-Martinez A   Phys. Rev. **B 40,** 1426 (1989)
[7] Kwok S H, Grahn H T, Ploog K, Merlin R   Phys. Rev. Lett. **69,** 973 (1992)
[8] LaPierre R R, Okada T, Robinson B J, Thompson D A, Weatherly G C   J. Cryst. Growth **155,** 1 (1995)
[9] Tsuchiya M, Gaines J M, Yan R H, Simes R J, Holtz P O, Coldren L A, Petroff P M, Phys. Rev. Lett. **62,** 466 (1989).
[10] Glass F   J. Appl. Phys. 62, 3201 (1987).

BAND LINEUP OF VAN DER WAALS-EPITAXY INTERFACES

R. SCHLAF, T. LÖHER, O. LANG, A. KLEIN, C. PETTENKOFER, W. JAEGERMANN
Hahn-Meitner-Institut, Abt. Grenzflächen, Glienicker Str.100, 14109 Berlin, Germany

ABSTRACT

Epitaxial lattice mismatched heterointerfaces between layered semiconductors and themselves and II-VI semiconductors (CdS, CdTe), respectively, have been prepared and their band lineup determined by photoemission. Different physical mechanisms, which govern the heterointerface formation, can be discriminated due to the specific properties of the van der Waals (vdW) surface. The interfaces between layered semiconductors mostly follow the electron affinity rule with a small but systematic deviation, which is assigned to the influence of interfacial quantum dipoles. However, the band lineup to the II-VI semiconductors shows a large interface dipole, which is related to a structural dipole from the polar, Cd terminated, face of the (111)- in case of Zinkblende CdTe- and the (0001)- in case of Wurtzite CdS- oriented overlayer film.

INTRODUCTION

Semiconductor heterocontacts are of great relevance for electronic devices. Hence, tailoring of valence and conduction band offset at the interface is one of the most essential parts in the design of modern semiconductor devices. Therefore heterojunction band offsets have been intensively investigated during the past four decades.[1] In addition, it has been attempted to predict the offsets from theoretical considerations either from *ab initio* calculations or from so called "linear models", which calculate the offsets from bulk properties of the involved semiconductors such as electron affinities, dielectric midgap energies (DME), charge neutrality levels or tight binding sp^3 energies.[1,2] The comparison of these models with experimental results are often complicated by experimental uncertainties in the determined offsets or the superposition of different physical effects.[1] Especially the formation of structural dipoles at the interfaces due to polar surfaces and possible lattice relaxation, the formation of defects, interdiffusion or chemical reactions at the interface can have significant impact on the magnitude of the offsets observed. Another important factor are electronic interface states, which lead to electronic interface dipoles and influence the band offsets.[3,4] Since both kinds of dipoles usually occur simultaneously at the commonly investigated heterosystems of III-V, IV, or II-VI semiconductors the detailed investigation of their relative magnitude and impact proved to be very difficult.

The investigation of heterojunction band offsets with layered chalcogenides may help to discriminate the different effects (Fig.1). In pure van der Waals heterointerfaces (Fig. 1a) each single sandwich layer of substrate and epilayer is chemically saturated and in the ideal case atomically abrupt. The sandwich layers are hold together by weak van der Waals type of interactions.[5] Due to these properties epitaxial growth of layered materials heterojunctions, named van der Waals epitaxy (vdWe), is almost unaffected by the structural problems indicated above.[6] This allows the detailed investigation of electronic dipoles without interference of structural or chemical dipoles. On the other hand structural interface dipoles are expected for quasi-van der Waals type interfaces, in which Zinkblende compounds are grown on layered chalcogenide substrates (Fig. 1b).[7]

We measured the band offsets of a number of vdWe layered chalcogenide heterojunctions and of quasi-vdWe interfaces to CdS and CdTe, which give information on the relative magnitude and polarity of the different interface dipoles. The details on the multi-step epitaxial growth of the overlayers are given elsewhere.[7,8] Also the photoemission (PES) results used for the determination of the band lineup will be published elsewhere.[9,10] Here we will only summarize the experimental results and discuss their implications.

Mat. Res. Soc. Symp. Proc. Vol. 448 © 1997 Materials Research Society

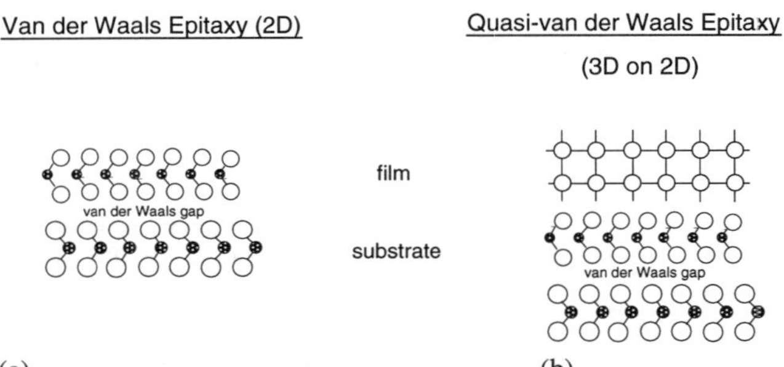

Fig.1: Van der Waals epitaxy (vdWe) of layered chalcogenide interfaces (a) and quasi-van der Waals epitaxy of three-dimensional semiconductors on layered chalcogenide substrates (b).

EXPERIMENTAL PROCEDURE

All presented experiments were performed in a photoelectron spectrometer (VG, ESCALAB, or ADES 500) connected to a MBE chamber. The experimental setup allows PES and electron diffraction (LEED) measurements in-situ whereas STM data and TEM data were obtained ex-situ. For details refer to Refs [7-10].

The valence band offsets ΔE_V were determined directly by PES measuring substrate-overlayer differences of core level binding energies, which were carried out after each step of the growth sequence. A detailed description of this method can be found in Ref. [11]. The reference spectra for the determination of the valence band maximum were obtained from single crystals.

The electron affinity χ is accessible from the experimentally determined work function (given by the secondary electron onset), the measured difference of the Fermi level to the valence band maximum (E_F-E_V) and the known band gaps (E_G=E_C-E_V) of the semiconductors. The interface dipole potential D across the interface is given from the difference of the electron affinity offset ΔE_C(EAR) between single crystal electron affinity of substrate and layer material assuming the validity of the electron affinity rule (EAR) and the experimentally deduced offset ΔE_C(exp) determined from the experimentally determined value ΔE_V(exp) and the bandgaps E_G.

RESULTS AND DISCUSSION

The goal of the experiments was the experimental determination of valence and conduction band offset of several heterojunctions and their comparison to the predictions of the electron affinity rule (EAR). The deviation gives a value for the electronic quantum dipole (QD) for each investigated interface. As determined from PES and LEED measurements all interfaces are non-reactive, and atomically abrupt. Epitaxial or strongly textured overlayer films are obtained.

The results of the evaluation of the valence band offsets of vdWe layered chalcogenide heterointerfaces versus the EAR predicted values are summarized in Figure 2 (additional ΔE_V values obtained from earlier experiments are added to the graph [12]). The dashed line represents the EAR limit. It is evident that the experimentally determined valence band offsets ΔE_V(exp) are close to the EAR rule as may be expected for vdWe heterointerfaces as the ideal vdW (0001) faces are known to be free of active surface states.[5] However, a close inspection of the experimental results suggest a small but systematic deviation from the predicted EAR values ΔE_V(EAR). The line fit through the data points indicates that the magnitude of the quantum dipoles depends on the EAR band offsets suggesting a linear correction term for the EAR. The value of the interface dipole D is given as difference of the experimental curve to the value predicted from EAR.

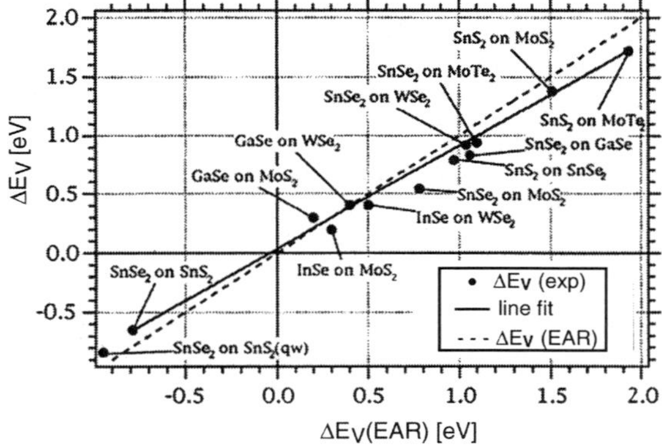

Fig.2: Experimentally determined valence band offsets ΔE$_V$(exp) versus values predicted by the electron affinity rule ΔE$_V$(EAR). The dashed line represents the EAR limit. The difference to the solid line gives the value of the interface dipole potential D.

The additional interface dipole D is related to the influence of electronic quantum dipoles at heterojunction interfaces as introduced by Tersoff.[3] These interfacial gap states result from tunneling of carriers from one side of the junction into the band gap of the other side as shown schematically in Fig.3. Due to the extension of the wave functions into the forbidden gap of the opposite side local charge transfer becomes possible as indicated by the "+" and "-" signs. In a straddled band lineup (Fig.3a) the transferred charges cancel each other (at least partially). In the case of a staggered lineup (Fig.3b) an electronic quantum dipole is formed since both of the offsets point in the same direction which leads to a net charge transfer across the interface. The resulting dipole potentials are shown at the bottom of Fig. 3.

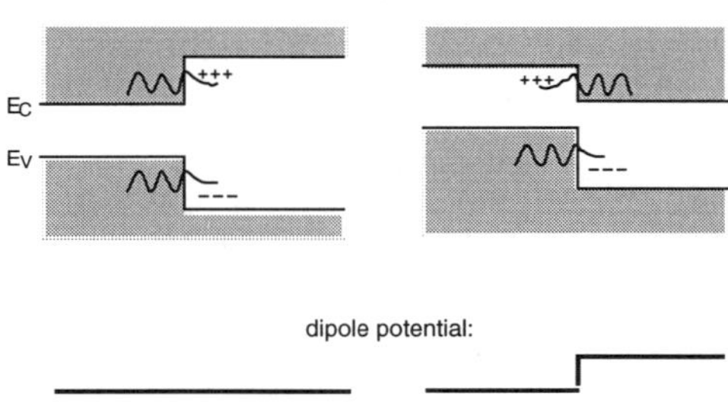

Fig.3: Schematic depiction of the origin of electronic quantum dipoles. The quantum dipole is induced by extending wavefunctions into the energy gaps of the opposite sides (after Reference [3]). Bottom: Resulting dipole potential for the two band lineup types shown above.

Following the concept of Ruan and Xing [4] the electronic quantum dipoles should depend in a first order approximation on the sum of the valence and conduction band offsets (using proper signs for "up" and "down" pointing offsets). Therefore, for a modified EAR the sum of the offsets are multiplied with a constant K to the original EAR-value. From

$$\Delta E_V = \Delta E_V(EAR) + K \cdot \left[\Delta E_V(EAR) + \Delta E_C(EAR) \right]$$
$$\Leftrightarrow \Delta E_V - \Delta E_V(EAR) = K \cdot \left[\Delta E_V(EAR) + \Delta E_C(EAR) \right] = eD \tag{1}$$

it follows immediately that the constant K can be determined by plotting the measured dipoles eD versus the sum of the EAR offsets ΔE$_V$(EAR)+ΔE$_C$(EAR). The constant K then equals the slope of a line fitted through the data points. The line fit yields a slope of K=-0.09. The negative sign of K means that the quantum dipoles always reduce the magnitude of the combined offsets. This correction term is smaller by a factor of 2 compared to the value for the three-dimensional semiconductors treated by Ruan and Xing.[4]

In contrast the band lineup of quasi-vdWe heterointerfaces (Fig. 1b) are expected to be influenced by structural dipoles. Therefore CdS and CdTe having different ionicities were grown on layered $MoTe_2$ and WSe_2 (0001) surfaces.[7,10] Strongly textured films are obtained oriented with the polar hexagonal plane ((0001) for Wurtzite CdS, (111) for Zinkblende CdTe) on the hexagonal nonpolar van der Waals (0001) plane, as deduced from LEED and TEM results. The band lineups as determined from photoemission results [10] are summarized in Table I.

Table I: Experimentally determined valence band offset ΔE_V and calculated conduction band offset ΔE_C and interface dipole D. Experimentally determined ionization potentials of substrate and film ($I_p{}^S$ and $I_p{}^F$) are also given in the table. For calculation of ΔE_C the bandgap values of $MoTe_2$ (1.1eV), WSe_2 (1.2eV), CdS (2.5eV) and CdTe (1.5eV) have been used.

Interface	ΔE_V	ΔE_C	D	$I_p{}^S$	$I_p{}^F$
CdS/MoTe$_2$	0.9	0.5	1.1	4.9	6.9
CdS/WSe$_2$	0.6	0.7	1.2	5.3	7.0
CdTe/MoTe$_2$	0.03	0.4	0.7	4.9	5.7
CdTe/WSe$_2$	-0.32	0.65	0.8	5.3	5.8

The important quantity is again the value of the dipole potential D across the interface given by the experimentally determined band lineup to the expected one based on the EAR (for the χ values of the overlayer the experimentally determined value of the polar face was taken). In contrast to the vdWe interfaces discussed above D is very large for the quasi vdWe interfaces and seems to be only dependent on the II-VI overlayer, showing a larger value to the more ionic CdS. This results suggest that the interface band lineup is mostly governed by a structural interface dipole as schematically shown in Fig.4. The direction of the interface dipole potential as well as the experimentally determined $I_p{}^F$ clearly indicates that the II-VI overlayer grows with the Cd temination towards the substrate and the group VI element towards vacuum. The magnitude of the surface dipole may be estimated from the oscillating potential typically present in polar directions of compound semiconductors.[13] However, for an estimate of the surface dipole surface reconstructions and relaxations must be considered, which occur to reduce the charging of the surface plane (compare also Fig.4). Unfortunately only very few systematic experimental studies of the orientation dependence of the ionization potential of compound semiconductors are given in the literature. With these uncertainties our estimate suggests that the large value of the interface dipole determined for the quasi-vdWe interfaces is dominated by the structural dipole related to the polar orientation of the overlayer growth. The experimental value has to be divided into an interface dipole at the phase boundary and a second one at the chalcogenide terminated polar II-VI surface to vacuum (in Fig.4 schematically shown for CdS (0001) on InSe (0001)). There may still be a small contribution of an electronic quantum dipole (QD), which we expect to be of negligible influence based on our results on the vdWe interfaces.

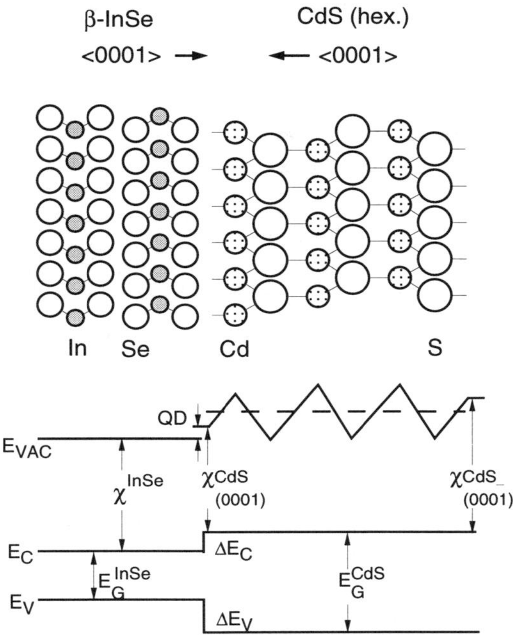

Fig.4: Orientation of the CdS overlayer on a layered chalcogenide (0001) surface (top). The related interface dipole is very large but smaller than the potential oscillations in the bulk (bottom).

SUMMARY AND CONCLUSION

The recently developed method of van der Waals epitaxy (vdWe) allows for a proper selection of semiconductors the detailed investigation of the relative contribution of physically different dipoles at semiconductor heterojunctions. In vdWe heterointerfaces of two-dimensional layered compound heterojunctions can be prepared which have no structural or chemical dipoles at the interface. The experimentally determined valence band offsets of these junctions show a small but significant deviation from the values predicted by the electron affinity rule indicating the presence of electronic quantum dipoles at the interfaces. These quantum dipoles could be quantified by adding a linear quantum dipole correction given by the combined offsets term to the original EAR.

In contrast quasi-vdWe interfaces between II-VI semiconductors and layered substrates show a large interface dipole D, which the band lineup is mostly determined by the structural dipole related to the polarity of the hexagonal growth plane ((111) for CdTe, (0001) for CdS).

ACKNOWLEDGEMENTS

Part of the work was supported by the BMFT, Germany. We are grateful to Herbert Sehnert and J. Lehmann for technical support, and to Yvonne Tomm for providing substrate crystals.

REFERENCES

1. F. Capasso and G. Margaritondo (eds), <u>Heterojunction Band Discontinuities</u>, North Holland, Amsterdam, 1987
2. H. Kroemer, in <u>Molecular Beam Epitaxy and Heterostructures,</u> edited by L. L. Ahngand and K. Ploog, Marinus Nijgoff Publishers, Dordrecht, 1985
3. J. Tersoff, Phys. Rev. B **30,** 4874 (1984)
4. Y. C. Ruan and W. C. Ching, J. Appl. Phys. **62**, 2885 (1987)
5. W. Jaegermann, in <u>Photoelectrochemistry and Photovoltaics of layered Semiconductors,</u> edited by A. Aruchamy, Kluwer, Dordrecht, 1992
6. A. Koma and K. Yoshimura, Surf. Sci. **174**, 556 (1986); A. Koma, Thin Solid Films **216**, 72 (1992)
7. T. Löher, Y. Tomm, C. Pettenkofer, and W. Jaegermann Appl. Phys. Lett. **65**, 555 (1994); T. Löher, Y. Tomm, A. Klein, D. Su, C. Pettenkofer, W. Jaegermann, J. Appl. Phys. **80**, 5718 (1996)
8. R. Schlaf, N. R. Armstrong, B. A. Parkinson, C. Pettenkofer, W. Jaegermann, Surf. Sci, (in press)

   R. Schlaf, D. Louder, O. Lang, C. Pettenkofer, W. Jaegermann, K. Nebesuy, P. Lee, B. A. Parkinson, N. R. Armstrong, J. Vac. Sci. Technol. A, 1761 (1995)
9. R. Schlaf, O. Lang, N. R. Armstrong, C. Pettenkofer, and W. Jaegermann, in preparation
10. T. Löher, A. Klein, C. Pettenkofer, W. Jaegermann, in preparation
11. J. R. Waldrop and R. W. Grant, Phys. Rev. Lett. 43, 1686 (1979)
12. O. Lang, Y. Tomm, R. Schlaf, C. Pettenkofer, W. Jaegermann J. Appl. Phys. 75, 781 (1994)
13. M. Lannoo and P. Friedel, <u>Atomic and Electronic Structure of Solid Surfaces</u>, Springer Verlag, Berlin 1991

Investigation of electric fields, interface charges, and conduction band offsets at ZnSe/GaAs heterojunctions with a novel photoreflectance technique

D. J. Dougherty, S.B.Fleischer, E. L. Warlick, J. L. House, G. S. Petrich, E. Ho, L. A. Kolodziejski, and E. P. Ippen

Research Laboratory of Electronics, MIT 36-325, 50 Vassar St., Cambridge, MA 02139

ABSTRACT

ZnSe/GaAs heterojunctions were investigated by contactless electroreflectance and photoreflectance techniques. Negative surface charge densities on the order of 10^{12} cm^{-2} were observed for films grown on n-type GaAs indicating a large contribution to the conduction band barrier between the materials due to band bending. The conduction band offset was also measured using a new photoreflectance technique involving a tunable pump laser.

INTRODUCTION

Because its lattice constant is well matched to ZnSe and it can be produced in high purity wafers, GaAs is a natural substrate material for ZnSe based opto-electronic devices. Parameters such as interface charges, electric fields, and even the conduction band offsets which determine the effective barrier height between the materials, and therefore transport properties, have either not been characterized or appear to depend on the growth conditions.

Surface second harmonic studies[1] for ZnSe grown on semi-insulating GaAs have shown a positive interface charge which causes band bending on the GaAs side of the junction leading to interface quantum well states and the absence of any effective conduction band barrier. X-ray photoemission experiments indicate that the valence band offset can vary from 1.2 eV to 0.6 eV with growth conditions[2]. Device designers have invoked up to 0.6 eV conduction band offsets to explain their measured I-V curves[3]. These results suggest that it would be desirable to have a technique for measuring effective barrier heights due to both the band offsets as well as band bending for understanding the effects of growth conditions. We have used contactless electroreflectance (CER) and photoreflectance (PR), which are powerful methods for measuring electric fields[4], to determine the band bending at the heterojunction. In addition, we have measured the conduction band offsets for several ZnSe films grown on n-type GaAs using a tunable pump PR technique which is similar to internal photemission (IPE) photoconductive methods[5].

ELECTROREFLECTANCE RESULTS

The ZnSe films used in this study were grown by MBE on semi-insulating and n-type GaAs buffer layers. The buffer layers were typically 2-3 μm thick and were grown

Mat. Res. Soc. Symp. Proc. Vol. 448 © 1997 Materials Research Society

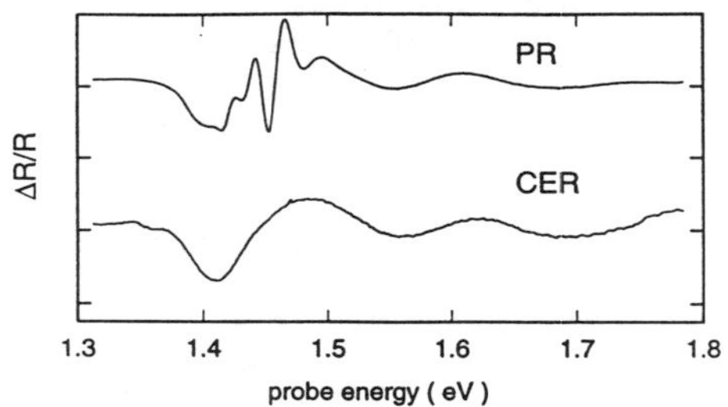

Figure 1

PR and CER data at the GaAs bandedge for a 0.12 μm ZnSe film grown on n-type GaAs

on either n-type or semi-insulating substrates. The nucleation of the ZnSe was done on GaAs c(4x4) reconstructed surfaces.

Figure 1 shows conventional CER and PR spectra of the GaAs bandedge for a 0.12 μm ZnSe film on a 7 x 10^{17} cm^{-3} n-type GaAs buffer layer. The long period Franz-Keldysh oscillations in the reflectivity spectrum indicate a electric field on the GaAs side of the junction. The shorter period signal seen in the PR trace is due to smaller fields at the junction between the n-type buffer layer and the semi-insulating substrate. The sign of the CER trace indicates that the field is pointing <u>towards</u> the interface. Both the direction and magnitude of this field are wrong for modulation doping effects. A fully depleted ZnSe layer would result in a maximum field of only 2 x 10^4 V/cm pointing <u>into</u> the GaAs. Thus, we attribute this large field to a 1.8 x 10^{12} cm^{-2} negative surface charge probably due to interface traps forming a depletion layer in the GaAs near the junction.

Using this interface charge and the doping levels, the conduction band bending was calculated and is shown in figure 2. A 200 meV conduction band offset was used here, but the band bending on the GaAs side adds another 250 meV to the effective barrier height. Similar band bending for ZnSe on n-type GaAs has been proposed by Ref [6] based on interpretation of Raman scattering experiments. Evidently, band bending has an important impact on the conduction band transport properties of the junction.

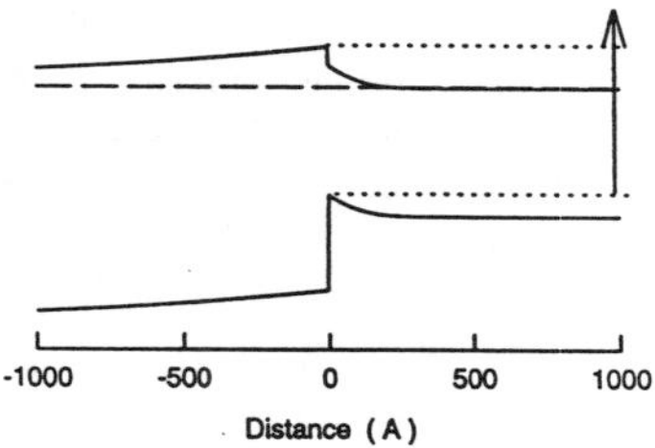

Figure 2

Calculated band bending at the heterojunction for a -1.8 x 10^{12} cm^{-2} interface charge

The existence of this barrier can be directly demonstrated by PR. Figure 3 shows conventional PR results for probe wavelengths in the blue spectral region. The top curve is for a Si doped GaAs substrate and shows that the signals at 2.9 and 3.2 eV originate from the GaAs layer. These are the transitions from the valence band to the L-valley. The bottom four traces are for the ZnSe/GaAs sample using different pump photon energies. For a UV pump, carriers are generated in both materials resulting in modulation at the ZnSe bandedge (2.67 eV) as well as in the GaAs. For pump photon energies intermediate to the two bandgaps, the GaAs signals are still present but the ZnSe signal is smaller and decreases as the pump energy is reduced. The fact that the ZnSe PR signal can be seen at all for pump energies below the ZnSe bandgap is significant because it indicates electron transport from the GaAs across the junction.

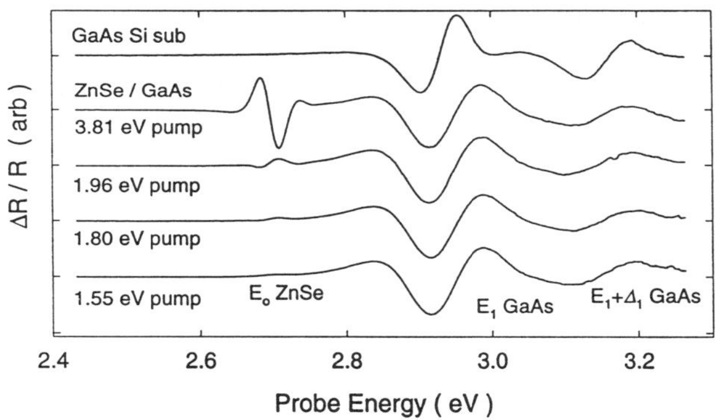

Figure 3
PR spectra taken with different pump photon energies.

TUNABLE PUMP PR

From 1-D electrostatics, photo-generated electron hole pairs confined to the GaAs cannot screen fields in the ZnSe. If electrons are created in the GaAs with enough excess energy to overcome the conduction band offset and enter the ZnSe; however, then fields in the ZnSe can be modulated. The dashed lines in figure 2 indicate the threshold energy for optically injected electrons to cross the junction. Because of the negative surface charge, the GaAs bands bend down. Thus, the threshold energy should be determined by the conduction band offset alone. By using a tunable pump laser and recording the size of the ZnSe PR signal with pump wavelength this threshold energy can be measured.

Several important points must be taken into account for this experiment to be successful. First, the ZnSe PR signal as seen in figure 2 rides on a background from the tail of the GaAs E_1 signal. Fortunately, the two signals have different time constants and by chopping at 400 Hz the ZnSe can be made approximately 30° out of phase with the pump and the E_1 signal. By setting the phase of the lock-in 90° to the pump, the ZnSe signal can be obtained background free. Second, the pump intensity must be low enough for the magnitude of the PR signal to be linear with the pump intensity. Finally, because the necessary tuning range is quite large (880 nm to 600 nm) several pump sources were used including a Ti:sapphire laser, a 688nm laser diode, and a HeNe laser. In order to ensure that the same intensity and spot size was maintained in switching sources, a 50/50 single mode fiber coupler was used for multiplexing. Light from the output of the coupler was defocused on the sample Power was measured at the fiber output in front of the sample to avoid the wavelength dependence of the coupler.

Results for both the ZnSe bandedge and GaAs E_1 PR signals are plotted in figure 4 versus pump energy shown as excess energy in the GaAs conduction band. The dark triangles are for the GaAs E_1 signal which is simply proportional to the absorbed pump intensity and is therefore a measure of the GaAs absorption spectrum. In contrast, the white circles representing the ZnSe signal show a dramatic quadratic increase with pump photon energy. This result is very similar to the behavior seen in IPE (internal photo-emission) spectroscopy where a photocurrent is measured as a function of pump photon energy[5]. Using the same analysis as in IPE, we extract a threshold, and thus conduction band offset as discussed above, of 150 ± 20 meV. Results for four other samples grown on n-type GaAs show band offsets between 150 and 250 meV suggesting variation with growth and nucleation conditions as Ref 2 has demonstrated.

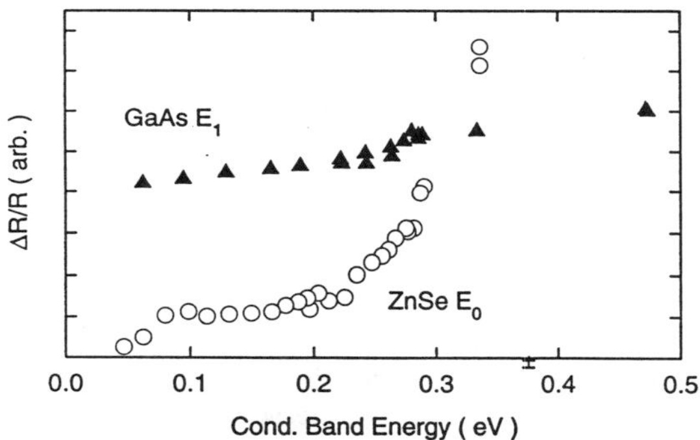

Figure 4
PR signals for the GaAs E_1 (triangles) and the ZnSe bandedge (circles) versus injected energy in the conduction band.

COMPARISON OF N-TYPE AND SEMI-INSULATING GaAs

To demonstrate that the existence of a conduction band barrier between the two materials depends on the GaAs doping, the left plot of figure 5 shows tunable pump results for samples of ZnSe grown on n-type (white circles) and semi-insulating (dark triangles) GaAs . The semi-insulating GaAs sample shows no pump energy dependence which is consistent with the absence of any barrier impeding electrons from entering the ZnSe. The n-type sample shows behavior similar to the sample of figure 4. Further evidence for the difference between these samples was found in the intensity dependence of the signals. The right plot of figure 5 shows the ZnSe PR signal size plotted against pump intensity. A HeNe laser was used as the pump. A factor of 300 increase in the saturation intensity for the n-type sample. Electrons created in the depletion region with sufficient energy to enter the ZnSe from the n-type GaAs must overcome the opposing electric field and must do so before they relax in energy to the bottom of the conduction band via LO phonon emission which occurs on a time scale of several hundred femtoseconds[7]. Both of these effects act to decrease the efficiency of the pump laser in saturating the ZnSe PR signal.

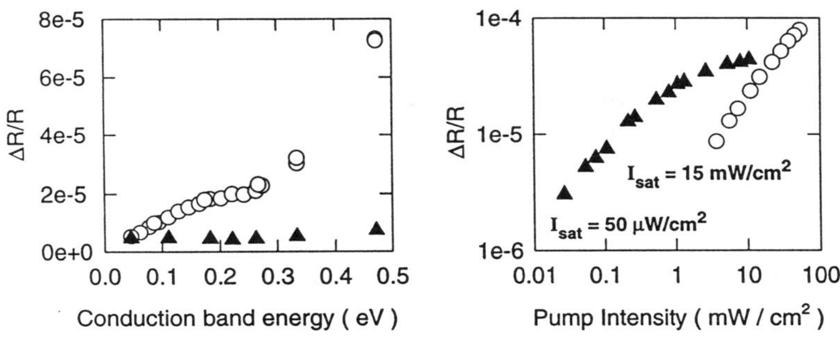

Figure 5
Left plot: ZnSe PR signals versus injected energy in the conduction band.
Right plot: Intensity dependence of the ZnSe PR signals using a HeNe laser.
N-type GaAs and semi-insulating GaAs are represented by open circles
and filled triangles respectively.

CONCLUSION

This study has demonstrated the importance of interfacial charges and the resulting band bending in understanding ZnSe/GaAs heterojunctions. The substrate doping appears to be crucial in determining the barrier height between the materials. The CER measurement of the GaAs electric fields leads directly to the band diagram of Figure 2. This picture immediately identifies the conduction band offset with the ZnSe PR threshold energy found in the tunable pump measurements. For all of the ZnSe/GaAs samples studied, a conduction band offset of between 150 and 250 meV was observed. Having established this framework of experiments and interpretation, a systematic investigation of the effect of doping and growth conditions on the electric fields and band offsets would appear to give very direct and useful information for device design.

ACKNOWLEDGMENTS

This research was supported in part by AFOSR, F49620-91-C-0091, by the Joint Services Electronics Program, and by DARPA through the National Center for Integrated Photonics Technology and the University Research Initiative Program under contract 284-25041.

REFERENCES

1.   M. S. Yeganeh, J. Qi, A. G. Yodh, ans M. C. Tamargo, Phys. Rev. Lett., **68,** 3761 (1992).

2.   A. Bonnani, L. Vanzetti, L. Sorba, A. Franciosi, M. Lomascolo, P. Prete, and R. Cingolani, Appl. Phys. Lett. **66**, 1092 (1995).

3.   J. Rennie, Y. Nishikawa, S. Saito, M. Onomura, and G. Hatakoshi, Appl. Phys. Lett. **68**, 2971 (1996).

4.   X. Yin, X. Guo, F. H. Pollak, G. D. Pettit, J. M. Woodall, and Eun-He Cirlin, SPIE Proceedings Vol. 1678, 168 (1992).

5.   H. K. Yow, P. A. Houston, and M. Hopkinson, Appl. Phy. Lett. **66**, 2852 (1995).

6.   O. Pages, M. A. Renucci, O Briot, and R. L. Aulombard, J. Appl. Phys, **80** 1128 (1996).

7.   R. G. Ulbrich, J. A. Kash, and J. C. Tsang, Phys. Rev. Lett. **62**, 949 (1989).

CONTACTLESS ELECTROREFLECTANCE STUDY OF $In_xGa_{1-x}As/InP$ MULTIPLE QUANTUM WELL STRUCTURES INCLUDING THE OBSERVATION OF SURFACE/INTERFACE ELECTRIC FIELDS

L.V. MALIKOVA *, J.Z. WAN *, FRED H. POLLAK *, J.G. SIMMONS ** and D.A. THOMPSON **
* Physics Department and New York State Center for Advanced Technology in Ultrafast Photonic Materials and Applications, Brooklyn College of the City University of New York, Brooklyn, New York 11210, USA
** Center for Electrophotonic Materials and Devices (CEMD), McMaster University, Hamilton, Ontario, L8S 4L7, Canada

ABSTRACT

Contactless electroreflectance measurements at 300 K were performed on two $In_xGa_{1-x}As/InP$ [$x = 0.53$ (lattice-matched) and 0.75] samples containing three quantum wells (QWs) grown by gas-source molecular beam epitaxy. The spectra consisted of two excitonic transitions (1e-1hh and 1e-1lh), corresponding to the fundamental conduction to heavy (h)- and light(l)- hole transitions, respectively, in the QW portion and a complicated Franz-Keldysh oscillation (FKO) pattern originating in the InP regions. Comparison between the experimental energies of 1e-1hh/1e-1lh and a theoretical envelope function calculation (including the effect of strain) made it possible to evaluate the conduction band offset parameters $Q_c = 0.34 \pm 0.03$ and 0.57 ± 0.03 for $x = 0.53$ and 0.75, respectively. The InP related FKO beat patterns were analyzed by a Fourier transform method. It was found that the FKO spectra were due to the simultaneous contribution of at least three different fields (106 kV/cm, 36 kV/cm, and 23 kV/cm), which originate in the various interfaces, i.e., substrate/buffer, cap layer/surface, and buffer/QW structure. Identification of the different fields has been accomplished by comparison of the Fourier-transformed spectra before and after sulfur passivation of the structure surface.

INTRODUCTION

Lattice-matched and strained $In_xGa_{1-x}As/InP$ quantum wells (QWs) and multiple QWs are especially useful for the design of optoelectronic communication devices working in the wavelength range of 1.3-1.55 μm [1]. Understanding of these $In_xGa_{1-x}As/InP$ QW structures is important from both fundamental and applied perspectives [2,3]. For example, the conduction band offset parameter is composition dependent in the $In_xGa_{1-x}As/InP$ heterostructure system [2,4] in contrast to GaAs/GaAlAs structures.[5] A knowledge of the energy band diagram of the structure is very important for device fabrication purposes. Also the presence of built-in electric fields at various surfaces/interfaces can influence device performance.

The modulation spectroscopy method of electromodulation (EM) has proven itself to be an extremely useful method for the study of a variety of semiconductor microstructures [6-9]. In EM measurements the optical constants of the sample are modified by the periodic variation of an externally applied electric field. The resultant spectra exhibit sharp, derivative-like features corresponding to intersubband transitions in QWs [6-8]. In addition, under certain conditions the signal exhibits oscillatory features above the band gap [Franz Keldysh oscillations (FKO)] which are a direct measure of the built-in electric field [6-9].

In this paper we have utilized contactless electroreflectance (CER) [6,7] at 300K to study two multiple quantum well (MQW) $In_xGa_{1-x}As/InP$ ($x = 0.53$ and 0.75) structures. The spectra

Mat. Res. Soc. Symp. Proc. Vol. 448 © 1997 Materials Research Society

consisted of two excitonic transitions (1e-1hh and 1e-1lh), corresponding to the fundamental conduction to heavy (h)- and light (l)- hole transitions, respectively, in the MQW portion and a complicated FKO pattern originating in the InP regions. Comparison between the experimental energies of 1e-1hh/1e-1lh and a theoretical envelope function calculation (including the effect of strain) [10,11] made it possible to evaluate the conduction band offset parameters $Q_c = 0.34 \pm 0.03$ and 0.57 ± 0.03 for $x = 0.53$ and 0.75, respectively. The InP related FKO beat patterns were analyzed by a fast Fourier transform (FFT) method [12,13]. It was found that the FKO spectra were due to the simultaneous contribution of at least three different fields (106 kV/cm, 36 kV/cm, and 23 kV/cm), which originate in the various interfaces, i.e., substrate/buffer, cap layer/surface, and buffer/QW structure. Identification of the different fields has been accomplished by comparison of the Fourier-transformed spectra before and after sulfur passivation of the structure surface [14].

EXPERIMENTAL DETAILS

Two $In_xGa_{1-x}As$/InP QW samples were studied in this paper: sample A contained three lattice-matched QWs ($x = 0.53$) while in sample B the three QWs were compressively strained ($x = 0.75$). The two materials were grown by gas-source molecular beam epitaxy (GSMBE) on (001) semi-insulating InP:Fe (001) substrates. Each sample was composed of three identical $In_xGa_{1-x}As$ QWs alternating between undoped InP barriers. The nominal well (L_W) and barrier widths are 30Å and 300Å, respectively, while the thicknesses of nominally undoped InP cap and buffer layers were 1500 Å and 2000 Å, respectively. The background doping level in the InP layers was $n \approx 2\times10^{15}$ cm^{-3}. The CER measurements were performed using a condenser-like system at room temperature. An *ac* modulating voltage (≈ 1 kV peak-to-peak) at a frequency of 200 Hz was employed. The value of the modulating voltage was sufficiently small so that all the measured line shapes were independent of the modulating voltage. The experimental details have been described previously [6,7]. The sulfur passivation procedure consisted of an ammonium sulfide ($(NH_4)_2S_x$) solution treatment of a previously etched sample, followed by drying in N_2 atmosphere in a glove box [14].

EXPERIMENTAL RESULTS

Displayed by the solid lines in Fig. 1(a) and 1(b) are the experimental CER spectra originating from the QW portion of samples A and B, respectively. The strong, lowest lying features are the fundamental intersubband electron to heavy-hole excitonic transition (1e-1hh), while the weaker high lying features are fundamental intersubband electron to light-hole excitonic transition (1e-1lh). The dashed lines in these figures are least-square fits to the first-derivative of a Lorentzian profile, which is appropriate for

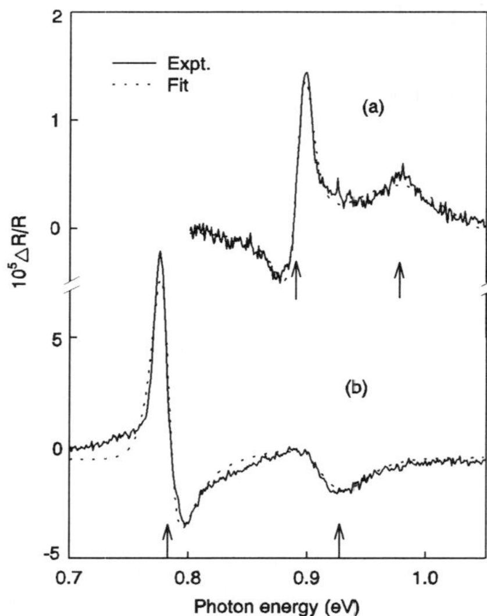

Fig. 1 Experimental CER spectra (solid lines) from the MQWs of samples (a) A and (b) B. The dotted lines are fits to a lineshape function.

excitonic transitions [6-8], The obtained energies of the 11H and 11L transitions are denoted by arrows.

In order to evaluate the conduction band offset parameter Q_c ($=\Delta E_c/\Delta E_g$, where ΔE_c is the conduction band discontinuity and ΔE_g the energy difference between the well and barrier materials, respectively) of our MQW structures we have compared the experimental exciton energies with an envelope function calculation [10] including the effects of strain. The electron, heavy- and light-hole masses and deformation potentials listed in Ref. [11] were employed. Exciton binding energies of $\approx$ 5 meV were deducted from the value of the theoretical transition energies. Any possible shifts related to the quantum-confined Stark effect due to the built-in fields were not taken into account.

We were not able to fit our experimental data, i.e., the energies of both 1e-1hh and 1e-1lh, to the theory for any value of Q_c if we used the nominal well width L_W = 30Å from the growth conditions. After several testing procedures, we determined that the energy positions for both the 1e-1hh and 1e-1lh exciton transitions could be accounted for simultaneously by assuming L_W = 36 Å with no changes in indium composition for either sample.

Figure 2 shows a comparison of the experimental results with the theoretical calculation with L_W = 36 Å for different values of Q_c. The dotted vertical lines are the experimental energies of the 1e-1hh and 1e-1lh exciton transitions while the shaded regions are the error bars. The solid lines indicates the calculated results as a function of Q_c. The best fit is obtained for Q_c = 0.34 $\pm$ 0.03 for the lattice-matched sample A (x = 0.53), as shown in Fig. 2a. A similar fit was made for sample B (x = 0.75) yielding Q_c = 0.57$\pm$0.03 (Fig. 2b).

The analysis of FKO is a very useful technique for the determination of built-in electric fields in a variety of semiconductor configurations [6-9]. Such fields are important for determining the energy band diagram of the semiconductor structure. In our MQW $In_xGa_{1-x}As$/InP structure there are several layers of the same material, e.g., InP. In general, the different built-in fields originating from charged localized surface/interface electronic states may simultaneously contribute to an EM spectrum. As a result one may observe a complicated spectrum with FKO exhibiting a beating pattern due to presence of a number of oscillations with different frequencies corresponding to the different fields as well as to transitions originating from the light- and heavy-hole valence bands [12,13,15,16].

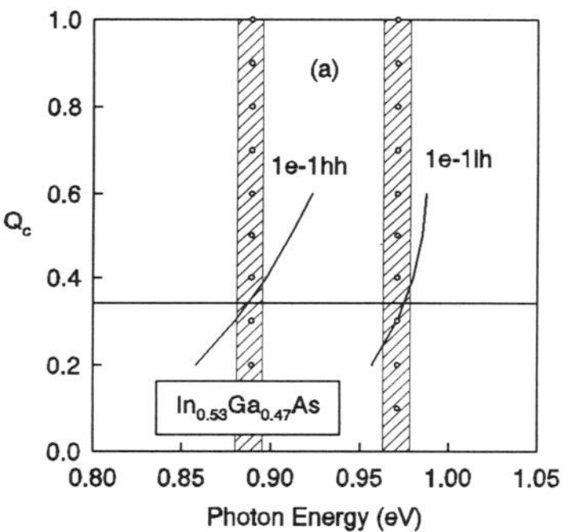

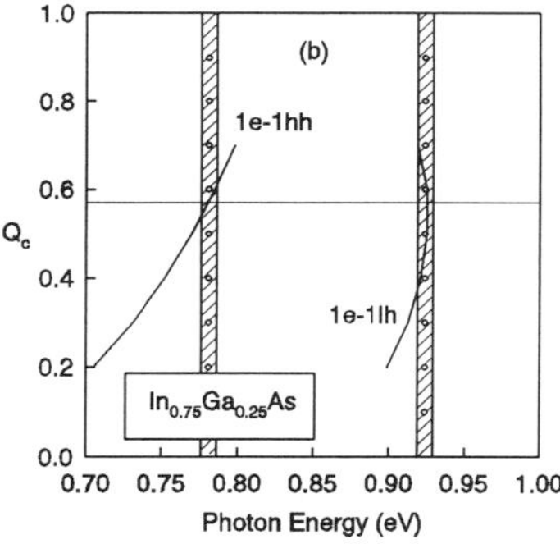

Fig. 2 Theoretical energies of the 1e-1hh and 1e-1lh transitions (solid lines) versus Q_c for samples (a) A and (b) B. The dotted lines are the experimental energies, error bars are shaded regions.

Shown in Fig. 3(a) is the CER data for sample A before sulfur passivation in the region of the InP band gap. The spectrum exhibits a complicated beat pattern indicating the contribution of several electric fields.

The asymptotic expression for FKO for $E > E_g$, where E and E_g are the photon energy and band gap energy, respectively, can be written as:[16]

$$E^2(E - E_g)\Delta R/R \sim \cos\{(4/3)[(E - E_g)^{3/2}(2\mu_\parallel)^{1/2}]/e\hbar F + \phi\} \tag{1}$$

where F is the electric field, $\mu_\parallel$ is the reduced interband effective mass in the direction of $\overline{F}$ and ϕ is an arbitrary phase factor. From Eq. (1) the extrema in the FKOs are given by:

$$n\pi = 4/3[(E_n - E_g)^{3/2}(2\mu_\parallel)^{1/2}/e\hbar F] + \phi \tag{2}$$

where n and E_n are the index number and energy of the nth extremum, respectively. A plot of $(4/3\pi)(E_n - E_g)^{3/2}$ versus index n yields a straight line whose slope is proportional to F.

Presented in Fig. 3(b) is the resultant FKO plot which has three different slopes. From this curve one cannot evaluate the corresponding electric fields, because each datapoint now represents the superposition of all effective electric fields in the structure.

To resolve the complex spectrum of Fig. 3(a) into its individual field contributions the FFT method was employed [12,13]. We first transformed the x-axis variable from E to $(E - E_g)^{3/2}$ and multiplied the spectrum by $E^2(E-E_g)$ in order to account for the inherent decay of the FKO [see Eq. (1)]. Shown in Fig. 4(a) is the FFT spectrum of Fig. 3(a). The transformed data clearly exhibit three peaks at frequencies of 26 eV$^{-3/2}$, 76 eV$^{-3/2}$ and 121 eV$^{-3/2}$, which we designate as f_1, f_2 and f_3, respectively. There is also a small feature at 53 eV$^{-3/2}$. There appears to be no evidence of a light hole transition contribution.

According to Eq. (1), the frequency (f) evaluated from the FFT is directly related to the magnitude of the electric field by:

$$f = 2(2\mu_\parallel)^{1/2}/3\pi e\hbar F \tag{3}$$

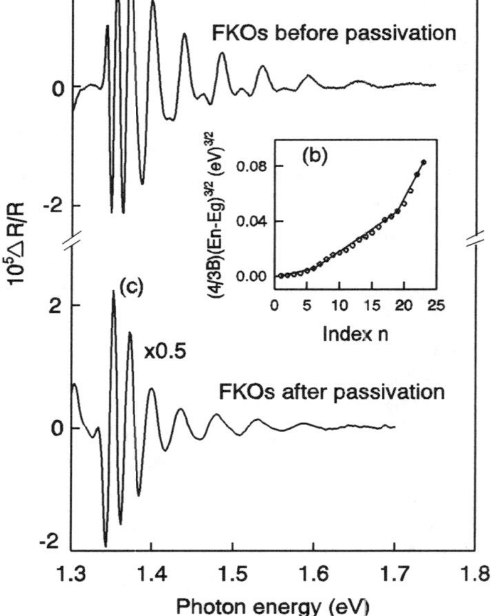

Fig. 3 CER spectra for sample A in the region of InP band gap before (a) and after (c) passivation; (b) is the resultant FKO plot before passivation.

From Eq. (3) the obtained electric fields corresponding to f_1, f_2 and f_3 are 106 kV/cm, 36 kV/cm and 22 kV/cm, respectively, while the field corresponding to the small feature equals 52 kV/cm. We used electron (0.079) and heavy-hole (0.45) effective masses (in units of the free electron mass) from Ref. [17].

To identify the origins of these various fields we have subjected the sample to a sulfurization treatment to suppress the surface contribution [14]. The CER spectrum in the region of the InP band gap and its FFT after passivation are displayed in Figs. 3(b) and 4(b), respectively. The passivation procedure has brought about a significant change in FKO spectrum. Comparing the FFT results before and after passivation demonstrates that this change is related mostly to the field associated with f_3 although there is also some influence on the f_2 structure. The f_1 resonance has remained essentially unchanged. We therefore conclude that the surface field is associated with f_3, i.e., 22 kV/cm, while f_1 and f_2 correspond to fields at the buried interfaces. We also observed that the passivation procedure did not effect the spectra originating from the MQWs, which indicates that there is only a small electric field within this region.

DISCUSSION

Our value for the conduction band offset parameter for the lattice-matched sample is in good agreement with the theoretical consideration of M. Hybertsen [18]. The variation of Q_c with indium composition is consistent with the results of Gershoni *et al.* [4]. The two monolayer deviation of the well width from the nominal

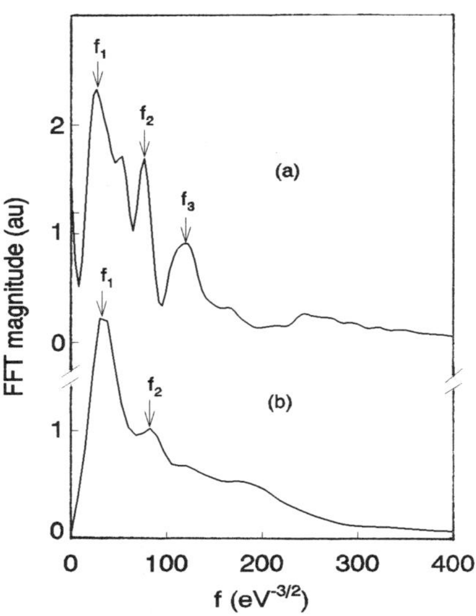

Fig. 4 FFT of the CER spectra in the region of the InP band gap before (a) and after (b) passivation for sample A.

value of 30Å can be explained by the existence of the interlayers between the well and barriers, so that the effective well width becomes broadened [19].

Given the low background doping level in the InP regions it is difficult to account for the observed electric fields based on space charge considerations. Therefore, in order to explain the results for the various electric fields we suggest that in our structures the Fermi level is pinned (a) at the top of the valence band at the substrate/buffer interface, probably by carbon [20], (b) at the top of the conduction band at the MQW/buffer layer interface and (c) at 0.3 V below the conduction band at the free surface [21]. Based on the dimensions of the cap layer (1500Å) the surface electric field is 20 kV/cm, in good agreement with the field associated with the f_3 structure. Also, based on the dimensions of the buffer layer (2000Å) and the pinning at the buffer/substrate and buffer/MQW interfaces the field in this region is calculated to be 70 kV/cm. This is somewhat smaller than the field associated with f_1 but there is reasonable agreement. The remaining field probably originates in the substrate.

The authors of Ref. [22] have observed a large built-in field (F = 39 kV/cm) near the interface formed by a semi-insulating InP:Fe substrate and an InGaAs epilayer. They considered the possibility that the Fe-impurities influenced the nature of the interface charges.

Also the fact that there is a small field in the MQW region is consistent with this picture.

ACKNOWLEDGEMENTS

The authors LVM, JZW and FHP gratefully acknowledge helpful discussion with Prof. D. L. Lile, Dr J. L. Freeouf and Dr. R. Iyer concerning the passivation techniques in this paper. This work was supported by the National Science Foundation grant #DMR-9414209,

PSC/BHE grant #666424 and the New York State Science and Technology Foundation through its Centers for Advanced Technology program.

REFERENCES

1. B. B. Elenkrig, D. A. Thompson, J. G. Simmons, D.M. Bruce, Y. Si, J. Zhao, J.D. Evans, and I.M. Templeton, Appl. Phys. Lett. **65**, 1239 (1994).
2. H. Temkin, D. Gershoni, and M. B. Panish, in Semiconductors and Semimetals, Vol.**40**, ed. A.C. Gossard (Academic Press, New York, 1994) p. 337.
3. J. Z. Wan, D. A. Thompson and J. G. Simmons, Nucl. Inst. & Methods in Phys. Res. **B 106**, 461 (1995).
4. D. Gershoni, H. Temkin, J. M. Vandenberg, S. N. G. Chu, R. A. Hamm, and M. B. Panish, Phys. Rev. Lett. **60**, 448 (1988).
5. See, for example, M. Missous in Properties of Aluminum Gallium Arsenide, ed. by S. Adachi (INSPEC, London, 1993) p. 73.
6. F. H. Pollak and H. Shen, Mater. Sci. Eng. **R10**, 275 (1993) and reference therein.
7. F. H. Pollak in Handbook on Semiconductors Vol. 2, Optical Properties of Semiconductors, ed. M. Balkanski (North Holland, Amsterdam, 1994) p.527.
8. O.J. Glembocki and B.V. Shanabrook in Semiconductors and Semimetals, Vol. **36**, ed. D.G. Seiler and C.L. Littler (Academic, New York, 1992) p. 221 and references therein.
9. H. Shen and M. Dutta, J. Appl. Phys. **78**, 2151 (1995).
10. G. Bastard and J. A. Brum, IEEE J. Quantum Electron., **QE-22**, 1625 (1986).
11. S. H. Pan, H. Shen, Z. Hang, F. H. Pollak, W. Zhuang, Q. Xu, A. P. Roth, R. Masut, C. LeCelle, and D. Morris, Phys. Rev. B **38**, 3375 (1988).
12. V. Alperovich, A. Jaroshevich, H. Scheibler and A. Terekhov, Solid State Electron., **37**, 657 (1994).
13. R. Holm, O. Glembocki and J. Tuchman, Mater. Res. Soc. Symp. Proc. **406**, 247 (1996).
14. R. Iyer, R.R. Chang, A. Dubey, and D.L. Lile, J. Vac. Sci. Technol. **B6**, 1174 (1988).
15. D. Wang and C. Chen, Appl. Phys. Lett. **67** (14), 2069 (1995).
16. D. Aspnes and A. Studna, Phys. Rev. B **7**, 4605 (1973).
17. Numerical Data and Functional Relationships in Science and Technology, ed. by O. Madelung, M. Schulz and H. Weiss, Landolt-Bornstein, New Series, Group III, Vol.17 (Springer, New York, 1982).
18. M.S. Hybertsen, Appl. Phys. Lett. **58**, 1759 (1991).
19. C. Orme, M. D. Johnson, J. L. Sudijono, K. T. Leung and B. G. Orr, Appl. Phys. Lett. **64**, 860 (1994).
20. M.L. Gray and F.H. Pollak, J. Appl. Phys. **74**, 3426(1993).
21. A. Ismail, A. Ben Brahim, L. Lassabatere and I. Lindau, J. Appl. Phys. **59**, 485 (1986).
22. W. Zhou, M. Dutta, H. Shen, and J. Pamulapati, J. Appl. Phys. **73**, 1266 (1993).

OPTICAL INVESTIGATIONS OF InAs GROWTH ON GaAs AND LASING IN SINGLY AND MULTIPLY STACKED ISLAND QUANTUM BOXES

A. KALBURGE, T. R. RAMACHANDRAN, R. HEITZ, Q. XIE, P. CHEN, and A. MADHUKAR
Photonic Materials and Devices Laboratory, University of Southern California, Los Angeles, CA 90089-0241

ABSTRACT

We report on the optical investigations of InAs growth on GaAs. *In-situ* STM/AFM studies show the presence of features 2-4 ML high, which we call quasi-3D (Q3D) clusters, well in advance of 3D island formation. Though the photoluminescence (PL) emission from these Q3D clusters is in the same wavelength regime as that from well developed 3D islands, they show characteristic differences in the PL excitation spectra and temperature dependence of PL, distinguishing them clearly from the 3D islands. Finally, we discuss the lasing observed from lasers containing single and five sets of InAs layers grown with conditions in which the *in-situ* STM/AFM studies show only 3D islands.

I. INTRODUCTION

The formation of coherent nm-scale islands in highly strained In(Ga)As on GaAs epitaxy has attracted a lot of interest in recent years [1-7]. Capped with GaAs, these islands act as high quality quantum boxes, dubbed quantum dots (QDs), providing strong carrier confinement with state splittings exceeding kT at room temperature[8]. These QDs have been assumed to be responsible for the lasing of injection diodes containing one or several InAs layers in the active region, which show the predicted [9] increase in the temperature stability of the threshold current density [10-16]. Photoluminescence (PL) spectra of samples with InAs depositions below the critical deposition for island formation (Θ_C) show strong PL in the spectral region around 850 nm, which is attributed to the InAs wetting layer (WL), whereas for depositions above Θ_C PL at wavelength greater than 900 nm is observed and attributed to 3D islands [3-7].

However, we demonstrated recently that for InAs on GaAs growth an intermediate surface state exists, where 2 to 4 ML high features are formed [7]. We dubbed them quasi-3D clusters in order to distinguish them from well developed 3D islands with heights of a few nm, which are the energetically favored surface features above Θ_C. The quasi-3D clusters are observed for coverages as low as 1.25 ML, well below Θ_C of 1.57 ML, and remarkably show a re-entrant behavior in the 2D to 3D morphology change depending on the kinetically controlled surface evolution. The quasi-3D clusters give rise to PL in the 900 to 1000 nm wavelength regime, in which most of the reported lasing from InAs/GaAs QD lasers is observed, thus raising a question whether these quasi-3D clusters may be involved in the lasing action.

In this paper, we report on investigations of the optical properties of these quasi-3D clusters using photoluminescence (PL) and PL excitation (PLE) spectroscopy. Though the PL wavelengths regimes for quasi-3D clusters and 3D islands are very similar, they show characteristic differences in the PLE behavior and the PL temperature dependence. We discuss the implications of these findings for the lasing observed from QD lasers fabricated from single and five stacks of InAs layers grown under conditions for which our STM investigations show *only* 3D islands.

Mat. Res. Soc. Symp. Proc. Vol. 448 © 1997 Materials Research Society

II. EXPERIMENTAL

Samples for optical investigation were grown via molecular beam epitaxy (MBE) on semi-insulating GaAs(001) ($\pm$ 0.1°) substrates. InAs was deposited at 500 °C on a 5000 Å GaAs buffer showing clear c(4×4) surface reconstruction at a growth rate of 0.22 ML/s and an As_4 partial pressure of 6×10^{-6} Torr. The RHEED pattern changed from streaky to spotty at 1.57 ML InAs deposition signifying the onset of 3D island formation. The samples were cooled down immediately after the InAs growth and capped with GaAs using migration enhanced epitaxy (MEE) at 400°C resulting in a high efficiency of the 3D island PL [4]. We report here on samples with InAs coverages of 1.25, 1.74 and 2.00 ML, given with respect to the GaAs surface density. Uncapped samples for in-situ STM/AFM studies were grown on n^+-substrates in an otherwise identical way [7,17]. For the laser structures, the active regions containing a single or five stacks of InAs layers separated by 36 ML GaAs are sandwiched between AlAs/GaAs superlattices forming a waveguide and n- and p- cladding layers [13].

III. QUASI-3D CLUSTERS

Recently, we reported detailed in-situ STM/AFM investigations of samples grown under identical conditions as used for this work with incremental increasing InAs deposition ranging from 0.87ML to 2.00 ML [7,18]. The results demonstrate that with increasing InAs deposition 3D islands are not spontaneously formed at Θ_C, but that they first coexist with 2D (1 ML high) and quasi-3D (2-4 ML high) clusters, the latter apparently acting as precursors to the islands. Only for coverage above 1.74 ML the 3D islands become the dominating surface feature with the disappearance of the quasi-3D clusters.

Fig. 1 gives STM images for the sample with 1.25 ML InAs deposition well below Θ_C showing that the InAs layer is far from being 2D at this deposition. In addition to small and large 2D clusters (denoted A and B) we find a significant density ($> 2\ \mu m^{-2}$) of quasi-3D clusters, which are either small (lateral extent $<\sim$ 20nm, denoted C') or large (lateral extent $>\sim$ 50nm, denoted C). Fig. 1(b) shows magnified image of a situation where a large quasi-3D cluster is topped by a small one resulting features upto 6ML high. Remarkably, for a slightly higher InAs deposition of 1.35ML, these quasi-3D clusters disappear completely before small quasi-3D clusters reappear with a much higher density (~140 μm^{-2}) for depositions above 1.45 ML, still well in advance of the regime of 3D island formation ($\Theta >$ 1.57 ML). Thus, the quasi-3D clusters establish a re-entrant behavior in the 2D to 3D morphology transition and show that in contrast to the common description of Stranski-Krastanow growth the

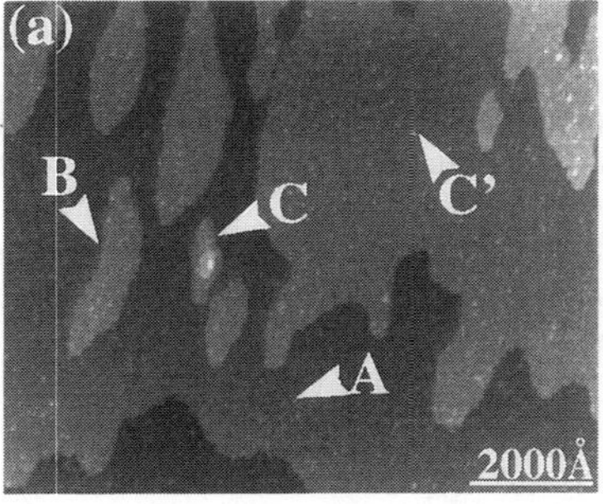

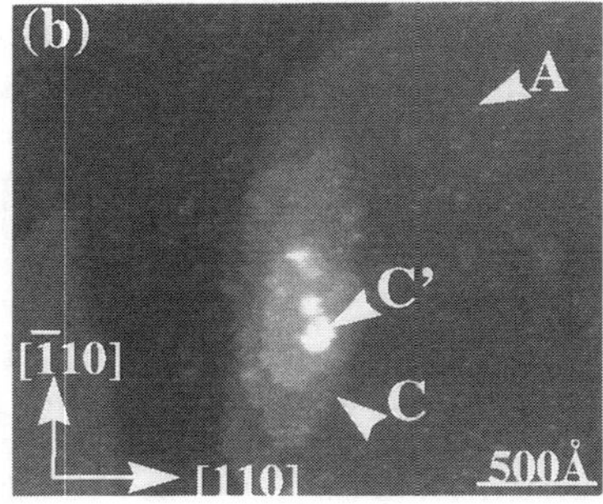

Fig. 1: In-situ STM images for an InAs deposition of 1.25 ML showing the presents of small and large quasi-3D clusters (C' and C).

strained InAs layer can assume an intermediate state.

The STM/AFM results show, that the used growth conditions allow us to study the optical properties of the quasi-3D clusters and the 3D islands separately, using samples with an InAs deposition of 1.25 ML and above 1.74 ML, respectively.

IV. OPTICAL PROPERTIES OF THE QUASI-3D CLUSTERS

Fig. 2 compares low temperature PL spectra excited with a density of 5 Wcm^{-2} at 514 nm for samples with InAs depositions of 1.25 and 2.00 ML. The 2.00 ML sample shows an almost Gaussian peak centered at 1020 nm, which we attribute to the ground state transition of 3D islands, which have a density of about 550 μm^{-2} and are the only 3D-feature observed in AFM studies on the equivalent uncapped samples. The PLE spectrum recorded at the PL peak maximum is represented by the dashed line. The strongly decreasing excitation efficiency with increasing excitation wavelength is typical for QDs formed from well developed 3D islands [4, 19]. It reflects the strong carrier localization and demonstrates the presence of excited states in the 3D island QDs. The sharp

excitation resonance at 857.8 nm indicates the 3D islands to sit on a high quality WL. From the spectral position we estimate a WL thickness near 1 ML.

Remarkably, the 1.25 ML sample, for which we find no 3D islands in corresponding uncapped samples, shows in addition to an intense wetting layer (WL) peak at 853 nm, a PL peak centered at 974 nm, which is about 500 times weaker than the 3D island PL in the 2.00 ML sample excited at 514 nm. In principle, this could be due to a very low density of 3D islands formed at this early stage. The PLE behavior (dashed spectrum) detected at 985 nm, however, differs completely from that of the 3D islands. The PL intensity increases about 100 times when exciting below the GaAs

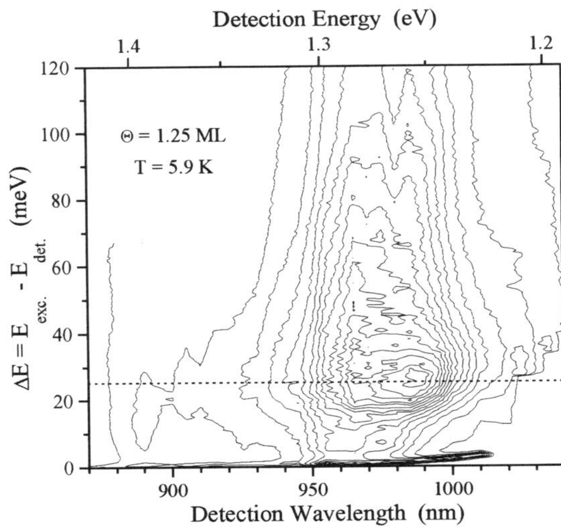

Fig. 2: PL excited at 514 nm and PLE spectra for samples with 1.25 and 2.00 ML InAs deposition. STM/AFM images show large quasi-3D clusters present for the 1.25 ML sample and 3D islands for the 2.00 ML sample.

Fig. 3: The PL intensity for the 1.25 ML sample in dependence of the detection wavelength and the difference ΔE between detection and excitation energy.

489

bandgap and reaches a maximum about at 965.5 nm near the detection as shown by the PLE spectrum in Fig. 2. This behavior indicates a high density of states until close above the luminescent state, indicating a QW-like structure, and the spectral position implies an InAs QW thickness above 2 ML, which is consistent with the large quasi-3D clusters with a lateral extension larger than the exciton Bohr diameter and which are present in a density of about 2 μm^{-2} at this delivery [7]. Fig. 3 shows the PL intensity dependence of the detection wavelength (energy) and the difference ΔE between detection and excitation energy. Independent of the detection wavelength the PL intensity has a maximum, exciting about 25 meV above the detection energy indicating that in

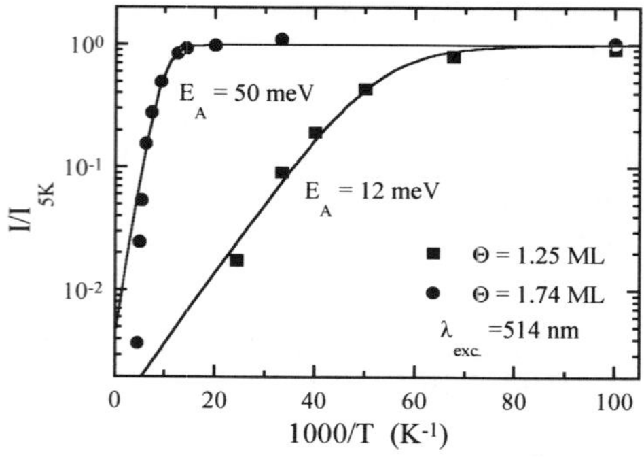

Fig. 4: Arrhenius plot of the normalized intensity of the 974 nm emission in the 1.25 ML sample and the emission of 3D island QDs in a 1.74 ML sample. Full lines correspond to a fit assuming a thermally-activated non-radiative recombination process.

fact localized states give rise to the observed PL. Comparing with the STM results we propose the small quasi-3D clusters to give rise to small QDs with a carrier localization of about 25 meV and to act as PL centers. The emission wavelength is determined by the properties of the local WL, i.e. the large quasi-3D clusters. The PLE spectra reveal no excitation resonance corresponding to the WL observed in PL indicating weak lateral diffusion.

Fig. 4 compares the temperature dependence of the PL intensity of the 974 nm peak in the 1.25 ML sample and that of the 3D island peak in a 1.74 ML sample. Whereas for 3D island QDs the intensity remains practically constant up to 70 K, reflecting the strong carrier localization, it starts to decrease above 15 K in the 1.25 ML sample. The full lines represent fits assuming a thermally-activated non-radiative recombination process, which yields an activation energy E_A of (12 ± 3) meV for the 1.25 ML sample. The activation energy indicates thermal evaporation of carriers out of the small quasi-3D clusters, providing a combined localization energy of 25 meV for electrons and holes, and efficient non-radiative recombination in the large quasi-3D clusters.

The presented results indicate that for the earliest stages of InAs on GaAs growth, two different structural features exist which give rise to PL at wavelength longer than 900 nm. One is 3D islands on a thin '1ML' WL as observed for the 2.00 ML sample and the other is small quasi-3D clusters sitting on locally thicker WL regions as provided by the large quasi-3D clusters. These features can be distinguished by their PLE behavior and result in a different temperature dependence of the PL intensity.

V. InAs/GaAs QUANTUM DOT LASERS

3D islands formed in highly strained epitaxy have been employed as active medium for injection lasing [10-16] due to their high optical quality combined with large carrier confinement and state splitting. The lasing occurs frequently on the shorter wavelength side of the PL peak leading to speculation on the role of excited states or the WL in the lasing process. However, in view of the present study quasi-3D clusters, providing optical transitions in the 900 to 1000 nm region, might contribute in the lasing in this spectral region.

Fig. 5 comprises results obtained for two laser structures, one with a single set and the other with a fivefold stack of InAs 3D islands as active medium. Details of the used laser structure are given elsewhere [13]. 2.00 ML InAs have been deposited in each single layer under the same conditions as used for the samples used in the above structural and optical studies yielding a 3D island density of about 550 μm^{-2}. The stack is grown with a 36 ML GaAs spacer between each layer, leading to an almost perfect vertical alignment of the 3D islands [20] without introducing significant electronic coupling. The PL spectra of both samples given as dotted lines in Fig. 5, show a PL peak centered at 1050 nm indicating the comparability of the formed QDs in both samples. PLE results show the WL to be around 860 nm.

The full lines in Fig. 5 show the 79K electroluminescence spectra for pulsed excitation and injection currents below and above the lasing threshold for lasers with cavity lengths of 2500 μm for the single set one and 2200 μm for the five set one. Sharp breaks in the respective L-I curves, shown in the insets, indicate threshold current densities (J_{th}) of 429 A/cm^2 and 310 A/cm^2, respectively. The L-I characteristics as well as the sharp

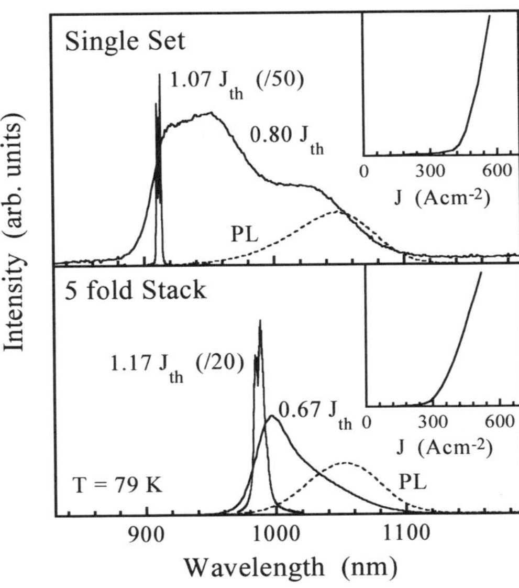

Fig. 5: Electroluminescence and lasing spectra of diode structures containing a single 2.00 ML InAs layer or a corresponding fivefold stack with 36 ML spacers as active layer. The dashed lines give normalized PL spectra of the laser structures for reference. The insets give the light output versus the injection current density.

spectral narrowing of the electroluminescence spectra above J_{th} indicate lasing. However, the lasing occurs at 912 and 988 nm for single set and the fivefold stacked samples, respectively, on the short wavelength side of the 3D island PL peak. Though, the lasing can not be attributed to the WL the spectral range corresponds to that of the quasi-3D clusters. However, our STM/AFM investigations show that under the used growth conditions and for an InAs deposition of 2.00 ML, in addition to the WL, only 3D islands are expected. Additionally, Fig. 4 shows that at 79 K, where the lasing is observed, only PL from the 3D islands remains, making it unlikely that the quasi-3D clusters can contribute to the lasing at this temperature. On the other hand, the PLE spectrum of the 2.00 ML sample shown in Fig. 2 reveals excitation resonances 66 and 98 meV above the detection energy indicating the presence of excited states in the 3D islands [4, 19].

Finally, we comment upon the shift in the lasing wavelength to the longer side along with the lower lasing threshold observed for the laser with five stacked of InAs layers as compared to laser with a single InAs layer. Shoji *et al* reported a similar jump in the lasing wavelengths for lasers made of single and three stacks of InGaAs layers [12]. They attributed this jump to lasing from excited states in the single stack laser but from the ground states in the three stack laser. Due to gain saturation [15] in the single set laser, lasing from ground states may be inhibited initiating lasing from the excited states. As the island density is increased in the stacked laser the mean occupancy of the states needed in QDs for lasing decreases [21]. Therefore, lasing threshold would decrease and the lasing wavelength would shift to longer values with increasing island density, as observed in the above lasers. It is suggested that the decrease in threshold with island density (i.e. number of stacks) is not monotonic but shows a minimum at an optimum value of island density [14, 21]. We therefore believe that under the growth conditions employed, such an optimum is not yet reached in our lasers. Further reduction in the threshold and lasing from ground states (i.e. at the wavelength corresponding to the PL maximum of the QDs) is possible by further increasing the number of stacks or by reducing the cavity loss by applying high-reflection coating on both facets [12].

VI. CONCLUSION

We have presented optical investigations of InAs growth on GaAs. *In-situ* STM/AFM studies have shown the presence of features 2-4 ML high, at an InAs deposition as low as 1.25 ML which is well in advance of 3D island formation. These Q3D clusters give photoluminescence (PL) emission in the same wavelength regime as well developed 3D islands thus raising a question about their possible role in lasing. They show a quantum well-like electronic structure in PLE and small localization energies for electrons and holes as indicated by temperature dependence of PL, revealing efficient thermally activated non-radiative recombination. We have presented we discuss the lasing observed from lasers containing single and five sets of InAs layers with nominal deposition of 2.00 ML. At this deposition our STM/AFM studies show that only 3D islands and the wetting layer are present. Therefore, the lasing observed is attributed to states of 3D island QDs. PLE spectra of 3D islands show the presence of excited states suggesting their role in lasing action.

This work was supported by U.S. AFOSR, ARO, and ONR.

REFERENCES

1. S. Guha, A. Madhukar, and K. C. Rajkumar, Appl. Phys. Lett. **57**, 2110 (1990)
2. D. Leonard, K. Ond, and P. M. Petroff, Phys. Rev. B **50**, 11687 (1994)
3. J. M. Gerard *et al*, J. Cryst. Growth **150**, 351 (1995)
4. Q. Xie *et al*, J. Crsyt. Growth **150**, 357 (1995)
5. A. Polimeni *et al*, Phys. Rev. B **53**, R4213 (1996)
6. D. Bimberg *et al*, Jpn. J. Appl. Phys. **35**, 1311 (1996)
7. T. R. Ramachandran *et al*, J. Cryst. Growth (to be published)
8. M. Grundmann *et al*, Phys. Rev. Ltt. **74**, 4043 (1995)
9. Y. Arakawa and H. Sakaki, Appl. Phys. Lett. **40**, 939 (1982)
10. N. Kirstaedter *et al*, Electron. Lett. **30**, 1416 (1994)
11. H. Shoji *et al*, IEEE Photon. Technol. Lett. **12**, 1385 (1995)
12. H. Shoji *et al*, Jpn. J. Appl. Phys. **35**, L903 (1996)
13. Q. Xie *et al*, IEEE Photon. Technol. Lett. **8**, 965 (1996)
14. O. Schmidt *et al*, Electron. Lett. **32**, 1302 (1996)
15. N. Kirstaedter *et al*, Appl. Phys. Lett. **69**, 1226 (1996)
16. K. Kamath *et al*, Electron. Lett. **32**, 1374 (1996); R. Mirin *et al*, Electron. Lett. **32**, 1732 (1996)
17. N. P. Kobayashi *et al*, Appl. Phys. Lett. **68**, 3299 (1996)
18. T. R. Ramachandran *et al*, Appl. Phys. Lett. (to be published)
19. R. Heitz *et al*, Appl. Phys. Lett. **68**, 361 (1996)
20. Q. Xie *et al*, Phys. Rev. Lett. **75**, 2542 (1995)
21. L. V. Asryan and R. A. Suris, Semicond. Sci. Technol. **11**, 554 (1996)

METROLOGY OF VERY THIN SILICON EPITAXIAL FILMS USING SPECTROSCOPIC ELLIPSOMETRY

Weize Chen*, Rafael Reif**
*Department of Materials Science and Engineering
**Department of Electrical Engineering and Computer Science
Massachusetts Institute of Technology, Cambridge, MA 02139

ABSTRACT

Non-destructive silicon epitaxial film thickness measurements have been performed using spectroscopic ellipsometry (SE) in the visible to near infrared range. The epitaxial films were grown by CVD at 700 – 900°C. It is shown that SE can simultaneously determine the epitaxial film thickness, the native oxide thickness, and the substrate dopant concentration. The epitaxial film thicknesses measured by SE are in excellent agreement with results of secondary ion mass spectrometry (SIMS). The dopant concentrations measured by SE also agree well with SIMS results for n-type substrates, but are consistently higher than SIMS values for p-type substrates. This study shows that spectroscopic ellipsometry could be used in semiconductor manufacturing environments as a viable non-destructive technique for sub-0.5 μm silicon epitaxial film thickness measurement.

INTRODUCTION

Silicon epitaxy is one of the most commonly practiced process technologies used in a wide variety of device structures, such as MOS, bipolar digital and analog devices, power devices, CCDs, BiCMOS, and HBTs. For most devices, the epitaxial layer thickness is a critical parameter that must be accurately measured and controlled. For example, several bipolar transistor device parameters such as breakdown voltage, junction capacitance, transistor current gain, and high frequency performance depend on the epitaxial layer thickness. CMOS latch-up, breakdown voltage, and drive-in time are also related to the thickness of the epitaxial films. As the dimension of modern devices is scaling down, the silicon epitaxial films used in integrated circuits will undoubtedly be decreasing. Consequently, advanced thin film deposition technologies and thin film metrology are receiving increasing attention. For example, a low temperature process is required to minimize both lateral and vertical diffusion and autodoping. Almost all applications that would use low temperature epitaxial silicon, however, would be less than 0.5 μm in thickness. This is below the detection limit of currently available non-destructive process monitor techniques such as Fourier transform infrared interferometry (FTIR) [1, 2]. In addition, the requirement in accuracy and precision will become more stringent as the overall thickness of the epitaxial layer decreases. From the economical point of view, as the wafer size increases, the cost of epitaxial wafers becomes substantial, making destructive measurement more and more costly. It is therefore necessary to extend the current non-destructive measurement capability into the sub-0.5 μm regime.

Spectroscopic ellipsometry (SE) has been widely used as an essential metrology tool for the semiconductor industry. This non-destructive optical technique offers fast measurement speed, high precision, and the ability of measuring very thin or multilayered films. Moreover, it is also suitable for on-line and *in-situ* monitoring and control in semiconductor manufacturing environments. The objective of this paper is to evaluate SE in the visible to near infrared range as a new technique to overcome the limitation mentioned above and extend

493

Mat. Res. Soc. Symp. Proc. Vol. 448 © 1997 Materials Research Society

the current measurement capability into the sub-0.5 μm regime.

EXPERIMENT

Undoped epitaxial films of various thickness were CVD grown on heavily boron or arsenic doped substrates at 700 – 900°C using commercial ASM reactors. For comparison and process monitoring, a few epitaxial films were also grown on lightly doped substrates under exactly the same deposition conditions. The idea of using low deposition temperatures is to minimize dopant depletion, outdiffusion and autodoping so that a sharp dopant profile across the metallurgical junction between the undoped epitaxial film and the heavily doped substrate can be obtained. Cross-sectional transmission electron microscopy (XTEM) and SIMS were used to characterize some of the samples.

Ellipsometric measurements were performed with a UVISEL spectroscopic phase modulated ellipsometer (SPME) made by Instrument SA, Inc. This ellipsometer is equipped with near infrared (NIR) extension and covers the spectral range 230 – 1700 nm. SPME measures the quantities I_s and I_c: [3]

$$I_s = sin2\psi sin\Delta \tag{1}$$

$$I_c = sin2\psi cos\Delta \tag{2}$$

The standard ellipsometry parameters ψ and Δ are defined by:

$$\rho = \frac{r_p}{r_s} = tan\psi e^{i\Delta} \tag{3}$$

where r_p and r_s are the Fresnel reflection coefficients for light polarized parallel and perpendicular to the plane of incidence, respectively.

Silicon is semi-transparent below the direct band edge, resulting in Δ being near 180° or 0°. Thus, for this particular application, SPME is superior to rotating analyzer ellipsometer (RAE) which measures the quantities $tan\psi$ and $cos\Delta$, but not $sin\Delta$ [4].

DATA ANALYSIS

The model of the sample structure used in this study consists of air, native oxide, undoped silicon epitaxial film, and heavily doped silicon substrate. The dielectric function of heavily doped silicon is modeled with the Drude theory:

$$\epsilon(\omega) = \epsilon_c(\omega) - \frac{4\pi Ne^2/m^*}{\omega^2 + i\omega/\tau} \tag{4}$$

where ω is the photon angular frequency; ϵ_c is the dielectric function of undoped crystalline silicon; N is the free carrier concentration; m^* is the free carrier effective mass, which is taken as $0.26m_0$ for electrons and $0.36m_0$ for holes, m_0 is the free electron mass; e is the electron charge; and τ is the free carrier relaxation time.

Optical constants of silicon dioxide are taken from ref. 5. Optical constants of undoped crystalline silicon are taken from ref. 5 for photon energies below 1.5 eV, and from ref. 6 for photon energies above 1.5 eV. For photon energies above 0.7 eV, the dielectric function of heavily doped silicon is insensitive to the free carrier relaxation time τ, thus, a value of 1.45×10^{-14} s is used in the calculation. This value is calculated with a boron concentration of 10^{19}cm^{-3} using the formula given in ref. 7. The native oxide thickness, epitaxial film

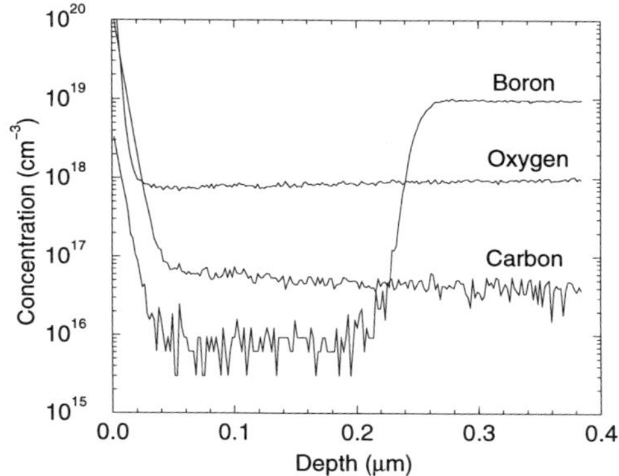

Figure 1: SIMS profiles of an undoped epitaxial film grown at 900°C on a heavily boron doped substrate.

thickness, and substrate free carrier concentration are then extracted from the measured ψ and Δ spectra through least-squares optimization using the Levenberg-Marquardt non-linear regression algorithm.

RESULTS

Figure 1 is the typical SIMS results of the samples measured at EVANS EAST, showing a good quality epitaxial film and a sharp dopant profile.

The defect density in the epitaxial film and at the epi/substrate interface is under the detection limit of high resolution XTEM. Results of SIMS and XTEM support the model of the sample structure used in this study.

Figure 2 plots measured ellipsometric data and model fits for four epitaxial films with different thickness and substrate dopant concentration. Three of them are boron doped, and the other is arsenic doped. Strong interference can be observed with an angle of incidence of 70° – 74°. However, as expected, no interference is detectable for those epitaxial films grown on lightly doped substrates. This observation proves that the interference shown in Figure 2 is brought about by the heavy doping in the substrates. Table 1 summarizes the 95% confidence intervals of the extracted parameters for these four samples. For comparison, SIMS results are also given in Table 1. As shown in Figure 2 and Table 1, excellent fits to the measured data are obtained using the Drude theory and the simple film structure described previously. The thicknesses measured by SPME are in excellent agreement with SIMS data. For n-type doping, the substrate dopant concentration agrees well with SIMS value. These results are in agreement with refs. 8 and 9. However, for p-type doping, the calculated substrate dopant concentrations are consistently higher than the SIMS results. The compressive strain in the epitaxial films is too small to account for this discrepancy. The discrepancy may be due to the fact that there are errors in SIMS results for boron, or the heavy doping causes the hole effective mass to decrease. However, van Driel's [10] theoretical calculation indicated that the hole effective mass increases with dopant concentration due to the nonparabolicity of the valence bands. Further study is needed for a plausible explanation.

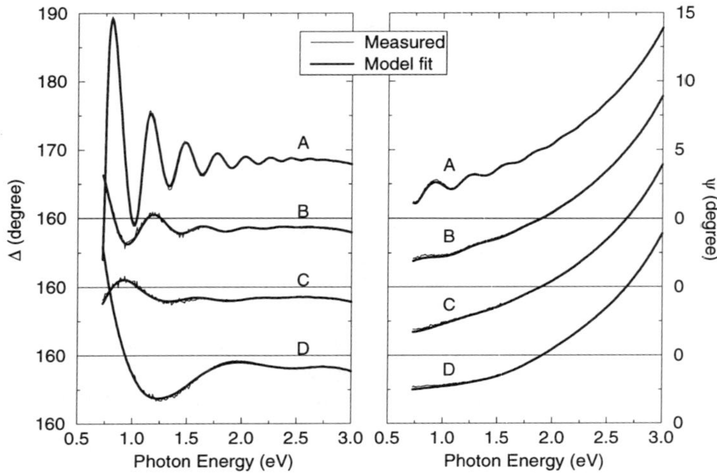

Figure 2: Measured *vs.* model fits at an angle of incidence $\phi = 73°$. The letters next to the plots are sample names. Results of parameter extraction and SIMS for these samples are summarized in Table 1.

Table 1: Results of parameter extraction and SIMS

Sample	Dopant	Resistivity (Ω-cm)	N ($\times10^{19}$cm^{-3}) Ellip.	SIMS	Epi film (Å) Ellip.	SIMS	Oxide (Å) Ellip.
A	Boron	0.001 − 0.002	8.61±0.07	5.02±0.07	5050±2	5200±180	17.3±0.1
B	Arsenic	0.001 − 0.005	1.60±0.05	1.61±0.04	3400±8	3600±220	17.1±0.1
C	Boron	0.005 − 0.010	1.43±0.07	0.95±0.02	2412±10	2500±220	17.4±0.1
D	Boron	0.001 − 0.002	9.53±0.11	5.47±0.08	1032±2	1100±160	17.7±0.1

CONCLUSIONS

The effect of heavy doping on the optical properties of crystalline silicon in the spectral range from 0.7 to 3 eV can be described by the Drude theory. Spectroscopic phase modulated ellipsometry in this spectral range is a viable non-destructive technique for sub-0.5 μm silicon epitaxial film thickness measurement. It can simultaneously determine the native oxide overlayer and the epitaxial film thicknesses, as well as the substrate dopant concentration.

ACKNOWLEDGMENTS

We would like to acknowledge helpful discussion with Dr. Victor Yakovlev from Instrument SA. We thank Dr. Wingra Fang from Stanford University and Dr. Richard Westhoff from Lawrence Semiconductor Research Laboratory for their help in sample preparation, and Dr. Michael Frost from EVANS EAST for performing SIMS measurements. This work was supported by SEMATECH and the Semiconductor Research Consortium under contract No. 94-SP-900.

REFERENCES

1. K. Krishnan and P. J. Stout, in <u>Practical Fourier Transform Infrared Spectroscopy</u>, edited by J. R. Ferraro and K. Krishnan (Academic, Orlando, 1990), Chap. 6, P. 285.

2. Z. H. Zhou, I. Yang, F. Yu, and R. Reif, J. Appl. Phys. **73**, 7331 (1993).

3. B. Drévillon, Prog. Cryst. Growth Charact. **27**, 1 (1993).

4. D. E. Aspnes and A. A. Studna, Appl. Opt. **14**, 220 (1975).

5. E. D. Palik, <u>Handbook of Optical Constants of Solids</u>, (Academic Press, Inc. New York, 1985), p. 749; p. 547.

6. G. E. Jellison Jr., Optical Materials **1**, 41 (1992).

7. W. E. Beadle, J. C. C. Tsai, and R. D. Plummer, <u>Quick Reference Manual for Silicon Integrated Circuit Technology</u>, (Wiley, New York, 1985), p. 2-68.

8. D. E. Aspnes, A. A. Studna, and E. Kinsbron, Phys. Rev. B **29**, 768 (1984).

9. G. E. Jellison Jr., S. P. Withrow, J. W. McCamy, J. D. Budai, D. Lubben, and M. J. Godbole, Phys. Rev. B **52**, 14607 (1995).

10. H. M. van Driel, Appl. Phys. Lett. **44**, 617 (1984).

AUTHOR INDEX

SUBJECT INDEX